Polymers: An Encyclopedic Sourcebook of Engineering Properties

Polymers: An Encyclopedic Sourcebook of Engineering Properties

ENCYCLOPEDIA REPRINT SERIES

Editor: Jacqueline I. Kroschwitz

A WILEY-INTERSCIENCE PUBLICATION

JOHN WILEY & SONS

NEW YORK · CHICHESTER · BRISBANE · TORONTO · SINGAPORE

Library of Congress Cataloging-in-Publication Data

Polymers: an encyclopedic sourcebook of engineering
  properties.

  (Encyclopedia reprint series)
  "A Wiley-Interscience publication."
  1. Polymers and polymerization—Mechanical
properties—Dictionaries.   I. Kroschwitz, Jacqueline I.
II. Title.

TA455.P58P695 1987   668.9   87-21693
ISBN 0-471-85652-5

Printed in the United States of America

10 9 8 7 6 5 4 3 2 1

CONTENTS

EDITORIAL STAFF

CONTRIBUTORS

F. J. Baltá-Calleja, *Institute for Structure of Matter CSIC, Madrid, Spain,* Hardness
Donald C. Clagett, *General Electric Company, Pittsfield, Massachusetts,* Engineering plastics
I. M. Daniel, *Illinois Institute of Technology, Chicago, Illinois,* Composites, testing
Ashok K. Dhingra, *E. I. du Pont de Nemours & Co., Inc., Wilmington, Delaware,* Fibers, engineering
Frank P. Gerstle, Jr., *Sandia National Laboratories, Albuquerque, New Mexico,* Composites

W. Brandt Goldsworthy, *Goldsworthy Engineering, Inc., Torrance, California,* Composites, fabrication

D. W. Hadley, *University of Reading, Reading, United Kingdom,* Mechanical properties

Paul M. Hergenrother, *NASA Langley Research Center, Hampton, Virginia,* Heat-resistant polymers

Richard W. Hertzberg, *Lehigh University, Bethlehem, Pennsylvania,* Fatigue

Michael Jaffe, *Celanese Research Company, Summit, New Jersey,* High modulus polymers

Roger P. Kambour, *General Electric Company, Schenectady, New York,* Crazing

H. H. Kausch, *École Polytechnique Fédérale de Lausanne, Lausanne, Switzerland,* Fracture

S. L. Kwolek, *E. I. du Pont de Nemours & Co., Inc., Wilmington, Delaware,* Liquid crystalline polymers

John K. Lancaster, *Royal Aircraft Establishment, Farnborough, Hants, United Kingdom,* Abrasion and wear

Herbert G. Lauterbach, *E. I. du Pont de Nemours & Co., Inc., Wilmington, Delaware,* Fibers, engineering

John A. Manson, *Lehigh University, Bethlehem, Pennsylvania,* Fatigue

Robert A. McCarthy, *Springborn Group, Inc., Enfield, Connecticut,* Chemically resistant polymers

Paul Morgan, *Consultant, Westchester, Pennsylvania,* Liquid crystalline polymers

Takayuki Murayama, *Monsanto Company, Springfield, Massachusetts,* Dynamic mechanical properties

D. R. Rueda, *Institute for Structure of Matter CSIC, Madrid, Spain,* Hardness

John Schaefgen, *Consultant, Wilmington, Delaware,* Liquid crystalline polymers

J. Martinez-Salazar, *Institute for Structure of Matter CSIC, Madrid, Spain,* Hardness

L. C. E. Struik, *Plastics and Rubber Research Institute TNO, Delft, The Netherlands,* Aging, physical

I. M. Ward, *University of Leeds, Leeds, United Kingdom,* Mechanical properties

J. G. Williams, *Imperial College of Science and Technology, London, United Kingdom,* Fracture

Albert F. Yee, *University of Michigan, Ann Arbor, Michigan,* Impact resistance

PREFACE

This volume is one of a series of carefully selected reprints from the world-renowned *Encyclopedia of Polymer Science and Engineering* designed to provide specific audiences with articles grouped by a central theme. Although the 19-volume Polymer Encyclopedia is widely available, many readers and users of this key reference tool have expressed interest in having selected articles in their specialty collected for handy desk reference or teaching purposes. In response to this need, we have chosen all of the original, complete articles related to engineering properties of polymers and composites to make up this new volume. The full texts, tables, figures, and reference materials from the original work have been reproduced here unchanged. All articles are by industrial or academic experts in their field and the final work represents the result of careful review by competent specialists and the thorough editorial processing of the professional Wiley staff. Introductory information from the Encyclopedia concerning nomenclature, SI units and conversion factors, and related information has been provided as a further guide to the contents for those concerned with the materials aspects of engineering resins and composites.

The contents of this volume include coverage of nearly every aspect of polymeric engineering materials and provide detailed information on methods of manufacture, properties, and uses. Alphabetical organization, extensive cross-references, and a complex index further enhance the utility of this Encyclopedia. The approximately 20 main entries in this Encyclopedia average 18,000 words in length and have been prepared by 30 leading authorities from industries, universities, and research institutes. The contents should be of interest to all those engaged in the manufacture and use of modern, lightweight, tough engineering materials for use in consumer goods, transportation, aerospace, communications, and related industrial activities. The contents include coverage of the newest high performance liquid crystalline materials, a wealth of physical and mechanical data, and standards and specifications for materials. The book should be an important research reference tool, desk-top information resource, and supplementary reading asset for teaching professionals and their students.

J. I. KROSCHWITZ

CONVERSION FACTORS, ABBREVIATIONS, AND UNIT SYMBOLS

SI Units (Adopted 1960)

A new system of measurement, the International System of Units (abbreviated SI), is being implemented throughout the world. This system is a modernized version of the MKSA (meter, kilogram, second, ampere) system, and its details are published and controlled by an international treaty organization (The International Bureau of Weights and Measures) (1).

SI units are divided into three classes:

Base Units

length	meter[†] (m)
mass[‡]	kilogram (kg)
time	second (s)
electric current	ampere (A)
thermodynamic temperature[§]	kelvin (K)
amount of substance	mole (mol)
luminous intensity	candela (cd)

Supplementary Units

plane angle	radian (rad)
solid angle	steradian (sr)

[†]The spellings "metre" and "litre" are preferred by ASTM; however, "-er" is used in the *Encyclopedia*.

[‡]"Weight" is the commonly used term for "mass."

[§]Wide use is made of "Celsius temperature" (*t*) defined by

$$t = T - T_0$$

where T is the thermodynamic temperature, expressed in kelvins, and $T_0 = 273.15$ K by definition. A temperature interval may be expressed in degrees Celsius as well as in kelvins.

Derived Units and Other Acceptable Units

These units are formed by combining base units, supplementary units, and other derived units (2–4). Those derived units having special names and symbols are marked with an asterisk in the list below:

Quantity	Unit	Symbol	Acceptable equivalent
*absorbed dose	gray	Gy	J/kg
acceleration	meter per second squared	m/s^2	
*activity (of ionizing radiation source)	becquerel	Bq	1/s
area	square kilometer	km^2	
	square hectometer	hm^2	ha (hectare)
	square meter	m^2	
*capacitance	farad	F	C/V
concentration (of amount of substance)	mole per cubic meter	mol/m^3	
*conductance	siemens	S	A/V
current density	ampere per square meter	A/m^2	
density, mass density	kilogram per cubic meter	kg/m^3	g/L; mg/cm^3
dipole moment (quantity)	coulomb meter	C·m	
*electric charge, quantity of electricity	coulomb	C	A·s
electric charge density	coulomb per cubic meter	C/m^3	
electric field strength	volt per meter	V/m	
electric flux density	coulomb per square meter	C/m^2	
*electric potential, potential difference, electromotive force	volt	V	W/A
*electric resistance	ohm	Ω	V/A
*energy, work, quantity of heat	megajoule	MJ	
	kilojoule	kJ	
	joule	J	N·m
	electronvolt[†]	$eV^†$	
	kilowatt hour[†]	$kW·h^†$	
energy density	joule per cubic meter	J/m^3	
*force	kilonewton	kN	
	newton	N	$kg·m/s^2$

[†]This non-SI unit is recognized by the CIPM as having to be retained because of practical importance or use in specialized fields (1).

Quantity	Unit	Symbol	Acceptable equivalent
*frequency	megahertz	MHz	
	hertz	Hz	1/s
heat capacity, entropy	joule per kelvin	J/K	
heat capacity (specific), specific entropy	joule per kilogram kelvin	J/(kg·K)	
heat transfer coefficient	watt per square meter kelvin	W/(m²·K)	
*illuminance	lux	lx	lm/m²
*inductance	henry	H	Wb/A
linear density	kilogram per meter	kg/m	
luminance	candela per square meter	cd/m²	
*luminous flux	lumen	lm	cd·sr
magnetic field strength	ampere per meter	A/m	
*magnetic flux	weber	Wb	V·s
*magnetic flux density	tesla	T	Wb/m²
molar energy	joule per mole	J/mol	
molar entropy, molar heat capacity	joule per mole kelvin	J/(mol·K)	
moment of force, torque	newton meter	N·m	
momentum	kilogram meter per second	kg·m/s	
permeability	henry per meter	H/m	
permittivity	farad per meter	F/m	
*power, heat flow rate, radiant flux	kilowatt	kW	
	watt	W	J/s
power density, heat flux density, irradiance	watt per square meter	W/m²	
*pressure, stress	megapascal	MPa	
	kilopascal	kPa	
	pascal	Pa	N/m²
sound level	decibel	dB	
specific energy	joule per kilogram	J/kg	
specific volume	cubic meter per kilogram	m³/kg	
surface tension	newton per meter	N/m	
thermal conductivity	watt per meter kelvin	W/(m·K)	
velocity	meter per second	m/s	
	kilometer per hour	km/h	
viscosity, dynamic	pascal second	Pa·s	
	millipascal second	mPa·s	
viscosity, kinematic	square meter per second	m²/s	
	square millimeter per second	mm²/s	

Quantity	Unit	Symbol	Acceptable equivalent
volume	cubic meter	m^3	
	cubic decimeter	dm^3	L(liter) (5)
	cubic centimeter	cm^3	mL
wave number	1 per meter	m^{-1}	
	1 per centimeter	cm^{-1}	

In addition, there are 16 prefixes used to indicate order of magnitude, as follows:

Multiplication factor	Prefix	Symbol	Note
10^{18}	exa	E	[a]Although hecto, deka, deci, and centi
10^{15}	peta	P	are SI prefixes, their use should be
10^{12}	tera	T	avoided except for SI unit-multiples
10^9	giga	G	for area and volume and
10^6	mega	M	nontechnical use of centimeter, as
10^3	kilo	k	for body and clothing
10^2	hecto	h[a]	measurement.
10	deka	da[a]	
10^{-1}	deci	d[a]	
10^{-2}	centi	c[a]	
10^{-3}	milli	m	
10^{-6}	micro	μ	
10^{-9}	nano	n	
10^{-12}	pico	p	
10^{-15}	femto	f	
10^{-18}	atto	a	

For a complete description of SI and its use, the reader is referred to ASTM E 380 (4).

A representative list of conversion factors from non-SI to SI units is presented herewith. Factors are given to four significant figures. Exact relationships are followed by a dagger. A more complete list is given in ASTM E 380–84 (4) and ANSI Z 210.1–1976 (6).

Conversion Factors to SI Units

To convert from	To	Multiply by
acre	square meter (m^2)	4.047×10^3
angstrom	meter (m)	1.0×10^{-10}[†]
are	square meter (m^2)	1.0×10^2[†]
astronomical unit	meter (m)	1.496×10^{11}
atmosphere	pascal (Pa)	1.013×10^5
bar	pascal (Pa)	1.0×10^5[†]
barn	square meter (m^2)	1.0×10^{-28}[†]

[†]Exact.

To convert from	To	Multiply by
barrel (42 U.S. liquid gallons)	cubic meter (m³)	0.1590
Bohr magneton (μ_β)	J/T	9.274×10^{-24}
Btu (International Table)	joule (J)	1.055×10^3
Btu (mean)	joule (J)	1.056×10^3
Btu (thermochemical)	joule (J)	1.054×10^3
bushel	cubic meter (m³)	3.524×10^{-2}
calorie (International Table)	joule (J)	4.187
calorie (mean)	joule (J)	4.1908
calorie (thermochemical)	joule (J)	4.184[†]
centipoise	pascal second (Pa·s)	1.0×10^{-3}[†]
centistokes	square millimeter per second (mm²/s)	1.0[†]
cfm (cubic foot per minute)	cubic meter per second (m³/s)	4.72×10^{-4}
cubic inch	cubic meter (m³)	1.639×10^{-5}
cubic foot	cubic meter (m³)	2.832×10^{-2}
cubic yard	cubic meter (m³)	0.7646
curie	becquerel (Bq)	3.70×10^{10}[†]
debye	coulomb·meter (C·m)	3.336×10^{-30}
degree (angle)	radian (rad)	1.745×10^{-2}
denier (international)	kilogram per meter (kg/m)	1.111×10^{-7}
	tex[‡]	0.1111
dram (apothecaries')	kilogram (kg)	3.888×10^{-3}
dram (avoirdupois)	kilogram (kg)	1.772×10^{-3}
dram (U.S. fluid)	cubic meter (m³)	3.697×10^{-6}
dyne	newton (N)	1.0×10^{-5}[†]
dyne/cm	newton per meter (N/m)	1.0×10^{-3}[†]
electron volt	joule (J)	1.602×10^{-19}
erg	joule (J)	1.0×10^{-7}[†]
fathom	meter (m)	1.829
fluid ounce (U.S.)	cubic meter (m³)	2.957×10^{-5}
foot	meter (m)	0.3048[†]
footcandle	lux (lx)	10.76
furlong	meter (m)	2.012×10^{-2}
gal	meter per second squared (m/s²)	1.0×10^{-2}[†]
gallon (U.S. dry)	cubic meter (m³)	4.405×10^{-3}
gallon (U.S. liquid)	cubic meter (m³)	3.785×10^{-3}
gallon per minute (gpm)	cubic meter per second (m³/s)	6.308×10^{-5}
	cubic meter per hour (m³/h)	0.2271
gauss	tesla (T)	1.0×10^{-4}
gilbert	ampere (A)	0.7958
gill (U.S.)	cubic meter (m³)	1.183×10^{-4}
grad	radian	1.571×10^{-2}
grain	kilogram (kg)	6.480×10^{-5}

[†]Exact.
[‡]See footnote on p. xii.

To convert from	*To*	*Multiply by*
gram-force per denier	newton per tex (N/tex)	8.826×10^{-2}
hectare	square meter (m²)	$1.0 \times 10^{4\dagger}$
horsepower (550 ft·lbf/s)	watt (W)	7.457×10^{2}
horsepower (boiler)	watt (W)	9.810×10^{3}
horsepower (electric)	watt (W)	$7.46 \times 10^{2\dagger}$
hundredweight (long)	kilogram (kg)	50.80
hundredweight (short)	kilogram (kg)	45.36
inch	meter (m)	$2.54 \times 10^{-2\dagger}$
inch of mercury (32°F)	pascal (Pa)	3.386×10^{3}
inch of water (39.2°F)	pascal (Pa)	2.491×10^{2}
kilogram-force	newton (N)	9.807
kilowatt hour	megajoule (MJ)	$3.6^{\dagger}$
kip	newton (N)	4.48×10^{3}
knot (international)	meter per second (m/s)	0.5144
lambert	candela per square meter (cd/m²)	3.183×10^{3}
league (British nautical)	meter (m)	5.559×10^{3}
league (statute)	meter (m)	4.828×10^{3}
light year	meter (m)	9.461×10^{15}
liter (for fluids only)	cubic meter (m³)	$1.0 \times 10^{-3\dagger}$
maxwell	weber (Wb)	$1.0 \times 10^{-8\dagger}$
micron	meter (m)	$1.0 \times 10^{-6\dagger}$
mil	meter (m)	$2.54 \times 10^{-5\dagger}$
mile (statute)	meter (m)	1.609×10^{3}
mile (U.S. nautical)	meter (m)	$1.852 \times 10^{3\dagger}$
mile per hour	meter per second (m/s)	0.4470
millibar	pascal (Pa)	1.0×10^{2}
millimeter of mercury (0°C)	pascal (Pa)	$1.333 \times 10^{2\dagger}$
minute (angular)	radian	2.909×10^{-4}
myriagram	kilogram (kg)	10
myriameter	kilometer (km)	10
oersted	ampere per meter (A/m)	79.58
ounce (avoirdupois)	kilogram (kg)	2.835×10^{-2}
ounce (troy)	kilogram (kg)	3.110×10^{-2}
ounce (U.S. fluid)	cubic meter (m³)	2.957×10^{5}
ounce-force	newton (N)	0.2780
peck (U.S.)	cubic meter (m³)	8.810×10^{-3}
pennyweight	kilogram (kg)	1.555×10^{-3}
pint (U.S. dry)	cubic meter (m³)	5.506×10^{-4}
pint (U.S. liquid)	cubic meter (m³)	4.732×10^{-4}
poise (absolute viscosity)	pascal second (Pa·s)	$0.10^{\dagger}$
pound (avoirdupois)	kilogram (kg)	0.4536
pound (troy)	kilogram (kg)	0.3732
poundal	newton (N)	0.1383
pound-force	newton (N)	4.448

†Exact.

To convert from	To	Multiply by
pound-force per square inch (psi)	pascal (Pa)	6.895×10^3
quart (U.S. dry)	cubic meter (m³)	1.101×10^{-3}
quart (U.S. liquid)	cubic meter (m³)	9.464×10^{-4}
quintal	kilogram (kg)	$1.0 \times 10^{2\dagger}$
rad	gray (Gy)	$1.0 \times 10^{-2\dagger}$
rod	meter (m)	5.029
roentgen	coulomb per kilogram (C/kg)	2.58×10^{-4}
second (angle)	radian (rad)	4.848×10^{-6}
section	square meter (m²)	2.590×10^6
slug	kilogram (kg)	14.59
spherical candle power	lumen (lm)	12.57
square inch	square meter (m²)	6.452×10^{-4}
square foot	square meter (m²)	9.290×10^{-2}
square mile	square meter (m²)	2.590×10^6
square yard	square meter (m²)	0.8361
stere	cubic meter (m³)	$1.0^\dagger$
stokes (kinematic viscosity)	square meter per second (m²/s)	$1.0 \times 10^{-4\dagger}$
tex	kilogram per meter (kg/m)	$1.0 \times 10^{-6\dagger}$
ton (long, 2240 pounds)	kilogram (kg)	1.016×10^3
ton (metric)	kilogram (kg)	$1.0 \times 10^{3\dagger}$
ton (short, 2000 pounds)	kilogram (kg)	9.072×10^2
torr	pascal (Pa)	1.333×10^2
unit pole	weber (Wb)	1.257×10^{-7}
yard	meter (m)	$0.9144^\dagger$

Abbreviations and Unit Symbols

Following is a list of commonly used abbreviations and unit symbols appropriate for use in the *Encyclopedia*. In general they agree with those listed in *American National Standard Abbreviations for Use on Drawings and in Text (ANSI Y1.1)* (6) and *American National Standard Letter Symbols for Units in Science and Technology (ANSI Y10)* (6). Also included is a list of acronyms for a number of private and government organizations as well as common industrial solvents, polymers, and other chemicals.

Rules for Writing Unit Symbols (4):

1. Unit symbols should be printed in upright letters (roman) regardless of the type style used in the surrounding text.

2. Unit symbols are unaltered in the plural.

3. Unit symbols are not followed by a period except when used as the end of a sentence.

4. Letter unit symbols are generally written in lowercase (eg, cd for candela) unless the unit name has been derived from a proper name, in which case the first letter of the symbol is capitalized (W, Pa). Prefix and unit symbols retain their prescribed form regardless of the surrounding typography.

†Exact.

5. In the complete expression for a quantity, a space should be left between the numerical value and the unit symbol. For example, write 2.37 lm, *not* 2.37lm, and 35 mm, *not* 35mm. When the quantity is used in an adjectival sense, a hyphen is often used, for example, 35-mm film. *Exception:* No space is left between the numerical value and the symbols for degree, minute, and second of plane angle, and degree Celsius.

6. No space is used between the prefix and unit symbols (eg, kg).

7. Symbols, not abbreviations, should be used for units. For example, use "A," not "amp," for ampere.

8. When multiplying unit symbols, use a raised dot:

$$N \cdot m \text{ for newton meter}$$

In the case of W·h, the dot may be omitted, thus:

$$Wh$$

An exception to this practice is made for computer printouts, automatic typewriter work, etc, where the raised dot is not possible, and a dot on the line may be used.

9. When dividing unit symbols use one of the following forms:

$$m/s \ or \ m \cdot s^{-1} \ or \ \frac{m}{s}$$

In no case should more than one slash be used in the same expression unless parentheses are inserted to avoid ambiguity. For example, write:

$$J/(mol \cdot K) \ or \ J \cdot mol^{-1} \cdot K^{-1} \ or \ (J/mol)/K$$

but *not*

$$J/mol/K$$

10. Do not mix symbols and unit names in the same expression. Write:

$$joules \ per \ kilogram \ or \ J/kg \ or \ J \cdot kg^{-1}$$

but *not*

$$joules/kilogram \ nor \ joules/kg \ nor \ joules \cdot kg^{-1}$$

Abbreviations and Units

A	ampere	*ac-*	alicyclic
A	anion (eg, H*A*); mass number	ACGIH	American Conference of Governmental Industrial Hygienists
a	atto (prefix for 10^{-18})		
AATCC	American Association of Textile Chemists and Colorists	ACS	American Chemical Society
		AGA	American Gas Association
		Ah	ampere hour
ABS	acrylonitrile–butadiene–styrene	AIChE	American Institute of Chemical Engineers
abs	absolute	AIME	American Institute of Mining, Metallurgical, and Petroleum Engineers
ac	alternating current, *n.*		
a-c	alternating current, *adj.*		

AIP	American Institute of Physics	bid	twice daily
AISI	American Iron and Steel Institute	Boc	t-butyloxycarbonyl
alc	alcohol(ic)	BOD	biochemical (biological) oxygen demand
Alk	alkyl	bp	boiling point
alk	alkaline (not alkali)	Bq	becquerel
-*alt*-	alternating as in alternating copolymer	C	coulomb
		°C	degree Celsius
amt	amount	C-	denoting attachment to carbon
amu	atomic mass unit	C_M	chain-transfer constant for monomer
ANSI	American National Standards Institute	C_P	chain-transfer constant for polymer
AO	atomic orbital	C_S	chain-transfer constant for solvent
AOAC	Association of Official Analytical Chemists		
AOCS	American Oil Chemists' Society	c	centi (prefix for 10^{-2})
		c	critical
APHA	American Public Health Association	ca	circa (approximately)
		cd	candela; current density; circular dichroism
API	American Petroleum Institute		
		CFR	Code of Federal Regulations
aq	aqueous		
Ar	aryl	cgs	centimeter-gram-second
ar-	aromatic	CI	Color Index
as-	asymmetric(al)	*cis*-	isomer in which substituted groups are on same side of double bond between C atoms
ASH-RAE	American Society of Heating, Refrigerating, and Air Conditioning Engineers		
		cl	carload
ASM	American Society for Metals	cm	centimeter
		cmil	circular mil
ASME	American Society of Mechanical Engineers	cmpd	compound
		CNRS	Centre National de la Recherche Scientifique
ASTM	American Society for Testing and Materials		
		CNS	central nervous system
at no.	atomic number	-*co*-	copolymerized with
at wt	atomic weight	CoA	coenzyme A
av(g)	average	COC	Cleveland open cup
AWS	American Welding Society	COD	chemical oxygen demand
b	bonding orbital	coml	commercial(ly)
bbl	barrel	conc	concentration
bcc	body-centered cubic	cp	chemically pure
bct	body-centered tetragonal	cph	close-packed hexagonal
Bé	Baumé	CPSC	Consumer Product Safety Commission
BET	Brunauer-Emmett-Teller (adsorption equation)	cryst	crystalline

cub	cubic	eng	engineering
D	Debye	EPA	Environmental Protection
D-	denoting configurational		Agency
	relationship	epr	electron paramagnetic
d	differential operator		resonance
d-	*dextro-*, dextrorotatory	ϵ	dielectric constant (unitless)
da	deka (prefix for 10^1)	eq.	equation
dB	decibel	esca	electron-spectroscopy for
dc	direct current, *n.*		chemical analysis
d-c	direct current, *adj.*	esp	especially
dec	decompose	esr	electron-spin resonance
detd	determined	est(d)	estimate(d)
detn	determination	estn	estimation
dia	diameter	esu	electrostatic unit
dil	dilute	η	viscosity
dl-; DL-	racemic	$[\eta]$	intrinsic viscosity
DMA	dimethylacetamide	η_{inh}	inherent viscosity
DMF	dimethylformamide	η_r	relative viscosity
DMG	dimethyl glyoxime	η_{red}	reduced viscosity
DMSO	dimethyl sulfoxide	η_{sp}	specific viscosity
DOD	Department of Defense	exp	experiment, experimental
DOE	Department of Energy	ext(d)	extract(ed)
DOT	Department of	F	farad (capacitance)
	Transportation	*F*	faraday (96,487 C); free
DP	degree of polymerization		energy
dp	dew point	f	femto (prefix for 10^{-15})
DPH	diamond pyramid hardness	FAO	Food and Agriculture
DS	degree of substitution		Organization (United
dsc	differential scanning		Nations)
	calorimetry	fcc	face-centered cubic
dstl(d)	distill(ed)	FDA	Food and Drug
dta	differential thermal		Administration
	analysis	FEA	Federal Energy
E	Young's modulus		Administration
(*E*)-	entgegen; opposed	FHSA	Federal Hazardous
e	polarity factor in Alfrey-		Substances Act
	Price equation	fob	free on board
e$^-$	electron	fp	freezing point
ECU	electrochemical unit	FPC	Federal Power Commission
ed.	edited, edition, editor	FRB	Federal Reserve Board
ED	effective dose	frz	freezing
EDTA	ethylenediaminetetraacetic	G	giga (prefix for 10^9)
	acid	*G*	gravitational constant =
em	electron microscopy		6.67×10^{11} N·m²/kg²;
emf	electromotive force		Gibb's free energy
emu	electronmagnetic unit	g	gram
en	ethylene diamine	(g)	gas, only as in $H_2O(g)$

g	gravitational acceleration	ir	infrared
-g-	graft as in graft copolymer	IRLG	Interagency Regulatory Liaison Group
gc	gas chromatography		
gem-	geminal	ISO	International Organization for Standardization
glc	gas-liquid chromatography		
g-mol wt; gmw	gram-molecular weight	IU	International Unit
		IUPAC	International Union of Pure and Applied Chemistry
GNP	gross national product	IV	iodine value
gpc	gel-permeation chromatography	iv	intravenous
		J	joule
GRAS	Generally Recognized as Safe	K	kelvin
		K	equilibrium constant
grd	ground	k	kilo (prefix for 10^3)
Gy	gray	k	reaction rate constant
H	henry	kg	kilogram
H	enthalpy	L	denoting configurational relationship
h	hour; hecto (prefix for 10^2)		
ha	hectare	L	liter (for fluids only) (5)
HB	Brinell hardness number	l-	$levo$-, levorotatory
Hb	hemoglobin	(l)	liquid, only as in $NH_3(l)$
hcp	hexagonal close-packed	LC_{50}	conc lethal to 50% of the animals tested
hex	hexagonal		
HK	Knoop hardness number	LCAO	linear combination of atomic orbitals
hplc	high-pressure liquid chromatography		
		LCD	liquid crystal display
HRC	Rockwell hardness (C scale)	lcl	less than carload lots
HV	Vickers hardness number	LD_{50}	dose lethal to 50% of the animals tested
hyd	hydrated, hydrous		
hyg	hygroscopic	LED	light-emitting diode
Hz	hertz	liq	liquid
^{i}(eg, Pr^i)	iso (eg, isopropyl)	lm	lumen
i-	inactive (eg, i-methionine)	ln	logarithm (natural)
		LNG	liquefied natural gas
IACS	International Annealed Copper Standard	log	logarithm (common)
		LPG	liquefied petroleum gas
ibp	initial boiling point	ltl	less than truckload lots
IC	inhibitory concentration	lx	lux
ICC	Interstate Commerce Commission	M	mega (prefix for 10^6); metal (as in MA)
ICT	International Critical Table	M	molar; actual mass
ID	inside diameter; infective dose	$\overline{M}_w$	weight-average mol wt
		$\overline{M}_n$	number-average mol wt
ip	intraperitoneal	$\overline{M}_v$	viscosity-average mol wt
IPS	iron pipe size	m	meter; milli (prefix for 10^{-3})
IPTS	International Practical Temperature Scale (NBS)		
		m	molal

m-	meta		neg	negative
max	maximum		NEMA	National Electrical
MCA	Chemical Manufacturers'			Manufacturer's
	Association (was			Association
	Manufacturing Chemists		NF	*National Formulary*
	Association)		NIH	National Institutes of
MEK	methyl ethyl ketone			Health
meq	milliequivalent		NIOSH	National Institute of
mfd	manufactured			Occupational Safety and
mfg	manufacturing			Health
mfr	manufacturer		nmr	nuclear magnetic resonance
MIBC	methyl isobutyl carbinol		NND	New and Nonofficial Drugs
MIBK	methyl isobutyl ketone			(AMA)
MIC	minimum inhibiting		no.	number
	concentration		NOI(BN)	not otherwise indexed (by
min	minute; minimum			name)
mL	milliliter		NOS	not otherwise specified
MLD	minimum lethal dose		nqr	nuclear quadruple
MO	molecular orbital			resonance
mo	month		NRC	Nuclear Regulatory
mol	mole			Commission; National
mol wt	molecular weight			Research Council
mp	melting point		NRI	New Ring Index
MR	molar refraction		NSF	National Science
ms	mass spectrum			Foundation
mxt	mixture		NTA	nitrilotriacetic acid
μ	micro (prefix for 10^{-6})		NTP	normal temperature and
N	newton (force)			pressure (25°C and 101.3
N	normal (concentration);			kPa or 1 atm)
	neutron number		NTSB	National Transportation
N-	denoting attachment to			Safety Board
	nitrogen		O-	denoting attachment to
n (as	index of refraction (for 20°C			oxygen
n_D^{20})	and sodium light)		o-	ortho
^{n} (as	normal (straight-chain		OD	outside diameter
Bu^n),	structure)		ω	frequency
n-			OPEC	Organization of Petroleum
n	neutron			Exporting Countries
n	nano (prefix for 10^9)		OSHA	Occupational Safety and
na	not available			Health Administration
NAS	National Academy of		owf	on weight of fiber
	Sciences		Ω	ohm
NASA	National Aeronautics and		P	peta (prefix for 10^{15})
	Space Administration		p	pico (prefix for 10^{-12})
nat	natural		p-	para
NBS	National Bureau of		p	proton
	Standards		p.	page

Pa	pascal (pressure)		RI	Ring Index
PAN	polyacrylonitrile		rms	root-mean square
pd	potential difference		rpm	rotations per minute
PE	polyethylene		rps	revolutions per second
pH	negative logarithm of the effective hydrogen ion concentration		RT	room temperature
			^{s} (eg, Bu^s); sec-	secondary (eg, secondary butyl)
phr	parts per hundred of resin (rubber)		S	siemens
π	osmotic pressure		(S)-	sinister (counterclockwise configuration)
p-i-n	positive-intrinsic-negative			
pmr	proton magnetic resonance		S	entropy
p-n	positive-negative		S-	denoting attachment to sulfur
po	per os (oral)			
POP	polyoxypropylene		s-	symmetric(al)
pos	positive		s	second
PP	polypropylene		(s)	solid, only as in $H_2O(s)$
pp.	pages		SAE	Society of Automotive Engineers
ppb	parts per billion (10^9)			
pph	parts per hundred		SAMPE	Society for the Advancement of Material and Process Engineering
ppm	parts per million (10^6)			
ppmv	parts per million by volume			
ppmwt	parts per million by weight		SAN	styrene–acrylonitrile
PPO	poly(phenyl oxide)		sat(d)	saturate(d)
ppt(d)	precipitate(d)		satn	saturation
pptn	precipitation		SBR	styrene–butadiene–rubber
Pr (no.)	foreign prototype (number)			
PS	polystyrene		sc	subcutaneous
pt	point; part		SCF	self-consistent field; standard cubic feet
PVC	poly(vinyl chloride)			
pwd	powder		Sch	Schultz number
py	pyridine		SFs	Saybolt Furol seconds
Q	reactivity of monomer in Alfrey-Price equation		SI	Le Système International d'Unités (International System of Units)
qv	quod vide (which see)			
R	univalent hydrocarbon radical		sl sol	slightly soluble
			SMC	sheet molding compound
(R)-	rectus (clockwise configuration)		sol	soluble
			soln	solution
r	precision of data		soly	solubility
rad	radian; radius		sp	specific; species
rds	rate determining step		SPE	Society of Plastics Engineers
Ref.	reference			
rf	radio frequency, n.		sp gr	specific gravity
r-f	radio frequency, adj.		SPI	Society of the Plastics Industry
rh	relative humidity			
ρ, d	density		sr	steradian

std	standard	TOC	Tagliabue (Tag) open cup
STP	standard temperature and pressure (0°C and 101.3 kPa)	*trans-*	isomer in which substituted groups are on opposite sides of double bond between C atoms
sub	sublime(s)		
SUs	Saybolt Universal seconds	TSCA	Toxic Substance Control Act
syn	synthetic	TWA	time-weighted average
^{t} (eg, Bu^t), *t-, tert-*	tertiary (eg, tertiary-butyl)	Twad	Twaddell
		UL	Underwriters' Laboratory
		USDA	United States Department of Agriculture
T	tera (prefix for 10^{12}); tesla (magnetic flux density)	USP	*United States Pharmacopeia*
		uv	ultraviolet
t	metric ton (tonne); temperature	V	volt (emf)
		var	variable
TAPPI	Technical Association of the Pulp and Paper Industry	*vic-*	vicinal
		vol	volume (not volatile)
τ	relaxation	vs	versus
TCC	Tagliabue (Tag) closed cup	v sol	very soluble
tex	tex (linear density)	W	watt
T_g	glass-transition temperature	Wb	weber
tga	thermogravimetric analysis	Wh	watt hour
θ	adjective of condition, solvent, temperature	WHO	World Health Organization (United Nations)
THF	tetrahydrofuran	wk	week
tlc	thin layer chromatography	yr	year
TLV	threshold limit value	(Z)-	zusammen; together; atomic number
TMS	tetramethylsilane		

Non-SI (Unacceptable and Obsolete) Units		*Use*
Å	angstrom	nm
at	atmosphere, technical	Pa
atm	atmosphere, standard	Pa
b	barn	cm^2
bar†	bar	Pa
bbl	barrel	m^3
bhp	brake horsepower	W
Btu	British thermal unit	J
bu	bushel	m^3; L
cal	calorie	J
cfm	cubic foot per minute	m^3/s
Ci	curie	Bq
cSt	centistokes	mm^2/s
c/s	cycle per second	Hz
cu	cubic	exponential form
D	debye	C·m

---

†Do not use bar (10^5Pa) or millibar (10^2Pa) because they are not SI units, and are accepted internationally only for a limited time in special fields because of existing usage.

Non-SI (Unacceptable and Obsolete) Units		Use
den	denier	tex
dr	dram	kg
dyn	dyne	N
dyn/cm	dyne per centimeter	mN/m
erg	erg	J
eu	entropy unit	J/K
°F	degree Fahrenheit	°C; K
fc	footcandle	lx
fl	footlambert	lx
fl oz	fluid ounce	m^3; L
ft	foot	m
ft·lbf	foot pound-force	J
gf/den	gram-force per denier	N/tex
G	gauss	T
Gal	gal	m/s^2
gal	gallon	m^3; L
Gb	gilbert	A
gpm	gallon per minute	(m^3/s); (m^3/h)
gr	grain	kg
hp	horsepower	W
ihp	indicated horsepower	W
in.	inch	m
in. Hg	inch of mercury	Pa
in. H_2O	inch of water	Pa
in.-lbf	inch pound-force	J
kcal	kilogram-calorie	J
kgf	kilogram-force	N
kilo	for kilogram	kg
L	lambert	lx
lb	pound	kg
lbf	pound-force	N
mho	mho	S
mi	mile	m
MM	million	M
mm Hg	millimeter of mercury	Pa
$m\mu$	millimicron	nm
mph	mile per hour	km/h
μ	micron	μm
Oe	oersted	A/m
oz	ounce	kg
ozf	ounce-force	N
η	poise	Pa·s
P	poise	Pa·s
ph	phot	lx
psi	pound-force per square inch	Pa

Non-SI (Unacceptable and Obsolete) Units		*Use*
psia	pound-force per square inch absolute	Pa
psig	pound-force per square inch gauge	Pa
qt	quart	m^3; L
°R	degree Rankine	K
rd	rad	Gy
sb	stilb	lx
SCF	standard cubic foot	m^3
sq	square	exponential form
thm	therm	J
yd	yard	m

BIBLIOGRAPHY

1. The International Bureau of Weights and Measures, BIPM (Parc de Saint-Cloud, France) is described on page 22 of Ref. 4. This bureau operates under the exclusive supervision of the International Committee of Weights and Measures (CIPM).
2. *Metric Editorial Guide (ANMC-78-1)* 3rd ed., American National Metric Council, 5410 Grosvenor Lane, Bethesda, Md. 20814, 1981.
3. *SI Units and Recommendations for the Use of Their Multiples and of Certain Other Units (ISO 1000–1981)*, American National Standards Institute, 1430 Broadway, New York, N.Y. 10018, 1981.
4. Based on *ASTM E 380–84 (Standard for Metric Practice)*, American Society for Testing and Materials, 1916 Race Street, Philadelphia, Pa. 19103, 1984.
5. *Fed. Regist.*, Dec. 10, 1976 (41 FR 36414).
6. For ANSI address, see Ref. 3.

R. P. LUKENS
American Society for Testing and Materials

ABRASION AND WEAR

The term abrasion is commonly used to refer to a process of wear in which there is displacement of material from a surface during relative motion against hard particles or protuberances (1). However, it is important to recognize that abrasion is only one among many other types of wear. In any given situation, the relative importance of all the different processes will depend on the type of material involved, eg, elastomers, plastics, plastics composites, and on the conditions of operation to which they are subjected, eg, sliding or rolling, speed, temperature, stress, etc. The subject of wear is thus one of considerable complexity and it is only within the past two or three decades that the essential scientific foundations have been laid.

Although the overall economic consequences of wear on a national scale are now well-appreciated (2,3), it is difficult to obtain quantitative data relating solely to polymers. At one extreme are circumstances in which excessive wear of a single component can jeopardize the viability of a complete assembly, eg, a dry-bearing in the main rotor assembly of a helicopter, while at the other, small amounts of wear distributed over a very large number of identical components can lead to an enormous cumulative wastage, eg, tires (qv). One report suggests that the cost of tire replacement in U.S. naval aircraft alone is on the order of 2×10^6/yr (4).

Surface Contact

Before proceeding to a detailed examination of abrasion and other wear processes, it is necessary to examine briefly the way in which surfaces come into contact. All surfaces prepared by conventional engineering techniques are rough and will touch only over a limited number of small, discrete areas. The real area

1

of contact is generally very much smaller than the apparent (geometric) area. The question then arises as to whether the localized contact deformation is likely to be elastic or plastic. A simple criterion to predict this relates the mechanical properties of a material to the limiting angle of slope of an asperity beyond which plastic flow will occur during flattening into the general plane of the surface (5). For this condition:

$$\tan \theta \,|\, _{\text{limit}} = k \frac{H}{E} (1 - \nu^2)$$

where H is the indentation hardness, E is Young's modulus, ν is Poisson's ratio, and $k = 2$ for full plasticity.

A more advanced criterion, the plasticity index, based on the computer analysis of real surface profiles has been devised (6):

$$\psi = \frac{H}{E} \left(\frac{\sigma}{\beta} \right)^{1/2}$$

where σ is the standard deviation of the asperity height distribution (Gaussian) and β is the average radius of curvature of the peaks.

Plastic deformation occurs only when $\psi \geqslant 1$. Some typical values of the limiting asperity angles for full plasticity and the plasticity indexes characteristic of a typical surface topography are given in Table 1 and show very clearly that polymers are much more likely to form elastic asperity contacts than are metals.

Table 1. Plasticity Criteria for Asperity Deformation

Material	θ_{crit}[a]	ψ[b]
polyethylene, low density	13	0.08
polytetrafluoroethylene	11.4	0.09
polyethylene, high density	8.6	0.12
nylon-6,6	6	0.17
poly(methyl methacrylate)	4.9	0.21
tool steel	3.9	1.1
low carbon steel	1.0	2.8
lead	0.25	4

[a] Critical angle of asperity for full plasticity.
[b] Plasticity index, assuming a typical roughness value of $(\sigma/\beta)^{1/2} = 10^{-2}$ and full plasticity.

Although the mode of deformation is not influenced by the magnitude of the applied load, the latter does, of course, affect the magnitude of the real area of contact A_r. For plastic deformation, A_r is proportional to the load L, irrespective of the particular shapes and sizes of the asperities in contact. However, for elastic deformation, shape and size become much more important. In the idealized case of a single contact between hemispherical asperities, elasticity (qv) theory leads to $A_r \propto L^{2/3}$ (7). Since the mechanical response of polymers to stress is usually viscoelastic rather than purely elastic, it is more realistic to expect a relationship $A_r \propto L^{2/m}$, where m lies somewhere between three for complete elasticity and two for complete plastic flow. For the deformation of real surfaces, it is necessary to introduce a statistical representation of their topographical features. Many surfaces prepared by conventional finishing techniques exhibit a Gaussian distribution of asperity heights (8), and for these it can be shown that $A_r \propto L$ irrespective of whether the mode of deformation is elastic, viscoelastic, or plastic (6). The essential feature here is that increasing the load brings more asperities into contact rather than increasing the size of the existing ones. Proportionality begins to break down, however, when the load increases to a value at which contact no longer remains restricted to the upper height levels of the Gaussian distribution. This point is reached when the surface asperities are becoming deformed into the general plane of the surface and from then the real contact area increases less rapidly than proportionately with the load (see MECHANICAL PROPERTIES; SURFACE PROPERTIES; VISCOELASTICITY).

Friction

Friction is the tangential force resisting motion when one surface is displaced relative to another under the influence of an external force and it generally arises from two types of contact. In one, the localized forces of adhesion (qv) must be overcome by shear, whereas in the other, interpenetrating asperities give rise to further deformation. The total frictional force F is thus given by the sum of two terms: $F = F_{adh} + F_{def}$.

The essential elements involved in this two-term model of friction have been summarized (9) and are shown in Figure 1. Deformation losses are most pronounced when sliding against rough surfaces where plastic flow may lead to microcutting and viscoelastic deformation can result in tearing, cracking, or fatigue (qv). At the other extreme, when sliding against smooth surfaces, frictional work is dissipated in a very narrow interfacial zone via the rupture of adhesive bonds. For polymers, the relevant adhesive forces are primarily of the van der Waals type, but sometimes supplemented by electrostatic interactions and/or hydrogen bonds. With the more rigid polymers, rupture of the adhesive junctions often leads to transfer and the formation of a layer of modified material on the polymer itself. These surface changes have been formalized into the concept of a third-body (10) and play a crucial role in subsequent friction and wear processes. With elastomers, however, sliding is usually restricted to the actual interface itself, leading to tearing, cracking, and fatigue or to energy dissipation via waves of detachment, Shallamach waves (11). Both bulk-deformation and interfacial sliding processes lead to heat generation which may ultimately result in surface oxidation, thermal degradation (qv), or actual melting.

Elastomers

For most practical sliding conditions it is difficult, and often impossible, to isolate the individual contributions of deformation and adhesion to friction. To examine the separate roles of each, it is therefore necessary to resort to idealized experiments. The simplest of these is the rolling of a hard sphere over a polymer plane where adhesion effects remain low, except for very low modulus elastomers (12). During rolling, each element of the polymer is compressed and then recovers

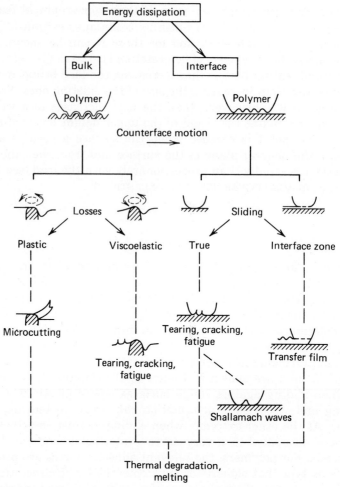

Fig. 1. Modes of energy dissipation and wear during sliding (9). Courtesy of Plenum Publishing Corp.

and energy is dissipated during this cycle by elastic hysteresis (13). The magnitude of the friction measured in this way agrees with theoretical predictions (14). As the load is increased, the coefficient of friction μ increases as $\mu \propto L^{1/3}$ and variations in friction with speed and temperature both show characteristic maxima which correlate closely with those obtained in mechanical loss measurements (15).

For elastomers sliding against hard counterfaces, the relative importance of the deformation and adhesion components in friction depends on the counterface topography; deformation is more important on rough surfaces and adhesion on smooth. Even in the latter case, however, viscoelastic energy losses still provide the main contribution to friction. This has been demonstrated very clearly on several elastomers in which the variations in friction with speed and temperature on three types of counterfaces were determined (16). All the separate variations with speed at different temperatures could be combined into a single master curve by a transformation of the Williams-Landel-Ferry type (17), and Figure 2 shows the results. On dusted silicon carbide, adhesion is minimized and the main contribution to the friction comes from the bulk-deformation losses. On nondusted silicon carbide, adhesion plays a more significant role and its importance increases still further when sliding against pin-head glass. Despite these variations in the relative contributions from adhesion and deformation, the curves all remain very similar.

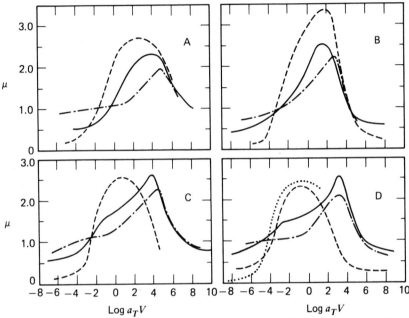

Fig. 2. Master curves for the variation in friction coefficient with speed for various rubbers sliding over different surfaces. —·— Dusted silicon carbide, — nondusted silicon carbide, ——— pin-head glass, ··· polished stainless steel. A, natural rubber, isomerized; B, butyl rubber; C, styrene–butadiene rubber; D, acrylate–butadiene rubber (16).

There have been numerous ideas about the detailed way in which energy dissipates during sliding when adhesion is the dominant factor in the friction mechanism, and the topic has been extensively reviewed (18–21). The behavior of the elastomer can be visualized as similar to that of a viscous liquid in which as the speed increases the rate of shear also increases, but the contact area decreases (22). On the other hand, it has been suggested that it is the strength of the individual adhesive bonds which increases with speed while the number of these bonds decreases (23). Both concepts lead to the prediction of a maximum in the friction–speed relationship, as observed experimentally. More recent ideas have emerged following the observation (11) that during sliding against smooth surfaces adhesion leads to localized surface buckling and the generation of a Shallamach wave moving across the contact area. During the propagation of this wave, each element of the elastomer is peeled from the counterface and then redeposited. The energy losses could thus result from either conventional deformation hysteresis or because more energy might be needed to detach the elastomer from its counterface than is liberated when it comes back into contact. The most recent experimental work supports the surface energy approach (24).

Plastics

Attempts to correlate deformation friction of the more rigid engineering plastics (qv) with viscoelastic properties have been reasonably successful during rolling (25), but somewhat less so during sliding (26). The main complication during sliding is the transfer that often occurs from the plastic to its counterface, usually metal, since the amount and type of transfer depend on the conditions of sliding imposed (load, speed, and temperature). On all but the very roughest counterfaces, adhesion effects predominate in the friction mechanism, and the main component of friction arises from shear within an interfacial zone of 50–200 nm in thickness (27). The frictional force F is then given by the product of the real area of contact A_r and the shear strength s of the junctions: $F = A_r s$. The relevant interfacial shear strength was originally equated to that of the bulk polymer, and by doing this, reasonable agreement was obtained between the predicted and measured trends in the coefficient of friction with temperature (28). Friction usually decreases with increasing temperature, as illustrated in Figure 3(**b**). More recent work, however, has shown that the interfacial shear strength is a linear function of the contact pressure, $S = S_0 + \alpha P$, where S_0 and α are material constants, and hence the coefficient of friction, $\mu = (A_r/L)S_0 + \alpha$ (29). In conditions of mixed elastic–plastic deformation, the real area of contact A_r increases less rapidly than proportionately with load, and the coefficient of friction thus tends to decrease with increasing load. This has been confirmed experimentally (30–32) and an example is shown in Figure 3(**c**) for polytetrafluoroethylene (PTFE).

The effects of speed on the friction of plastics are more complex than those of load. For many thermoplastics, the coefficient of friction varies only slightly with speed until the associated rise in temperature causes appreciable thermal softening, leading to an increase in the real contact area and a rise in friction (33). Under very severe conditions of sliding (high loads), the coefficient of friction can then decrease again at high speeds as the surfaces become effectively lubricated by a molten polymer film (34). When examined over a very wide range of speeds, some polymers also exhibit shallow maxima in friction, which can be attributed to viscoelastic losses within the bulk-deformation component (35). These maxima are most pronounced with crystalline thermoplastics. A further complication can also arise with some crystalline thermoplastics: when the time involved in localized shear becomes commensurate with the molecular relaxation time, the mechanism of transfer changes from one involving thin films to the deposition of relatively large lumps (36). PTFE is the outstanding example of this type of behavior and Figure 3(**a**) shows a synthesis from various sources of the trend in friction with speed.

Polytetrafluoroethylene (PTFE)

Although the coefficient of friction of PTFE is not always very low, as Figure 3 shows, it can nevertheless become lower than that of any other polymer in

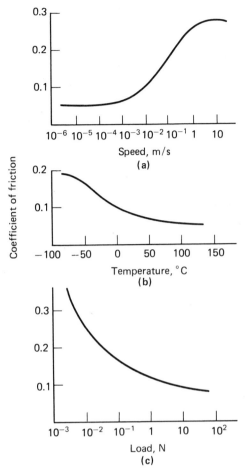

Fig. 3. General trends between the coefficient of friction of PTFE on steel and **(a)**, speed; **(b)**, temperature; and **(c)**, load. To convert N to kgf (kilogram force), divide by 9.807. British Crown Copyright.

certain conditions of sliding. Values of 0.05 or less are obtainable at low speeds, heavy loads, or high temperatures, and when PTFE is in the form of a thin film supported by a harder substrate, that is, the classical, thin film, solid lubrication process (37). The original explanation for the low friction of PTFE was that it arose as a consequence of the chain morphology (38). The charge on the carbon atoms is screened by the larger fluorine atoms (39), resulting in low interchain forces and thus enabling them to slip easily over each other. Evidence was subsequently obtained, however, indicating that the frictional behavior of PTFE could be related to its bulk rather than its molecular structure (40). Within the bulk structure are thin crystalline bands separated by amorphous or disordered regions (41). Easy slip was postulated to occur within these disordered regions, enabling the crystalline bands to be drawn out as long, oriented films or fibrils. This explanation has been criticized on the grounds that marked changes to the bulk structure produced by different heat treatments make little difference to the frictional behavior (41), and in a return to the original molecular ideas, it is now postulated that the critical factor permitting easy interchain slip is the smooth molecular profile of the chains. Four items of evidence support this mechanism: (1) Low friction is obtained only after the early stages of motion have oriented the chains parallel to the direction of sliding. (2) Modifications to the chain profile by replacing some of the fluorines with larger chlorine atoms (poly-chlorotrifluoroethylene) or with CF_3 groups (perfluorinated poly(ethylene-co-propylene)), lead to appreciably higher friction. (3) Neutron irradiation to induce cross-linking also increases the coefficient of friction (42). (4) Frictional behavior somewhat analogous to that of PTFE is also exhibited by high density polyethylene, which again has a relatively smooth molecular profile. The behavior of low density polyethylene, wherein the chains contain numerous side groups, is very different (36) (see CHLOROTRIFLUOROETHYLENE POLYMERS; CROSS-LINKING BY RADIATION; ETHYLENE POLYMERS; TETRAFLUOROETHYLENE POLYMERS).

Wear Testing

Terminology

The concept of broadly classifying wear processes in terms of bulk-deformation and interfacial sliding (Fig. 1) is comparatively recent and it is therefore necessary to examine how it is related to the older and more conventional terminology. *Abrasion,* as already defined, is essentially a deformation wear process. For the more rigid plastics, microcutting is the most important element, whereas for elastomers, tearing, crack-propagation, and fatigue are more relevant. Abrasion is commonly subdivided into two-body abrasion, where one of the two surfaces in contact is hard and rough, and three-body abrasion, where the abrasive material is present in the contact zone as loose particles. When abrasive particles are dispersed in a moving fluid and impinge against a solid surface, the resulting process is known as *erosion.* Abrasive erosion relates to the condition where particle movement is almost parallel to the surface, whereas impact- or impingement-erosion occurs when the particles strike the surface almost normally. Wear as a consequence of fatigue (qv) can arise in both bulk-deformation and interfacial sliding, and is significant for both elastomers and plastics. *Adhesion* (qv), on the other hand, is confined mainly to the regime of interfacial sliding. With plastics, adhesion leads to the transfer of material from one surface to the other, whereas with elastomers it influences the localized stress distributions and may lead directly to tearing or indirectly affect fatigue processes. Finally, *corrosion,* in the broad sense, is relevant to both deformation and interfacial wear when the localized surface temperatures become sufficiently high to cause oxidation and change the mechanical properties of the surface layer. The corrosion of metal counterfaces, against which polymers often slide, can also be important if the severity is such as to change the metal surface composition and topography.

Apparatus

A very wide variety of devices for assessing wear has been and continues to be used. In general, these fall into three levels of complexity. At the lowest level is simple laboratory apparatus designed specifically to study the basic mechanisms by which surface damage is initiated and propagated; typical examples are rigid cones, spheres or needles slowly traversing polymer surfaces. At the other extreme is complex custom-built equipment, which is intended to simulate the conditions of operation existing in a particular application as closely as possible. Information obtained from such equipment, although of considerable value to the user, is seldom of more general relevance to other applications or problems. Between these two extremes is a range of laboratory apparatus designed to examine not only the relative ranking of materials in different regimes of wear, but also the ways in which wear varies with the conditions of sliding, such as load, speed, temperature, environment, etc. A selection of the more widely-used equipment of the latter type is shown schematically in Figure 4. For experimental convenience, polymer samples which are much smaller in area than the coun-

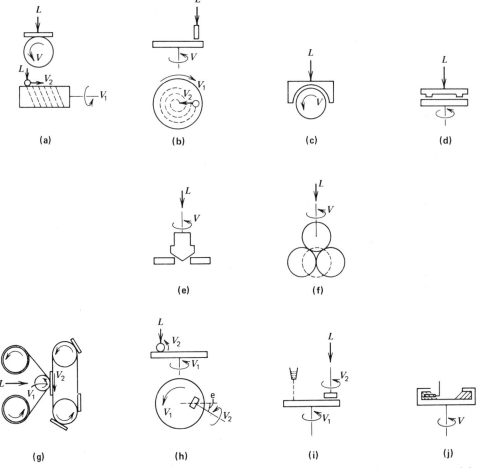

Fig. 4. Schematic arrangements of various types of wear-testing apparatus: **(a)**, pin on ring; **(b)**, pin on disk; **(c)**, block on ring; **(d)**, thrust washers; **(e)**, cone and washer; **(f)**, four ball; **(g)**, tape abrader (Armstrong); **(h)**, rolling with slip (Taber); **(i)** pin and disk—loose abrasive (Olsen); **(j)**, abrasive erosion (NBS). British Crown Copyright.

terface against which they slide are usually employed. In addition, the surfaces of the more rigid polymers may be shaped to enable the volume of wear to be determined from microscope measurements of a wear scar. When this is not feasible, as with elastomers, the usual method of wear measurement is by weight loss.

Apparatus of the pin-on-ring or pin-on-disk type (Fig. 4(a) and (b)) is particularly suitable for wear measurements of plastics sliding against metals. When the metal surface is smooth and the location of the polymer pin remains fixed, wear is of the interfacial type, as in a bearing. Deformation wear can be induced, however, by making the metal surface rough or covering it with abrasive paper or cloth. Usually the polymer then traverses the rotating counterface in a helical or spiral track. A well-known example of this type of apparatus for studying the abrasive wear of elastomers is the DuPont Abrader (43). Figures 4(c) and 4(d) represent the conforming-geometry analogues of (a) and (b) and more closely simulate journal and thrust bearing applications, respectively. The essential difference in the bearing arrangements is that it is much harder for the wear debris to escape from the contact zone; this can modify the structure and thickness of the transfer films generated on the counterface. For assessing three-body abrasive wear, loose particles may be introduced into conforming contacts such as (c) or (d), either in a continuous stream or dispersed in a carrier fluid.

Configurations (e) and (f) have been used extensively for fundamental studies at relatively high stresses; the conical element in (e) and the top ball in (f) is usually steel (44). Device (f) has also been used in reverse with three steel balls and one polymer ball, or with the polymer ball replaced by a cone. The remaining configurations, (g)–(j), are primarily, although not exclusively, intended for abrasion/erosion studies on elastomers. In (g), flat samples are worn against abrasive paper or cloth (45), whereas in (h) one or more rubber wheels driven against a rotating disk simulate certain aspects of tire wear. A similar geometrical arrangement, with idling abrasive wheels and a driven disk, is the well-known Taber Abraser, widely used for routine wear tests on plastics, textiles, and coated fabrics. Three-body abrasive wear is simulated in (i), where the polymer pad can be either stationary or rotating. An element of impact abrasion can also be introduced by arranging for the pad to rise and fall repetitively. Finally, configuration (j) is able to assess erosive wear of a wide range of materials, either as shown or with the specimen itself rotating in a stationary container of abrasive particles. Alternative schemes for assessing erosion (not illustrated) include air-driven abrasive jets impacting polymer flats (46), abrasive powder falling under gravity onto a material inclined at 45° (47), or, for rain erosion, samples rotating at high speed in an artificial rain field (48).

More detailed descriptions of all the above tests, together with their different variants, are given in review articles (49–52). The American Society of Lubrication Engineers has also produced a comprehensive compilation of friction and wear devices, although not specifically concerned with polymer wear (53). Many types of apparatus have been incorporated within standardized test procedures devised by various national and international organizations, and a complete list of all current ASTM standards relevant to polymer friction and wear is given in Table 2.

Table 2. ASTM Standards on Abrasion and Wear Testing of Polymeric Materials, 1981

ASTM number	Description
G 6-77	abrasion resistance of pipeline coatings
F 510-81	resistance to abrasion of resilient floor coverings using the Taber Abraser with the Frick grit-feed method
D 658-44	abrasion resistance of coatings of paint, varnish, lacquer, and related products with the air-blast abrasion tester
D 673-70	mar-resistance of plastics
F 735-81	abrasion resistance of transparent plastics and coatings using the oscillating sand method
D 821-47	evaluating degree of abrasion, erosion, or a combination of both, in road service tests of traffic paints
D 968-51	abrasion resistance of coatings of paint, varnish, lacquers, and related products with the falling sand method
D 1044-78	resistance of transparent plastics to abrasion
D 1175-80	abrasion resistance of textile fabrics (oscillatory cylinder and uniform abrasion method)
D 1242-56	resistance of plastics materials to abrasion
D 1395-58	abrasion resistance of clear floor coatings
D 1630-61	rubber property—abrasion resistance (NBS Abrader)
D 2228-69	rubber property—abrasion resistance (Pico Abrader)
D 2625-69	endurance (wear) life and load-carrying capacity of solid film lubricants (Falex method)
D 2716-71	coefficient of friction and wear characteristics of dry solid film lubricants in vacuum and other controlled atmospheres
D 2981-71	wear of solid film lubricants in oscillatory motion
D 3108-76	standard method for coefficient of friction (yarn to metal)
D 3181-80	standard practice for conducting wear testing on textile garments
D 3389-75	coated fabrics—abrasion resistance (rotary platform, double-head abrader)
D 3412-78	standard method for coefficient of friction (yarn to yarn)
D 3702-78	wear rate of materials in self-lubricated rubbing contact using a thrust-washer testing machine
D 3884-80	abrasion resistance of textile fabrics (rotary platform, double-head method)
D 3885-80	abrasion resistance of textile fabrics (flexing and abrasion method)
D 3886-80	abrasion resistance of textile fabrics (inflated diaphragm method)

Wear Indexes

The basic parameter of interest in all wear investigations is the *wear rate* and a variety of units are used to specify this quantity. When the most serious consequence of wear is the loss in fit of a component, such as a bearing, it is common to use a depth wear rate, $W_h = h/d$, where h is the thickness removed and d is the distance of sliding. The use of distance rather than time is preferable in order to eliminate the influence of speed. A more fundamental parameter, which arises from several simple theories of wear, is a volumetric wear rate expressed as $W_v = V/dL$, where V is the wear volume and L is the load applied. This quantity is referred to as a specific wear rate, K-factor, wear coefficient, or abrasion factor, and has units of m^2/N. The units are sometimes expanded to $mm^3/(N \cdot m)$ ($in.^3/cal$) to confer greater physical insight into what is actually being represented. The terms *wear resistance* and *abrasion resistance* are also widely used and correspond to the reciprocal of the wear rate, in whatever units are being used. In routine materials testing, wear resistance is frequently compared with that of a standard reference material, and the ratio expressed as a relative wear resistance. A further parameter, mostly encountered in the USSR, is an *energy index of wear,* which is defined as the volume of material removed per unit of work expended during sliding, $W_I = V/\mu dL$, where μ is the coefficient of friction. This quantity corresponds physically to the ratio of the specific wear rate to the coefficient of friction and, for elastomers, has also been called *abradability*. The inverse of the energy index or abradability is also used in the USSR and referred to as the coefficient of abrasion resistance.

Deformation Wear

Abrasion

The simplest theory of abrasive (cutting) wear can be derived by considering the penetration of hard irregularities, such as cones, into a softer material. If the base angle θ of the cone is sufficiently large to cause full plasticity, then it may readily be shown from geometrical considerations that the specific wear rate is given by

$$\frac{V}{Ld} = \frac{K \tan \theta}{\pi P_m}$$

where L is the applied load, d is the sliding distance, P_m is the flow pressure (approximately equal to the indentation hardness), and K is the proportion of material displaced which is removed as debris.

The wear rate is thus a product of two parameters: a material property, typified by P_m and a geometrical factor given in this case by $\tan \theta/\pi$. The latter will vary with the particular shape assumed for the penetrating asperities.

Plastics.   The above expression has been repeatedly verified for metals (54) and an example showing wear rate proportional to $\tan \theta$ is given in the top curve, A, of Figure 5 for the idealized experiment of a conical indenter sliding against tin. In the equivalent experiment with a polymer, however, as shown in curve

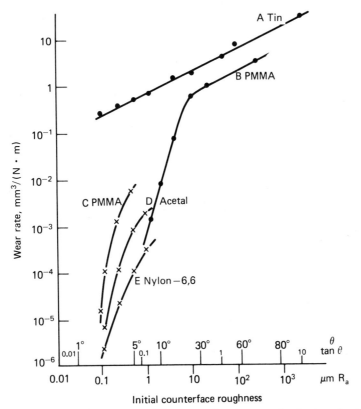

Fig. 5. Variation of wear rate with counterface topography during single traversals; A,B, conical indentors; C,D,E, steel surfaces. To convert mm³/(N·m) to in.³/cal, multiply by 0.00255. British Crown Copyright.

B for poly(methyl methacrylate), proportionality is obtained only when θ exceeds about 20°. This result is broadly consistent with the deformation criterion mentioned earlier and reflects the fact that elastic deformation becomes increasingly important at lower asperity slope angles. For the more realistic arrangement of polymers sliding against metal counterfaces, it can be shown that the center-line-average roughness R_a of abraded metal surfaces is approximately proportional to $\tan^2\theta$ (55). Equivalent roughnesses are given along the abscissa of Figure 5 and demonstrate that abrasive (cutting) wear of plastics is only likely to be obtained against surfaces rougher than about 10 μm R_a . Such roughnesses are rarely encountered on metals prepared by conventional finishing treatments, where the typical range is 0.1 to 2 μm. Wear measurements for a number of plastics sliding in single traversals over metals with roughnesses within this range are given in Figure 5 (curves C–E), and show that the wear rates decrease very rapidly with decreasing roughness, in agreement with the idealized experiment using a conical indenter. One interpretation of this steep decrease is that the wear process is a varying mixture of microcutting (plastic deformation) and fatigue (elastic deformation).

There has been a considerable amount of work in the USSR on the wear of polymers against abrasive papers. One study shows that the wear rate tends to increase with increasing grit size but the relative wear resistance of different polymers remains the same (56). Wear rates are also usually proportional to the applied load. A simple theory (57) of polymer wear against abrasive papers considers that three stages are involved in the production of a wear particle: plastic deformation produces a real contact area the size of which is inversely proportional to hardness H; relative motion is opposed by a frictional force $F = \mu L$; and disruption of the polymer over the real contact area requires an expenditure of work equivalent to the work to fracture. The latter is approximately equal to the product of the breaking stress s and the elongation to break e. Since these three processes occur sequentially, the total wear may be regarded as proportional to the probability of completion of each stage, ie, the wear rate $= \text{const}\,[(\mu L)/(Hse)]$. Figure 6 shows the general way in which the individual parameters in this expression, μ, H, s, and e, and wear rate vary with temperature for an amorphous

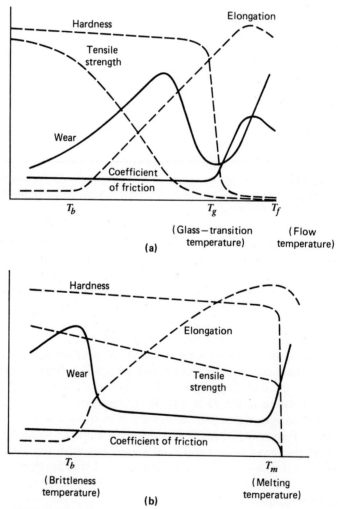

Fig. 6.   Derivation of the wear rate–temperature variation for **(a)**, amorphous polymers; **(b)**, crystalline polymers (20).

and a crystalline polymer. Complete experimental verification of these trends does not appear to have been made for wear against abrasive papers, but certain elements have been confirmed in experiments against rough metals, notably the existence of the minimum in wear at a particular temperature (59).

It has also been confirmed that a general correlation exists between the abrasive wear rates of many polymers and the reciprocal of the product *se,* as determined from conventional tensile tests (60). This is shown in Figure 7 for 18 different plastics.

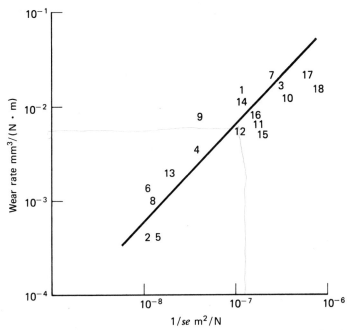

Fig. 7.   Variation of wear rate with the reciprocal of the product breaking strength × elongation to break for various polymers during single traversals over rough steel (1.2 μm R$_a$). 1, Poly(methyl methacrylate) (PMMA); 2, low density polyethylene (LDPE); 3, polystyrene; 4, acetal copolymer; 5, nylon-6,6; 6, polypropylene; 7, epoxy; 8, polytetrafluoroethylene; 9, PMMA—acrylonitrile copolymer; 10, polyester; 11, polychlorotrifluoroethylene; 12, polycarbonate; 13, nylon-11; 14, acrylonitrile–butadiene–styrene (ABS); 15, poly(phenylene oxide); 16, polysulfone; 17, poly(vinyl chloride); 18, poly(vinylidene chloride). To convert mm^3/(N·m) to in.^{3}/cal, multiply by 0.00255. British Crown Copyright.

The product *se* is related to the impact strength or fracture toughness of a material, and correlations have also been sought between these parameters and abrasive wear. One study considers that the notched impact strength is most relevant because the associated strain rates then approach those of the wear process more closely (56). However, experiments on abrasive papers show only a nonlinear relationship between wear and notched impact strength and show considerable scatter. More fundamentally, it should also be possible to relate abrasive wear rates of polymers to their cohesive energies and this has been confirmed (61).

Elastomers.   When elastomers slide over hard, rough surfaces, the probability of abrasion occurring via microcutting is small. An examination of the model situation of a blunt needle sliding over a rubber sheet has shown that the most common phenomenon is the formation of an elastic bulge in front of the needle (62). Tensile stresses are produced at the rear of the contact and failure ultimately occurs by tearing in a direction transverse to that of the maximum stress. The bulge in front of the needle may eventually elongate into a lip, which folds back on itself after tearing has occurred to constitute a raised surface imperfection and, hence, an area of preferential contact during a subsequent traversal of the needle. During repeated sliding against rough surfaces, a sequence of individual processes of this type gradually leads to the formation of abrasion patterns on the elastomer, that is, a pattern of ridges transverse to the direction of motion (63). These ridges somewhat resemble sawteeth pointing against the direction of motion of the abrading surface. A comprehensive review of the factors influencing the formation of abrasion patterns has been given (64).

There have been numerous attempts to relate the wear of elastomers on abrasive papers to various combinations of mechanical properties. For example, the empirical relationship wear rate $= k_1 - k_2 H_s - k_3 \sigma$, where $k_{1,2,3}$ are constants, H is the Shore hardness, and σ is the tensile strength, has been proposed (58). Alternatively, it has been suggested that the wear rate is proportional to $\mu(100 - D)/\sigma$, where μ is the coefficient of friction, D is the percentage rebound resilience (characterizing the hysteresis losses), and σ is again the tensile strength (65). Such relationships tend to have only limited validity for particular test conditions and over a restricted range of materials. One of the main problems in attempting to relate wear of elastomers to conventional mechanical properties is that the relevant properties are not those of the bulk elastomer, but of a thin surface layer which has been modified by the sliding process itself. A strain-induced reduction in stiffness of the surface layer of carbon black-filled rubbers is a common phenomenon (66), and temperature increases may also result in oxidation and/or thermal degradation of the molecular structure (67) (see DEGRADATION).

In view of the considerable success achieved in correlating speed and temperature effects on the friction of elastomers by means of WLF-type transforms, attempts have been made to apply similar treatments to wear. Figure 8 shows some results for a butadiene–acrylonitrile rubber sliding over abrasive paper (68). Hysteresis could influence wear through its effect on both the coefficient of friction and the tensile strength, but whereas the latter increases monotonically with speed, the former exhibits a maximum (Fig. 2) (69). The shape of the wear curve in Figure 8 thus implies that tensile strength variations are more relevant to the wear process than friction variations.

Three-body abrasive wear, by loose particles within a polymer–polymer or polymer–metal contact, has received little attention, but the general features are likely to be similar to those already discussed for two-body abrasion. One additional feature, however, is that for polymers sliding against metals, abrasive particles may damage the metal surface and this could then affect the wear of the polymer. Of relevance here is the extent to which abrasive particles can be temporarily or permanently embedded into polymer surfaces during three-body abrasion. Experiments have shown that the amount of metallic wear resulting

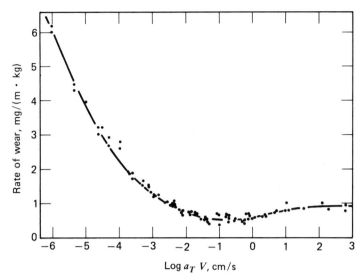

Fig. 8. Composite master curve, reduced to 20°C; of wear rate against speed for a butadiene–acrylonitrile rubber sliding against 180 grade silicon carbide paper. Individual points obtained at 11 temperatures over the range −10° to 96°C (68).

from embedded abrasive in different elastomers increases very rapidly with increasing elastic modulus or Shore hardness (wear proportional to $H_s^{10.2}$) (70). Metallic wear during sliding against elastomers has also been reported to occur as a consequence of corrosion by free radicals liberated during degradation of the elastomer, but the magnitude of this effect is relatively small (71).

Erosion

When impacting abrasive particles have an appreciable velocity component parallel to the surface under attack, the mechanisms of wear and the relationships between wear and material properties are very similar to those already discussed for two-body abrasion. In general, the greater the impact strength, toughness, or product *se*, the greater the erosion resistance. For the more rigid plastics, *se* seldom exceeds about 1 MN/m² (145 psi), whereas values for metals can be much higher, eg, 10 MN/m² for mild steel. The erosion resistance of plastics is therefore likely to be significantly less than that of most metals and this has been confirmed experimentally (72). Erosion resistance increases markedly, however, for low modulus elastomers and such materials are widely used for erosion-resistant coatings in pumps, or pipelines. The selection of the most resistant elastomer for particular applications is usually based either on previous experience or on empirical tests, eg, ASTM G 6-77. There are few general guidelines beyond the facts that natural rubber is often more erosion-resistant than synthetics (73), and that unfilled elastomers are usually superior to filled ones (74).

One especially significant aspect of erosion in the aerospace environment is that of damage by raindrop impact to forward-facing surfaces on aircraft during high-speed flight (75). Plastic-based radomes are particularly susceptible. The usual method for assessing rain erosion resistance is by whirling-arm tests in a simulated rain field, and a general idea of the susceptibility of various plastics materials to damage produced in this way is given in Table 3. A threshold velocity of impact exists below which damage remains negligible, and for polymers this is around 100 m/s (328 ft/s). Above this threshold, there is an induction period during which damage is also negligible, but beyond this, rapid erosion begins at a rate characteristic of the material concerned. The appropriate relationship is $R = K(V - V_T)^x$, where R is the rate of erosion, V is the threshold velocity, K is a material constant, and the exponent x is also material-dependent, but roughly of the order of three. The erosion rates of the various polymeric materials given in the right-hand column of Table 3 cover a wide range. Glass-reinforced thermosetting composites for radomes usually require protection by coatings of polyurethanes or fluoroelastomers at least 0.3 mm thick. Similar protection is also essential for lightweight radome constructions, such as honeycomb and polyurethane or syntactic foams. It may also be noted from Table 3 that the rain-erosion resistance of polymers is generally inferior to that of metals, although that for the recently developed poly(ether ether ketone) (PEEK) is approaching a similar level of resistance.

Table 3. Comparative Rain-erosion Data on Various Materials[a]

Material	Time to "unacceptable" damage, min	Erosion rate, μm/min
honeycomb laminate	1	
paint, acrylic	4	
paint, epoxy	7	
polyester–glass fiber laminate	10	1000
0.3 mm polyurethane coating on above	100–300	
poly(methyl methacrylate)		200
polytetrafluorethylene		40
poly(ethylene terephthalate)		10
polyamide 6		3
polychloroprene		2
poly(ether ether ketone)		0.8
aluminum		1.2
copper		0.25
nickel		0.1

[a] Whirling-arm tests. 225 m/s (738 ft/s), 25 mm/h rain (ca 1 in./h) intensity, ca 2 mm diameter mean drop size.

Fatigue

Recognition of fatigue (qv) as a main mechanism of wear for polymers has arisen mainly from work in the USSR, which was reviewed in English in 1967 (44). The simplest theoretical treatment assumes indentation of a rigid hemispherical asperity of radius r into a smooth polymer plane under an applied load L. If n repeated contacts are required to produce a debris particle of dimensions comparable to the diameter of the contact spot a, then the volume removed per unit sliding distance V/d, $\propto a^2/n$. The value of n can be derived from the Wöhler relationship characterizing the fatigue behavior of the polymer:

$$n \propto \left(\frac{\sigma_0}{\sigma}\right)^t$$

where σ is the applied stress, σ_0 is the failure stress, and t is a material constant. From elasticity (qv) theory:

$$a \propto \left(\frac{rL}{E}\right)^{1/3} \text{ and } \sigma \propto \left(\frac{E}{r}\right)^{2/3} L^{1/3}$$

and the wear rate is thus given by

$$\frac{V}{d} \propto \sigma_0^{-t} E^{(t-1)} r^{-(t-1)} L^{(t+2)}$$

More advanced theories based on different and more realistic distributions of asperities have been devised and lead to slightly different power relationships (76,77). However, the essential elements remain the same in that the wear rate is the product of three groups of parameters: mechanical properties (σ_0, E, and t), topography (r), and applied stress (L).

The magnitude of the exponent t derived from conventional fatigue data ranges from about 1.5 to 3.5 for elastomers, and from about 3 to 10 for the more rigid plastics and their composites. Confirmation of the relevance of the analysis has been sought from measurements of the wear of several polymers during single traversals over steel surfaces of varying roughness, and the values of t derived are generally of the same order as those obtained from conventional fatigue tests (78). The agreement thus provides strong circumstantial evidence supporting the idea of a fatigue wear process for polymers sliding against surfaces insufficiently rough to cause microcutting.

One arrangement widely used in USSR work to examine fatigue wear of polymers is to slide against a fine wire gauze in order to obtain repeated stress cycles with defined macroasperities and avoid complications from transfer. Using this method, it has been shown that, in general, wear rates are proportional to L^α, where in terms of the simple theory above $\alpha = \frac{1}{3}(t + 2)$ (57). Values of t obtained from these experiments range from 2 to 7 for various thermoplastics. For elastomers, t increases with increasing molecular polarity, filler content, and cohesive energy density, but decreases when the rubber swells following immersion in a fluid (79).

Another view of fatigue wear considers that the wear rate is proportional to the rate at which bonds in the polymer chain are broken, and that the activation energy for bond rupture is reduced by the applied stress (80). This concept suggests that it might be possible to reduce polymer wear by introducing stabilizing additives of the type commonly used to inhibit oxidation or thermal degradation. In confirmation, experiments show that the addition of N,N'-bis (β-naphthyl)-p-phenylenediamine to nylon-6,8 reduces the rate of wear (against wire gauze) by a factor of about fifty. One complication is that the amount by which wear decreases is load-dependent. The introduction of additives tends to increase the exponent α, and since the wear rate is proportional to L^α, their wear-reducing effect thus decreases with increasing load. Similar effects of oxidation inhibitors on the fatigue wear of elastomers sliding over rough metals have been reported (81). Additions of stabilizers to polymers are largely ineffective in reducing abrasive (cutting) wear, however, because they have no opportunity to function when failure occurs in a single contact event (see STABILIZATION).

Interfacial Wear

The main characteristics of interfacial wear are that energy dissipation occurs within a narrow region adjacent to the sliding interface, and adhesion plays a principal role. This regime is mainly associated with the more rigid plastics sliding against smooth metals, and breaking of the interfacial bonds often leads to transfer to the metal counterface, localized high temperatures, and structural or chemical degradation of the plastic. Before discussing these processes, it is necessary to examine briefly an intermediate situation between deformation and interfacial wear which arises when elastomers slide against smooth metals. Whereas the adhesive forces involved here can be large, deformation still provides the main contribution to wear. The high tractive forces involved in overcoming adhesion lead to stretching of the elastomer and ultimately to the formation of rolls. Roll formation results in characteristic surface markings on the elastomer, somewhat similar to the abrasion patterns produced during sliding against rough surfaces, but they are on a smaller scale and disappear when the interface is lubricated and adhesion is reduced. Roll formation has also been observed with some plastics, although on a very small scale of size (82). Its precise relevance to the overall wear of plastics is still uncertain.

Surface Modifications

When polymers slide repeatedly over smooth metal surfaces, their wear rates seldom, if ever, correlate directly with their bulk mechanical properties because surfaces of both polymer and counterface are modified by the sliding process to produce surface layers or third-bodies. The properties of the third-bodies can be very different from those in bulk.

Insofar as changes to polymer surfaces are concerned, the molecules in the surface layer frequently undergo chain scission (83) with a consequent reduction in molecular weight (84,85). These lower molecular weight constituents may then lead to a surface layer of lower elastic modulus, similar to those produced on elastomers during abrasive wear (62). At high sliding speeds, when interfacial temperatures increase, thermal or oxidative degradation occurs; for thermosetting polymers, this may reduce the ductility of the surface layer. Elevated temperatures also lead to structural modifications, and the disappearance of the spherulites in the surface layers of a number of crystalline thermoplastics has been noted after wear (33). With PTFE and high density polyethylene, sliding is also known to induce a preferred orientation of the molecular chains (36,40). Another factor modifying polymer surfaces during sliding is the incorporation of material transferred from the counterface (86). Transfer of metal fragments occurs when the subsurface strength is lower than that of the interfacial adhesive bonds and this can arise when conventional finishing treatments leave localized areas of the surface in a weakened state (70).

The two most important modifications to metal counterfaces occurring during sliding against polymers are transfer and abrasion. The successive deposition of transferred fragments gradually builds up a more-or-less uniform film on the counterface which can vary in thickness from a few nm to a few μm, depending upon the type of polymer, the initial counterface roughness, and the conditions of sliding. However, not all polymers transfer; brittle polymers below their glass-transition temperatures and highly cross-linked structures usually tend to shear at the original sliding interface. Most thermoplastics do exhibit transfer above their T_gs, although it should be noted that the relevant T_g may differ from that of the bulk material because of the high stresses and shear rates involved in the sliding process (87). The fragments transferred from crystalline thermoplastics are usually commensurate in size with the individual areas of real contact, but there are two notable exceptions. PTFE and high density polyethylene, which have already been noted as having low friction because of their smooth molecular profiles, transfer as extremely thin highly oriented films, <10 nm thick. This anomalous behavior, however, is restricted to a comparatively narrow range of sliding conditions and particularly to low speeds of sliding. At high speeds, molecular relaxation to permit chain orientation is inhibited and the transfer behavior becomes more normal, ie, lumpy particles rather than thin films.

The adhesion of transferred polymer fragments to a metal counterface is usually relatively weak (ie, van der Waals forces), but can increase appreciably when the sliding conditions initiate polymer degradation. The latter aspect has been extensively studied and becomes very significant for filled and reinforced polymers (88). There may also be an element of mechanical keying of the transferred fragments into the surface depressions on the counterface. Reductions in wear associated with transfer film formation result from two main causes: (1) The sliding interface ultimately becomes that of polymer against polymer, which reduces the work of adhesion leading to lower adhesive forces. This is particularly important for PTFE and high density polyethylene when there is preferential chain orientation. (2) Filling of the counterface depressions by transfer reduces the surface roughness and, in turn, the magnitude of the local contact stresses. Steady-state wear rates thus become much less sensitive to the initial counterface roughness.

The second principal process of counterface modification when sliding against polymers is abrasion. In view of the marked influence of counterface topography on wear (Fig. 5), it is clear that any abrasion that improves the initial surface finish will considerably reduce the wear rate. Pure polymers do not normally abrade metal surfaces, unless metal pickup occurs, as already discussed. By far the greatest contribution to counterface abrasion arises from the fillers or reinforcing fibers in polymer composites, which will be discussed later.

Load, Speed, and Temperature

In contrast to the rather complex ways in which deformation wear depends on load, speed, and temperature via viscoelastic effects, the corresponding relationships for interfacial wear tend to be simpler. The wear rates of thermoplastics sliding against smooth metals are generally proportional to the applied load, until the temperature rise due to frictional heating becomes sufficient to cause thermal softening and, ultimately, extrusion and flow (89). As might be expected, the critical load at which rapid wear begins owing to thermal softening increases when the polymer is cross-linked by γ-radiation (90,91). The influence of speed on the wear rate (per unit distance of sliding) is also relatively slight until thermal softening occurs. There is good agreement between the calculated asperity contact temperatures at the onset of rapid wear and the crystalline melting point of the polymer involved (92). An example for acetal on steel is shown in Figure 9.

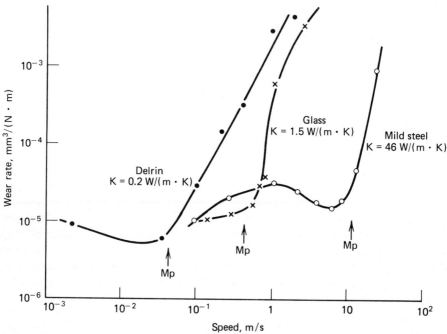

Fig. 9.   Variation of wear rate with speed for acetal resin, (Delrin), sliding against counterfaces of differing thermal conductivity. The arrows correspond to the calculated speed for which the total asperity contact temperatures reach the melting point (176°C). British Crown Copyright.

Furthermore, the variation in critical speed with counterface thermal conductivity, as predicted from contact temperature theory, also agrees with experimental observation (93). The very marked reduction in critical speed with decreasing counterface conductivity is a consequence of the fact that most of the frictionally generated heat is dissipated by conduction through the counterface. These results explain why polymer–polymer combinations are often found to exhibit wear rates very much greater than those of polymers against metals in standardized test conditions (94). Figure 9 demonstrates that at very low speeds, where no surface melting occurs, the differences are no longer very large.

Composites

The primary reason for adding fillers or reinforcing fibers to polymers is to improve their mechanical properties, but the effects on wear are not invariably beneficial. The most important instance of this occurs during deformation (abrasive) wear against rough surfaces where the relevant mechanical property controlling the wear rate is the product *se*, as discussed earlier. Whereas reinforcements increase the breaking strength *s*, they usually greatly decrease the elongation to break *e*, and hence the product *se* may become smaller than that of the unreinforced polymer. In consequence, reinforcement frequently increases abrasive deformation wear rates, as shown by the results in Table 4. Similar considerations apply to the erosion of composites, either by solid particles or by liquid impact. In complete contrast, the effects of fillers and reinforcements on interfacial wear processes are much more favorable. Table 4 shows that the addition of graphite fibers greatly reduces wear during sliding against smooth steel. The magnitude of these reductions is far greater than can be accounted for solely by the changes in mechanical properties; other mechanisms must therefore contribute. These are summarized in Figure 10.

Mechanical/Thermal Properties

One reason for the large reductions in interfacial wear achieved by fillers is that they support the load preferentially by virtue of their greater strength and stiffness (95). The extent of preferential load support is influenced by particle size (smaller particles tend to be buried within the polymer more easily than larger ones), shape (large aspect ratios are most favorable (96)), and orientation (for fibrous reinforcements). The introduction of fillers into PTFE is particularly successful in reducing wear and here an additional process comes into play. The interfacial wear mechanism for PTFE involves the drawing out of long fibrils or ribbons from within the crystalline bands, and filler particles in the surface greatly impede this extrusion process (97) (see THERMAL PROPERTIES).

Transfer

It has long been established that adding copper and lead, or their oxides, to PTFE promotes transfer-film formation on a metal counterface and greatly reduces wear (98). Adhesion of the transfer film is believed to be increased by degradation of the PTFE during sliding and the formation of fluoride ions (99). It has further been suggested that copper catalyzes a reaction between the polymer degradation products and lead (100). Additions of copper to other polymers can also reduce wear if the metal is present in a sufficiently finely divided form. This can be achieved, for example, by incorporating copper compounds, such as formate or oxalate, which decompose at relatively low temperatures during sliding, leaving the metal in a highly reactive, colloidal state (101). Such additions are even more effective for wear reduction when sliding in the presence of certain fluids, notably glycerol, where a process of selective transfer occurs (102); both the polymer and its counterface become plated by a very thin film of copper.

Table 4. Influence on Wear of Reinforcement by 30% Wt Graphite Fibers

	Deformation wear[a]		Interfacial wear[b]	
	Wear rate (nonreinforced), 10^{-5} mm³/N·m	Wear rate ratio, reinforced/nonreinforced	Wear rate (nonreinforced), 10^{-5} mm³/N·m	Wear rate ratio, reinforced/nonreinforced
polypropylene	810	1.9	25	0.008
acetal copolymer	615	1.3	0.3	0.23
polycarbonate	490	1.8	4	0.03
polytetrafluoroethylene	405	3	45	0.004
polyethylene, low density	82	1.3	0.4	0.23
nylon-6,6	32	2.7	0.5	0.5

[a] Single traversals over rough (1.2 μm R_a) steel.
[b] Steady-state repetitive sliding on smooth (0.1 μm R_a) steel.

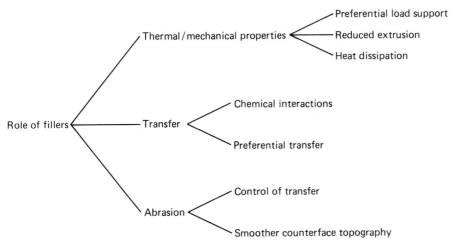

Fig. 10. Various ways in which fillers influence the wear processes of polymers. British Crown Copyright.

Apart from the smooth molecular profile polymers, PTFE and high density or ultrahigh molecular weight polyethylene, most other materials benefit from the addition of lubricating fillers, which reduce the coefficient of friction. Such fillers, PTFE, MoS_2, and graphite, tend to transfer preferentially to the counterface and thereby reduce wear as well as friction. PTFE is a common additive to a wide variety of thermoplastics and its tendency to reduce strength and stiffness can be offset by fiber-reinforcement (103). Graphite and MoS_2 are generally less effective than PTFE, at least in concentrations less than about 15% by volume, but graphite has the additional merit of providing some degree of reinforcement. Orientation of the graphite particles also influences friction and wear; lowest friction but highest wear are obtained when the basal planes are oriented parallel to the sliding interface (104).

Abrasion by Fillers

Most of the fillers or fibers commonly added to polymers to improve mechanical properties are abrasive in some degree to metal counterfaces. Table 5 shows some comparative data on filler abrasiveness obtained from an arbitrary test in which the wear of a bronze ball was measured during sliding against various filled polymers (105). The difference in abrasiveness between the two types of carbon fibers, high modulus (graphite) and high strength (nongraphite) is particularly interesting and is reflected in the corresponding wear behavior of PTFE reinforced with these fibers (see CARBON FIBERS). Figure 11 shows that the wear of graphite fiber composites is much more dependent on the initial counterface roughness than that of composites containing the more abrasive nongraphitic fibers. The latter are able to polish the counterface to a constant topography during the early stages of sliding. For composites containing largely nonabrasive fillers, PTFE or organic fibers, the deliberate introduction of small amounts of conventional abrasives can often be beneficial (106). The optimum concentration needed for minimum wear, however, is not easy to predict because the amount of abrasion depends on a whole range of factors: particle size and shape, hardness relative to the metal counterface, and the type of polymer into which they are incorporated. The importance of the latter is evident from the data in Table 5, which show that the amount of abrasion caused by 25% volume of glass fiber depends greatly on the polymer to which it is added. Two aspects are important here: the elastic modulus of the polymer, which influences the level of exposure of the abrasive filler, and the extent to which polymer deformation and flow can encapsulate the filler particles.

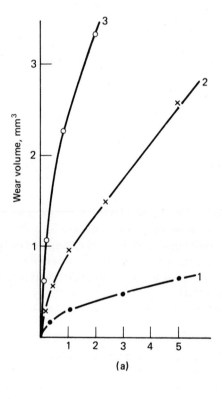

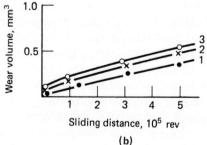

Fig. 11. Influence of initial counterface roughness on the wear of carbon fiber-reinforced PTFE. **(a),** High modulus fibers; **(b),** high strength fibers. Counterface—low carbon steel of roughness: 1, 0.04 μm R_a; 2, 0.18 μm R_a; 3, 0.5 μm R_a. British Crown Copyright.

Table 5. Comparative Abrasiveness of Polymers with Various Fillers and Fibers

Material	Abrasion rate,[a] 10^{-6} mm³/Nm
Polymers alone	
nylon-6,6–acetal copolymer	<0.01
polyethylene, polytetrafluoroethylene	<0.01
polyester	<0.01
PTFE-composites	
asbestos fiber, 25% wt	10
high-strength carbon fiber, 25% wt	8
flake mica, 30% wt	3.1
coke, 25% wt	1.7
high modulus carbon fiber, 25% wt	0.2
bronze, 40% wt	0.08
graphite, 33% wt	0.05
Glass-reinforced polymers, 30% wt	
polyester	146
polytetrafluoroethylene	62
acetal copolymer	15
nylon-6,6	7.3

[a] Abrasion of a bronze ball, H = 185 VPN (Vicker's Pyramid Number), sliding in epicyclic motion at a load of 10 N over the polymer composite surface.

Lubrication

The ideal function of a fluid lubricant is to provide complete separation between two moving surfaces so that resistance to motion only arises from viscous shear within the fluid itself. This is hydrodynamic or elastohydrodynamic lubrication. The low elastic moduli of polymers, compared with metals, facilitate the development of this lubrication regime because elastic deflections under load can become sufficient to generate the appropriate geometrical surface contours on either a macro- or micro-scale (107). A further advantage of polymers, again because of their low elastic moduli, is that the contact stresses are unable to increase enough to cause significant increases in fluid viscosity and therefore the viscous resistance to motion always remains low (108). Fluid-film lubrication plays an important role in some applications, eg, elastomeric, dynamic seals, but in others it is an unwelcome feature; with footware and tires, the aim is to maintain adhesion and high friction in the presence of fluids rather than to reduce it. A good review of the main features associated with the formation and collapse of elastohydrodynamic films during the sliding or rolling of elastomers has been published (24).

For the more rigid plastics and composites, the regime of elastohydrodynamic lubrication becomes less important than for elastomers, but still more so than for metals. With heavy loads and at low-to-moderate speeds of sliding, boundary or mixed lubrication processes predominate; the load is supported mainly by solid–solid contact through adsorbed monolayers of lubricant molecules. When plastics slide against themselves, reductions in friction by these adsorbed boundary lubricants, eg, fatty acids, are relatively small. This arises because of the inability to form close-packed monolayers on polymers (109) and because the structure and shear strength of fatty acids are not very different from those of polymers themselves. However, when polymers slide against metals, reductions in friction by boundary lubricants can be more significant because of the interaction between the lubricant and the metal (110).

A particularly interesting feature of some polymer–fluid combinations is their ability to maintain low friction against metals for very long periods, even when the fluid is present only in trace amounts (111,112). Some results illustrating behavior of this type are given in Figure 12(a), which shows the duration of the low friction period produced by a single 0.5 μL drop of mineral oil for a number of polymers sliding against stainless steel. There are marked differences between the various polymers and Figure 12(b) shows that, for acetal, there are also marked differences between various types of lubricants. The reasons for such behavior are not yet fully understood. Wetting (113) and plasticization (114) are both likely to play a role, and it has also been suggested that effective lubrication occurs when the interaction energy between the fluid molecules and the metal

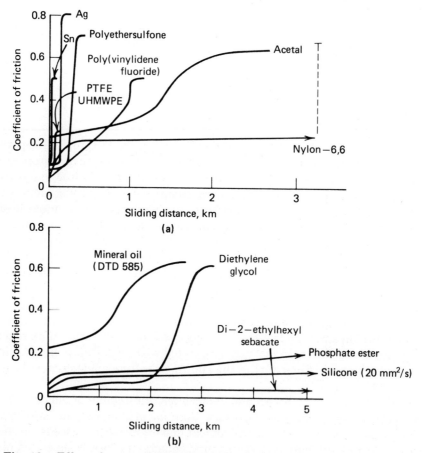

Fig. 12.   Effect of traces of fluid on the friction of polymers. (a) Various polymers with mineral oil, UHMWPE is ultra high molecular weight polyethylene; (b), acetal resin with various fluids. Counterface—smooth stainless steel (0.15 μm R_a) with 0.5 μl of fluid added following initial dry sliding. British Crown Copyright.

is greater than that between the polymer and the metal (115). The phenomenon has been exploited practically by deliberately incorporating small pockets of fluid within the bulk structure of polymers during fabrication (116,117). The fluid is then gradually released at the sliding interface during wear.

The effects of fluids on the wear of polymers in the boundary lubrication regime, ie, in conditions of essentially solid–solid contact, are generally much greater than those on friction. Three general types of behavior can be identified, as illustrated in Figure 13, which shows the wear rates of three materials plotted

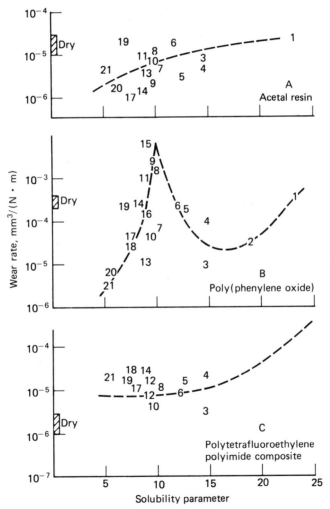

Fig. 13. Effects of fluids on the wear of polymers sliding against smooth stainless steel (0.15 μm R_a). 1, Water; 2, formamide; 3, ethylene glycol; 4, methanol; 5, N,N-dimethylformamide; 6, n-propanol; 7, n-octanol; 8, acetone; 9, benzene; 10, di-2-ethylhexyl sebacate; 11, carbon tetrachloride; 12, toluene; 13, tricresyl phosphate; 14, cyclohexane; 15, methylisopropyl ketone; 16, decalin; 17, n-hexadecane; 18, n-heptane; 19, n-hexane; 20, perfluorodimethylcyclohexane; 21, polydimethylsiloxane (1.5 mm^2/s). To convert mm^3/ (N·m) to in.^{3}/cal, multiply by 0.00255. British Crown Copyright.

against the solubility parameters of the fluids in which they were worn. For polymers that do not transfer to a metal counterface, fluids tend to reduce the rate of wear, as shown in Figure 13(**a**). Complications can arise with those polymers that are subject to enhanced stress crazing or cracking in particular fluids, eg, poly(phenylene oxide) in benzene, and Figure 13(**b**) shows that in these circumstances the wear rate may increase well above its unlubricated value. Finally, Figure 13(**c**) shows that for polymer composite systems that rely on transfer film formation to achieve low wear, the presence of fluids is invariably deleterious and leads to increased wear. The main reason for this is that adsorption of the fluid onto debris particles prevents them from adhering to the metal counterface or to themselves, and thus inhibits the formation of third-body films (118). Water is particularly active in this respect and thus problems often arise in applications where polymer–metal combinations are exposed to the outside environment. In these circumstances, low wear in both wet and dry conditions can only be ensured by incorporating mildly abrasive fillers into the polymer composite which polish the metal counterface to a very smooth topography (106).

Applications

Because of the very broad range of tribological applications of polymers, it is only possibly to summarize a selected few and to draw attention to the relevant literature in which more details can be found. Uses of polymers in bearings, both dry and lubricated, are widespread and the materials available have been reviewed (112,119). A particularly successful composite construction is that of a porous bronze layer on a steel backing infiltrated with a mixture of PTFE and lead for dry operation, or acetal for lubricated conditions (112). Other types of composites intended for high stress and low speed conditions, such as those found in airframe control bearings, comprise fabric-reinforced thermosetting resin liners with PTFE incorporated either as a dispersion throughout the resin, or as a fiber interwoven into the reinforcing fabric (120). The conventional way of rating the relative performance of these various dry-bearing materials is in terms of their pressure–velocity (Pv) properties, either as a limiting Pv factor, above which wear or friction increases rapidly, or as a Pv diagram corresponding to a particular rate of wear. Examples of the latter are shown in Figure 14 for a number of materials sliding against smooth steel and a constant rate of wear of 25 μm/100 h. One problem with the method of characterization is that it ignores all effects of temperature, and to obviate this a modified approach to materials selection has been devised (121). In essence, this involves measuring the rate of wear of a material sliding against smooth steel at light loads and low speeds, where temperature effects remain negligible, and then applying correction factors to this basic value to take into account the temperature, pressure, speed, type of motion, and counterface roughness relevant to the application. A general idea of the range of basic wear rates characteristic of different groups of polymeric dry-bearing materials is given in Table 6.

As an alternative to fabricating bearings from bulk materials, extensive use is made of thin surface films bonded to metallic substrates. PTFE films can be deposited by powder-spraying, electrophoresis (qv) or radio-frequency sput-

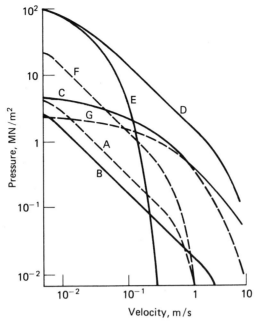

Fig. 14. Pressure–velocity limits for various polymeric dry-bearing materials corresponding to a wear rate of 25 µm/100 h. A, unfilled thermoplastics; B, PTFE; C, filled PTFE; D, porous bronze–PTFE–lead; E, woven PTFE–glass fiber–thermosetting resin; F, reinforced thermoset–MoS$_2$; G, thermosetting resin–carbon/graphite–PTFE. To convert MN/m^2 to psi, multiply by 145. British Crown Copyright.

tering, and self-adhesive PTFE tape is also available for light-duty applications (see COATING METHODS; COATINGS, ELECTRODEPOSITION). To avoid the necessity for high temperature sintering, PTFE is sometimes used in conjunction with thermosetting resins which cure at low temperatures. Other types of solid lubricant coatings containing graphite or MoS$_2$ also incorporate thermosetting resins, such as phenolics, epoxies, acrylics, or polyimides, in proportions usually

Table 6. Comparative Wear Rates of Various Groups of Polymers and Composites[a]

Material	Wear rate, mm^3/Nm
unfilled PTFE	ca 10^{-4}
unfilled thermoplastics	3×10^{-7}–10^{-5}
filled–reinforced thermoplastics	10^{-7}–10^{-5}
reinforced thermosets and solid lubricants	3×10^{-7}–3×10^{-6}
filled PTFE formulations	10^{-7}–10^{-6}
polyethylene, ultrahigh mol wt	10^{-7}–5×10^{-7}
woven PTFE fiber–resin thin layers	3×10^{-8}–5×10^{-7}
porous bronze–PTFE lead	3×10^{-8}–2×10^{-7}

[a] Sliding against smooth steel; frictional heating negligible.

ranging from 25% to 50% volume (122). The performance of MoS_2-resin bonded coatings depends critically on the particular type of resin chosen (123), but so far, it has proved impossible to define the optimum resin properties needed; development continues to be largely empirical. Another method of thin film solid lubrication, which is becoming increasingly important, is transfer lubrication, where a solid film is deposited continuously on one or both of the contacting surfaces by transfer from a third component. The main application of this technique is in rolling-element bearings, where the retainers are made from a lubricating composite such as PTFE–MoS_2–glass fiber (124). Design information is now available to relate bearing life to load, speed, temperature, and bearing size (125). Polymeric retainers in rolling bearings are also being widely used in conjunction with conventional fluid lubrication; materials of interest include fabric-reinforced thermosetting resins, acetal, and porous polymers such as polyamides (qv), polyimides (qv), and polyamide–imides.

Applications of polymers for gears are less widespread than for bearings, although fabric-reinforced thermosetting resins have been used for many years in light-duty conditions to reduce noise and vibration. More recently, thermoplastics, such as polyamides and acetal resins, have emerged for unlubricated applications and a wide range of suitable filled and unfilled grades is commercially available (126). The chief factor affecting gear life in unlubricated conditions is usually wear, but in the presence of lubricants, which reduce wear, tooth fracture via fatigue becomes more significant (127). Design information on gear-tooth design, strength, and fatigue properties has been published (128), but data on wear are usually derived from tests in other conditions; there have been few direct measurements of wear on polymer gears.

Information on abrasion and wear of polymers in many other applications is widely scattered throughout the technical literature and tends to be highly specialized. With the notable exception of tire wear (62), there have been few reviews. References for discussions of abrasion and wear for other applications follow: brake materials (129), piston rings (130), elastomeric seals (131), dental restoratives (132), flooring materials (qv) (52), textile fibers (qv) and fabrics (133), paints (134), wiper blades (24), face seals (134), machine tool slideways (135), water-lubricated bearings (136), marine applications (qv) (137), prosthetic devices (138), conveyor belting (qv), and rolling bearings (139) (see also COATINGS; DENTAL APPLICATIONS; FABRICS, COATED; SEALANTS; TEXTILE RESINS).

BIBLIOGRAPHY

"Abrasion Resistance" in *EPST*, 1st ed., Vol. 1, pp. 7–41, by E. Everett Auer, Mobil Chemical Company.

1. *Friction, Wear and Lubrication—Tribology. Glossary of Terms and Definitions,* Organisation for Economic Cooperation & Development, (OECD), Paris, 1968.
2. *Lubrication (Tribology) Education and Research,* (Jost Report), Her Majesty's Stationery Office, London, 1966.
3. *Strategy for Energy Conservation through Tribology,* ASME, New York, Nov. 1977.
4. M. J. Devine, *Proceedings of a Workshop on Wear Control to Achieve Product Durability,* Office of Technology Assessment, Washington, D.C., Feb., 1976.
5. J. S. Halliday, *Proc. Inst. Mech. Eng. London* **169,** 777 (1955).
6. J. A. Greenwood and J. B. P. Williamson, *Proc. R. Soc. London Ser. A* **295,** 300 (1966).

7. S. Timoshenko and N. N. Goodier, *Theory of Elasticity*, 2nd ed. McGraw-Hill, New York, 1951.
8. J. B. P. Williamson in P. M. Ku, ed., *Interdisciplinary Approach to Friction and Wear*, NASA SP-181, Washington, D.C., 1968, p. 85.
9. B. J. Briscoe in K. L. Mittal, ed., *Physicochemical Aspects of Polymer Surfaces*, Plenum Publishing, New York, 1982, p. 387.
10. M. Godet and D. Play, *Polymères et Lubrification, Coll. Int. du CNRS, No. 23*, CNRS, Paris, 1974, p. 361.
11. A. Shallamach, *Wear* **17,** 301 (1971).
12. K. Kendall, *Wear* **33,** 351 (1975).
13. D. Tabor, *Proc. R. Soc. London, Ser. A* **229,** 198 (1955).
14. J. A. Greenwood, H. Minshall, and D. Tabor, *Proc. R. Soc. London Ser. A* **259,** 480 (1961).
15. D. G. Flom, *J. Appl. Phys.* **31,** 306 (1960).
16. K. A. Grosch, *Proc. R. Soc. London Ser. A* **274,** 21 (1963).
17. M. L. Williams, R. F. Landel, and J. D. Ferry, *J. Am. Chem. Soc* **77,** 3701 (1955).
18. D. F. Moore and W. A. Geyer, *Wear* **22,** 113 (1972).
19. A. Shallamach, *Rubber Chem. Techol.* **41,** 209 (1968).
20. G. M. Bartenev and V. V. Lavrentev, *Friction and Wear of Polymers*, Elsevier North-Holland, Amsterdam, 1981.
21. D. F. Moore, *Friction and Lubrication of Elastomers*, Pergamon Press, Oxford, UK, 1972.
22. G. M. Bartenev and A. I. Elkin, *Wear* **8,** 8 (1965).
23. A. Shallamach, *Wear* **6,** 375 (1963).
24. A. D. Roberts, *Prog. Rubber Technol.* **41,** 121 (1978).
25. D. G. Flom, *Anal. Chem* **32,** 1550 (1960).
26. K. C. Ludema and D. Tabor, *Wear* **9,** 329 (1966).
27. B. J. Briscoe in D. Dowson and co-eds., *Friction and Traction*, IPC Science & Technology Press, Guildford, UK, 1981, p. 81.
28. R. F. King and D. Tabor, *Proc. Phys. Soc. London Sect. B* **66,** 728 (1953).
29. B. J. Briscoe, B. Scruton, and R. F. Willis, *Proc. R. Soc. London Ser. A* **333,** 99 (1973).
30. M. W. Pascoe and D. Tabor, *Proc. R. Soc. London Ser. A* **235,** 210 (1956).
31. D. Tabor and D. E. Wynne Williams, *Wear* **4,** 391 (1961).
32. A. J. G. Allen, *Lub Eng* **14,** 211 (1958).
33. K. Tanaka and Y. Uchiyama in L.-H. Lee, ed., *Advances in Polymer Friction and Wear*, Vol. 5B of *Polymer Science and Technology Symposium Series*, Plenum Publishing, New York, 1974, p. 499.
34. M. Watanabe, M. Karasawa, and K. Matsubara, *Wear* **12,** 185 (1968).
35. K. G. McLaren and D. Tabor, *Proceedings of the 1st Lubrication and Wear Convention*, Institution of Mechanical Engineers, 1963, p. 210.
36. C. M. Pooley and D. Tabor, *Proc. R. Soc. London Ser. A* **329,** 251 (1972).
37. F. P. Bowden and D. Tabor, *The Friction and Lubrication of Solids*, Clarendon Press, Oxford, UK, 1954.
38. K. V. Shooter and P. H. Thomas, *Research (London)* **2,** 533 (1949).
39. W. E. Hanford and R. M. Joyce, *J. Am. Chem. Soc.* **68,** 2082 (1946).
40. K. R. Makinson and D. Tabor, *Proc. R. Soc. London Ser. A* **281,** 49 (1964).
41. C. J. Speerschneider and C. H. Li, *J. Appl. Phys.* **33,** 1871 (1962).
42. K. G. McLaren and D. Tabor, *Wear* **8,** 3 (1965).
43. *ASTM D 394, ASTM Standards*, Part 28, American Society for Testing and Materials, Philadelphia, Pa., 1966, p. 187. Now discontinued.
44. D. I. James, ed., *Abrasion of Rubber*, MacLaren & Sons, London, 1967.
45. F. M. Gavan, *ASTM Bull.* **238,** 44 (1959).
46. A. G. Roberts, W. A. Crouse, and R. S. Pizer, *ASTM Bull.* **208,** 36 (1955).
47. L. Boor, *ASTM Bull* **244,** 43 (1960).
48. A. A. Fyall, R. B. King, and R. N. C. Strain, *J. R. Aeronaut. Soc.* **66,** 447 (1962).
49. J. M. Buist, *Trans. Inst. Rubber Ind.* **26,** 192 (1950).
50. A. E. Lever and J. Rhys, *The Properties and Testing of Plastics Materials*, Chemical Publishing, New York, (1958).
51. F. C. Harper, *Wear* **4,** 461 (1961).
52. F. M. Gavan in J. V. Schmitz and W. E. Brown, eds., *Testing of Polymers*, Vol. 3, Interscience Publishers, a division of John Wiley & Sons, Inc., New York, 1966, p. 138.

53. R. Benzing and co-workers, *Friction and Wear Devices,* 2nd ed., American Society of Lubrication Engineers, Park Ridge, Ill., 1976.
54. M. A. Moore in D. Scott, ed., *Treatise on Materials Science and Technology,* Vol. 13, Academic Press, London, 1979, p. 217.
55. J. K. Lancaster, *Proceedings of 2nd Lubrication & Wear Group Convention.* Institution of Mechanical Engineers, London, 1964, p. 190.
56. S. B. Ratner in ref. 44, p. 23.
57. S. B. Ratner, I. I. Farverova, O. V. Radyukevich, and E. G. Lure in ref. 44, p. 145.
58. J. M. Buist and O. L. Davies, *Trans. Inst. Rubber Ind.* **22**(2), 68 (1946).
59. J. K. Lancaster, *Proc. Inst. Mech. Eng. London* **183**(3P), 100 (1969).
60. J. K. Lancaster, *Plast. & Polym.* **41**, 297 (1973).
61. J. P. Giltrow, *Wear* **15**, 71 (1970).
62. A. Shallamach, *J. Polym. Sci.* **9**, 385 (1952).
63. A. Shallamach in L. Bateman, ed., *The Chemistry and Physics of Rubber-like Substances,* MacLaren and Sons, London, 1963, Chapt. 13.
64. E. Southern and A. G. Thomas, *Proceedings of the 3rd Leeds-Lyon Symposium on Tribology,* Mechanical Engineering Publications, London, 1978, p. 157.
65. S. B. Ratner and G. S. Klitenik, *Zavod. Lab.* **25**, 1375 (1959).
66. A. Shallamach, *Proc. Phys. Soc. London Sect. B* **66**, 817 (1953).
67. A. Shallamach, *Rubber Chem. Technol.* **30**, 1097 (1957).
68. K. A. Grosch and L. Mullins, *Rev. Gen. Caoutch. Plast.* **39**, 1781 (1962).
69. A. Shallamach, *Wear* **1**, 384 (1958).
70. R. B. King and J. K. Lancaster, *Wear* **61**, 341 (1980).
71. A. N. Gent and C. T. R. Pulford, *J. Mater. Sci.* **14**, 1301 (1979).
72. G. R. Davies, *I. Mech. E. Paper AD P9/66,* Institution of Mechanical Engineers, London, 1966.
73. M. Clement, *Wear* **4**, 450 (1961).
74. A. I. Marei and P. V. Izvozchikov in ref. 44, p. 274.
75. A. A. Fyall in J. D. Walton, ed., *Radome Engineering Handbook,* Marcel Dekker, New York, 1970, Chapt. 8.
76. M. M. Reznikovskii in ref. 44, p. 119.
77. I. V. Kragelskii and E. F. Nepomnyashchii, *Wear* **8**, 303 (1965).
78. A. E. Hollander and J. K. Lancaster, *Wear* **25**, 155 (1973).
79. G. S. Klitenik and S. B. Ratner in ref. 44, p. 64.
80. S. B. Ratner and E. G. Lure in ref. 44, p. 155.
81. G. I. Brodskii and M. M. Reznikovskii in ref. 44, p. 81.
82. S. M. Aharoni, *Wear* **25**, 309 (1973).
83. V. A. Belyi and A. I. Sviridyonok in ref. 33, p. 745.
84. S. B. Ratner and E. G. Lure in ref. 44, p. 161.
85. B. C. Arkles and M. J. Schireson, *Wear* **39**, 177 (1976).
86. W. A. Brainard and D. H. Buckley, *Wear* **26**, 75 (1973).
87. B. J. Briscoe, *Wear of Materials,* ASME, New York, 1981, pt. 3, p. 7.
88. V. A. Belyi, A. I. Sviridenok, M. I. Petrokovets, and V. G. Savkin, *Friction and Wear in Polymer-based Materials,* Pergamon Press, Oxford, UK, 1982, p. 195.
89. J. F. Archard, *Wear* **2**, 438 (1958).
90. K. Matsubara and M. Watanabe, *Wear* **10**, 214 (1967).
91. C. Shen and J. H. Dumbleton, *Wear* **30**, 349 (1974).
92. J. K. Lancaster, *Tribology* **4**, 82 (1971).
93. J. C. Jaeger, *Proc. R. Soc. N.S.W.* **76**, 203 (1943).
94. J. E. Theberge, B. C. Arkles, and P. Cloud, *Mach. Des.*, 60 (Oct. 1974).
95. J. K. Lancaster, *J. Phys. D.* **1**, 549 (1968).
96. K. Tanaka and S. Kawakami, *Wear* **79**, 221 (1982).
97. B. C. Arkles, S. Gerakaris, and R. Goodhue, in ref. 33, p. 663.
98. D. C. Mitchell and G. C. Pratt, *Proceedings of the Lubrication & Wear Conference, London, 1957,* The Institution of Mechanical Engineers, London, 1957, p. 416.
99. D. H. Buckley and R. L. Johnson, *Tech. Note D-2073,* NASA, Washington, D.C., 1963.
100. G. C. Pratt, *Plastics Institute (London), Trans. J.* **32**, 255 (1964).
101. V. A. Belyi, I. V. Kraghelskii, V. G. Savin, and A. I. Sviridyonok, *Trans. ASME, J. Lub. Tech.* **100**, 185 (1978).

102. V. A. Belyi and co-workers, *Am. Chem. Soc. Div. Org. Coat. Plast. Chem. Pap.* **33,** 227 (1973).
103. J. E. Theberge and B. C. Arkles, in ref. 102, p. 227.
104. N. P. Istomin and M. M. Krushchov, *Mech. Sci.* **1,** 97 (1973).
105. J. K. Lancaster, *Tribology* **1,** 240 (1968).
106. D. C. Evans, *Proceedings of the 2nd Conference on Solid Lubricants,* ASLE SP-6, American Society of Lubrication Engineers, Park Ridge, Ill., 1978, p. 202.
107. C. J. Cudworth and G. R. Higginson, *Wear* **37,** 299 (1976).
108. C. J. Hooke and J. P. O'Donohugue, *J. Mech. Eng. Sci.* **14,** 34 (1972).
109. T. Fort, *J. Phys. Chem* **66,** 1136 (1962).
110. G. A. Vinogradov and M. D. Mezborodko, *Wear* **5,** 467 (1962).
111. E. R. Booser, E. H. Scott, and D. F. Wilcock, in ref. 98, p. 366.
112. G. C. Pratt in E. R. Braithwaite, ed., *Lubrication and Lubricants,* Elsevier North-Holland, Amsterdam, 1967, Chapt. 7.
113. J. M. Senior and G. H. West, *Wear* **18,** 311 (1971).
114. W. L. Skelcher, T. F. J. Quinn, and J. K. Lancaster, *ASLE Trans.* **25,** 391 (1982).
115. R. Butterfield, D. Farmer, and E. M. Scurr, *Wear* **18,** 243 (1971).
116. B. C. Arkles and J. E. Theberge, *Lubr. Eng.* **29,** 552 (1973).
117. M. N. Abouelwafa, D. Dowson, and J. R. Atkinson, *Proceedings of the 3rd Leeds-Lyon Symposium on Tribology,* Mechanical Engineering Publications, London, 1978, p. 56.
118. R. W. Bramham, R. B. King, and J. K. Lancaster, *ASLE Trans.* **24,** 479 (1981).
119. J. K. Lancaster, *Tribology* **6,** 219 (1973).
120. F. J. Williams, *SAMPE Q.* **8,** 30 (1977).
121. *A Guide on the Design and Selection of Dry Rubbing Bearings,* Engineering Sciences Data Unit, London, 1976, Item 76029.
122. W. E. Campbell in F. F. Ling, E. E. Klaus, and R. S. Fein, eds., *Boundary Lubrication,* ASME, New York 1969, Chapt. 10.
123. B. C. Stupp, *Lubr. Eng.* **14,** 159 (1958).
124. D. Sitch, *Tribology* **6,** 262 (1973).
125. *Self-lubricating Bearings—A Performance Guide,* National Centre of Tribology, Risley, UK, 1977.
126. J. E. Theberge, *Proceedings of the International Conference on Solid Lubricants,* ASLE SP-3, American Society of Lubrication Engineers, Park Ridge, Ill., 1971, p. 166.
127. J. Target and J. E. Nightingale, *Conference on Gearing. Proceedings Institution of Mechanical Engineers* **184,** 30 (1970).
128. R. Shanley and L. Lamond, *Mech. Des.,* 125 (Dec. 9, 1976); 75 (Jan. 6, 1977).
129. A. E. Anderson in M. B. Peterson and W. O. Winer, eds., *Wear Control Handbook,* ASME, New York, 1980, p. 843.
130. J. H. Fuchsluger and V. L. Vandusen, in ref. 129, p. 667.
131. J. M. Brown, *ASLE Trans.* **20,** 161 (1977).
132. H. L. Lee, J. A. Orlowski, P. D. Kidd, R. W. Glace, and E. Enabe in ref. 33, p. 705.
133. A. Hepworth and J. Sikorski, *Proceedings of the 3rd Leeds-Lyon Symposium on Wear of Non-metallic Materials,* Mechanical Engineering Publications, London, 1978, p. 165.
134. E. Mayer, *Mechanical Seals,* 2nd ed., Iliffe, London, 1972.
135. C. P. Hemingray, *Proceedings of the Machine Tool Design and Research Conference,* Macmillan, New York, 1973, p. 99.
136. S. Hother-Lushington, *Tribology* **9,** 257 (1976).
137. W. V. Smith in J. V. Schmitz and W. E. Brown, eds., *Testing of Polymers,* Vol. 3, John Wiley & Sons, Inc., New York, 1967, p. 221.
138. J. H. Dumbleton, *Tribology of Natural and Artificial Joints,* Elsevier North-Holland, Amsterdam 1981.
139. W. H. Wilson, *Proceedings of the 3rd Leeds-Lyon Symposium on Wear of Nonmetallic Materials,* Mechanical Engineering Publications, London, 1978, p. 269.

JOHN K. LANCASTER
Royal Aircraft Establishment

AGING, PHYSICAL

Materials are said to age when their properties change with time; usually the change is adverse. For polymers, such aging may be due to a large number of causes: oxidation, with or without the stimulation of uv light; a gradual loss of plasticizers or other low molecular weight additives; after-curing; hydrolysis; etc. Such chemical or physicochemical degradation processes are not considered here; only aging processes of a physical nature will be treated (see DEGRADATION).

There are many kinds of physical aging processes: secondary crystallization; recrystallization; static or dynamic fatigue (qv), that is, degradation due to the prolonged action of mechanical stresses; and so forth. This article is confined to the very general and important type of aging that is due to the inherent instability of the amorphous glassy state (see AMORPHOUS POLYMERS). The process is also known as volume recovery, structural relaxation, stabilization (qv), annealing (qv), etc, and can be characterized as follows: Products of glassy polymers are usually made by shaping them in the molten (rubbery) state, followed by a fixation of the shape by rapid cooling to below the glass transition (qv) temperature (T_g). During this cooling, the material solidifies and stiffens, but the period is too short for this process to be completed. Solidification and stiffening continue during the service life of the product; thus, many material properties change with time. For example, the elastic modulus and the yield stress increase, the ductility may decrease, the creep (qv) and stress relaxation rates decrease, the dielectric relaxation (qv) is delayed, the propensity for crazing (qv), and shear banding is enhanced, etc. The origin, most important aspects, and practical results of this aging process are reviewed herein, and some consequences for materials other than purely amorphous glassy polymers are discussed. The implications of aging for understanding the nonlinear mechanical behavior of plastics are also summarized.

The Origin of Aging

The general picture of aging can be deduced from the results on time–temperature superposition obtained in the 1950s (1–4). At that time it was established that for the temperature range just above T_g, a change in temperature T causes a shift of the mechanical or dielectric relaxation curves along the logarithmic time scale (see Fig. 1). Shifting occurs without appreciable change in shape, and this forms the basis for time–temperature superposition procedures. The whole effect of temperature upon relaxation can be described by the shift function log $a(T)$, and this function applies to the mechanical relaxation of small strains, to dielectric relaxation, as well as to the mechanical behavior at large strains, eg, fracture (qv) behavior of rubbery materials (5). Since temperature only affects the position of the response curve on the time scale, we can use a time parameter τ instead of the shift factor a. In Figure 1, τ is arbitrarily defined as the time at the midpoint of the relaxational transition.

Experiments show a tremendous change in τ over the temperature range between T_g and 50–100°C above T_g. This change of roughly 10 orders of magnitude is schematically illustrated in Figure 2. Williams, Landel, and Ferry (1) have proposed that the variations in τ are not primarily due to thermal activation

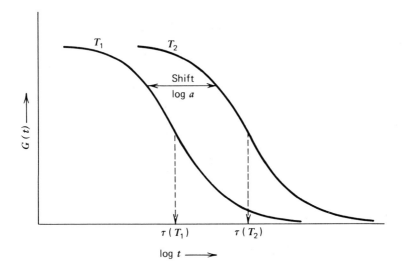

Fig. 1. Effect of temperature on mechanical relaxation in the temperature range just above T_g; $G(t)$ denotes the stress relaxation modulus.

but to thermal expansion, ie, to the expansion of free volume v_f with increasing temperature, and using an equation proposed by Doolittle, these authors derived the famous WLF equation.

Next, the material is rapidly cooled through the T_g-range and is subsequently kept at a constant temperature $T_a < T_g$. At T_g, the retardation times become so long that the changes in v_f can no longer keep up with the cooling

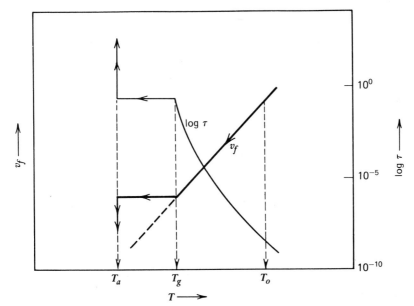

Fig. 2. Change in free volume v_f and relaxation time τ during rapid cooling from $T_o > T_g$ to $T_a < T_g$ and subsequent isothermal aging at T_a. The time units on the τ-scale are arbitrary; the dashed line is the supposed equilibrium line for v_f below T_g.

process. To a first approximation the value of v_f at T_g is frozen-in and v_f no longer decreases with temperature during further cooling from T_g to T_a (Fig. 2). If the WLF ideas can be applied to the temperature range below T_g, we can expect a similar course for τ vs temperature, ie, a sharp increase in τ during cooling above T_g and a constant value of τ during further cooling from T_g to T_a. In Figure 2, the direct thermal activation effect on τ is neglected, and thus it is assumed that constancy of v_f during cooling from T_g to T_a results in a similar constancy of τ. This is, of course, an approximation.

Next, the isothermal processes at the aging temperature T_a are considered. Several studies show that at T_a, specific volume v slowly decreases with time (6–10). The usual interpretation of this volume relaxation or aging effect is that free volume v_f slowly relaxes towards its equilibrium value, the dashed line in Figure 2. So, if the WLF ideas can be applied in the temperature range below T_g, the decrease in v_f will be accompanied by an increase in τ (Fig. 2). Consequently, during isothermal aging below T_g, the mechanical response curves will be affected in the same way as during cooling in the temperature range above T_g. That is, with increasing age, the response curves will shift along the logarithmic time scale; the shifts reflect a delay in relaxation and occur without appreciable change in the shape of the curve.

These effects were found in the 1960s (11–18) and have been reviewed comprehensively (19–20). A striking illustration of the aging effect is shown in Figure 3, which is from a study of the amorphous polymer poly(vinyl chloride) (PVC), and shows that for a change in aging time t_e from 0.03 to 1000 days, creep is retarded by a factor of 10^4–10^5. This article principally discusses the effects of

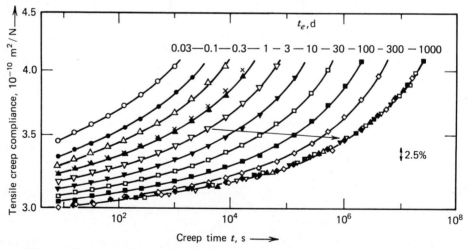

Fig. 3. Small-strain tensile creep curves of rigid PVC quenched from 90°C (ie, ca 10°C above T_g) to 20°C and kept at 20 ± 0.1°C for a period of 4 yr. The different curves were measured for various values of the time t_e, in days, elapsed after the quench. The master curve at t_e = 1000 d gives the result of a superposition by shifts which were almost horizontal; the shifting direction is indicated by the arrow. The crosses (×) on the t_e = 1 d curve were found when, after 1000 d of aging, the sample was reheated to 90°C, requenched to 20°C, and remeasured for a t_e of 1 d; they show that the aging is thermoreversible. Courtesy of Elsevier.

aging on mechanical properties such as creep. The underlying volume relaxation process is not treated in detail; for this, see references 10, 21–22.

Typical Aspects of Aging

Aging is a consequence of the instability of the glassy state and should be found in all glassy materials. This has been confirmed experimentally, and effects similar to those shown in Figure 3 have been found in many glassy polymers (19), low molecular weight organic glasses (13,16,19), substances such as bitumen (19), cured epoxies (19,23–25), inorganic glasses (26–28), and glassy metals (29–30).

Reversibility

In contrast to chemical aging or degradation, the physical aging shown in Figures 2 and 3 is a reversible process, ie, by reheating the aged material to $T_o > T_g$, the original state of thermodynamic equilibrium is recovered and a renewed cooling to T_a will induce the same aging effects as before. The aging effects can be reproduced an arbitrary number of times on the same sample; this repeatability is shown by the crosses in Figure 3.

Aging Primarily Affects Retardation Times

According to Figure 2, aging induces a gradual increase in τ. Therefore, the creep curves shift progressively along the logarithmic time scale with increasing age. This shifting without appreciable change in shape is shown in Figure 3 and has been found in all amorphous polymers studied (19). The shifting effect has also been found for stress relaxation curves (31–33) and for dielectric relaxation curves (14–15,34–35). Sometimes the horizontal shifts are accompanied by (small) vertical shifts, the origin of which has not yet been determined.

The shifting effect has implications for isochronal values of several properties. For example, Figure 4 shows the effects of aging on the creep compliance at a fixed creep time or the dynamic modulus at a fixed frequency. These properties change with age: The creep compliance decreases, the modulus increases, and the isochronal dielectric constant decreases just as the creep compliance. It should, however, be realized that these changes are not the result of intrinsic changes in stiffness or polarizability (dielectric constant); the changes in the isochronal values are merely reflections of the delay in relaxation with increasing age.

The relative change in the isochronal value depends on the magnitude of the shift due to aging, but also on the slope of the creep or modulus curve on the $\log t$ or $\log \omega$ scale. Just below T_g, this slope increases with increasing time and decreasing frequency (compare the changes at t_1 and t_2 or ω_1 and ω_2 in Fig. 4). Consequently, dynamic tests at high frequencies are not the proper means to investigate the aging effects; the much slower creep or stress relaxation tests are more appropriate.

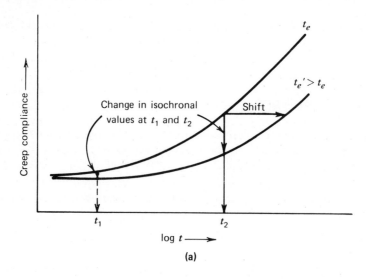

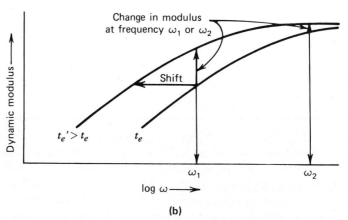

Fig. 4. Effect of aging on the isochronal values (constant loading time t or frequently ω) of the creep compliance (**a**) or the dynamic modulus (**b**).

Relationship of Retardation Time to Aging Time and Temperature

Retardation time τ is proportional to aging time t_e and the temperature difference, $T_g - T$, according to the formula:

$$\tau \cong k\, t_e\, (T_g - T)$$

where k is a constant.

According to Figure 2, the aging process is of a self-delaying nature. The more v_f decreases, the higher τ will be, and the process will continue with more difficulty (more slowly). For such self-delaying processes a general law could be derived (19), stating that τ will increase proportionally with t_e. Thus, each tenfold increase in t_e will induce a tenfold increase in τ, ie, a shift of the creep curve by one decade; this is clearly visible in Figure 3. This proportionality between τ and

t_e has been found for the small-strain creep of a large number of amorphous polymers (19), for the creep of a low molecular weight organic glass (16), for the dielectric relaxation of polymeric glasses (14–15,34–35), for the changes in the viscosity of inorganic glasses (27–28), and for the aging of glassy metals (29–30).

The above equation further implies that τ will vary proportionally with $T_g - T$; this has been derived and compared with experimental data (19), and a full discussion about the equation is given in ref. 20. An important conclusion follows when the variations in t_e and $T_g - T$, which may occur under practical conditions, are compared. For plastic products, t_e may vary over many orders of magnitude; for example, in Figure 3 there are 4.3 decades between 1 d and 50 yr after processing. The changes in $T_g - T$ are much more limited. Normally, plastic products are not used at temperatures close to T_g, which means that under practical conditions $T_g - T$ will be $\geq 10°C$. On the other hand, the upper limit of $T_g - T$ is ca 100°C, so that the maximum variation in $T_g - T$ is approximately a factor of 10 in contrast to a variation of ca 10^4 in t_e. Consequently, for characterizing the mechanical properties of glassy polymers under practical conditions, t_e is a more important parameter than temperature. Experimental data demonstrating this statement are shown in Figure 5; by a proper choice of t_e, creep at 0°C can be faster than creep at 40°C.

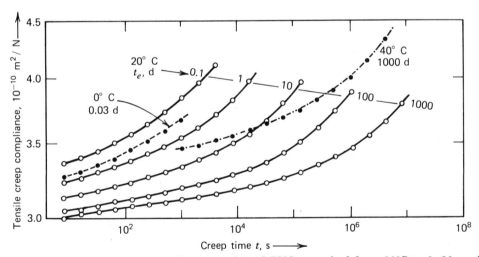

Fig. 5.   Small-strain tensile creep of rigid PVC quenched from 90°C to 0, 20, and 40°C and tested at the aging temperature after various values of t_e. For 20°C, the data are given for t_e's of 0.1–1000 d, for 40°C for a t_e of 1000 d only, and for 0°C for a t_e of 0.03 d only. Notice that creep at 40°C may be slower than at 0°C. Courtesy of Elsevier.

In addition, the retardation-time equation has implications other than practical ones. For example, it shows that τ does not vary exponentially with T (or $1/T$), but rather varies linearly. This means that in comparing states at different temperatures but of equal age (t_e), the whole concept of Arrhenius-type thermal activation does not apply (20). Therefore, care should be taken with experiments for which the interpretation is completely based on thermal activation theory (T-jump experiments (36–37) and thermally stimulated discharge (TSD) and thermally stimulated current (TSC) tests (38–39).

Aging Occurs Between T_g and T_β

The upper limit of the aging range follows from the discussion of the origin of aging. It equals T_g, since above T_g the material will be in thermodynamic equilibrium. The lower limit is T_β. That is, experiments as well as theoretical arguments show that the aging phenomena disappear at temperatures below the first (highest) secondary transition T_β (19). So, the aging range runs from T_β to T_g, eg, for PVC from ca -50 to 80°C and for polycarbonate (PC) from -120 to 150°C.

Deaging by Partial Heating

So far, the phenomena discussed pertain to isothermal aging after a quench (rapid cooling) from above to below T_g. Under such circumstances, aging always runs in the same direction (increasing τ). Peculiar phenomena occur, however, when after a period of aging at temperature T_1, the material is heated to a final temperature T_x between T_1 and T_g ($T_1 < T_x < T_g$). An example for PVC ($T_g \cong$ 80°C) is shown in Figure 6. Two samples are compared for which the aging times at $T_1 = 20$°C differ by a factor of 18. When tested at 20°C, their creep properties differ in the usual way. However, when the two samples, aged at 20°C, are heated to and tested at 50°C, their creep properties are identical. Obviously, the heating to 50°C has erased the differences in aging history at 20°C.

The origin of the erasure of aging history can be understood from the volume relaxation peaks (7). An example for poly(vinyl acetate) (PVAc), which has a T_g of 32°C, is shown in Figure 7. A direct quench from 40 to 30°C results in a monotonic contraction process (curve A). A quench to $T_1 = 10, 15,$ or 25°C followed

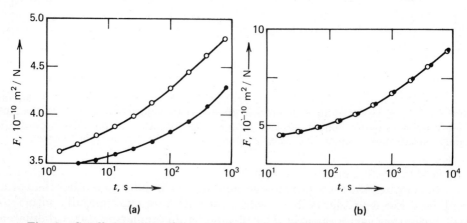

(a)   (b)

Fig. 6.   Small-strain tensile creep curves of rigid PVC quenched from 90 to 20°C, and aged at 20°C for ⅓ h (○, sample 1) and 6 h (●, sample 2). After these two aging histories at 20°C, the samples were tested at 20°C (**a**) and at 50°C (**b**). The equilibration time at 50°C was ½ h. Courtesy of Elsevier.

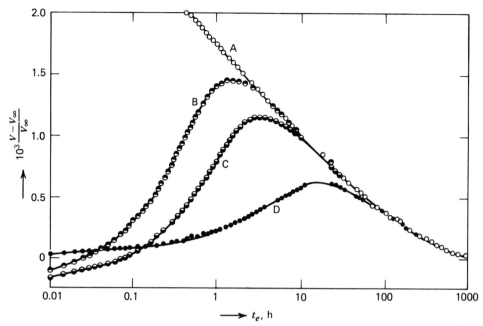

Fig. 7. Isothermal volume relaxation of PVAc at $T_\infty = 30°C$ after the following thermal histories: A, quench from equilibrium at 40°C to 30°C; B, quench from equilibrium at 40°C to $T_1 = 10°C$ followed by an aging period t_1 of 160 h at T_1 and an upquench to $T_\infty = 30°C$ (note: 40 (∞) ↓ 10 (160 h) ↑ 30°C); C, same conditions as B, but now 40 (∞) ↓ 15 (140 h) ↑ 30°C; D, same conditions as B, but now 40 (∞) ↓ 25 (90 h) ↑ 30°C. Courtesy of *Fortschritte der Hochpolymeren-Forschung.*

by an aging period at T_1 of 90–160 h, results in volume relaxation curves showing a maximum after heating to $T_\infty = 30°C$. This is a general effect; it is well understood and found in many polymers (19). The important point here is that after passing the maximum, all volume relaxation curves merge with the one measured after a direct quench from 40 to 30°C. This implies that for times larger than that of the volume relaxation maximum, the aging history at T_1 has been erased. This reasoning explains the PVC data of Figure 6 (20).

The erasing or deaging phenomenon is also well-known from dsc studies of organic (40–42) and metallic glasses (43). The behavior can be summarized schematically as shown in Figure 8 and can be described formally by the multiparameter theory for volume and enthalpy relaxation developed by a number of authors (19,21–22,41–46). The physical interpretation of the effect has been discussed (19,43).

Practical Consequences of Aging

The facts about aging discussed so far have a number of practical consequences. First, data such as that in Figures 3 and 5 show that aging time t_e should be included in the specification of material properties, particularly for low frequency properties such as creep and stress relaxation (Fig. 4). In fact, creep curves with unspecified values of t_e are almost useless.

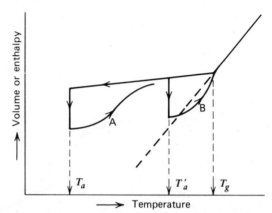

Fig. 8. Deaging during heating after aging at temperature T_a or $T_a' > T_a$ respectively as shown by the schematic course of volume or enthalpy vs temperature (19,40–42). After aging at T_a, deaging occurs below T_g (curve A), and plots of thermal expansivity (dv/dT) or heat capacity (dH/dT) show peaks below T_g. If the aging temperature approaches T_g (T_a', curve B), the endothermic C_p peak changes into the well-known overshoot peak at T_g. The recovery curve A does not match the original cooling curve; the reason is that new aging occurs during heating from T_a to T_g. Courtesy of Hanser.

The second point is that the deaging phenomena should be taken into account in experimental studies. The data, shown in Figure 9, are from an investigation of the effect of annealing at 60°C on the yield stress at 35.5°C. In contrast to usual thinking, annealing first causes a decrease in σ_y; only after some time does σ_y increase as expected. The origin of this effect is the deaging at 60°C which follows the aging at room temperature for several months. This is clearly demonstrated by the dsc diagrams, included in Figure 9; the original endotherm owing

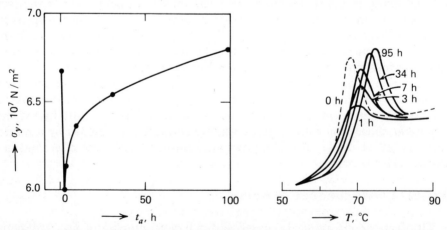

Fig. 9. Effect of annealing (aging) at 60°C for various times t_a on tensile yield stress σ_y at 35.5°C (strain rate 10%/min) and on the dsc diagrams measured at a heating rate of 32°C/min. The material was a sample machined from a commercial (extruded) sheet of rigid PVC that had been stored under room temperature conditions for several months. Courtesy of Elsevier.

to the aging at room temperature disappears, and only after some time, a new endothermic peak, related to the aging at 60°C, develops.

A third consequence of aging concerns the prediction of long-term behavior from short-time tests. In a long-term mechanical deformation process such as creep, two processes occur simultaneously: first the creep process itself, and secondly, the aging process that continuously stiffens the material and delays the creep. Methods for predicting long-term behavior which take these simultaneous effects into account have been worked out and apply to creep at small strains, to stress relaxation, and to dimensional instabilities such as shrinkage, ie, creep under the action of internal stresses (19). The methods allow for extrapolations over two to three decades on the time scale, and extrapolation errors have been analyzed theoretically.

Finally, attempts to accelerate creep by performing tests at higher temperatures are generally useless for predicting the long-term creep at the working temperature. In the first place, heating changes the state of the material, as shown for example in Figure 6, and the properties measured at the higher temperature are not representative for those at the working temperature. Secondly, time–temperature superposition methods do not apply for long-term creep because it is affected by simultaneous aging, the rate of which is different at different temperatures (19–20). As a result, long-term creep curves measured at different temperatures cannot be superimposed by horizontal (and vertical) shifts, even when creep data are plotted in double-logarithmic diagrams.

Aging in Materials Other Than Amorphous Glassy Polymers

Aging is an inherent property of the glassy state and therefore is found in all glasses, irrespective of their chemical nature (synthetic amorphous polymers, cured thermosets, low molecular weight organic glasses, inorganic glasses, glassy metals, etc). From a practical point of view, one might ask what effect the internal stresses and molecular orientations, which always arise when plastic products are made by injection molding, extrusion, etc, have on aging (47). The answer is simple; since internal stresses and orientations do not remove the glass–rubber transition phenomenon, they also cannot remove aging. This has been shown experimentally for injection moldings as well as for hot-drawn PVC samples with a stretch factor of ca 2.2 (19). Of course, internal stresses and orientations will influence the physical properties of the glass, but their effect on the way in which these properties change with aging time t_e is remarkably small.

Semicrystalline polymers are considered as two-phase composites: one phase amorphous, the other crystalline. The amorphous phase possesses a glass transition, and, therefore, aging effects are to be expected below the T_g of the semicrystalline material. Such effects have indeed been found for a large number of semicrystalline polymers; an example is reproduced in Figure 10 (19).

Remarkably, semicrystalline polymers also age above their T_g. An example for polypropylene (PP), is shown in Figure 11. Polypropylene ages strongly at temperatures of at least 40°C above T_g (ca 0°C); the aging effects are just as important as they are in amorphous polymers, and creep at 40°C can be slower than that at 20°C by a proper choice of t_e (Fig. 5). Similar effects were found in

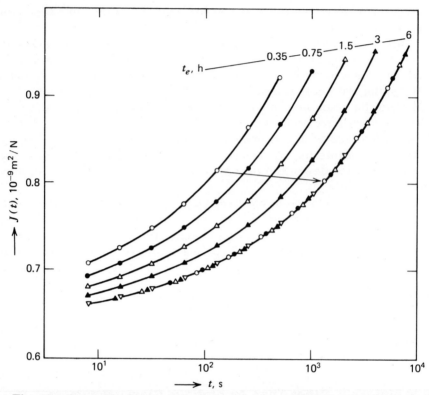

Fig. 10. Aging of polypropylene at temperatures below the glass transition; the material was quenched from 120 to $-20°C$ and tested (small-strain torsional creep) at different times t_e elapsed after the quench. The master curve was obtained by horizontal and (small) vertical shifts; the arrow denotes the direction of shifting. Courtesy of Elsevier.

all semicrystalline polymers investigated: polyethylene (LDPE and HDPE), nylon-6, nylon-12, and poly(ethylene terephthalate) (PET).

There is a strong similarity between the aging of semicrystalline polymers above and below their T_g; in fact, as far as aging and volume relaxation are concerned, there is complete continuity at T_g. Consequently, aging above and below T_g has been explained (19,48) on the basis of the same mechanism, that is, from phenomena in the amorphous phase (49). Below T_g, the aging is due to the instability of the glassy state. If the same mechanism applies to the temperature range above T_g, we must assume that part of the amorphous state is still glassy at temperatures above the conventional T_g of a semicrystalline polymer. This effect can be explained by the idea that crystallites in semicrystalline polymers have an effect similar to that of carbon black particles in filled rubbers (50–52). The particles disturb the amorphous phase and reduce segmental mobility. This reduction is at a maximum in the immediate vicinity of the particles (crystals), and the properties of the amorphous phase become equal to those of the bulk amorphous material only at large distances from the crystals or particles. The consequence of this immobilization is that the glass transition is extended towards the high temperature side as shown in Figure 12. Above the T_g of the

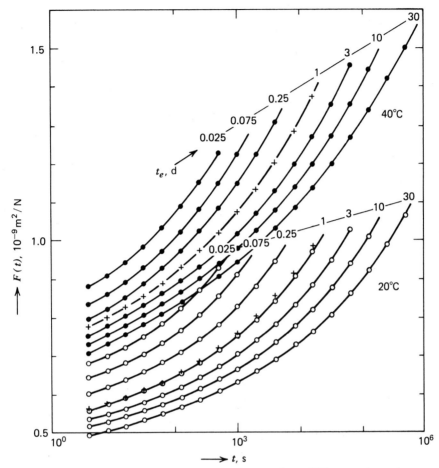

Fig. 11. Small-strain tensile creep of polypropylene (PP) at various times t_e after quenches from 120 to 20 and 40°C. The crosses (+) refer to duplicate tests on the same specimen after completion of the first sequence of tests. In these duplicate tests, the creep was measured for a t_e of 1 d only. These duplicate tests show that the aging is thermo-reversible and spontaneous (no mechanical enhancement). Courtesy of *Plastics and Rubber Processing and Applications.*

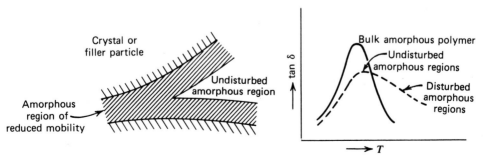

Fig. 12. Extended glass transition in semicrystalline polymers and filled rubbers. Courtesy of Elsevier.

bulk amorphous material, some parts of the amorphous phase are rubbery, other parts are glassy, and still other parts are just passing their glass transition; the nonrubbery parts are responsible for the strong aging and thermal history effects.

Determining whether filled rubber systems show the same aging effects as semicrystalline polymers above their T_g was crucial in developing this model. As shown in Figure 13, this is indeed the case; moreover, when the filler is removed, the aging effects almost completely disappear. The aging effects become more pronounced for finer carbon blacks (see CARBON; FILLERS).

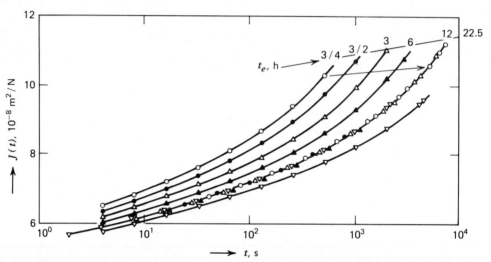

Fig. 13. Aging of styrene–butadiene rubber filled with HAF black (SBR–HAF) quenched from 70 to 20°C. Before quenching, the sample had been kept at 70°C for 16 h. The sample was tested (small-strain torsional creep) at different time t_e elapsed after the quench. The master curve was obtained by horizontal and (small) vertical shifts; the arrow denotes the direction of shifting. Courtesy of Elsevier.

The model of Figure 12 appears to work well. However, it conflicts with some current views of the mechanical and dielectric relaxation of semicrystalline polymers. For example, the gradual softening of materials such as PE and PP, which occurs in a wide temperature range above T_g, is often explained by relaxations taking place in the crystalline phase; the amorphous phase is assumed to be entirely rubbery and incapable of giving rise to strong relaxations. In view of the model of Figure 12, the latter assumption is premature; the gradual softening of PE and PP may be due, at least in part, to the broadened glass transition of the amorphous phase.

Although the model of Figure 12 works satisfactorily, there is still room for doubt because of the aging effects found in polycrystalline metals. An example, for lead, is shown in Figure 14, and these effects for duralumin and stainless steel have been reported (20). For these polycrystalline metals, aging as well as creep behavior are similar to that of amorphous polymers (19), but the molecular mechanisms must be different. Different molecular mechanisms can thus result in very similar aging and creep phenomena. The same conclusion follows from the aging effects found in margarine (53), a thixotropic material.

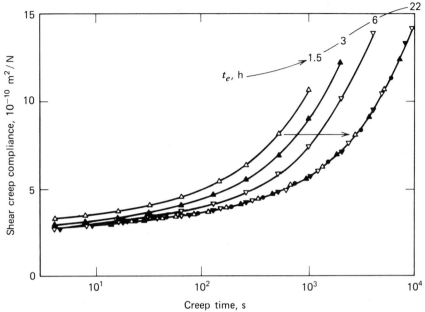

Fig. 14.   Small-strain torsional creep at 180°C of polycrystalline lead at various times t_e after heating from 20 to 180°C. The master curve was obtained by shifting; the arrow denotes the shifting direction. The creep strains were < 0.004 %. Courtesy of Elsevier.

In view of these results, the idea cannot be excluded that for the semicrystalline polymers, the very similar aging phenomena above and below T_g are owing to different molecular mechanisms, eg, a glasslike aging below T_g and secondary crystallization above T_g. Moreover, above T_g there may be a combination of secondary crystallization and aging according to the model of Figure 12.

Natural polymers are just as sensitive to aging as their synthetic counterparts as has been shown for shellac and cheese (19), elastin (54), the semisynthetic polymer cellulose acetobutyrate (55), and wool (qv) (56–58). The work on wool is particularly interesting because it has some relevance to the area of paints and lacquers. The aging of wool was studied after wetting followed by rapid drying. Plasticization by wetting has a similar effect to heating above T_g, and rapid drying is similar to quenching from above to below T_g. The aging behavior of wool turned out to be identical to that of amorphous polymers, ie, the stress relaxation is retarded by a factor of 10 for each tenfold increase in t_e (time elapsed after drying). Phenomena similar to those in wool are to be expected in paints and lacquers which will age physically after evaporation of the solvent. The same will occur in other systems that stiffen by evaporation of a solvent (eg, adhesives). In paints, the actual situation is much more complex because the stiffening process is also caused by chemical reactions. The same is true for the curing (qv) process of thermosets. These chemical reactions are influenced, however, by the physical aging process that starts during curing, when the T_g of the material is raised to above the curing temperature. Physical aging determines segmental mobility and in this way the rate constants of the chemical reactions. An example of an experiment where physical aging and chemical cross-linking (qv) reactions occur simultaneously is shown in Figure 15. Notice that the creep curves shift faster than by a decade per tenfold increase in t_e; eg, between t_e = 1.5 and 12 h, the curves of Figure 15 shift by about 2 decades. The final stages of the hardening process of paints, lacquers, adhesives, and thermosets are influenced by physical aging.

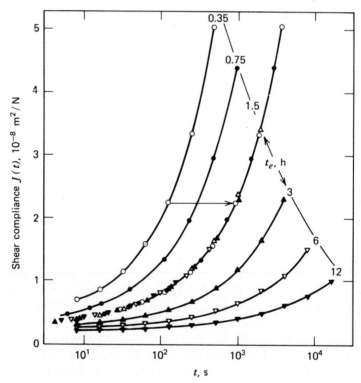

Fig. 15. Isothermal changes in the small-strain torsional creep properties during curing of a methyl methacrylate and allyl methacrylate copolymer (90:10). The sample was prepolymerized at 50°C during 42 h, next heated to 80°C, and tested at various times, t_e at 80°C. The master curve for 1.5 h was obtained by almost horizontal shifts (see the arrow). Courtesy of Elsevier.

Nonlinear Viscoelasticity, Yielding, and Fracture

Deaging effects can be brought about by high mechanical stresses as well as by partial heating (19). In the extreme of yielding and necking, the deaging is complete and the material behaves as if it were actually heated to above T_g (40).

Aging plays a key role in the nonlinear viscoelastic behavior of plastics. In fact, such materials behave as thixotropic solids such as margarine. When left at rest and at constant temperature, there is continuous stiffening. This aging for margarine is due to the gradual formation and perfection of a network of fat crystals. However, when the aged material is slightly heated or mechanically deformed, it is deaged and softened. For margarine, this is due to the breakdown of the network of crystals. The nonlinear viscoelastic behavior can thus be viewed as a competition between structure formation due to aging and structure destruction due to deformation. The implications of this view on nonlinear phenomena such as crazing, shear banding, yielding, and necking are discussed in ref. 20.

BIBLIOGRAPHY

1. M. L. Williams, R. F. Landel, and J. D. Ferry, *J. Am. Chem. Soc.* **77**, 3701 (1955).
2. J. D. Ferry, *Viscoelastic Properties of Polymers*, 2nd ed., John Wiley & Sons, Inc., New York, 1970.
3. F. R. Schwarzl and A. J. Staverman, *J. Appl. Phys.* **23**, 838 (1952).
4. A. J. Staverman and F. R. Schwarzl in H. A. Stuart, ed., *Die Physik der Hochpolymeren*, Vol. IV, Springer Berlin, 1956, Chapt. 5.
5. F. R. Eirich and T. L. Smith in H. Liebovitz, ed., *Fracture: An Advanced Treatise*, Vol. VII, Academic Press, New York, 1972, Chapt. 7.
6. E. Jenckel, *Z. Elektrochemie* **43**, 769 (1937); **45**, 202 (1939).
7. A. J. Kovacs, Thesis, Fac. Sci., Paris, France, 1954.
8. A. J. Kovacs, *J. Polym. Sci.* **30**, 131 (1958).
9. A. J. Kovacs, *Trans. Soc. Rheol.* **5**, 285 (1961).
10. A. J. Kovacs, *Fortschr. Hochpolym. Forsch.* **3**, 394 (1964).
11. A. J. Kovacs, R. A. Stratton, and J. D. Ferry, *J. Phys. Chem.* **67**, 152 (1963).
12. H. H. Meyer, P. M. F. Mangin, and J. D. Ferry, *J. Polym. Sci.* **A3**, 1785 (1965).
13. H. H. Meyer and J. D. Ferry, *Trans. Soc. Rheol.* **9**(2), 343 (1965).
14. S. Kästner and M. Dittmer, *Kolloid-Z* **204**, 74 (1965).
15. S. Kästner, *J. Polym. Sci.* **C16**, 4121 (1968).
16. D. Plazek and J. H. Magill, *J. Chem. Phys.* **45**, 3038 (1966).
17. S. Turner, *Br. Plast.* 682 (Dec. 1964).
18. L. C. E. Struik, *Rheol. Acta* **5**, 303 (1966).
19. L. C. E. Struik, *Physical Aging of Amorphous Polymers and other Materials*, Elsevier, Amsterdam, The Netherlands, 1978.
20. L. C. E. Struik in W. Brostow and R. D. Corneliussen, eds., *Mechanical Failure of Plastics*, Hanser, Munich, FRG, 1983, Chapt. 15.
21. J. M. Hutchinson, J. J. Aklonis, and A. J. Kovacs, *A. C. S. Polym. Preprints* **16**(2), 94 (1975).
22. A. J. Kovacs, J. J. Aklonis, J. M. Hutchinson, and A. R. Ramos, *J. Polym. Sci. (Phys.)* **17**, 1097 (1979).
23. E. S. W. Kong, *J. Appl. Phys.* **52**, 5921 (1981).
24. E. S. W. Kong, *A. C. S. Org. Coat. Appl. Polym.* **46**, 568 (1982).
25. T. D. Chang, S. H. Carr, and J. O. Brittain, *Polym. Eng. Sci.* **22**, 1221, 1228 (1982).
26. H. R. Lillie, *J. Am. Ceram. Soc.* **16**, 619 (1933).
27. A. L. Zijlstra, *Phys. Chem. Glass.* **4**, 143 (1963).
28. M. Prod'homme, *Rheol. Acta* **12**, 337 (1973).
29. P. M. Anderson and A. E. Lord, *Mater. Sci. Eng.* **44**, 279 (1980).
30. A. I. Taub and F. Spaepen, *J. Mater. Sci.* **16**, 3087 (1981).
31. F. A. Myers and S. S. Sternstein in C. Klason and J. Kubét, eds., *Proceedings VIIth International Congress on Rheology*, Göteborg, Sweden, 1976, p. 260.
32. S. S. Sternstein, *A. C. S. Polym. Preprints* **17**, 136 (1976).
33. M. Cizmecioglu, R. F. Fedors, S. D. Hong, and J. Moacanin, *Polym. Eng. Sci.* **21**, 940 (1981).
34. G. E. Johnson, S. Matsuoka, and H. E. Bair, *A. C. S. Org. Coat. Plast. Preprints* **38**, 350 (1978).
35. M. Uchidoi, K. Adachi, and Y. Ishida, *Rep. Progr. Polym. Phys. Japan* **18**, 421 (1975); *Polym. J.* **10**, 161 (1978).
36. N. G. McCrum and M. Pizzoli, *J. Mater. Sci. (Letters)* **12**, 1920 (1977).
37. C. K. Chai and N. G. McCrum, *Polymer* **23**, 589 (1982).
38. J. v. Turnhout, *Thermally Stimulated Discharge of Polymer Electrets*, Elsevier, Amsterdam, The Netherlands, 1975.

39. C. K. Chai and N. G. McCrum, *Polymer* **21**, 706 (1980); **23**, 589 (1982).
40. A. R. Berens and I. M. Hodge, *Macromolecules* **15**, 756 (1982).
41. I. M. Hodge and A. R. Berens, *Macromolecules* **15**, 762 (1982).
42. I. M. Hodge and G. S. Huvard, *Macromolecules* **16**, 371 (1983).
43. H. S. Chen, *J. Non. Cryst. Solids* **46**, 289 (1981); *J. Appl. Phys.* **52**, 1868 (1981).
44. R. Gardon and O. S. Narayanaswamy, *J. Am. Ceram. Soc.* **53**, 380 (1970).
45. O. S. Narayanaswamy, *J. Am. Ceram. Soc.* **54**, 491 (1971).
46. M. A. de Bolt, A. J. Easteal, P. B. Macedo, and C. T. Moynihan, *J. Am. Ceram. Soc.* **59**, 16 (1976).
47. L. C. E. Struik, *Polym. Eng. Sci.* **18**, 799 (1978).
48. L. C. E. Struik, *Plast. Rubber Process. Appl.* **2**, 41 (1982).
49. M. K. Agarwal and J. M. Schultz, *Polym. Eng. Sci.* **21**, 776 (1981).
50. P. P. A. Smit, *Rheol. Acta* **5**, 277 (1966).
51. F. R. Schwarzl and co-workers, *Rheol. Acta* **5**, 270 (1966).
52. G. Kraus, *Adv. Polym. Sci.* **8**, 155 (1971).
53. L. C. E. Struik, *Rheol. Acta* **19**, 111 (1980).
54. G. Pezzin and L. C. E. Struik, *Biopolymers* **19**, 1667 (1980).
55. G. Levita and L. C. E. Struik, *Polymer* **24**, 1071 (1983).
56. B. M. Chapman, *J. Text. Inst.* **64**, 667, 729 (1973).
57. B..M. Chapman, *J. Text. Inst.* **66**, 339, 343 (1975).
58. B. M. Chapman, *Rheol. Acta* **14**, 466 (1975).

L. C. E. Struik
Plastics and Rubber Research Institute TNO

CHEMICALLY RESISTANT POLYMERS

Some polymeric materials have outstanding chemical resistance. Most polymers in commercial use are resistant in that they do not deteriorate by an electrochemical mechanism, as metals do. However, the resistance of a polymer is often affected by the material or the environment. Generalizations about chemical resistance may be adequate for simple comparison, but only analytical data obtained by standard test procedures are acceptable. Polymer selection must be based on laboratory data and confirmed by testing under simulated conditions of use.

The chemical resistance of a polymeric material is its ability to withstand chemical attack with minimal change in appearance, dimensions, mechanical properties, and weight over a period of time. Test conditions include the length of exposure, concentration, temperature, and internal stress. The final classification as chemically resistant depends on the application.

Mechanisms

Polymers can absorb, react chemically, become plasticized, dissolve, or be stress-cracked by chemical environments. Chemical change is stimulated by heat and concentration. If the chemical exposure is limited or infrequent, the service life of the polymeric part may still be long, but some change in properties usually occurs.

Absorption, solvation, and plasticization may be tolerated under certain conditions. Polymers, eg, polyolefins, may absorb chemicals, such as hydrocarbons, and swell. Mechanical properties are reduced and dimensions changed, but functionality may remain adequate. Dissolution may result in loss of weight and mechanical strength, whereas plasticization may weaken cohesive forces and reduce mechanical properties, especially creep resistance. Stress-cracking is related to these mechanisms (1). It is a surface phenomenon intensified by stress and can be caused by selective absorption of the chemicals from the environment, eg, in the amorphous regions of semicrystalline polymers. Softening with lowering of glass transition and weakening can lead to local failure, such as crack propagation. Selective dissolution of low molecular weight portions of a polymer can weaken and crack the surface. Reaction at active sites, similar to chemical attack at boiler rivets, can destroy polymer crystallinity and result in loss of strength, failure, and cracking.

Fabrication, compounding, and application affect chemical resistance. The choice of polymer is aided by the following general rules (2):

(1) Solubility is reduced and resistance enhanced by increasing molecular weight.

(2) Susceptibility to oxidation increases with unsaturation.

(3) Solubility is favored by chemical similarity of solute and solvent.

(4) The solubility of a polymer is reduced by chain branching and cross-linking (qv) (see also SOLUBILITY OF POLYMERS).

Thermosetting Plastics

Most thermosetting plastics are resistant to aqueous salt solutions and moisture. The largest volume thermosets, phenol–formaldehyde, melamine–formaldehyde, and urea–formaldehyde resist nonpolar solvents, but not strong oxidants and alkali (see AMINO RESINS; PHENOLIC RESINS). Thermosetting silicone resins have good solvent resistance, but are attacked by strong acids and alkali (see SILICONES). Polyurethanes (qv) are attacked by hot water, strong acids and bases, and polar solvents; they are not used if chemical resistance is required. Epoxy resins (qv) have good chemical resistance that partially depends on the curing technique; they are used in coatings in the chemical process industry. They resist weak acid and alkali and hydrocarbons, but are attacked by strong acids, oxidizing agents, and some nonpolar solvents.

Unsaturated polyesters are the polymers most widely used by the chemical processing industry, particularly reinforced with glass and fabricated into large parts. They resist weak acid and alkali, but are attacked by strong alkali, solvents, and oxidizing agents (see POLYESTERS, UNSATURATED). Chlorinated polyesters, alkyd resins (qv), and vinyl esters have better solvent resistance. Chemical resistance of alkyd resins is similar to that of polyesters. Thermosetting polyimides have high heat resistance and are unaffected by dilute acids, hydrocarbons, esters, ethers, alcohols, freons, hydraulic fluids, aviation fuel, and kerosene. They are attacked by dilute alkali and concentrated inorganic acids. Their high cost is often justified by their resistance to heat and chemical attack.

The low volume thermosetting furan resins have very high chemical resistance (see FURAN POLYMERS). They withstand mineral acids, alkali, and organic solvents, but, like most thermosetting resins, are attacked by oxidizing agents and organic acids such as formic and acetic. The widely used epoxy and polyester resins have the best combination of properties in this group.

Thermosetting Elastomers

Thermosetting elastomers resist aqueous salt solutions, alkali, moisture, and nonoxidizing acids. They are not highly resistant to nonpolar solvents and oxidizing acids. Resistance to mineral and vegetable oils, jet fuel, and gasoline varies. Styrene–butadiene (SBR), natural rubber, and polyisoprene are attacked by oils and solvents. Urethane rubbers are attacked by caustic and steam. Polyacrylic rubber has excellent resistance to petroleum products and vegetable oils, but swells in aromatic hydrocarbons and is attacked by alkali and steam.

Moderate chemical resistance is shown by silicone, ethylene–propylene, and butyl rubbers. However, they are not resistant to solvents, and silicone rubber is not resistant to steam.

Good chemical resistance is shown by polychloroprene (Neoprene), chlorosulfonated polyethylene (Hypalon), and polysulfide rubber. Neoprene is resistant to aliphatic hydrocarbons, alcohols, and glycols, but is attacked by chlorinated and aromatic hydrocarbons, phenols, and oxidizing acids. Hypalon has excellent resistance to most chemicals, oils, greases, aqueous salt solutions, alcohols, bases, and concentrated sulfuric acid. It is attacked by hydrocarbons, gasoline, and chlorinated solvents. Polysulfide rubbers resist polar solvents, gasoline, and oils; solvent swelling is reduced with increasing sulfur content (3). They are less resistant to nonpolar solvents, alkali, and oxidizing acids (see also CHLOROPRENE POLYMERS; ETHYLENE POLYMERS; POLYSULFIDE POLYMERS).

Of the widely used thermosetting elastomers, nitrile rubbers (acrylonitrile–butadiene) have the best chemical resistance. They have good resistance to oils, solvents, alkali, and aqueous salt solutions. In aliphatic hydrocarbons, fatty acids, alcohols, and glycols, they swell slightly, but without significant physical deterioration. They are attacked by strong oxidizing agents, ketones, ethers and esters, but are useful when gasoline and oil resistance are required. Coatings made from solutions of nitrile, polysulfide, polychloroprene, chlorosulfonated polyethylene, and butyl rubbers resist chemical attack.

Chemical resistance of fluoroelastomers is higher than that of nitrile rubbers, but they are more expensive. There are five types which vary in chemical resistance: copolymer of chlorotrifluoroethylene and vinylidene fluoride (Fluorel); copolymer of hexafluoropropylene and vinylidene fluoride (Viton and Fluorel); perfluorobutyl acrylate (Kalrez); phosphonitrilic polymers (PNF); and trifluoropropyl siloxane polymers. Viton and Fluorel have excellent resistance to hot oils, synthetic lubricants, gasoline, jet fuels, dilute mineral acids, aqueous salt solutions, alkali, and chlorinated solvents; resistance to strong acids at room temperature is good. Kalrez resists almost all environments, and is attacked only by fluorine-containing compounds and molten alkali metals. Limited in use, PNF exhibits excellent resistance to oils and gasoline, and solvent resistance is comparable to that of Viton and Fluorel (see POLYPHOSPHAZENES). The automotive industry is making increased use of fluoroelastomers in hose and tubing. Silastic fluorosilicones have very good resistance to acids, alkali, solvents, oils, and fuel, as well as good resistance to alcohols, esters, and ketones (see CHLOROTRI-FLUOROETHYLENE POLYMERS; FLUORINATED ELASTOMERS; and VINYL FLUORIDE POLYMERS).

Thermoplastics

The thermoplastic polymers are the most widely used. They are not cross-linked, ie, cured, and are usually low in chemical resistance. Poly(vinyl alcohol)s, polycarbonates, cellulosic resins (acetate, acetate–butyrate, and nitrate), polyesters, and polyurethanes have poor alkali resistance. Poor solvent resistance, especially to aromatic hydrocarbons, is common with polystyrene, styrene–acrylonitrile (SAN), acrylonitrile–butadiene–styrene (ABS), silicone polymers, poly(vinyl acetate), and poly(vinyl butyral).

Chemical resistance is somewhat better in acetal resins (qv), polyamides, and polyethylenes. Acetals differ in resistance to alkali; the homopolymer is attacked by strong bases, whereas the copolymer is resistant. Both have excellent resistance to organic solvents, alcohols, glycols, ketones, and vegetable and mineral oils, but they are attacked by strong and warm, weak acids.

Nylons have good solvent resistance, but are attacked by strong mineral acids, oxidizing agents, and some salts. Nylon-6,6 has the best chemical resistance of the polyamides (qv). Polyolefins, including polypropylene, polybutylene, all forms of polyethylene, and ionomers, are resistant to acids and alkali, but are attacked by oxidizing acids. Solvent resistance varies; chlorinated solvents, hydrocarbons, certain esters, and oils weaken these polymers and may cause stress-cracking.

Unplasticized poly(vinyl chloride), the second largest volume thermoplastic, has very good chemical resistance and is widely used in chemical processing equipment. Rigid PVC resists alkali, salt solutions, and mineral acids, is attacked by organic acids, especially acetic, but resists oxidizing acids, an unusual property for thermoplastics. Alcohols, oils, and aliphatic hydrocarbons have little effect on PVC, which is, however, attacked by chlorinated and aromatic hydrocarbons. Chlorinated PVC and poly(vinylidene chloride)–PVC copolymer have higher heat resistance, but poorer solvent resistance. These materials are used in pipes, fittings, pump liners, and ducts (see VINYL CHLORIDE POLYMERS).

In recent years, the use of so-called high performance or engineering plastics (qv) has increased. These materials, usually thermoplastics, are characterized by good heat and chemical resistance. Usage is limited because of higher prices. Some engineering plastics, such as poly(phenylene oxide), blended with high impact polystyrene in Noryl, have limited chemical resistance and are attacked by hydrocarbons to which polystyrene is susceptible (see POLYOXYPHENYLENES).

Modified poly(phenylene ether), Prevex, resembles Noryl; resistance of Cadon or Dylark impact-modified styrene–maleic anhydride terpolymer is superior to modified poly(phenylene oxide), polycarbonate, and polycarbonate–ABS alloys.

Poly(4-methyl-1-pentene) (TPX) resembles the polyolefins in that it is attacked by strong oxidizing agents and swelled by light hydrocarbons and chlorinated solvent. The polyarylates (see POLYESTERS, AROMATIC) are subject to environmental stress-cracking, as are the other amorphous polymers. Their chemical resistance is limited to hydrocarbons, esters, ketones, and chlorinated solvents, just as is the case for polycarbonate.

Among the more resistant engineering plastics are the polysulfones (qv) and poly(phenylene sulfide). Polysulfone has good resistance to mineral acids, alkali, salt solutions, and aliphatic hydrocarbons, but is attacked by polar organic solvents and chlorinated and aromatic hydrocarbons. Poly(phenylene sulfide) has good chemical resistance except to strong oxidizing agents, sodium hypochlorite, and mineral acids (see POLY(ARYLENE SULFIDE)S).

The chemical resistance of thermoplastic polyimides (qv) is less outstanding than that of the thermosetting grades. The polyetherimide Ultem is resistant to salt solutions, dilute bases, gasoline, aliphatic hydrocarbons, and alcohol. It is particularly resistant to mineral acids, but partially halogenated solvents such as methylene chloride or trichloroethane and strong bases attack it. Other thermoplastic polyimides are resistant to most organic solvents and dilute acids, but are soluble in highly polar solvents and weakened by prolonged exposure to caustic solutions. The poly(amide-imide) Torlon is resistant to hydrocarbons, halogenated solvents, and acids and bases, but is attacked by caustic at elevated temperatures, steam, some amino compounds, and some oxidizing agents. Aqueous systems at elevated temperatures attack the polyamide component.

The poly(ether sulfone) Victrex and the newer polyetheretherketone (qv), Victrex PEEK, are aromatic polyethers with high chemical and heat resistance; poly(ether sulfone) has good resistance to most inorganic chemicals, oils, greases, aliphatic hydrocarbons, and gasolines. It is attacked by esters, ketones, polar aromatic solvents, and methylene chloride (see also POLYSULFONES). PEEK is reported to withstand most solvents and 60% sulfuric acid or 40% sodium hydroxide at elevated temperatures. It is dissolved by concentrated sulfuric acid and attacked by other mineral acids, such as nitric, but only a limited number of solvents cause stress-crazing, and then only under severe stress conditions.

Polytetrafluoroethylene (PTFE) (Teflon, Halon, Fluon), the largest volume fluoroplastic, is not melt processible, and almost chemically inert. Fluorinated ethylene–propylene copolymer (FEP) and perfluoroalkoxy resin (PFA) are melt processible, but differ little from PTFE in other respects. They are attacked by fluorine at elevated temperatures, chlorine trifluoride, and molten alkali metals, but resist less extreme conditions. Ethylene–tetrafluoroethylene copolymer (ETFE) (Tefzel) is the highest melting fluoroplastic and has slightly poorer solvent resistance than PTFE (see TETRAFLUOROETHYLENE POLYMERS).

Poly(vinylidene fluoride) (PVDF) (Kynar) is used for chemical equipment linings. It resists most acids, but not fuming sulfuric or chlorosulfonic acid. Its resistance to inorganic bases is excellent to 120°C, and it resists most organic bases except some primary amines. Solvent resistance is excellent except to acetone and some other highly polar solvents. It is resistant to halogens, oxidizing acids, and inorganic salts (see VINYLIDENE FLUORIDE POLYMERS).

Table 1. Chemical Resistance of Polymers[a]

Polymers	Acids	Alkali	Oxidizing agents	Oils, greases, petroleum, and fuels	Hydrocarbons	Average value
PTFE, FEP, PFA[b]	10	10	10	10	10	10
Kalrez fluoroelastomer	10	10	10	10	10	10
ETFE, ECTFE[c]	10	10	10	10	8	9.6
PVDF[d]	10	10	10	10	7	9.4
other fluoroelastomers	9	7	10	9	8	8.8
PCTFE, PVF[e]	10	10	10	8	5	8.6
PVC—unplasticized PVC, PVDC–PVC copolymer, CPVC[f]	9	10	10	9	5	8.6
furan resins	10	10	1	10	10	8.2
poly(phenylene sulfide)	6	8	4	10	10	7.6
thermosetting polyimides	6	6	8	8	9	7.4
poly(ether sulfone), polyarylketone	8	9	6	8	6	7.4
epoxy resins	7	8	3	9	9	7.4
polysulfone	8	8	7	6	6	7.0
nitrile rubber	7	8	4	10	5	6.8
acetal resins	5	10	4	8	9	6.8
chloroprene polymers	9	8	4	8	5	6.8
unsaturated polyester, FRP[g]	7	5	5	8	8	6.6
polyolefins[h]	8	9	7	6	3	6.6
polyacrylic rubber	6	6	6	9	6	6.6
chlorosulfonated rubber	10	7	7	5	4	6.6
polysulfide rubber	4	8	4	7	9	6.4
silicone rubber	7	9	7	5	2	6.2
poly(amide-imide), polyetherimide, thermoplastic polyimides	8	5	2	8	8	6.2
polypropylene, ethylene–propylene elastomers	9	9	6	5	2	6.2
thermoplastic polyesters	6	5	4	8	8	6.2
butyl rubber	10	10	4	4	2	6.0
nylon	3	8	3	8	8	6.0

[a] Based on an arbitrary scale of increasing resistance of 0 to 10.
[b] PTFE = polytetrafluoroethylene; FEP = fluorinated ethylene–propylene copolymer; PFA = perfluoroalkoxy resins.
[c] ETFE = ethylene–tetrafluoroethylene copolymer; ECTFE = ethylene–chlorotrifluoroethylene copolymer.
[d] PVDF = poly(vinylidene fluoride).
[e] PCTFE = polychlorotrifluoroethylene; PVF = poly(vinyl fluoride).
[f] PVC = poly(vinyl chloride); PVDC = poly(vinylidene chloride); CPVC = chlorinated poly(vinyl chloride).
[g] Fiber-reinforced plastics.
[h] Except polypropylene.

Polychlorotrifluoroethylene (PCTFE) (Kel-F), sold as Aclar film, has excellent barrier properties to gases (see also BARRIER POLYMERS). It is resistant to strong bases and acids, including oxidizing acids, but solvent resistance is poor at 100°C. Ethylene–chlorotrifluoroethylene copolymer (ECTFE) (Halar) has excellent chemical resistance to organic solvents, strong acids, and alkali. It is attacked by hot amines such as aniline and dimethylamine. ECTFE absorbs chlorinated solvents, ketones, esters, ethers, and common polymer solvents, although with some deterioration of properties.

Poly(vinyl fluoride) (PVF) is available as Tedlar film and used as a weather-resistant laminate. It resists strong acids and bases and most common solvents, including hydrocarbons and chlorinated solvents, at room temperature. It is partially soluble in a few highly polar solvents at temperatures above 150°C (see VINYL FLUORIDE POLYMERS).

The chemical resistance of some polymers is given in Table 1. Average values have little meaning if only one environment is of interest; however, the table is intended to show a composite chemical resistance of outstandingly chemical-resistant polymers ranked in order of excellence based on available information.

Test Methods and Standards

Test methods are given in ASTM D 543 (4) and ASTM C 581 (5). Standard specimens are immersed in reagents for 7 d at elevated or room temperature, and changes in weight, dimensions, appearance, and mechanical properties are determined. Exposure of 30–50 d is suggested. A classification of results from ASTM D 543 is given in Table 2 (6).

Method ASTM D 1693 (7) is used for evaluating polyethylene. A U-shaped strip is immersed in a chemical environment. This test is restricted to flexible materials, but ASTM D 2990 is more widely applicable. Here, a sample tensile bar is placed under static tension in a chemical environment at elevated temperature or stress-cracking coating is applied to the midpoint of the bar at room temperature. These are creep-rupture tests.

Table 2. Classification of Test Results[a]

| | Change in, % | | | |
Category	Weight	Dimension	Tensile and flexural strengths	Change in appearance
significant data	>1.0	>0.5	>20	softened, crazed, warped, distorted optically
significant data, but not conclusive	0.5–1.0	0.2–0.5	10–20	discolored
not significant	<0.5	<0.2	<10	slightly discolored

[a] Ref. 6. Courtesy of McGraw-Hill, Inc.

In stress-cracking tests, a strip of blotting paper that has been immersed in the test chemical is placed on a cantilevered strip with a weight clamped to one end. The strip covers the length of the polymer bar; stress-cracking is more severe in the areas of the greatest stress. Stress is highest at the fixed end, and lowest at the free weighted end. The weight is selected to produce cracking at the midpoint of the bar. This simple test lacks reproducibility, but is useful for rigid and semirigid materials for which ASTM D 1693 cannot be used and when equipment for D 2990 is unavailable; 8–12 h are required for this test.

In the bent-strip test, the sample is bent over an elliptical shape and clamped at each end. A blotting-paper strip soaked with the chemical to be tested is placed on the sample. This test is unsatisfactory for rigid materials and is limited in stress development to the shape of the ellipse. Stress relaxation may occur, and results are doubtful.

Prototype testing is more valuable for establishing acceptability. Ultimately, there is no substitute for testing under actual use conditions.

Applications

Applications of chemically resistant polymers are given in Table 3.

Table 3. Applications of Chemically Resistant Polymers

Polymers	Applications
high fluorine-content fluorocarbons, PTFE, FEP, PFA[a]	bearings, gaskets, packaging, O-rings, seals, tank linings, valve seats, siphons and jets for pickling tanks, piston rings, compression joints, chemical tubing, film, and coatings
fluoroelastomers, Kalrez	O-rings, seals, backup rings, diaphragms, pump and valve linings, hose and tubing, and coated fabrics and sheet goods
ETFE, ECTFE[b]	containers, diaphragms, linings and coatings for tanks, pumps, valves, pipes, scrubbing towers, reactors, agitators, thermocouple wells, centrifuge components, metering exchangers, ducting, unsupported pipe and tubing, tower packings, valve seats, filters, dust collectors, mist eliminators, and closures
PVDF[c]	solid and lined pipe and fittings, solid, lined, and coated valves and pump components, seals, gaskets, coated and lined process vessels, agitators and tanks, hose and tubing for transfer, tubing for roll covers, monofilament for column packing and filter cloths, and containers and liners for shipping
PNF[d], Viton, Fluorel, fluorosilicones	PNF is suited for oil seals, gaskets, diaphragms, and hoses; Fluorel and Viton for O-rings, seals, fuel cells, flexiliners for pumps, seals and diaphragms for metering pumps, V-rings, valve liners, hose, pipe couplings, packing, coated fabrics, and precision-molded parts; fluorosilicones for sealants and bonding, and sealing or caulking applications where resistance to swelling effects of fuels, oils, and solvents is required
PCTFE, PVF, Kel-F, Tedlar[e]	PVF (Tedlar) film is used for pipe insulation in aggressive chemical environments; Kel-F resin in V-rings, pump seals and linings, joints, and gasketing

Table 3. *(Continued)*

Polymers	Applications
rigid PVC, CPVC, PVDC[f]	pipe valves, tubing, battery cases, drums, containers, tanks, cladding, linings, ducts, and pipe insulation
furan resins	coatings
poly(phenylene sulfide), Ryton	ball valve seats, battery components, chemical etch-bath trays, filters, seals, pump vanes, pump housings, valve disks, and coatings
polyimide thermosetting resin, Kapton film, Vespel parts	coatings, bearings, seals, valve seats, gaskets, and bushings
poly(ether sulfone), polyetheretherketone	Victrex is clear and is used for eyeglasses and pump housings; Victrex PEEK for diaphragms
epoxy resins	protective coatings, interiors of pipe and metal cans, and maintenance coatings
polysulfone	process pipe and tubing, tower packings, food-processing equipment, pumps, filter modules and support plates, and membranes
nitrile rubber	hose, O-rings, packings, gaskets, oil seals, diaphragms, blow tubes for foundry cores, oil-well parts, fuel cell liners, fuel lines, adhesives, rolls, packing, coated fabric, and tubing
polychloroprene, neoprene	hose, O-rings, packings, gaskets, tank liners, (flexible) diaphragms, coated fabrics, diaphragms, adhesives, caulks, sealants, bushings, and sealings
unsaturated polyesters, glass-reinforced	pipe, couplings, tanks, tank liners, wires, troughs, scrubbers, ducts, railings, gratings, ladders, ventilators, filters, manholes and manhole covers, fans, baffles, and pumps
polyolefins[g]	low density polyethylene is used for pipe and tubing; high density polyethylene for industrial pipe, containers, and chemical tanks; ultrahigh molecular weight polyethylene in down-hole tubing in oil fields, food- and beverage-processing machinery parts, and in chemical processing ducts; ionomers as packaging materials; polybutylene for gas piping; cross-linked polyethylene in chemical fertilizer tanks
polyacrylic rubber	automotive seals, O-rings, packings, gaskets, spark-plug covers, hose, molded parts for antipollution devices, seals for valve steams, and lip, shaft, piston and transmission seals
chlorosulfonated polyethylene, Hypalon	similar uses as for Neoprene, although resistance to ketones and aromatic solvents is poorer; hoses, seals, liners for pumps and valves, pond liners for chemical wastes, expansion joints, coatings, flexible fuel lines, and coated fabrics
polysulfide rubbers	paint spray, lacquer, fuel hose, gaskets, sealants for aircraft wing tanks and fuselages, cements, and putties
silicone rubber	sealants, formed-in-place gaskets, seals, hoses, O-rings, automotive seals, spark-plug boots, adhesives, and pastes
polyimides	Torlon poly(amide-imide) is used for bushings and thrust-washers in aircraft jet and stationary engine parts, and in seals; Ultem polyetherimide for automotive and aircraft applications in oil and fuel exposure; thermoplastic polyimide for bearings, bushings, seals, thrust-washers, and piston rings in hot and corrosive environments; fibers made from this material are used for filtering corrosive gases and liquids

Table 3. (Continued)

Polymers	Applications
PP, EP elastomers[h]	for highest resistance to chemicals in pipes and valves, PTFE is used, followed by PVDF, PVDC, and polypropylene. Other PP uses are in containers for food and household chemicals and in pump components. EP rubber is used for chemical-transfer hoses, pump seals, and gaskets, ie, except for oil or hydrocarbon solvent exposure
thermoplastic polyesters	poly(butylene terephthalate) (PBT) is used in under-the-hood automotive applications, and poly(ethylene terephthalate) as packaging material and tape
butyl rubber	hose, seals, gaskets, and uses not requiring exposure to petroleum solvents and oils
nylon	mechanical parts functioning in solvent and hydrocarbon environments; under-the-hood automotive applications

[a] PTFE = polytetrafluoroethylene; FEP = fluorinated ethylene–propylene copolymer; PFA = perfluoroalkoxy resins.
[b] ETFE = ethylene–tetrafluoroethylene copolymer; ECTFE = ethylene–chlorotrifluoroethylene copolymer.
[c] PVDF = poly(vinylidene fluoride).
[d] PNF = phosphonitrilic polymers.
[e] PCTFE = polychlorotrifluoroethylene; PVF = poly(vinyl fluoride).
[f] PVC = poly(vinyl chloride); CPVC = chlorinated poly(vinyl chloride); PVDC = poly(vinylidene chloride).
[g] Except polypropylene.
[h] PP = polypropylene; EP = ethylene–propylene.

BIBLIOGRAPHY

"Chemically Resistant Polymers" in *EPST* 1st ed., Vol. 3, pp. 665–683, by Henry Z. Friedlander, Dorr-Oliver, Inc.

1. J. Agranoff, ed., *Modern Plastics Encyclopedia 1983–1984,* McGraw-Hill, Inc., New York, 1983, pp. 412 and 413.
2. B. Golding, *Polymers and Resins—Their Chemistry and Chemical Engineering,* D. Van Nostrand Co., Inc., Princeton, N.J., 1959, p. 126.
3. C. Harper, ed., *Handbook of Plastics and Elastomers,* McGraw-Hill, Inc., New York, 1975, Chapt. 4, p. 4–63.
4. *ASTM D 543, Resistance of Plastics to Chemical Reagents,* American Society for Testing and Materials, Philadelphia, Pa., 1984.
5. *ASTM C 581, Chemical Resistance of Thermosetting Resins Used in Glass Fiber-Reinforced Structures,* American Society for Testing and Materials, Philadelphia, Pa., 1983.
6. Ref. 1, p. 414.
7. *ASTM D 1693, Environmental Stress-Cracking of Ethylene Plastics,* American Society for Testing and Materials, Philadelphia, Pa., 1970.

General References

J. Agranoff, ed., *Modern Plastics Encyclopedia 1984–1985,* McGraw-Hill, Inc., New York, 1984, pp. 423–428.
Rubber Plast. News **XIII,** 489 (Dec. 1983).
R. B. Seymour, *Plastics vs. Corrosives (Society of Plastics Engineers Monographs),* Wiley-Interscience, New York, 1982, Chapt. 14, pp. 147–262.

ROBERT A. MCCARTHY
Springborn Group, Inc.

COMPOSITES

Combining and orienting materials to achieve superior properties is an old and well-proven concept; examples of this synergism abound in nature (1). Wood contains an oriented hard phase which provides strength and stiffness, and a softer phase for toughness. Other natural composites are found in teeth, bones, bird feathers, and plant leaves. Synthetic composites are found in artifacts and recorded history. The use of chopped straw by the Israelites to control residual cracking in bricks is one example (2). More representative of modern structural composites are Mongolian bows, which were laminates of wood, animal tendons, and silk (2), and Japanese samurai swords, formed by repeated folding of a steel bar back upon itself. The resulting structure contains as many as 2^{15} alternating layers of hard oxide and tough, ductile steel.

A composite can be defined as the material created when two or more distinct components are combined, but this definition is too broad to be useful; even if limited to polymers, it would include copolymers and blends, reinforced plastics (qv), and materials such as carbon-black-filled rubber (qv). In this article, composites are those materials formed by aligning extremely strong and stiff continuous fibers in a polymer resin matrix or binder. The materials in this class have exceptional mechanical properties and are often termed advanced composites to distinguish them from chopped-fiber or otherwise-filled polymers. Indeed, the term reinforced plastic is a misnomer when applied to advanced composites, because the fibers are what make them useful; the fibers are typically 50 times stronger and 20–150 times stiffer than the matrix polymers. The role of the matrix is primarily that of a glue or binder which enables the fibers to support the applied loads. Advanced composites are more accurately described as bonded-fiber materials than as reinforced plastics (2).

The fibers that dominate the field of advanced composites are, in order of chronological development (3), S-glass, boron on a tungsten filament core, graphite or carbon, and aromatic polyamides (Kevlar). They possess the desirable properties of low density (1.44–2.7 g/cm³) and extremely high strengths (3–4.5 GPa) and moduli (80–550 GPa).

When combined with a resin binder and laminated to support the applied loads, these fibers provide mechanical properties equal to or exceeding those of most metals. Typical fiber and unidirectional composite properties are given in Table 1. Their low density produces specific mechanical properties (normalized by density) that considerably exceed those of metals. A comparison of specific stiffness and tensile strength of unidirectional advanced composites and three common structural metals is shown in Figure 1. Such plots are misleading as guides to material selection. They compare tensile properties of the composites in one direction with properties that the isotropic metals have in all directions. It is preferable to compare quasi-isotropic properties obtained by arranging fibers to produce planar isotropy. Also, there are other factors to consider in choosing a material for an application (see MATERIALS SELECTION).

Three factors affect the competition of advanced composites with traditional engineering materials: cost, reliability, and complexity (4). The cost barrier has been reduced by mass production, and fiber prices have either fallen or remained constant. Fibers exhibit linear elastic behavior to failure at strains of only 1–3%.

Table 1. Comparative Fiber and Unidirectional Composite Properties

Fiber/composite	Elastic modulus, GPa[a]	Tensile strength, GPa[a]	Density, g/cm³	Specific stiffness, MJ/kg	Specific strength, MJ/kg
E-glass fiber	72.4	2.4	2.54	28.5	0.95
epoxy composite	45	1.1	2.1	21.4	0.52
S-glass fiber	85.5	4.5	2.49	34.3	1.8
epoxy composite	55	2.0	2.0	27.5	1.0
boron fiber	400	3.5	2.45	163	1.43
epoxy composite	207	1.6	2.1	99	0.76
high strength graphite fiber	253	4.5	1.8	140	2.5
epoxy composite	145	2.3	1.6	90.6	1.42
high modulus graphite fiber	520	2.4	1.85	281	1.3
epoxy composite	290	1.0	1.63	178	0.61
aramid[b] fiber	124	3.6	1.44	86	2.5
epoxy composite	80	2.0	1.38	58	1.45

[a] To convert GPa to psi, multiply by 145,000.
[b] Aromatic polyamide.

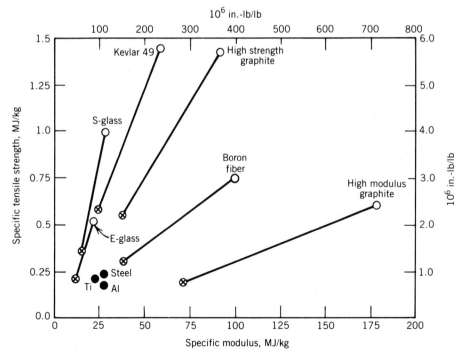

Fig. 1. Specific strength vs specific modulus of unidirectional and quasi-isotropic materials; fibers are in an epoxy matrix. ○, unidirectional composite; ●, bulk material; ⊗, quasi-isotropic composite. To convert MJ/kg to 10⁶ in.-lb/lb, multiply by 4.023.

Engineers accustomed to designing to yield with a built-in safety margin of 10% or more strain due to plastic flow have been skeptical about using these materials as primary structures. Other reliability concerns include reported variability in properties and lack of information about response to fatigue, sustained loading, and different environments. Complexity stems from the anisotropy and inhomogeneity (on the microscale) of advanced composites. Besides the obvious directional dependence of their thermoelastic properties, composites exhibit a variety of strengths and failure modes. Thus, direct substitution of a composite for an existing metal using traditional isotropic material design, manufacturing, and joining practices has led to unsuccessful designs (see also COMPOSITES, FABRICATION).

The first significant use of advanced composites came in the aerospace industry, where weight is at such a premium that new lightweight alloys and reductions of materials by improved stress and failure analyses are sought continuously. Advanced composites were first used in filament-wound rocket-motor casings and in laminated fiber–epoxy composites for airframes. Successes have been achieved by cooperation among design engineers, structural analysts, materials scientists, process engineers, and quality-assurance engineers. In the sporting-goods industry, traditional materials are being supplanted by advanced composites; tennis rackets, fishing rods, golf clubs, and kayaks are common examples of this. Fiber composites are also used in pipes (see PIPE).

Principles of Composite-property Improvement

Structural materials require strength, stiffness, and toughness. Strength measures the maximum stress a material can withstand; stiffness, strain under a given stress; and toughness, resistance to stress concentrations and the propagation of flaws. Other properties, such as resistance to corrosion, creep, fatigue (qv), temperature, or moisture, are also needed in most structural materials. However, these properties are important because of their effect on strength, stiffness (dimensional stability), and toughness.

Fiber Role. Bulk polymers typically have room-temperature strengths, 30–130 MPa, which are low compared with steel or aluminum alloys, and their low moduli, ca 2–4 GPa, would result in unacceptably large deformations in structures of any size. The first principle of continuous-fiber composites is that the fibers must support all main loads and limit deformations acceptably. A striking example of the latter requirement is the now-famous drawing of an aircraft in which the low modulus wings are strained 1.6% (Fig. 2) (5). Even though materials can strain to this extent without failing, the function of the aircraft is seriously impaired. Ideally, then, fibers are oriented so that the matrix experiences only negligible stresses, and overall deformation is controlled by the fiber modulus.

The material must contain a sufficiently high volume fraction of fibers aligned in directions as needed to support the applied loads. In contrast to chopped fibers in reinforced plastics, continuous fibers, eg, boron monofilaments or yarns and woven cloths of graphite, aramid, or glass, can be precisely placed in a structure

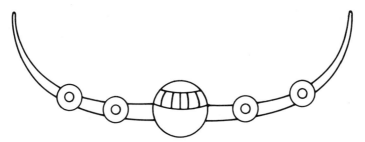

Fig. 2. Aircraft with wings strained 1.6% (5). Courtesy of Penguin Books, Ltd.

and provide good filament packing and high fiber volume. Continuous-fiber composites routinely have strengths two to three times greater than discontinuous-fiber composites made from the same constituents.

Predictions of composite strength and stiffness as a function of constituent properties are often controversial (1). However, to a first approximation, composite tensile properties in the fiber direction (L) can be estimated by a rule of mixtures based on fiber (f) properties, and neglecting the matrix contribution. Thus,

$$E_L \cong E_f v_f \tag{1}$$

$$F_L \cong F_f v_f \tag{2}$$

where E, F, and v_f refer to modulus, strength, and fiber volume fraction, respectively. The need for a large fiber volume is obvious from these expressions.

Strength and stiffness in the direction normal to the fibers are much less. Advanced composites are anisotropic, with strength, stiffness, and toughness functions of direction relative to that of the fiber. Anisotropy constitutes one of the chief differences between advanced composites and isotropic structural metals.

Matrix Role. Although properties of the matrix do not appear in equations 1 and 2, the matrix does play an important role. It maintains the desired fiber orientations and spacings, transmits shear loads between layers of fibers so that they resist bending and compression, and protects the fiber from surface damage. The matrix role is more subtle and complex in the areas of strength and toughness. Glass, graphite, and boron fibers are linear elastic, brittle solids. They fail catastrophically, without plastic flow, whenever the stress on them is sufficient to cause unstable flaw growth. Flaws are typically surface notches or steps from handling or processing.

The tendency toward brittle-fiber fracture is mitigated in several ways. The fibers are quite small and of high perfection; any flaws they contain are therefore small. Because the fibers are separated by a layer of matrix, the stress concentration and thereby the tendency to transmit the crack from the broken fiber to an intact one are reduced. The interfacial bond between fiber and matrix may fail in either transverse tension or shear ahead of the crack and cause splitting along the interface, normal to the crack, effectively blunting the crack. Fibers are not all equally strong; when a weaker fiber breaks, the crack does not propagate if the neighboring fibers are sufficiently strong to withstand the resulting stress concentration. When the material is stressed enough so that some fiber failures do occur, the variation in fiber strengths produces multiple breaks in such a way that the fibers do not all fail in the same plane. More crack surface is formed and some fibers are pulled out, each process dissipating energy.

Thus, the combination of a brittle fiber in a brittle matrix produces a material that is quite tough, indeed much tougher than either of them alone. This synergism is achieved by a combination of mechanisms that tends to keep cracks small, isolated, and blunted and that dissipates energy (2). These mechanisms, based in part on the heterogeneity of composites, constitute another difference from structural metals. Toughness in advanced composites is achieved without the large-scale plastic flow seen in tough metals.

The matrix also transfers load into the fibers. Besides the need to transmit shear between layers, tensile loads must be reintroduced into a fiber that has broken or otherwise ends within the material. This problem is studied by shear-lag analysis (6–8). A shear stress is present between the matrix and fiber at its end. The fiber tension gradually increases until it equals that of its neighbors, and since normal strain is equal in fiber and matrix, the shear stress becomes zero. The length of fiber necessary to reach full tension is the critical or ineffective length; typically, it is on the order of 2 to 200 fiber diameters. The absolute value of this length is more important to discontinuous-fiber-reinforced plastics than to advanced composites. However, the concept is important to both, because it explains why a broken fiber carries the same load over most of its length as an unbroken one.

High values of matrix and interfacial shear strength are required to keep the ineffective fiber length small. This requirement, together with the higher density of stress concentrations in a discontinuous- or chopped-fiber material, explains the lower strengths measured in such materials. This phenomenon has been analyzed (9); for comparable fiber volume fraction, a chopped-fiber composite with carefully aligned fibers of large aspect ratio can approach continuous-fiber properties in stiffness but not in strength. Experimental studies on carefully made E-glass–epoxy (10,11) show that the modulus of a discontinuous-fiber composite nearly equals that of the continuous-fiber material, but its strength is only 59% of the latter's. In short, it is easy to stiffen a reinforced plastic, but difficult to strengthen it.

The role of the matrix in advanced composites is more complex than that of the fiber. Compromises are necessary depending upon the stress state and application. In applications where the fibers are loaded only in tension, the matrix and interface should be as weak as possible to take full advantage of the toughening that that condition produces. This has been demonstrated for Kevlar–epoxy pressure vessels (12). By applying a silicone release agent (qv) to the fibers before they are wet-wound into a vessel, a 20% increase in burst pressure over vessels made with conventional, ie, strong, interfacial bonding was obtained. However, structures supporting only in-plane tensile loads are not usual. The most common structural elements are beams, plates, and shells, all of which must withstand bending and torsion moments that give rise to combinations of tensile, compressive, and shear stresses. Because of their small diameters, fibers of graphite, glass and Kevlar are not stable in compression unless well bonded to the matrix. Common failure modes are interfacial cracking or microbuckling of the fibers into resin-rich or void regions. To support shear and compressive loading, the composite must contain a consistent and strong interface, and be relatively void-free. Materials that satisfy all the requirements of a matrix and interface do not exist, and the failure-limiting constituent in advanced composites is often the resin matrix.

Prediction and Measurement of Lamina Properties

The simplest representative form of advanced composites is the unidirectional ply, or lamina, which consists of fibers oriented in a single direction and bonded together by a resin. If the properties of this assemblage are known, it is possible to calculate most thermoelastic and transport properties of laminates containing arbitrary numbers and orientations of laminae. The unidirectional lamina may be considered the basic building block of more complex laminates (qv).

A discussion of lamina properties must treat inhomogeneity, anisotropy, and the commonly invoked assumptions of micromechanics. A fiber–resin composite is heterogeneous on the microscale. However, if the volume of material used to measure or calculate a particular property is large compared with fiber and matrix–interfiber dimensions, the composite may be assumed homogeneous on the macroscale. Thus, heterogeneity of fiber and matrix does not affect modulus and thermal conductivity, which depend only on the direction (not on the position) within the lamina. Directional dependence of properties is the basis for anisotropy, which is large because of the large differences between properties parallel and transverse to the fibers. Anisotropy creates complexities; in the worst case, 21 elastic constants must be specified for the fourth-rank tensor, which relates the six components of strain and stress by Hooke's law (13):

$$
\begin{Bmatrix} \epsilon_1 \\ \epsilon_2 \\ \epsilon_3 \\ \gamma_{23} \\ \gamma_{31} \\ \gamma_{12} \end{Bmatrix}
=
\begin{bmatrix}
S_{11} & S_{12} & S_{13} & S_{14} & S_{15} & S_{16} \\
 & S_{22} & S_{23} & S_{24} & S_{25} & S_{26} \\
 & & S_{33} & S_{34} & S_{35} & S_{36} \\
 & & & S_{44} & S_{45} & S_{46} \\
 & \text{symmetric} & & & S_{55} & S_{56} \\
 & & & & & S_{66}
\end{bmatrix}
\begin{Bmatrix} \sigma_1 \\ \sigma_2 \\ \sigma_3 \\ \tau_{23} \\ \tau_{31} \\ \tau_{12} \end{Bmatrix}
\tag{3}
$$

Here, ϵ_i and γ_{ij} are normal and engineering shear strains, respectively; σ_i and τ_{ij} are normal and shear stresses, respectively; and S_{ij} represents the components of the elasticity compliance matrix. Since a lamina has three planes of material symmetry, equation 3 may be simplified if the coordinate system is aligned with the principal directions of the material:

$$
\begin{Bmatrix} \epsilon_1 \\ \epsilon_2 \\ \epsilon_3 \\ \gamma_{23} \\ \gamma_{31} \\ \gamma_{12} \end{Bmatrix}
=
\begin{bmatrix}
S_{11} & S_{12} & S_{13} & 0 & 0 & 0 \\
 & S_{22} & S_{23} & 0 & 0 & 0 \\
 & & S_{33} & 0 & 0 & 0 \\
 & & & S_{44} & 0 & 0 \\
 & \text{symmetric} & & & S_{55} & 0 \\
 & & & & & S_{66}
\end{bmatrix}
\begin{Bmatrix} \sigma_1 \\ \sigma_2 \\ \sigma_3 \\ \tau_{23} \\ \tau_{31} \\ \tau_{12} \end{Bmatrix}
\tag{4}
$$

The resulting material, with nine elastic constants, is called orthotropic. Note that in equation 4 the normal-shear coupling terms (S_{ij}, $i = 1, 2, 3, j = 4, 5, 6$) vanish, which occurs only when the loading coincides with principal material axes. Ordinarily, such coupling is present and can produce indeterminate stress fields in testing and unexpected failures in applications.

A further simplification is possible if the fiber array is the same through the ply thickness as it is across the ply. The ply then has a plane with isotropic properties lying transverse to the fiber direction. This is the 2–3 plane in Figure 3. Designating the 1 direction as L and the 2 and 3 directions as T, the number of elastic constants needed is five.

$$
\begin{Bmatrix} \epsilon_1 \\ \epsilon_2 \\ \epsilon_3 \\ \gamma_{23} \\ \gamma_{31} \\ \gamma_{12} \end{Bmatrix} =
\begin{bmatrix}
\dfrac{1}{E_L} & \dfrac{-\nu_{LT}}{E_L} & \dfrac{-\nu_{LT}}{E_L} & 0 & 0 & 0 \\
 & \dfrac{1}{E_T} & \dfrac{-\nu_{TT}}{E_T} & 0 & 0 & 0 \\
 & & \dfrac{1}{E_T} & 0 & 0 & 0 \\
 & & & \dfrac{1}{G_{TT}} & 0 & 0 \\
 & & & & \dfrac{1}{G_{LT}} & 0 \\
 & \text{symmetric} & & & & \dfrac{1}{G_{LT}}
\end{bmatrix}
\begin{Bmatrix} \sigma_1 \\ \sigma_2 \\ \sigma_3 \\ \tau_{23} \\ \tau_{31} \\ \tau_{12} \end{Bmatrix}
\tag{5}
$$

In equation 5, the elastic compliances are written in terms of the more familiar and readily measured engineering constants E, ν, and G. Also, although there appear to be six, only five are independent, since $E_T = 2G_{TT}(1 + \nu_{TT})$.

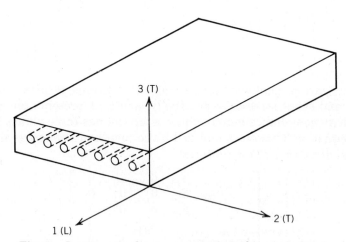

Fig. 3. Lamina coordinate system; 2–3 plane is isotropic.

Transverse isotropy also reduces the number of entries in the tensors, which comprise the thermophysical properties of the lamina, a partial list of which includes:

Elastic: E_L, E_T, ν_{LT}, G_{LT}, ν_{TT}

Thermal expansion: α_L, α_T

Thermal conductivity: k_L, k_T

Diffusivity: D_L, D_T

Moisture expansion: β_L, β_T

To these can be added viscoelastic properties, electrical conductivities, and dielectric constants. Also required are limiting properties, such as strength, maximum strain, or dielectric strength, which differ from constants relating linear material response to applied fields. Full characterization of mechanical strength requires measurement of longitudinal and transverse failure stresses in both tension and compression, as well as interlaminar and out-of-plane shear.

Micromechanical Analysis

Even with the simplification afforded by transverse isotropy, there is still a formidable array of properties. Because of the large number of measurements needed to characterize a lamina fully, techniques for estimating lamina properties from properties of the constituents have been developed. This field, known as composite micromechanics, utilizes many of the principles of continuum (macro) mechanics to estimate properties and provide an understanding of composite behavior on the microscale; this enables design and fabrication of composites having desired physical properties (13) (see also MICROMECHANICAL MEASUREMENT).

Elasticity Constants.   Because of its importance to stress analysis, the most intensely analyzed property has been elastic stiffness. The analyses range in complexity from simple but unrealistic "netting" analysis to three-dimensional elasticity approaches which can only be solved numerically. Most are based on certain common assumptions about constituent behavior: both fiber and matrix are assumed to be elastic, homogeneous solids, completely bonded in a void-free assemblage and subjected to small strains. The lamina itself is usually assumed to be transversely isotropic with fibers regularly spaced and exactly aligned so that the lamina properties are statistically homogeneous.

When a lamina is subjected to a uniaxial tensile or compressive stress in the fiber direction, the material deforms, away from the grip region, with approximately equal strains in fiber and matrix. This occurs because both constituents are continuous between load points, and the applied load is divided between fiber and matrix in proportion to their relative moduli. The resulting longitudinal modulus and Poisson's ratio are given by

$$E_L = E_f v_f + E_m v_m \tag{6}$$

$$\nu_{LT} = \nu_f v_f + \nu_m v_m \tag{7}$$

where m = matrix. These rule-of-mixtures expressions usually agree quite well with data (13,14), ie, to within the limits of measurement of E_f and ν_f. Equation 6 is an upper bound for E_L (1) and is approximated by equation 1 for $E_f \gg E_m$.

The remaining elastic properties, E_T, G_{LT}, and ν_{TT}, are more elusive. When a stack of unidirectional plies is bonded into a laminate and then loaded in transverse tension or compression, the matrix phase is continuous but stiffened to some extent by the fibers. The assumption of equal stress in each constituent leads to the inverse rule of mixtures, ie,

$$\frac{E_T}{E_m} = \frac{1}{v_m + \dfrac{E_m}{E_f} v_f} \tag{8}$$

$$\frac{G_{LT}}{G_m} = \frac{1}{v_m + \dfrac{G_m}{G_f} v_f} \tag{9}$$

Equations 8 and 9 predict values that are a factor of 2 below experimental results at v_f of interest (12,13) because the model does not account for the out-of-plane, longitudinal stiffening that the fibers provide when stressed transversely. Also, many fibers are anisotropic. However, the necessary transverse elastic properties have not been measured directly because of experimental problems.

Another predictive approach is the self-consistent model (15,16) based on the case of a fiber in an infinite medium with the same macroscopic properties as the composite. It has proven accurate only for low values of v_f. The variational approach using energy minima to establish upper and lower bounds for moduli has been used (17,18). Although better than those predicted by equations 8 and 9, lower bound transverse properties predicted by this method fall well below measured data. Other work using exact theory of elasticity improved accuracy somewhat, but involves numerical solutions too cumbersome for design use. It is also possible to incorporate anisotropic fiber properties into these analyses. However, obtaining direct measurement of such properties as E_T and G_{LT} of the fiber itself has proven very difficult.

The best compromise appears to be the Halpin-Tsai equation (19), which provides an adjustable constant set empirically for the material being analyzed. The general form for transverse properties is

$$\frac{M}{M_m} = \frac{1 + \xi \eta v_f}{1 - \eta v_f} \tag{10}$$

where

$$\eta = \frac{\dfrac{M_f}{M_m} - 1}{\dfrac{M_f}{M_m} + \xi}$$

and M = corresponding lamina property E_T, G_{LT}, ν_{TT}; M_f = corresponding fiber property E_f, G_f, ν_f; and M_m = corresponding matrix property E_m, G_m, ν_m. ξ is a measure of fiber reinforcement, which includes fiber geometry and packing, and boundary conditions. Typically, $\xi = 2$ for $M = E_T$ and $\xi = 1$ for $M = G_{LT}$. The semiempirical equation provides a means for extrapolating to other lamina properties or constituent combinations that would be costly to measure. In a comparison (20) of predictive techniques for E_T for graphite and Kevlar composites with three different values of E_m, the Halpin-Tsai equation gave predictions within experimental error.

The equation reduces to a rule of mixtures (eqs. 6 and 7) as $\xi \to \infty$ and to an inverse rule of mixtures (eqs. 8 and 9) for $\xi = 0$ (13). Figure 4 is a plot of upper and lower bounds and equation 10 for $\xi = 2$. The large difference in E_T/E_m at $v_f = 0.5$–0.6 is readily seen in the plot.

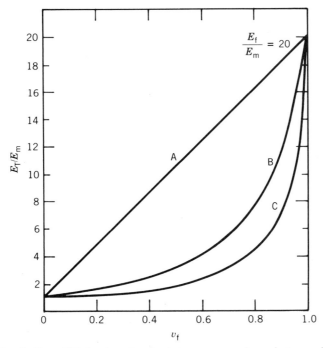

Fig. 4.   Prediction of E_T by A, rule of mixtures, upper bound, $\xi = \infty$; B, Halpin-Tsai equation, $\xi = 2$; C, inverse rule of mixtures, lower bound, $\xi = 0$.

Thermophysical Properties.   Transport properties such as thermal conductivity, diffusivity, and electrical conductivity are simpler to analyze for the transversely isotropic unidirectional lamina than elastic moduli because they are second- rather than fourth-rank tensors. An excellent summary of techniques used to predict bounds for this class of properties has been given (14). Longitudinal values are approximated quite well by the rule of mixtures,

$$k_L = k_f v_f + k_m v_m \qquad (11)$$

Transverse values estimated by the inverse rule of mixtures are, like their elastic counterparts, too low. The Halpin-Tsai equation can be adapted to predict these properties (21,22), as can expressions for dielectric constant (14):

$$\frac{k_T}{k_m} = 1 + \frac{v_f}{\dfrac{1}{k_f - k_m} + \dfrac{v_m}{2k_m}} \tag{12}$$

Properties such as the coefficient of thermal expansion α_i and moisture swelling coefficient β_i must include elastic as well as expansion properties of the constituents. Expansion coefficients are important to composite structural design because stresses and deformations due to temperature or moisture can cause failure in the lamina or produce unacceptable distortions during fabrication. Thermal and moisture expansion may be included in a generalized Hooke's law as

$$\{\epsilon_i\} = [S_{ij}]\{\sigma_j\} + \{\alpha_i\}\Delta T + \{\beta_i\}\Delta M \tag{13}$$

where $i,j = 1, 2, 3$ and where ΔT and ΔM are changes in temperature and moisture content, respectively. Thermal expansion and moisture swelling are second-rank tensors. If the strain axes are aligned with material axes, the expansion is dilatational only. Isotropy transverse to the fibers reduces the number of independent constants in the expansion tensors to two, ie, α_L and α_T. A simple relationship equating lamina properties to constituent properties and volume fraction has been derived (1,14,23). The most generally used expressions (24) are

$$\alpha_L = \frac{E_f v_f \alpha_f + E_m v_m \alpha_m}{E_f v_f + E_m v_m} \tag{14}$$

$$\alpha_T = (1 + v_f)\alpha_f v_f + (1 + v_m)\alpha_m v_m - (v_f v_f + v_m v_m)\alpha_L \tag{15}$$

These equations were obtained from thermoelastic energy considerations, and have been corroborated experimentally on several composites. As in the case of elastic constants, inaccuracies owing to fiber anisotropy and deficiencies in the needed elastic properties make it difficult to verify the equations completely. Equations 14 and 15 should be valid for all forms of dilatational swelling (25). This is true if the amount of moisture absorbed is unaffected by the presence of fibers and if the dilatation is totally elastic (14).

Strength and Failure Criteria

A unidirectional lamina exhibits strengths that are a function of both fiber orientation and loading direction. If the lamina is transversely isotropic, it exhibits six distinct ultimate strengths: longitudinal tensile (F_L) and compressive (F'_L); transverse tensile (F_T) and compressive (F'_T); interfiber shear (F_{LT}) and shear through the fiber (F_S). Predicting these from constituent properties has proven difficult. Many of the simplifying assumptions used in predicting elastic and thermophysical properties do not apply to strength analysis. Mechanical failure is usually caused by fiber defects, weak interfacial bonds, voids, or residual stresses. For failures controlled by the matrix and interface, eg, transverse tension and interfiber shear, there is considerable localized inelastic deformation before final fracture. In fiber-dominated failure, eg, longitudinal tension, differences in fiber behavior owing to defects or variation in diameter or modulus complicate analysis. In general, predictions of values or bounds of lamina strengths have not been as successful as thermoelastic micromechanics analyses.

Longitudinal Tension. Longitudinal tensile strength, a key property for most composite applications, has been the most thoroughly analyzed. The simplest approach assumes a uniform strain in fiber and matrix, and uniform strength F_f in all fibers; for this case, the strength is

$$F_L = F_f v_f + F_m^* v_m \tag{16}$$

where F_m^* is the matrix stress at the strain where fiber failure occurs. Since $F_m^* \ll F_f$, equation 16 reduces to a simple mixture rule given by equation 2. For most polymer matrices, the neat resin failure strain is greater than the fiber's, but the *in situ* failure strain is less because of axial and transverse residual stresses set up in the resin by shrinkage and differing thermal contractions; resin strain is further aggravated by the constraints of dissimilar Poisson contraction. Thus, considerable resin cracking generally occurs before fiber fracture. Equation 16 is valid only for material with fiber-controlled fracture, ie, for reasonably large v_f. A number of analyses have been advanced to predict critical v_f; typically, these result in very small values, eg, 2.6% for graphite–epoxy and 0.6% for glass–polyester (26). Thus, for practical structural composites, equation 2 is a good approximation.

The simplicity of equation 2 is complicated by the lack of a simple value for F_f. Like all material properties, F_f is a statistical variable, with a range of values, but unlike most properties, it is not well described by a mean value. The lower values of fiber strength play a disproportionately large role in the tensile strength exhibited by a fibrous composite.

Fiber strength is usually measured by gauge lengths from 2.5 to 250 mm. The shorter lengths exhibit higher mean strengths but larger coefficients of variations (CV). This has been attributed to the decreased likelihood of finding a large flaw in a small volume of material. In modern fiber production, another likely cause is the variation in fiber diameter along its length; this is difficult to measure, but significantly affects calculated strength. When n fibers are tested as an untwisted (parallel) bundle, the strength is always less than the average fiber strength for that gauge length, with a lower CV. In this configuration, once a fiber fails, it is removed from the load train at every point along its length. All other fibers are subjected to an overstress at all points. Failure is thus a cumulative process governed by the weakest fibers. In a twisted bundle of fibers, the broken fiber can support load away from the break because of shear stress due to fiber–fiber friction. The mean strength of a slightly twisted bundle is higher than an untwisted bundle because the weak links do not play such a dominant role. Adding resin matrix to an untwisted bundle yields the same result as twisting without introducing stress gradients. The impregnated bundle, called a strand, most nearly represents an actual lamina, and exhibits the highest strength, typically higher than the average F_f for a fiber of the same gauge length. A fiber break in this configuration is restricted to a small region, called the ineffective length, which is smaller than the affected length in the twisted bundle because fiber–matrix friction and bond strength are greater than fiber–fiber friction.

The issues central to estimating composite strengths are inherent strength variation at a given length, strength–length dependence, stress concentration, and ineffective length. Data from the first two have been modeled with some success by an extreme value distribution of the Weibull type (27):

$$P(F) = 1 - \exp\left[- \frac{\ell}{\ell_0} \left(\frac{F}{F_0}\right)^m \right] \tag{17}$$

where F_0 is the Weibull scale parameter, m is the Weibull shape parameter, and ℓ_0 is the reference gauge length. Equation 17 is plotted in Figure 5 for three different gauge lengths ($m = 10$). If, as shown, F_0 does not vary with ℓ/ℓ_0, Figure 5 demonstrates the usual observation that shorter gauge-length fibers exhibit higher strengths. The shape parameter is usually not the same for all gauge lengths, since it is inversely proportional to the coefficient of variation of the data ($\cong 1.2/CV$). Typically, m is less, ca 5–10, for short gauge lengths and greater, ie, ca 10–25 for large ones. The effect of reduced m is to flatten $P(F)$ over a large range of F/F_0, indicative of greater scatter.

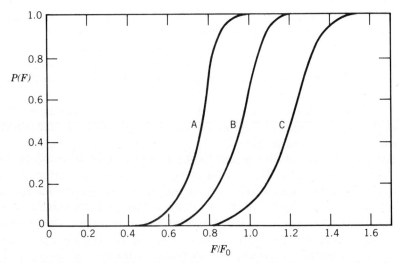

Fig. 5. Weibull probability plot of fiber strength for three gauge lengths ℓ/ℓ_0: A, $\ell/\ell_0 = 10$; B, $\ell/\ell_0 = 1$; C, $\ell/\ell_0 = 0.1$. $P(F) = 1 - \exp\left[- \frac{\ell}{\ell_0}\left(\frac{F}{F_0}\right)^m \right]$ and $m = 10$.

Equation 17 suffers from two deficiencies. It predicts that some very long fibers have strengths approaching zero. Experiments on optical communication fibers several kilometers long have shown no such trend. Also, it does not contain a limiting upper bound for the strength of very short lengths, which cannot exceed the theoretical strength of the unflawed material.

When a fiber breaks, the surrounding material is subjected to a stress concentration and the broken fiber is unloaded for a distance, called the ineffective or critical length. The simplest stress-concentration treatment is to assume all other fibers in the plane share the load equally, ie, they experience $n/(n - 1)$ times the applied stress. This approach has been used to predict failure in an unimpregnated fiber bundle, where the ineffective length is infinite (28), and has been applied to continuous-fiber composites modeled as a chain of bundles, the length of which was equal to the ineffective or critical length (29). The latter cumulative weakening model predicts a considerably higher strength than the former does, but it does not agree well with composite data, since it tends to overestimate observed strengths.

A more realistic analysis (30) of strength using calculated (31) local stress concentrations around a broken fiber predicts rather low strengths. Although these analyses (29,30) provide bounds on strength, they are too far apart to be of much practical use. More accurate analysis of the effect of local stress concentration has been presented (27).

Each of the composite statistical strength models relies on knowledge of ineffective or critical length as an integral part of the analysis. Analyzed first in 1952 (6), and subsequently (7,8,32), the ineffective length has proven an elusive parameter to compute or measure. It depends upon a large number of fiber, matrix, and interface properties, some of which cannot be measured directly. Estimates vary from a few fiber diameters for strongly bonded materials to 10^4 fiber diameters for poorly bonded or large strain-to-failure fibers. The situation is even more complex for Kevlar fibers, which do not fracture straight across. The definitive, experimentally corroborated determination of critical length has not yet been done. Until this occurs, it is justifiable to use equations 16 or 2 with a fiber strength corresponding to a reasonable, ie, 25–250 mm, gauge length.

Transverse, Compressive, and Shear Strengths. The tensile strength of a unidirectional lamina transverse to the fiber direction is quite low, less than the strength of the parent resin. Instead of reinforcing the resin, the fibers appear to act as stress concentrators in this direction. Transverse tensile strength is also reduced by voids, weak interfacial bonds, and residual stresses owing to cure shrinkage and thermal-expansion mismatch. Unlike longitudinal strength, transverse strength is quite sensitive to fiber spacing; closely spaced fibers produce severe stress or strain amplification. If the fiber–matrix bond is poor, analysis (33) that treats the fibers as holes applies. The effect of voids has been analyzed with reasonable experimental agreement (34). Estimates for F_T are 40% and 20% of matrix strength for well- and poorly bonded fibers, respectively. Attempts to increase F_T by use of thin coatings or rubberized or ductile resins have not shown much promise, perhaps because at 60 vol % fiber there is too little volume of resin left for any significant inelastic deformation.

Since failure in transverse tension does not seriously affect fiber load-carrying capacity, it is not usually a concern. This is not the case for longitudinal compressive strength, which can constitute a primary failure mode. Unlike longitudinal tensile strength, which is dominated by fiber strength, F'_L depends upon various factors: these include fiber alignment and spacing, matrix shear modulus, interfacial bond strength, and void content. Compressive-strength testing is complicated by the occurrence of a number of failure modes: longitudinal splitting due to Poisson effects, fiber microbuckling, and shear or kinkband formation.

A comprehensive theory does not exist, but certain general observations can be made. Well-made composites with large-diameter, nearly isotropic fibers (such as boron) can have F'_L values that considerably exceed F_L. Smaller-diameter isotropic fibers such as glass are usually stronger in compression than tension. Anisotropic fibers such as Kevlar are usually much poorer in compression than tension. Early models of compression failure tended to grossly overestimate compressive strengths in anisotropic fibers. Typically, F'_L is larger if the failure is controlled by shear rather than fiber microbuckling (26). In the best graphite composites, ie, well-bonded, void-free, F'_L approaches F_L. In aramid composites, typically having poor interfacial bonds and fiber that readily shears in compression, F'_L is about 20% of F_L.

Interfiber shear strength F_{LT}, also termed intralaminar shear, depends upon many of the same factors as F_T and F'_L do. The stress concentration due to fibers becomes acute at high fiber volume fraction (35) or, equivalently, when fibers are bunched together. Composites with lower modulus, toughened resins achieve a higher fraction of the neat resin shear strength than do those with brittle matrices. The limiting value of F_{LT} seems to be the matrix shear strength, which is usually about 60% of its tensile strength.

Failure due to transverse compression F'_T and shear through the fiber F_s has not been thoroughly studied or often measured. Each requires that fibers fail under loads applied normal to the fiber direction. Consequently, the test conditions needed for a unique failure are difficult to achieve. The values measured tend to be very large, ie, on the order of $F_L/2$, and many times greater than F_T or F_{LT}. These latter two are often exceeded in testing for F'_T.

Phenomenological Failure Criteria.   Complexity of failure processes and the concurrent difficulty in modeling have led to the adoption of several types of empirical criteria. The most common are maximum stress, maximum strain, and various polynomials that define a failure surface. The maximum stress criterion states that failure occurs when the maximum stress in a particular direction exceeds the corresponding lamina material strength, independent of any other stresses present in the body. The maximum strain criterion is analogous; the appropriate strengths and ultimate strains must be determined for each material and failure mode of interest. The maximum stress and strain criteria are simple to apply and have the added advantage of indicating which type of failure has occurred.

Most of the interactive polynomial failure criteria is similar to the Von Mises yield surface, modified to account for material anisotropy. Of these, the most common are the Hill-Tsai or Tsai-Wu polynomials (36), which account for the coupling of one stress on failure in another direction. They are nonspecific about failure mode and weigh failure in transverse tension or shear equally with longitudinal tension, and may provide a more accurate representation of the angular dependence of strength than the maximum stress or strain criteria. In particular, the maximum stress criterion appears to predict that longitudinal tensile strength is greater when the lamina is loaded at a nonzero angle to the fiber direction. This would lead to a physically implausible situation in which poor fiber alignment would increase tensile strength. Several studies (37,38), however, refute this criticism, which is based in part on a misunderstanding of the fiber stresses present in a tension test. There appear to be no strong arguments for selecting polynomial criteria over the simpler maximum stress or strain criteria. One survey (39) shows the latter two to be the most popular among composite structural designers.

Toughness.   In most structures, failure originates at a stress concentration necessary for design function, ie, bolt or rivet hole, joint, cutout, etc, or at a flaw, ie, manufacturing defect, in-service damage, etc. A useful structural material must tolerate reasonable stress concentrations and flaws without failing. The material property that quantifies this tolerance is toughness.

In isotropic structural materials, the theoretical and experimental bases for fracture toughness are well established and have been successfully used by designers for many years. The key concept of linear elastic fracture mechanics is that, in the vicinity of a crack tip, the stress field can be described by a stress intensity factor K, which depends only upon the crack geometry and the magnitude and direction of the applied stress (40) (see also FRACTURE). For simple tension on a plate containing a small crack of length $2a$ in its center,

$$K_I = \sigma\sqrt{\pi a} \tag{18}$$

When K_I reaches a critical value, called the fracture toughness K_{IC}, the material fails. K_{IC} is a measure of the flaw sensitivity of a material. Toughness may also be described in terms of the energy consumed or absorbed by the material during fracture, which is equivalent to the amount of work that must be done to extend the crack. In isotropic solids, this critical fracture surface energy $\mathcal{G}_{IC}$ is related to K_{IC}:

$$\mathcal{G}_{IC} = \frac{K_{IC}^2}{E} \tag{19}$$

Typically, K_{IC} is more useful to the designer, and has been especially valuable in predicting fatigue life of alloys. $\mathcal{G}_{IC}$ is more readily related to microscopic fracture mechanisms (41).

Simple relationships such as equations 18 and 19 do not exist for fiber composites because of their anisotropy and inhomogeneity. Anisotropy itself is not a great obstacle: rolled aluminum exhibits higher resistance to cracks perpendicular to the rolling direction than those running parallel to it. Aluminum, however, is not heterogeneous on a scale that violates the basic assumptions of fracture mechanics. Composites exhibit heterogeneity, large damage zones, and many different failure modes during fracture. There is no set of invariant, test-independent parameters to measure a composite's resistance to fracture by an arbitrarily oriented crack; it is unlikely that such a set of properties exists. However, with certain restrictions, fracture mechanics can be applied to fiber composites; this has been explained (42).

For fracture parallel to the fibers, the crack tends to run either in the matrix or along the interface, and fracture mechanics has been used to predict crack growth under both transverse and shear loading (43). The measured fracture toughness and critical surface energy values are comparable to the neat resin (41).

For the technologically important case of cracking across fibers, the situation is more complicated. In this case, the crack must traverse fiber, interface, and matrix. To measure fracture toughness in metals, introduction of a sharp crack that grows colinearly across the specimen is necessary. It is practically impossible to grow a sharp crack in a unidirectional resin matrix composite, because of notch-tip stresses exceeding the interfiber shear or transverse tensile strength. Instead of growing colinearly, a transverse notch turns 90° and runs parallel to the fiber. Such splitting results in an apparent notch insensitivity, ie, an immeasurable fracture toughness, but is in fact an artifact of the test geometry and loading rate available in a typical testing machine. This was demonstrated in an elegant experiment (44) in which an unnotched tensile specimen was photographed during fracture; it showed that the main crack completely traverses and fractures the specimen before any interfiber splitting occurs. Crack blunting or branching by delamination does not occur during rapid fracture. Thus, toughness for trans-fiber fracture is a valid concept under certain loading conditions, but it is not readily measured in a unidirectional material. Nor has an analytical method of predicting K_{IC} from constituent properties been developed.

In longitudinal tension, composites do exhibit some notch sensitivity, but much less than would be expected from an assemblage of two brittle materials. Cracks shorter than the ineffective length of the material do not affect static strength (41). For longer cracks or those forming when fibers fall under load, a composite absorbs energy principally by a combination of debonding and fiber pullout.

Fracture surface energies have been measured for a variety of composites and are much higher than would be predicted by simple mixture rules (41). Typical values range from 10^3 to over 10^4 J/m²; these exceed those of some structural metals.

The most promising fracture toughness models account for the heterogeneity of the composite. A shear-lag analysis has been used to calculate the stress concentration factors in fibers adjacent to single and multiple breaks (31,45). For the simple case of a well-bonded plane of fibers, the local stress concentration owing to r broken fibers is given by

$$K_r = \frac{4 \cdot 6 \cdot 8 \ldots (2r + 2)}{3 \cdot 5 \cdot 7 \ldots (2r + 1)} \tag{20}$$

This analysis was extended to include matrix yield and plastic flow in a well-bonded unidirectional boron–aluminum composite (46). However, a general treatment for the more difficult case of a poorly bonded resin matrix composite has not been done. Toughness is known to increase with decreasing interfacial bond strength; indeed, the least notch-sensitive fibrous structure is a dry bundle. The stress concentration K_r for r broken fibers in this case is only $n/(n - r)$, less than that predicted by equation 20. The value of K_r for a partially debonded material lies between these two values. As mentioned, a weak interfacial bond reduces tensile strength and, more critically, ruins compressive and shear strengths. The optimum interfacial bonding for all possible loadings has not yet been, and perhaps cannot be, determined.

Toughness and strength place conflicting demands on the bond required. In general, fracture mechanics is valid for composites if the crack direction starts and remains parallel to one of the principal material directions, and if the damage zone is small compared with the crack dimensions. Fracture surface energy $\mathcal{G}_{IC}$ may be more useful for screening composites for energy-absorption and crack-deflection capability. It also yields information on which fracture mechanisms operate and what their relative (energy) contributions are.

Time-dependent Deformation and Failures

A comprehensive treatment of viscoelastic deformation in composites has been presented (47–49) and shows that for thermorheologically simple materials, creep and stress relaxation can be predicted in composites if the orthotropic viscoelastic constants are known. This is accomplished by adapting the correspondence principle of mechanics to anisotropic materials. Most continuous-fiber composites do not exhibit large viscoelastic strains in the fiber direction, since the fibers are nearly elastic at temperatures that the resin can withstand. Significant viscoelastic deformations can occur in the resin-dominated modes transverse to the fiber. These affect the residual stresses resulting from the cure cycle. A viscoelasticity approach has been used to calculate optimum cool-down cycles to minimize residual stress (50).

Whereas the effect of time on fiber-controlled deformation is not large, its effect on strength is pronounced. Glass fiber is especially susceptible to static fatigue, a loss of strength under sustained load. This effect is accelerated by moisture, which increases the crack-growth rate in stressed glass. An analysis combining the effects of slow crack-growth and fiber-strength statistics to predict failure time of glass–resin strands gives good experimental agreement (51).

Kevlar fibers also exhibit static fatigue (52), but the degradation mechanism is not well understood. Considerable simplification and improvement in life prediction was made possible when it was discovered that failure life varied significantly between specimens made from different spools of yarn (53). Failure time was determined to be a power-law function of stress, just as it is in glass, ie,

$$t = t_0 \left(\frac{\sigma}{\sigma_0}\right)^{-c} \tag{21}$$

where t_0, σ_0, and c are empirically determined constants. However, since Kevlar fibers do not fail in a simple tensile manner, it is not known if the functional form has any physical basis.

Preliminary studies on graphite-fiber composites show that they too exhibit static fatigue. The failure mechanism is not yet characterized (see also CARBON FIBERS).

Cyclic fatigue in composites has been reviewed (54). Prediction techniques suffer from a lack of well-developed models to describe crack growth as a function of stress and number of cycles. In stress states and failure modes for which fracture mechanics can be applied, it is possible to estimate fatigue life using the same approach used for metals. This approach can be used to predict fatigue crack growth in matrix-dominated failure modes.

Longitudinal fatigue properties of most advanced composites are excellent, provided the controlling fatigue mode is tensile fiber fracture. Only glass composites exhibit a steep S–N (stress vs number of cycles to failure) slope, probably because of stress corrosion cracking due to moisture. Fatigue resistance in composites probably stems from the nearly perfect elastic deformation of the fibers. Absence of hysteresis in the fiber indicates lack of plastic flow or lack of flaw growth over most of its strain cycle. Thus, the loss mechanisms ultimately causing fatigue failure in metals and polymers are not present in graphite or Kevlar fibers. Their fatigue limits are quite high, in excess of 60% of the static strength.

Test Methods

It is generally possible to measure thermal and elastic properties of composites by using test techniques similar to those that the ASTM has published for homogeneous isotropic materials. The specimen must be designed to allow for material anisotropy, particularly low shear modulus and strength. For example, the minimum span-to-depth ratio for valid determination of flexural modulus and strength has been calculated (55). Anisotropic elastic properties can give rise to large shear stresses in constrained specimens (56). This point has been elaborated upon (13); the importance of careful determination of the stress and strain field whenever the load does not coincide with material principal axes has been shown.

Grip effects in tensile testing of unidirectional composites continue to be a problem. It is difficult to break extremely strong, anisotropic materials away from the grip region. Grip failures can be avoided by using tabs with properties graded to match the composite elastic properties. A method to treat the effect of grip failure on fiber-strength statistics has been developed (57) (see COMPOSITES, TESTING).

Constituent and Composite Properties

Advanced composites derive their impressive structural properties primarily from high performance continuous fibers, which have been commercially available since World War II. The most important of these have been glass, boron, graphite, and aramid (Kevlar). Other fibers of interest include SiC and Al_2O_3, as well as several polymer fibers not yet in mass production. Graphite fibers are now the most widely used in advanced composites, followed by aramid, and then glass. Boron usage has lagged because of its continuing high cost.

Resins used as matrix materials for advanced composites are usually thermosets, eg, epoxy, polyester, phenolic, vinyl ester, polyimide, and silicone; epoxy is by far the most common. Some thermoplastics are in use, including thermoplastic polyimide, polysulfone, and polyetheretherketone (PEEK). The thermosets provide good chemical environmental durability, dimensional stability, and reasonable high temperature properties. However, they tend to be rather brittle, have limited shelf life, and are time-consuming to process. Thermoplastic matrices are not as stable dimensionally, but they are tougher, and simpler to process; no chemical handling or mixing is required.

Thermosetting Resins.   Epoxy resins (qv) are compatible with all fibers used in advanced composites and are the most commonly used. They adhere well to most fibers, cure with relatively little shrinkage compared with polyesters and phenolics, have no condensate on cure, are reasonably thermally stable, and are moderately priced. Disadvantages include brittleness, considerable moisture absorption, and a tedious processing cycle. Any molecule containing the oxirane ring is classified as an epoxy; only two are used extensively in advanced composites. They are the diglycidyl ether of bisphenol A (DGEBPA) and tetra-glycidylmethylenedianiline (TGMDA), ie, N,N,N',N'-tetraglycidyl-4,4'-diamino-diphenylmethane (58).

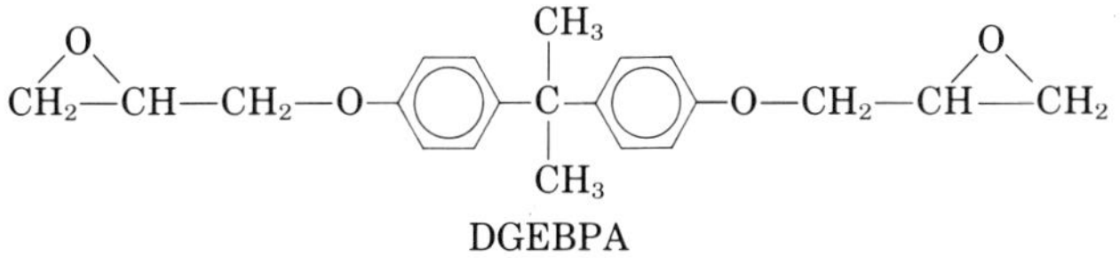

DGEBPA

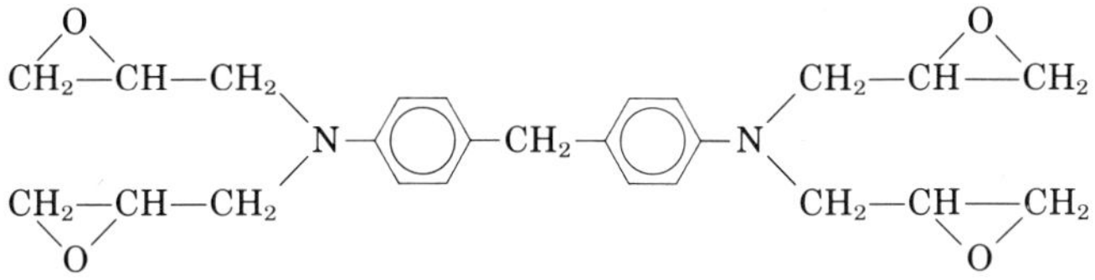

TGMDA

DGEBPA forms the basis for most filament winding (qv), which is typically a liquid resin process. TGMDA is used mainly as a partially polymerized, B-staged, material preimpregnated (prepregged) into the fibers and then cured in an autoclave or hot press. DGEBPA resins are also used as prepregs.

Epoxies react with various other molecules to form stiff three-dimensional networks. The usual curing agents are primary or secondary amines or anhydrides. Amines are more inert chemically and environmentally and therefore are more widely used. Depending upon the degree of cross-linking, the resulting material can have a low glass-transition temperature (ca 100°C) and behave as a hard elastomer, or be an organic glass with a high glass-transition temperature (ca 225°C). Maximum service temperatures for unmodified epoxies are ca 175°C (dry) and ca 125°C (wet) (59). Moisture drastically lowers the glass transition, and thus the service temperature. Typical neat resin properties are listed in Table 2.

Table 2. Properties of Matrix Resins[a]

Property	Epoxy	Polyimides, thermosetting
elastic modulus, GPa[b]	2.8–4.2	3.2
tensile strength, MPa[c]	55–130	55.8
compressive strength, MPa[c]	140	187
density, g/cm^3	1.15–1.2	1.43
thermal expansion, 10^{-6}/°C	45–65	50.4
thermal conductivity, W/(m·K)	0.17–0.21	0.36
glass-transition temperature, °C	130–250	370

[a] At RT.
[b] To convert GPa to psi, multiply by 145,000.
[c] To convert MPa to psi, multiply by 145.

Phenolics are too brittle and difficult to process to be used in pure form as matrix materials (see PHENOLIC RESINS). However, they can be copolymerized with an epoxy to form a structure with cross-link density greater than epoxies, with a maximum service temperature of ca 225°C. These resins are termed epoxy–novolacs.

Polyester resins form the basis for the chopped E-glass-fiber-reinforced composite industry. They exhibit high shrinkage during cure and do not have the environmental durability of epoxies. They are never used in advanced composites (59). Vinyl esters have been used successfully with Kevlar fibers. These resins show better moisture resistance than epoxy does when both are cured at room temperature (59).

Thermosetting polyimides were first synthesized at NASA Lewis Research Center. The material, known as PMR-15, is polymerized by an addition reaction and has a maximum service temperature of ca 350°C (60). It is brittle and requires extended high temperature, high pressure cure cycles. See Table 2.

Thermoplastic Resins.   Polyimide can also be made in an uncross-linked thermoplastic form by a condensation reaction. This material, DuPont's NR-150, is less expensive than the corresponding thermoset and easier to process. However, the volatile water condensate typically causes considerable, ie, 5–10%, porosity. Maximum use temperature, ie, from 300° to 325°C, is less than the polyimide thermoset. Mechanical properties are also somewhat reduced by the porosity.

A new class of thermoplastics, polyetheretherketone (PEEK) (qv) was introduced several years ago by ICI, Ltd. in the UK. Compatible with both glass and graphite, this material has better toughness and moisture resistance than epoxy. As a thermoplastic, it has an indefinite shelf life and can be formed rapidly into desired structural shapes. Its disadvantages are much higher cost ($55/kg) and lower use temperature (ca 150°C) than epoxy. The dimensional stability of PEEK under sustained load has not yet been determined.

Strong Fibers.   There are several features common to most fibers suitable for use in advanced composites. High strength requires an atomic structure with a closely spaced, continuous network of strong bonds. The theoretical strength of these fibers is on the order of $10^{-1} E$. None of the fibers exhibit any ductility on the macroscale; in tension, they behave as linear elastic brittle solids. The strengths they exhibit are largely owing to the reduction of stress-concentrating flaws. Most strong fibers are sensitive to surface damage caused by handling. The stress at the tip of a semielliptical crack of length a and a tip radius r subjected to a uniform stress σ_0 is (61)

$$\sigma_t = \sigma_0\left(1 + 2\sqrt{\frac{a}{r}}\right) \tag{22}$$

For brittle solids with atomically sharp cracks, r is ca 0.2 nm; for a crack only one micrometer long, $\sigma_t \cong 100\,\sigma_0$. Thus, if σ_t is the theoretical strength, the stress needed to fracture a specimen would be only $10^{-3} E$. This is typical of bulk specimens made from strongly bonded brittle materials.

Most strong fibers usually cannot be formed to produce a reduced cross section for strength measurements. Grip failures, along with surface-sensitive strengths, make it difficult to obtain accurate, reproducible mechanical test results. Consequently, strength, and to a lesser extent, modulus data in the literature may reflect scatter greater than what is actually in the material. Strength statistics are confused by the strength–length effect (see also ENGINEERING FIBERS).

Glass Fibers and Composites. Glass fibers are the oldest form of strong fibers used in composite structural materials. Continuous fibers are made by a gravity extrusion process. Molten glass is placed into a drawing furnace and is formed into fibers by passing through small-diameter orifices. After cooling, a protective, lubricating finish or size is applied to protect the fibers from self-abrasion and improve handleability (62). The fibers are collected into either an untwisted strand or twisted yarn and then wound onto a spool or "cake." The fiber is available either in this form or woven into a wide variety of cloths.

Many types of glass fibers have been made, but only two are used in structural composites: E (for electrical) and S (for strength). E-glass is a low alkali, aluminum–borosilicate composition with excellent electrical properties and reasonable strength and modulus. In chopped form, it is the staple of the fiber-glass molding-compound industry. S-glass is a magnesium–aluminosilicate composition with considerably higher strength and modulus (Table 3). S-glass has been used in filament-winding applications.

Adherence between glass fibers and most resins is poor. The protective sizing is usually formulated to promote wetting by the resin as well as bonding to the hydroxyl groups on the glass surface. The most common sizes contain poly(vinyl acetate) as a film-former and either a methacrylic chromic chloride complex or an organosilane coupling agent. The coupling agent (qv) acts as a chemical bridge between fiber and matrix (62).

Epoxy and polyester are the most common matrix materials used with glass fibers. Properties of unidirectional glass fiber–epoxy composites are given in Table 4. Elastic properties follow the mixture rules (eqs. 6 and 7) and the Halpin-Tsai equation (eq. 10). Agreement is likely since the fiber is isotropic. Longitudinal tensile strengths follow the mixture rules (eq. 2), provided the fiber is carefully handled during composite fabrication. S-glass fiber composites are strong, tough, and have good tolerance for high temperatures, but relatively low moduli. The value of 55 GPa for E_L for S-glass–epoxy in Table 4 is not low; however, when laminated into a useful structure such as a plate or shell, the effective modulus is considerably below that of structural metals. Glass fibers are also susceptible to stress corrosion cracking under sustained load and have the poorest fatigue resistance of any strong fiber. These limitations have slowed the growth of S-glass applications. E-glass is suitable for many less stringent structural applications, especially in cloth form.

Principal applications for S-glass composites are pressure bottles, rocket-motor cases, and missile shells. E-glass resin composites are used in automobile bodies, for tubing and pipes, and in sporting-goods equipment, eg, surfboards, skis, and fishing rods. Electrically insulating structural members, such as the booms on cherry pickers, are made of E-glass–epoxy.

Graphite Fibers and Composites. The evolution of graphite or carbon fibers (qv) has been reviewed (63). Although a misnomer, graphite has become a common designation for this class of fibers, even though carbon is more accurate. (Carbon fiber is used preferentially in the *Encyclopedia*.) The higher modulus fibers, $E \geq 500$ GPa, are more graphitic than the medium modulus fibers, $E \approx 250$ GPa.

Graphite fibers have been made from a number of precursors, but are now mainly produced from polyacrylonitrile (PAN), pitch, and rayon. They derive

Table 3. Fiber Properties[a]

Property	E-glass	S-glass	HS[b] graphite	HM[c] graphite	Kevlar 29	Kevlar 49	Boron, W core
diameter, µm	3–20[d]	9	6–8	7–9	12.1	11.9	100, 140, 200
density, g/cm³	2.54	2.49	1.7–1.8	1.85	1.44	1.44	2.65, 2.45, 2.38
tensile strength, GPa[e]	2.4	4.5	3–4.5	2.4	3.5	3.6	3.5
elastic modulus, GPa[e]	72.4	85.5	234–253	345–520	59	124	386–400
thermal expansion, 10^{-6}/°C	5.0	5.6	−0.5(a)[f] 7(r)[f]	−1.2(a)[f] 12(r)[f]	−2(a)[f] 58(r)[f]	−2(a)[f] 59(r)[f]	5.4
thermal conductivity, W/(m·K)	1.86	2.55	8–25(a)[f]	105(a)[f]		3.1	2.7
cost, $/kg	1.1	22–33	66–110	220–660	11–22	22–33	330–440

[a] Data culled from many sources; should not be used for design.
[b] High strength.
[c] High modulus.
[d] Most common roving sizes are 9, 10, and 13 µm.
[e] To convert GPa to psi, multiply by 145,000.
[f] (a) = axial; (r) = radial.

Table 4. Glass Fiber Unidirectional Lamina Properties

Property	E-glass–epoxy	S-glass–epoxy
Elastic moduli		
E_L, GPa[a]	45	55
E_T, GPa[a]	12	16
G_{LT}, GPa[a]	5.5	7.6
ν_{LT}	0.20	0.26
Strengths		
F_L, MPa[b]	1100	1600–2000
F_T, MPa[b]	40	40
F_L', MPa[b]	620	690
F_T', MPa[b]	140	140
F_{LT}, MPa[b]	70	80
Thermophysical		
ρ, g/cm^3	2.1	2.0
α_L, 10^{-6}/°C	6.3	3.5
α_T, 10^{-6}/°C	30	29
k_L, W/(m·K)	1.26	1.58
k_T, W/(m·K)	0.59	0.57

[a] To convert GPa to psi, multiply by 145,000.
[b] To convert MPa to psi, multiply by 145.

their enormous strength and modulus from the alignment of covalently bonded basal planes of carbon atoms along the axis of the fiber. Modulus and strength of these planes are very high, ca 1000 GPa and ca 100 GPa, respectively. Production fibers do not attain these values, but either the highest modulus (>550 GPa) or the highest strength (ca 5 GPa) of any commercial fiber can be obtained with graphite fibers. Typical properties are given in Table 3.

The conversion of PAN to carbon fiber is a multistep process involving spinning and stretching of the precursor, stabilization in air at 220°C, carbonization at ca 1500°C, and graphitization at ca 3000°C in inert atmospheres (63). Perfection in the final product depends partly on the precursor and partly on process control. The strength of PAN-based graphite fibers has steadily risen over the years.

Pitch-based fibers are made by a liquid-crystal spinning process; high modulus has been achieved, but only modest strengths. Rayon-based fibers are processed much like PAN. The yield is low, however, and the fibers tend to be more expensive (>$660/kg). Graphite fiber, like glass, is available as strand or yarn and on individual spools, or as woven cloth. A large number of companies produce PAN fibers, at costs ranging from ca $66/kg for medium strength material to over $220/kg for very high strength or high modulus fiber.

Graphite fibers are most often used with epoxy resins, although they can be used with polyimides and various thermoplastics as well. They adhere reasonably well to some resins without special surface treatments. However, sizings or surface treatments are usually added. The interface can be tailored to improve interlaminar shear strength (64). Most carbon-fiber applications involve the use of prepregs, usually with TGMDA epoxy as the chief resin constituent. An unusual property of graphite fibers is their negative axial coefficient of expansion, which results from the extreme anisotropy induced by pyrolysis and graphitization. Designers have used this property to produce structures with excellent dimensional stability.

Unidirectional graphite–epoxy properties for high strength and high modulus fiber are given in Table 5. Transverse and shear elastic properties do not agree as well with predicted values as do those of glass because of fiber anisotropy. Longitudinal compressive strength is quite high, although it can be reduced substantially if the matrix or interface is degraded. Since graphite fibers are conductors, both thermal and electrical conductivities of graphite–resin composites are much higher than for other advanced composites.

Table 5. Carbon Fiber Unidirectional Lamina Properties

Property	High strength graphite–epoxy	High modulus graphite–epoxy
Elastic moduli		
E_L, GPa[a]	145	220–290
E_T, GPa[a]	10	6.2–6.9
G_{LT}, GPa[a]	4.8	4.8
ν_{LT}	0.25	0.25
Strengths		
F_L, MPa[b]	1240–2300	900–1200
F_T, MPa[b]	41	21
F'_L, MPa[b]	1200	620
F'_T, MPa[b]	170	170
F_{LT}, MPa[b]	80	60–70
Thermophysical		
ρ, g/cm^3	1.5–1.6	1.63
α_L, 10^{-6}/°C	−0.1	−0.5 to −0.8
α_T, 10^{-6}/°C	25	27–30
k_L, W/(m·K)	10–17	48–130
k_T, W/(m·K)	0.7	0.8–1.0

[a] To convert GPa to psi, multiply by 145,000.
[b] To convert MPa to psi, multiply by 145.

An outstanding feature of graphite fiber is its resistance to fatigue and corrosion. Graphite fibers are inert to temperature and humidity effects over the range of most resin service temperatures. Although moisture is absorbed by the resin, the effects are reversible unless the interfacial bond is degraded. Cyclic fatigue resistance in fiber-dominated failure modes is exceptional. The S–N curve for unidirectional material is very flat (59), and is superior to structural alloys. Endurance limits (fatigue strength as a percentage of static strength) exceed 70%. Static fatigue of graphite has not received much attention.

One problem in joining graphite composites to metals is galvanic corrosion. Because of fiber conductivity, fiber ends in a drilled plate serve as excellent contacts to create a galvanic cell. This can be avoided if the metal fastener is insulated or if the joint is designed to preclude fiber–metal contact.

Applications for graphite composites are mainly in the aerospace and sporting-goods industries. Graphite–epoxy has been used in military aircraft for both primary and secondary structural members for some years. Nineteen percent (by weight) of the AV-8B fighter airframe is carbon-fiber composite (Fig. 6). Commercial aircraft such as the Boeing 757 and 767 employ graphite–epoxy as secondary structural members. Boeing uses 1800 kg per plane, with a net savings of 900 kg. Various sporting-goods manufacturers are now marketing tennis rackets, fishing rods, golf-club shafts, and skis. Besides the properties already mentioned, these applications exploit the damping of the composite, which is superior to metal. Industrial applications will account for a large use of graphite composites (65); any machine with a inertia-limited production rate can employ graphite composites. If fiber prices drop to $10/kg or less or if the fiber properties continue to increase toward theoretical limits, graphite composites will continue to replace metals. Their combination of strength, stiffness, fatigue resistance, density, and toughness is unique in the advanced-composites field.

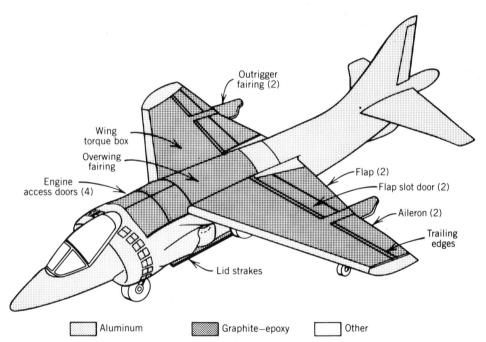

Fig. 6. AV-8B airframe. Structural weight: aluminum, 55%; graphite–epoxy, 19%; other, 26%.

Aramid Fibers and Composites.   Covalently bonded organics offer the potential for high strength and stiffness fibers. The ultimate strength of aligned organics has been estimated between 20 and 50 GPa, and a theoretical modulus of about 200 GPa has been calculated (66,67). Although these values should be attainable in any organic in which chains are closely packed and precisely aligned, the problems in processing such materials are formidable. Laboratory quantities of polyethylene fibers with strengths over 3.5 GPa and moduli of 85 GPa have been made by flash spinning and drawing from solution (68). However, the upper use temperature, ie, <100°C, of polyethylene limits it as a structural material. The only materials approaching their theoretical potential in commercial quantities are the aromatic polyamides, or aramids, marketed by DuPont under the trade name Kevlar (68).

Kevlar fibers are the result of many years of study of spinning of fibers from liquid crystals in solution. As predicted by Flory (68), liquid crystal solutions exhibit a reduced viscosity with increasing solids content as the solution becomes anisotropic. This property was exploited to extrude strong fibers of poly(p-benzamide) directly from solution (69). Current Kevlar fibers are made of poly(p-phenyleneterephthalamide) or PPTA. PPTA dope is the product of a condensation reaction of p-phenylenediamine and terephthaloyl chloride; this is dissolved in hot, concentrated sulfuric acid until a liquid-crystal solids concentration of ca 20 wt % is reached. The PPTA–H$_2$SO$_4$ is extruded through dry spinnerettes into a coagulating bath (dry jet-wet spin), after which the fiber is neutralized with NaOH and then washed. The resulting material, called Kevlar or Kevlar 29, has properties shown in Table 3. By drawing this material slightly at elevated temperature, the modulus may be increased from 59 to 124 GPa, the value for Kevlar 49. Kevlar is available as both medium and high modulus material in either yarn or cloth form.

The structure of these fibers is highly crystalline, with a molecular arrangement thought to be that shown in Figure 7. The para-arrangement of the main chain bonds provides stiffness; the polar nature of the amide bond results in hydrogen bonds between adjacent chains and probably improves interchain packing. Kevlar's density of 1.44 g/cm³ is among the highest for organics. Molecular weight of the PPTA molecules is considered rather low. Although some details of the crystal structure of Kevlar have been uncovered by x-ray studies, a comprehensive model does not exist. Available studies have been reviewed (70); crystallites in Kevlar are slightly misaligned with respect to the fiber axis, and can be oriented by a tensile load. From creep tests at room temperature, the moduli of both Kevlar 29 and 49 have been measured at over 175 GPa, considerably above the initial Kevlar 49 value. Evidence of alternating regions in Kevlar with and without defects has been found (71); the individual Kevlar molecules probably span a number of these defect zones, which contribute to the low compressive strength of Kevlar. Figure 8 shows the collapsed zones on the compressive side of the knotted Kevlar fiber.

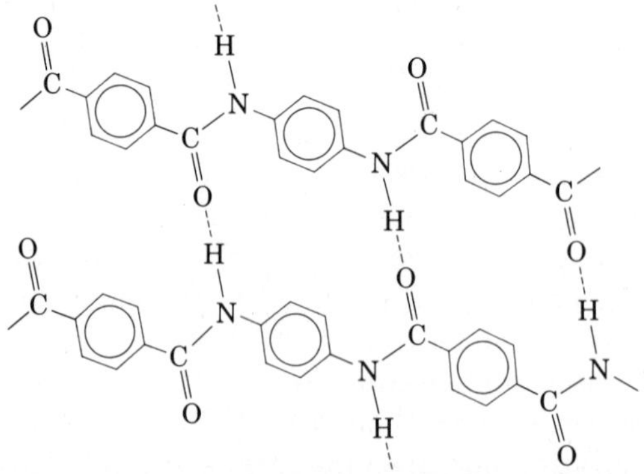

Fig. 7.   Chemical structure of Kevlar (PPTA).

500×

Fig. 8. Knotted Kevlar fiber, showing compressive failure. Courtesy of Albany International Research Co.

The fibrillar nature of Kevlar, shown schematically in Figure 9, is very evident in the surface of a failed Kevlar fiber (Fig. 10). Instead of breaking straight across, as do most other high strength fibers, Kevlar splits axially over considerable lengths. To fracture the fiber, a crack must therefore traverse a number of different planes; the fiber behaves as though it were a composite in

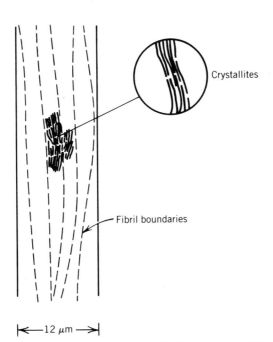

Fig. 9. A schematic representation of the fibrillar nature of Kevlar. The crystallites are ca 5 × 50 nm. Courtesy of Richard H. Ericksen, Sandia National Laboratories.

Fig. 10. Kevlar fiber failed in tension and showing fibrillation. Courtesy of Stephanie C. Kunz, Sandia National Laboratories.

microscale. Fibrillar fracture explains Kevlar's resistance to cutting either by machine tools or by projectiles. It is difficult to machine cleanly or to penetrate with a bullet. When axial splitting of the fiber during fracture was first studied (72), extremely long axial splits in fatigued specimens were observed. Splits extended almost 500 fiber diameters, ie, ca 6 mm. Such splitting contributes greatly to material toughness by preventing local stress concentrations.

Unidirectional properties of Kevlar 49–epoxy are given in Table 6. The tensile strength is noteworthy; because of low density, this fiber has the highest specific strength of any engineering material. Like graphite, Kevlar has a negative axial expansion coefficient, which results in near-zero values of α_L. Less desirable properties are compressive, transverse, and shear strengths. Compressive strength appears to be limited by the availability of shear planes in the fiber. Transverse tensile and shear strengths are functions of weak interfacial bonds and fiber anisotropy. Because of its structure, Kevlar has few primary bond sites on its surface, and no sizing appears to modify it. Deposition of amine groups on the surface by plasma activation has produced improved interfacial bonds (73). This process also decreases the considerable moisture uptake of Kevlar. Like all polymers with amide bonds, Kevlar is degraded by ultraviolet light.

Long-term and fatigue properties of Kevlar 49–epoxy are considerably better than those of glass composites. Although creep rupture does occur, it is predictable if the variability in long-term life between spools is accounted for (53). Fatigue resistance is not quite as good as that of graphite composites; endurance limits appear to fall between 0.5 and 0.6 of F_L, depending upon load cycle.

Kevlar fibers were developed primarily as a tire cord; this and cabling constitute the principal applications of the low modulus material. Kevlar 49

**Table 6. Kevlar Unidirectional
Lamina Properties**

Property	Kevlar 49–epoxy
Elastic moduli	
E_L, GPa[a]	80
E_T, GPa[a]	5.5
G_{LT}, GPa[a]	2.1
ν_{LT}	0.31
Strengths	
F_L, MPa[b]	2000
F_T, MPa[b]	20
F'_L, MPa[b]	280
F'_T, MPa[b]	140
F_{LT}, MPa[b]	40
Thermophysical	
ρ, g/cm^3	1.38
α_L, 10^{-6}/°C	−4.0
α_T, 10^{-6}/°C	60
k_L, W/(m·K)	1.7–3.2
k_T, W/(m·K)	0.15–0.35

[a] To convert GPa to psi, multiply by 145,000.
[b] To convert MPa to psi, multiply by 145.

composites are used for filament-wound rocket-motor cases and pressure vessels. The space shuttle has 17 metal-lined Kevlar overwrapped vessels. Kevlar composites are used in interior cabin structures in aircraft and also in some secondary exterior fairings. The military has adopted a Kevlar–epoxy helmet. Kevlar components are also used in boat and kayak hulls. Its strength, toughness, density, and cost, ie, $30/kg, are the bases for its selection and popularity.

Ceramic Fibers and Composites. Boron was the first high strength, high modulus fiber to be produced. It is made by the reduction of BCl_3 in a chemical vapor-deposition process onto a hot tungsten or graphite filament. The resulting amorphous boron has excellent properties, but the process is very costly. Current prices are about $330/kg. The fiber is produced in several diameters, 0.10, 0.14, and 0.20 mm. Properties for tungsten core material are listed in Table 3. Graphite core material is potentially cheaper, but its properties are not as consistent.

Boron–epoxy composites have excellent properties and have been used in a few military aircraft, but cannot compete with graphite composites for most applications. Boron composites have been used in some sporting-goods equipment, eg, tennis rackets and fishing rods.

Silicon carbide and aluminum oxide filaments have been developed. SiC fibers can be produced in a manner similar to that of boron, ie, vapor deposition of an organosilane onto a heated substrate. Strength and modulus are similar to boron's, with slightly higher, (10%), density. Al_2O_3 fibers were developed at DuPont from a slurry casting technique. Strength properties are rather low, ca 2 GPa, compared with other high performance fibers. Both SiC and Al_2O_3 fibers are used more for metal and ceramic composites than for resin matrix materials. Their use in resin matrix composites may increase if they are coupled with Kevlar to improve the compressive strength of composites made from that fiber (see also FIBERS, INORGANIC; PRECERAMIC FIBERS).

Laminate Design, Analysis, and Properties

Unidirectional composites have properties suitable for one-dimensional stress states, such as might be present in a leaf spring or fishing rod. But even these apparently simple structures must withstand both bending and torsional moments. Indeed, most structures must support a variety of loads. The very low values of F_T and F_{LT} for unidirectional composites usually require that fibers be oriented to support all principal tensile and shear loads. This is done by bonding two or more lamina together to form a laminate (see also LAMINATES).

Traditionally, structural design has focused on well-characterized, tough metals such as steel, aluminum, and titanium. Application of metal-design precepts to advanced composites can be catastrophic. Composite design requires an integrated approach involving structural, materials, and manufacturing engineering, cost and value analysis, and quality assurance and certification.

Structural Mechanics.   The heart of the composite-design process is structural mechanics, which leads to selection and specification of material, properties, geometry, and lamination sequence. It usually includes both analytical and experimental studies. An accurate analysis requires specification of the complete service environment and expected load history of the part, ie, loads, temperatures, service life, moisture, radiation, gases, chemical agents, freeze–thaw, etc. There must also be an accurate characterization of a unidirectional ply, the basic composite building block. From this analysis, the designer may specify a laminated structure that meets design requirements. The stress state must be completely evaluated, because off-axis strength allowables are quite low and may be exceeded. The designer should then identify and reinforce possible trouble spots caused by low material ductility, eg, joints, cutouts, stress-risers, etc, and estimate the reduction in properties caused by fatigue, sustained loading, temperature, or environment. Experimental verification of these estimates is required, usually by material property tests.

Structural analysis consists of two parts: stress analysis and failure analysis. It is the tool permitting a designer to specify a lamination sequence and part geometry that can meet design requirements. Governing equations and assumptions are the same as used for classical boundary value problems except that the constitutive relations are anisotropic. Also, the solution must be more detailed than for isotropic materials, because stresses other than principal values may exceed design allowables. Another reason for more detail is that safety margins in metals are usually based on yield strength (rather than on ultimate values as in composites), which results in built-in conservatism in the form of stress redistribution by plastic flow. The field of composite structural mechanics has been heavily studied for the last 15 years, resulting in the solution of numerous boundary value problems both analytically and by powerful finite difference and finite element techniques (74). These analyses, termed macromechanics to distinguish them from the micromechanics previously discussed, usually assume the unidirectional ply to be a homogeneous anisotropic continuum. Analyses pertinent to a number of important structural geometries, ie, bars, plates, shells, sandwich structures, have been reviewed (75).

Linear Elastic Deformations. This is by far the best-developed branch of composite-materials mechanics, because composites loaded parallel to a fiber direction at moderate temperatures exhibit approximately linear elastic behavior up to the time of failure. Techniques for treating anisotropic elastic solids and solutions to many common problems have been given (76,77). This topic has been developed in detail for laminated structures (13,22,75,78). The stress and strain fields for composite laminates subjected to moderate mechanical, thermal, or moisture loading can be estimated accurately, provided ply properties are known. The ply-property tensors required include stiffness or compliance, thermal-expansion coefficient, thermal conductivity, moisture-swelling coefficient, and moisture diffusivity.

The basic method outlined in the references cited is that of classical lamination theory, which uses tensor transformations to rotate the ply tensor into the appropriate orientation for calculating laminate properties. This method is for thin, flat plates; similar techniques are available for shells (75). The stress–strain relations (eq. 3) can be simplified by considering the case of a thin plate in plane stress (out-of-plane stresses $= 0$). This is not unduly restrictive, since many engineering structures are approximated by plates with principal stress only in the plane of the plate. The stress–strain relations for an orthotropic lamina in plane stress loaded parallel to principal material axes are (13)

$$\begin{Bmatrix} \sigma_L \\ \sigma_T \\ \tau_{LT} \end{Bmatrix} = \begin{bmatrix} Q_{LL} & Q_{LT} & 0 \\ Q_{LT} & Q_{TT} & 0 \\ 0 & 0 & Q_{SS} \end{bmatrix} \begin{Bmatrix} \epsilon_L \\ \epsilon_T \\ \gamma_{LT} \end{Bmatrix} \tag{23}$$

where Q_{ij} are the plane stress (reduced) stiffnesses of the ply, defined in terms of lamina engineering constants as

$$\begin{aligned} Q_{LL} &= E_L/\lambda & Q_{TT} &= E_T/\lambda \\ Q_{LT} &= Q_{TL} = \nu_{LT}Q_{TT} & Q_{SS} &= G_{LT} \end{aligned} \tag{24}$$

where $\lambda = 1 - \nu_{LT}\nu_{TL} = 1 - \nu_{LT}^2 E_T/E_L$.

When stresses are applied along axes rotated (about the z axis) at an arbitrary angle θ to the L–T axes, all of the Q_{ij} matrix components are nonzero:

$$\begin{Bmatrix} \sigma_x \\ \sigma_y \\ \tau_{xy} \end{Bmatrix} = \begin{bmatrix} \overline{Q}_{11} & \overline{Q}_{12} & \overline{Q}_{16} \\ \overline{Q}_{12} & \overline{Q}_{22} & \overline{Q}_{26} \\ \overline{Q}_{16} & \overline{Q}_{26} & \overline{Q}_{66} \end{bmatrix} \begin{Bmatrix} \epsilon_x \\ \epsilon_y \\ \gamma_{xy} \end{Bmatrix} \tag{25}$$

where $\overline{Q}_{ij}$ are the components of Q_{ij} transformed into the xy axes:

$$\begin{Bmatrix} \overline{Q}_{11} \\ \overline{Q}_{12} \\ \overline{Q}_{22} \\ \overline{Q}_{66} \\ \overline{Q}_{16} \\ \overline{Q}_{26} \end{Bmatrix} = \begin{bmatrix} m^4 & 2m^2n^2 & n^4 & 4m^2n^2 \\ m^2n^2 & m^4 + n^4 & m^2n^2 & -4m^2n^2 \\ n^4 & 2m^2n^2 & m^4 & 4m^2n^2 \\ m^2n^2 & -2m^2n^2 & m^2n^2 & (m^2-n^2)^2 \\ m^3n & -mn(m^2-n^2) & -mn^3 & -2mn(m^2-n^2) \\ mn^3 & mn(m^2-n^2) & -m^3n & 2mn(m^2-n^2) \end{bmatrix} \begin{Bmatrix} Q_{LL} \\ Q_{LT} \\ Q_{TT} \\ Q_{SS} \end{Bmatrix} \tag{26}$$

with $m = \cos\theta$ and $n = \sin\theta$ (13,78). Equations 26 show the strong dependence of stiffness upon ply orientation. It is important to note that $\overline{Q}_{ij}$ has entries in all nine positions relating stress to strain, in contrast to zeros for the normal-shear coupling coefficients in the Q_{ij} matrix. Thus, the simplification for an orthotropic solid no longer applies, and a normal stress generally produces a shear strain (and *vice versa*).

The second issue to be addressed in calculating response of laminated plates is the position of the ply relative to the midplane of the plate. For plates satisfying the Kirchoff-Love hypothesis, strain in the kth lamina is the sum of the mid-surface strain plus the mid-surface curvature κ times its distance z from the mid-surface (Fig. 11):

$$\left\{\begin{array}{c}\epsilon_x \\ \epsilon_y \\ \gamma_{xy}\end{array}\right\}_k = \left\{\begin{array}{c}\epsilon_x^0 \\ \epsilon_y^0 \\ \gamma_{xy}^0\end{array}\right\} + z\left\{\begin{array}{c}\kappa_x \\ \kappa_y \\ \kappa_{xy}\end{array}\right\} \tag{27}$$

Here, κ_x and κ_y are bending curvatures and κ_{xy} is the twist curvature of the middle surface. Substituting equation 27 into equation 25 shows that in the most general case, application of a normal stress gives rise to both normal and shear strains as well as bending and twisting curvatures in the plate. This is usually undesirable and can be counteracted by the judicious selection of angles and positioning of plies relative to the mid-surface. To do this, analysts use resultant or averaged stresses and moments and three averaged stiffness matrices that characterize overall laminate response. For a plate of thickness h, the stress $\{N\}$ and moment $\{M\}$ resultants are given by

$$\left\{\begin{array}{c}N_x \\ N_y \\ N_{xy}\end{array}\right\} = \int_{-h/2}^{h/2}\left\{\begin{array}{c}\sigma_x \\ \sigma_y \\ \tau_{xy}\end{array}\right\}_k dz = \sum_{k=1}^{N}\int_{z_{k-1}}^{z_k}\left\{\begin{array}{c}\sigma_x \\ \sigma_y \\ \tau_{xy}\end{array}\right\}_k dz \tag{28}$$

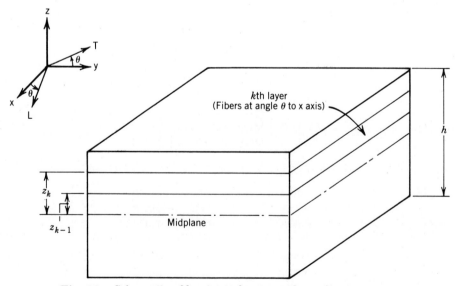

Fig. 11. Schematic of laminate lay-up and coordinate system.

$$\begin{Bmatrix} M_x \\ M_y \\ M_{xy} \end{Bmatrix} = \int_{-h/2}^{h/2} \begin{Bmatrix} \sigma_x \\ \sigma_y \\ \tau_{xy} \end{Bmatrix}_k zdz = \sum_{k=1}^{N} \int_{z_{k-1}}^{z_k} \begin{Bmatrix} \sigma_x \\ \sigma_y \\ \tau_{xy} \end{Bmatrix}_k zdz \qquad (29)$$

where k designates the kth layer in a laminate of N layers and z its distance from the mid-surface (Fig. 11). The three stiffness matrices are termed extensional A_{ij}, bending-stretching coupling B_{ij}, and bending D_{ij}. These are defined

$$A_{ij} = \sum_{k=1}^{N} (\overline{Q}_{ij})_k (z_k - z_{k-1}) \qquad (30)$$

$$B_{ij} = \frac{1}{2} \sum_{k=1}^{N} (\overline{Q}_{ij})_k (z_k^2 - z_{k-1}^2) \qquad (31)$$

$$D_{ij} = \frac{1}{3} \sum_{k=1}^{N} (\overline{Q}_{ij})_k (z_k^3 - z_{k-1}^3) \qquad (32)$$

The constitutive equations for the most general laminated plate are given by

$$\begin{Bmatrix} N_x \\ N_y \\ N_{xy} \\ \hline M_x \\ M_y \\ M_{xy} \end{Bmatrix} = \begin{bmatrix} A_{11} & A_{12} & A_{16} & B_{11} & B_{12} & B_{16} \\ A_{12} & A_{22} & A_{26} & B_{12} & B_{22} & B_{26} \\ A_{16} & A_{26} & A_{66} & B_{16} & B_{26} & B_{66} \\ \hline B_{11} & B_{12} & B_{16} & D_{11} & D_{12} & D_{16} \\ B_{12} & B_{22} & B_{26} & D_{12} & D_{22} & D_{26} \\ B_{16} & B_{26} & B_{66} & D_{16} & D_{26} & D_{66} \end{bmatrix} \begin{Bmatrix} \epsilon_x \\ \epsilon_y \\ \gamma_{xy} \\ \hline \kappa_x \\ \kappa_y \\ \kappa_{xy} \end{Bmatrix} \qquad (33)$$

For this most general case, all stress and strain terms are coupled, which is unfortunate. For a laminate with constant thickness plies, eliminating certain undesirable components is possible by proper selection of the stacking sequence. The bending-stretching coupling matrix B_{ij} is troublesome, causing warpage and twisting. Since B_{ij} is an even function of z, it may be made identically zero if the ply stacking sequence is symmetric about the mid-surface. Examples are $(+\theta, 0, +\theta)$, $(+\theta, -\theta, -\theta, +\theta)$, and $(0, +\theta, -\theta, -\theta, +\theta, 0)$. A second problem is normal-shear coupling, the 16, 26 terms in the A_{ij} and D_{ij} matrices. A_{16} and A_{26} may be eliminated by balancing the number of $+\theta$ plies on one side of the mid-surface with an equal number of $-\theta$ plies on the other. Thus A_{16} and A_{26} are zero for laminates such as $(+\theta, -\theta)$ and $(+\theta, -\theta, -\theta, +\theta)$. To eliminate D_{16} and D_{26}, the $+\theta$ and balancing $-\theta$ plies must be placed at the same distance (z_k) from the midplane. D_{16} and D_{26} are zero for a $(+\theta, -\theta)$ laminate, but not for $(+\theta, -\theta, -\theta, +\theta)$. It is not possible (except for $0°$, $90°$ cross-ply laminates), for both B_{ij} and all the 16, 26 terms to be identically zero. However, for symmetric laminates with a large number of alternating layers, the D_{16}, D_{26} terms become quite small (22,78).

A laminate of considerable practical importance can be formed by orienting the plies in such a way as to obtain elastic isotropy in the plane of the fibers. Plies in the quasi-isotropic laminate must be of uniform thickness and properties and be oriented at an angle θ_k with respect to a reference direction and satisfy the relation (79)

$$\theta_k = \pi(k - 1)/N \qquad (34)$$

where $k = 1, \ldots N$ and $N \geq 3$. Examples are $(0°, 60°, 120°)$ for $N = 3$ and $(0°, 45°, 90°, 135°)$ for $N = 4$. Such laminates are isotropic in extension $(A_{11} = A_{22}, 2A_{66} = A_{11} - A_{12})$, but not in bending. The apparent Young's modulus for a quasi-isotropic laminate is given by (78)

$$\overline{E} = \frac{A_{11}}{h}\left(1 - \left(\frac{A_{12}}{A_{11}}\right)^2\right)$$
(35)

where

$$\frac{A_{11}}{h} = \tfrac{1}{8}[3(Q_{LL} + Q_{TT}) + 2(Q_{LT} + 2Q_{SS})]$$

$$\frac{A_{12}}{h} = \tfrac{1}{8}[6Q_{LT} + Q_{LL} + Q_{TT} - 4Q_{SS}]$$

For highly anisotropic lamina, $E_L \gg E_T$, $Q_{LL} \cong E_L$, and $Q_{TT} \cong E_T$ and equation 35 may be approximated

$$\overline{E} = \tfrac{3}{8}E_L + \tfrac{5}{8}E_T$$
(36)

Lamination reduces the unidirectional modulus of the composite to about 40% of E_L in the quasi-isotropic case. Test results on various laminates confirm the validity of classical plate theory.

There are obviously many laminations available for any given composite. The quasi-isotropic lay-up provides a useful structure with which to compare metal structural materials. The comparison is more accurate than for lamina, since the quasi-isotropic laminate approximates the elastic isotropy of metals. Representative quasi-isotropic properties for various composites are given in Table 7, along with those of an aircraft aluminum. The composite specific properties are considerably higher than for metals, but the absolute values are similar.

Table 7. Properties of Quasi-isotropic Laminates

Property	E-glass–epoxy	S-glass–epoxy	High strength graphite–epoxy	Kevlar 49–epoxy	2024 aluminum
elastic modulus, GPa[a]	24	30	60	33	69
tensile strength, GPa[a]	0.45	0.7	0.5–0.9	0.8	0.43
fracture toughness, MPa[b]·m$^{1/2}$	15–30	20–30	11–25	30–38	32
allowable flaw for 400 MPa[b] stress, mm	0.45–1.8	0.8–1.8	0.2–1.2	1.8–2.9	2.0
density, g/cm^3	2.1	2.0	1.6	1.38	2.7
thermal expansion, 10^{-6}/°C	12	10.8	1–3	−0.9 to 0.9	23
thermal conductivity, W/(m·K)	0.9	1.1	5	0.9	189

[a] To convert GPa to psi, multiply by 145,000.
[b] To convert MPa to psi, multiply by 145.

Analysis of dilatation owing to temperature or moisture can be included in equations 33 (13). The technique involves adding a set of terms for thermal expansion and moisture swelling to each of the equations.

With this formalism in place, a large number of lay-ups and materials can be evaluated by computer simulation. Most of the necessary properties exist, and the analytical approach has proven quite accurate. When thermal or moisture effects are involved, it is necessary to account for gradients in temperature or moisture in nonequilibrated structures. However, the basic approach remains the same.

Strength and Fracture Analysis. The well-developed formulation of stress analysis by classical plate theory for laminates has no counterpart in failure analysis. There is no generally agreed-upon technique or criterion. The best approach is to calculate overall laminate response to a particular set of boundary conditions, and then calculate the stresses in individual plies. These stresses may be compared with the respective lamina strengths to determine if failure stress levels have been exceeded. It is usually better to analyze strength and failure on the ply level. Laminates are susceptible to an additional failure mode: interlaminar shear.

On the ply level, the various strength criteria previously discussed for lamina can be employed. Prior to failures involving fiber fracture, it is usual for some plies to fail in transverse tension or shear. These are ordinarily constrained by adjacent plies oriented at an angle to the break. For example, in a (0°, ±45°, 90°) quasi-isotropic laminate stressed in the 0° direction, the 90° plies fail first, followed by shear failure in the 45° layers. Fiber fracture in the 0° plies occurs last. Microcracking in 90° plies has little effect on laminate ultimate strength, and is usually neglected in failure analysis. Shear failure may or may not be critical, depending upon the application.

Since strength is a statistical variable, guidelines for selecting safe, working allowables have been developed. An A allowable is the value above which 99% of the strengths lie within a 95% confidence level. A B allowable is the value above which 90% of the strengths lie within a 95% confidence level (80). A allowables are used in critical applications where failure of that member would result in loss of structural integrity. B allowables are used in less critical elements where load redistribution would occur after failure.

In angle-ply laminates, failure often begins at the edges of plates or at holes because of the buildup of interlaminar shear and normal stresses at a free surface. In the first analyses of these stresses, it was found that out-of-plane normal stresses σ_z could be minimized for certain laminates by proper stacking sequences (81,82). The complete problem has yet to be solved. The quasi-isotropic laminate is useful as a yardstick for strength and toughness of composite plates. Strengths for typical tape lay-up advanced composites are given in Table 7. The strength of a composite with fibers randomly oriented in a plane has been estimated as $F_L/3$ (6). Quasi-isotropic plate strengths, ie, when tested in a fiber direction, are higher, usually about 40% of F_L.

Fracture toughness has been measured on notched quasi-isotropic plates from various materials (83). Such tests are possible, and may be valid because the crack is constrained to run colinearly across fibers owing to constraint provided by adjacent angled layers (84). Values of K_{IC} for advanced composites are given in Table 7. Also listed are critical flaw lengths, assuming a center notch geometry. Both the K_{IC} and critical flaw values are competitive with those of structural aluminum. Although made of brittle constituents, advanced composites have considerable tolerance to flaws, but the designer must be aware of the different techniques used to minimize stress concentrations around fasteners and joints (85).

Fatigue strength of laminates is controlled by the behavior of those plies where fibers support the principal loads (54). Delaminations during fatigue do not appear to affect tensile strength, but do reduce compressive strength. Microscopic examination of fatigue failure in graphite–epoxy laminates reveals little difference from static failure modes (54).

Analysis of failure in advanced-composite laminates usually is an iterative process which heavily relies on a knowledge of the stress state in each ply. Material or geometry modifications are best done on an interactive computer.

Fabrication.   The differences between metal and composite structural design become especially evident during the manufacturing process. For advanced composites, the design and process engineers must interact closely, because the mechanical properties of the structure are developed as the part is being made. Continuing dialogue, especially regarding the limitations of each fabrication technique, can identify and prevent potential problems early on.

Techniques used to fabricate advanced composites are filament winding (qv), pultrusion (qv), and lamination (see LAMINATES). These are shown schematically in Figure 12 (86). Many of the procedures used in fabricating advanced composites were originally developed for glass fibers. Processing information needed for design has been summarized (87,88). For further discussion of processing, see COMPOSITES, FABRICATION.

Current processing techniques produce good-quality, thin (5–10 mm thick) laminates from resins with cure cycles up to 200°C. For thick, complex parts or unusual cure cycles, the reject rate increases, usually because of excessive void content or delaminations.

Advanced composite materials are inherently more expensive as raw materials, but offer manufacturing processes that are less costly than metal fabrication processes. Several factors contribute to this savings: fewer parts in the assembled structure, less energy required to produce the structure, and use of automated manufacturing methods.

A large source of information on methods of manufacture for composite materials and on the relative costs of manufacture is available from more than 310 government-sponsored programs in fabrication and testing of composite components. It is contained in the *DOD/NASA Structural Composites Fabrication Guide* (89) and stored in a computerized data-management system that can be accessed for historical data on costs. Figure 13 provides examples of historical information on the time required to fabricate composite parts as a function of the number of parts produced, the area of the part, and the method of fabrication, respectively.

Quality Assurance and Certification.   Quality assurance includes inspection and control of incoming materials, processing specifications for fabrication steps, and destructive and nondestructive inspection. In general, the instruments

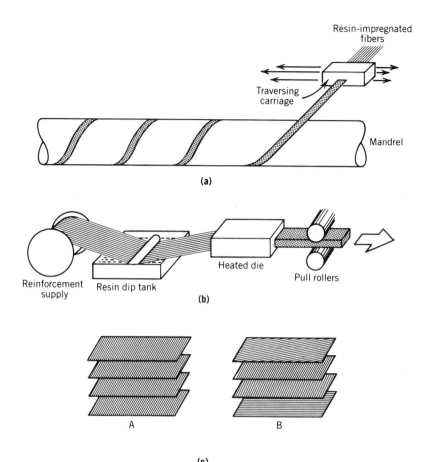

Fig. 12.   Fabrication methods for advanced composites. Continuous fiber-composites can be produced by filament winding, pultrusion, or lamination. (**a**) Filament-wound products are produced by winding resin-impregnated fibers or tape onto a mandrel. Here, a traversing carriage lays down impregnated fiber on a revolving, lathelike mandrel. (**b**) Pultrusion is used to produce constant cross-sectional shapes such as rods and I beams. The fibers pass through a resin tank and are then pulled through a heated die to cure and form the final product. (**c**) Laminated composites are formed by laying up a number of prepreg mats or prepreg tapes; the prepreg laminate is then cured in an oven. Lamination: A, unidirectional lamination consists of fiber-impregnated tapes laminated with fibers running in the same direction for strength along one axis; B, quasi-isotropic consists of impregnated tapes or mats laminated in three, four, or more directions to produce isotropic properties in the plane of the fibers (86). Courtesy of SRI International.

needed to inspect the material at each stage of the process are available. However, it is not always possible to determine if a particular out-of-specification property, procedure, or flaw constitutes an unacceptable final part. For example, common nondestructive inspection techniques can detect most flaws owing to voids or delaminations (see NONDESTRUCTIVE TESTING). However, unless they affect the primary failure mode, such flaws are often tolerable. Gross flaws such as badly misaligned or missing plies are always cause for rejection. Although techniques

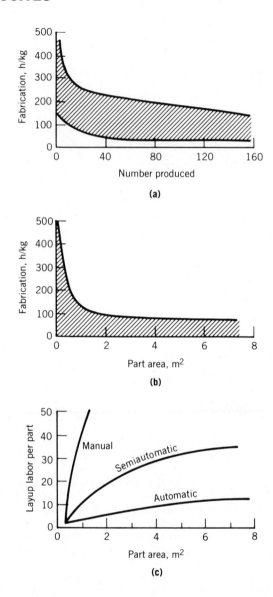

Fig. 13.   Examples of the time required to fabricate composite parts. (**a**), Lay-up; reduction in fabrication time vs number of parts produced; (**b**), all parts; dependence of fabrication time on part area; (**c**), lay-up ply-on-ply; reduction in labor by automation (89). Courtesy of Air Force Wright Aeronautics Laboratory.

to detect flaws in composites exist, the problem lies in the ability to interpret their effect on strength or life of the structure. As fracture mechanics concepts and test techniques become better established for composites, this situation should change; a real benefit would be the establishment of a firm basis for proof testing.

In metal structures, proof testing is based on the tenet that flaws greater than a certain size will grow catastrophically under a certain, ie, greater than design, load. The proof test thus ensures that no flaw greater than an acceptable size is present in the structure. Fracture mechanics, on which the concept of proof testing is based, requires that the flaw grow in a colinear, self-similar manner. In composites, crack-branching and multiple cracking may occur before final failure. Typical flaws in composites include matrix voids, matrix cracking, delamination between plies, and fiber breakage (especially around machined holes and cutouts). Proof testing should work for any laminate with a fracture toughness value effectively independent of the specimen used to measure it. This seems to be the case for quasi-isotropic laminates loaded with a principal stress axis coincident with a ply direction. However, it has yet to be determined if proof testing can be used to guarantee the performance of an arbitrarily oriented laminate.

Future Directions

Improvements in Fiber Properties.   Fiber manufacturers now have greater understanding and control of their products. Ten years ago, graphite and Kevlar fibers had guaranteed mean strengths of about 2.5 and 2.75 GPa, respectively. Today, graphite can be obtained with strengths of about 4.5 GPa, and Kevlar with 3.8 GPa. Since neither value approaches the theoretical strength (63,68,90), further increases can be anticipated.

A melt-spinnable aramid fiber has been discussed for some time, but there is no commercial process. The current solution process for Kevlar is sufficiently complex that no cost reductions appear likely. Similarly, the cost of high quality precursors has kept graphite prices high. Increased volume might reduce the cost of these materials, and if fiber properties continue to rise, the cost per unit performance will certainly decrease (86).

Development of Matrix Resins.   The ideal matrix is an inexpensive, tough material that readily bonds to the fiber and can be processed rapidly and reproducibly at low temperatures into the final structure. Such a combination of properties probably does not exist in one material. Thermally stable, moisture-resistant thermoplastics offer an enormous processing advantage over thermosets. Thermoplastics remove the burden of handling and mixing chemicals from the composite fabricator (90). Rapid mass production of large-area plate and shell structures such as those needed by automobile and truck manufacturers would be possible if an inexpensive version of PEEK could be produced (91). It might also be feasible to develop joining processes analogous to welding in metals.

Three-dimensional Reinforcement.   For advanced composites to gain a share of the auto-industry market, they must become tough, forgiving materials. Composite laminates exhibit in-plane toughnesses comparable to aluminum alloys. Values for transverse toughness, tensile strength, and interlaminar shear are quite low, on the order of glassy polymers. Some improvement can be made through use of tougher thermoplastic resins, but a more immediate remedy is three-dimensional reinforcement. Early attempts at 3-D fiber arrays were not very successful because the addition of the third direction diluted in-plane stiffness and strength. Also, the techniques utilized were clumsy and hard to automate. Newer attempts with very strong graphite or Kevlar and some novel weaving techniques, such as interlocking plies, are much more promising. A few volume percent fiber oriented in the out-of-plane direction can significantly increase transverse toughness and strength without seriously reducing in-plane properties. The aircraft industry is evaluating 3-D composites to help solve the problem of delamination and loss of compressive properties owing to lateral impact loading.

Standardization and Automation.   The most commonly used shapes and lay-ups could be produced more economically if they were standardized (59). The analogues are plate, channel, and I beams for mild steel; plate, rod, and bar stock for alloy steel, aluminum, and titanium; and standardized boards and plywood sheets for wood products. Composite sheets with fibers in cloth form in a thermoplastic matrix are likely candidates for standardization. Braided tubes can be made in any cross section having an axis of symmetry. The fiber-composite field would benefit greatly if designers could specify a composite structural member as confidently as they would a piece of 1040 steel.

Improved and Novel Processing.   Much current composite processing is still empirical. Models of the cure process based on resin-reaction kinetics, gelation, and vitrification are needed.

Another processing-related question requiring further study is near-net-shape final product. For example, with interactive computers, it may be possible to choose a lamination sequence that would bend to the desired shape upon cooldown without using expensive curved tooling. Asymmetric or unbalanced laminates may also be incorporated in clever structural and process designs (92).

***In Situ* Reinforcements.**   An exciting possibility for advanced composites is the potential for growing a rigid rod molecule from a matrix of the same material. The metal analogue of a directionally solidified eutectic does not achieve the extreme strength and stiffness increases possible with an oriented polymer chain. Research sponsored by the Air Force on poly(phenylene benzobisthiazole) and poly(phenylene benzobisoxazole) polymers has shown considerable promise (93) (see POLYBENZOTHIAZOLES). Other materials and processes for integral reinforcement are being explored.

Long-term Behavior of Fibers.   The origin of flaws during manufacture and their growth owing to stress and environment require further study. The hypothesis that high static strength ensures long fatigue life is intuitively appealing, but has not been established, and such matters as how fiber fractures interact and propagate to adjoining fibers are still unresolved (90).

If long-term failure processes can be resolved, diagnostic techniques can be developed that permit determination of the degree to which a structure has degraded and an estimate of the remaining safe life. Field repair, as an inexpensive alternative to replacement, also then becomes possible (59).

Nomenclature

A_{ij}	= extensional stiffness matrix
B_{ij}	= bending-stretching coupling stiffness matrix
D_i	= diffusivity
D_{ij}	= bending stiffness matrix
E_i	= elastic modulus
F_i	= strength
G_{ij}	= shear modulus
K	= stress intensity factor
K_r	= stress concentration
K_{IC}	= fracture toughness
M_{ij}	= bending or twisting moment
M_i	= elastic property
ΔM	= change in moisture content
N_{ij}	= stress resultant

N = number of plies
P = probability
Q_{ij} = plane stress stiffness matrix
S_{ij} = elastic compliance matrix
ΔT = change in temperature
$\mathcal{G}_{IC}$ = energy-release rate
a = crack length
c = power-law exponent
h = plate thickness
k_i = thermal conductivity
ℓ = gauge length of fiber
m = Weibull shape parameter; also, $\cos \theta$
n = number of fibers; also, $\sin \theta$
r = crack tip radius; also, number of
       broken fibers
t = time
v = volume fraction
x,y,z = coordinates
z = the distance of the kth lamina from
       the mid-surface
α_i = coefficient of thermal expansion
β_i = coefficient of moisture expansion
γ_{ij} = engineering shear strain
ϵ_i = normal strain
η = factor in Halpin-Tsai equation
θ = ply angle
κ_{ij} = curvature
λ = Poisson factor
ν_{ij} = Poisson ratio
σ_i = normal stress
τ_{ij} = shear stress
ξ = factor in Halpin-Tsai equation

Subscripts
f   = fiber
m   = matrix
L   = longitudinal, ie, parallel to fiber
T   = transverse, ie, perpendicular to fiber

BIBLIOGRAPHY

1. C. W. Bert and P. H. Francis, *Trans. N. Y. Acad. Sci. Ser. II* **36**(7), 663 (1974).
2. J. E. Gordon, *The New Science of Strong Materials or Why You Don't Fall Through the Floor*, 2nd ed., Penguin Books, Ltd., Harmondsworth, UK, 1976. An excellent historical perspective on structural materials and a fine primer on stiffness, strength, and toughness.
3. D. V. Rosato in G. Lubin, ed., *Handbook of Composites*, Van Nostrand Reinhold Co., New York, 1982, pp. 1–14.
4. K. M. Prewo in M. A. Myers, ed., *Frontiers in Materials Technology*, Elsevier, Amsterdam, 1985.
5. Ref. 2, p. 40.
6. H. L. Cox, *Br. J. Appl. Phys.* **3**, 72 (1952).
7. A. H. Cottrell, *Proc. R. Soc. London Ser. A* **282A**, 2 (1964).
8. A. Kelly, *Strong Solids,* 2nd ed., Clarendon Press, Oxford, UK, 1973.
9. P. E. Chen, *Polym. Eng. Sci.* **11**, 51 (1971).
10. J. L. Kardos, J. C. Halpin, and S. L. Chang in G. Astarita, G. Marrucci, and L. Nicolais, eds., *Rheology*, Plenum Publishing Corp., New York, 1980, pp. 255–260.
11. J. L. Kardos, E. Masoumy, and L. Kacir in L. Nicolais and J. C. Seferis, eds., *Role of the Polymeric Matrix in the Processing and Structural Properties of Composite Materials*, Plenum Publishing Corp., New York, 1983, pp. 407–423.
12. N. A. Munford, P. C. Hopkins, and B. A. Lloyd, *Matrix/Fiber Interface Effects on Kevlar Pressure*

Vessel Performance, AIAA/SAE/ASME 18th Joint Propulsion Conference, June 21–23, 1982, Cleveland, Ohio, Paper 82-1069.

13. R. M. Jones, *Mechanics of Composite Materials*, Script Book Co., Washington, D. C., 1975. Contains excellent treatments of anisotropic elasticity and laminated plate theory.

14. D. K. Hale, *J. Mater. Sci.* **11**, 2105 (1976).

15. R. Hill, *J. Mech. Phys. Solids* **12**, 199 (1964).

16. *Ibid.*, **13**, 189 (1965).

17. Z. Hashin, *Appl. Mech. Rev.* **17**, pp. 1 (1964).

18. Z. Hashin, *Theory of Fiber Reinforced Materials, NASA CR-1974*, NASA, Washington, D.C., 1972.

19. J. A. Halpin and S. W. Tsai, *Environmental Factors in Composite Materials Design, AFML TR 67-423*, Air Force Materials Laboratory, Dayton, Ohio, 1969.

20. R. E. Allred and F. P. Gerstle, Jr., *The Effect of Resin Properties on the Transverse Mechanical Behavior of High Performance Composites, 30th Anniversary Technical Conference*, Reinforced Plastics/Composites Institute, Society of the Plastics Industry, Inc., New York, 1975, Sect. 9-B, pp. 1–6.

21. S. G. Springer and S. W. Tsai, *J. Compos. Mater.* **1**, 166 (1967).

22. J. E. Ashton, J. C. Halpin, and P. H. Pettit, *Primer on Composite Materials: Analysis*, Technomic Publishing Co., Inc., Stamford, Conn., 1969.

23. V. M. Levin, *Mech. Solids USSR* **2**, 58 (1967).

24. R. A. Schapery, *J. Compos. Mater.* **2**, 380 (1968).

25. J. C. Halpin and N. J. Pagano, *Consequences of Environmentally Induced Dilatation in Solids, AFML TR 68-395*, Air Force Materials Laboratory, Dayton, Ohio, 1969.

26. D. Hull, *An Introduction to Composite Materials*, Cambridge University Press, Cambridge, UK, 1981. Contains an excellent treatment of composite-strength theories for various modes of failure.

27. H. D. Wagner, S. L. Phoenix, and P. S. Schwartz, *J. Compos. Mater.* **18**(4), 312 (1984); S. L. Phoenix, *Int. J. Fract.* **14**(3), 327 (1978).

28. B. D. Coleman, *J. Mech. Phys. Solids* **7**, 60 (1958).

29. B. W. Rosen, *Fiber Composite Materials*, American Society for Metals, Metals Park, Ohio, 1965, pp. 37–75.

30. C. Zweben, *AIAA J.* **6**(12), 2325 (1968).

31. J. M. Hedgepeth, *Stress Concentrations in Filamentary Structures, NASA TN D-882*, NASA, Washington, D.C., May 1961.

32. G. S. Holister and C. Thomas, *Fiber Reinforced Materials*, Elsevier, New York, 1966.

33. J. Tirosh, E. Kutz, G. Lifschwtz, and A. S. Tetelman, *Eng. Fract. Mech.* **12**, 267 (1979).

34. L. B. Greszczuk, *AGARD Conf. Proc.* **163**, Paper 12 (1974).

35. D. F. Adams and D. R. Doner, *J. Compos. Mater.* **1**, 4 (1967).

36. S. W. Tsai and H. T. Hahn in C. T. Herakovich, ed., *Inelastic Behavior of Composite Materials*, ASME Publ. AMD-Vol. 13, American Society of Mechanical Engineers, New York, 1975, pp. 73–96.

37. J. W. S. Hearle, *Fibre Sci. Technol.* **3**, 321 (1971).

38. F. P. Gerstle, Jr. and E. D. Reedy, Jr., *J. Compos. Mater.* **19**(5), (Sept. 1985).

39. R. C. Burk, *Astronaut. Aeronaut.* 58 (June 1983).

40. P. H. Francis and C. W. Bert, *Fibre Sci. Technol.* **8**, 1 (1975). An excellent survey of composite strength and toughness with an extensive reference list.

41. D. C. Phillips and A. S. Tetelman, *Composites* **4**, 216 (1972). The classic work on toughness in composites from a materials-science viewpoint.

42. E. M. Wu in S. W. Tsai, J. C. Halpin, and N. J. Pagano, eds., *Composite Materials Workshop*, Technomic Publishing Co., Inc., Stamford, Conn., 1968, pp. 20–43.

43. D. J. Wilkens, J. R. Eisenmann, R. A. Canin, W. S. Margolis, and R. A. Benson in K. Reifsnider, ed., *ASTM STP 775, Damage in Composite Materials*, American Society for Testing and Materials, Philadelphia, Pa., 1982, pp. 168–183.

44. C. A. Berg and A. Rinsky, *Fibre. Sci. Technol.* **3**, 295 (1971).

45. J. M. Hedgepeth and P. Van Dyke, *J. Compos. Mater.* **1**, 294 (1967).

46. E. D. Reedy, Jr., *J. Mech. Phys. Solids* **28**, 265 (1980).

47. R. A. Schapery in Ref. 42, pp. 153–192.

48. R. A. Schapery in G. P. Sendecky, ed., *Composite Materials*, Vol. 2, (*Mechanics of Composite Materials*), Academic Press, Inc., New York, 1974, pp. 85–168.

49. R. A. Schapery in Ref. 36, pp. 127–155.

50. Y. Weitsmann, *J. Appl. Mech.* **47**(1), 35 (1980).
51. A. Kelly and L. N. McCartney, *Proc. R. Soc. London Ser. A* **374**, 475 (1981).
52. T. T. Chiao, J. E. Wells, R. L. Moore, and M. A. Hamstad in *ASTM STP 546, Composite Materials: Testing and Design (Third Conference),* American Society for Testing and Materials, Philadelphia, Pa., 1974, pp. 209–224.
53. F. P. Gerstle, Jr. and S. C. Kunz in T. Kevin O'Brien, ed., *ASTM STP 813, Long Term Behavior in Composites,* American Society for Testing and Materials, Philadelphia, Pa., 1983, pp. 263–292.
54. H. T. Hahn in S. W. Tsai, ed., *ASTM STP 674, Composite Materials: Testing and Design (Fifth Conference),* American Society for Testing and Materials, Philadelphia, Pa., 1979, pp. 383–417.
55. C. W. Zweben, W. S. Smith, and M. W. Wardle in S. W. Tsai, ed., *ASTM STP 674, Composite Materials: Testing and Design, (Fifth Conference),* American Society for Testing and Materials, Philadelphia, Pa., 1979, pp. 228–262.
56. N. J. Pagano and J. C. Halpin, *J. Compos. Mater.* **2**, 18 (1968).
57. S. L. Phoenix and R. G. Sexsmith, *J. Compos. Mater.* **6**, 322 (1972).
58. L. S. Penn and T. T. Chiao in Ref. 3, pp. 57–88.
59. A. A. Watts, ed., *ASTM STP 704, Commercial Opportunities for Advanced Composites,* American Society for Testing and Materials, Philadelphia, Pa., 1980.
60. T. T. Serafini in Ref. 3, pp. 89–114.
61. B. A. Proctor, *Composites* **3**, 85 (June 1971).
62. C. E. Knox in Ref. 3, pp. 136–159.
63. D. M. Riggs, R. J. Shuford, and R. W. Lewis in Ref. 3, pp. 196–271. An excellent, comprehensive treatment of graphite fibers and composites containing 389 references.
64. L. T. Drzal, M. J. Rich, and P. F. Lloyd, *J. Adhesion* **16**(1), (1983).
65. E. M. Trewin in B. Noton, R. Signorelli, K. Street, and L. Phillips, eds., *Proceedings of the 1978 International Conference on Composite Materials,* American Institute of Mining, Metallurgical and Petroleum Engineers, New York, 1978, pp. 1462–1473.
66. F. C. Frank, *Proc. R. Soc. London Ser. A* **319**, 127 (1970).
67. H. F. Mark in A. V. Tobalsky and H. F. Mark, eds., *Polymer Science and Materials,* John Wiley & Sons, Inc., New York, 1971, pp. 231–246.
68. E. E. Magat, *Philos. Trans. R. Soc. London Ser. A* **294**, 463 (1980).
69. S. L. Kwolek, *Polym. Prepr. Am. Chem. Soc. Div. Polym. Chem.* **21**(1), 12 (1980).
70. R. H. Ericksen, *Polymer* **26**(5), 733 (1985).
71. P. Avakian, R. C. Blume, T. D. Gierke, H. H. Yang, and M. Panar, *Polym. Prepr. Am. Chem. Soc. Div. Polym. Chem.* **21**(1), 8 (Mar. 1980).
72. L. Kanopasek and J. W. S. Hearle, *J. Appl. Polym. Sci.* **21**, 2791 (1977).
73. R. E. Allred, E. W. Merrill, and D. K. Roylance, *Surface Chemistry and Bonding of Plasma-Aminated Polyaramid Filaments,* Proceedings of the ACS Symposium on Composites, Interfaces, Seattle, Wash., Mar. 21–25, 1983.
74. D. W. Purdy in C. C. Chamis, ed., *Composite Materials,* Vol. 8 (*Structural Design and Analysis Part II*), Academic Press, Inc., New York, 1975, pp. 1–31.
75. C. W. Bert and P. H. Francis, *AIAA J.* **12**(9), 1173 (1974).
76. S. G. Lekhnitskii, *Theory of Elasticity of an Anisotropic Body* (translated by P. Fern), Holden-Day, Inc., San Francisco, Calif., 1963.
77. S. G. Lekhnitskii, *Anisotropic Plates* (translated by S. W. Tsai and T. Cheron), Gordon & Breach Publishers, Inc., New York, 1968.
78. C. W. Bert in Ref. 74, Vol. 7 (*Structural Design and Analysis Part I*), pp. 149–206.
79. F. Werren and C. B. Norris, *Mechanical Properties of Laminate Designed to be Isotropic, Report 1841,* Forest Products Laboratory, Madison, Wisc., 1953.
80. B. H. Jones in Ref. 74, pp. 34–72.
81. N. J. Pagano and R. B. Pipes, *J. Compos. Mater.* **5**, 50 (1971).
82. N. J. Pagano and R. B. Pipes, *Int. J. Mech. Sci.* **15**, 679 (1973).
83. G. Caprino, J. C. Halpin, and L. Nicolais, *Composites* **11**, 223 (Oct. 1979).
84. R. G. Hamilton and C. A. Berg, *Fibre Sci. Technol.* **6**, 55 (1973).
85. G. C. Grimes and L. F. Greimann in Ref. 74, pp. 135–230.
86. N. K. Hiester, *Trends in Advanced Composites, Report No. 700,* SRI International, Menlo Park, Calif., spring 1984, p. 12.
87. E. K. Welhart, *Mater. Eng.* **8**, 50 (1977).
88. G. Lubin, *Handbook of Fiberglass and Advanced Plastics Composites,* Van Nostrand Reinhold Co., New York, 1969, Chapts. 13–15 and 17–20.

89. U.S. Dept. of Defense, *DOD/NASA Structural Composites Fabrication Guide,* 2nd ed., prepared by Lockheed-Georgia Co., under contract no. F33615-77-C-5256 for the Air Force Wright Aeronautics Laboratory Manufacturing Technology Division, Dayton, Ohio, May 1979.
90. "Report of Research Briefing Panel on High-Performance Polymer Composites," *Research Briefings 1984,* National Academy Press, Washington, D.C., 1984, pp. 45–56.
91. M. M. Schwartz, *Composite Materials Handbook,* McGraw-Hill Book Company, New York, 1984, Chapt. 8, pp. 8.1–8.27.
92. S. W. Tsai, "A Matured Technology," *J. Compos. Mater.* **15**(1), Frontispiece editorial, no page number, (Jan. 1981).
93. T. E. Helminiak and R. C. Evers in *Polymers for Fibers and Elastomers,* American Chemical Society, Washington, D.C., 1984, pp. 414–420.

FRANK P. GERSTLE, JR.
Sandia National Laboratories

COMPOSITES, FABRICATION

Composite fabrication techniques have evolved from early manual labor to microprocessor-controlled systems. Process and materials developers have balanced competition and cooperation to maintain steady growth.

Early pioneers combined raw materials and formed them into a finished structure by hand. Tooling consisted mainly of jigs and plaster molds, although a few production-oriented companies utilized metal molds heated and used in standard presses (1).

An early development was the invention of preform machines that sprayed chopped-glass strands onto a three-dimensional air screen that approximated the shape to be molded. The preformed glass mat was placed in a matched metal die in a compression molding press, and liquid resin was added to form the product. Preform–die-molding processes have been used to mass-produce automobile bodies, truck cabs, furniture, boxes, trays, reusable concrete forms, and small boat shells (1).

The impression that fiber-reinforced plastic–composite products are unsuitable for automated production is false. Once the proper sequence of necessary process steps is known, automatic machinery can be designed by straightforward engineering.

For convenience, terms frequently encountered in composite fabrication are defined below:

Angle of winding (wind angle)	The angle the roving band is laid with respect to the mandrel axis of rotation.
Closed-end vessels	Parts that have much smaller diameters or totally closed domes at the ends.
Debulking	A process in which air is squeezed out of a prepreg laminate in order to promote adhesion.
Doff	Roving package.
Fiber payout	Transfer of rovings from delivery system to product line.
Overtravel	The additional carriage or eye travel beyond the ends of the part mandrel that is necessary to provide laydown of the fiber on the mandrel.

Mat	Chopped or random reinforcement material cut to the contour of a mold, usually impregnated with resin just before or during the molding process.
Peel ply	The outside expendable layer of a laminate, which is removed to improve bonding of additional layers.
Prepregs	Reinforcements in the form of a cloth impregnated with thermosetting resin advanced in cure only through the B-stage. The term also covers fabrics such as jute and rayon, which have been impregnated with a thermoplastic resin, eg, vinyl or ABS.
Roving	A form of reinforcing fiber glass comprised of 8–120 (and usually 60) single strands gathered into a bundle, and treated with a coupling agent to promote adhesion of the glass to the plastic matrix.
Tape	A form of reinforcement similar to mat and broad goods, cut in 7.5–30 cm widths, and of flexible consistency for laying onto a mold. Tapes are generally prepregs and are supplied with backing paper to prevent loss of integrity prior to use.
Tow	Term used instead of roving to designate fiber strands when referring to graphite or boron reinforcements in the fiber strand form.
Wet systems	Composite fabrication processes and equipment that incorporate reinforcement impregnation as part of the process, just before raw material die entry or mold contact.

Product and Machine Design

Product and process optimization are achieved through cooperative design, planning, and manufacturing. Traditional procedures, particularly in the aircraft industry, have allowed each discipline to proceed independently, with little consultation with manufacturing; the result has too often been equipment that did not operate efficiently.

An integrated approach is necessary to achieve the best results in product automation and economics. Cooperative effort among all disciplines is required from the start of a project.

The first step during the preliminary stage is product design (qv). Although it is not necessary for the machine designer to be familiar with every detail of manufacture, it is important to have certain basic data, such as environment, load paths, throughput rates, economic constraints, and configuration information.

Structural analysis to determine property requirements is recommended. Hybrid systems sometimes offer advantages with respect to weight, properties, and cost (2).

The design factor with the greatest impact on cost and performance is fiber orientation control. Strength properties depend on fiber orientation, and departures from the optimum can result in drastic property reduction. The difficulty of making complex arrangements by means of automated equipment is the primary reason why hand lay-up and other labor-intensive processes are still prevalent in the composite industry. This is not to say that complex arrangements are impossible; fiber orientation in nearly every conceivable configuration has been achieved on automated equipment (see Figures 9–27). Appropriate design and arrangement of supply spools and folding tooling for pultrusion equipment, multiaxis configuration and controls on filament-winding and tape-placement equipment, and other devices make complex laminate schedules feasible.

The most common composite-reinforcing materials are fiber glass, carbon fibers, aromatic polyamide (aramid), and boron. The term composite is used throughout to include glass fibers and other so-called advanced composite materials (see COMPOSITES). Reinforcement selection is based on intended usage, and property values established for these materials can be used as guidelines. Although most machines accommodate all reinforcements, designers who know the type of reinforcement to be used can take unique characteristics such as high modulus into consideration.

In addition to the common roving strands, raw material is available in many other forms, such as preimpregnated and wet systems, mats (paper and nonwovens), woven specialties with knit and braided reinforcement, multiple-ply broad goods, core materials as they apply to automation, fillers, extenders, and additives (films, coatings, plating, etc). The characteristics of the reinforcement must be familiar to the machine designer.

Matrix selection is another important design consideration; resin chemistry can accommodate nearly every conceivable product application. Running speed, cure rates, and final properties vary widely with resin, catalyst, additive, and fillers. Advance consultation with the supplier is recommended.

In addition to wet systems, where reinforcements are resin-coated during the process, more convenient prepregs are available that can be used for many applications. Suppliers assist in the selection of preimpregnated materials, which are available as unidirectional, woven and nonwoven tape, tow, mat, and so forth.

Although thermosets are the standard composite matrices, recent advances in thermoplastic technology may change this situation.

The diversity of products that can be fabricated on automated equipment is fast expanding as suppliers increase the flexibility of raw materials, and designers and manufacturers gain experience with integrated designs.

Methods and Processes

This discussion omits the more common forms of processing used to produce nonstructural components, such as SMC (sheet-molding compound (qv)), BMC (bulk-molding compound), preform molding, injection molding (qv), and spray-up. These methods have already shown their economic advantages in high and low volume production. The machines and processes described here are used in the manufacture of structural components, ie, those using continuous fiber reinforcements.

Composite fabrication is expanding into a whole array of specialty machinery for composite aircraft, automobiles, and other products.

Pultrusion.   The pultrusion (qv) process makes structural profiles of any shape continuously from composite materials (3–6), just as they would be extruded in aluminum. It is a one-step process that converts raw stock into finished product at rates up to 4.5 m/min.

Reinforcements are drawn into the system, impregnated with resin, and simultaneously formed and cured in a heated die; radio frequency or induction energy can be used to supplement curing (7–9).

The graphs shown in Figures 1–4 (10) illustrate the heat–pressure–time (line speed) interactions taking place inside a 90-cm long die during curing. The temperature profiles of Figure 1 show that the die temperature must be lower at the inlet to prevent pregel, and also at the exit to prevent hot cracking. Preheating permits a more gradual die heat profile and avoids skinning.

As can be seen from Figure 2, superimposing material temperature on the die heat profile shows inversion of the heat-flux relationship. First, the material absorbs heat from the die, then the die cools the cured profile. In Figure 3, material

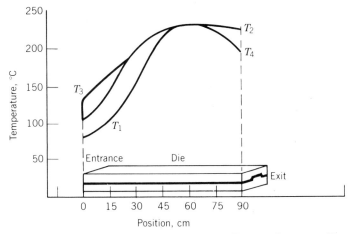

Fig. 1. Die temperature profiles. T_1, start-up; T_2, steady-state; T_3, with preheat; and T_4, cool-down.

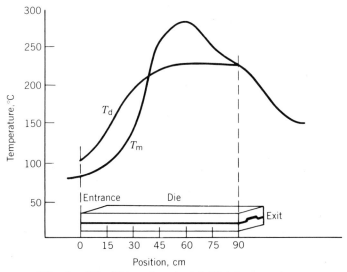

Fig. 2. Die (T_d) and material (T_m) temperatures.

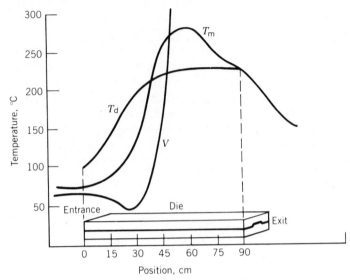

Fig. 3.   Viscosity and temperature. V, viscosity; T_d, die temperature; and T_m, material temperature.

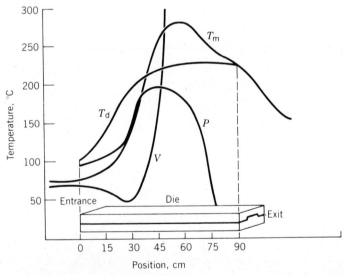

Fig. 4.   Pressure, viscosity, and temperature. P, pressure; V, viscosity; T_d, die temperature; and T_m, material temperature.

viscosity is superimposed on temperature curves. Viscosity at the die inlet must be high enough to prevent porosity formation inside the die.

Pultrusion Machines.   A typical pultrusion machine is shown in Figure 5. Reinforcements are impregnated by one of the available devices, in this case a pass-through-type wet-out tank, before passing through squeeze-out bushings. All the energy required to cure the resin system can be supplied by a radio frequency (r-f) generator. The process is run fast enough to bring the material

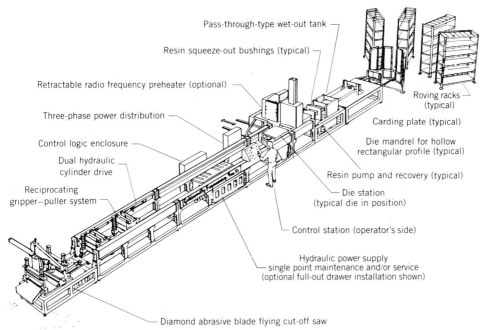

Pass-through-type wet-out tank

Resin squeeze-out bushings (typical)

Retractable radio frequency preheater (optional)

Three-phase power distribution

Control logic enclosure

Dual hydraulic cylinder drive

Reciprocating gripper–puller system

Roving racks (typical)

Carding plate (typical)

Die mandrel for hollow rectangular profile (typical)

Resin pump and recovery (typical)

Die station (typical die in position)

Control station (operator's side)

Hydraulic power supply single point maintenance and/or service (optional full-out drawer installation shown)

Diamond abrasive blade flying cut-off saw

Fig. 5.   Typical pultrusion machine.

into the forming die just before gelation. The die, 75–120 cm long, is heated to prevent it from becoming a heat sink and drawing from the package. The cured product leaving the die proceeds downstream through the pullers, which gives rise to the term pultrusion. Unlike extrusion (qv), in which the material is forced through a die with pressure, pultrusion pulls material through the die. As the stock leaves the pulling mechanism, a flying saw cuts it to length.

A recent pultrusion machine, built in modular sections, is shown in Figure 6.

Fig. 6.   Pulmaster pultrusion machine model PM 8-24-0. Courtesy of Goldsworthy Engineering, Inc.

Reciprocating clamp-puller systems with multistream capability, such as the equipment shown in Figure 7, are commonly employed.

Fig. 7.   Reciprocating clamp-puller system. Courtesy of Pultrusion Technology, Inc.

Product center-line control is an often overlooked but necessary feature in pultrusion processing. This control can be conveniently provided by hydraulic jackscrews, which can change the height of the bottom platen (see Fig. 8). Center-line control can also be obtained by shimming of tools on the bottom platen. Center-line height must be compatible with the height of the cutoff saw.

Fig. 8.   Product center-line control. Courtesy of Pultrusion Technology, Inc.

Capacity.   The capacity of pultrusion equipment is steadily increasing and the future market for pultruded product may be in profiles much larger than anything possible with traditional materials such as aluminum or rolled steel sizes. The Goldsworthy Pulmaster produces a profile 60 cm wide and up to 20 cm high. A similar machine produces a profile 90 cm wide and 45 cm high. Another machine is being designed for 90 × 60-cm profiles.

The potential of automated composite fabrication is further illustrated by the pultrusion of residential building panels up to 2.6 m wide and 10–15 cm deep, with internal ribbing. A full-length panel can be tilted up on a building in any desired length, limited only by available means of transportation. Such large panels can be made because pultrusion is not a pressure operation, and size is limited only by the size of the surrounding structure. Competitive processes, such as the steel rolling and aluminum extrusion, require very high pressures, and sizes are correspondingly limited.

Very large profiles are difficult to cut off. To reach the center of a 90 × 60-cm profile, a diamond saw, 19 m in diameter, would be required. For the 90 × 45-cm capacity machine, two saws are used.

Fiber Orientation.   Contrary to common belief, pultrusion is not limited to unidirectional reinforcements, but can bring reinforcing fibers into the profile in any needed orientation.

Equipment such as that shown in Figure 9 increases compression and burst strengths of continuously pulled profiles by adding circumferential winding to simple profiles on a continuous basis.

Fig. 9.   Roving winders for circumferentially wound pultrusions. Courtesy of Pultrusion Technology, Inc.

Square tubing is made on the equipment shown in Figure 10, which includes longitudinal reinforcement of pure roving, three plies of random continuous fiber mat for omnidirectional plies, and a circumferential overwind. Random mat is fed into a folding shoe that convolutely folds the mat around the mandrel, while longitudinal fiber is brought in from supply racks and the circumferential ply is applied by the rotating wheel.

Nearly any construction can be achieved. Tooling fabricated for round tubing with 6.25-mm wall starts with an inside ply of polyester wall mat, a ply of random glass fiber, and a ply of longitudinal fibers. Because of the thickness of the wall,

Fig. 10. Pultrusion machine tooled for square tubing. Courtesy of Goldsworthy Engineering, Inc.

this package is taken through an impregnating bushing; two additional plies of random glass mat are added, and the package wrapped with plies of pure circumferentials by a winding wheel. Another ply of random mat is added, following which a $+45°$ ply and a $-45°$ ply are added by large winding wheels. Another impregnating bushing wets out this last group of plies. The package then enters the r-f system, where the energy is introduced. More plies can be added. The product proceeds through the pullers in the usual manner.

Production Rate. Throughput speed depends largely on the resin system. A vinyl ester system at 1.2–1.5 m/min is used for the round tubing. Polyester and vinyl ester systems have been developed that permit a production rate of ca 4.5 m/min with pure unidirectional reinforcement. However, when an off-axis reinforcement such as mat is added, viscous shear forces increase exponentially with velocity, and at speeds of 2–3 m/min the reinforcements tear.

Radio-frequency heat is recommended only for glass-reinforced stock. For graphite or conductive fiber reinforcement, inductive heating is applied. The mass can be instantly heated through by one of these techniques.

Curved Configuration. Since there is no way to bend pultrusions after curing, curved pultrusion must be made in the desired configuration. The roving is drawn off supply racks in the usual manner, impregnated, passed through an r-f generator, and brought between a fixed and a rotating die. The preheated stock is pulled through the system by a curved die located on the face of a rotating curved quadrant, pulling the part continuously through the orifice formed by the fixed and rotating dies (11). Goldsworthy Engineering developed this curved pultrusion process and built prototype equipment for NASA for making graphite hat-section stiffener rings for the space shuttle.

Filament Winding. Filament winding (qv) allows control of fiber volume and orientation, utilizing continuous filaments, thereby maximizing laminate strength. The reinforcement, its alignment, and the resin matrix are determined during the design, according to process requirements. The winding process parameters can be calculated from the product geometry and the wind angle for each layer. The angle of rotation required through certain portions of the mandrel is calculated, and circuit pattern, band width, and the number of circuits needed are determined. This is an iterative process; several variables are interdependent. After test winds, winding parameters are often changed; this process is repeated to give full optimization.

Nongeodesic winding, though more complex, utilizes the same theory. A friction factor is introduced into the fiber-path motion function, allowing the winding to be located out of its geodesic pattern; the matrix resists side slip. Tension and resin viscosity greatly influence the degree and speed of winding such patterns without resorting to mechanical resistance to side slippage.

Traditionally, the pattern was based on the fiber tension; winding proceeded along geodesic paths irrespective of structure. Innovative filament-placement techniques, however, deviate widely from the geodesic path, and may lead to winding of very complex structures.

This innovative technique was made possible by two developments: computer control and the winding on nonuniform mandrel surfaces. For instance, it would be possible in one operation to wind an aircraft fuselage with interruptions to the pure geodesic path such as doublers around doors and transparencies.

Filament winding was originally used to fabricate tanks and pipes; early equipment consisted of little more than converted lathes, with a reciprocating feed eye winding a helical pattern onto a rotating mandrel. This created the impression that helical winding of pipes and tanks is necessary, when in fact, other winding patterns are equally effective.

Winding can be accomplished on a horizontal (Figs. 11 and 12) or vertical (Fig. 13) tooling face. Employment of mechanical or computer-controlled filament-winding machines depends on equipment cost, part geometry, and usage, ie, dedicated to a single task or adaptable for a broad range (12). Capital equipment costs are significantly lower for mechanical machines, which, however, require time-consuming calculations and hardware changes.

If large quantities of simple parts are to be produced, a mechanical winding

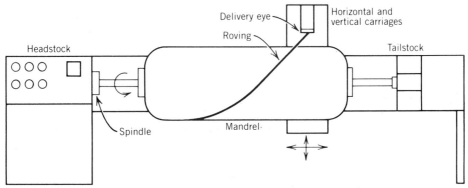

Fig. 11. Horizontal filament winding machine. Courtesy of Engineering Technology, Inc.

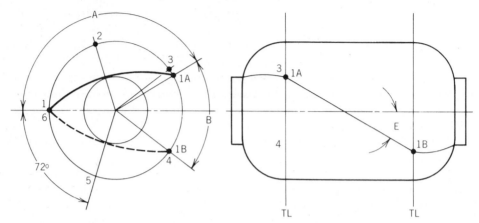

Fig. 12.   Helical winding pattern. Courtesy of Engineering Technology, Inc.

machine offers advantages. Programming consists of coordinating the motion between spindle and carriage, by selecting a drive-gear ratio and determining the carriage drive sprocket and chain length.

Computerized programmable equipment is usually more cost effective where a variety of parts are to be produced or where part configuration is complex, such as a wind turbine blade produced by Hamilton Standard Division of United Technologies. This process employs a large winding machine with five control axes, a sophisticated controller, and data-processing systems including five computers. Each circuit or cycle of the fiber payout system follows a unique path to position the fibers on the mandrel properly (12).

Sophisticated equipment such as Engineering Technology's FiberGrafix color graphic programming station assists in the development of such complex winding patterns. Most of the important aspects of a pattern can be examined and modified through this station (13). With a pattern–time reference, winding time can be minimized subject to the limitations of the winding machine and pattern smoothness. Insertion, deletion, replacement, and smoothing of the stored pattern positions are possible via an interactive color graphic station. A band advance function allows the winding pattern to be easily modified, and the winding material can be properly distributed on the mandrel. The pattern can also be modified by scaling, offsetting, and rotation (see also PROCESSING, COMPUTER CONTROL).

Equipment.   A typical early filament-winding machine with rudimentary computer controls is the lathe-type winder illustrated in Figure 14, utilizing rotating mandrel and traversing carriage.

Another type is known as the orbital, or racetrack winder, where the winding heads move completely around the mandrel, as shown in Figure 15. It is used primarily for rocket motor cases and where polar winding is desired; high speed is attainable.

In the tumble-type winder (Fig. 16), the mandrel is tumbled end over end during a polar wind, whereas the feed eye is traversed and the mandrel rotated in the normal lathe-type fashion for helical or circumferential winds. This type is widely used for high volume commercial products such as water-softener tanks, pool filter tanks, etc.

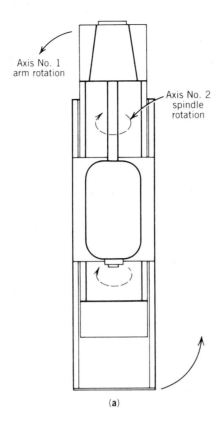

(a)

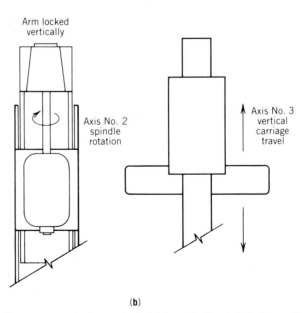

(b)

Fig. 13. Winding on a vertical mandrel: (**a**) longitudinal; (**b**) circumferential. Courtesy of Engineering Technology, Inc.

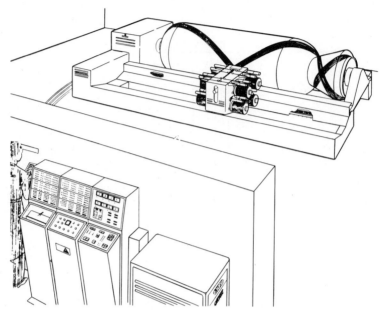

Fig. 14.   Typical lathe-type filament-winding machine.

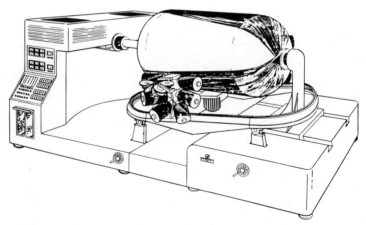

Fig. 15.   Typical racetrack-type filament-winding machine.

Whirling-arm filament winding is illustrated in Figures 17 and 18. Polar wind is accomplished by means of the C-shaped arm, which supplies filaments from feed eyes at both extremities. While whirling on its horizontal axis, the arm winds on the polar axis of the mandrel, which is circumferentially indexed at appropriate intervals. During polar wind, the horizontal winding arm is retracted; when the polar wind is completed, the horizontal arm extends and reciprocates vertically to effect helical or circumferential winds while the mandrel rotates.

Although filament winding is most commonly employed on curved surfaces of rotation, it is also possible to wind on flat-sided surfaces. A flat-sided whirling-arm winding machine was built in the early 1960s for winding high strength

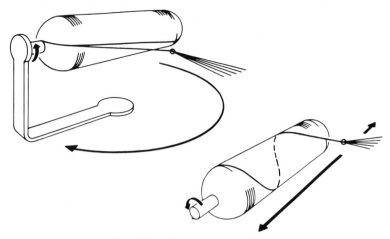

Fig. 16. Tumble-type winder.

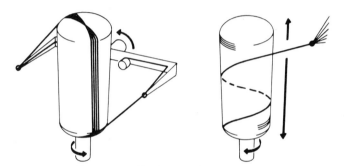

Fig. 17. Whirling-arm-type winder.

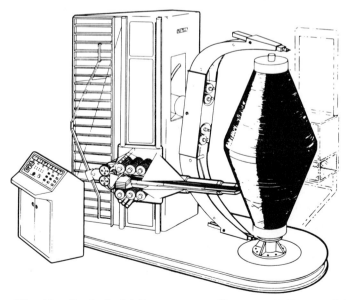

Fig. 18. Typical whirling-arm-type filament-winding machine.

instrument cases (14). All six faces of the tool had to be wound, and rather than use normal support mandrels, the tool was flipped into position by means of hydraulically operated tables.

Spherical winding is most difficult (Fig. 19). Before the advent of the computer, most machines were digitally controlled.

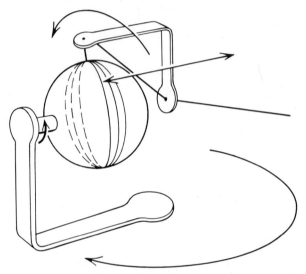

Fig. 19.   Spherical winder.

Contact-type balls, such as basketballs, volleyballs, and soccer balls, are filament wound. Although a totally random wind must be effected, it must be controlled; because it is necessary to wind all over the ball, there is no way of holding onto it or drive it. The required controlled random pattern can be obtained by suspending the rubber bladder on an air column and changing product rotational speed and thereby direction of wind by air jetting the column.

In a ring winder (Figs. 20 and 21), the winding head rotates around and traverses a passive mandrel. This type originated for the filament winding of massive structures, for example, a 45-m windmill blade. It is clearly a formidable task to support and rotate such a structure. The ring-winder design allows the mandrel to remain passive while the rotating ring–feed-eye assembly reciprocates along its length, laying a helical pattern. Steady rests stationed along the length of the part are hydraulically lowered by the carriage and raised behind it.

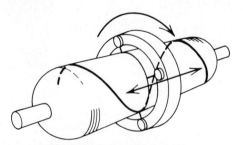

Fig. 20.   Ring-type winder.

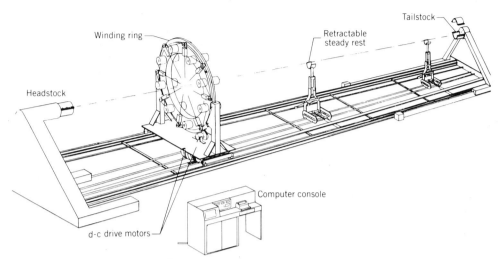

Fig. 21. Ring-type filament-winding machine. Winding ring can mount up to 14-kg glass packages. Drive motors are 11 kW (15 hp). Ring length is 152 cm; maximum part length is 9 m. Courtesy of Goldsworthy Engineering, Inc.

Tape Placement. Tape-placement equipment utilizes an automated tape-laying head to place the tape contiguously on a pattern table or mold in such a way that each tape lies within a specified distance of any previous tape course with no overlap. Precise sequential-layer orientation (cross-plying) is mandatory for strength; taper can be achieved by step-back buildup (15). The raw materials used at present are graphite, aramid fiber, or glass reinforcements in the form of unidirectional or woven tape. Widths range from 2.5–30 cm. The tape is usually impregnated with an epoxy-based resin system.

Early automated tape placement imitated manual operations to lay 7.5-cm unidirectional tape on a flat table. All manufacturers at first produced 7.5-cm tape, and tape-placement machines were built accordingly. Tape sizes were then extended to 15 and 30 cm, then back to 2.5 cm. Today most machines have two heads: one for 2.5-cm, and another for 15-cm tape.

This progress resulted from the fact that if tape of any width is laid over a compound curvature, one edge of the tape travels farther than the other. Hand lay-up in the manufacture of aircraft structures established the tradition of using unidirectional tape for flat patterns, which in most cases were draped in the tooling to form simple curved parts. As the need to produce parts with compound curvature became apparent, manual methods were adapted to machinery, although they usually incorporate guided adjustments that the machine does not imitate, leaving ripples at the long edge.

Satisfactory automated lay-up of compound curvatures will undoubtedly require the replacement of tape by tow-placement machines, with individual tows formed into a tape at the lay-down shoe.

A typical early tape-placement head is illustrated by Figure 22 (15). As this device was built without computer control, optical devices were utilized to sense the edge of the tool, accelerate machine motion, set shear to the right angle, and trim the tape. At that time, the head was usually built separately and then

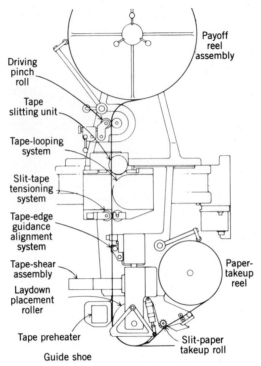

Payoff reel assembly

Driving pinch roll

Tape slitting unit

Tape-looping system

Slit-tape tensioning system

Tape-edge guidance alignment system

Tape-shear assembly

Laydown placement roller

Tape preheater

Guide shoe

Paper-takeup reel

Slit-paper takeup roll

Fig. 22. Typical early tape-placement head.

attached to an existing machine. For example, a head built by Goldsworthy Engineering was used by Lockheed with a profiler as the host machine; the lay-up is trimmed after the machine is in place.

The first machine to use 30-cm tape fabricated weapon bay doors for the B1 bomber. A government contract limited the expenditure by the prime contractor (Rockwell), but without any sacrifice in accuracy. Requirements were met by a manually operated machine with automatic width indexing, with other motions initiated by the operator.

Computer Control. The Army's desire to test an all-purpose, multiaxis tape placement machine led to the next step forward, an automatic machine built in 1974 for helicopter rotor blades for the Army Aviation Systems Command (AVSCOM) (17). This was the first six-axis minicomputer-controlled tape-placement machine. The entire gantry, including the Y-axis structure, was built from composites. At first, army personnel strenuously objected because of the established use of iron machine tools. However, an essential element of tape-placement machines is rapid X-axis travel, and it was vital to control the structure at the desired speeds. The composite gantry provided this capability.

The machine was actually an eight-axis machine, with two steady rests to support the blade tooling. Extreme compound curvatures existed in the transition area between the air-foil section of the blade and the hub section. For conformity with these extreme compound curvatures, the tape head had the capability of splitting the 7.5-cm tape into 24 individually tensioned 0.31-cm tapes.

In the first composite blade to be laid up, the spar was built up on the machine, honeycomb put in place to form the section of the blade aft of the spar, and the skin wound over it. In effect, this structure became a mandrel in place.

This machine was an excellent tool for developing the techniques of spar making, skin lay-ups, and attachment of helicopter rotor blades; it pointed the way to specialty production machinery (see Fig. 23).

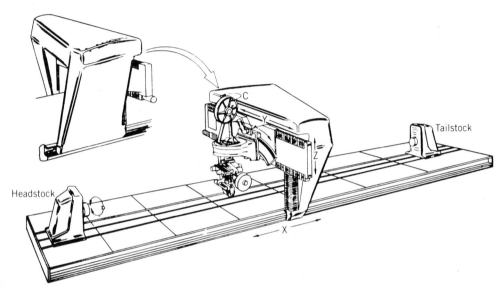

Fig. 23. Original eight-axis automated tape lay-up system (ATLAS). Courtesy of Goldsworthy Engineering, Inc.

The Cincinnati Milacron seven-axis computer numerical control (CNC) machine illustrated in Figures 24 and 25 has the inherent capability for following and uniformly applying 7.5-cm wide tape to compound curvatures that rise and fall at angles up to 15°. A wide variety of outersurface airframe parts can be manufactured, such as ailerons, stabilizers, flaps, rudders, fins, and wing skins. The machine head consists of the tape-feed roll and drive, a backing-paper windup spool and drive, a floating rubber shoe for laying down and compacting the tape, and a stylus for cutting. It is mounted to a saddle by mechanisms that allow for computer-controlled rotation about a vertical axis (C), and for sensor-controlled pivoting about a point on the center line of the floating rubber shoe (A axis). Combining the computer-controlled rotation of the head (C axis) with the longitudinal movement of the gantry (X axis) and the cross-movement of the carrier (Y axis) permits tape laying in large work areas (18). Two computer-controlled axes are built into the tape head to control the functions of the cutter. The W axis moves the cutter across the tape width; it is synchronized with head speed for correct angle of cut. The D axis provides angular positioning of the cutter to assure alignment along the direction of the cut.

Composite tape can also be wrapped around a part by specially designed tape-wrapping equipment. A tape wrapping head is illustrated in Figure 26.

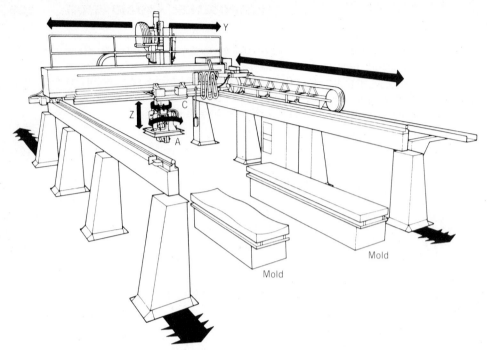

Fig. 24. Cincinnati Milacron seven-axis computer numerical control (CNC) machine. The X-axis travel can be extended at any time beyond the basic 75 cm in 30-cm increments. Courtesy of Cincinnati Milacron, Inc.

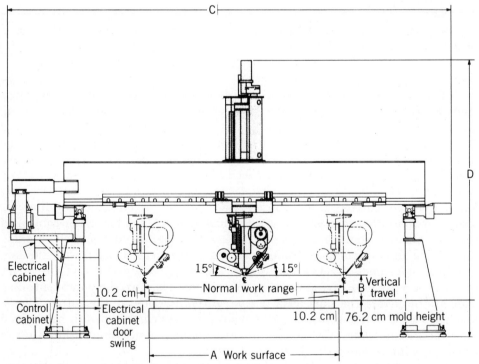

Fig. 25. Line drawing of seven-axis CNC tape-placement machine. Courtesy of Cincinnati Milacron, Inc.

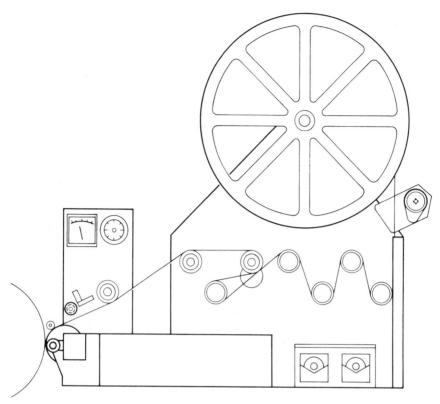

Fig. 26.   Tape-wrapping head. Courtesy of Cincinnati Milacron, Inc.

Production Design.   These machines were adequate in the airframe industry for prototypes and testing, but large structural components require production
tape-placement machines. These presented certain problems, of which the most
pressing was that the tape, lacking integrity, must be guided by the backing
paper. If tape and paper are separated, tape control is therefore lost. It follows
that a device is required that will shear the tape to the proper trim angles without
shearing the backing paper. A number of devices were developed to perform this
difficult task; their success rate has been about 93%. Unfortunately, a 7% failure
rate is disastrous since even one uncut filament destroys the integrity of the
entire system by pulling it up when the head lifts; the damage must be repaired
manually before resuming production. In other words, even a low failure rate
destroys the automated functioning of the machine.

   High angle cuts present a similar problem. The long tapering tail on one
side of the cut needs to be pressed down, and the other side picked up.

   To overcome these and other difficulties, Goldsworthy Engineering developed a two-phase system (Fig. 27): tape preparation is accomplished in phase I,
and placement in phase II.

   In the preparation phase designated ACCESS (Advanced Composite Cassette Edit–Shear System), the supply spool is interfaced with CAD system soft-
ware specific to the part to be produced. The material from the supply spool is
measured, cut, angled on both ends, and transferred to new backing paper in

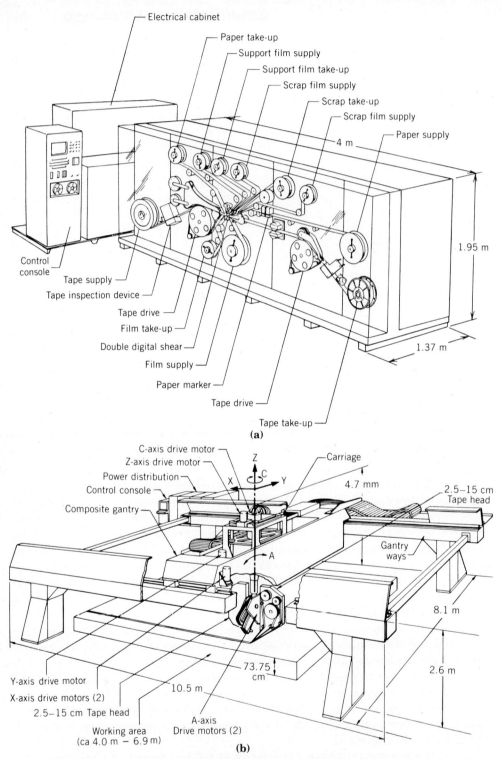

Fig. 27. Two-phase tape-placement system. (**a**) Phase I: tape preparation machine, ACCESS (Advanced Composite Cassette Edit–Shear System); (**b**) Phase II: tape placement machine, ATLAS II (Automated Tape Lay-up System, second version). Travel: X, 7.28 m; Y, 4.08 m; Z, 4.8 m; C, ±200°; and A, ±30°. Courtesy of Goldsworthy Engineering, Inc.

order to be spaced to compensate for long tails. Defective material is removed. Voids, cut failure, and high angle problems during laydown are thus eliminated. The cut tape on its new backing paper is rewound on a new supply spool, and the "composite cassette" installed on the tape-placement machine.

The second phase is termed ATLAS II (Automated Tape Lay-up System, second version). This system offers the following advantages: 100% shear dependability; properly spaced tails; preinspected and edited tape; and elimination of all waste motion from the tape placement machine. Higher production rates result.

Pulforming. The curved-pultrusion process developed for the NASA shuttle stiffener rings led to so-called pulforming. In this process, as in pultrusion, primary reinforcing fibers are drawn through a tank for resin impregnation and then through a die. Pulforming may be curved or straight with machines that produce profiles of changing volume and changing shape or constant volume and changing shape.

Changing Volume and Shape. An example is the composite hammer handle made continuously on straight pulforming equipment (19). Volume is changed as follows: a bulk molding compound (BMC) in the form of extruded rope on a supply spool is fed through a monitoring device that measures length to provide the desired volume. The length is cut and crimped around a single traveling roving strand at uniform intervals. The result resembles a string of beads proceeding downstream. A blivet press traveling at line speed forms the BMC pieces into the shape of the changing volume. Further downstream, all primary reinforcing fibers come together to encapsulate the center fiber with the formed BMC. Film is convolutely folded and ultrasonically welded around the package, creating a roll, which passes through a shrink tunnel; this tightens the film to produce a very controlled package.

The package next enters a split female die, which travels at line speed downstream and is closed with a C-press, which injects the die into a belt-clamping system that butts a whole stream of the dies together. The dies are heated as they traverse the clamping area; when they reach the end, the dies open, releasing cured stock, and return to the upstream C-press end of the die section. Operation is thus effected continuously. The part is cut to length with a flying saw.

Constant Volume and Changing Shape. The automatic machine pictured in Figure 28 (20,21) was developed to make automotive leaf springs. Reinforcing materials from supply racks pass through an impregnating tank, are heated in an r-f generator, and enter the die cavity, which is at the convergence of a stainless-steel belt and the rotating die. The die cavity is in the shape of the spring to be produced. The stainless-steel belt is clamped to the die face and runs continuously with it, thereby closing the fourth side of the die. Curing of the part is finished along that belt; it is then peeled out of the die cavity and guided into the path of the flying saw. At present such machines are controlled by standard switch and relay logic; future machines will utilize programmable controllers. Production rate, depending on specific configuration, is approximately two springs per minute. Springs produced by pulforming are much cheaper than steel springs.

Specialty Machines. Machinery can usually be designed to accommodate any reinforcing fiber and resin system on the market today. Since the goal is the lowest possible cost, the industry usually employs roving or tow as the cheapest

Fig. 28. Pulformer curved pultrusion machine; constant volume–changing shape. Courtesy of Goldsworthy Engineering, Inc.

raw material, and thermoset polyesters and vinyl ester resins, rather than epoxies, for the highest running rates.

Most of the machines described here are flexible as to fibrous form; however, very high modulus fibers, such as carbon graphite, can create problems. If designers know that high modulus fibers are to be used on prospective equipment, they can plan accordingly.

The specialty machinery described below was designed for a variety of applications and demonstrates the flexibility of automated composite production equipment.

Plywood.   Composite-faced plywood equipment for the truck and container industry continuously applies a bidirectional glass facing to both sides of standard plywood (3 × 1.2 m), producing an endless laminate-faced plywood panel 3 m wide. An automatic saw cuts it to the desired length, the standard being 12 × 3 m (22–24).

The plywood is stacked at the entrance of the machine, where it is automatically elevated, separated, and butted at the edges, then driven into the system. Cross-ply glass is applied with a large winding wheel, holding about 80 doffs of roving, which rotates around the plywood as it proceeds at a controlled rate. Longitudinal rovings are added by feeding roving through carding racks above and below the product stream just ahead of an opposed-belt laminator. Resin is poured onto the upper and lower surfaces of the product just upstream of the longitudinal roving contact point, wetting cross and longitudinal fiber as it enters the opposed belt laminator. The package proceeds through the laminator and is cured, trimmed, and cut to length.

Sporting Goods.   A continuous-taper tube-winding machine is used to manufacture graphite–epoxy golf clubs (25). Stock is filament wound using ten stacked winding wheels; speed is individually controlled in such a way that angle of wind and tape can be varied for production of flexible or stiff shafts. Steel mandrels are centerless ground to the taper requirement; the small end of one mandrel is indexed into the large end of the other, and the mandrels can pass continuously through the machine.

Insulation.   A reinforced foam machine designed for McDonnell Douglas may be the most complex automated composite fabrication equipment ever built (26). Its purpose was to produce a 70×22.5 cm three-dimensionally reinforced, urethane foam log continuously, to be used for insulating tankers transporting liquefied natural gas. This equipment probably holds a record number of ends of reinforcing material handled on a single machine. A supply rack feeds 3500 ends of glass fibers into the system for X and Y reinforcement; additional creels in the center section supply 1500 ends of Z-direction fiber.

Military Equipment.   Filament-winding machines are used to wind wet filament on 155-mm artillery projectiles, five shells at a time on a three-minute cycle. The steel casings are automatically loaded horizontally onto a continuous chain- and spud-conveyor system. The casings are oriented vertically under a row of winding heads, which move into position for winding and then withdraw. At the next station the filament is severed, while the conveyor brings a new bank of shells into position. The wound shells are cured in an r-f oven and removed at the end of the machine.

During the Vietnam war a machine was designed to transport water, gas, and jet fuels to the front. Rolled steel or aluminum tanks were bulky, awkward, and required seam welding, a skilled operation. A pultrusion machine was installed in a standard 12-m highway trailer, and a composite tank constructed on-site (27).

After the foundation was poured, a continuously pultruded tongue-and-groove profile resembling hardwood flooring was produced in the length required for the tank wall (ca 1560 m). An unusual feature was that two streams of two-sided bondable Tedlar were fed in with the glass roving reinforcements, to create an impermeable barrier. The profile was fed into an erection machine that wound and zippered it continuously together while injecting epoxy bonding. With unskilled labor, the prototype tank (6 m diameter by 3 m high) was erected in about four and a half hours and then grouted into the foundation.

Under the auspices of the Air Force Materials Laboratory, Air Force Wright Aeronautical Laboratories (AFWAL), several primary airframe manufacturers such as Grumman Aerospace, Northrop Aircraft, North American/Rockwell, and General Dynamics have spent several years in research and development of automated integrated composite laminating centers, commonly termed "factory of the future."

Grumman's integrated laminating center (ILC), for example, fills the need for more fully automated fabrication of gently contoured composite structures from unidirectional tape (28). Elimination of manual operation significantly cut costs of airframe components such as the 56-ply covers of the F-14A horizontal stabilizer (utilizing 7.5-cm wide boron–epoxy tape), and hybrid boron graphite–epoxy B-1 stabilizers. In the late 1970s, Grumman teamed with AFWAL to develop mechanized equipment for fabricating severely contoured, integrally stiffened, complex structures for high performance aircraft such as the ATS. For the Automated Integrated Manufacturing System (AIMS) (Fig. 29), three modules were integrated with the ILC, ie, a contour ply handler, broad-goods prepreg dispenser, and translaminar stitcher. Computerized air-passage contour data base; computerized detail design of flat ply patterns, templates, and curing tools; and computer-generated nesting pattern for automatic laser trimming are included.

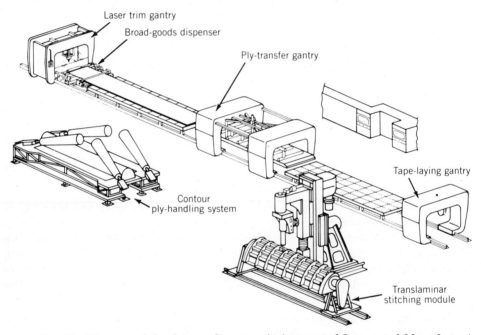

Fig. 29.   Factory of the future: Grumman's Automated Integrated Manufacturing System (AIMS). Courtesy of Grumman Aerospace Corp.

Automotive Applications.   Another example of low cost production is provided by a machine continuously manufacturing automotive drive shafts. A hybrid drive shaft is produced with glass fiber in the $\pm 45°$ directions to take torsional loads and graphite as longitudinal fiber for stiffness. An endless stream of mandrels is fed through the machine; steel end fittings are inserted and later welded to universal joint spiders. Winding wheels apply glass fiber, and longitudinal graphite fibers are fed in at the next station. Circumferential fibers are added last. The mass is cured by induction heating.

Other automotive applications (qv) include pulformed leaf springs, bumpers, door frames, and others.

Aircraft.   A "pin winder" built for Bell Helicopter for making blade spars is shown in Figure 30. Carrying impregnated S glass, the carriage moves along

Fig. 30. Helicopter spar "pin winder." Courtesy of Goldsworthy Engineering, Inc.

the tooling bed, rotating 180° at each end. The beam is computer-controlled for vertical Z-axis motion; the feed eye moves in a plane but any path is possible.

Interest has recently grown in the fabrication of structural aircraft components from composites, but it must be understood that neither economic nor performance benefits are realized by designing a metal airplane and building it from composites.

A 2.4-m full-scale test section of a geodesic aircraft fuselage is shown in Figure 31. The entire fuselage section is filament wound. Although during World War II thousands of Wellington bombers were built with geodesic structures, they are troublesome with metal. However, a geodesic pattern is probably the simplest design for composite aircraft structures. By a straightforward procedure, the structure is wound first, and then the skin. The test section shown in Figure

Fig. 31. Geodesic aircraft fuselage. Courtesy of Goldsworthy Engineering, Inc.

31 was for a pressurized airplane, and therefore required bulkheads at both ends; these usually create severe stress problems. Problems also arise in the filament winding of end sections. They were solved by winding across the pressure bulkhead and keeping it in tension with the structure being stiffened. The wound skin was cured in a female die to form aerodynamic surfaces. A peel ply was wound with the skin in the geodesic pattern, leaving a clean surface after stripping. Both diamond and round windows were wound into the test section, the former because they were compatible with the geodesic structure.

Composites in Orbit. A ribbon-forming machine was developed for the NASA SSPS (satellite solar power system) program (Fig. 32). The $0+45-45-0$ graphite–thermoplastic matrix ribbon formed would be installed in a cassette in a machine in the cargo bay of the space shuttle. In orbit the machine would process the ribbon into a triangular truss beam (see Fig. 33). The graphite-reinforced, polysulfone matrix ribbon is formed and welded into the triangular closed beam. The intercostals are continuously wound or welded on the beam to form a truss. Prototype equipment was built in the Goldsworthy Engineering laboratory, and a triangular truss beam is on display at NASA Huntsville, Ala.

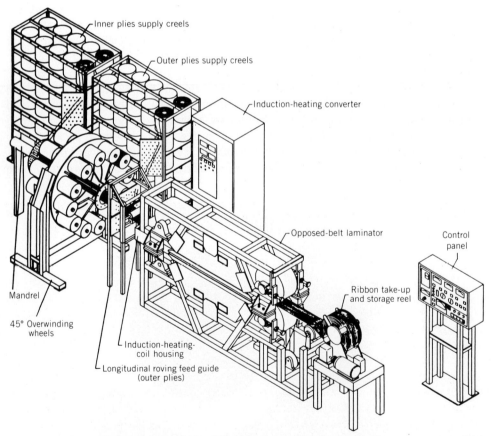

Fig. 32.   Earthbound ribbon-forming beam machine; thermoplastic matrix-graphite fiber $(0+45-45-0)$.

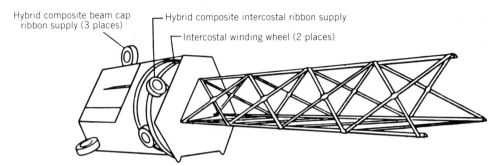

Hybrid composite beam cap ribbon supply (3 places)

Hybrid composite intercostal ribbon supply

Intercostal winding wheel (2 places)

Fig. 33.   Beam machine: continuously wound and welded intercostal version.

Thermoplastics.   Thermoplastic matrix composites represent a new and intensively researched area. They offer promise in significantly improved properties and faster processing.

Thermoplastic matrices, by definition, require no cure cycle. They are heated until they flow or permanently deform, pushed into their new shape, and cooled. Elimination of cure cycles reduces processing cycle times.

Compression molding and filament winding are already commercially viable and production processes and equipment are in limited service. However, application must justify the higher materials cost. Engineering thermoplastic matrices are currently 1.5–4 times the cost of thermosets, and the equipment requires adaptations (see also COMPRESSION AND TRANSFER MOLDING).

Continuous processes such as pultrusion still represent a challenge to processors because the very high melt viscosities of thermoplastic matrices, compared to the low thermoset viscosities, limit the throughput rate severely because of high hydraulic shear forces generated in the dies and forming tooling. These forces damage the reinforcement and generate high pull loads unless line speeds are kept to a few centimeters per minute.

However, fluidized bed powder matrix application and derivative processes may solve these problems. Ideally, processors would like to see a material with the properties of poly(ether ether ketones) (qv), poly(arylene sulfides) (qv), polysulfones (qv), and similar engineering thermoplastics with melt viscosities below 1 Pa·s ($=$ cP).

Other processes like tape laying also show great promise in the application of thermoplastic matrices. Present tape-laying operations, even when machine performed, require several "debulk" cycles plus the normal multihour cure cycle. Capability for heating under pressure and debulking may make it possible to have a complete trim-ready wing skin at the end of the lay-up operation, without debulk or cure cycles.

Clearly the potential here, both in the tougher, higher temperature physical properties available from thermoplastics, and in the near instantaneous processing, will increase research and investment in thermoplastic matrix-processing of composites. It is possible that in 10–20 years thermoplastics will be the dominant form of matrix. Much depends both on the progress of research and applications and the competitive response of the thermoset manufacturers.

Standards and Specifications

Specifications and standards pertaining to reinforced plastics–composites fall into one of the following classifications:

1. Standards developed and published primarily by ASTM and ANSI committees.
2. Codes set by users such as building officials, American Petroleum Institute, American Water Works Association, etc.
3. Design data generated by commercial and government research.

The codes and standards under the first two classifications relate to product rather than process and equipment; no standards have been written to cover processes and equipment. Design data are available which document procedures developed by research agencies and fabricators.

In addition to the ASTM standards given below, the ASTM D 20.23 Committee has 17 standards in various stages of development. Trade associations and government agencies supply code information (29).

List of Standards and Specifications

ANSI Standard A14.5, Safety Requirements for Portable Reinforced Plastic Ladders, 1981.

ASTM Specification F 711, FRP Rod and Tube for Live Line Tools, 1981.

ASTM Standard D 3647-84, Practice for Classifying Reinforced Plastic Pultruded Shapes According to Composition, 1985.

ASTM Standard D 3914-84, Test Method for In-plane Shear Strength of Pultruded Glass Reinforced Plastic Rod, 1985.

ASTM Standard D 3916-84, Test Method for Tensile Properties of Pultruded Glass Fiber-Reinforced Plastic Rod, 1985.

ASTM Standard C 581, Chemical Resistance of Resins, 1984.

ASTM Standard C 582, Laminates, Contact-molded, for C-R Equipment, 1984.

ASTM Standard D 1598, (joint with F-17) Time-to-failure, Constant Internal Pressure, 1984.

ASTM Standard D 1599, (joint with F-17) Rupture Strength, Short Time, 1984.

ASTM Standard D 1694, Threads, RTRP, 1984.

ASTM Standard D 2105, Longitudinal Tensile Properties, RTRP, 1984.

ASTM Standard D 2143, Cyclic Pressure Strength, RTRP, 1984.

ASTM Standard D 2310, Pipe, Classification of RTRP, 1984.

ASTM Standard D 2412, Parallel Plate Loading, 1984.

ASTM Standard D 2517-81, Pipe, Glass Pressure, Epoxy Resin, 1984.

ASTM Standard D 2924, External Pressure Resistance, 1984.

ASTM Standard D 2925, Beam Deflection, RTRP, Full Bore Flow, 1984.

ASTM Standard D 2992, Hydrostatic Design Basis for RTRP and Fittings, 1984.

ASTM Standard D 2996-81, Pipe, Filament-wound RTR, 1984.

ASTM Standard D 2997-77, Pipe, Centrifugally-cast, 1984.

ASTM Standard D 3262-82, Pipe, RPR Sewer, 1984.

ASTM Standard D 3299-82, Tanks, Filament-wound, 1984.

ASTM Standard D 3517-81, Pipe, RPM Pressure, 1984.

ASTM Standard D 3567, Dimensions, RTRP and Fittings, 1984.

ASTM Standard D 3615, Chemical Resistance Thermoset Compounds, 1984.

ASTM Standard D 3681, Chemical Resistance of RTRP in Deflected Con- dition, 1984.

ASTM Standard D 3753-81, Manholes, 1984.

ASTM Standard D 3754-81a, Pipe, RPM Sewer and Industrial Pressure, 1984.

ASTM Standard D 3839, Underground Installation of RTRP and RPM, 1984.

ASTM Standard D 3840-81, Fittings, RPM Pipe for Non-pressure Appli- cations, 1984.

ASTM Standard D 3982-81, Hood Specification, 1984.

ASTM Standard D 4021-81, Tanks, Underground Petroleum Storage, 1984.

ASTM Standard D 4024-81, Flanges, RTR Contact-molded, 1984.

ASTM Standard D 4097-82, Tanks, Contact-molded, 1984.

ASTM Standard D 4160-82, Fittings, RTRP for Non-pressure Applications, 1984.

ASTM Standard D 4161-82, Joints, RTRP Bell-and-spigot, Using Elasto- meric Seals, 1984.

ASTM Standard D 4162-82, Pipe, RTRP Sewer and Industrial Pressure, 1984.

ASTM Standard D 4163-82, Pipe, RTRP Pressure, 1984.

ASTM Standard D 4166-82, Measurement of Thickness via Digital Magnetic Intensity, 1984.

ASTM Standard D 4167-82, Fans and Blowers, 1984.

ASTM Standard D 4184-82, Pipe, RTRP Sewer, 1984.

ASTM Standard D 4398-84, Chemical Resistance by On-side Panel Expo- sure, Test Method, 1984.

Economic Aspects

The composites industry has usually been perceived as providing specialty parts at premium prices. Today, however, composites are being reclassified from "specialty" to "commodity" industry. The growth of composite product develop- ment has resulted in lower prices without loss of properties, such as light weight, high stiffness, corrosion resistance, etc. Composites are now cheaper than tra- ditional materials, and may become the economic barometer of the country, sim- ilar to steel in an earlier era.

This change was accelerated by the 1974 energy crisis and the resulting sharp price increase of traditional materials, such as aluminum and steel. Com- posites rose less in price and became increasingly competitive.

Energy requirements are given in Table 1. In general, those for composite materials and processing are lower than those for steel and aluminum.

Table 1. Energy Requirements for Traditional Raw Materials and Composites[a]

Material	MJ/kg[b]
Traditional metals	
steel, sheet	29
aluminum	
sheet	175
die cast	90
Glass fiber	
roving	43
chopped strand mat	52
chopped strands	44
woven roving	46
yarn	63
Resins	
polyester	92
epoxy	167
phenolic	82
polyamide	190
polycarbonate	146
Composite compounding	
BMC[c], 18% glass–52% filler	44
SMC[d], 30% glass–35% filler	54
polyamide–30% glass	153
polycarbonate–30% glass	126
polypropylene–30% glass	107
Composite products	
contact-molded	82
sprayed-up	80
hot-press-molded	88
SMC	60
pultruded	58
filament-wound	80

[a] Ref. 30.
[b] To convert J to cal, divide by 4.184.
[c] Bulk-molding compound.
[d] Sheet-molding compound.

Cost Savings Example.   Leaf springs provide a clear example of cost savings that are possible with composites. Composite springs weigh only a fifth as much as steel and perform far better. In a fatigue test, a steel spring fails at 35,000–50,000 cycles, whereas the composite spring survives 500,000 or even 5×10^6 cycles with no sign of failure. The cost savings extend beyond material cost to weight reduction and every aspect of production and performance. For example, the steel spring requires skilled labor in a multiple-step process, whereas the composite spring is made on automatic, single-stage machinery.

A study conducted by General Motors, based on an annual capacity of three million (10^6) units, produced the following cost figures for spring manufacture:

	Steel	*Composite*
floor space, m²	9000	540
capital equipment, $	6×10^6	400×10^3

Environmental Advantages.   With respect to expensive safety and pollution-control devices, composites have a marked advantage. A steel-spring plant presents a variety of hazards and serious pollution problems, whereas composites fabrication is conducted at a console. Under current EPA regulations, the cost of pollution control may exceed the cost of the basic capital equipment.

Quality Assurance Techniques

Quality assurance techniques for composite products fall into two categories: process control and product evaluation.

Raw materials control is, of course, critical to all manufacturing processes, and is assumed in this article.

This type of manufacturing confronts problems not found elsewhere, such as process variables and determining the size and degree of a fault area in large composite pieces. Still another problem is that faults in composite materials do not propagate as in more traditional materials and are harder to locate. As a result, in-process quality assurance is of much greater importance than in traditional manufacturing. Some defects can be corrected while the process is under way; others can be detected only in the end product, and therefore corrective action must wait until another cycle (see also COMPOSITES, TESTING).

Tape-placement Process.   The automated tape-placement process assumes the correct size, angle of cut, material constituency, and order of progression on the roll of the prepreg tape supplied by a programmed tape magazine. It can be termed an intermediate process, since the product is moved to another process center for thermal development of properties.

Defects in orientation, tape width, length, ply thickness, angle of cut, or smoothness are found mainly by visual inspection. An extremely high accuracy and control are required of the tape supply magazine and its programming.

Detection of defects is easier after thermal processing. Pressure and temperature must be monitored closely, usually by self-correcting mechanisms and without manual adjustment. After thermal processing, the part may be subjected to appropriate nondestructive testing (qv), as described later.

The size, complexity, and value of the processed part may limit testing to "on-site" techniques. Thorough visual inspection usually reveals significant surface defects, such as porosity, scratches, deficiency or surplus of resin, tape misorientation, and wrinkles.

Hardness of peripheral regions should be tested. The impressions or indentations made by instruments such as the Barcol tester are considered destructive testing, although many materials recover fully.

Separate specimens are used for destructive analysis and evaluation with regard to fiber fraction, void content, porosity, and mechanical properties.

Pultrusion.   In contrast to tape placement, pultrusion requires almost continuous monitoring to ensure continuous flow and optimum product properties. Curing of the resin matrix is critical, ie, gelation must be reached before the stock leaves the curing die. The degree of cure is checked by dielectric inspection. As marginal conditions are discovered, heating corrections to ensure gelation and continuity of flow through the dies can be made. This is true for polyesters and even more for epoxy resins, which require more time to gel.

The mandatory visual inspection can detect discoloration, crazing, surface fractures, pitting, bubbles, porosity, voids, finish inconsistencies, and surplus or deficiency of resin.

The pultruded part is evaluated by ultrasonic C-scan or infrared scanning. Uniform fiber dispersion and even coating by the matrix resin are the main requirements.

Pulforming.   Pulforming resembles pultrusion, except that the products can be straight or curved and may have constant volume and changing shape or changing volume and shape.

During pulforming, complications due to changes in cross-sectional shape and volume of the developing part require that close attention be paid to shrink rate and surface uniformity. The proper timing of resin gelation with respect to entry into the pulforming dies is determined by dielectric inspection of the degree of cure.

The surface is inspected visually during processing, but as in pultrusion, ultrasonic C-scan or infrared scanning are valuable techniques. Uniform dispersion of the fibers and evenness of the resin coating are essential.

Both pultrusion and pulforming may benefit from inspection by flash X ray or cineradiography, with associated good visual display.

Filament Winding.   In filament winding, many factors influence the product, including mandrel stiffness and surface finish, winding speed, resin precure, filament alignment and tension and temperature control.

The three winding methods include wet winding, in which the roving passes from a spool through the impregnating bath and onto the mandrel; dry winding with preimpregnated B-staged roving fed directly onto a heated mandrel or first through a softening oven; and postimpregnation, in which the roving is dry wound onto the mandrel and the resin applied by brushing, pressure, or vacuum. All methods require close control of fiber and resin properties.

The filaments are applied to the mandrel by circumferential or helical winding; both require uniform filament tension, even resin coating, correct filament patterns, and temperature control throughout.

The importance of inspection during winding may be reduced by computerized control. Visual inspection can detect filament out of position and uneven resin distribution, indicating variation in tension.

With the completion of filament winding, the testing of the wound and cured part encounters other difficulties. The correlation of standard test results with final performance poses a major problem. Since the filaments are wound in a curve, tensile tests of detached specimens are not directly relatable to material requirements. Recently, special test specimens have been developed for use with standard ASTM methods.

Nondestructive tests applicable to filament-wound parts include ultrasonic C-scan and radiographic and visual inspection.

Advantages and disadvantages of nondestructive methods are given in Table 2.

Outlook.   Continuous-flow processes, eg, for composite pipe, are adequately controlled by continuous scanning. With the improvement of such systems as infrared scanning and cineradiography and self-correcting loop-feedback controls, continuous processing machines will gain in scope. For noncontinuous processes, functional design criteria obtained from field use may facilitate the control of

Table 2. Nondestructive Test Methods for Wound Filaments

Method	Advantages	Disadvantages
penetrant inspection	good detection of defects or discontinuities open to the surface familiarity and reliability	no graphic record in normal operation not always portable
holographic interferometry	good definition of material integrity, especially subsurface good graphic record	expensive not always applicable not portable
radiographic inspection	good definition of material integrity, surface and subsurface good graphic record detects almost every type of defect	radiation hazard time-consuming expensive not always portable
ultrasonic C-scan	good definition of material integrity, surface and subsurface choice of techniques for specific applications	expensive requires liquid coupling agent that is not always practical not always portable
visual, aided or unaided	convenient inexpensive wide scope	no graphic record limited to surface conditions
infrared scanning	as accurate as ultrasonic techniques good graphic record portable no coupling agent required	still under development as standard equipment
dielectric inspection	can determine degree of cure on moving material	still under development as standard equipment

low volume composite products. Automated composite-fabrication systems and their quality control methods are at the frontiers of composite technology.

BIBLIOGRAPHY

1. E. E. Hardesty, "Forming Fiber Reinforced Products," *Automation Cleveland,* (Sept. 1970).
2. B. H. Jones, *Analysis of an Energy Absorbing Pultruded Beam,* final report of a contract carried out for Ford Motor Co. by the Composite Products Technology Center of Goldsworthy Engineering, Inc., Torrance, Calif., Nov. 1973.
3. U.S. Pat. 2,871,911 (Feb. 3, 1959), W. B. Goldsworthy (to Glastrusions, Inc.).
4. U.S. Pat. 2,990,091 (June 27, 1961), W. B. Goldsworthy (to Glastrusions, Inc.).
5. U.S. Pat. 3,556,888 (Jan. 19, 1971), W. B. Goldsworthy (to Glastrusions, Inc.).
6. U.S. Pat. 3,684,622 (Aug. 15, 1972), W. B. Goldsworthy (to Glastrusions, Inc.).
7. U.S. Pat. 3,674,601 (July 4, 1972), W. B. Goldsworthy (to Glastrusions, Inc.).
8. U.S. Pat. 3,793,108 (Feb. 19, 1974), W. B. Goldsworthy (to Glastrusions, Inc.).
9. U.S. Pat. 3,960,629 (June 1, 1976), W. B. Goldsworthy (to Glastrusions, Inc.).
10. J. E. Sumerak and J. D. Martin, "It's Time We Really Understood Pultrusion Process Variables," *Plast. Technol.,* (Feb. 1983).

11. U.S. Pat. 3,873,399 (Mar. 25, 1975), W. B. Goldsworthy (to Goldsworthy Engineering, Inc.).
12. C. D. Hermansen and R. R. Roser, "Filament Winding Machines: Which Type is Best for Your Application?" Paper presented at the *36th Annual Conference of the SPI,* RP/C Institute, Feb. 16–20, 1981, Society of Plastics Industry, New York, 1981.
13. R. R. Roser, *Computer Graphics Streamline the Programming of the Filament Winding Machine,* Engineering Technology, Inc., Salt Lake City, Utah.
14. U.S. Pats. 3,701,489 (Oct. 31, 1972), 3,740,285 (June 19, 1973), and 3,738,637 (June 12, 1973), W. B. Goldsworthy (to Goldsworthy Engineering, Inc.).
15. E. E. Hardesty, "Advanced Composite, High-modulus Tape Placement Machines," *Paper presented at the SAMPE 15th National Conference,* Los Angeles, Calif., Apr. 1969, Society of Aerospace Materials and Process Engineers, Covina, Calif., 1969.
16. W. B. Goldsworthy, "Thermoplastic Composites: The New Structurals," *Plast. World,* (Aug. 1984).
17. E. E. Hardesty, "Design and Construction of a Large, Fully Automated Tape Placement Machine for Aircraft Structures," *Composites,* (Nov. 1972).
18. *Multi-axis Composite Tape-laying CNC Machine,* Pamphlet, Cincinnati Milacron, Inc., Cincinnati, Ohio.
19. U.S. Pat. 4,462,946 (July 31, 1984), W. B. Goldsworthy (to Goldsworthy Engineering, Inc.).
20. U.S. Pat. 4,440,593 (Apr. 3, 1984) and 4,469,541 (Sept. 4, 1984), W. B. Goldsworthy (to Goldsworthy Engineering, Inc.).
21. U.S. Trademark 1,187,389 (Pulformer) (Jan. 26, 1982), W. B. Goldsworthy (to Goldsworthy Engineering, Inc.).
22. U.S. Pat. 3,801,407 (Apr. 2, 1974), W. B. Goldsworthy (to Goldsworthy Engineering, Inc.).
23. U.S. Pat. 4,402,778 (Sept. 6, 1983), W. B. Goldsworthy (to Goldsworthy Engineering, Inc.).
24. U.S. Pat. 4,420,359 (Dec. 13, 1983), W. B. Goldsworthy (to Goldsworthy Engineering, Inc.).
25. U.S. Pat. 4,125,423 (Nov. 14, 1978), W. B. Goldsworthy (to Goldsworthy Engineering, Inc.).
26. U.S. Pat. 4,032,383 (June 28, 1977), W. B. Goldsworthy and H. E. Karlson (to McDonnell Douglas).
27. U.S. Pat. 3,966,533 (June 29, 1976), W. B. Goldsworthy (to Goldsworthy Engineering, Inc.).
28. *Automated Integrated Manufacturing System: Cost-cutting Composite Fabrication for the '80s,* Materials Laboratory, Air Force Wright Aeronautical Laboratories (AFWAL), Dayton, Ohio, and Gruman Aerospace Corp., Midgeville, Ga.
29. J. McDermott, "Codes, Standards, and Design Data on Reinforced Plastics," *Paper presented at the 36th Annual Conference of the SPI,* RP/C Institute, Feb. 16–20, 1981, The Society of the Plastics Industry, Inc., New York, 1981.
30. *Energy Content of Reinforced Plastics Materials,* International Reinforced Plastics Industry (IRPI), London, Nov. 1981.

General References

H. S. Katz, and J. V. Milewski, eds., *Handbook of Fillers and Reinforcements for Plastics,* Van Nostrand Reinhold Co., Inc., New York, 1978.
B. H. Jones, *Probablistic Design and Reliability,* Academic Press, Inc., New York, 1974.
G. Lubin, ed., *Handbook of Composites,* Van Nostrand Reinhold Co., Inc., New York, 1982.
R. J. Roark, *Formulas for Stress and Strain,* McGraw-Hill, Inc., New York, 1965.
M. M. Schwartz, *Composite Materials Handbook,* McGraw-Hill, Inc., New York, 1984.
L. R. Whittington, *Whittington's Dictionary of Plastics,* Technomic Publishing Co., Inc., Wesport, Conn., 1968.
J. Agranoff, ed., *Modern Plastics Encyclopedia,* McGraw-Hill, Inc., New York. Published annually.
Advanced Composites Design Guide, 3rd ed. rev., Vols. III, IV, and V, prepared under Contract No. F33615-74-C-5075 by Los Angeles Aircraft Division of Rockwell International, Los Angeles, Calif., for Air Force Systems Command (Structures Division, Air Force Flight Dynamics Laboratory), Wright-Patterson Air Force Base, Dayton, Ohio, Jan. 1977.
Development of an Automated Layup System for Monofilament Fiber Structures, final report prepared under Contract No. DAAJ01-70-C-0049(P1G) by Boeing Vertol Co. for Army Aviation Systems Command, St. Louis, Mo.
J. Delmonte, *Technology of Carbon and Graphite Fiber Composites.*
Improved Automated Tape Laying Machine, Technical Report AFML-TR-73-307, Vought Systems Division, Vought Corp., Dallas, Texas, Feb. 1974.

W. B. Goldsworthy, *Volume Component Manufacturing,* short course presented to UCLA Extension, Composite Materials, Sept. 17–21, 1984.

J. E. Johnson, J. W. Lee, and R. M. DuPuis, "Reinforced Plastic Beam Shapes," *Paper presented to ASCE Conference on Selection, Design and Fabrication of Composite Structures,* Nov. 1972, American Society of Civil Engineers, 1972.

SAMPE Proceedings, 2020 Vision in Materials for 2000, Vol. 15, Society of Aerospace Material and Process Engineers, Azusa, Calif., 1983.

Journals

Journal of Composite Materials and *International Reinforced Plastics Industry* (UK) are good general references.

K. L. Stone, *SAMPE Q.* **15,** (1984).

SAMPE Q. **15,** (1983).

SAMPE Q. **15,** (1984).

SAMPE J. **20,** (1984).

Mater. Eval. **43,** 1084 (1984).

Adv. Astronaut. Sci. **53,** (1983).

Patents

Filament winding:

U.S. Pat. 4,251,036 (Feb. 17, 1981), P. H. McLain (to Shakespeare Co.).

U.S. Pat. 3,397,850 (1968), J. C. Anderson (to Engineering Technology, Inc.).

Laminates:

U.S. Pat. 4,277,531 (July 7, 1981), C. E. Picone (to PPG Industries).

Pulforming:

U.S. Pat. 4,422,680 (Dec. 27, 1983), M. Goupy (to Regie Nationale des Usines Renault, France).

Pultrusion:

U.S. 4,419,400 (Dec. 6, 1983), R. R. Hindersinn (to Occidental Chemical Corp.).

U.S. 4,249,980 (Feb. 10, 1981), S. M. Shobert and E. B. Fish (to Plas/Steel Products).

Tape placement:

U.S. Pat. 4,516,561 (May 14, 1985), A. Schaeffer (to Cincinnati Milacron, Inc.).

U.S. Pat. Appl., D. Peterson (to Cincinnati Milacron, Inc.). Adjustable tape chute.

U.S. Pat. Appl., Grone and co-workers (to Cincinnati Milacron, Inc.). Composite tape-laying machine and method.

U.S. Pat. Appl., J. Wisbey (to Cincinnati Milacron, Inc.). Tape laminator.

W. Brandt Goldsworthy
Goldsworthy Engineering, Inc.

COMPOSITES, TESTING

The design and analysis of composite structures relies heavily on experimental data. Testing of composites serves a variety of purposes such as:

1. Characterization of constituent materials, ie, fiber and matrix, for micromechanics analysis in order to predict the behavior of the lamina and hence of the laminates and structures.

2. Verification of micromechanics analysis, including effects of curing stresses, temperature, and moisture.

3. Qualification and characterization of the basic unidirectional lamina, which forms the building block of all laminated structures. This characterization,

especially the determination of irreversible and ultimate properties, is essential in providing data for subsequent stress analysis using lamination theory.

4. Experimental stress analysis by measuring strain distributions and using the appropriate constitutive relations.

5. Fracture characterization, including identification of fracture mechanisms and modes, initiation, and propagation.

6. Testing under special conditions of loading, eg, fatigue, multiaxial, and high-rate testing.

7. Assessment of structural integrity and reliability using nondestructive evaluation methods.

8. Life prediction through accelerated testing.

The various experimental methods include photoelastic methods, strain gauges, moiré, holographic and interferometric methods, ultrasonics, acoustic emission, x radiography, and thermography. Test methods applicable to composite materials have been reviewed in the literature (1–8).

Characterization of Constituent Materials

Tensile properties of the fiber and matrix are of particular interest. Single fibers are tested in tension to determine modulus, strength, and ultimate strain. The method is discussed in detail in ASTM D 3379-75 for fibers of modulus higher than 21 GPa (3×10^6 psi). For matrix resin systems that can be processed in thick sheets, a standard specimen is used as described in ASTM specification D 638-72. Resin systems that cannot be fabricated in thick sheets are cast into thin sheet or film form. Thin strips can be tested in tension as described in ASTM D 882-73.

Experimental Micromechanics

Stress distributions on the fiber scale can be determined experimentally by means of photoelastic models (4). Stress concentrations in the matrix of composites have been determined using two- and three-dimensional photoelastic models. Fracture mechanisms and crack propagation on the micromechanical scale have been studied by means of two-dimensional dynamic photoelastic models (see also FRACTURE; MICROMECHANICAL MEASUREMENTS).

Physical Properties

Physical properties of interest are the density, fiber/volume fraction, and thermal and moisture expansions.

The density of a composite sample is usually determined by the displacement method (ASTM D 792-66). The sample is first weighed in air; then it is suspended by a wire and weighed while totally immersed in a liquid of known density and good wettability, such as alcohol. Finally, the suspended wire is weighed immersed in the liquid to the same depth as during the sample weighing. Similar procedures are used for measuring the density of the polymeric resin matrix and the fiber separately.

The fiber/volume ratio (FVR) can be determined by two methods. In the first, the matrix is dissolved in a liquid solvent. In the case of nonburning fibers, the matrix is burned and the weight of the remaining fibers and the densities of the fiber and composite are used in the determination of the FVR. The second method is gravimetric and uses the densities of the constituent and composite materials as follows:

$$\mathrm{FVR} = V_f = \frac{\rho_c - \rho_m}{\rho_f - \rho_m}$$

where ρ_c, ρ_m, and ρ_f = densities of composite, matrix, and fiber, respectively.

The coefficient of thermal expansion is determined by measuring thermal strains in a composite specimen as a function of temperature. Commercial strain gauges are suitable for measuring small thermal strains in composites. However, the readings must be corrected for the purely thermal output of the gauge (7). Thermal strains measured in unidirectional graphite–epoxy, Kevlar 49–epoxy, and S glass–epoxy specimens are shown in Figure 1. Both Kevlar 49–epoxy and graphite–epoxy exhibit negative thermal strains in the longitudinal (fiber) direction. Coefficients of thermal expansion are obtained as the slopes of these curves. The coefficients in the fiber and transverse to the fiber directions are sufficient for the computation of thermal deformations in any direction and for any laminate configuration.

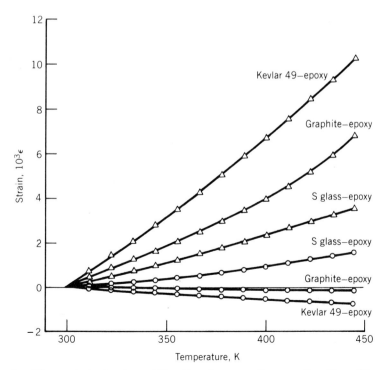

Fig. 1.   Thermal strains in unidirectional composites as a function of temperature.
—O— = longitudinal strain, ϵ_{11} and —△— = transverse strain, ϵ_{22}.

The coefficients of moisture expansion are determined by measuring specimen deformations over a range of moisture weight gains (7). In the case of unidirectional laminates most of the swelling occurs in the direction transverse to the fiber.

Mechanical Properties of Unidirectional Composites

Some mechanical properties (qv) are obtained by so-called qualification testing. The usually 15-ply thick unidirectional specimens are subjected to three-point bending. Flexural test specimens are 10.2 cm long and 1.27 cm wide with a 6.4-cm span length. Interlaminar shear-strength specimens are 1.5 cm long and 0.64 cm wide with a 1-cm span length. Standard-beam formulas are used to determine the flexural strength and interlaminar shear strength. Results can be regarded only as qualitative. The flexural strength reflects the mixed behavior of tension and compression and the effects of stress gradient. The short-beam test produces a nonuniform interlaminar shear stress superimposed on the longitudinal normal stress.

Tensile Tests.   Uniaxial tensile tests are performed on unidirectional laminae to determine the following properties:

$$E_{11} = \text{longitudinal Young's modulus}$$
$$E_{22} = \text{transverse Young's modulus}$$
$$\nu_{12} = \text{major Poisson's ratio}$$
$$\nu_{21} = \text{minor Poisson's ratio}$$
$$F_{1T} = \text{longitudinal tensile strength}$$
$$F_{2T} = \text{transverse tensile strength}$$
$$\epsilon_{11T}^{u} = \text{ultimate longitudinal tensile strain}$$
$$\epsilon_{22T}^{u} = \text{ultimate tranverse tensile strain}$$

Tensile specimens are straight-sided and of constant cross section with bonded glass–epoxy tabs. They are equipped with strain gauges and loaded in tension to failure. Typical stress–strain curves for 0° and 90° graphite–epoxy specimens are shown in Figures 2 and 3.

Compression Tests.   Compression testing of composites is difficult because of the tendency for premature failure due to crushing or buckling. Compression test methods can be classified into three broad categories (7). In the first one, Type I, specimens of very short but unsupported gauge length are used. In the so-called Celanese test, long glass–epoxy tabs are attached to the specimens; the gauge length is approximately 1.27 cm. Load is introduced through friction by means of split conical collet grips that fit into matching sleeves, which in turn fit into a snugly fitting cylindrical shell (ASTM D 3410-75). This method requires a perfect cone-to-cone contact, which is not normally achieved. A modification is represented by the Illinois Institute of Technology Research Institute (IITRI) test fixture, in which the conical grips are replaced with trapezoidal wedges (9) (Fig. 4). This eliminates the problem of line contact, since surface-to-surface contact can be attained at all positions of the wedges.

In the second category of compression test methods, Type II, a relatively

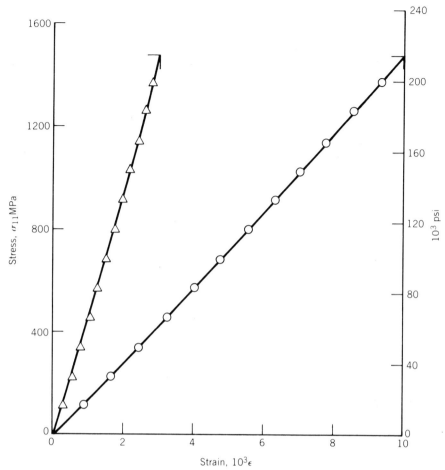

Fig. 2. Strains in 0° unidirectional specimen under uniaxial tensile loading (SP286T300 graphite–epoxy). —○— = Longitudinal strain, ϵ_{11} and —△— = transverse strain, $-\epsilon_{22}$. Results: E_{11} = 145 GPa (21.0 × 10⁶ psi); ν_{12} = 0.31; F_{1T} = 1477 MPa (214 × 10³ psi); and ϵ_{11T}^u = 0.0100.

long, fully supported specimen is used which tends to give lower values due to some premature buckling, despite the lateral support.

The third type of compression specimen, Type III, is a sandwich beam made by bonding the composite laminate to a honeycomb core. However, this method suffers from transverse stress and interfacial shear-stress interactions.

Shear Properties. In-plane shear properties are determined by several methods. The first one utilizes a $[\pm 45]_{2s}$ specimen under uniaxial tensile loading (10). The in-plane shear stress and in-plane shear strain referred to the fiber direction in each lamina are given by

$$\sigma_{12} = \frac{\sigma_x}{2}$$

$$\epsilon_{12} = \frac{\epsilon_x - \epsilon_y}{2}$$

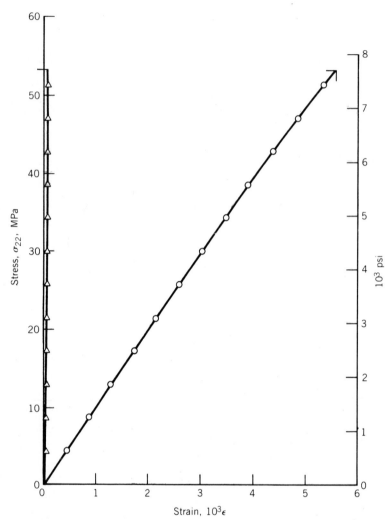

Fig. 3. Strains in 90° unidirectional specimen under uniaxial tensile loading (SP286T300 graphite–epoxy). —O— = Axial strain, ϵ_{22} and —△— = transverse strain, $-\epsilon_{11}$. Results: E_{22} = 9.9 GPa (1.43 × 10^6 psi); ν_{21} = 0.014; F_{2T} = 53 MPa (7.7 × 10^3 psi); and ϵ_{22T}^u = 0.0056.

where σ_x is the applied uniaxial stress and ϵ_x and ϵ_y are the axial and transverse strains measured with two-gauge rosettes. The in-plane (or intralaminar) shear modulus of the unidirectional lamina is obtained from the slope of the σ_{12} vs ϵ_{12} curve as

$$G_{12} = \frac{\sigma_x}{2(\epsilon_x - \epsilon_y)}$$

This method tends to overestimate the shear strength because of the constraint imposed on the lamina by the adjacent plies.

The second test method is the 10° off-axis test (11,12). The fibers in the 6-ply unidirectional specimen are oriented at 10° with the loading axis. The tabs

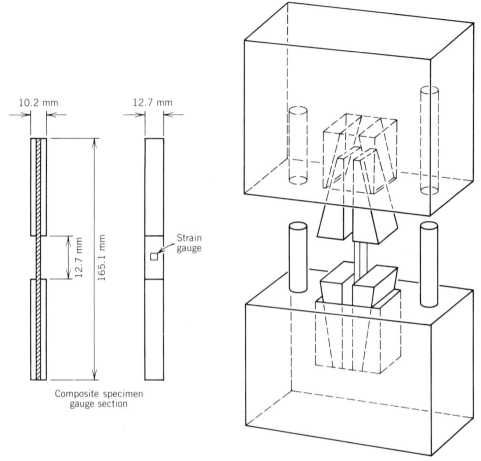

Fig. 4. IITRI compression test specimen and fixture (9). Courtesy of the *Journal of Testing and Evaluation*.

are tapered, and a three-gauge rosette is attached on each side. The in-plane shear stress and shear strain, respectively, are given by

$$\sigma_{12} = \sigma_x \sin \theta \cos \theta = 0.171\sigma_x$$

$$\epsilon_{12} = \frac{\epsilon_x - \epsilon_y}{2} \sin 2\theta + \left(\epsilon_{45} - \frac{\epsilon_x + \epsilon_y}{2} \right) \cos 2\theta$$

where $\theta = 10°$; and ϵ_x, ϵ_y, and ϵ_{45} are the axial, transverse, and 45° strains, respectively. The in-plane shear modulus is obtained from the slope of the σ_{12} vs ϵ_{12} curve as

$$G_{12} = \frac{\sigma_{12}}{2\epsilon_{12}}$$

The ultimate values of σ_{12} and ϵ_{12} determine the shear strength and ultimate shear strain. This method tends to underestimate these ultimate properties due to interaction of the tensile stress across the fibers.

The third method of determining shear properties is the rail shear test (two-

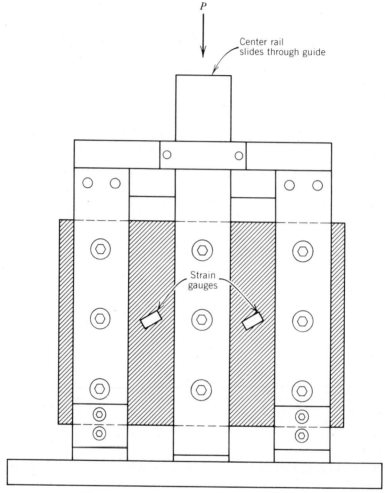

Fig. 5.   Three-rail shear fixture; P = load.

or three-rail test). In the three-rail test a rectangular specimen is clamped between three parallel pairs of rails (Fig. 5). The load is applied to one end of the
middle rails and reacts at the opposite ends of the two outer pairs of rails. These
tests do not produce a uniform shear stress in the gauge section.

The fourth method is the torsion method, utilizing a solid rod or a hollow
tubular specimen subjected to torque. A shear stress–shear strain curve can be
obtained by plotting torque versus angle of twist or shear strain measured directly
with strain gauges. Such specimens are difficult to make and load. The solid rod
is less desirable because of the shear-stress gradient across the section.

Biaxial Testing

A variety of specimen types and techniques has been proposed and used for
biaxial laminate characterization. They include the off-axis coupon, crossbeam

sandwich specimen, rectangular plate loaded in biaxial tension, and thin-wall tubular specimen.

The off-axis unidirectional specimen has been used successfully in coupon and ring forms. It is loaded uniaxially at an angle to the fiber direction. The main limitations of this method are that biaxial normal stresses are always of the same sign and it is not possible to vary the three stress components independently.

The crossbeam sandwich specimen consists of two intersecting beams subjected to pure bending. The laminate is used as one skin of a honeycomb-core sandwich. However, the state of stress in the gauge section is not uniform and cannot be calculated without knowledge of the material properties.

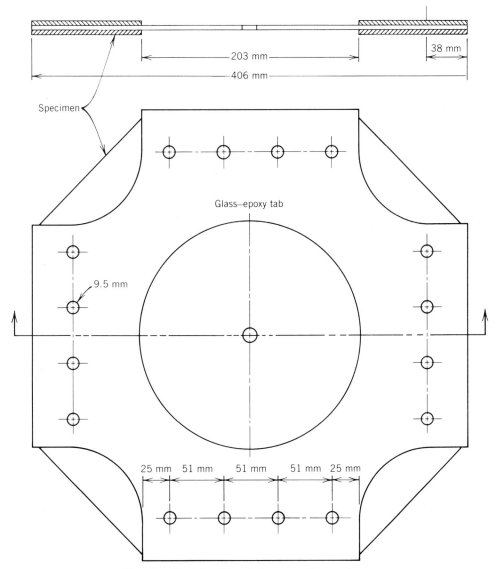

Fig. 6.   Sketch of biaxial composite specimen.

The flat-plate specimen is usually a square plate subjected to tension–tension loading on its sides through glass–epoxy tabs. Nonproportional loading is possible to some degree. The specimen is most suitable for biaxial testing of laminates with stress concentrations (13). A sketch of a biaxial plate specimen is shown in Figure 6. This test is primarily limited to tension–tension loading. In the case of unnotched specimens, it is very difficult to induce failure in the test section.

The tubular specimen is the most versatile. Any desired biaxial state of stress can be achieved by the independent application of axial loads, internal or external pressure, and torque. A typical tubular specimen geometry is illustrated in Figure 7.

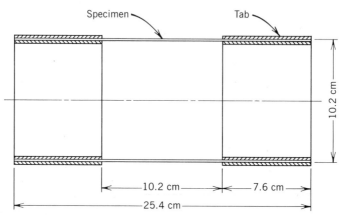

Fig. 7. Thin-wall tubular specimen for biaxial testing of composites.

Nondestructive Evaluation

A variety of nondestructive evaluation (nde) techniques is used for evaluating the integrity of composite materials. They include radiographic (x ray, neutron), optical (moiré, birefringent coatings, holography, and interferometry), thermographic, acoustic (acoustic wave, acoustic emission), embedded sensor, and ultrasonic techniques (14) (see also NONDESTRUCTIVE TESTING).

Flaws are introduced into composites during fabrication or in service. They include voids, inclusions, incorrect fiber orientation, matrix cracking, delaminations, and fiber breaks. Different methods are best suited for detecting different types of flaws.

The most widely used methods are ultrasonics and x radiography. The former is based on the attenuation of an ultrasonic pulse passing through a defect. It is best suited for detecting delaminations, inclusions, and large cracks. An example of an ultrasonic record depicting a film intraply inclusion is shown in Figure 8. X radiography, enhanced with x-ray-opaque penetrants, is best suited for detecting matrix cracks and delaminations. Optical methods are more sensitive to near-surface defects. Thermographic methods are sensitive to defects but do not yield a clear definition of them.

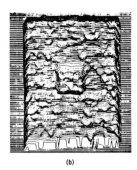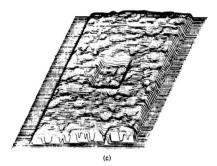

(a) (b) (c)

Fig. 8. Ultrasonic C-scans of $[(0/\pm45/90)_s]_2$ graphite–epoxy specimen with a film patch. (**a**) Pen-lift scan; (**b**) analogue scan (normal); and (**c**) analogue scan (perspective).

BIBLIOGRAPHY

1. S. P. Prosen, *Composite Materials: Testing and Design, ASTM STP 460,* American Society for Testing and Materials, Philadelphia, Pa., 1969, pp. 5–12.
2. E. M. Lenoe, *J. Eng. Mech. Div.* **96,** EM6, 809 (1970).
3. C. W. Bert in C. C. Chamis, ed., *Structural Design and Analysis,* in L. J. Broutman and R. M. Krock, eds., *Composite Materials,* Vol. 8, Academic Press, Inc., New York, 1975, Pt. II, pp. 73–133.
4. I. M. Daniel in G. P. Sendeckyj, ed., *Mechanics of Composites,* in Ref. 3, Vol. 2, 1974, pp. 433–489.
5. T. T. Chiao and M. A. Hamstad, in *Proceedings of the 1975 International Conference of Composite Materials,* Vol. 2, The Metallurgical Society of AIME, Warrendale, Pa., Geneva, Switzerland, and Boston, Mass., 1976, pp. 884–915.
6. B. D. Agarwal and L. J. Broutman, *Analysis and Performance of Fiber Composites,* John Wiley & Sons, Inc., New York, 1980, Chapt. 8.
7. J. M. Whitney, I. M. Daniel, and R. Byron Pipes, *Experimental Mechanics of Fiber Reinforced Composite Materials,* Monograph No. 4, Society for Experimental Stress Analysis, Brookfield Center, Conn., 1982.
8. I. M. Daniel in Z. Hashin and C. T. Herakovich, eds., *Mechanics of Composite Materials,* Pergamon Press, Elmsford, N.Y., 1982, pp. 473–496.
9. K. E. Hofer and P. N. Rao, *J. Test. Eval.* **5,** 278 (1977).
10. B. W. Rosen, *J. Compos. Mater.* **6**(4), 552 (1972).
11. I. M. Daniel and T. Liber, *NASA CR-134826,* National Technical Information Service, Springfield, Va., 1975.
12. C. C. Chamis and J. H. Sinclair, *Exp. Mech.* **17,** 339 (1977).
13. I. M. Daniel, *Exp. Mech.* **20**(1), 1 (1980).
14. I. M. Daniel and T. Liber in W. W. Bradshaw, ed., *Proceedings of the 12th Symposium on Nondestructive Evaluation,* San Antonio, Texas, April 1979, Southwest Research Institute, San Antonio, Texas, 1979.

I. M. DANIEL
Illinois Institute of Technology

CRAZING

In glassy polymers such as polystyrene (PS), the appearance of what seem to be many small cracks has long been termed crazing. Misnamed, since they are not true cracks, these entities, now called crazes, usually grow normal to the direction of largest tensile stress, reflect light, and eventually fracture if stressed sufficiently. In these ways, crazes resemble cracks.

Unlike cracks, however, these entities are load-bearing as a result of a web of stretched microfibrils that spans the space between the walls of what would otherwise be called cracks. As a result of this ability to bear a substantial load, a craze growing near another passes by as though the other were not there. In contrast, a crack tends to veer and terminate perpendicularly in the flank of a neighboring crack: thus the appearance of so-called mud-flat cracking. This difference often provides the quickest way to judge whether the entity is a crack or a craze.

Crazes are important for several reasons. First, they are the defects in which true cracks initiate. Second, as the crack propagates, more crazes are produced at the crack tip, thus constituting a plastic zone which blunts the crack. Third, their formation and deformation is the principal source of energy dissipation, ie, toughness, in many rubber-toughened plastics. Fourth, crazing, or a cavitational process akin to crazing, occurs in the shoulder of the neck formed during the cold-drawing of many semicrystalline plastics. Lastly, craze formation and breakdown in amorphous plastics is often markedly facilitated by small-molecule organic agents, and this susceptibility is the single greatest vulnerability of amorphous thermoplastics.

In the last three decades, crazing has come to be appreciated as a novel deformation process found in essentially all thermoplastics, but no other materials. Simultaneously exhibiting some of the characteristics of shear flow and cracking, the process is scientifically unique. Crazing is capable of raising the toughness and crack resistance of thermoplastics relative to thermosets by two or three orders of magnitude. However, shear flow, the alternative deformation process available in thermoplastics, is a superior mechanism of energy absorption for two reasons. First, unlike crazing, shear flow is essentially a process continuous in space, ie, one that may spread through a much greater volume fraction of the stressed body and thus consume much more energy in total. Second, shear flow is much less sensitive to environmental effects. In short, shear deformation is better than crazing, but crazing is better than no deformation at all (see also FRACTURE).

Crazing is an important source of toughness in rubber-modified thermoplastics. In some cases, eg, polystyrene, it is the only source. In other cases, eg, polycarbonate modified with polyethylene, crazing is of minor importance compared with shear flow. This subject is comprehensively addressed in a recent monograph (1).

Composition and Structure

A craze can be described as a layer of polymer a nanometer to a few micrometers thick, which has undergone plastic deformation approximately in the direction normal to the craze plane as a response to tension applied in this direction. In contrast to the stretching of a rubber band or the conventional drawing of a ductile material such as copper, the process occurs without lateral contraction (2,3). As a result, the polymer volume fraction in the craze is proportional to λ^{-1}, where λ is the draw ratio in the craze. The reduction in density occurs on such a small scale that the refractive index is markedly reduced, which accounts for the reflectivity of the craze (Fig. 1). Indeed, refractive-index measurements provide the simplest means of calculating craze polymer content, which varies widely according to conditions, for example, as low as 10 vol % under stress, 50 vol % when unstressed, and >90 vol % upon annealing without stress.

The polymer content consists of a web of microfibrils, which, under stress, is oriented in the major principal stress direction (Fig. 2) (3,4). The fibrils consist of a plastically oriented polymer that, in the case of glassy resins, is amorphous. Upon annealing an unstressed specimen above its T_g, the long-known plastic-memory effect is observed, ie, the fibrils retract and thicken, the density and refractive index of the craze returning to values very nearly equal to those of the normal resin. (Since true cracks lack an internal retractive mechanism, annealing provides another simple means, if there is no residual stress, of differentiating between a craze and a crack.) The polymer volume fraction of the craze also depends on temperature, time under stress, and degree of swelling by a plasticizing agent (4,5). Increase in any of these factors tends to produce a coarser craze structure through the breaking and retraction of fibrils, leaving the structure ever more voided and weak (Fig. 3). Indeed, in many cases, crazes gradually

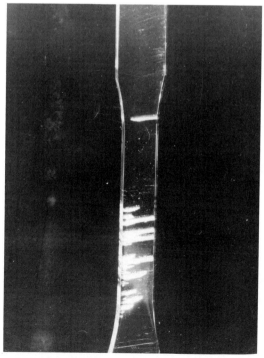

Fig. 1.   A bar of bisphenol A-polycarbonate containing several unusually large crazes that have run through the entire cross section without specimen failure, demonstrating that crazes are load-bearing (3).

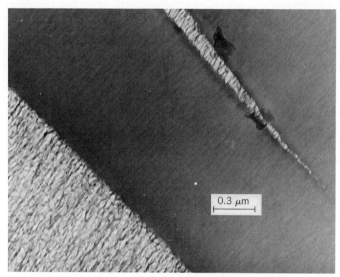

Fig. 2. Electron micrograph of ultrasection through craze tip and part of another craze in poly(2,6-dimethyl-1,4-phenylene oxide). Fibril orientation, sharp boundary between craze and bulk polymer, and uniformity of structure except at youngest part of tip can be seen (3).

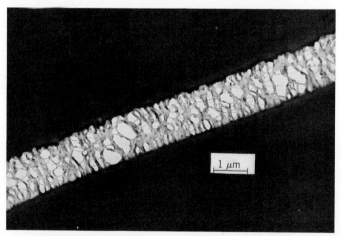

Fig. 3. Polystyrene craze with coarse weak structure grown in plasticizing fluid near T_g. Dark spots on fibril appear to be the residues of fibrils that broke and retracted (3).

evolve into cracks by such a mechanism. Resin molecular weight is also important; the higher the molecular weight, the more resistant the craze fibrils to breakdown.

The remainder of the craze, ie, the void content, usually consists of the gaseous or liquid environment in which the craze was formed (if, in the latter case, the liquid is a wetting liquid). Organic liquids and even water solutions penetrate crazes more or less rapidly owing to capillary forces, the continuous morphology of the craze void content, and its connection to the exterior of the specimen (3).

When formed under conditions such that the T_g of the craze material is far above ambient temperature, ie, low test temperature or weak organic agent, the composition seems reasonably uniform throughout, ie, from very near its tip (Fig. 3) backward and its midplane to the craze–normal polymer interface. Specifically, density is uniform and the fibril diameters are constant (3,4). These observations show that older portions of the craze thicken as it grows forward in the plane by the transformation of solid to craze at the craze–bulk polymer interfaces. Polystyrene crazes, however, seem to have a central layer of lower density that forms at the craze tip, where the stresses are highest (4).

From electron-microscopy (qv) and x-ray scattering studies, craze-fibril diameters are 0.6 to approximately 30 nm in most crazes formed far below T_g. The uniformity in diameters throughout a given craze shows that strain-hardening, ie, an increasing resistance to drawing with increasing λ, is a key factor in stabilizing the microstructure in a way analogous to the stabilization of so-called necks in the cold-drawing of most ductile glassy thermoplastics. The strain-hardening rate, ie, the increase in the resistance with strain, has been correlated with chain entanglements (4); the higher the chain-entanglement density, the higher the rate of hardening. The strain-hardening rate also increases as the difference between T_g and T_{test} increases. Thus, fibril volume fractions increase with these parameters, resulting in higher craze strength. It may sometimes exceed the yield stress for shear failure, and a neck may actually be grown through the surrounding bulk polymer without altering the craze (3). When stretched taut, the aligned fibrils confer nearly linear elastic stress–strain behavior on the craze (3). The elastic modulus of the stretched craze is low, eg, 25% of that of the bulk resin, only because the volume fraction of polymer is low. However, when the fibrils are originally unstressed and therefore almost randomly oriented, loading causes them to reorient slowly at first and then more rapidly. This response often confers a yield stress and large elongation on such a craze before the fibrils become sufficiently aligned for the modulus to approach its final high value (Fig. 4).

Stresses in and Around the Craze

Characterizing the stresses around crazes is important for understanding their morphology, growth mechanism, microstructure, and the way they alter stresses at crack tips. Clearly, the force field in a specimen must be altered by the formation of a craze more or less surrounded by undeformed resin. The boundaries of the craze are further apart than they were before the cavitation and drawing took place by the distance $\Delta t = t\varepsilon/(1 + \varepsilon)$, where t is the craze thickness

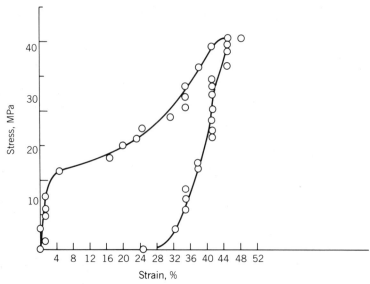

Fig. 4. Cyclic stress–strain curve for craze in bisphenol A polycarbonate after aging at low stress for several months (3). To convert MPa to psi, multiply by 145.

and ε the strain. However, the fibrils can bear a finite load, the magnitude of which depends on their number, stiffness, and some positional and geometric factors. Obviously, if the craze is large enough and passes entirely through the specimen cross section, the force lines pass through the craze uniformly and the fibrils must bear all the load. On the other hand, if the craze is small, low compliance results in a lower-than-average stress being transmitted by the fibrils in the craze body and an excess stress being transmitted by the bulk polymer at the tip.

A number of quantitative analyses of craze-stress distributions have been made; the most elegant, detailed, and complete is based on electron-microscopy results and stress analysis (4) (see also MICROMECHANICAL MEASUREMENTS). Thin polymer films are caused to adhere to copper microscope grids which are stretched plastically; this causes a measurable uniform strain in the film. With some resins, eg, PS, that are much more prone to crazing than to shear flow, this strain is sufficient to cause craze formation. Thickness, ie, distance between the two craze–bulk polymer interfaces, and strain are measured point by point along the craze. The strain is determined from the mass thickness of the craze relative to that of the surrounding bulk resin. From these measurements, the craze-displacement distribution is calculated. Then, by two independent analytical methods, the stress distribution at the craze–bulk polymer boundary is calculated.

The results show that dry crazes and crazes formed in weakly plasticizing liquids can be divided into two regions: a tip region roughly 20 μm long, where the stress is a maximum, and a craze body behind the tip where the stress has a fairly constant lower value that approximates the applied or far-field stress (Fig. 5). The maximum stress at the craze tip is between 1.2 and 1.7 times the far-field stress, depending on the resin, environment, and details of the calculations. In highly plasticizing liquids, ie, in which T_g is lowered nearly to ambient

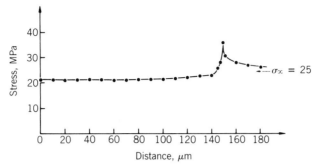

Fig. 5.   Stress distribution in the bulk polymer at the boundary of a craze in polystyrene of M_w = 200,000 and M_w/M_n = 1.06 (4). To convert MPa to psi, multiply by 145. Courtesy of E. J. Kramer and Springer-Verlag.

temperature, the load-bearing capability of the craze body is diminished and a higher stress concentration exists at the tip (Fig. 6) (5).

A concentration in the major principal stress σ_y adjacent to the craze tip implies similar concentrations in the minor principal stress σ_x, ie, that stress operating parallel to the craze-growth direction, and in the resolved shear stress. In thin films of relatively craze-resistant polymers, the maximum resolved shear stress can be large enough to stop crazing by the production of shear-deformation zones growing in the craze-growth direction in front of the tip (Fig. 7a) (3). In the center of a thick specimen, the direction of maximum shear stress is on a diagonal in the xy plane and shear zones occasionally emanate in these directions (Fig. 7b); two crazes growing toward each other but on slightly different planes frequently stop when their tips reach positions where these maximum shear stresses overlap and shear flow occurs between the two tips.

Mechanics and Kinetics of Craze Initiation

Shear flow, craze initiation, and craze growth bear a strong resemblance to each other in certain respects. Like shear flow, craze initiation and growth are stress-biased, temperature-activated processes. In addition, prior orientation makes both shear yielding and crazing more difficult if the stress is applied in the orientation direction, and easier if applied in the transverse direction (3). These qualitative similarities are not surprising given the plastic deformation that must be part of craze-fibril formation. However, experimental quantities such as activation energies differ quantitatively, and seemingly small changes in test rates and temperatures or degrees of preorientation or relative stress direction often cause a preempting of crazing by shear flow. In some cases, for example, poly(methyl methacrylate), the change from crazing to shear flow tends to occur when test rate or stress is lowered, temperature raised, or molecular orientation increased, but only if the testing stress is applied parallel to the orientation direction. There are many examples, however, that contradict rate and temperature generalizations at least in part. For example, most resins that fail by shear flow in conventional tensile tests, eg, test time < ½ h, develop crazes if held for several hours at stresses below those at which shear flow begins.

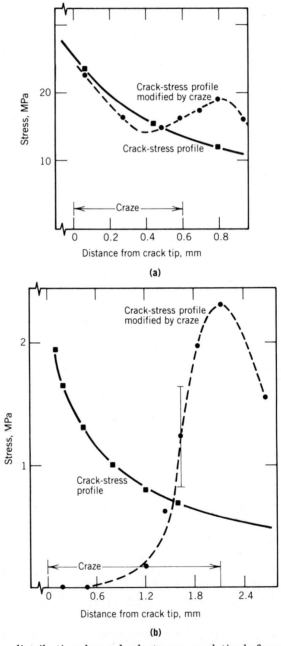

Fig. 6. Stress distributions beyond polystyrene crack tips before and after formation of environmentally induced crazes: (**a**) weakly plasticizing agent, ie, methanol; (**b**) strongly plasticizing liquid, ie, heptane (5). Much lower stress is borne by craze body in (**b**). To convert MPa to psi, multiply by 145. Courtesy of E. J. Kramer and Applied Science Publishers, Ltd.

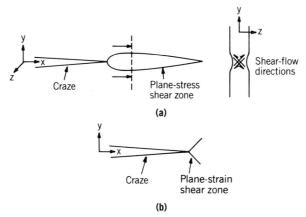

Fig. 7. Shear zones in front of craze tips in (**a**) thin film and (**b**) center of thick specimen.

Isochronal Mechanical Criteria. Crazing and shear flow are affected differently by shear stress and by hydrostatic stress. Shear flow at a given temperature and given test rate occurs when the resolved shear stress, more exactly the so-called octahedral shear stress τ_{oct}, is brought to a critical level τ^*_{oct} which varies slightly with the magnitude of the hydrostatic tension P (6). Thus,

$$\tau^*_{oct} = A - \mu P \tag{1}$$

where

$$\tau_{oct} = \frac{1}{3}[(\sigma_x - \sigma_y)^2 + (\sigma_y - \sigma_z)^2 + (\sigma_z - \sigma_x)^2]^{1/2} \tag{2}$$

and

$$P = \frac{1}{3}(\sigma_x + \sigma_y + \sigma_z) \tag{3}$$

and the coefficient μ = ca 0.1; A is usually much larger than μP. Thus, shear flow can occur when P is negative, ie, compressive, albeit not quite as easily as under a positive P.

By contrast, crazing is believed never to occur when P is negative. Moreover, under combinations of two stresses applied orthogonally (the so-called two-dimensional stress field since the third stress $\tau_z = 0$ by design) the shape of the locus of craze initiation differs qualitatively from that of the locus for the onset of shear flow (Fig. 8). The crazing locus, first characterized for poly(methyl methacrylate) (6,7), was fitted by the following expression:

$$\sigma^*_y - \sigma^*_x = \frac{A}{(\sigma^*_y + \sigma^*_x)} + B \tag{4}$$

where σ^*_y and σ^*_x are the critical values of the major and minor principal stresses, and A and B empirical time- and temperature-dependent constants. Since the maximum shear stress in the xy plane τ_{xy} is $(\sigma_y - \sigma_x)/2$ and the hydrostatic

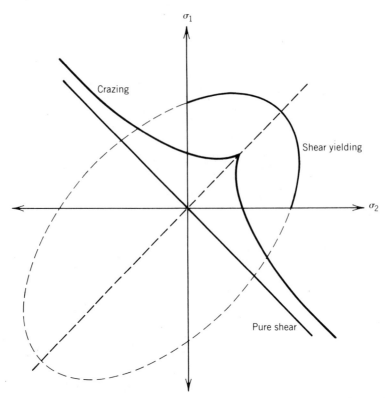

Fig. 8. Biaxial stress-field loci for shear flow and craze initiation in poly(methyl methacrylate) at 60°C (6,7). Courtesy of S. Sternstein and co-workers; The American Chemical Society; and *Journal of Macromolecular Science-Physics*.

tension P is $\frac{1}{3}(\sigma_x + \sigma_y)$, equation 2 can be rewritten:

$$\tau^*_{xy} = A'/P^* + B' \tag{5}$$

Under uniaxial tensile conditions, A'/P^* usually constitutes a large fraction of the right side in this equation, whereas in the shear-yielding equation, μP is small compared with A. This contrast highlights the much greater prominence of hydrostatic tension in crazing and the converse prominence of shear stress in shear yielding. On the basis of these results, it was suggested (6,7) that craze initiation required a shear stress the size of which was reduced by increasing hydrostatic tension. Since that time, other studies (8–12) have resulted in somewhat different mathematical formulations and physical interpretations of stress-field crazing loci. Slightly obscuring the fundamental significance of most of these studies is the fact that their specimens had machined surfaces subsequently polished with abrasives to produce uniform finishes with grooves, each of which was fractions of a micrometer wide; moreover, abrasion frequently produces small degrees of shear deformation under the abrading particle. The abrasion question was avoided by studying crazing at the interfaces of rubber and glass spheres that are caused to adhere well to polystyrene matrices (10,12). Under tensile loading in the polar direction, the crazes initiate in one case at the equator

($\Theta = 90°$) (rubber sphere) and in the other at the poles ($\Theta = 0°$) (glass sphere). Advantage is taken of the fact that the various stress, strain, and strain energies reach maxima at different values of Θ. Moreover, these locations shift radically with the hardness of the sphere. On this basis, it has been suggested that, of the simple criteria, the major principal stress σ_y and the dilation Δ both come closest to being the criterion for crazing (12). However, if the true criterion is a complex one such as has been suggested (6), it is recognized that imbedded-sphere studies are unable to define this criterion.

Nevertheless, in spite of these questions, the fact that crazing is inherently a cavitational process prompts most investigators to believe that crazing never occurs when the true local values of P or Δ are negative, in other words, compressive. Thus, whatever the ultimate three-dimensional stress-field criteria for craze initiation turns out to be, it seems highly likely that hydrostatic tension or dilation will be a prominent, but not the only, ingredient in such criteria.

Kinetic-mechanical Criteria for Initiation Under Uniaxial Loading. Craze initiation under uniaxial loading occurs more rapidly the higher the applied stress or strain. Frequently, constant stress tests show a linear dependence of log craze-initiation time on applied stress. This dependence, in combination with temperature-acceleration effects, leads to a fitting of initiation kinetics into an Eyring stress-biased activation framework,

$$t_i^{-1} = \alpha\exp\left(-E_a + \beta\sigma/RT\right) \tag{6}$$

where t_i is the initiation time, E_a the apparent activation energy, α and β constants, and σ the applied stress. For constant strain tests, plots of applied strain vs log time are usually nonlinear; the initiating time appears to approach infinity as a lower strain threshold is approached. These results suggest that a critical applied strain exists for the given experimental conditions, ie, a strain below which craze initiation never occurs. Whether or not this is always strictly true under dry conditions, it is certainly observed frequently in swelling environments.

In recent years, another kinetic-mechanical criterion has been proposed, ie, that crazing occurs in constant stress or strain experiments when the mechanically dissipated energy per unit volume, ie, that part of the total deformation energy not recoverable even on a delayed-time basis, reaches a critical value ω_s, which is a material constant (13). The time and temperature dependence is then considered to be that involved in the energy-dissipation mechanism. The few values reported to date for ω_s vary substantially with the material. The applicability of this approach over wide ranges of temperature has yet to be demonstrated. Moreover, the effects of stress state on ω_s remains an important question: if the stress-field results described above (eqs. 1–4) are valid, it should be expected that as the shear component of stress is increased relative to the hydrostatic tension, ω_s goes to infinity.

Effects of Polymer Orientation. For nearly four decades, molecular orientation (qv) has been known to affect craze initiation profoundly (3). When the crazing stress is applied parallel to the orientation direction, the resistance of polystyrene appears to rise linearly with the orientation birefringence. In the extreme, the resistance may exceed the ductile failure stress. (This effect made possible the development of so-called stretched acrylic airplane canopies in World War II and the use of biaxially stretched polystyrene for tear-resistant commercial packaging materials (qv).) Conversely, crazing occurs more readily if the stress is applied perpendicular to the orientation direction. Orientation effects are probably to be understood in terms of the relative number of chain segments aligned in the stress direction, ie, the greater their number density, the more strain-hardening becomes effective in raising flow resistance. A study of stretched polystyrene showed that the craze-fibril volume fraction was nine times greater when the stress paralleled the orientation direction than when it was perpendicular to it (14). Thus, true stresses are much higher in the former case, both on each fibril in the body of the craze and at the craze tip itself.

The modulus E of an oriented polymer is greatest in the orientation direction ($\Theta = 0°$) and least in the perpendicular direction ($\Theta = 90°$). When the stress direction Θ_σ lies between 0 and 90° for such a specimen, the direction of maximum elastic strain Θ_ε differs from Θ_σ by an amount $\Delta\Theta$ that depends both on the ratio $E(0°)/E(90°)$ and on Θ_σ. Under these conditions, the planes of the crazes formed are normal no longer to Θ_σ, but rather lie normal to the estimated Θ_ε. On this basis, it has been suggested that major principal strain is the correct stress-field criterion for initiation in isotropic as well as oriented specimens (15). However, the lack of data for modulus vs Θ prevented calculation of the precise direction of major principal strain in the cited study.

Preexistent shear bands, which usually have high orientation levels, can stop craze growth, probably owing to the fact that their moduli in the loading direction are substantially increased relative to the surrounding isotropic resin (Fig. 9).

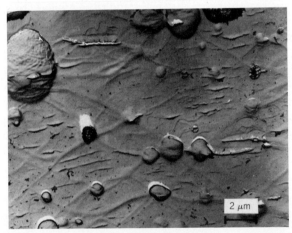

Fig. 9.   Crazes (nearly horizontal, short dark lines) stopped by shear bands (long, diagonal, diffuse bands) in a blend of rubber-modified polystyrene and poly(2,6-dimethyl-1,4-phenylene oxide) (1). Courtesy of C. B. Bucknall and Applied Science Publishers, Ltd.

Molecular Effects.   Molecular weight does not influence craze-initiation resistance, at least above a certain minimum molecular weight, ie, $2M_c$, the chain-entanglement molecular weight (3).

On the other hand, initiation resistance is a strong function of intermolecular cohesive forces (16); this can be seen in Figure 10, where constant-strain

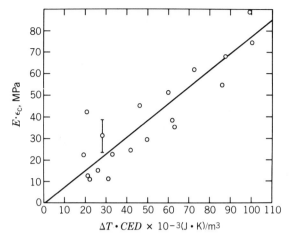

Fig. 10. Modulus times critical applied strain for craze initiation in air of a wide variety of resins vs $(T_g - T_{test})$ times cohesive energy density CED (16). To convert MPa to psi, multiply by 145; to convert J to cal, divide by 4.184 (16).

crazing resistance in the form of modulus times critical strain, ie, approximately a critical stress, is displayed vs resin cohesive energy density CED multiplied by ΔT, the difference between test temperature and T_g (the temperature term is introduced in order to compensate for the thermal effects on craze initiation discussed earlier). The dependence is reasonably linear up to a value of $(CED \cdot \Delta T)$ where the corresponding applied strain produces measurable plastic (shear) deformation (17). The connection between this correlation and the cavitational aspect of crazing is obvious.

This correlation and the molecular weight limitations cited above help in understanding why crazing is a phenomenon found only in thermoplastics, and not in other structural solids. The ratio of yield-point stress to CED is ca 3700 for bisphenol A polycarbonate but only 300 or so for steel, as examples of two solids susceptible to plastic deformation. In other words, the ratio of the cavitational resistance to the shear resistance is ten times smaller in organic solids than in metals. Only the ability to strain-harden upon drawing confers stability on the craze, which thus rules out crazing in low molecular weight organic solids. Consequently, crazing appears confined to intrinsically ductile, low CED, strain-hardening solids, ie, organic thermoplastics.

Shear-deformation Effects. As noted above, uniform shear flow can retard or eliminate crazing when, subsequent to the flow, the crazing stress is applied in the orientation direction. Localized shear deformation can have other profound effects on initiation that vary greatly with geometry. For example, when a sharply notched specimen is compressed uniaxially, sharply bounded shear bands grow from the notch tip. When the stress is removed, the nonuniform elastic recovery of the shear band and the surrounding resin produce local tensile stresses in the band that can cause crazes to grow perpendicular to the shear-band planes (18).

Another more important and complex situation exists with notched specimens of materials such as polycarbonate that are ductile in uniaxial tensile tests over wide temperature and test-rate ranges. With these materials, notched specimens usually show an abrupt change in fracture behavior, from ductile to apparently brittle failure, as any of a number of variables is changed, eg, increasing specimen thickness (19), decreasing notch radius (20), decreasing temperature (20), increasing test rate (19), or thermal aging below T_g (21,22). At the heart of the transition lie the effects of the notch on shear deformation (23): in the outer layers of the specimen, the constraining effect of the notch on the ease of shear deformation is relatively low, ie, the ratio of τ_{oct} to the applied stress is relatively high. Moreover, the constraint decreases rapidly in front of the notch, ie, in the x direction in Figures 11a and c. Consequently, plane-stress shear flow spreads further into the interior of the specimen with increasing load and increasing distance x.

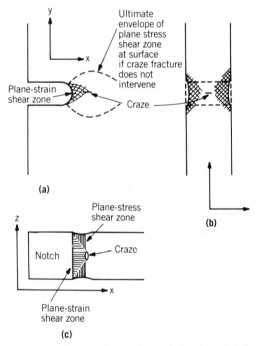

(a)

(b)

(c)

Fig. 11.   Shear zones and craze beyond notch in ductile plastic: (**a**), plane-strain shear zone and craze in midplane of specimen; (**b**), section through specimen at craze showing invading plane-stress shear zone; (**c**), section through specimen at notch midplane showing craze and both shear zones.

In the interior of the specimen near the notch, root contraction in the z direction is restricted (Fig. 11a and c); consequently, a zone of so-called plane-strain shear flow grows into the specimen in the x direction. The constraint of the notch causes tensile stress σ_x in the x direction to build up at the zone tip as the notch causes the tensile stress σ_x in the x direction to build up at the zone tip as the zone grows. As a result (see eq. 2), the level of the major principal stress σ_y at the zone tip necessary to keep τ_{oct} equal to τ_{oct}^* increases. The growth of both σ_y and σ_x also results in an increase in P (see eq. 2). These stresses may become great enough to cause craze initiation at the zone tip (24–26) (see Fig.

12). According to theory,

$$\sigma_y = \tau^*_{oct}[1 + \ln(1 + x/\rho)] \qquad (7)$$

$$P = \tau^*_{oct}[\tfrac{1}{2} + \ln(1 + x/\rho)] \qquad (8)$$

at the zone tip, where x is the distance from the notch root and ρ is the notch radius. Thus, if τ^*_{oct} is decreased, eg, owing to temperature increase, or if P is increased, the critical position x^* at which σ_y or P would become large enough for craze initiation, if the specimen had been sufficiently thick, reaches the edge of the plane-stress zone. The lower stresses in the latter zone prevent craze initiation entirely. With the brittle failure mechanisms preempted, shear flow proceeds through the whole net section below the notch with the consumption of large amounts of energy; hence, the brittle-to-ductile transition.

Fig. 12. Thin section (xy plane) in interior of Trogamid glassy nylon specimen showing zone of plane-strain shear flow emanating from notch root and two crazes originated at shear-zone tip. Transmitted polarized monochromatic light. After Mills (24).

Aging (qv) below T_g causes primarily an increase in the shear-flow resistance (21), which thus causes the critical level of stress for craze breakdown to be reached earlier in the growth of the plane-strain shear zone (25). The effect of molecular weight is thought to be mostly in the lowering of the craze strength, and secondarily in affecting the thermal aging rate (21).

Since craze initiation and breakdown are both time- and temperature-dependent, the rate-temperature dependence of the ductile–brittle transition involves a complex competition between these processes and shear flow, all mediated by specimen geometry. This ductile–brittle phenomenon in notched specimens is limited to resins in which the craze-fracture stress exceeds the plane-strain yield stress, since otherwise initiation and fracture occur from the notch root before shear flow begins.

Craze Growth

Shape.   After initiation, the craze tends to grow in all three dimensions. Surface-initiated crazes tend to grow more rapidly along the surface and less rapidly into the specimen, and frequently a very elongated half-moon shape develops to the craze area. There are probably several reasons for such shape, even in the absence of environmental agents. First, the surface may be abraded or degraded in some way, eg, photodegradation. Second, the surface may have a skin stressed during the molding process. Third, if the specimen is bent, the surface layers are under the highest stress. The older parts of the craze also thicken as the tip advances, albeit at a decreasing rate, and therefore the craze takes on a shape like the profile of a sewing needle, ie, a tip region having the shape of a wedge with internal angle of no more than 5°, and further back, a region with parallel sides. The length of the tip region varies from micrometers to a millimeter, probably depending on the stress distribution along the craze.

Kinetics.   Under constant applied tensile stress, crazes, once initiated, grow in lateral dimension either at a constant rate, eg, polystyrene, or at a rate decreasing linearly with inverse time under load, eg, poly(methyl methacrylate) containing a plasticizer and polycarbonate at high temperature. Moreover, as a rough rule, the final size attained by a craze is smaller the greater the critical strain for initiation. At least one reason for this difference in kinetics seems connected with competition from homogeneous (shear) deformation: matrix creep is expected in materials requiring a high applied strain for craze growth, and the resultant plastic strain may suppress growth (1% plastic strain in ductile plastics is known to prevent subsequent craze initiation completely (17)).

It has also been suggested that so-called stress-aging effects may occur in the matrix at the craze interface or in the fibrils themselves (27). Aging would tend to produce an increase with time in the resistance to further fibril pullout. Such an effect would tend to reduce the tip-stress concentration with time and thus lower growth rates.

As expected, orientation affects eventual sizes. When the crazing stress is normal to the orientation direction, larger crazes are usually obtained than when the stress parallels the orientation. Molecular weight also affects craze sizes. This is probably so because of increases in the strain hardening rates and strengths of craze fibrils with molecular weight; these increases will tend to reduce craze tip stresses. In injection moldings, another indirect effect exists: residual orientation in the flow direction tends to be higher with higher molecular weight resins, and therefore craze initiation and growth resistance is higher when the flow direction and subsequent applied stress direction coincide.

A detailed study has been made of craze growth on finely abraded surfaces of polystyrene under various combinations of macroscopic stress (28). Ultimate growth rates correlated with the magnitude of the maximum principal tensile stress, and the existence of varying levels of shear stress seemingly made no difference.

Models of Craze Initiation and Growth

In crystalline solids, there are a number of easily identified defects that act as stress-concentration points and weak spots for initiation of dislocations, ie, the basic elements of shear flow in these materials, and eventually of cracks. In glassy polymers of reasonable purity, it has been difficult to identify craze-initiating defects when surface roughness can be ruled out. The key term may be reasonable purity, however. No commercial resins are totally free of foreign particles, ie, 50 nm or greater, for example. Here, foreign particle can be defined as having an elastic modulus different from the resin. Thus, it is highly likely that every cubic millimeter of solid contains many weak spots equal to or smaller than 0.1 μm. Since even spherical particles have stress concentrations on the order of two at their boundaries, it seems probable that in all studies to date, craze initiation has involved what may be called heterogeneous nucleation, ie, nucleation at a foreign particle or at a scratch, etc.

Initiation. Craze initiation may involve three stages according to one model (11). First, a region of microporosity develops, perhaps as a result of inhomogeneous shear flow. Second, when the void content in this region is large enough, a spontaneous reorganization takes place; the voids agglomerate into a large stable void, allowing an elastic unloading of the surrounding, stressed material. Third, an extension process occurs, resulting in the craze taking on the familiar planar form. The first step is proposed to be governed kinetically by the thermally activated shear process. This model has been criticized for ignoring both surface-tension and chain-entanglement effects (4). In a modification of the shear-patch idea, it was suggested that crazes may initiate at the intersection of two shear bands (9). With both models, it is difficult to see how local shear bands can form inside resins such as polystyrene so far below the macroscopic shear-flow stress unless they emanate from a defect. (The critical tensile stress for crazing in PS is 10 MPa (1450 psi), whereas the tensile stress for shear yielding is about eight times greater.) But if foreign particles are the sites of initiation adhesive failure of the resin/particle interface as the beginning of void formation has to be considered.

In another nucleation model, flaws produce a local hydrostatic stress concentration which results in such an increase in free volume that T_g is reduced to ambient temperature and cavitation develops in the rubbery material (29). Just what flaws exist that could generate the necessary high stress concentrations is unclear. Moreover, shear-flow mechanisms invoking such a T_g reduction have not proved useful.

Although the stress-concentration factor at a spherical particle boundary is two, independent of particle diameter d, the factor falls off with distance into the matrix at a rate related to this size, and at $d/10$ this factor falls to less than 1.5. Therefore, if it is assumed that a critically sized nucleus for initiation exists, particles effective in nucleating crazes must be large enough to give a sufficiently high stress over a region larger than the critical nucleus. This size may be ca 75 nm, which is three or four fibril spacings (4). Until rigorously clean resins can be made and well-defined foreign particles 50–500 nm in diameter introduced into these materials, craze-nucleation mechanisms will remain obscure.

Growth. In the earliest model of craze-tip growth, it was assumed that each increment of advance occurred by the separate nucleation of a void beyond the existing craze tip (30). This model was found unsatisfactory for several theoretical reasons (28). Instead, a model was proposed with features that are accepted today by many investigators: the so-called Taylor meniscus instability model. Meniscus instabilities occur when the boundary between a less viscous and a more viscous fluid recedes in the direction of the more viscous fluid owing to a driving force. Beyond a certain velocity that depends on several factors, including the interfacial tension and the plastic-flow resistance, the boundary departs more and more from linearity and eventually develops a series of deep indentations (Fig. 13). The material between each pair of indentations gets pinched off repeatedly from the receding viscous liquid front. Such instabilities in polymeric systems are most readily demonstrated when a soggy piece of adhesive tape is peeled off the skin slowly (air displaces elastomer, the more viscous fluid). At the junction between tape and skin, the failing elastomer is seen to consist of separate fibers stretched out between the two.

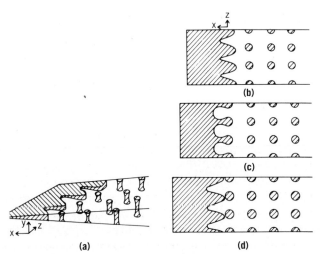

Fig. 13.   Argon-Salama meniscus instability model of craze growth (4). Courtesy of E. J. Kramer and Applied Science Publishers, Ltd.

In this model, it is assumed that the plastic has begun to undergo plane strain shear flow in a zone at the growing craze tip, ie, it is fluidized by stress. The meniscus instability then develops as a result of the surface tension of the fluid. Just why the thickness of the craze tip should be limited to ca 10 nm is not clear. Moreover, the diameters of the fibrils predicted from the model seem too large, ie, 100 nm vs ca 20 nm found by x-ray scattering for, admittedly, the bulk of the craze and not its tip. Finally, the mechanism by which each finger or web of polymer breaks up into discrete fibers has not been made clear. Nevertheless, the qualitative validity of this view of craze-tip advance remains unchallenged.

A variant of this model has been developed for the thickening of the craze (4), ie, the transformation of bulk polymer to fibril at the craze–bulk interface. Again, it is postulated that a layer of bulk polymer at the interface is fluidized by stress and that the greater the surface tension and the smaller the spacing between the fibrils, the greater the applied force necessary to produce further transformation. The fibril diameter itself is a function of the strain-hardening rate, which, in turn, is believed to be controlled by chain entanglement. The smaller the chain-entanglement molecular weight, the more rapidly strain-hardening increases, the larger the diameter of the mature fibril, and presumably the greater the craze strength (Fig. 14).

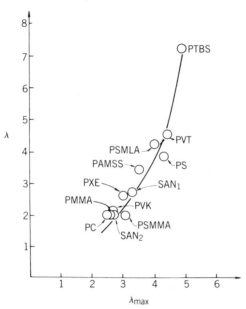

Fig. 14. Dependence of craze fibril extension λ on theoretical maximum possible extension λ_{max} of a single isolated chain for several different polymers (4). $\lambda_{max} = l_e/k(M_e)^{1/2}$ where l_e is the chain-contour length, k a constant peculiar to each resin, and M_e the molecular weight between entanglements. Key: PTBS = poly(*tert*-butylstyrene); PVT = polyvinyltoluene; PSMLA = poly(styrene-*co*-maleic anhydride); PS = polystyrene; PAMSS = poly(α-methylstyrene-*co*-styrene); SAN = poly(styrene-*co*-acrylonitrile); PXE = poly(xylenyl ether); PVK = polyvinylcarbazole; PMMA = poly(methyl methacrylate); PC = polycarbonate; and PSMMA = poly(styrene-*co*-methyl methacrylate). Courtesy of E. J. Kramer and Springer-Verlag.

Craze Breakdown and Crack Initiation

In glassy polymers crazes constitute the defects in which brittle cracks initiate. Moreover as the crack propagates, more crazes are produced at its tip, thus tending to resist its advance. Resistance to breakdown varies with time, stress, stress history, temperature, molecular weight, chain structure, and environment.

The craze breaks down in at least two different ways. Sometimes localized cracks initiate at certain weak spots in a craze that otherwise seems undamaged (Fig. 15). These internal cracks grow slowly and eventually link up. This mechanism is found primarily in high molecular weight resins at temperatures well below the T_g of the craze. Closer to the T_g and in lower molecular weight resins,

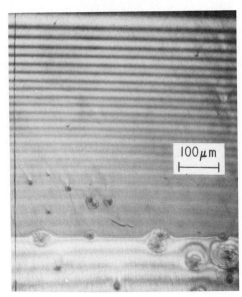

Fig. 15.   Set of interference fringes from craze at polystyrene crack tip (3). Craze contains several circular dark spots and sets of interference rings arising from independently nucleated cracks. Monochromatic reflected light.

a craze may deteriorate by a process of fibril breakdown throughout, which results in a coarsening of the structure and general reduction in craze strength, as illustrated in Figure 3. Cyclic stress in some cases seems to produce a generalized weakening as well; this results in a phenomenon called discontinuous crack growth (DCG). Beyond a critical point in the slow growth of internal cavities an abrupt change occurs and the crack jumps forward, rapidly producing one or more new crazes at its tip. This phenomenon apparently plays an important role in DCG. The kinetics of craze breakdown, like those of shear flow and of failure in general, are described by the Eyring stress-biased thermal-activation expressions. At their root, the kinetics must derive from the mechanism of fibril breakdown. The stress on the fibril is determined by $\sigma_0\lambda$ where σ_0 is the stress on the craze and λ is the extension ratio of the fibril. In turn, λ reflects the strain-hardening characteristics of the particular resin (4). Under stress, the fibril breaks down as the more highly stressed chain segments break with time. Unfortunately, little is known about the time–temperature behavior of strain-hardening and subsequent creep in plastics.

The importance assigned to chain entanglements is strongly supported by two pieces of evidence. First, the time to failure under constant load is an extreme function of molecular weight M, eg, M^{11}, in one case (31). Second, in contrast to craze initiation, which is independent of M over wide ranges, craze strength drops rapidly toward zero in the neighborhood of M_c to $2M_c$ (4). Breakdown sensitivity to molecular weight seems to extend to much higher molecular weights under cyclic loading (32) than under static loadings, probably as a result somehow of the very large motions, eg, roughly 50–300% strain that the craze structure undergoes internally upon cycling from zero to some high level of stress.

Environmental Effects

A wide variety of environments act as so-called stress-cracking agents for plastics. They fall into two classes: chemical corrosion agents that produce chain scission, ie, that break primary bonds (for example, nitric acid on polypropylene); and small-molecule solvents and swelling agents, ie, agents that interfere with the cohesive forces between chains. Although the porosity of the craze and the small size of its fibrils makes the craze vulnerable to both classes of agents, the second class causes the most difficulty. Because of this prominence and because of a variety of subtleties and complications this second class, loosely termed solvent stress crazing and cracking, has been studied extensively.

Solvent Stress Crazing and Cracking.   Although these problems are seen in crystalline plastics, particularly polyolefins stressed in detergents (33), glassy polymers are generally far more susceptible. The agents causing these effects dissolve in the polymer at hand to a degree sufficient to alter molecular mobility (3). For the most part, these agents are relatively small-molecule organic liquids and vapors. In addition, gases, eg, CO_2 or argon, at high enough pressure or low enough temperatures can dissolve sufficiently to induce crazing under stress (34,35) (see also PLASTICIZERS).

Craze Initiation.   As with crazing in air, the lower the applied stress or strain, the greater the initiation time. However, an abrupt halt to this behavior occurs as time proceeds, and below a certain applied stress or strain called the critical stress σ_c or strain ε_c, no crazing or cracking is seen. The time t_c at which this change occurs varies greatly from one agent to another; t_c heavily depends on rates of absorption. As a rule, the higher the viscosity of the fluid, the longer the time to attain σ_c or ε_c.

Although reduction in true surface energy has been postulated often as a fundamental role of crazing liquids, the preponderance of evidence favors the view that the fundamental roles played by the agent are connected with its absorption in the resin on a molecular scale (3,5). Of greatest importance by far is the reduction in T_g by the plasticizing action of the agent. This effect is basically the same as that caused by raising the test temperature. A second role played by the dissolved agent appears to be that of a so-called antiplasticizer. In other words, the diluent, even in low concentrations, tends to raise modulus and shear-flow resistance above those values that would have obtained had T_{test} been raised rather than T_g lowered (3,4) (such an effect seems to occur through a suppression of the secondary losses caused by localized segmental motions). Although this change is relatively subtle compared with T_g reduction, it may be responsible for the switch from shear flow to crazing that is seen when a plasticlike polycarbonate, which is ductile in inert environments from far below room temperature up to T_g, is stressed in weakly absorbing liquids, eg, decane, which swells polycarbonate to ca 0.8% by volume.

The amount of agent absorbed depends on the heat and entropy of mixing. The first can be crudely predicted on the basis of the difference between the so-called solubility parameter of polymer δ_p and solvent δ_s. Consequently, craze-initiation resistance in a wide spectrum of small-molecular organic liquids can often be correlated with δ_s roughly, as seen in Figure 16 (3). Since the entropy of mixing has a large combinatorial contribution, the size of which decreases

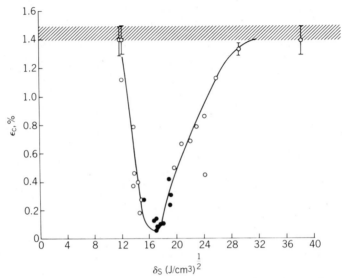

Fig. 16. Critical strain for solvent craze initiation of poly(2,6-dimethyl-1,4-phen-ylene oxide) vs solubility parameter δ_s of a wide variety of small-molecule organic liquids: ○, crazing agents; ●, cracking agents (3). To convert J to cal, divide by 4.184.

with the volume of the absorbing agent, the degree of swelling of a set of liquids, eg, the normal alkanes, of nearly the same δ_s decreases markedly with molar volume of the absorbing agent (36).

Rates of absorption of small-molecule agents are functions of many variables, including molecular size and stress. The effect of tension (probably the hydrostatic stress component) is to accelerate absorption rates, sometimes several times (37). The sensitivity of rate to stress multiplication seems to increase with diffusant molecular size. The effect is probably highest with agents having low equilibrium solubilities, eg, 1–3%. The practical result is that stress-free plastic parts, which are seemingly unswollen by a certain organic liquid, quickly undergo crazing and cracking when stressed in contact with the agent.

A final effect of the diluent worth mentioning is that its absorption is attended by mechanisms of stress relaxation which compete with crazing (3). The first of these is volume increase: each 1% of swelling increases the linear dimensions of a volume element by an average of ca 0.3%, which thus tends to negate an equal elastic strain. Thus, when a specimen is exposed for a time to a swelling agent at strains below ε_c, a swollen and somewhat relaxed or even compressed surface layer develops because the part is now in fact inhomogeneous. A much higher deformation must be applied to raise the surface layer strain up to ε_c. The second mechanism is homogeneous stress relaxation, which is undoubtedly accelerated by the reduction in T_g owing to plasticization.

Craze Growth.   Environmentally induced crazes grow much more rapidly and to much larger sizes (Fig. 1) than those grown in inert environments. The main reason for these differences is probably the localized effect of the diluent on the mechanical properties of the fibrils and the immediately surrounding bulk resin. First, the fibrils are softer and weaker; a greater stress-concentration factor therefore exists at the craze tip (5). Second, the immediately adjacent layer of bulk polymer is plasticized, and less stress is required to complete its necessary fluidization for fibril formation. By contrast, bulk polymer further removed is still unswollen and thus still hard and creep resistant, ie, fully capable of transmitting stresses to the craze tip.

However, crazes induced by liquid environments can penetrate the specimen at a linear rate dl/dt no faster than liquids can flow through the craze structure and keep up with the tip. The driving force for liquid flow is capillary pressure: $p = 2\gamma \cos \Theta/r$ where γ is the liquid–vapor surface tension, Θ is the wetting angle of the fluid on the polymer (Θ is approximately 90° for most organic fluids), and r is the effective capillary radius (roughly one-half the spacing between fibrils); typically, $p = 3.5$ MN/m² (507.5 psi) for a craze. The maximum flow rate in the craze is easily shown to be approximately $dl/dt = \gamma \cos \Theta/4\ \eta l$, where η is the viscosity of the fluid and l is depth to which the liquid has penetrated the craze. By integration, it is seen that the depth to which a solvent craze penetrates a specimen tends to be proportional to $t^{1/2}$ (Fig. 17). It is also clear that the velocities of liquid-induced cracks are no higher than the rate at which the liquid can flow through the zone of crazes at the crack tip (Fig. 18) (38). This limitation sometimes causes serious engineering-design problems: short-term, high stress tests in the presence of high viscosity fluids often fail to identify liquids that are potent long-term cracking agents. Given the high viscosities of many common organic liquids, eg, greases, detecting craze and crack growth may require weeks or months of laboratory testing.

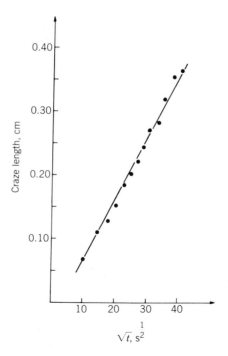

Fig. 17. Dependence of craze length on time$^{1/2}$ for poly(methyl methacrylate) in methanol (5). Courtesy of E. J. Kramer and Applied Science Publishers, Ltd.

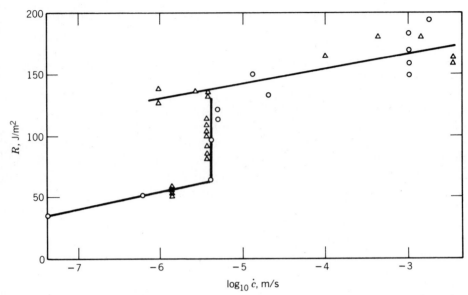

Fig. 18.   Crack resistance R vs crack velocity $\dot{c}$ for ABS terpolymer under crazing agent butyl oleate. High velocity values of R are typical of those found in air; those at low velocity are true environmental values. Different symbols indicate different specimens (38). To convert J/m^2 to $ftlbf/in.^2$, divide by 2100. Courtesy of *Polymer Engineering and Science.*

Crazing in Other Resins

Crazing also occurs in crystalline plastics such as polypropylene and nylons, where it is less evident because of their lack of transparency. Moreover, organic environmental agents, on the average, seem to cause fewer cracking problems than in glassy polymers; an important exception is the detergent cracking of polyolefins. Another reason is probably that with most of these resins, the ratio of shear-flow resistance to crazing resistance in thin sections is lower and failure is more often ductile in nature. A recent review treats this subject in great detail (39).

Cross-linking (qv) seems to bring about the total disappearance of crazing at sufficiently high cross-link densities, eg, well-cured epoxies (40), even though shear ductility as seen in compression tests or in tension testing of rubber-toughened epoxies is still present.

Crack Propagation

Crack propagation in thermoplastics, as in metals, takes two forms. First, so-called plane-stress failure tends to occur in thin sections; in more ductile, craze-resistant materials; at higher temperatures; and at lower speeds. Shear flow in directions that result in a thinning of the region in front of the crack tip is responsible for plane-stress crack toughness. Second, so-called plane-strain fracture tends to occur in thick sections; in less ductile, more easily crazed materials; and at lower temperatures and higher speeds. Crack toughness in this case arises from the continual generation of one or more crazes at the crack tip and sometimes a degree of plane-strain shear flow (3,41,42). The coexistence of shear flow with crazing in the plastic zone at the crack tip is apparently found when the ratio of the plane-strain shear-yield stress to the craze fracture stress is less than unity.

The number of crazes at the crack tip, and thus the crack resistance, varies greatly with conditions and resin. Evidently, the number, from 1 to more than 50, depends primarily on the ratio of the breaking stress σ_b, which thus depends on craze-fibril volume fraction and fibril strength, to the craze-initiation stress σ_c. The development of shear flow at stresses between σ_c and σ_b evidently serves to alter the stress distribution in the plastic zone in such a way as to reduce stress buildup on the crazes at the crack root and thus delay their failure to higher loads. A recent review covers the mechanics of crazes at crack tips in depth (43) (see also FRACTURE).

BIBLIOGRAPHY

"Crazing" in *EPST* 1st ed., Vol. 4, pp. 294–307, by R. W. Raetz, Catalin Corporation.

1. C. B. Bucknall, *Toughened Plastics,* Applied Science Publishers, Ltd., London, 1977.
2. S. Rabinowitz and P. Beardmore in E. Baer, ed., *Critical Reviews In Macromolecular Science,* Vol. 1, CRC Press, Cleveland, Ohio, Mar. 1972, p. 1.
3. R. P. Kambour, *Macromol. Rev.* **7,** 1 (1973).
4. E. J. Kramer in H. H. Kausch, ed., *Crazing in Polymers, Advances in Polymer Science,* Vol. 52/53, Springer-Verlag, Berlin, 1983, Chapt. 1.
5. E. J. Kramer in E. H. Andrews, ed., *Developments in Polymer Fracture* Vol. 1, Applied Science Publishers, Ltd., London, 1979, Chapt. 3.
6. S. S. Sternstein and L. Ongchin, *Polym. Prepr. Am. Chem. Soc. Div. Polym. Chem.* **10,** 1117 (1969).
7. S. S. Sternstein and F. A. Meyers, *J. Macromol. Sci. Phys.* **8,** 539 (1973).
8. R. J. Oxborough and P. B. Bowden, *Philos. Mag.* **28,** 547 (1973).
9. M. Kawagoe and M. Kitagawa, *J. Polym. Sci. Polym. Phys. Ed.* **19,** 1423 (1981).
10. T. T. Wang, M. Matsuo, and T. K. Kwei, *J. Appl. Phys.* **42,** 4188 (1971).
11. A. S. Argon and J. G. Hanoosh, *Philos. Mag.* **36,** 1195 (1977).
12. M. E. J. Dekkers and D. Heikens, *J. Mater. Sci.* **18,** 3281 (1983).
13. O. L. Brueller, *Polym. Eng. Sci.* **23,** 844 (1983).
14. N. R. Farrar and E. J. Kramer, *Polymer* **22,** 691 (1981).
15. P. Beardmore and S. Rabinowitz, *J. Mater. Sci.* **10,** 1763 (1975).
16. R. P. Kambour, *Polym. Commun.* **24,** 292 (1983).
17. *Ibid.,* **25,** 130 (1984).
18. C. C. Chau and J. C. M. Li, *J. Mater. Sci.* **18,** 3047 (1983).
19. A. S. Morecroft and T. L. Smith, *J. Polym. Sci. Part C* **32,** 269 (1971).
20. G. Allen, D. C. W. Morley, and T. Williams, *J. Mater. Sci.* **8,** 1449 (1973).
21. D. G. LeGrand, *J. Appl. Polym. Sci.* **13,** 2129 (1969).
22. M. Ishikawa and I. Narisawa, *J. Mater. Sci.* **18,** 2826 (1983).
23. A. S. Tetelman and A. J. McEvily, Jr., *Fracture of Structural Materials,* John Wiley & Sons, Inc., New York, 1967, Chapts. 2 and 8.
24. N. J. Mills, *J. Mater. Sci.* **11,** 363 (1976).
25. M. Ishikawa, I. Narisawa, and H. Ogawa, *J. Polym. Sci. Polym. Phys. Ed.* **15,** 1791 (1977).
26. M. Ishikawa, H. Ogawa, and I. Narisawa, *J. Macromol. Sci. Phys.* **19,** 421 (1981).
27. N. Verheulpen-Heymans, *Polymer* **21,** 97 (1980).
28. A. S. Argon and M. M. Salama, *Philos. Mag.* **36,** 1217 (1977).
29. A. N. Gent, *J. Mater. Sci.* **5,** 925 (1970).
30. A. S. Argon, *J. Macromol. Sci. Phys.* **8,** 573 (1973).

31. J. F. Rudd, *J. Polym. Sci. Part B* **1,** 1 (1963).
32. R. W. Hertzberg and J. A. Manson, *Fatigue of Engineering Plastics*, Academic Press, Inc., New York, 1980, especially p. 119.
33. Y. Ohde and H. Okamoto, *J. Mater. Sci.* **15,** 1539 (1980).
34. N. Brown and S. Fischer, *J. Polym. Sci. Polym. Phys. Ed.* **13,** 1315 (1975).
35. W. V. Wang and E. J. Kramer, *Polymer* **23,** 1667 (1982).
36. C. H. M. Jacques and M. G. Wyzgoski, *J. Appl. Polym. Sci.* **23,** 1153 (1979).
37. T. L. Smith and R. E. Adam, *Polymer* **22,** 299 (1981).
38. R. P. Kambour and A. F. Yee, *Polym. Eng. Sci.* **21,** 218 (1981).
39. K. Friedrich in Ref. 4, Chapt. 5.
40. A. J. Kinloch, S. J. Shaw, D. A. Tod, and D. L. Hunston, *Polymer* **24,** 1341 (1983).
41. R. P. Kambour and S. A. Smith, *J. Polym. Sci. Polym. Phys. Ed.* **20,** 2069 (1977).
42. M. T. Takemori, R. P. Kambour, and D. S. Matsumoto, *Polym. Commun.* **24,** 297 (1983).
43. W. Doll in Ref. 4, Chapt. 3.

R. P. KAMBOUR
General Electric Company

DYNAMIC MECHANICAL PROPERTIES

Dynamic mechanical properties are the mechanical properties of materials as they are deformed under periodic forces. The dynamic modulus, the loss modulus, and a mechanical damping or internal friction express these properties. The dynamic modulus indicates the inherent stiffness of material under dynamic loading conditions. It may be a shear, a tensile, or a flexural modulus, depending upon the investigating technique. Dynamic modulus is especially important for engineering structural applications of plastics and composites since the variation of properties is easily determined as a function of temperature and frequency (or time). The polymeric materials under use conditions are often in the dynamic stress and strain field, such as parts of automobiles, aircraft, and aerospace structures. Knowledge of the dynamic moduli and properties of polymeric materials is indispensable for design of these materials.

The mechanical damping or internal friction indicates the amount of energy dissipated as heat during the deformation of the material. The internal friction of material is important as a property index and for environmental and industrial applications. Since noise is radiated by vibration, especially of metallic materials with small internal friction (0.001–0.004), the application of damping materials to the vibrating surface converts the energy into heat, which is dissipated within the damping materials rather than being radiated as airborne noise. Many viscoelastic polymers are good damping materials with high internal friction (0.1–0.3). High damping or internal friction is essential in decreasing the effect of undesirable vibration and in reducing the amplitude of resonance vibrations to safe limits, as well as in playing a key role in all kinds of structures from airplanes to buildings.

The investigation of the dynamic modulus and internal friction over a wide range of temperatures and frequencies has proved to be very useful in studying the structure of high polymers and the variations of properties in relation to performance (1–6). These dynamic parameters have been used to determine the glass-transition region, relaxation spectra, degree of crystallinity, molecular orientation, cross-linking, phase separation, structural or morphological changes resulting from processing, and chemical composition of polymer blends, graft polymers, and copolymers. In addition to these structure–property relationships, dynamic mechanical studies have been extended to material composites (qv) and structural systems.

Dynamic mechanical properties, such as the dynamic modulus and internal friction, are the most basic of all mechanical properties (qv). The dynamic modulus indicates stiffness of polymeric material under the dynamic stress and strain condition. It varies from 100 kPa to 100 GPa (10^6–10^{12} dyn/cm^2) and depends upon the type of polymer, temperature, and frequency. This value is related to the Young's modulus of polymers. The dynamic loss modulus, or internal friction, is sensitive to many kinds of molecular motion, transitions, relaxation processes, structural heterogeneities, and the morphology of multiphase systems (crystalline polymers, polymer blends, and copolymers). Therefore, interpretations of the dynamic mechanical properties at the molecular level are of great scientific and practical importance in understanding the mechanical behavior of polymers.

Much research has been done on the relationships between molecular parameters and dynamic mechanical properties. A comprehensive review of dynamic mechanical spectroscopy in relation to the molecular structure of polymers is given in Ref. 7. A number of useful generalizations and interpretations on this subject are found in Refs. 3 and 4; new interpretations on the relaxation spectrum have been proposed, and the molecular motions in amorphous and semicrystalline polymers with respect to the anelastic spectra have been reviewed (8,9). This has been the most active area of study of the dynamic mechanical properties of polymers.

The effects of molecular parameters on the dynamic mechanical properties are based on internal structure. However, many polymeric materials are combined as structural units, eg, laminated plastics, fiber-reinforced composites, tires, and fabrics. Dynamic mechanical properties are useful in determining the properties of these structural units and the effects of the structural geometry on dynamic properties. The ultimate performance of these structural assemblies is influenced by the inherent properties of the components and the external structural factors of the systems.

The damping characteristics of structural systems have been the subject of investigation for many years. In the past several decades, engineers have expressed considerable interest in the damping (internal friction) properties of structures at stress levels encountered in engineering design. They have contended that resonant structural vibrations could be controlled by the effective use of this property.

In addition to the internal damping properties of structural materials, attention has been paid to structural fabrications involving Coulomb-slip and viscoelastic-shear damping mechanisms. High-speed space vehicles require degrees of structural damping higher than those provided by the structural material itself. Investigators thus turned to structural damping techniques such as interfacial Coulomb-slip damping, viscoelastic laminate damping, and Coulomb- or viscoelastic-junction damping. Detailed studies of structural damping can be found in Refs. 10 and 11.

Theory

Dynamic mechanical properties are determined by various instruments measuring the response (deformation) of polymers and their assemblies to periodic forces. Thus, the vibrational parameters—amplitude, frequency, and type of oscillation and wave propagation—become important in this analysis.

In studies of the response of a material to vibrational forces, stress, strain, and frequency are the key variables. The stress applied to a viscoelastic body results in a linear or nonlinear dynamic response. For example, if the stress varies sinusoidally with time at a given frequency, the strain varies cyclically at the same frequency. If the amplitude of the stress is small enough, the strain is also sinusoidal with time, and the amplitude of the strain is proportional to the amplitude of the stress at a given temperature and frequency, ie, the behavior is linear. However, if the amplitude of the stress is large enough, the strain varies with the same frequency but is nonsinusoidal; it can have high harmonics, showing nonlinear viscoelastic behavior.

The method of sinusoidal excitation and response is very useful for determination of the dynamic mechanical properties of polymeric materials (3,12,13). The applied force and the resulting deformation both vary sinusoidally with time; the rate is usually specified by the frequency f in Hz of $\omega = 2\pi f$ in rad/s. For linear viscoelastic behavior, the strain alternates sinusoidally but is out of phase with the stress. This phase lag results from the time necessary for molecular rearrangements and is associated with relaxation phenomena (7). The stress σ and strain ϵ can be expressed as follows:

$$\sigma = \sigma_0 \sin (\omega t + \delta) \tag{1}$$

$$\epsilon = \epsilon_0 \sin \omega t \tag{2}$$

where ω is the angular frequency and δ is the phase angle. Then,

$$\sigma = \sigma_0 \sin \omega t \cos \delta + \sigma_0 \cos \omega t \sin \delta \tag{3}$$

the stress can be considered to consist of two components, one in phase with the strain ($\sigma_0 \cos \delta$) and the other 90° out of phase ($\sigma_0 \sin \delta$). When these are divided by the strain, the modulus can be separated into an in-phase (real) and out-of-phase (imaginary) component.

The relationships are:

$$\sigma = \epsilon_0 E' \sin \omega t + \epsilon_0 E'' \cos \omega t \tag{4}$$

$$E' = \frac{\sigma_0}{\epsilon_0} \cos \delta \text{ and } E'' = \frac{\sigma_0}{\epsilon_0} \sin \delta \tag{5}$$

where E' is the real part of the modulus and E'' the imaginary part. The complex representation for the modulus can be expressed as follows:

$$\epsilon = \epsilon_0 \exp i\omega t \tag{6}$$

$$\sigma = \sigma_0 \exp i(\omega t + \delta) \tag{7}$$

Then,

$$\frac{\sigma}{\epsilon} = E^* = \frac{\sigma_0}{\epsilon_0} e^{i\delta} = \frac{\sigma_0}{\epsilon_0} (\cos \delta + i \sin \delta) = E' + i E'' \tag{8}$$

and similarly for the other deformation types. For example, the ratio of the peak stress to peak strain for shear is

$$|G^*| = G'^2 + G''^2 \tag{9}$$

where G^* is the shear complex modulus, G' is the real part of the modulus, and G'' the imaginary part; the phase angle δ is given by

$$\tan \delta = G''/G' \tag{10}$$
$$G' = |G^*| \cos \delta \text{ and } G'' = |G^*| \sin \delta \tag{11}$$

The real parts of the moduli E' and G', are called the storage moduli, because they are related to the storage of energy as potential energy and its release in the periodic deformation. The imaginary parts of the moduli, E'' and G'', are called the loss moduli and are associated with the dissipation of energy as heat when the materials are deformed.

These dynamic moduli can also be expressed in terms of a complex compliance

$$J^* = 1/G^* = J' - i J'' \tag{12}$$

where J' is the storage compliance and J'' is the loss compliance.

The loss tangent $\tan \delta$, called internal friction or damping, is the ratio of energy dissipated per cycle to the maximum potential energy stored during a cycle.

These viscoelastic behaviors can be explored theoretically using several models (3,13). The earliest models consisted of a spring and dashpot in series (Maxwell element), a spring and dashpot in parallel (Voigt element), and simple combinations of these basic elements (1).

Each Maxwell or Voigt element is characterized by a relaxation or retardation time defined as $\tau = \eta/E$, which is the viscous damping constant; E is the real modulus. When subjected to a constant stress, the Maxwell element exhibits instantaneous elasticity followed by linear creep. If the loading input is an instantaneous constant strain, the stress-relaxation behavior is an instantaneous elastic response which then decays exponentially. The Voigt element responds to a constant stress input with asymptotically increasing strain. If the stress is removed, the deformation decays as the strain asymptotically approaches zero. Combining these two elements produces a model which exhibits the combined characteristics of the previous two, ie, instantaneous elasticity, linear creep, and delayed elasticity. These simple models provide insight into the understanding of such phenomena as creep, relaxation, and memory (see also ELASTICITY; POLYMER CREEP).

Coupling Maxwell elements in parallel or Voigt elements in series adapts these basic elements to the description of a wider range of phenomena and has led to the concept of a distribution of relaxation times. It assumes that the behavior of a real material can be represented mathematically by such models, if they are generalized to contain an infinite number of springs and dashpots in such a manner that the relaxation times τ of the various individual combinations of springs and dashpots are distributed continuously throughout the system. Both of these generalized models have been used for analyzing the mechanical behavior of polymeric materials (3).

Analysis of the behavior of these generalized systems under periodic loading has in turn led to the concept of the complex modulus for characterization of the dynamic mechanical properties of viscoelastic materials. For example, the stress–strain relationship for a generalized Maxwell model containing an infinite number of basic elements subject to a harmonic input of the form $\epsilon = \epsilon_0^{i\omega t}$, with ω constant, can be written in the form

$$\sigma = (E' + i\,E'')\,\epsilon_0 e^{i\omega t} \tag{13}$$

where

$$E' = \int_0^\infty \frac{\omega^2 \tau^2}{1 + \omega^2 \tau^2}\,\phi(\tau)\,d\tau \tag{14}$$

$$E'' = \int_0^\infty \frac{\omega \tau}{1 + \omega^2 \tau^2}\,\phi(\tau)\,d\tau \tag{15}$$

and $\phi(\tau)$ is the continuous distribution of relaxation times for the material in question (1).

From equation 13 it is possible to define a complex modulus for this model as

$$E^* = \frac{\partial \sigma}{\partial \epsilon} = E' + i\,E'' \tag{16}$$

The individual components of this complex modulus represent the contributions of each Maxwell element in the generalized model summed over the relaxation-time distribution, as indicated in equations 14 and 15. This complex formulation indicates that the resulting stress is out of phase with the imposed strain.

By rearranging equation 13, it is possible to identify an elastic modulus and viscous damping constant in terms of an effective single Voigt element for a viscoelastic material at a fixed frequency, ie,

$$\sigma = E'\epsilon_0 e^{i\omega t} + \frac{E''}{\omega}\frac{d}{dt}(\epsilon_0 e^{i\omega t}) \tag{17}$$

$$\sigma = E'\epsilon + \eta\frac{d\epsilon}{dt} \tag{18}$$

where $\eta = E''/\omega$. Thus the real part of the complex modulus relates directly to the time-dependent elastic behavior of the material, whereas the complex component represents the time-dependent dissipative properties. Both properties are, however, a function of frequency.

If the mechanical behavior of viscoelastic material is desired for some loading condition other than constant stress (creep test), constant strain (relaxation test), or harmonic strain variation (complex modulus measurement), the Boltzmann superposition principle is applied. It states that the total strain of a sample is an additive function of the individual strains resulting from each of a sequence of stress applications. Each stress application in the sequence can be handled in terms of one of the simple loading conditions and the particular material properties associated with it.

The stress or strain amplitude and the frequency are related in the heat generation in the polymeric materials during oscillation (4).

$$Q = \pi\,E''\,\epsilon_0^2 \tag{19}$$

where Q is the heat dissipated per cycle, ϵ_0 the maximum value of the strain amplitude during a cycle, and E'' the loss modulus.

Thus, the heat generation in a material is proportional to the square of the amplitude, the frequency, and the loss modulus. If a polymeric material is subjected to high-frequency, high-amplitude oscillation, the temperature of the specimen is increased. For example, energy losses in the rubber compounds and fabric of tires under dynamic loading result in a drag component manifesting itself in internal heat generation which is readily observed as an increase in temperature (14). Internal heat generation by oscillation is also a cause of structural changes in polymers. Therefore, in the determination of dynamic mechanical properties, low levels of strain amplitude and low frequencies are often used. This low strain amplitude is in the linear region of the stress–strain curve. When large strain or stress amplitude is applied to the viscoelastic material, it exhibits increased internal heat generation and nonlinear viscoelastic behavior. Analyses of dynamic response in nonlinear viscoelasticity are generally more complex.

In the nonlinear region, the polymers are changed permanently in part of their structure, as in crazing or the formation of microscopic cracks or voids. High strain or stress amplitude at high frequencies can also result in the breaking of bonds, fatigue, and rupture.

Primary and Secondary Transitions

The dynamic mechanical properties of polymers are usually studied over a wide temperature range (-150–$300°C$). Typical dynamic mechanical properties of polymers as a function of temperature are illustrated in Figure 1.

In the region where the dynamic-modulus–temperature curve has an inflection point, the internal friction (tan δ) curve goes through a maximum. This dispersion is called the glass-transition region where the dynamic modulus E' changes from approximately 1 GPa (10^{10} dyn/cm^2) in the glassy state to about 1 MPa (10^7 dyn/cm^2) in the soft rubbery state.

In the transition region, the damping is high owing to the initiation of micro-Brownian motion in molecular chains. Some of the molecular chain segments are free to move, others are not. A frozen-in segment can store much more energy for a given deformation than a free-to-move rubbery segment. Thus, every time a stressed, frozen-in segment becomes free to move, its excess energy is dissipated as heat. Micro-Brownian motion is concerned with the cooperative diffusional motion of main-chain segments. This transition is so conspicuous that it is called the primary dispersion (the α-peak).

The loss modulus E'' goes through a peak at a slightly lower temperature than the internal friction E''/E'. The maximum heat dissipation per unit deformation occurs at the temperature where E'' is maximum; at 1 Hz, this temperature is very close to the value of the glass-transition temperature as determined by volume–temperature measurements.

Other relaxation transitions can be found in a glassy state on the lower temperature side of the primary dispersion. These are called secondary dispersions and are usually designated β, γ, etc, in order of decreasing temperature. The β-dispersion, for example, has been shown by a combination of comparative nmr and dielectric measurement studies on similar polymers to be associated

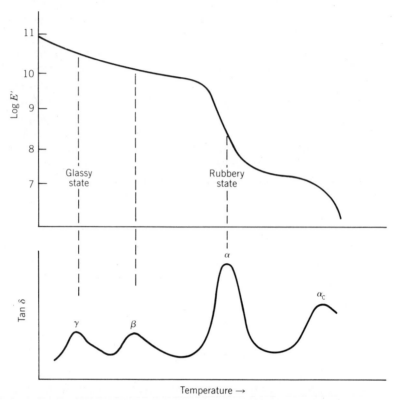

Fig. 1. Typical dynamic mechanical properties of polymers.

with side-chain motion of the ester group (15–18). The γ-dispersion involves motion of the methyl groups attached to the main chain. The activation energy of these secondary dispersions is generally lower than that of the α-dispersion, since it corresponds to segment motion which is smaller than the micro-Brownian motion of the main chain.

The primary relaxation of a crystalline polymer (α-peak), corresponding to the glass transition (T_g), results from the initiation of micro-Brownian motion of the amorphous chains. In this glass-transition region, the dynamic modulus of the glassy state, 1–10 GPa (10^{10}–10^{11} dyn/cm^2) decreases to the leathery-state modulus of around 100 MPa (10^9 dyn/cm^2), about 100 times the modulus in the rubbery region. The appearance of the leathery state is caused by the micro-Brownian motion of the noncrystalline region under the structural restraint of the neighboring molecular chains in the crystalline regions. For this reason, even after crystallizing a given polymer, the α-peak tends to shift to higher temperature and broaden in comparison with the amorphous state.

Highly crystalline polymers, such as high density polyethylene and isotactic polypropylene, exhibit another dispersion between the α-dispersion and melting temperatures. This relaxation, called the α_c- or α'-dispersion, has been attributed to molecular motion within the crystalline phase (6). This crystalline dispersion is due to the frictional viscosity among specific crystalline planes or molecules inside the crystals.

Molecular Motion and Loss Factor

Motions of long chain segments in the polymer structure have a profound effect on the loss factors (tan δ and loss modulus) of the dynamic mechanical properties. The tan δ value at the α-peak (the glass transition) is of greater magnitude than at the dispersion peaks for lower temperatures. It is accompanied by the greatest decrease in dynamic modulus with increasing temperature.

The loss factors plotted against temperature (Fig. 1) often are called relaxation or anelastic spectra (7,19); they contain several loss peaks. Analyses of these spectra are useful for determining molecular motions in combination with dielectric and nmr measurements. Interpretations of the spectra are found in Refs. 4–6, 8, 9.

The loss factors are most sensitive to molecular motion. For example, in the glassy region the dynamic modulus shows a small degree of change, yet the loss factors exhibit damping peaks.

The largest loss peak is generally the α-peak, which is associated with the glass-transition temperature. This transition occurs in the amorphous regions of the polymer with the initiation of the micro-Brownian motion of the molecular chains. The magnitude of the α-peak in the amorphous polymer is much greater than in the semicrystalline polymer, because the chain segments of the amorphous polymer are freer from restraints imposed by crystalline polymers in the glass-transition region.

The glass-transition temperature is affected by a number of chemical and molecular structures. Flexibility of the molecular chain, bulkiness of the side groups attached to the backbone, and molecular polarity affect the value of T_g. The peak height and temperature at which the α-peak occurs are closely related to the glass-transition phenomenon. Therefore, similar interpretations are used for both the α-peak and T_g (see also GLASS TRANSITION).

The magnitude of the β-peak is small in comparison with the α-peak. The temperature at the β-peak is about 0.75 T_g. This dispersion is associated with an amorphous-phase relaxation in most polymers. It is resolved into a double or triple peak and appears to involve local motion in the chain. In the case of amorphous polymers, the β-peak is very broad and may appear as a shoulder of the α-peak. It is associated with motion about the chain backbone of a relatively small number of monomer units or with motion of side groups (8).

The γ-peak temperature is below the β-peak temperature. In this region, the chain segments are frozen in, and side-group motion is made possible by defects in packing or configuration in the glassy and crystalline states. This γ-peak of polymers can be related to the following general classifications: side-group motions in the amorphous and crystalline phases, end-group rotation, crystalline defects, backbone-chain motions of short segments or groups, and phase separation of impurities or diluents (3,8).

The γ-relaxation in both amorphous and crystalline polymers could in many cases be attributed to a restricted motion of the main chain, which requires at least four —CH_2 groups in succession on a linear part of the chain (20). This is the crankshaft mechanism (21), (22), (23). It has been proposed that this mechanism is relevant to the γ-peak in polyethylene, polyamides, polyoxymethylene, and poly(propylene oxide).

The α' (α_c) peak occurs between T_g and the melt temperature. This process relates rotation about the chain axis of chain-folded polymers to a second-order expansion of lattice parameters perpendicular to the chain direction in the crystalline phase.

In the case of amorphous polymers, the α' dispersion, often designated the T_{ll} relaxation, involves a melting process. The T_{ll} relaxation is exhibited in polystyrene (24,25), poly(vinyl chloride) (26), and poly(methyl methacrylate) (27). It has been extensively studied in synthetic rubber blends of the polybutadiene type (28). In this study, it was found that T_{ll} dispersion is related to the motion of the low molecular weight components of a polymer with a broad molecular-weight distribution.

The relaxation spectra of nylon-6 and -6,6 are shown in Figures 2 and 3, respectively. The results for tan δ on nylon-6 are for three samples of different crystallinity (6). The least crystalline specimen was obtained by quenching from the melt to $-78°C$, the sample of intermediate crystallinity was prepared by crystallizing at $200°C$, and for the preparation of the sample of highest crystallinity residual monomer (ϵ-caprolactam) was removed. The nylon-6,6 sample was prepared by quenching rapidly after annealing unoriented nylon-6,6 film at $266°C$ (29).

The three loss peaks observed in Figures 2 and 3 have been labeled α, β, and γ occurring at approximately 90 to $110°C$, -70 to $-40°C$, and -120 to $-140°C$, respectively. In addition to these three peaks, a further relaxation region in nylon-6 is indicated by the shoulder on the high-temperature side of the α-peak. Since this shoulder is most prominent for the specimen of highest crystallinity, it has been attributed to a relaxation process involving the crystalline regions. This relaxation might be related to a crystal-structure change from monoclinic to hexagonal, apparent from x-ray and specific-volume/temperature data. An additional crystalline relaxation mechanism might also be operative in the region of $190°C$; the detection of this relaxation might be hindered by the relatively low crystalline content of the polyamides.

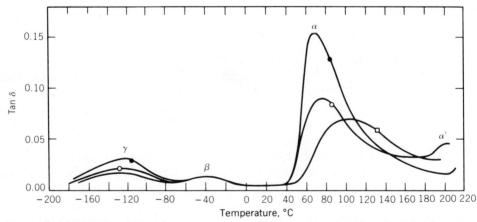

Fig. 2   Tan δ vs temperature for three nylon-6 samples prepared by different methods; data at 100 Hz: □, Caprolactam removed; ●, $-78°C$ quenched; and ○, $200°C$ crystallization (6).

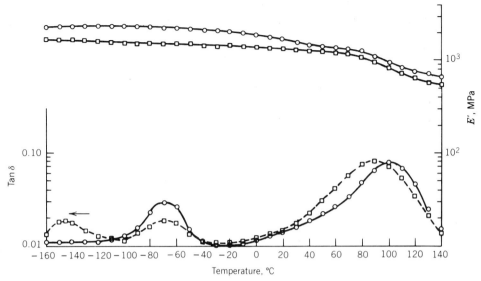

Fig. 3. E' and tan δ vs temperature for nylon-6,6: $\bigcirc$, Nylon-6,6 film annealed at 266°C and cooled slowly (Form II); $\square$, nylon-6,6 film annealed at 266°C and quenched rapidly (Form I).

It can be seen in Figure 2 that the α -peak for nylon-6 decreases in height with increasing crystallinity; this suggests that the α -process involves motions within the amorphous phase. The α -damping-peak magnitude and temperature for nylon-6 and -6,6 are sensitive to both the chemical composition of the nylon and the effect of moisture. The β -damping peak is attributed to the carbonyl group of the polyamide-forming hydrogen bonds. The temperature of this β -peak is influenced by the moisture content of nylon (29). Since nylons absorb some moisture from environmental humidity, moisture absorption characteristics must be considered in prospective application. The low temperature γ transition is attributed to the crankshaft-like rotation of four or more methylene carbon atom chains between the amide groups.

Determination of the three relaxation spectra of polymers is useful for understanding the molecular motions in the structure, as well as practical applications. The α -dispersion is associated with the glass-transition temperature and is accompanied by the greatest change in the stiffness of materials. Nearly all tough ductile glassy polymers and those with high impact strength have prominent secondary transitions (β - and γ -peaks) (4).

Crystallinity and Molecular Orientation

Many polymers, eg, poly(ethylene terephthalate) and nylon, crystallize when cooled from the melt, and chain-folding of the molecules in lamellar crystals takes place. The crystalline aggregates are known as spherulites.

The degree of crystallinity and molecular orientation are affected by heat treatment and stress in the polymer processes. These structural changes have profound effects on the dynamic mechanical properties.

The dynamic modulus generally increases with increasing degree of crystallinity. This effect is clearly seen in the modulus above T_g. The following expressions have been derived for the relationships between the modulus above T_g and the degree of crystallinity w_c up to 70% (30):

$$\log_{10} G = 6.763 + 4.77\ w_c \tag{20}$$

$$\log G = w_c \log G_c + (1 - w_c) \log G_a \tag{21}$$

where G is the dynamic shear modulus of semicrystalline polymers, G_c is the modulus of the pure crystal, and G_a is the modulus of the amorphous polymer above T_g. This equation implies that $G_c = 34$ GPa (3.4×10^{11} dyn/cm^2) for polymer crystals, and $G_a = 570$ kPa (5.7×10^6 dyn/cm^2) for the rubbery phase. Equation 21 is called the logarithmic rule of mixtures, and is known to hold for many types of two-phase systems in which both phases are continuous (4).

Crystallinity has a great effect on α-dispersion. Poly(ethylene terephthalate) (PET) is a good example of the α-peak changes due to crystallinity and molecular orientation. Controlling these parameters led to the introduction of a number of new products with a variety of mechanical properties. Details on PET follow.

Effect of Annealing Temperature on Crystallinity and Crystal Size of PET. The effect of structural changes on the height and position of the α-dispersion in PET has been extensively studied (31–35). The dynamic mechanical properties of PET measured by nonresonant forced vibration imply that the α-transition shifts from 80 to 125°C with increasing crystallinity (36). An investigation by a cantilever vibration method showed that, although the α-transition initially shifted to higher temperature with increasing crystallinity, above a certain crystallinity the α-transition began to shift back (37). This shifting back was also found (38), when a Rheovibron dynamic viscoelastometer was used, in addition, to a linear relation between the height of the α-dispersion peak and crystallinity. As the crystalline T_g increased, the height of the α-peak decreased.

The intensive study of the dynamic properties of unoriented PET films of different degrees of crystallinity by a torsion-pendulum method showed that the α-peak shifted to higher temperatures for crystallinities up to 30%. At higher crystallinities it shifted toward lower temperatures (39).

This behavior was attributed to the effect of crystal size on the amorphous regions. At low to medium crystallinities, many small crystallites act like crosslinks and inhibit segment motion in the amorphous regions; at high crystallinities, the crystallites are fewer and larger, and allow more freedom to the segments in the amorphous regions.

The degree of crystallinity is generally determined by x-ray diffraction (40) or by measuring the density (41,42) (see CRYSTALLINITY DETERMINATION). Morphological structure, including the length of chains between folds in crystals and spherulitic structure, is determined by light scattering (qv) (43–47), small-angle x-ray scattering (48–51), and electron microscopy (qv) (52,53).

Unoriented PET fibers of low crystallinity (2–3%) were studied by a combination of x-ray diffraction (qv) (54) and mechanical dynamic methods (55). Samples of different degrees of crystallinity were prepared by heating freely hanging skeins of yard in an oven for 6 h within a nitrogen atmosphere. During this time period, essentially all the crystallization was completed.

The effect of annealing temperature on crystallinity of unoriented PET fibers is shown in Table 1. No crystallization was detected until an annealing temper-

Table 1. X-ray Crystallinity and Vibron Data for Unoriented Annealed PET Fibers

Annealing temperature, °C	Temperature at which maximum occurs in E'', °C	E'' max, MPa	Degree of crystallinity w_c, %	Width of E'' peak at $E_{max}/\sqrt{2}$
85	83	250	0	6
97	84	220	0	8.5
98.5	87	210	1.5	6
100	90	175	11.5	8
101	95	162	18.5	14
102	97	150	25.5	20
110	98	140	31.0	21
125	100	135	35.3	25
155	100	125	42.2	26
176	100	123	49.0	26
200	98	125	49.7	26
228	95	120	48.0	20
245	92	123	48.2	21

ature of 98°C was reached, even though the glass transition of amorphous PET is 67°C (56). The results show that between 67 and 90°C, parallelism of the benzene rings can occur without the formation of crystallites (57). The degree of crystallinity of PET increases rapidly between 98 and 170°C. This behavior is a consequence of two crystallization processes: the formation of nuclei and their subsequent growth. The first process is dominant at low annealing temperatures, the second at high annealing temperatures. At intermediate temperatures both processes can occur. At low annealing temperatures, crystallization proceeds via the formation of new crystallites rather than by the growth of existing crystallites. Thus, the crystal size should be almost independent of temperature in this region. At high temperatures the crystallization takes place by a diffusion process in which crystallites grow in size with little or no new nucleation (see also AN-NEALING; KINETICS OF CRYSTALLIZATION).

The long period d obtained from small-angle x-ray scattering measurements is obtained as a function of annealing temperature. At low annealing temperature it changes slowly with temperature, whereas for annealing temperatures near the melting point it increases rapidly. These results show that the increase in d is due to increase in crystal size.

Crystal sizes obtained from measurements of the ⟨010⟩ reflection are shown in Table 2. The crystal sizes of PET increase with increasing annealing temperature between 102 and 245°C. These crystal sizes are weight-average sizes which employ integral breadths obtained by wide-angle x-ray measurements. The true number-average size of crystals can be related to the weight-average size, if the correction for lattice distortion is known. A variety of measurements indicate that the lattice distortion should be approximately 1%. The volume of a crystallite as a function of annealing temperature is calculated from the crystal size, as shown in Table 2, where it was assumed that the crystal volume is equal to the cube of the crystal size obtained from the ⟨010⟩ reflection. Ideally, crystal sizes should be obtained from several reflections, giving the crystal shape. In the absence of such information it is reasonable to assume the shape to be a cube.

Table 2. Crystal-Size Data for Unoriented PET Fibers

Annealing temperature, °C	Integral breadth, c	Crystal size, nm	Crystal volume, nm³	Amorphous volume/crystal,[a] nm³	X/V,[b] 10^{-6} nm^{-3}	Long period, nm
85						
97						
98.5	c			17850	55	
100	c			2100	422	9.45
101	c			1200	680	9.48
102	−0239	6.49	272	795	937	9.48
110	−0237	6.50	274	610	1130	9.62
125	−0233	6.63	290	531	1220	9.76
155	−02175	7.09	355	486	1190	10.22
176	−02145	7.18	369	385	1330	10.40
200	−0201	7.69	452	457	1100	10.84
228	−0182	8.48	608	659	789	11.62
245	−0160	9.66	900	967	535	12.85

[a] Number of crystals.
[b] Number of crystals/volume per unit.
[c] Peak too weak to measure. Crystal size assumed equal to that of 102° sample.

Relationships between Crystallinity and Dynamic Mechanical Properties.

The dynamic mechanical properties of the same series of unoriented annealed PET fibers have been examined with a Rheovibron (6) operated at a frequency of 11 Hz. Samples were heated at 1°C/min in a nitrogen atmosphere under relaxed conditions, and the tensile modulus E and the loss factor tan δ were measured at increments of 5°C, except in the transition region, where smaller increments were used. Samples were allowed to equilibrate at temperature for 5 min before measurements were made. Measurements were continued up to the temperature at which the sample had been annealed. The annealing temperature was not exceeded, because this would have caused additional crystallization.

The temperature dependence of the real part of the modulus E', for this series of PET is shown in Figure 4. Changes in crystallinity only slightly alter the room-temperature value of E', but have a great effect on the temperature dependence of E'.

The values of the maximum loss modulus, E''_{max}, and the temperature at which the peak occurs are given in Table 1. Also included are the half-power widths of the peak in the loss modulus. As crystallinity increased, E''_{max} decreased. The width of the loss-modulus peak generally increased as crystallinity increased, although at higher crystallinities it decreased somewhat. The increase in width is taken to represent the presence of an increased range of order. The temperature at which the maximum in E'' occurs is plotted versus crystallinity. The intensity of tan δ decreases with increasing degree of crystallinity. The intensity of the damping peak (tan δ) is given by equation 22 (58)

$$\tan \delta = w_c (\tan \delta)_c + (1 - w_c)(\tan \delta)_a \tag{22}$$

where w_c is the degree of crystallinity, and the subscripts a and c refer to the contributions of the amorphous and pure crystalline phases, respectively. Since

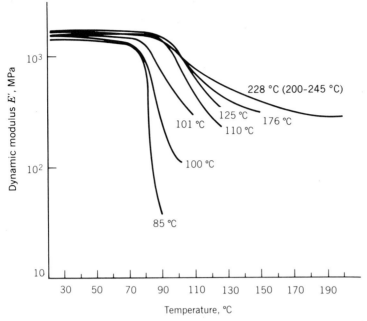

Fig. 4. Dynamic modulus E' as a function of temperature for PET of various crystallinities (11 Hz). Above 200°C, the curve remains the same.

the loss tangent (tan δ), in general, is mostly due to the amorphous phase, the tan δ equation for the maximum in the tan δ peak simplifies to

$$\tan \delta \simeq (1 - w_c)(\tan \delta)_a \tag{23}$$

This equation shows that intensity decreases with increasing degree of crystallinity.

Effect of Molecular Orientation on the α-Peak Position. When a polymer is stretched, the molecules may be preferentially aligned along the stretch direction. In order to determine the effect of molecular orientation, a series of PET yarns were stretched, with draw ratios from 2× to 5.5×, by drawing over a hot pin at 90°C. For these samples, dynamic measurements using a Rheovibron were made at temperatures up to 160°C. The results are given in Table 3. The shift in E''_{max} for the drawn series is much greater than for heat-crystallized unoriented fibers. For example, the sample of 5.5× draw ratio has a crystallinity of 42%,

Table 3. Vibron and X-ray Data for Drawn PET Fibers

Draw ratio	Position of maximum in E'', °C	E''_{max}, MPa	Degree of crystallinity, %	Width of E'' peak at $E_{max}/\sqrt{2}$
1	84	520	0	5
2	90	480	5.5	10
3	95	420	13.5	20
4	118	490	21.0	46
5	128	460	33.0	52
5.5	132	400	42.0	53

which would indicate a change in position of E''_{max} from 80 to ca 100°C. However, the maximum occurs at 132°C. The results cannot be treated in the same way as for the undrawn series, since the drawn samples were not thermally conditioned. Some additional crystallization occurs on heating in the Rheovibron, as well as some loss of orientation as the samples are heated under relaxed conditions. These changes would tend to decrease the shift in E''_{max}.

For the drawn samples, two effects may contribute to the greater shift of E''_{max}. The crystals in a drawn sample may be smaller than those in an annealed sample of the same crystallinity. Furthermore, in the drawing process the amorphous regions become oriented, which greatly reduces the free volume. It cannot be said unequivocally that the greater changes in the positions of E''_{max} are due to orientation alone. It has been shown that the combined effect of crystallinity and orientation is significantly higher for PET than for polystyrene (59).

The variation of dynamic modulus E' with temperature is obtained. The modulus is one order of magnitude higher for the drawn series than for the heat-crystallized samples. There is also a greater change in E' within the drawn series.

The effects of the orientation in the crystalline and noncrystalline regions on the α-transition have been studied by using x-ray and birefringence on PET yarns drawn $3\times$ and $4.25\times$ over a hot pin at 80°C (60). The samples were heated for 6 h up to 240°C *in vacuo* under conditions which allowed the samples to shrink. All samples were subsequently boiled in a relaxed condition in water for 1 h.

Dynamic measurements were made with a Rheovibron operated at a frequency of 11 Hz. Crystallinity measurements were made with a Norelco x-ray diffractometer, which was also used to obtain the crystallite orientation from scans on the $\langle 105 \rangle$ reflection. The orientation parameter employed was f_c given by (61)

$$f_c = \tfrac{1}{2}\,(3[\cos^2 \theta] - 1) \qquad\qquad (24)$$

where $[\cos^2 \theta]$ is the mean-square cosine of the angle θ between a $\langle \bar{1}05 \rangle$-plane normal and the fiber axis. The orientation in the noncrystalline regions, f_{am}, was obtained from a combination of birefringence, crystallinity, and f_c (62). Crystal-size measurements along the fiber direction were made from small-angle data collected with a Kratky camera (63).

X-ray and Rheovibron data for these drawn samples are given in Table 3. The crystallinity increases with annealing temperature, with higher crystallinities for the $4.25\times$ than for the $3\times$ yarns, except at high annealing temperatures, where the crystallinities are similar. The orientation of the crystalline regions is almost the same for both draw ratios and is independent of annealing temperature. The orientation of the noncrystalline regions is also independent of annealing temperature, but is higher for the $4.25\times$ yarns than for $3\times$ yarns. The orientation of the noncrystalline regions is considerably lower than that of the crystalline regions. The long period in the small-angle scattering increases with annealing temperature and is higher for the $4.25\times$ yarn series except at the highest annealing temperature. Crystal sizes calculated from the small-angle data increase with increased annealing temperature and are again somewhat higher for the $4.25\times$ yarns except at the highest annealing temperature.

These results show that drawn crystalline PET yarns exhibit a shift in the position of the α-transition to lower temperatures for high annealing temperature;

this effect has also been observed for unoriented crystalline PET. The main difference between the results from the drawn and unoriented yarns lies in the position of the α-transition on the temperature scale; the α-transition occurs at higher temperatures for higher draw ratios. The shift in the position of the α-transition due to annealing is interpreted in terms of the number of crystals, whereas the shift due to drawing is explained by orientation in the amorphous regions (see also ORIENTATION; ORIENTATION PROCESSES).

Molecular Weight and Cross-linking

Effect of Molecular Weight.   Increasing molecular weight has a large effect in the glass-transition range, transforming the behavior from viscous flow to a plateau range of rubber-like behavior. The length of the rubbery plateau region increases with increasing molecular weight.

The loss tangent (tan δ) above T_g is also strongly dependent upon molecular weight. The value of the tan δ at the minimum above T_g decreases as the molecular weight increases (3,64–67).

The relationship between the dynamic mechanical properties and molecular weight for nylon-6,6 is discussed in Ref. 68. In order to prepare high-molecular-weight nylon-6,6 film and fiber, thermally induced solid-state polycondensation (SSP) is applied (69,70). Briefly, the process is carried out by heating polyamide fibers, such as poly(hexamethylene adipamide) fibers, with or without the usual additives (eg, TiO_2, copper, potassium salts, phosphorus-containing compounds) in the absence of oxygen to temperatures below the melting point (120–230°C) for 30 min–24 h. These temperatures are high enough to allow the amine and acid ends of the polymer molecules to react, thus increasing the length of the molecules, and consequently the molecular weight. Time and temperature are dependent on each other inasmuch as higher temperatures require a shorter period of time to effect the desired result. The exclusion of oxygen and moisture is most important; a blanket of an inert gas, such as nitrogen, or a vacuum may be employed.

In accordance with the patents (69,70), films and oriented fibers composed of essentially linear polyamide molecules are produced. These materials are soluble in formic acid at 25°C. Table 4 shows the conditions of the thermally induced

Table 4. Effects of Thermally Induced Solid-State Polycondensation Conditions on Nylon-6,6 and Molecular Weight

Sample	Conditions	Intrinsic viscosity η, dL/g	Average molecular weight $\overline{M}_v$
film 1	control	1.6	45,000
2	extruded from resin	2.0	60,000
3	174°C, 24 h[a]	3.5	120,000
fiber 4	control	1.3	35,000
5	140°C, 18 h[a]	1.6	45,000
6	160°C, 8 h[a]	1.7	48,500
7	160°C, 18 h[a]	1.8	52,000

[a] At 1.06 kPa (8 mm Hg).

solid-state polycondensation process for the test samples. The density did not change during treatment (1.14 g/cm³); the fiber sample was wound on perforated metal bobbins with both ends of the threadline tied down to reduce relaxation.

The loss tangent at 0% rh as a function of temperature for a control and for a typical thermally induced solid-state polycondensed nylon-6,6 film and fiber are shown in Figure 5. The temperature of the maximum of the tan δ peak (T_g), the maximum tan δ value, and the width of the tan δ peak (ΔT) for these samples are summarized in Table 5.

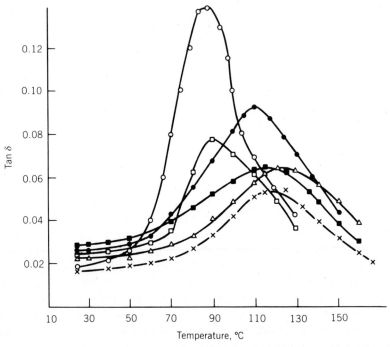

Fig. 5. Loss tangent of high molecular weight nylon-6,6 film and fiber as a function of temperature at 0% rh. Nylon-6,6 film [η], dL/g: ○, control 1.6; □, sample 2, 2.0; and ×, sample 3, 3.5. Fiber [η], dL/g: ●, control 1.3; △, sample 6, 1.7; and ■, sample 7, 1.8.

Table 5. Peak Parameters of Nylon-6,6 Made by Thermally Induced Solid-State Polycondensation

Sample	0% Rh			30% Rh		
	α-Peak temperature, °C	Tan δ_{max}	ΔT^a	α-Peak temperature, °C	Tan δ_{max}	ΔT^a
film 1	88	0.138	23	60	0.104	57
2	90	0.080	40	72	0.076	42
3	120	0.054	48	98	0.082	62
fiber 4	110	0.094	34	90	0.084	40
5	118	0.062	49	110	0.062	46
6	121	0.063	64	103	0.064	41
7	123	0.062	62	105	0.065	50

[a] Width of α-peak at (tan δ)$_{max}$/$\sqrt{2}$.

Intrinsic viscosities were measured in 90% formic acid at 20°C. The peak temperatures of tan δ for the treated films and fibers are shifted 8–32°C higher than those of the corresponding control nylon-6,6 (film and fiber), whereas the maximum height of the tan δ peaks decreases.

The occurrence of the α-transition, indicated by the maximum in tan δ at 110°C, for the highly oriented nylon-6,6 fiber is typical of the values obtained by other workers for similarly prepared nylon-6,6 (71–73). The effect of annealing and drawing on the α-peak of nylon-6,6 has been reported (73). The α-transition is little affected by the annealing treatment, but an increase in orientation causes the transition to move to a higher temperature. The results on dynamic mechanical properties of high-molecular-weight nylon-6,6 film and fiber clearly show that the α-transition is shifted to higher temperatures. The clearest and most dramatic effect of molecular weight on tan δ is shown in the unoriented films, whose intrinsic viscosities are 1.6, 2.0, and 3.5 dL/g. In these curves, the loss of orientation occurring during SSP with its attendant effects in viscoelastic properties is not present to obscure the basic effects of molecular-weight increases. Even the film with $\eta = 3.5$ dL/g was free of cross-linking (69) and the change in film properties can be attributed to linear growth of molecules. The magnitude of the change in α-peak temperature, as the intrinsic viscosity of the unoriented film is increased to 3.5 dL/g, becomes apparent when one notes that this α-peak temperature becomes higher than that found for oriented nylon of normal molecular weight. A shift of the α-peak to high temperatures is usually a symptom of increasing crystallinity or more closely packed morphology. However, the study of the dynamic mechanical behavior of nylon-6,6 (73) indicated that crystallinity is little affected between 100 and 180°C (46.4–55% crystallinity). The α-peak in high-molecular-weight nylon-6,6 may lead to the formation of the additional interlamellar connections (74).

Effect of Cross-linking. The effect of cross-linking on the dynamic properties of the vulcanized rubber is evident above T_g (4). The dynamic modulus of a cross-linked rubber does not decrease with increasing temperature. The loss tangent is a sensitive indicator of cross-linking. At temperatures well above T_g, damping decreases with increasing degree of cross-linking. Epoxies and highly cross-linked phenol-formaldehyde resins are examples of polymers with a much higher degree of cross-linking (qv) than normal rubbers.

The effect of a high degree of cross-linking on dynamic mechanical properties is shown in Figures 6 and 7 (75,76). The dynamic modulus above T_g increases dramatically, and the α-peak (T_g) shifts to higher temperatures with increasing cross-linking. These epoxy samples were Epon 828 cured with various amounts of methylenediamine (MDA); the amount of amine was 6–102% of the stoichiometric amount. The amine was melted and mixed with the diepoxy at 100°C. A sheet was then cast on a glass plate. After curing (qv) in an air over 45 min at 60°C, 30 min at 80°C, and finally 2½ h at 150°C, the samples were allowed to cool slowly to room temperature.

A plot of tan δ as a function of temperature is shown in Figure 6. A greater than stoichiometric amount of amine decreases the temperature and increases the height of the loss peak. An excess of amine causes a significant shift in T_g, whereas a moderate excess of epoxy does not. This is reasonable, considering that excess amine increases the molecular weight between cross-links, M_c (77), whereas

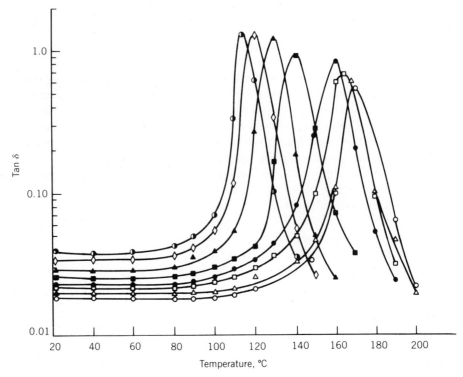

Fig. 6. Loss tangent as a function of temperature. Excess epoxy, %: ○, 2.41; △, 5.96; excess MDA, %: □, 11.64; ●, 26.68; ■, 46.20; ▲, 65.61; ◇, 84.04; and ◑, 98.57.

the M_c for all epoxy-excess polymers should be about the same; the polymers differ only in regard to the number of large cross-linked molecules in the system and the number of dangling epoxy ends.

The elastic component E' of the complex modulus for these epoxy samples is shown as a function of temperature in Figure 7. Again, there is a shift in the transition temperature as the stoichiometric excess of amine is increased. The order of the transition temperatures is the same as the order of the moduli at room temperature; the highest glass-transition temperature is associated with the highest room-temperature modulus. The curves have the shape characteristic of cross-linked amorphous polymers in the transition region.

Theory (78) requires that high cross-linking produces a higher value of E' after the transition than low cross-linking. In the present case, it would be expected that samples near zero excess amine would have a higher degree of cross-linking than the samples near 100%, and that they therefore would have a higher value of E' at $T_g + 40°C$. This is observed in Figure 7.

The shear modulus G of cross-linked rubbers is given by (79)

$$G = \frac{\bar{r}_i^2}{\bar{r}_f^2} \frac{dRT}{M_c} \left(1 - \frac{2M_c}{\bar{M}_n} \right) \tag{25}$$

where $\bar{r}_i^2/\bar{r}_f^2$ is the ratio of the mean-square end-to-end distance of the polymer chain in the sample to the same quantity in a randomly coiled chain (79) (here

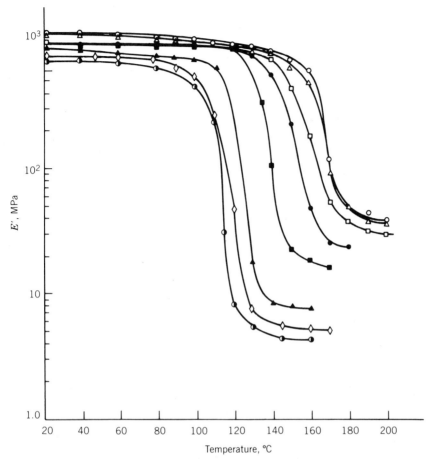

Fig. 7. Dynamic modulus E' as a function of temperature: Excess epoxy, %: ○, 2.41; △, 5.96; excess MDA, %: □, 11.64; ●, 26.68; ■, 46.20; ▲, 65.61; ◇, 85.04; and ◖, 98.57.

this ratio is assumed to be 1.0); d is the polymer density, R the gas constant, T the absolute temperature, M_c the molecular weight between cross-links, and $\overline{M}_n$ the chain-backbone molecular weight. The factor in parentheses is a correction for chain ends and becomes negligible when $\overline{M}_n$ is very large relative to M_c, as it is in the present case. Thus,

$$G = \frac{dRT}{M_c} \tag{26}$$

G must be measured in the rubber plateau region above T_g. For present purposes, E' was measured at $T_g + 40°C$. The shear modulus G can be taken as $E'/3$ (68).

The values for M_c in equation 26 have been obtained (75). The resulting values of G agree reasonably well with the reaction-stoichiometry values at low values of M_c, but are too high for samples with stoichiometric excess amine greater than 60%.

A number of studies have been made which relate the distance between cross-links to the shift in T_g (80–84). A rough estimate of M_c can be obtained from

$$M_c = \frac{3.9 \times 10^4}{T_g - T_{go}} \tag{27}$$

where T_{go} is the transition temperature of a polymer not cross-linked. To estimate T_{go}, a polymer was made by partial reaction of MDA and Epon 828 in such a way that almost all of the primary amine groups, but very few of the secondary amine groups, were consumed; the result approximated the linear polymer. The transition temperature was estimated by heating rapidly and noting the apparent softening temperature under stress with a Vibron. The transition was found in the region 75–80°C. The M_c values calculated from the T_g shift are consistent with the stoichiometry (75).

Polymer Blends and Copolymers

The dynamic mechanical properties of polymer blends (qv) are determined primarily by the mutual solubility of the two homopolymers. If two polymers are compatible and soluble in one another, the properties of the blend are nearly the same as those of a random copolymer of the same composition (see COMPATIBILITY). However, many polymer mixtures form two phases, due to the insolubility of the components. In this case, the damping-temperature curve shows two peaks; each peak is characteristic of the glass-transition temperature of one of the components. The two steps in the dynamic-modulus–temperature curves are characteristic of an immiscible two-phase system (4). The dynamic mechanical properties of an ABS resin (85) are shown in Figure 8, where polybutadiene (PBD) particles are dispersed in a phase of styrene–acrylonitrile copolymer (SAN) with the formation of graft bonds on its boundary; the temperature dependence of the dynamic modulus E' and loss modulus E'' is shown in Figure 8. The two blend ratios, 77/23 (open circles) and 56/44 (full circles), are expressed as volume fractions of SAN/PBD; the dynamic properties of PBD film and SAN copolymer at 138 Hz are the dashed lines in Figure 8 (6). The two transition peaks observed are the dispersion at 115°C, the α-peak of SAN copolymer, whereas the α-peak of PBD is at -82°C. Both of these polymer blends exhibit dispersions at 115 and -82°C. However, the intensities of the peaks depend on the SAN or PBD concentrations in the polymer blends. The magnitude of each peak is a characteristic of the relative concentrations of the two components. This behavior is found whether the phases are dispersed or continuous (86–88) (see also ACRYLONITRILE-BUTADIENE-STYRENE POLYMERS).

Using these results, the theoretical moduli of an SAN/PBD blend can be calculated from the equation of the mechanical equivalent model of two-phase mixtures (6,89).

The complex modulus E^* of this equivalent model is expressed by

$$E^* = \left[\frac{\phi}{(1 - \lambda)E_1^* + \lambda E_2^*} + \frac{1 - \phi}{E_2^*} \right]^{-1} \tag{28}$$

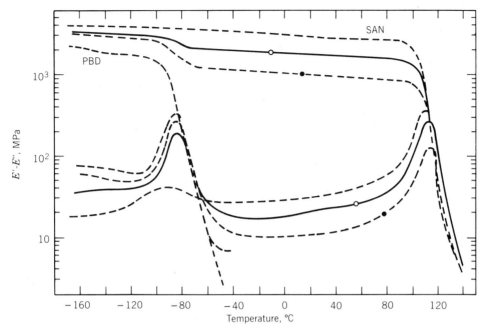

Fig. 8. E'-E'' at 138 Hz vs temperature for polybutadiene (PBD) in styrene–acrylonitrile copolymer (SAN); ----- experimental; and —— calculated: (SAN)/PBD: ○, 77/23; and ●, 56/44 (6).

where E_1^* and E_2^* are the complex moduli of the medium and the dispersed particles, respectively, and λ and ϕ are parameters representing the mixing state; they are so related that $\lambda\phi$ is equal to the volume fraction of the particles.

In the case of SAN/PBD blends, the parameters in equation 28 are as follows: E_1^* is the value for PBD; E_2^* is the value for SAN copolymer; for the volume fraction of PBD, $vR = \lambda\phi = 0.23$, $\lambda = 0.600$, $\phi = 0.380$; and for SAN, $vR = 0.44$, $\lambda = 0.775$, $\phi = 0.568$. The calculated moduli are illustrated by the solid lines in Figure 8. The observed and calculated curves are in satisfactory agreement, except for a slight deviation in the dispersion region.

The dynamic mechanical properties of copolymers are strongly influenced by the characteristics of the homopolymers and the copolymer composition. In a homogeneous copolymer all molecules have the same chemical composition, and hence should have a single sharp α-transition (T_g). Many copolymers are heterogeneous, that is, different molecules have different chemical composition. This heterogeneity of chemical composition comes about because one monomer tends to be more reactive than the other. At the beginning of polymerization, one comonomer tends to concentrate in the copolymer more than the other. During the course of the polymerization the more reactive monomer is depleted first and the copolymer is enriched with the slower-reacting monomer later. Thus, the heterogeneous copolymer contains molecules with widely varying T_g (4). Therefore, the damping curve of the dynamic properties of the heterogeneous copolymer reflects the broadening of the transition region caused by the increase in heterogeneity (90). The broadening of the α-peak is related to the mutual solubility of the two homopolymers. If the homopolymers tend to be insoluble in one another, heterogeneous copolymers have broad transitions. If the molecular interaction between the two polymers is high, the effect of interchain chemical heterogeneity is small, and the peak in the transition region is sharp (4).

The relative concentration of the two copolymers or polymer blends has a dramatic effect on the primary dispersion (6,91). The dependence of dynamic mechanical properties on the amount of components is shown in Figure 9. The samples of different compositions were prepared by vaporizing a tetrahydrofuran solution of PVC and butadiene–acrylonitrile copolymer rubber (NBR). The dynamic modulus and loss modulus are shifted on the temperature scale in proportion to the relative concentration of the two components. The damping peaks in the intermediate compositions are broader in the transition region. This suggests the existence of microheterogeneity due to the different segmental environments. In the case of homogeneous copolymers, the shape of the dispersion

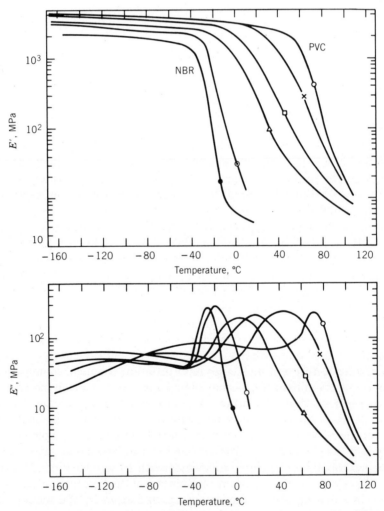

Fig. 9. E' and E'' vs temperature for blended system of PVC and NBR of various compositions (PVC/NBR) at 138 Hz: ○, 100/0; ×, 83/17; □, 67/33; △, 50/50; ⊙, 20/80; and ●, 0/100 (6).

does not change, and only the position of the dispersion shifts with composition (92). Similar effects on dynamic properties are observed with plasticizers (93).

Copolymerization and polymer blends are useful for modifying the properties of polymers and improving the processability of the molten polymers (94).

Composite Structures

Composite materials consist of two or more components and phases (see COMPOSITES). The structure of a composite depends on the structure and arrangement of these components. Filled composites have a continuous matrix phase and a discontinuous filler phase made up of discrete particles. These include polymer blends, filled poly(vinyl chloride) formulations (used for floor tiles and wire coating), filled rubbers, and glass fiber-filled plastics (95).

Many composites contain fibers as the filler phase, for example, glass fibers in polyesters and epoxies, glass fabric laminates, asbestos fibers in phenolic resins, and chopped glass fibers in polystyrene and nylons. Recent developments are boron and graphite fibers in epoxy resins or polyimides, and whiskers (alumina or silicon carbide) in resins for aerospace applications (96). The importance of fiber-filled composites arises largely from their unusually high strength and stiffness (modulus) per unit weight (97).

The dynamic mechanical properties of composite materials are determined by the properties of the components, the morphology of the system, and the nature of the interface between the phases. Therefore, a great variety of properties can be obtained by varying the structure of the system or the interface properties. An important property of the interface is the degree of the adhesive bonding between the phases.

A study of the effect of rigid fillers on dynamic mechanical properties of filled rubber indicates that the dynamic shear modulus G at 1 Hz for a filled polyurethane rubber increases with increasing amounts of sodium chloride filler (98,99). Fillers have a larger effect on the modulus above T_g than below. The most pronounced effect of rigid fillers on the damping G''/G' is a broadening of the transition region by high concentration of fillers.

Fillers also decrease the damping as expressed by G''/G'. In this case the damping can be approximated by (100,101)

$$G''/G' = (G''/G')_1\phi_1 + (G''/G')_2\phi_2 \qquad (29)$$

where ϕ_1 and ϕ_2 are the volume fractions of the filler and matrix, and (G''/G') and (G''/G') are the loss tangents of the filler and matrix, respectively. However, fillers increase the damping of a mica-filled phenoxy resin (97), which may be the result of particle-polymer friction. In some cases the filler shifts the loss-tangent peak and T_g to higher temperatures (100–103). The shift in T_g is related to the surface area of filler. The effect should increase with concentration and the size of the particles.

Polymer blends and block and graft copolymers have two polymeric phases which are composite structures. The dynamic mechanical properties of these polymers show two glass transitions, characteristic of the material in each phase

(Fig. 8). The loss-tangent curves show two peaks (104–106). The relative height of the two loss peaks is determined by the concentration of the two components and the structure of the system (see BLOCK COPOLYMERS; GRAFT COPOLYMERS).

Many fiber-filled composites are anisotropic (uniaxially oriented), with tremendous strength and modulus in one direction due to alignment of the reinforcing fibers. These composite structures are made by a number of different techniques (107–110). Examples are laminates (qv) of sheets and woven fabrics, injection-molded polymers containing short fibers, and impregnated fiber mats.

The dynamic modulus and loss factors in fiber-reinforced composites are influenced by the fiber orientations. The dynamic modulus and damping have been determined (111) by studying the free and forced transverse vibrations of cantilever beams made of the material in which the fibers were uniformly oriented in the plane normal to the direction of vibration, making angles with the beam longitudinal axis of 0, 22.5, 45, and 90°. The results are given in Table 6. The dynamic moduli decrease with decreasing fiber angle, whereas the damping factors increase. The effects of fiber orientation on the dynamic properties are determined by ultrasonic techniques (111–114).

Table 6. Dynamic Properties of Fiber-Reinforced Composites at Different Fiber Orientations

Beam material	Fiber orientation, deg	Dynamic modulus E', GPa	Damping ratio, 10^{-2}
aluminum		74.5	0.08
epoxy		2.9	1.17
composite	0	35.5	0.16
	22.5	19.1	0.65
	45	11.6	0.92
	90	10.2	0.85

A number of equations have been proposed for estimating static or dynamic moduli (115–117). The longitudinal modulus E_L in tension on the uniaxially oriented system is given by

$$E_L = E_1\phi_1 + E_2\phi_2 \tag{30}$$

where E_1 and E_2 are the moduli of matrix and fibers, respectively, and the corresponding volume fractions of the two phases are ϕ_1 and ϕ_2.

The transverse modulus E_T has been estimated (118,119):

$$E_T E_1 = \frac{1 + AB\phi_2}{1 - B\psi\phi_2} \tag{31}$$

where

$$A = 0.5 \tag{32}$$

$$B = \frac{E_2/E_1 - 1}{E_2/E_1 + A}$$

and

$$\psi = 1 + \left[\frac{1 - \phi_m}{\phi_m^2} \right] \phi_2 \tag{33}$$

where ϕ_m is the maximum packing fraction of the fibers ($\phi_m = 0.785 \approx 0.907$).

The damping factor (tan δ) of the composite can be estimated (1). Structural damping is expressed (11) as a structural loss coefficient η as follows:

$$\eta = \frac{D}{2\pi W} \tag{34}$$

where D is the energy dissipated in the composite system per cycle of vibration and W is the vibration energy of the system.

The different composite structures are designed to improve stiffness, strength, dimensional stability, impact strength, mechanical damping, and similar properties. Tires (qv), for example, are fiber-reinforced cord-rubber composites of three basic designs: bias-ply, radial-ply, and bias-belted. These structural designs affect the dynamic behavior of tire composites (120). The dynamic moduli of rubbers and tire cords (nylon, polyester, rayon, and steel) influence the stiffness of a pneumatic tire (120,122), whereas the loss modulus and tan δ of these components are closely related to the amount of heat generated (123–125).

Methods of Determination

An accurate measurement of the response of a polymeric material to vibrational forces requires many instrumentational considerations. A large number of techniques has been developed by using different vibrational principles at various frequencies. The methods are free vibration, resonance-forced vibration, off-resonance forced vibration, and acoustic and wave-pulse propagation techniques. The frequency for each method has a limited range; for instance, the frequencies of forced vibration are about 10^{-3}–10^2 Hz.

Since polymeric materials exhibit a wide variety of shapes and moduli (polymer melts, rubbers, plastics, and fibers), the instrument is often specialized for one type of material. No instrument is capable of characterizing all types of polymers over a wide range of frequency and temperature. Many instruments are described in Refs. 1–6, 126, 127.

In recent years, a wide variety of commercial dynamic testing instruments has become available for operation over a range of testing frequencies. Representative instruments include torsion pendulum, torsional-braid analyzer (TBA), Rheometrics Dynamic Spectrometer, Dynastat, the PL-dynamic mechanical thermal analyzer, B&K complex modulus apparatus, DuPont 981 Dynamic Mechanical Analyzer (DMA), and the Rheovibron and Autovibron. In 1982, the Perkin-Elmer Company introduced a new DMA module for the TMS-2 thermomechanical analyzer.

The torsion pendulum is one of the first and still frequently used methods to determine dynamic mechanical response. A specimen (cylindrical rod, rectangular block or film) is rigidly clamped at one end, while the other end is attached to an inertia disk which is free to oscillate. After the disk is set in oscillation, the polymer sample causes the amplitude of the oscillation to decay. The tangent of phase lag is determined from the ratios of amplitudes of two consecutive peaks. Frequencies of disk oscillation are typically between 0.01 and 50 Hz. ASTM D2236-64T describes the procedure.

Torsional-braid analysis (TBA) is an extension of the torsion-pendulum method (128). This is the only torsion pendulum method for measuring properties of solids or melts on samples less than 100 mg. The sample specimen is prepared by impregnating a multifilament glass-braid substrate with a solution of the material. Changes in the mechanical properties of the composite are attributed to changes in the polymer. Specimen preparation is generally simplified by deposition from a melt or solution onto the substrate. By using a support, a material can be investigated through various regions of the mechanical spectrum, including those in which the material may not be capable of supporting even its own weight. This provides an opportunity for studying resin-forming reactions of thermoplastic polymers, both of which often start in the liquid state and proceed through gel, rubbery, and glassy states (129,130).

The mechanical spectrometer is a rotational rheometer for measurement of the basic rheological properties, such as shear viscosity, normal stresses, dynamic moduli, and melt elasticity in fluids and solids. The Weissenberg Rheogoniometer and Rheometrics Dynamic Spectrometer are examples of the commercial instruments. During the past several years these instruments have been equipped with electronic circuits, improved transducers, signal-receiving systems, and computerized controlling systems.

A newer commercial mechanical spectroscope is the DuPont 981 Dynamic Mechanical Analyzer (DMA). This instrument measures the resonant frequency and energy dissipation without tension on a wide range of samples as a function of temperature (or time). The resonant frequency is related to the modulus by the sample geometry. Energy dissipation relates to impact resistance, brittleness, and noise abatement (131–132). The analyzer contains the mechanical oscillator, driver, and digital display.

The standard Rheovibron Viscoelastometer DDV-11(6) was designed to read tan δ directly, greatly simplifying and speeding the characterization of a material. A number of improvements in this instrument have been proposed by users, eg, a mechanical model and modified drive expression for the moduli to take into account system compliance, sample yielding within the tensile grips, and system inertia (133,134). These effects require a correction to the raw data for glassy polymer samples in the tensile mode. In addition to the standard tensile grips, a new sample holder has been introduced which adapts the Rheovibron to the flexural and shear mode for investigation of the rheological properties of styrene–ethylene oxide block copolymers (135) and polyurethane elastomer (1). These modifications of grips have been extended to the compression (136) mode and bending mode for measuring anisotropic dynamic mechanical properties. Recently, the Rheovibron has been modified to measure the dynamic viscoelastic properties of polymer melts and liquids (137).

In addition, the model DDV-III-C and the automation system have been introduced, with larger stress capability (5 kg) and a hydraulic driving system. The Autovibron provides essentially a continuous monitoring of the changes in storage modulus, loss modulus, and tan δ as a function of temperature. Improved accuracy and sensitivity are provided by a combination of multiprogrammer, programmable calculator, and lock-in analyzer. The Autovibron has expanded capability for measuring the dynamic mechanical properties of materials as well as being a thermal analyzer.

Rheooptical studies of semicrystalline polymers (138,139) have utilized electromagnetic radiation in studying the deformation and flow of polymers. In this study, optical as well as mechanical responses of the material are investigated simultaneously against mechanical excitation. Multichannel narrow sector technique for dynamic x-ray diffraction measurements was developed. This technique can be applied in principle to oscillatory measurement of every rheooptical response, not only to the dynamic wide-angle x-ray diffraction but also to dynamic birefringence, dynamic infrared absorption, dynamic wide- and small-angle light scattering, and dynamic small-angle x-ray scattering.

BIBLIOGRAPHY

1. T. Murayama, *Dynamic Mechanical Analysis of Polymeric Materials,* Elsevier Scientific Publishing Co., Amsterdam, 1978; 2nd ed., 1982.
2. B. E. Read and G. D. Dean, *The Determination of Dynamic Properties of Polymers and Composites,* John Wiley & Sons, Inc., New York, 1978.
3. J. D. Ferry, *Viscoelastic Properties of Polymers,* 3rd ed., John Wiley & Sons, Inc., New York, 1980.
4. L. E. Nielsen, *Mechanical Properties of Polymers and Composites,* Marcel Dekker, Inc., New York, 1974.
5. I. M. Ward, *Mechanical Properties of Solid Polymers,* John Wiley & Sons, Inc., New York, 1971.
6. M. Takayanagi, *Viscoelastic Properties of Crystalline Polymers,* Vol. 23, No. 1, Memorandum of the Faculty of Engineering, Kyushu University, 1963.
7. N. G. McCrum, B. E. Read, and G. Williams, *Anelastic and Dielectric Effects in Polymer Solids,* John Wiley & Sons, Inc., London, 1967.
8. R. F. Boyer in E. Bear, and S. V. Radcliffe, eds., *Polymeric Materials: Relationships Between Structure and Mechanical Behavior,* American Society for Metals, Metals Park Ohio, 1974, Chapt. 6, pp. 227–368.
9. R. F. Boyer, *Polymer* **17,** 996 (1976).
10. J. E. Ruzicka, *Structural Damping,* American Society of Mechanical Engineers, New York, 1959.
11. E. M. Kerwin, Jr., *Internal Friction, Damping and Cyclic Plasticity, Special Technical Publication 378,* American Society for Testing Materials, Philadelphia, Pa., 1964.
12. H. Kolsky, *Stress Waves in Solids,* Clarendon Press, Oxford, 1953.
13. H. Leaderman, *Physics of High Polymers,* Tokyo Institute of Technology, Tokyo, 1957.
14. S. K. Clark, ed., *Mechanics of Pneumatic Tires, Nat. Bur. Stand. Monogr.,* 122, 1971.
15. K. Deutsch, E. A. W. Hoff, and W. Reddish, *J. Polym. Sci.* **13,** 365 (1954).
16. J. G. Powles, B. I. Hunt, and D. J. H. Sandiford, *Polymer* **5,** 505 (1964).
17. K. M. Sinnot, *J. Polym. Sci.* **42,** 3 (1960).
18. J. Heijboer, *Physics of Non-Crystalline Solids,* North-Holland, Amsterdam, 1965, p. 231.
19. P. Hedvig, *Dielectric Spectroscopy of Polymers,* John Wiley & Sons, Inc., New York, 1977.
20. A. H. Willbourn, *Trans. Faraday Soc.* **54,** 717 (1958).
21. T. F. Shatzki, *J. Polym. Sci.* **57,** 496 (1962).
22. R. F. Boyer, *Rubber Rev.* **34,** 1303 (1963).
23. E. Helfand, *Science* **226,** 4675 (1984).

24. E. Helfand, *Rubber Chem. Technol.* **36,** 1303 (1963).
25. E. Helfand, *J. Polym. Sci. Part C* **14,** 267 (1966).
26. L. Utracki, *J. Macromol. Sci. Phys.* **B10**(3), 477 (1974).
27. K. Ueberreiter and J. Naghazadeh, *Kolloid-Z.Z.-Polym.* **250,** 927 (1972).
28. E. A. Sidorovitch, A. I. Marei, and N. S. Gashto'd, *Rubber Chem. Technol.* **44,** 166 (1971).
29. J. P. Bell and T. Murayama, *J. Polym. Sci. Part A-2* **7,** 1059 (1969).
30. L. E. Nielsen, *J. Appl. Polym. Sci.* **2,** 351 (1959).
31. G. Farrow, J. McIntosh, and I. W. Ward, *Makromol. Chem.* **38,** 147 (1960).
32. D. E. Kline and J. A. Sauer, *Polymer* **2,** 401 (1961).
33. R. Meridith and H. Bay-Sung, *J. Polym. Sci.* **61,** 271 (1962).
34. P. R. Pinnock and I. M. Ward, *Proc. Phys. Soc.* **81,** 260 (1963).
35. P. R. Pinnock and I. M. Ward, *Polymer* **7,** 255 (1966).
36. A. B. Thompson and D. W. Woods, *Trans. Faraday Soc.* **52,** 1383 (1956).
37. T. Kawaguchi, *J. Polym. Sci.* **32,** 417 (1958).
38. M. Takayanagi, M. Yoshino, and S. Minami, *J. Polym. Sci.* **61,** 171 (1962).
39. K. H. Illers and H. Breuer, *J. Colloid Sci.* **18,** I (1963).
40. R. L. Miller in R. A. V. Raff and K. W. Doak, eds., *Crystalline Olefin Polymers,* Wiley-Interscience, New York, 1965, Pt. 1, p. 577.
41. F. P. Price, *J. Chem. Phys.* **19,** 973 (1951).
42. B. Wunderlich, *Macromolecular Physics 1980,* Vol. 1, Academic Press, New York, 1980.
43. M. B. Rhodes and R. S. Stein, *J. Appl. Phys.* **32,** 2344 (1961).
44. R. S. Stein, *Polym. Eng. Sci.* **9,** 320 (1966).
45. R. S. Stein and F. H. Norris, *J. Polym. Sci.* **21,** 381 (1956).
46. R. S. Stein and M. B. Rhodes, *J. Polym. Sci. Part A* **27,** 1539 (1969).
47. R. S. Stein, *Polym. Prepr. Am. Chem. Soc. Div. Polym. Chem.* **24**(1), 180 (1983).
48. G. Porod., *Adv. Polym. Sci.* **2**(3), 363 (1961).
49. T. Jeffrey, Koberstein, and R. S. Stein, *Polymer* **25,** 171 (1984).
50. R. J. Fredericks, A. J. Melveger, and L. J. Dolegiewitz, *J. Polym. Sci. Polym. Phys. Ed.* **22,** 57 (1984).
51. H. Kawai, *Polym. Prepr. Am. Chem. Soc. Div. Polym. Chem.* **24**(1), 181 (1983).
52. P. H. Geil, *Polymer Single Crystals,* Wiley-Interscience, New York, 1963.
53. A. Keller, *Philos. Mag.* **2,** 1171 (1957).
54. J. H. Dumbleton and T. Murayama, *Kolloid-Z.Z.-Polym.* **220,** 41 (1967).
55. J. H. Dumbleton and B. B. Bowles, *J. Polym. Sci.* **4,** 951 (1966).
56. H. P. Klug and L. E. Alexander, *X-ray Diffraction Procedures,* John Wiley & Sons, Inc., New York, 1954.
57. H. G. Kiliam, H. Halboth, and E. Jenckel, *Kolloid.-Z.Z.-Polym.* **172,** 166 (1960).
58. R. W. Gray and N. G. McCrum, *J. Polym. Sci. Part A-2* **7,** 1329 (1969).
59. S. Newman and W. P. Cox, *J. Polym. Sci.* **46,** 29 (1960).
60. J. H. Dumbleton, T. Murayama, and J. P. Bell, *Kolloid-Z.Z.-Polym.* **228,** 54 (1968).
61. R. J. Samuels in O. J. Sweeting, ed., *Science and Technology of Polymer Films,* Wiley-Interscience, New York, 1968, Chapt. 7.
62. J. H. Dumbleton, *J. Polym. Sci. Part A-7* **7,** 667 (1969).
63. D. Tsvankim, *Ya. Polym. Sci. USSR* **6,** 2304 (1964).
64. W. P. Cox, R. A. Isaksen, and E. H. Merz, *J. Polym. Sci.* **44,** 149 (1960).
65. G. Pezzin, G. Ajroldi, and C. Garbuglio, *Rheol. Acta* **8,** 304 (1969).
66. T. Q. Nguyen and H. H. Kausch, *J. Appl. Polym. Sci.* **29,** 455 (1984).
67. R. Kimmich, *Polymer* **25,** 187 (1984).
68. T. Murayama and B. Silverman, *J. Polym. Sci. Polym. Phys. Ed.* **11,** 1873 (1973).
69. U.S. Pat. 3,562,206 (Feb. 9, 1971), B. Silverman and L. E. Stewart (to Monsanto Co.).
70. U.S. Pat. 3,548,584 (Dec. 22, 1970), B. Silverman and L. E. Stewart (to Monsanto Co.).
71. A. E. Woodward, J. M. Crissman, and J. A. Sauer, *J. Polym. Sci.* **44,** 23 (1960).
72. J. P. Bell and T. Murayama, *J. Polym. Sci. Part A-2* **7,** 1059 (1969).
73. J. H. Dumbleton and T. Murayama, *Kolloid-Z.Z.-Polym.* **238,** 410 (1970).
74. A. Miyagi and Wunderlich, *Bull. Am. Phys. Soc. Ser.* **11,** 16(3), 410 (1970).
75. T. Murayama and J. P. Bell, *J. Polym. Sci. Part A-2* **8,** 437 (1970).

76. M. F. Drumm, C. W. H. Dodge, and L. E. Nielsen, *Ind. Eng. Chem.* **48,** 76 (1956).
77. J. P. Bell, *J. Polym. Sci. Part A-2* **8,** 417 (1970).
78. P. J. Flory, *Principles of Polymer Chemistry,* Cornell University Press, Ithaca, N.Y., 1953, Chapt. II.
79. A. V. Tobolsky, D. W. Carison, and N. Indictor, *J. Polym. Sci.* **54,** 175 (1961).
80. K. Ueberreiter and G. Kanig, *J. Chem. Phys.* **18,** 399 (1950).
81. T. G. Fox and S. Loshack, *J. Polym. Sci.* **15,** 371, 391 (1955).
82. G. M. Martin and L. Mandelkern, *J. Res. Nat. Bur. Stand.* **62,** 141 (1959).
83. H. D. Heinze, K. Schmieder, G. Schnell, and K. A. Wolf, *Kautschuk Gummi* **7,** 208 (1961); *Rubber Chem. Technol.* **35,** 776 (1962).
84. E. A. DiMarzio, *J. Res. Nat. Bur. Stand.* **68A,** 611 (1964).
85. N. E. Davenport, I. W. Hubbard, and M. R. Putti, *Br. Plast.,* 549 (1959).
86. T. Miyamoto, K. Kodama, and K. Shibayama, *J. Polym. Sci. Part A-2* **9,** 2131 (1971).
87. R. Buchdahl and L. E. Nielsen, *J. Appl. Phys.* **21,** 482 (1950).
88. L. E. Nielsen, *Appl. Polym. Symp.* **12,** 249 (1969).
89. M. Takaynagi, S. Uemura, and S. Minami, *J. Polym. Sci. Part C* **5,** 113 (1964).
90. L. E. Neilsen, *J. Am. Chem. Soc.* **75,** 1435 (1953).
91. D. I. Livingston and J. E. Brown, Jr., in S. Onogi, ed., *Proceedings 5th International Congress on Rheology,* Vol. 4, University of Tokyo Press, Tokyo, 1970, p. 25.
92. A. V. Tobolsky, *Properties and Structures of Polymers,* John Wiley & Sons, Inc., New York, 1960, p. 79.
93. K. Schmeider and K. Wolf, *Kolloid-Z.Z.-Polym.* **127,** 65 (1952).
94. D. R. Paul and S. Newman, *Polymer Blends,* Vol. 1, Academic Press, New York, 1978.
95. L. Holliday, ed., *Composite Materials,* Elsevier, Amsterdam, 1966.
96. S. W. Tsai, J. C. Halpin, and N. J. Pagano, *Composite Materials Workshop, Technomic,* 1968.
97. L. E. Neilsen, *J. Macromol. Sci.* **C3,** 69 (1969).
98. F. R. Schwarzl, H. W. Bree, C. J. Nederveen, G. A. Schwippert, L. C. E. Struik, and C. W. Van der Wal, *Rheol. Acta* **5,** 270 (1966).
99. F. R. Schwarzl, H. W. Bree, G. A. Schwippert, L. C. E. Struik, and C. W. Van der Wal in S. Onogi, ed., *Proceedings 5th International Congress on Rheology,* Vol. 3, University of Tokyo Press, Tokyo, 1970, p. 3.
100. L. E. Neilsen, R. A. Wall, and P. G. Richmond, *SPE J.* **11,** 22 (1955).
101. R. F. Landel and T. L. Smith, *ARS J.* **31,** (1961).
102. G. Kraus and J. T. Gruver, *J. Polym. Sci. Part A-2* **8,** 571 (1970).
103. S. A. Schreiner, P. I. Zubov, T. A. Volkova, and I. I. Vokulonskaya, *Colloid J. USSR* (Engl. Transl.), **26**(5) 541 (1965).
104. T. Miyamoto, K. Kodama, and K. Shibayama, *J. Polym. Sci. Part A-2* **8,** 2095 (1970).
105. R. Buchdahl and L. E. Nielsen, *J. Appl. Phys.* **21,** 482 (1950).
106. E. Perry, *J. Appl. Polym. Sci.* **8,** 2605 (1964).
107. L. J. Broutman and R. H. Krock, *Modern Composite Materials,* Addison-Wesley, Reading, MA, 1967.
108. P. Morgan, *Glass Reinforced Plastics,* Iliffe, London, 1961.
109. S. S. Oleesky and J. G. Mohr, *Handbook of Reinforced Plastics,* Reinhold, New York, 1964.
110. D. J. Duffin, *Laminated Plastics,* Reinhold, New York, 1966.
111. E. Behrens, *Textile Res. J.* **38,** 1075 (1968).
112. J. M. Whitney, *J. Compos. Mater.* **1,** 188 (1967).
113. A. W. Schultz, *Mater. Res. Stand.* **7,** 341 (1967).
114. J. M. Whitney and M. B. Riley, *AIAA J.* **4,** 1537 (1966).
115. S. W. Tsai, *Structural Behavior of Composite Materials,* NASA Rept. CR-71, 1964.
116. J. J. Hermans, *Proc. Koninkl. Nederl. Adad. Wet.* **70B,** 1 (1967).
117. J. C. Halpin, *J. Compos. Mater.* **3,** 732 (1969).
118. L. E. Nielsen, *J. Appl. Phys.* **41,** 4626 (1970).
119. T. B. Lewis and L. E. Nielsen, *J. Appl. Polym. Sci.* **14,** (1970).
120. S. D. Gehman, *Mechanics of Pneumatic Tires, Nat. Bur. Stand. Monogr.,* 1981.
121. P. Kainradl and G. Kaufmann, *Rubber Chem. Technol.* **49,** 823 (1976).
122. P. R. Willett, *J. Appl. Polym. Sci.* **19,** 2005 (1975).

123. P. R. Willett, *Rubber Chem. Technol.* **46**, 425 (1973).
124. G. E. R. Lamb, H. D. Weigmann, and B. C. Goswami, *Rubber Chem. Technol.* **46**, 425 (1973).
125. G. E. R. Lamb, H. D. Weigmann, and B. C. Goswami, *Rubber Chem. Technol.* **49**, 1145 (1976).
126. *Polym. Prepr. Am. Chem. Soc. Div. Polym. Chem.* **22**(1), 251 (1981).
127. T. Murayama, *Rheology*, Vol. 3, Plenum Publishing Corp., New York, 1980, pp. 397–408.
128. J. K. Gilham, *Techniques and Methods of Polymer Evaluation*, Vol. 2, Marcel Dekker, Inc., New York, 1970, p. 225.
129. L. T. Manzione, U. C. Paek, C. F. Tu, and J. K. Gillham, *J. Appl. Polym. Sci.* **27**, 1301 (1982).
130. R. F. Boyer, *Polymer* **17**, 996 (1976).
131. J. Karger-Kocsis and V. N. Kuleznev, *Polymer* **23**, 699 (1982).
132. R. J. Keeling, *Preprints Society Plastic Engineers, Technical Paper XXVII*, 1982, pp. 176–179.
133. D. J. Massa, *J. Appl. Phys.* **44**, 2595 (1973).
134. A. R. Wegewood and J. C. Seferis, *Polymer* **22**, 966 (1981).
135. P. F. Erhardt, J. J. O'Malley, and R. G. Crystal in S. L. Aggarwal, ed., *Block Copolymers*, Plenum Publishing Corp., New York, 1970, pp. 195–213.
136. T. Murayama, *Polym. Eng. Rev.* **2**(1), 83 (1982).
137. T. Murayama, *J. Appl. Polym. Sci.* **27**, 89 (1982).
138. R. S. Stein and T. Oda, *J. Polym. Sci. B* **9**, 543 (1971).
139. S. Suehiro, T. Yamada, H. Inagaki, and H. Kawai, *Polym. J.* **10**(3), 315 (1978).

Takayuki Murayama
Monsanto Company

ENGINEERING PLASTICS

The *Dictionary of Scientific and Technical Terms* defines engineering plastics as those "plastics which lend themselves to use for engineering design, such as gears and structural members" (1). This implies that engineering plastics can substitute for traditional materials of construction, particularly metals. The terms engineering plastics and engineering thermoplastics can be used interchangeably (2). In fact, it is convenient to restrict the term to plastics that serve engineering purposes and are processed by injection and extrusion methods. This excludes the so-called specialty plastics, eg, fluorocarbon polymers and infusible film products such as Kapton and Upilex polyimide film, and thermosets including phenolics, epoxies, urea–formaldehydes, and silicones (qv), some of which have been termed engineering plastics by other authors (3) (see FLUOROCARBON ELASTOMERS; TETRAFLUOROETHYLENE POLYMERS; PHENOLIC RESINS; EPOXY RESINS; AMINO RESINS). Many other polymers can be processed by injection molding and extrusion, but do not exhibit enough toughness, temperature resistance, or dimensional stability for engineering use. For the purposes of this article, the following definition, which is in accord with that of the *Kirk–Othmer Encyclopedia of Chemical Technology*, is offered (4): engineering plastics are thermoplastics that maintain dimensional stability and most mechanical properties above 100°C and below 0°C. This definition encompasses plastics which can be formed into functional parts that can bear loads and withstand abuse in temperature environments commonly experienced by the traditional engineering materials: wood, metals, glass, and ceramics. Generic resins falling within the scope of this definition include acetal resins (qv), polyamides (qv) (nylons), polyimides (qv), polyetherimides, polyesters (qv), polycarbonates (qv), polyethers, polysulfide polymers (qv), polysulfones (qv), blends or alloys of the foregoing resins, and some examples from other resin types (see also POLYETHERETHERKETONES). Thermoplastic elastomers and commodity resins, such as most styrenic resins, acrylics, polyolefins, polyurethanes (qv), poly(vinyl chloride)s, and related chlorinated polyolefins, lose mechanical properties below 100°C (see also ELASTOMERS, THERMOPLASTIC). Exceptions include certain acrylonitrile–butadiene–styrene polymers (ABS) (qv) which maintain engineering properties in the vicinity of 100°C; some unreinforced nylons exhibit creep under load.

The development of engineering thermoplastic resins (3,5) began with the pioneering work of Carothers on nylons at E. I. du Pont de Nemours & Co., Inc. in the 1930s (6) and has continued with the recent introduction by General Electric Co. of its polyetherimide Ultem resin (7). Certain trends are discernible. First, the introduction of new polymers has slowed. Several new varieties of established polymer resins were introduced in the 1960s and 1970s, but polyetherimide resins are the only new generic polymers that have been commercialized in the 1980s. This is not surprising in light of the high risk associated with the introduction of new materials costing millions of dollars and years of development. However, many new blends of engineering materials, often tailored to specific applications, have been tried in the last few years. For these, only 2–4 yr are required for commercialization. Plant and equipment investment costs are minimal since compounding and blending equipment is used for manufacture.

Plastics are either crystalline or amorphous (8). Acetals, most nylons, some polyesters, poly(aromatic sulfide), and polyetherketones are crystalline resins. In general, crystalline polymers do not transmit light, whereas amorphous polymers do. The mode of processing or treatment can alter the morphological form. However, a polymer is considered crystalline if that is the morphology which develops upon cooling the melt to ambient temperature. Amorphous polymers are defined similarly (see also AMORPHOUS POLYMERS; POLYMERS, CRYSTALLINE).

By far the most distinguishing properties between amorphous and crystalline polymers are organic solvent resistance and light transmission (Table 1). In applications where solvent resistance is important, such as in a gear box exposed to oils and greases, a crystalline polymer is the likely choice. On the other hand, for filter bowls, glazing, etc, where clarity is required, an amorphous polymer might be preferred. When treatment of polymers results in morphological change, properties can change as well. For example, poly(ethylene terephthalate) (PET) normally solidifies from a melt as a partially crystalline, opaque solid. As a biaxially oriented film, the polymer is even more crystalline owing to heat-setting under restraint, yet it also is transparent.

Table 1. Morphology Effects on Properties

Property	Crystalline	Amorphous
organic solvent resistance	high	low
light transmission T	none to low	high
lubricity	high	low
dimensional stability	high[a]	moderate
mold shrinkage	high	low

[a] With a high degree of crystallinity and no water absorption.

Lubricity of crystalline polymers is usually higher than that of amorphous polymers. Excellent machinery parts are made from nylon-6,6 resins, eg, gears, cams, wedges, and other components not requiring lubrication. Gears made of polyimide resin, on the other hand, do not exhibit this feature.

Crystalline polymers tend to be more stable dimensionally than amorphous materials. The closer packing and regular physical interactions of the crystalline species would be expected to yield less creep and susceptibility to deformation. Crystalline nylon expands and contracts with changes in ambient moisture (9). Thus, although crystalline polymers are dimensionally stable at a given rh, changes in humidity can result in warpage. This is important for large parts where dimensional change of a few micrometers per centimeter can result in serious deformation. In addition, crystalline polymers exhibit high mold shrinkage compared with amorphous materials. Mold shrinkages of most amorphous resins are ca 0.005 mm/mm, whereas crystalline resins shrink ca 0.015 mm/mm (10).

Polymers with differing morphologies respond differently to fillers (qv) and reinforcement (qv). In crystalline resins, heat-distortion temperature (HDT) increases as the aspect ratio and amount of filler and reinforcement are increased. In fact, glass reinforcement can result in the HDT approaching the melting point. Addition of fillers, however, interrupts the physical interactions, and certain properties, such as impact strength, are reduced. Amorphous polymers are much less affected.

Engineering thermoplastics are priced between the very expensive resins and the high volume, low priced commodities, eg, poly(phenylene oxide)–polystyrene alloys at $2/kg to polyetheretherketones at $60/kg (Fig. 1). Many specialty and commodity resins are addition polymers. Their prices are primarily governed by raw material costs. Thus, the raw material for fluoropolymers is much more expensive than the olefins used for polyethylene (PE), polystyrenes (PS), acrylics, vinyl chloride polymers, etc. Engineering polymers tend to be composed of aromatic monomers, except for acetals and nylons, linked by con-

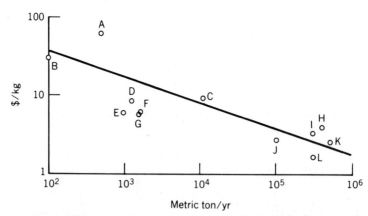

Fig. 1. Engineering resins cost vs annual volume (HDT, °C): A, polyetheretherketone (288); B, polyamideimide (>270); C, polyarylether sulfone (170–> 200); D, polyimide (190); E, amorphous nylons (124); F, poly(phenylene sulfide) (>260); G, polyarylates (170); H, crystalline nylons (90–220); I, polycarbonate (130); J, midrange poly(phenylene oxide) alloy (107–150); K, polyphthalate esters (180–260); L, acetal resins (110–140).

densation (ester, carbonate, amide, imide), substitution (sulfide), or oxidative coupling (ether). Monomer costs and complexity of polymerizations account for higher prices (11). Larger volumes of production take advantage of economics of scale. Representative producers and trademarks of engineering plastics are given in Table 2.

Processing

Injection Molding. Injection molding (qv) is the process most commonly used for engineering thermoplastics (12). Typically, the resin is processed as a pellet ca 2 mm in diameter and 3 mm long, made by chopping an extruded strand and drying in a forced-draft oven (see also PELLETIZING). The pellets are introduced into a hopper which feeds into a screw- or plunger-actuated melt chamber. The melted plastic is injected into a mold in which it solidifies to the finished article. The size and complexity of the part capable of being produced are determined by the flow characteristics of the molten resin at the processing temperature. Typically, the viscosity of a resin should be ca 10,000 Pa·s (1×10^5 P) at a shear rate of 100 s^{-1} for convenient injection molding or extrusion (see Fig. 2).

A variation of conventional injection molding is foam-injection molding (13,14). The molten resin, under pressure in the extruder melt chamber, is injected with an inert gas or an *in situ* heat-activated chemical blowing agent (qv) is decomposed to produce a gas. When the resin is injected into the mold, it leaves the high pressure melt chamber and enters the lower pressure mold cavity. The hot gases expand to form a honeycomb structure in the solidified foam plastic part. Larger and more complex parts are produced by foam molding than by conventional injection molding. Parts are internally reinforced by the honeycomb structure; however, part walls are thick.

Blow Molding. In blow molding (qv), the molten resin is extruded as a hollow, cylindrical parison (15–17). The mold shell is closed over the parison which is blown against the mold, resulting in a hollow part. The resin should exhibit melt strength under the low shear conditions on the parison but good flow during the high shear blowing process. Some engineering resins are specifically supplied for blow molding, eg, slightly branched polycarbonate (PC) resin by Bayer, Mobay, and General Electric.

Extrusion. Sheet, film, and profiled articles are made by extrusion (qv) (18). The resin is melted and forced through a die plate or head. Variations include multilayer and blown-film applications. In multilayer coextrusions, different combinations of plastics are separately but concurrently extruded to form layered sheet or film. In the packaging industry, specialty resins such as high barrier ethylene–vinyl alcohol copolymers are combined with heat- and impact-resistant thermoplastics for food packages. The properties of each resin layer are added, as opposed to the "averaging of property" in blends. Multilayers are also used for blow-molded containers, films, and sheet products (see also FILMS, MULTILAYER).

Blown film is made similarly to blow-molded articles; the molten resin is extruded continuously as a parison, which is blown to form a continuous film cylinder. This cylinder can be rolled up as a two-layer structure or slit to form one large or two smaller single layers.

Table 2. Producers of Engineering Plastics and Their Trademarks

Company	Abbreviation	Trademark	Resin
E. I. du Pont de Nemours & Co., Inc.	DuPont	Kapton	polyimide
		Rynite	polyester
		Selar	nylon–polyolefin blend
		Zytel	nylon
		Delrin	acetal
		Minlon	nylon
Ube Industries, Ltd.	Ube	Upilex	polyimide
Imperial Chemical Industries, Ltd.	ICI	Victrex	PEEK
Amoco Chemicals Corp.	AMOCO	Torlon	polyamideimide
Union Carbide Corp.	Union Carbide	Udel	polysulfone
		Radel	polysulfone
		Ardel	polyacrylate
Minnesota Mining and Manufacturing Co.	3M	Astrel	polysulfone
Rohm & Haas Co.	Rohm & Haas	Kamax	polyimide
Phillips Petroleum Co.	Phillips	Ryton	poly(phenylene sulfide)
Unitika Chemicals, Ltd.	Unitika	Unitika	polyarylate
General Electric Co.	GE	Ultem	polyetherimide
		Valox	polyester
		Lexan	polycarbonate
		Noryl	PPO–PS blend
		PPO	poly(phenylene oxide)
		PPC	polyester carbonate
		Xenoy	PBT–PC blend
Celanese Plastics Co.	Celanese	Durel	polyarylate
		Celcon	acetal
		LCP	polyarylate
Ticona Polymerwerke AG	Ticona	Hostaform	acetal
Polyplastics Co.	Polyplastics	Duracon	acetal copolymer
Bayer AG	Bayer	Durethan	nylon–PU
		Macrolon	polycarbonate
Mobay Chemical Corp.	Mobay	Merlon	polycarbonate
		Bayblend	ABS–PC blend
Teijin, Inc.	Teijin	Panlite	PBT
Mitsubishi Gas and Chemical Co., Ltd.	MGC	Iupilon	polycarbonate
		Iupiace	modified poly(phenylene ether)
Dynamit Nobel AG	Dynamit Nobel	Trogamid T	nylon
Emser Werke AG	Emser	Amidel	nylon
Allied Chemical Corp.	Allied	Capron	nylon
Monsanto Polymer Products Co.	Monsanto	Vydyne	polyamide
Badische Anilin und Soda Fabrik AG	BASF	Ultramid	nylon
Borg-Warner Chemicals, Inc.	Borg-Warner	Prevex	PPO–PS blend
Eastman Kodak Co.	Eastman	Kodar	polyester
Asahi	Asahi	Xyron	modified poly(phenylene ether)
Dartco, Inc.	Dart	Xydar	polyarylate
Carborundum Co.	Carborundum	Ekcel	polyarylate

Thermoforming (qv) of extruded sheet is achieved by warming the sheet to its glass-transition point and collapsing it over a mold shape (19). The collapse can be effected by vacuum suction or applied pressures on the sheet above the mold. This technique is used to make trays and profiled advertising signs. Both single and multilayer sheets can be used.

Extruded engineering thermoplastic stock can be treated like other building material and can be machined, cut, and fastened. However, none of the engineering plastics can be considered a one-for-one substitute for metals or wood. For example, impact resistance must be considered, and glues, paints, etc, must be screened for chemical aggressiveness.

Reaction-injection Molding; Reactive Casting. Reaction-injection molding (qv) (RIM) (20) and reactive casting (21) have been demonstrated on nylon-6, which is polymerized by catalytic ring opening and linear recondensation of ϵ-caprolactam (22):

$$n \underset{(CH_2)_5}{\overset{O \atop \|}{\overbrace{C}}} NH \xrightarrow[\text{catalyst}]{\text{heat}} \left(NH-(CH_2)_5-\overset{O \atop \|}{C} \right)_n$$

Engineering thermoplastics have also been used in preimpregnated constructions. The thermoplastic is thoroughly dispersed as a continuous phase in glass, other resin, carbon fibers (qv), or other reinforcement. Articles can be produced from these constructions using thermoforming techniques. The aerospace industry uses polyetheretherketone (PEEK) in woven carbon-fiber tapes (23). Experimental uses of other composite constructions have been reported (24) (see also COMPOSITES).

Properties

For physical, thermal, electrical, and mechanical properties, ASTM test methods are employed (25). Flammability (qv) limitations are based on UL standards (26). Underwriters Laboratories' flammability ratings given in this article are not intended to reflect the hazards presented by the resins under use conditions. Typical properties are given in Table 3. More details and additional properties are given in Refs. 5 and 27–29.

Physical Properties. Physical properties include specific gravity, water absorption, mold shrinkage, and transmittance, haze, and refractive index. Specific gravity affects performance and has commercial implications. The price of the material divided by the specific gravity gives the yield in cost per unit volume. Comparison of yields gives an evaluation of raw-material costs.

If water is absorbed, predrying is required before injection molding or extrusion. At high processing temperatures, the absorbed water is converted to superheated steam. Failure to remove absorbed water may result in degradation of the material and poor performance. For example, molded PET containing 0.1% water exhibits splay marks on the surface. In PC, 0.1% water causes backbone degradation. Extrusion at 290°C of undried linear PC with an intrinsic viscosity

Table 3. Units of Resin Properties

Property	ASTM method	Unit English	SI
Physical			
density (specific gravity)	D 792		g/cm³
water absorption, 24 h at 23°C	D 570	wt %	wt %
mold shrinkage	D 955	mil/in.	mm/mm
transmittance	D 1003	%	%
haze	D 1003	%	%
refractive index		dimensionless	
yellowness index (YI)	D 1925	dimensionless	
Thermal			
deflection temperature	D 648		
at 0.46 MN/mm² [a]		°F	°C
at 1.82 MN/mm² [a]		°F	°C
specific heat		Btu/(lb·°F)	J/(kg·K)
thermal conductivity		Btu·in./(h·ft²·°F)	W/(m·K)
coefficient of thermal expansion	D 696	per °F	per °C
flammability ratings[b], UL Standard 94			
1.6-mm thickness		various codes	
3.2-mm thickness		various codes	
Electrical			
dielectric strength, short time,	D 149	V/mil	kV/mm
3.2-mm thickness			
dielectric constant	D 150		
60 Hz		dimensionless	
10⁶ Hz		dimensionless	
dissipation factor	D 150		
60 Hz		dimensionless	
10⁶ Hz		dimensionless	
volume resistivity at 23°C, dry	D 150	Ω·cm	Ω·cm
Mechanical			
tensile strength	D 638		
yield		psi	N/mm² [a]
ultimate		psi	N/mm² [a]
elongation	D 638	%	%
flexural strength	D 790	psi	N/mm² [a]
flexural modulus	D 790	psi	N/mm² [a]
shear strength	D 732		
yield		psi	N/mm² [a]
ultimate		psi	N/mm² [a]
Izod impact strength, 3.2-mm thickness	D 256		
notched		ftlbf/in.	J/m
unnotched		ftlbf/in.	J/m
tensile impact strength S	D 1822	ftlbf/in.²	kJ/m²
falling-dart impact strength,		ftlbf	J
3.2-mm thickness			
Rockwell hardness	D 785		
M		dimensionless	
R		dimensionless	
Rheological			
viscosity			
dynamic		poise	Pa·s
intrinsic, [η]			dL/g
kinematic			mm²/s
spiral flow, 3.2-mm thickness		in.	cm
melt index	D 1238		g

[a] N/mm² = MPa.

[b] These ratings are not intended to reflect the hazards presented by any material under actual use conditions.

$[\eta]$ of 0.69 dL/g reduces $[\eta]$ to 0.50 dL/g. Absorbed water can thus greatly affect final properties. In nylons, the amide linkages accept water molecules through hydrogen-bonding interactions. The absorbed water content varies with humidity, and marked changes in properties are observed. For example, the volume electrical resistivity of nylon-6 is reduced by a factor of 10^7 between anhydrous and 100% rh conditions (30); mechanical property changes occur also. When dried at 23°C, the notched Izod impact strength is 40 J/m (0.75 ftlbf/in.). With 3% water absorption, the impact increases to 150 J/m (2.8 ftlbf/in.) (31). Water absorption after molding can result in warpage. Thus, plastics that tend to absorb water are not suitable for making large parts which must maintain dimensional stability. Dimensional changes can be prevented by compounding or blending with mineral fillers or reinforcing fibers.

Mold shrinkage of unreinforced crystalline plastics is high. This is critical in part and mold design, especially for large parts.

Almost all amorphous engineering thermoplastics, except PC and some polyester carbonates, are colored. Even polycarbonates have yellowness indexes (YI) (32) of 0.1 to 5.0. Colorless material is produced from these resins by compounding with complementary blue dyes which reduce transmission. Haze in amorphous resins is an indication of particulates. Haze reduces optical clarity and transmission.

Thermal Properties. Thermal properties (qv) include heat-deflection temperature (HDT), specific heat, thermal conductivity, coefficient of thermal expansion, and flammability ratings. Heat-deflection temperature is a measure of the minimum temperature that results in a specified deformation of a plastic beam under loads of 1.82 or 0.46 N/mm^2 (264 or 67 psi, respectively). For an unreinforced plastic, this is typically ca 20°C below the glass-transition temperature T_g, at which the molecular mobility is altered. Sometimes confused with HDT is the UL Thermal Index, which Underwriters Laboratories established as a safe continuous operation temperature for apparatus made of plastics (33). Typically, UL temperature indexes are significantly lower than HDTs. Specific heat and thermal conductivity relate to insulating properties. The coefficient of thermal expansion is an important component of mold shrinkage and must be considered when designing composite structures.

The UL flammability ratings describe the relative combustibility of plastics. Tests include the measurement of flame propagation, time to self-extinguish, melt and drip with and without flame, and oxygen indexes. Some engineering plastics, eg, polyetherimides, are inherently nonflammable. Others can be made nonflammable by compounding with flame retardants such as bromine compounds, antimony oxide, phosphates, and the like. No statements about flammability (qv) can be made without knowledge of ingredients. The UL ratings of each product must be considered.

Electrical and Mechanical Properties. Electrical properties (qv), including dielectric strength, dielectric constant, dissipation factors, and volume resistivity, are self-descriptive in their application; these properties can change with temperature and absorbed water.

Mechanical properties (qv), like electrical properties, are self-explanatory, and are also subject to thermal and water-content changes. For a part to exhibit structural stiffness, flexural moduli should be above 2000 N/mm^2 (290,000 psi). Notched Izod impact values should be determined at different thicknesses. Some plastics exhibit different notch sensitivities. For example, PC, 3.2 mm thick, has a notched Izod impact of 800 J/m (15 ftlbf/in.) which drops to 100 J/m (1.9 ftlbf/in.) at 6.4-mm thickness. On the other hand, a 100(50/50)0 bisphenol-A phthalate-based polyarylate resin maintains a 250-J/m (4.7-ftlbf/in.) notched Izod impact at both thicknesses. Toughness depends on the structure of the part under consideration as well as the plastic employed to make the part.

Chemical Resistance.   Chemical resistance is less commonly defined than the previously discussed properties. Measurement methods include immersion in selected vapors or liquid, and various surface treatments, including painting. Elevated temperature, temperature cycling, wet–dry cycling, application of stress, and posttreatments are often employed. Parts tested are typically ASTM flexural, tensile, or impact specimens, or a commercial part. Resistance to chemical treatment is rated by comparing mechanical and optical properties before and after. Organic chemicals, aqueous acids, bases, salts, and buffers, light of various wavelengths, and combinations of the foregoing are agents affecting chemical resistance.

In many commercial brochures, chemical resistance is indicated as excellent, good, fair, or poor. Although the test method is usually outlined, wide interpretation is possible. Immersion tests are usually described in this manner.

A common semiquantitative chemical immersion test incorporates stress and uses Bergen strain jigs (34). An ASTM flex bar is bent over a shallow arc of metal during the immersion. The arc varies from almost no bend on one end to pronounced curvature at the other. The stress on the part therefore increases along the arc. Crazing (qv) or cracking indicates failure. Results are reported in terms of time to failure and the distance along the arc where failure initiates. This method is used when testing resistance to gasoline, oils, brake fluids, and cleaning solvents.

Hydrolytic stability is tested by salt-spray cycling or autoclave cycling.

In a clear amorphous plastic, effects on light transmission are easily measured. Retention of surface gloss or mechanical properties such as impact can also be used (see also CHEMICALLY RESISTANT POLYMERS).

Susceptibility to photochemical factors is important for glazing and outdoor applications (35). Many engineering plastics contain aromatic molecules and absorb uv and visible light, particularly light with wavelengths between 2500 and 3500 nm. Laboratory tests use artificial light sources; outdoors, the part is exposed to solar radiation. Optical, gloss, and mechanical property changes are used to measure effects (see also WEATHERING).

Rheological Properties.   Rheological properties of engineering thermoplastics are determined in both solid and molten states. Melt-flow behavior is important for the design of plastic parts and molds and affects formation of knit lines, locations of internal stresses, mold fill, melt strength, and the conditions necessary for extrusion and molding. Useful rheological information is gained from viscosity vs shear-rate curves. In Figure 2, for example, polymer A has a higher apparent viscosity than polymer B at low shear, and a steeper reduction in viscosity as injection and blow-molding shear rates (100 s^{-1}) are reached. This suggests that A is less Newtonian than B in its flow properties and exhibits more melt strength at low shear and greater fluidity at high shear. Polymer A would therefore make a better candidate for blow molding than B because the molten parison would sag less after extrusion. On the other hand, B should be extruded and injection molded at lower shear with greater facility.

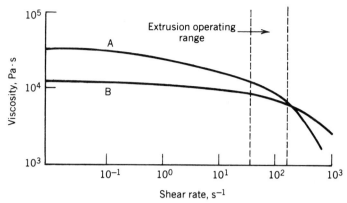

Fig. 2.   Shear sensitivity of polycarbonate resins at 320°C (1 Pa·s = 10 P). Polymers A and B are discussed in the text.

When a complete extrusion rheometric profile is not available, resin processibility can be estimated in a number of ways. Knowledge of the kind of plastic, its filler level, and intrinsic viscosity $[\eta]$ is often sufficient for an experienced plastics technologist. Comparative melt-flow data are obtained with one-point extrusion rheometer measurements taken at a reference temperature at fixed process conditions. Data are usually based on the extrusion time or the amount of plastic that is extruded in a given time. For example, using an extrusion plastometer (Tinius Olson Testing Machine Co.), the relative time to extrude a Lexan PC resin at 300°C vs a Unitika polyarylate resin is 6,000 vs 66,000 units. This suggests that the Unitika polyarylate is stiffer than PC at this temperature. A variation of this method measures the spiral flow of resins by injection molding. The longer the resin travels within the spiral, the higher the melt flow. In the case of the PC vs polyarylate above, the PC would be expected to have a much greater flow length than the polyarylate.

Complete property profiles including rheological data are not available for all grades of engineering thermoplastics. However, the situation is changing as a result of analytical efforts by the resin suppliers. These data are essential for CAD and CAE (computer-aided design and computer-aided engineering), which are being applied to engineering plastic part and mold designs (36,37) (see also RHEOLOGY).

Resins

Polyetheretherketone Resin (PEEK)

The PEEK resin was commercialized by Imperial Chemical Industries, Ltd. (ICI) in the late 1970s. It is now produced in both the UK and the United States under the trade name Victrex PEEK resin (38,39) via the displacement reaction of 4,4′-difluorodiphenyl ketone by the potassium salt of hydroquinone:

PEEK is a difficult resin to manufacture because of batch procedures and insolubility.

The PEEK resin is gray, crystalline, and has outstanding solvent resistance; T_g is ca 185°C, and it melts at 288°C. The unfilled resin has an HDT of 165°C, which can be increased to near its melting point by incorporating glass filler. The resin is thermally stable, and maintains ductility for over 1 wk after being heated to 320°C; it can be kept for years at 200°C. Hydrolytic stability is excellent. The resin is flame retardant, has low smoke emission, and can be processed at 340–400°C. Recycled material can be safely processed. Properties are given in Table 4.

Table 4. Properties of PEEK Resins

| | | | Reinforced Victrex PEEK[b] | |
| | | Victrex | | |
Property	ASTM method	PEEK resin[a]	30% glass	30% carbon
HDT at 1.82 N/mm² [c], °C	D 648	165	282	282
notched Izod at 20°C, 3.2-mm thickness, J/m[d]	D 256	150	110	70
flex modulus at 20°C, N/mm² [c]	D 790	2000	7700	1540
specific gravity	D 792	1.32	1.44	1.32
flammability rating, UL 94		V-O	V-O	V-O

[a] Ref. 38.
[b] Ref. 40.
[c] To convert N/mm² to psi, multiply by 145.
[d] To convert J/m to ftlbf/in., divide by 53.38.

The PEEK resin is marketed as neat or filled pellets for injection molding, as powder for coatings, or as preimpregnated fiber sheet and tapes. Applications include parts that are exposed to high temperature, radiation, or aggressive chemical environments. Aerospace and military uses are prominent. At present, polyamideimide (PAI) resin and poly(arylene sulfide)s (qv) are the main competitors for applications requiring service temperatures of 280°C. At lower temperatures, polyethersulfones, amorphous nylons, and polyetherimides (PEI) can be considered.

Production of PEEK resin is less than 500 metric tons/yr worldwide (41). Growth is not expected to exceed 8–10%/yr. Because of chemical and manufacturing constraints, the price, ca $60/kg (40), is likely to remain high and unaffected by economies of scale (8) (see also POLYETHERETHERKETONE).

Polyamideimide Resin (PAI)

AMOCO introduced the first commercial PAI resin, Torlon, in the early 1970s (42,43). It is produced by the solution condensation of trimellitic trichloride with methylenedianiline (44):

In an alternative route, aromatic diisocyanates are treated with an N-methyl-pyrrolidinone–xylene mixture. This route is used to make solution-borne PAI coating materials. Rhone-Poulenc may produce its coating products by this method.

Torlon resin is an amber-to-gray amorphous polymer with an HDT of 274°C, flexural modulus of 4600 MN/mm² (667,000 psi), and good impact resistance. Although it contains amide linkages, it is dimensionally much more stable under various humidity conditions than the standard crystalline nylons. Like aromatic polyamides and imides, Torlon is flame retardant and has low smoke emission. Creep under load is low. Electrical, chemical, hydrolytic, and radiation stability are excellent. Because of these properties, PAI replaces metal in gears, connectors, and other mechanical parts. It is also used in electrical applications where high temperature and chemical resistance are required. Properties are given in Table 5.

Table 5. Properties of Torlon Polyamideimide (PAI) Resin

Property	ASTM method	Torlon 4203L[a]	Torlon 7130[b,c]
HDT at 1.82 N/mm²[d], °C	D 648	274	278
notched Izod at 20°C, 3.2-mm thickness, J/m[e]	D 256	135	45
flex modulus at 20°C, N/mm²[d]	D 790	5,110	20,000
specific gravity	D 792	1.38	1.42
flammability rating, UL 94		V-O	V-O

[a] Ref. 45.
[b] 30% graphite-reinforced.
[c] Ref. 46.
[d] To convert N/mm² to psi, multiply by 145.
[e] To convert J/m to ftlbf/in., divide by 53.38.

The PAI resin is utilized for small, simple parts. Because of its stiff melt flow, high pressure and temperature are required for injection molding. Mold shrinkage and potential for warpage after molding require special attention; annealing is recommended (43).

The PAI resin is available as injection-moldable pellets. Grades include neat, filled, and glass-reinforced resins. Because of high prices and specific gravities above 1.3, the market potential is limited. World production is ca 90 t/yr, with a growth rate of ca 10%. Resins competing with PAI include PEI, polyethersulfones, fluoropolymer, and thermoset resins.

Polysulfone Resins

Commercially important polysulfones (qv) are aromatic, ie, in the generalized formula for the repeating unit

$$R-\overset{\overset{\displaystyle O}{\|}}{\underset{\underset{\displaystyle O}{\|}}{S}}-R'$$

R and R' both contain aromatic rings. They all possess ether linkages as well, so that use of the designations polysulfone, polyarylsulfone, polyarylethersulfone, and polyethersulfone is somewhat arbitrary.

Union Carbide introduced the first commercial polyarylsulfone resin, Udel, in 1966. Four other products have since been marketed by BASF, Union Carbide, Minnesota Mining and Manufacturing (3M), and ICI (47).

The resins are made by batch processes employing Friedel–Crafts reactions or nucleophilic aromatic substitution. Udel resin and Radel resin are produced by the nucleophilic displacement of chloride on 4,4'-dichlorodiphenylsulfone by the potassium salts of bisphenol A and 4,4'-biphenol, respectively (48):

Victrex resin and possibly Radel A400 resin are made by other displacement reactions (49).

Astrel 360 resin is made by the Friedel–Crafts reaction of biphenyl with 4,4'-dichlorosulfonyldiphenyl ether (50). This product may no longer be commercially available.

Polyarylethersulfones are amber-to-yellow transparent amorphous resins. Because of high HDTs, resistance to thermooxidative conditions, and hydrolytic stability, they are suitable for appliances, electrical and electronic components, aircraft interior parts, and biological and medical devices requiring steam sterilization; flammability and smoke generation are low. Udel resin has excellent microwave properties and is employed in dishware suitable for microwave ovens. These resins, however, are subject to attack by organic solvents. Melt flow is stiff, which limits the size and complexity of injection-molded parts. The most serious problem with these resins is uv light instability; for outdoor use, coating is recommended. Properties are given in Table 6. The resins are sold primarily for injection-molding applications. Products include neat, filled, reinforced, pigmented, and modified pellets.

Table 6. Properties of Polyarylethersulfone Resins

Property	ASTM method	Udel[a,b]	Victrex[a,c]	Astrel 360[d]
HDT at 1.82 N/mm²[e], °C	D 648	174	203	274
notched Izod at 20°C, 3.2-mm thickness, J/m[f]	D 256	64	90	100–200
flex modulus at 20°C, N/mm²[e]	D 790	2689	2600	2720
specific gravity	D 792	1.24	1.37	1.36
flammability rating, UL 94		V-O	V-O	V-O
light transmission T, %		85	0	85
organic solvent resistance		fair	excellent	fair

[a] Neat.
[b] Ref. 51.
[c] Ref. 52.
[d] Ref. 53.
[e] To convert N/mm² to psi, multiply by 145.
[f] To convert J/m to ftlbf/in., divide by 53.38.

Polyarylethersulfones compete with a number of other resins, eg, PEEK, PAI, and PEI resins. Total worldwide yearly sales volumes are 8000–9000 t/yr, with Udel resin accounting for the main share. Growth is expected to be 15%/yr (54,55). Unless uv stability and, hence, weatherability are improved, primary applications will continue to be indoor parts.

Polyimide Resins

Polyimide resins are used when heat and environmental resistance is required. Kapton resin (56) and, more recently, Upilex resin (57) are sold as films, partially polymerized resin coating solutions, and sinterable powders.

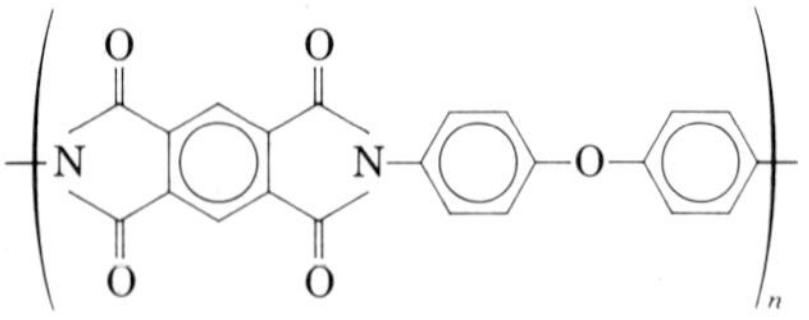

Kapton resin

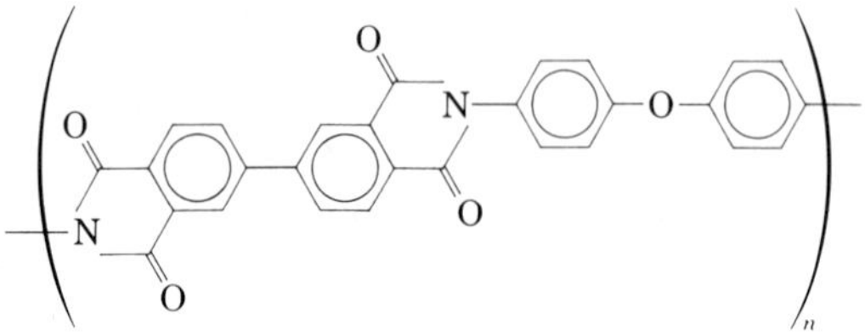

Upilex R resin

These resins have extensive applications in the high performance insulation markets (traction motor, wire and cable) and flexible printed circuits, but are not melt reprocessible.

Upjohn's Polyimide 2080 resin (58), Rhone-Poulenc's Kerimide resin (59), Mitsubishi's BT resins (60), and Toshiba's Imidaloy resin (61) are based on bis-maleimide (3) technology. Maleic anhydride reacts with a diamine to produce a diimide oligomer. Further polymerization by additional diamine and heat produces the molding resins (62):

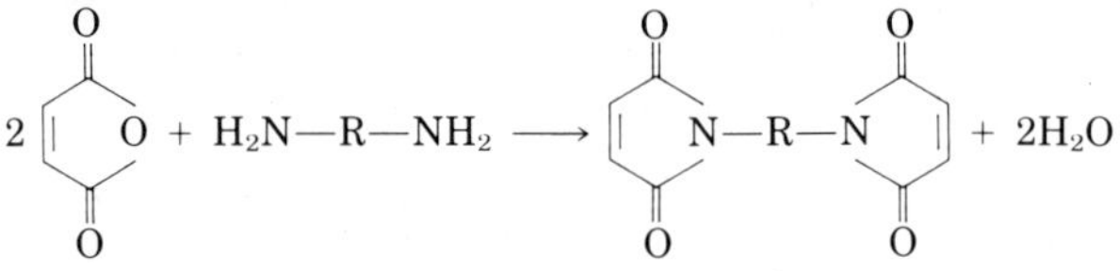

(3)

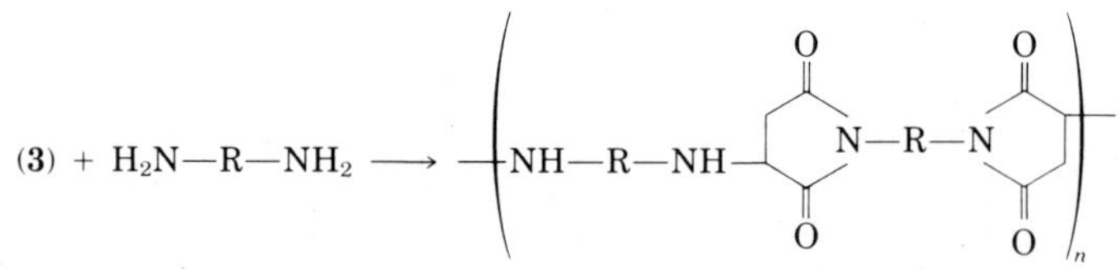

Although these resins have been on the market for a number of years, volumes and applications remain low in spite of favorable properties including high temperature resistance, excellent wear and friction properties, good electrical properties, chemical inertness, and inherent nonflammability. Processibility, price, and thermal performance may be the limiting factors.

In 1979, Rohm & Haas introduced Kamax resin, which was thought to be an N-methylamine imidization product from poly(methyl methacrylate) (63). This crystalline resin had excellent nonflammability, flex modulus, chemical resistance, and a surprisingly high notched Izod impact. However, thermal-service temperatures were not much better than those of PC resin, which limited application. The product has since been withdrawn.

At present, there appear to be no clearly identified engineering applications for thermoplastic polyimide resins. Current sales volumes are less than 400 t/yr; growth prospects are uncertain (see also POLYIMIDES).

Polyetherimide Resins (PEI)

Polyetherimide resins (PEI) are the newest generic engineering plastics (7). They are produced by an unusual nucleophilic substitution process:

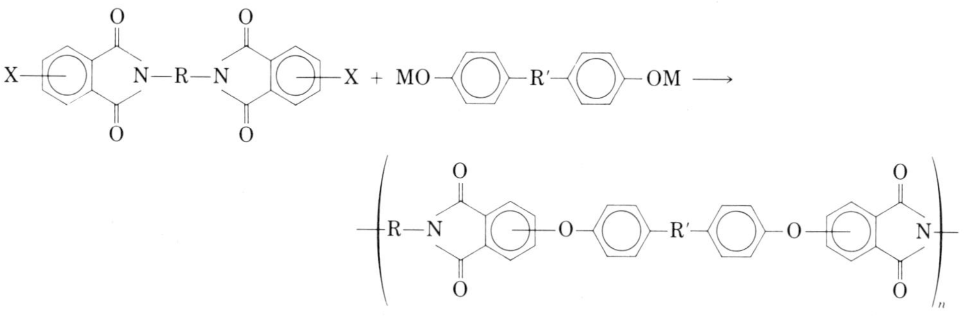

where M = metal

or by a conventional condensation of diamines and dianhydrides:

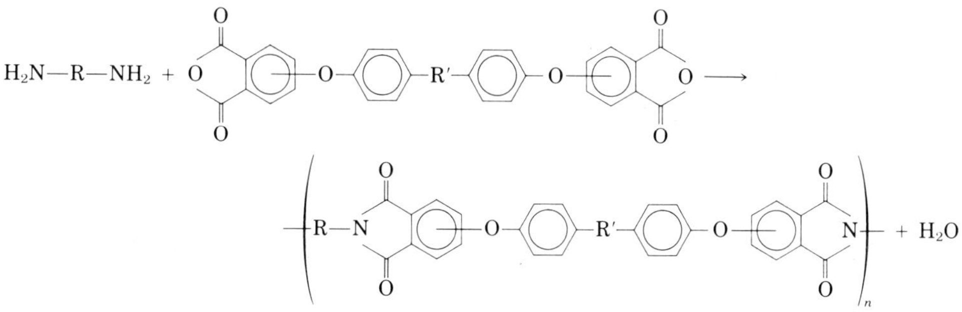

Many different resins can be made by varying R and R′, as in Ultem resins.

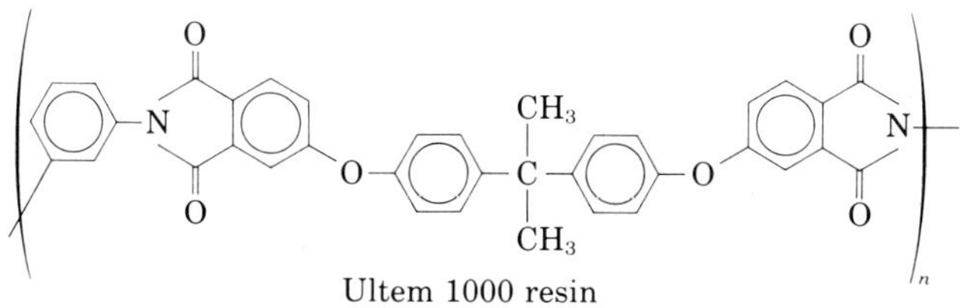

Ultem 1000 resin

Ultem PEI resins are amber and amorphous, with heat-distortion temperatures between polyarylate resins and high temperature crystalline polymers such as PEEK or PAI resins. Ultem resins exhibit high modulus and are stiff yet ductile. Light transmission is low. In spite of the high use temperature, they are processible by injection molding, structural foam molding, or extrusion techniques at moderate pressures between 340 and 425°C. They are inherently flame retardant and generate little smoke; dimensional stabilities are excellent. Large flat parts such as circuit boards or hard disks for computers can be injection-molded to maintain critical dimensions. Creep properties are excellent; for example, Ultem 1000 maintains its apparent flexural modulus for thousands of hours at a loading of 34 N/mm^2 (4930 psi). Unlike PAI resins, annealing (qv) is not required. Chemical resistance is high for an amorphous resin. Ultem resins are unaffected by automotive organic chemicals and aromatic, chlorocarbon, and ester solvents; hydrolytic stability is excellent. Properties are given in Table 7.

Table 7. Properties of Polyetherimide (PEI) Resin[a]

Property	ASTM method	Ultem 6200[b]	Ultem 1000[c]
HDT at 1.82 N/mm^2 [d], °C	D 648	223	200
notched Izod at 20°C, 3.2-mm thickness, J/m[e]	D 256	80	50
flex modulus at 20°C, N/mm^2 [d]	D 790	6550	3240
specific gravity	D 792	1.43	1.7
flammability rating, UL 94		V-O	V-O

[a] Ref. 64.
[b] 20% glass-reinforced.
[c] Neat.
[d] To convert N/mm^2 to psi, multiply by 145.
[e] To convert J/m to ftlbf/in., divide by 53.38.

Ultem PEI resins are used in numerous engineering applications where a combination of heat, chemical, and flame resistance is needed. Typical uses include rigid memory disks, fuse bodies, gears, fans for computer and business equipment, sterilizable surgical appliances, hot combs, brushes, iron skirts, microwave-oven components, aircraft interior components, and high temperature switches, connectors, and circuit boards for industrial and military devices. Recently, Ultem 1000 resin has been used as the external rigid layer in multilayer sheet extrusion which can be converted to food packaging for microwave and convection ovens.

The Ultem PEI resins compete with PAI, polyarylethersulfone, nylon, and polyester resins in certain markets. General Electric Co. is the sole U.S. manufacturer of PEI resins. Growth of high technology markets is anticipated to be good. The Japanese market, for example, is predicted to grow by 15%/yr (65).

Poly(phenylene sulfide) Resins (PPS)

Poly(phenyl sulfide) resin (66) is manufactured from p-dichlorobenzene and sodium sulfide in a dipolar aprotic solvent (67).

$$\text{Cl}-\bigcirc-\text{Cl} + \text{Na}_2\text{S} \longrightarrow \left(\bigcirc-\text{S}\right)_n + \text{NaCl}$$

It was commercially introduced under the trade name Ryton in 1970 by Phillips Petroleum Co., which is the only supplier. The polymer is crystalline with a melting point of 287°C. It is a flame retardant and has excellent hydrolytic and chemical resistance. Like most crystalline polymers, the neat material exhibits high mold shrinkage and a tendency for molded parts to warp, which can be corrected by fillers and reinforcement. Although molded parts maintain their mechanical properties during prolonged heating, under some conditions the resin continues to polymerize. This reduces the thermoplastic character of the resin. Recycling should be carefully considered. Properties are given in Table 8.

Table 8. Properties of Poly(phenylene sulfide) (PPS) Resin[a]

Property	ASTM method	PPS resin[b,c]	Ryton R-4[d,e]
HDT at 1.82 N/mm^2[f], °C	D 698	111	241
notched Izod at 20°C, 3.2-mm thickness, J/m[g]	D 656	11	63
flex modulus at 20°C, N/mm^2[f]	D 790	3,845	11,000
specific gravity	D 792	1.36	1.6
flammability rating, UL 94		V-O	V-O

[a] Annealed; properties are influenced by the degree of crystallinity.
[b] Neat.
[c] Ref. 68.
[d] 40% glass-reinforced.
[e] Ref. 69.
[f] To convert N/mm^2 to psi, multiply by 145.
[g] To convert J/m to ftlbf/in., divide by 53.38.

PPS resins are chiefly used for injection molding. The melt flow of the glass-filled resins is very stiff, and high injection pressures are required. Mold surface wear is heavier than for most other engineering plastics. Molding melt temperatures are near 330°C; for optimum surface gloss and impact strength, mold temperatures of 130°C should be used. The resins are brown to brown–black.

Injection-molding pellets are available in a number of grades including neat, pigmented, mineral-filled, and glass-reinforced. Applications include chemical-resistant parts for pumps and processing equipment such as impellers. The high heat resistance of glass-reinforced grades allows application in large exterior light reflectors and automotive engine components.

PPS resins have to compete with PEI, terephthalate polyester, nylon, phenolic, and polytetrafluoroethylene (PTFE) resins. Currently, over 9000 t is sold annually worldwide (70). Annual growth may reach 10%/yr. Competitive PPS resins are expected to be marketed by Japanese and European firms.

Polyester Resins

The general formula of engineering thermoplastic polyester resins (71) is shown below:

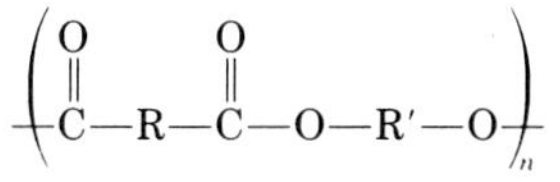

$$\left(\begin{array}{c} O \quad\quad O \\ \parallel \quad\quad \parallel \\ \text{C—R—C—O—R}'\text{—O} \end{array} \right)_n$$

Most polyesters (qv) are based on phthalates. They are referred to as aromatic–aliphatic or aromatic according to the copolymerized diol. Thus, poly(ethylene terephthalate) (PET), poly(butylene terephthalate) (PBT), and related polymers are termed aromatic–aliphatic polyester resins, whereas poly(bisphenol A phthalate)s are called aromatic polyester resins or polyarylates; PET and PBT resins are the largest volume aromatic–aliphatic products. Other aromatic–aliphatic polyesters (72) include Eastman Kodak's Kodar resin, which is a PET resin modified with isophthalate and dimethylolcyclohexane. Polyarylate resins are lower volume specialty resins (see POLYESTERS, AROMATIC).

Aromatic–Aliphatic Polyester Resins. Unlike most other classes of engineering plastics, which are made by only a few manufacturers, aromatic–aliphatic polyester resins are produced and compounded by several dozen firms (see Tables 9 and 10) (73). The aliphatic polyester resin marketplace is charac-

Table 9. Suppliers of Poly(ethylene terephthalate) (PET) Resin

Company	Bottle	Film	Engineering	Country
		Trademark[a,b]		
American Hoechst Corp.		Hostaphan		United States
Bemis Co., Inc.		Esterphane		United States
E. I. du Pont de Nemours &		Cronar		United States
Co., Inc.		Mylar	Rynite	United States
Eastman Kodak Co.	Kodapak	Estar		United States
ICI America		Melinex		United States
Minnesota Mining and	X	Scotchpac		United States
Manufacturing Co.		Scotchpar		United States
Goodyear Tire and	Cleartuf			United States
Rubber Co.				
Rohm & Haas Co.	Carodel			United States
Hoechst AG	X	X		FRG
Akzo Plastics BV	X		Arnite	Netherlands
ICI Holland BV	X	X		Netherlands
ICI	X	X		UK
Japan Unipet Co.	X			Japan
Kendo Synthetic Fibers	X			Japan
Mitsui PET Resin Co.	X			Japan
Nippon Ester Co., Ltd.	X			Japan
Teijin, Ltd.	X		X	Japan
Agfa-Gevaert NV		X		Belgium
Kodak-Pathe SA				France
Rhone-Poulenc Films SA		X	Techster	France
3M Italia SpA		X		Italy
duPont de Nemours SA		X		Luxembourg
Allied Corp.			Petra	United States
Mobay Chemical Corp.			Petlon	United States
ATO CHEM SA			Orgater	France
Bayer AG			Pocan	FRG
Chemische Werke Huels			Vestodur	FRG
CIBA-GEIGY Marienberg			Crastine	FRG
BIP Chemicals, Ltd.			Beetle PET	UK
duPont Japan, Ltd.			X	Japan
Mitsubishi Rayon, Ltd.			X	Japan
Toyobo Co., Ltd.			X	Japan
Unitika, Ltd.			X	Japan

[a] Ref. 71.
[b] X = trademark not available.

Table 10. Suppliers of Poly(butylene terephthalate) (PBT) Resin[a]

Company	Engineering-resin trademark[b]	Country
Celanese Corp.	Celanex	United States
GAF Corp.	Gafite	United States
	Gaftuf	United States
General Electric Co.	Valox	United States
ATO CHEM SA	Orgater	France
Rhone-Poulenc Chemie SA	Techster	France
BASF AG	Ultradur	FRG
Bayer AG	Pocan	FRG
Chemische Werke Huels	Vestodur	FRG
CIBA-GEIGY GmbH	Crastine	FRG
Montepolimari	Pibiter	Italy
Akzo Plastics BV	Arnite	Netherlands
Dianippon Ink & Chemicals, Inc.	X	Japan
Engineering Plastics, Ltd.	Valox	Japan
Mitsubishi Chemical Industries, Ltd.	X	Japan
Mitsubishi Rayon Co., Ltd.	X	Japan
Mitsui Petrochemical Industries, Ltd.	X	Japan
Polyplastics Co., Ltd.	Duranex	Japan
Teijin, Ltd.	X	Japan
Toray Industries, Inc.	X	Japan
Toyobo Co., Ltd.	X	Japan

[a] Ref. 72.
[b] X = trademark not available.

terized by wide product differentiation and competition. Some firms make only a few hundred tons per year and presumably still retain profitability because of the availability of low cost monomer and the simplicity of the processes employed. Low investment and low manufacturing costs are possible even for small volume operations.

Low molecular weight PET and PBT resins are made by melt processes. For higher molecular weight resins, both melt processes or solid-state polymerization are used. Although terephthalic acid can be directly esterified, the most common process involves transesterification of dimethyl terephthalate with ethylene glycol or 1,4-butanediol in the presence of trace amounts of metal ion cataylsts (74,75).

The methanol is collected overhead and the neat resin is extruded from the reactor in a batch or continuous process. The larger volume firms employ continuous operations with outputs per line of 11,000–14,000 t/yr (75).

The engineering applications of PET resins include blow-molded bottles, films, molding, and extrusion. Resins made for the latter two uses and related purposes are called molding resins in this article. The huge volumes of PET resin used for textile filaments and industrial fibers, eg, tire cord, are not included here. The PBT resins are mainly used for molding and related applications.

Bottle-grade PET resin has an $[\eta]$ range of 0.72–0.85 dL/g. Gels and particulates are required for low intrinsic viscosity, and the manufacturing process is different than that for film or molding resins. Instead of continuing to heat and stir the low molecular weight resin to build a higher molecular weight, the resin is extruded as pellets at an $[\eta]$ of 0.6 dL/g. The pellets are heated in a hot nitrogen gas stream to promote further polymerization in the solid state. The temperatures used are lower than those required for melting, and shear effects are eliminated (73).

In 1977, consumption of PET resins in bottle applications was raised when the FDA banned acrylonitrile resins owing to toxicity considerations (recently rescinded) (76) and when the 2-L bottle was accepted for beverage sale worldwide (77). The carbon dioxide barrier properties of PET are sufficient to provide the 6-mo shelf life necessary for carbonated beverages (see also BARRIER POLYMERS).

Most PET bottles are produced by injection blow molding (78) the resin over a steel-core rod. The neck of the bottle is formed with the proper shape to receive closures and resin is provided around the temperature-conditioned rod for the blowing step. The rod with the resin is indexed to the mold, and the resin is blown away from the rod against the mold walls, where it cools to form the transparent bottle. The finished bottle is ejected and the rod is moved again to the injection-molding station. This process is favored for single cylindrical bottles, but cannot be used for more complex shapes such as bottles with handles.

The PET resins used for film and sheet applications are exceptionally clean resins with $[\eta]$ ranging from 0.6 to 1.0 dL/g. Most of the large volume general-purpose packaging film is resin of the lower (textile-fiber range) viscosity range. General-purpose as well as high tensile strength films are made by stretching the film longitudinally or laterally with timed periods of heating and cooling. Films stretched along two axes are termed biaxially oriented. The higher the extent of stretching, the higher the tensile strength and the lower the elongation to break. Stretching increases crystallinity, whereas heating anneals the stretched film and confers dimensional stability between -70 and 150°C. For heat sealing, the film is often coated with poly(vinylidene chloride) (PVDC) (79). Applications of biaxially oriented films include packaging films, heat-shrink film, magnetic-coated tapes, and computer floppy disks (79) (see also ORIENTATION).

Both PET and PBT resins are used for moldable and extrudable engineering resins; intrinsic viscosity is typically in excess of 0.7 dL/g. Firms manufacturing molding resins either produce a certain molding-grade resin or use resins for compounding. This is especially true for PET resin products. The PBT resin was the first to be commercialized for injection molding in 1970, because PET crystallized slowly. The crystallinity of molded PET parts varied from molding to molding and changed over time, which often resulted in cracking. In 1978, DuPont introduced the PET molding compounds in its Rynite line, using nucleating agents which enhanced the crystallization rates (80).

Both PET and PBT molding resins are available as neat resins or compounded with other resins (including each other), pigments, fillers, flame retardants, and reinforcements. Filled and reinforced resins are most commonly used for large parts and those with flat surfaces. Fillers retard mold shrinkage and the warpage characteristic of neat resins. However, fillers increase notched impact sensitivity and some grades contain additional impact modifiers. Another advantage of the filled resins is the increase in heat-deflection temperature from the glass-transition temperature to values approaching the melting point. The properties given in Table 11 are for cream-colored resins with good-to-excellent solvent resistance and good hydrolytic stability, but no clarity.

Table 11. Properties of Aromatic–Aliphatic Polyester Resins

Property	ASTM method	Valox 325[a,b]	Valox DR-51[b,c]	Rynite 530[d,e]
HDT at 1.82 N/mm^2 [f], °C	D 648	130	160	224
notched Izod at 20°C, 3.2-mm thickness, J/m[g]	D 256	50	55	90
flex modulus at 20°C, N/mm^2 [f]	D 790	2300	4530	6050
specific gravity	D 792	1.31	1.41	1.56
flammability rating, UL 94			V-O	HB

[a] Neat PBT resin.
[b] Ref. 81.
[c] PBT resin, 15% glass-reinforced.
[d] PET resin, 30% glass-reinforced.
[e] Ref. 82.
[f] To convert N/mm^2 to psi, multiply by 145.
[g] To convert J/m to ftlbf/in., divide by 53.38.

Both PET and PBT resins often compete for the same applications. The PET resins are considered in cases where higher heat-deflection temperatures and greater rigidity are desired. The PBT resins have the advantages of faster crystallization and reduced moisture absorption, which allows faster molding cycles and reduced drying time. Typical applications include automotive parts such as distributor caps, air-blower deflectors, connectors, and housings, and exterior parts, eg, cowls and fender extensions. The solvent resistance and high heat-deflection temperatures of PBT and PET resins permit replacement of automotive metal parts by these resins. Blends containing PBT resins are used for auto bumpers. Highly ductile, specially formulated polyester resins will soon compete for other large exterior parts, including doors, fenders, and quarter panels. Other markets are electrical and electronic parts (which use flame-retarded UL 94 V-O grades), home appliances, plumbing valves, pump housings, brackets, medical components, and parts for sports equipment.

Processing is similar to other engineering plastic resins. Drying is necessary before extrusion or molding. Special drying precautions are required for PET products to prevent degradation and splay. For injection-molding PET resins, molds must be maintained at ca 10°C higher than molds used for PBT resins (71).

Both PET and PBT resins compete with other plastics. Nylon, unsaturated polyester fiber glass, phenolic, PC, and polyarylate resins are sometimes considered for the same molded part. Bottle-grade PET resin with marginal gas-barrier properties competes with PC resins which can be coextrusion blow-molded with specialty barrier resins. PEI film affords higher heat-deflection temperature and dimensional stability required by small electronic circuit constructions. Polyolefin films are less expensive. Nonetheless, combined PET and PBT polyester resins are sold at a rate of 800,000 t/yr worldwide (Table 12) and are expected to have an average annual volume growth rate of 7–12% (55,73); in Europe, bottle-grade PET is growing at a rate of 23% (73). Growth rates for molding resins could increase if usage for large exterior automotive parts increases (83).

Table 12. Terephthalate Resin Sales, 10^5 t/yr

Area	PET[a]		PET and PBT molding[b]
	Bottle	Film	
Western hemisphere	2.55	2.15	0.55
Europe	0.50	0.94	0.64
Asia	0.27	0.93	
Total	*3.32*	*4.02*	*1.19[c]*

[a] Ref. 73.
[b] Ref. 84.
[c] 45,000 t/yr is PBT (71); estimates are as high as 1.63×10^5 t/yr (55), but it is not clear what is included in this figure.

Aromatic Polyester Resins.   Commercial aromatic polyester resins or poly-arylates are a combination of bisphenol A with isophthalic acid or terephthalic acid (85). The resins are made commercially by solution polymerization

or melt transesterification (86) such as

Blends of iso- and terephthalates give amorphous, clear resins, mostly yellow in color. Heat-deflection temperatures are higher than those of 100% PC resins and depend on the iso- to terephthalate ratio. For example, a resin with a 1:1 ratio has a value of 160°C. These resins are flame retardant; mechanical and electrical properties are similar to those of PC resins. The notched Izod impacts are lower at 250–300 J/m (4.7–5.6 ftlbf/in.), even in thick sections. Long-term uv radiation stabilities are excellent, but yellowness increases during initial exposure owing to photo-Fries rearrangements (87); o-hydroxybenzophenone units are produced along the polymer chains.

The chromophores are yellow and act as uv screening agents. The reactions occur at surfaces and uv penetration is avoided. Mechanical properties are thereby protected (86).

Polyarylates are sensitive to heat. Although mechanical properties are not much affected, colors darken. Properties are given in Table 13. Hydrolytic stability and resistance to organic solvents are fair.

Table 13. Properties of Aromatic Polyester Resins[a]

Property	ASTM method	Ardel D-100	Durel 400
HDT at 1.82 N/mm^2 [b], °C	D 648	174	170
notched Izod at 20°C, 3.2-mm thickness, J/m[c]	D 256	210	275
flex modulus at 20°C, N/mm^2 [b]	D 790	2100	2300
specific gravity	D 792	1.21	1.21
flammability rating, UL 94		V-O	V-O
color		yellow	colorless
light transmission T, %		75	84–88

[a] Ref. 86.
[b] To convert N/mm^2 to psi, multiply by 145.
[c] To convert J/m to ftlbf/in., divide by 53.38.

Polyarylates have been employed for electrical and electronic components, fireman helmets, and applications requiring higher heat-deflection temperatures than PC resins. Polyarylates were first used for lighting, especially for small automotive lenses and sodium light outdoor lamps. However, the yellow color and heat-induced darkening have limited sales.

Although several companies have produced polyarylates, today Unitika in Japan and Celanese are the only ones. Unitika resin is marketed in the United States by Union Carbide under the Ardel trademark. Celanese announced plans to market Durel polyarylate resin, which they acquired from Occidental Petroleum Co. The reason for this decline may be the competition of PC, polyester carbonates, and polysulfone resins. Liquid-crystal polyarylate specialty resins include Dart's Xydar (formerly Carborundum Ekcel) and Celanese's LCP resins (85) (see also LIQUID CRYSTALLINE POLYMERS). Worldwide polyarylate sales are probably less than 1000 t/yr; growth prospects are about the same as the GNP performance (see POLYESTERS, AROMATIC).

Polycarbonate (PC) Resins

Polycarbonates (qv) based on bisphenol A are sold in large quantities. Other bisphenols can be incorporated, but do not give the same favorable combination of properties and cost (88). Small quantities of PC based on tetramethylbisphenol A are used as blending resins (89) and polyester carbonate copolymers are used for applications requiring heat-deflection temperatures above those of standard PC resins (86).

Bisphenol-A Polycarbonate Resins. These resins are manufactured by interfacial polymerization (qv) (90).

The terminal R groups can be aromatic or aliphatic. Typically, they are derivatives of monohydric phenolic compounds including phenol and alkylated phenols, eg, t-butylphenol. In interfacial polymerization (qv), bisphenol A and a monofunctional terminator are dissolved in aqueous caustic. Methylene chloride containing a phase-transfer catalyst is added. The two-phase system is stirred and phosgene is added. The bisphenol-A salt reacts with the phosgene at the interface of the two solutions and the polymer extends into the methylene chloride, and the sodium chloride by-product enters the aqueous phase. Chain length is controlled by the amount of monohydric terminator. The methylene chloride–polymer solution is separated from the aqueous brine-laden by-products. The facile separation of a pure polymer solution is the key to the interfacial process. The methylene chloride solvent is removed, and the polymer is isolated in the form of pellets, powder, or slurries.

Although some 12 firms produce PC resin worldwide, Bayer AG along with its subsidiary Mobay Chemical Co., and General Electric are the principal producers.

PC resins are glassy, amorphous polymers with little color and excellent optical properties. The high molecular weight resins are tough, with notched Izod impact values of 600–800 J/m (11.2–15.0 ftlbf/in.). Heat-deflection temperatures of unfilled grades are near 130°C; glass reinforcement increases these values by 10°C. Flex moduli above 2400 N/mm^2 (348,000 psi) allow PC resins to be formed into large rigid parts. Below 100°C, creep under load is low, as is predictable from the high glass-transition temperatures. PC resin products have good hydrolytic stability but should not be used for applications requiring repeated autoclaving. Like most amorphous resins, they are susceptible to attack by some organic solvents. Although PC resins have good resistance to fruit juices, aliphatic hydrocarbons, aqueous alcohol, and battery acids, they dissolve in chlorocarbon solvents and stress-crack in contact with ketonic solvents such as acetone and methyl ethyl ketone. Properties are given in Table 14.

Table 14. Properties of Polycarbonate (PC) Resins

Property	ASTM method	Lexan 101[a]	Merlon 8320[b,c]	Lexan 4701[a,d]
HDT at 1.82 N/mm^2[e], °C	D 648	132	142	163
notched Izod at 20°C, 3.2-mm thickness, J/m[f]	D 256	600	100	500
flex modulus at 20°C, N/mm^2[e]	D 790	2300	3380	2300
specific gravity	D 792	1.21	1.27	1.2
flammability rating, UL 94		V-2	V-1	HB
light transmission T, %		85	opaque	85

[a] Ref. 91.
[b] 20% glass-reinforced.
[c] Ref. 92.
[d] Polyester carbonate.
[e] To convert N/mm^2 to psi, multiply by 145.
[f] To convert J/m to ftlbf/in., divide by 53.38.

PC resin materials are sold mainly as injection-moldable pellets available in transparent, flame-retarded, mineral-filled, reinforced, and impact-modified grades. PC resins are also used for foam molding. Some grades contain blowing agents (qv) activated by heat; others receive the blowing agents by on-line gas or volatile liquid injection during foam molding. Sheet products are available in forms ranging from uv-stabilized construction grades to multilayer laminates designed for bullet-resistant glazing and lightweight twin-wall extrusions. Both blown and extruded film are produced. Some branched PC resins are made specifically for blow-molding applications by introducing polyfunctional monomers such as trimellityl trichloride during polymerization; branching increases melt strength.

Processing PC resins by extrusion or injection-molding methods requires melt temperatures of 290–320°C and is easy to control. High melt viscosity at low shear rates prevents mold flash and drool. At injection shear rates, apparent viscosities decrease, and easy melt flow allows manufacture of large, complex parts.

PC resins find application in many areas. Sheet products are used for glazing and thermoformed products such as signs, cases, and furniture. PC resin film is suitable for silk-screen printing and finds uses in illuminated consoles and soft-touch controls. PC resins are utilized for packaging as components of multilayer extruded sheet and film and in coextruded blow-molded containers. Injection molding consumes 80% of production. Applications include electrical and electronic parts such as housings, connectors, and fuses, appliance parts, power-tool housings, and parts for the transportation industries ranging from automobile dashboards to headlamps. Sports equipment requiring high impact and other uses might be categorized under metal or glass replacement. PC blow-molding resins are used for containers including large water bottles. Blending trades off on toughness and impact resistance with properties of other resins such as solvent resistance.

Current worldwide PC resin sales are ca 290,000 t/yr (76,93); annual growth rates are expected to be near 10% (94).

Polyester Carbonate Copolymers. Polyester carbonate resins have been commercialized during the past decade. The molecular structures are composed of iso- and terephthalate units in conjunction with the standard bisphenol-A PC moieties.

A convenient method of compositional designation uses molar percentages of ester and carbonate linkages coupled with the molar percentages of iso- and terephthalate units in the polymer. A 70% ester, 30% carbonate polyester carbonate with 60 parts of isophthalate and 40 parts of terephthalate is designated 70(60/40)30. Similarly, a standard PC resin is 0(0/0)100 and a polyarylate resin composed of a 1:1 molar ratio of iso- to terephthalate units esterified with bisphenol A is designated 100(50/50)0.

Polyester carbonate resins are made by the interfacial process described for standard PC resins. The phthalate units are introduced by reaction of the appropriate phthaloyl dichlorides concurrent with the phosgenation. At present, Bayer, Dow, GE, and Mobay produce polyester carbonate resins (86); sales volume is low, probably ca 100 t/yr. Polyester carbonates are used primarily in applications requiring 5–25°C higher heat-deflection temperature and better hydrolytic performance than provided by standard PC resins. Properties are given in Table 14.

Acetal Resins

Acetal resins (qv) are poly(methylene oxide) (polyformaldehyde) homopolymers and formaldehyde copolymerized with aliphatic oxides such as ethylene oxide (95,96). The homopolymer resin is produced by the anionic catalytic polymerization of formaldehyde. For thermal stability, the resin is endcapped with an acyl or alkyl function.

$$H_2C{=}O \xrightarrow{R} {-}(CH_2{-}O)_n{-}$$

The copolymers are produced by the cationic catalyzed reaction of 1,3,5-trioxane and an ethylene oxide.

About six firms, not including their subsidiaries, make acetal resins; Celanese, DuPont, and their subsidiaries are the principal producers.

Polyacetals are translucent white and crystalline. They are solvent resistant and self-lubricating. Unlike nylons, they have low moisture absorption and maintain dimensional stability under varying humidity. Melt-flow characteristics are excellent, allowing for short mold cycles. Good fatigue resistance allows applications such as ski-binding parts. On the other hand, impact resistance is low, which affects toughness. Mold design must allow for shrinkage above 25% for unfilled resins. Specific gravities are high. Effective fire retardants have not been developed. For these reasons, applications are limited to small parts or those for which mold warpage, flammability, and toughness are not important. The choice of homopolymer versus copolymer resin involves some trade-offs. For example, homopolymers have higher HDTs but lower unloaded maximum use temperatures and dielectric strengths. These could be important considerations for electrical and electronic applications. Properties are given in Table 15.

Table 15. Properties of Acetal Resins

Property	ASTM method	Delrin 500 homopolymer resin[a]	Celcon M90 copolymer resin[b,c]
HDT at 1.82 N/mm^2 [d], °C	D 698	136	110
notched Izod at 20°C, 3.2-mm thickness, J/m[e]	D 256	75	70
flex modulus at 20°C, N/mm^2 [d]	D 790	2830	2625
specific gravity	D 792	1.42	1.41
flammability rating, UL 94		HB	HB

[a] Ref. 97.
[b] Oxymethylene and oxyethylene groups.
[c] Ref. 98.
[d] To convert N/mm^2 to psi, multiply by 145.
[e] To convert J/m to ftlbf/in., divide by 53.38.

Acetal resins are processed mainly by injection molding. Most resin products are virgin or pigmented neat resins; uv-light stabilized, glass-reinforced, and plateable grades are available. Applications are in consumer, transportation, and industrial and construction industries. Consumer articles include cassette cartridges, zippers, and container valves. Automotive parts include mechanical components such as cranks, gears, and bearings, etc. Although the parts are small, the number of applications are large and a typical American automobile contains ca 1 kg of polyacetal. In the industrial and construction areas, valves and connectors are made of acetal resins. Because they are easily chrome-plated, polyacetals are used in plumbing fixtures.

Worldwide acetal resin capacity is ca 228,000 t/yr (86). Applications are tied to the consumer, construction, and automotive markets and will probably grow only as fast as the world economy. Growth rates of 2–10% are projected for various areas (96).

Nylon Resins

Nylon engineering thermoplastic resins have the polyamide structures shown below.

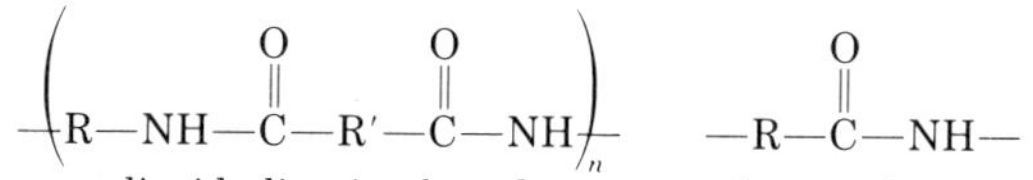

$$\left(\text{R}-\text{NH}-\overset{\overset{\displaystyle O}{\|}}{\text{C}}-\text{R}'-\overset{\overset{\displaystyle O}{\|}}{\text{C}}-\text{NH}\right)_n \qquad -\text{R}-\overset{\overset{\displaystyle O}{\|}}{\text{C}}-\text{NH}-$$

diacid, diamine-based   lactam-based

The straight-chain aliphatic nylons are commonly designated by the chain lengths of the monomer constituents. For example, the largest volume nylon, polyhexamethyleneadipamide, has hexamethylenediamine and adipic acid as monomer constituents and is called nylon-6,6; the diamine chain length is noted first. Where a lactam is the precursor monomer as in polydodecanoamide, its chain-length number is used, and the polymer is called nylon-12. When aromatic monomers are used, a letter is used as designation. Polyhexamethyleneisophthalamide can be designated nylon-6,I. This system of nomenclature breaks down when a nylon is composed of three or more monomers. However, since most commercial nylons are made with one or two aliphatic monomers, the system works well in practice (see also POLYAMIDES).

Commercial engineering thermoplastic nylons are mainly crystalline resins. Nylon-6,6 is the largest volume resin, followed by nylon-6 (99). Other commercially available but much lower volume crystalline nylons are -6,9, -6,10, -6,12, -11, and -12. The crystallinity of the molded part decreases with chain size (100). A few truly amorphous commercial nylon resins contain both aromatic and aliphatic monomer constituents (101). For example, Trogamid T resin is made from a mixture of 2,2,4- and 2,4,4-trimethylhexamethylenediamines and terephthalic acid (102).

In the United States, 11 companies manufacture nylon resins; DuPont, Monsanto, Celanese, and Allied are the leaders. In Europe, there are 22 manufacturers and in Japan six (103).

Nylon resins are made by numerous methods (104) ranging from ester amidation (105) to the Schotten–Baumann synthesis (106). The most commonly used method for making nylon-6,6 and related resins is the heat-induced condensation of monomeric salt complexes (107). In this process, stoichiometric amounts of diacid and diamine react in water to form salts.

$$n \; H_2N\!-\!(CH_2)_6 NH_2 \; + \; n \; HO\!-\!\overset{\overset{O}{\|}}{C}\!-\!(CH_2)_4\overset{\overset{O}{\|}}{C}\!-\!OH \; \xrightarrow{H_2O}$$

$$n \; [H_3\overset{+}{N}\!-\!(CH_2)_6\overset{+}{N}H_3][\; {}^-O\!-\!\overset{\overset{O}{\|}}{C}\!-\!(CH_2)_4\overset{\overset{O}{\|}}{C}\!-\!O\;]^- \; \xrightarrow{heat}$$

$$-\!(CH_2)_6 NH\!-\!\overset{\overset{O}{\|}}{C}\!-\!(CH_2)_4\overset{\overset{O}{\|}}{C}\!-\!NH\!+_n \; + \; H_2O$$

Water is removed and further heating converts the carboxylate functions to amide linkages. Chain lengths are controlled by small amounts of monofunctional reagents. The molten finished nylon resin can be directly extruded to pellets.

The preparation of nylon resins from lactam precursors involves ring opening, which is facilitated by a controlled amount of water in the reaction mixture. The salt complex condenses internally to produce the polyamide (108). The synthesis of nylon-6 from ε-caprolactam is given below.

$$H_2O \; + \; (CH_2)_5\!\underset{NH}{\overset{C\overset{\displaystyle O}{\nearrow}}{}} \; \longrightarrow \; (CH_2)_5\!\underset{\overset{+}{NH_3}}{\overset{C\overset{\displaystyle O}{\nearrow}{\!-\!O^-}}{}} \; \xrightarrow{heat} \; -\!(CH_2)_5\overset{\overset{O}{\|}}{C}\!-\!NH\!+_n \; + \; H_2O$$

Physical properties of commercial nylons are shown in Table 16. Crystalline nylons are white and chemically and hydrolytically stable. Flammability, as

Table 16. Properties of Nylon Resin

Property	ASTM method	Nylon-6,6 Zytel 101[a,b]	Nylon-6 Zytel 211[a,b]	Trogamid T[c,d]
HDT at 1.82 N/mm[2][e], °C	D 648	194	129	124
melting point, °C	D 2117	270	228	
notched Izod at 20°C, 3.2-mm thickness, J/m[f]	D 256	53	80	65
flex modulus at 20°C, N/mm[2][e]	D 790	540–2800	1000	2700
specific gravity	D 792	1.14	1.13	1.12
flammability rating, UL 94		V-2	HB	V-O

[a] As molded; crystalline.
[b] Ref. 109.
[c] Amorphous yellow resin; light transmission 85–90%.
[d] Ref. 110.
[e] To convert N/mm^2 to psi, multiply by 145.
[f] To convert J/m to ftlbf/in., divide by 53.38.

measured by laboratory tests, varies with composition: the higher the aliphatic carbon content, the more flammable the resin. The filled resin is a lubricant. Mineral- and glass-filled crystalline nylons are creep resistant; heat-deflection temperatures approach the crystalline melting points in dry resin. With good melt flow and rapid crystallization rates, nylons are easy to process. On the other hand, they exhibit high mold shrinkage, and dimensions and properties vary with the amount of absorbed water, which can be as high as 10% by weight; an increase in water content reduces the glass-transition temperatures. The lower the glass-transition temperature, the greater creep under load. Mineral filling reduces but does not eliminate these problems. Thus, part and mold designs have to compensate for dimensional changes and warpage.

Amorphous nylons are transparent. Heat-deflection temperatures are lower than those of filled crystalline nylon resins, and melt flow is stiffer; hence, they are more difficult to process. Mold shrinkage is lower and they absorb less water. Warpage is reduced and dimensional stability less of a problem than with crystalline products. Chemical and hydrolytic stability are excellent. Amorphous nylons can be made by using monomer combinations that give structures which crystallize with difficulty or by adding crystallization inhibitors to crystalline resins such as nylon-6 (111).

Crystalline nylons are processed by injection molding and extrusion. The extrusion products are mostly films. Coated or laminated with moisture-barrier resins, eg, PVDC or PE polymers, they are used for meat wrappings and shaped food containers. For injection moldings, extruded pellets are used, available in neat, mineral-filled, glass-reinforced, pigmented, and toughened grades. Notched Izod impact resistance can be increased from the usual 100–150-J/m (1.9–2.8-ftlbf/in.) (under ambient moisture conditions) to the 1000-J/m (19-ftlbf/in.) range (112). Injection-molded parts are used for small mechanical, electrical, and construction articles. In the automotive field, nylon resins are used for interior and exterior applications and under the hood. Parts include filter bowls (amorphous nylon resins), window cranks, electrical connectors, fuse boxes, and speedometer gears. The electrical and electronic industries use various nylon resins for connectors and clamps. Mechanical applications include metal replacement in parts such as gears, sprockets, and wedges. The construction industry uses injection-molded parts for fasteners, hardware, and tool parts. In many applications, nylons compete with polyester and PC resins.

Over 565,000 t/yr of nonfiber crystalline nylons is sold worldwide (113). Since markets are controlled by the economy, a modest growth of 5–8%/yr is expected, unless new markets are developed such as large exterior automotive parts for which nylon blends are being tested. Although currently only ca 900 t/yr of amorphous nylons is sold worldwide (114), a growth rate of 10% is expected because of increased research activity. Currently, the amorphous nylon resins compete with PEI and polyesters in many applications (see also POLYAMIDES, PLASTICS).

Poly(phenylene ether) Resin Alloys

Poly(phenylene ether) resins (115), composed of phenolic monomers, have a very high T_g. The commercial resins are based on 2,6-dimethylphenol. The resin is produced by oxidative polymerization (qv) in toluene solution over an amine catalyst (see also POLYOXYPHENYLENES).

With the glass-transition temperatures above 260°C, neat poly(phenylene ether) resins are very difficult to process; melting does not begin below 255°C. At extrusion or injection-molding temperatures of 300–350°C, melt flow is very stiff and special precautions must be taken to minimize oxidation. Poly(phenylene ether) resins are soluble in all proportions with PS resin and other styrenic polymers, which allows blends or alloys to be made. The higher the PS content, the lower the heat-deflection temperature and the easier is alloy processing. Without further modifications, however, these alloys are brittle. Rubber-impact modifiers improve toughness. Quality and cost are controlled by blending resins with different characteristics. In practice, the poly(phenylene ether) resin–PS resin alloys are compounded with many ingredients, eg, stabilizers and flame retardants. Thus, although the alloys are clear and amorphous, the compounded alloys are opaque.

With the recent expiration of General Electric's worldwide patents (116) on blends of poly(phenylene ether) resins with polystyrenes, other companies are entering the marketplace.

Poly(phenylene ether) resin–styrene resins are cream-colored and are characterized by excellent melt flow. They are easily foamed either by an internal, heat-activated blowing agent or by in-line injection of a volatile material. Flame retardance is conferred by halogen-free compounds. Electrical properties are outstanding and moisture absorbance is low. With excellent dimensional stabilities, the alloys can be made into very large parts. Although hydrolytic stability is very good, organic solvent resistance is poor. This has limited applications such as large automotive parts. Final finishing must take chemical resistance into consideration. Another problem area is color stability to uv radiation. Properties are given in Table 17.

Table 17. Properties of Poly(phenylene ether)-based Resins

Property	ASTM method	Noryl SE100[a]	Prevex VQA[b]	Noryl GFN3[a,c]
HDT at 1.82 N/mm² [d], °C	D 698	100	129	135
notched Izod at 20°C, 3.2-mm thickness, J/m[e]	D 256	250	268	120
flex modulus at 20°C, N/mm² [d]	D 790	2500	2600	7700
specific gravity	D 792	1.10	1.06	1.36
flammability rating, UL 94		SE-1	V-1	SE-O

[a] Ref. 117.
[b] Ref. 118.
[c] Glass-reinforced.
[d] To convert N/mm² to psi, multiply by 145.
[e] To convert J/m to ftlbf/in., divide by 53.38.

Resins other than styrene are being investigated for blending. For example, poly(phenylene ether) resins blended with polyamide resins give dimensionally stable, solvent-resistant materials (119). This illustrates the use of compatibilization to overcome insolubility (see COMPATIBILITY). The materials are very promising and could have an impact on large-part injection-molding and blow-molding technologies.

Pellets for injection-molding, extrusion, and blow-molding applications include mineral-filled, glass-filled, pigmented, and flame-retarded grades. The alloys can be processed into parts requiring hydrolytic and autoclave stability, such as desalinization filter frames and medical devices. Foamed products are used in cabinets for business machines, computers, and copiers. Automotive seat backs are made by blow molding and replacements for metal parts by injection molding.

Worldwide sales of poly(phenylene ether)–styrene resin alloys are 100×10^3–160×10^3 t/yr (86,120); annual growth rates are ca 9%. Other resins, particularly acrylonitrile–butadiene–styrene (ABS) polymers (qv) and blends of these resins with PC resins, compete for similar applications.

Blends

Engineering resin blends, sometimes called engineering resin alloys, are mixtures of engineering resins or mixtures of engineering resins with commodity resins (121). In the list below, asterisks indicate sales above 900 t/yr.

Engineering resin with engineering resin
poly(butylene terephthalate)–poly(ethylene terephthalate)*
polycarbonate–poly(butylene terephthalate)
polycarbonate–poly(ethylene terephthalate)*
polycarbonate–polyester carbonate
polysulfone–poly(ethylene terephthalate)
polyarylate–nylon
poly(phenylene oxide)–nylon

Engineering resin with other resins

polysulfone–ABS
modified acetal
modified nylon*
modified poly(butylene terephthalate)
polycarbonate–ABS*
polycarbonate–styrene maleic anhydride
poly(phenylene oxide)–polystyrene*

Companies produce polymer blends (qv) in order to extend their existing resin capacity and product lines without high capital investment and to achieve property profiles required for specific applications.

For blending, resin compatibility must be considered. For example, PET–PC blends can be either compatible or noncompatible, depending on stoichiometry. Blends of these two resins in which the PC resin concentration is less than 30 wt % have glass-transition temperatures that are not characteristic of either neat component. This is generally viewed as indicating compatibility (qv) (miscibility or solubility are terms often used as well). On the other hand, a 50–50 mixture of the two resins, although seemingly soluble, retains the glass-transition temperatures of each of the neat resin components. In this compositional range, the resins are incompatible (122).

Incompatibility is more common than compatibility. Blends of incompatible resins can be unstable unless compatibilization techniques, which are frequently proprietary, are employed. Compatibilization provides chemical linkages between the resins. This need not result in a near-homogeneous mixture, and, in the cases of controlled morphology blends such as the DuPont Selar resins, laminar inclusions of one resin in another is, in fact, sought (123). The linkages are effected by grafting, surfactant technology, direct covalent bonding, ionic bonding, and filler penetration.

Grafting is employed when one of the resins can react to produce end groups or side-chain functionalities which are soluble or provide a bonding interaction with the other blend resin. The grafted resin can be a primary blend resin, but more often it is used as a blend stabilizer. Grafted PET is an example. Graft points are typically copolymerized acrylate esters. The grafted resins thus have reactive vinyl terminating groups which react with styrenic, rubber, and similar moieties.

Heat-stable, nonvolatile surfactants facilitate processing. Resins with surfactant properties, such as styrene maleic anhydride resins, have sometimes been employed (119).

Direct covalent bonding or ionic bonding can be achieved by inducing chain rupture and chemical combination through shear forces in blending extruders or by the addition of peroxide cross-linking agents. This method resembles grafting. Fillers (qv) stabilize blends by retarding segregation.

The properties of PBT and PC resins and of a blend of these two resins are given in Table 18. The chemical resistance of crystalline PBT is reduced, but that of amorphous PC is increased. Hydrolytic stability is good throughout. Impact performance is lower than that of the components. It can be improved by

Table 18. Properties of Polycarbonate–Poly(butylene terephthalate) Blends

Property	ASTM method	PBT	50–50 blend	PC
HDT at 1.82 N/mm^2 [a], °C	D 648	130	99	132
notched Izod at 20°C, 3.2-mm thickness, J/m[b]	D 256	50	25	800
flex modulus at 20°C, N/mm^2 [a]	D 790	2300	2000	2300
specific gravity	D 792	1.31	1.25	1.21
organic solvent resistance		good	fair	poor

[a] To convert N/mm^2 to psi, multiply by 145.
[b] To convert J/m to ftlbf/in., divide by 53.38.

modifiers. A commercial example of this type of resin blend is the General Electric Xenoy resin which is used in automotive bumpers.

Modified nylons are blends of nylon resins and specially grafted nylon resins. In the DuPont family of Zytel resins, certain blends have been designated Supertough to emphasize the improvement in impact that blends provide over standard resins.

Blends of PC and ABS resins are being used more and more. Sufficient PC is used to raise the heat-distortion temperature of the ABS above 100°C. Impact resistance is reduced without sacrificing utility, and cost savings can be considerable.

Currently, over 110,000 t/yr of engineering resin blends are consumed worldwide, primarily in the transportation, business-machine, hardware, electrical, and appliance industries. Annual growth is projected to be ca 17%/yr. New blends based on PC, terephthalate, and nylon resins are experiencing the greatest expansion (124). These projections could be surpassed if large volume metal applications such as automotive panels are replaced by engineering resin blends which are currently being field-tested.

Outlook

Applications of engineering resins in industry are likely to increase if the conservative estimates of 8–8.5% worldwide annual growth rates are realistic (55). The current consumption of 2×10^6 t/yr should double by 1990, mainly because of the promotion of resins tailored for specific uses. Current resins will most likely be modified, as opposed to the marketing of new resins owing to investment considerations. Modification of resin molecular weights, branching and endcapping, compounding with impact modifiers, flame retardants, stabilizers, and reinforcements, and blending can be expected (125). Nylons (126) will offer chemical and heat resistance, modulus, and toughness in order to replace metals in large parts. Blends will afford engineering properties at lower prices. Liquid-crystal (85), advanced-polyimide, and amorphous-nylon resin technologies (110) will be applied.

Processing will continue to employ injection molding, foam injection molding, and extrusion to produce profiled stock, sheet, and film; blow molding, coextrusion blow molding, and multilayer extrusion will be utilized. Large parts, such as those for the fast-growing automotive exterior markets (127), are likely applications. Graphite-reinforced engineering plastic composites are commercially available in PEEK and polysulfone (23). Additional uses are expected in applications now being served by reinforced thermoset composite materials (128). Much activity will be seen in reaction-injection-molded nylons, with large parts being obvious targets (129). Conductive plastics are being developed especially for parts requiring inherent r-f shielding capability (130) (see also ELECTROMAGNETIC INTERFERENCE; and CONDUCTIVE PLASTICS in Supplement).

Engineering-resin suppliers will work with manufacturers using computer-aided design, engineering, and manufacturing methods (131). The time between product conception and marketing should be dramatically reduced.

Consult Ref. 132 for more information on the future of engineering resins.

BIBLIOGRAPHY

1. D. N. Lapedes, ed., *Dictionary of Scientific and Technical Terms,* 2nd ed., McGraw-Hill, Inc., New York, 1978, p. 542.
2. D. W. Fox and E. N. Peters in R. W. Tess and G. W. Poehlein, eds., *Applied Polymer Science,* 2nd ed., The American Chemical Society, Washington, D.C., 1985, pp. 495–514.
3. G. F. Foy, *Engineering Plastics and Their Commercial Development,* in R. F. Gould, ed., *Advances in Chemistry Series,* The American Chemical Society, Washington, D.C., 1969, pp. 3 and 4.
4. M. Bakker in M. Grayson, ed., *Kirk–Othmer Encyclopedia of Chemical Technology,* 3rd ed., Vol. 9, Wiley–Interscience, New York, 1980, p. 118.
5. O. H. Fenner in C. A. Harper, ed., *Handbook of Plastics and Elastomers,* McGraw-Hill, Inc., New York, 1975, pp. 4-1 to 4-34.
6. H. Mark and G. S. Whitby, eds., *Collected Papers of W. H. Carothers,* Interscience Publishers, New York, 1940.
7. U.S. Pat. 3,875,116 (Apr. 1, 1975), D. R. Heath and J. G. Wirth (to General Electric Co.).
8. C. W. Bunn in R. Hill, ed., *Fibers from Synthetic Polymers,* Elsevier Publishing Co., Amsterdam, 1953, pp. 240–256.
9. E. M. Lacey in M. I. Kohan, ed., *Nylon Plastics,* John Wiley & Sons, Inc., New York, 1973, p. 268.
10. J. Brandrup and E. H. Immergut, eds., *Polymer Handbook,* 2nd ed., John Wiley & Sons, Inc., New York, 1975, p. VIII-5.
11. R. D. Deanin in Ref. 3, pp. 9–12, 18.
12. H. Faig in J. Agranoff, ed., *Modern Plastics Encyclopedia 1982–1983,* Vol. 60, No. 10A, McGraw-Hill, Inc., New York, 1983, pp. 248–262, 268.
13. F. R. Nissel in Ref. 12, pp. 234–237.
14. K. Alex in Ref. 12, pp. 237 and 238.
15. M. Yahr in Ref. 12, pp. 178–188.
16. R. W. Haviland in Ref. 12, pp. 188–190.
17. S. L. Belcher in Ref. 12, pp. 190 and 191.
18. E. Steward in Ref. 12, pp. 214–231.
19. W. F. Kent in Ref. 12, pp. 322–327.
20. C. M. Hall in Ref. 12, p. 311.
21. P. E. Gage in Ref. 12, pp. 193 and 194.
22. R. W. Lenz, *Organic Chemistry of Synthetic High Polymers,* Interscience Publishers, New York, 1967, p. 562.
23. R. B. Rigby, *VICTREX® Polyethersulfone (PES) and VICTREX® Polyetherethersulfone (PEEK),* SPE-RETEC, Mississauga, Ontario, Canada, Oct. 17, 1984, pp. 6 and 7.
24. S. G. Hill, E. E. House, and J. T. Haggatt, *Final Report Advanced Thermoplastic Composite Development Contact N 00019-77-C-0561,* Naval Air Systems Command, U.S. Department of the Navy, Washington, D.C., May 1979, pp. 7–9.
25. *Annual Book of ASTM Standards,* The American Society for Testing and Materials, Philadelphia, Pa., 1984, Index.

26. *Tests for Flammability of Plastic Materials for Parts in Devices and Appliances, UL 94,* 3rd ed., Underwriters Laboratories, Inc., Northbrook, Ill., 1980.
27. C. A. Harper in Ref. 5, Chapt. 2.
28. G. E. Rudd and R. W. Sampson in Ref. 5, Chapt. 3.
29. R. S. Lenk, *Polymer Rheology,* Applied Science Publishers, Ltd., London, 1978.
30. R. M. Bonner, M. I. Kohan, E. M. Lacey, P. N. Richardson, T. M. Roder, and L. T. Sherwood, Jr. in Ref. 9, p. 337.
31. C. A. Harper in Ref. 5, p. 347.
32. Ref. 25, D 1925, 08.02.
33. *Polymeric Materials—Long Term Property Evaluations, UL 746B,* 2nd ed., Underwriters Laboratories, Inc., Northbrook, Ill., 1979, pp. 6–26.
34. R. L. Bergen, Jr., *SPE J.* **18,** 667 (1962).
35. P. D. Gabriche, J. R. Geib, R. M. Iannusis, and W. J. Reid, "Photochemical Degradation and Biological Defacement of Polymers," in D. P. Garner and G. A. Stahl, eds., *The Effects of Hostile Environments on Coatings and Plastics,* The American Chemical Society, Washington, D.C., 1983, pp. 317–330.
36. S. Collins, *Plast. Des. Forum* **8**(2), 17 (1983).
37. M. Huecker and S. Blackson, *CAE,* 56 (1984).
38. Ref. 23, pp. 107.
39. H. J. Dillon in Ref. 12, p. 54.
40. *Mod. Plast. Int.* **11**(11), 76 (1981).
41. *Chem. Week* **12,** 90 (June 27, 1984).
42. *TORLON®, Brochure D0881,* Amoco Chemicals Corp., Chicago, Ill.
43. *TORLON®, Technical Data and Applications Information, Brochure 1117-7.5M-480,* Amoco Chemicals Corp., Chicago, Ill.
44. E. Helmes, E. M. Connolly, L. Waddams, and J. Shimosato, *Chemical Economics Handbook,* SRI International, Menlo Park, Calif., 1984, p. 580.1401 G.
45. *Amoco TORLON® Engineering Resins, Data Sheet Grade 4203L,* Code Number AT-5, Chicago, Ill., Feb. 1982.
46. *Amoco TORLON® Engineering Resins, Data Sheet Grade 7130,* Code Number AT-14, Chicago, Ill., Feb. 1982.
47. R. L. Maffly, O. Kamatari, and J. Bakker in Ref. 44, 1982, p. 580.0600 J-P.
48. R. N. Johnson, A. G. Farnham, R. A. Clendinning, W. F. Hale, and C. N. Merriam, *J. Appl. Polym. Sci. Part A-1* **5,** 2375 (1967).
49. N. J. Ballintyn in Ref. 4, Vol. 18, 1982, p. 843.
50. E. J. Goethals in N. M. Bikales, ed., *Encyclopedia of Polymer Science and Technology,* 1st ed., Vol. 13, Wiley–Interscience, New York, 1969, pp. 448–477.
51. *Polysulfone Design Engineering Data Handbook, Lit. #F-47178,* Union Carbide Corp., New York, Mar. 1979, 4th rev.
52. N. J. Ballintyn in Ref. 4, Vol. 18, 1982, p. 844.
53. R. N. Johnson in Ref. 50, pp. 447–463.
54. R. L. Maffly, O. Kamatari, and J. Bakker in Ref. 44, 1982, p. 580.0600 F.
55. J. Fredos, *Plast. Focus* **16**(49), 1, 2 (1985).
56. Ref. 44, p. 580.1401E.
57. *Mod. Plast. Int.* **14**(4), 81 (Apr. 1984).
58. D. Bradley, *Plast. World* **40,** 94 (Apr. 1982).
59. *Resine Polyimide pour Composites Haute Temperatures, Brochure CF H711,* Rhone-Poulenc, Paris (in French).
60. *Plast. Ind. News* **27**(11), 164 (Nov. 1981).
61. Ref. 44, p. 480.1401 V.
62. Ref. 44, pp. 580.1401 F–G.
63. *KAMAX™ 201, Engineering Thermoplastic Resin, Product Brochure,* Rohm & Haas Co., Philadelphia, Pa., 1979.
64. *ULTEM®, Polyetherimide Properties Guide,* General Electric Co., Pittsfield, Mass., Mar. 1985.
65. Ref. 44, p. 580.1401 Z.
66. N. J. Ballintyn in Ref. 4, Vol. 18, 1982, pp. 794–813.
67. U.S. Pat. 3,354,129 (Nov. 21, 1967), J. T. Edmonds, Jr. and H. W. Hill, Jr. (to Phillips Petroleum Co.).

68. N. J. Ballintyn in Ref. 4, Vol. 18, 1982, pp. 810–811.

69. *RYTON® Polyphenylenesulfide Physical, Chemical, Thermal, and Electrical Properties, TSM-266*, Phillips Chemical Co., Bartlesville, Okla., Apr. 1981, p. 2.

70. Ref. 55, p. 1.

71. D. B. G. Jaquiss, W. F. H. Borman, and R. W. Campbell in Ref. 4, Vol. 18, 1982, pp. 549–574.

72. C. Nitschke in Ref. 12, pp. 46 and 47.

73. C. S. Hughes, J. Bakker, and K. Tsuchiya in Ref. 44, pp. 580.1172 I–J, T, W, Z, 580.1173 Q, T, V, W, 580.1174 C, D, F.

74. W. Griehl and G. Schnock, *J. Polym. Sci.* **30**, 413 (1958).

75. C. S. Hughes, J. Bakker, and K. Tsuchiya in Ref. 44, pp. 580.1172 D, F.

76. J. Fredos, *Plast. Focus* **16**(32), 1 (1984).

77. C. S. Hughes, J. Bakker, and K. Tsuchiya in Ref. 44, pp. 580.1172 K–L.

78. J. Antonopoulos in Ref. 12, Vol. 61, No. 10A, 1984, pp. 192–194.

79. C. S. Hughes, J. Bakker, and K. Tsuchiya in Ref. 44, pp. 580.1172 B, F–G, Q.

80. C. S. Hughes, J. Bakker, and K. Tsuchiya in Ref. 44, pp. 580.1172 C–D, G–I.

81. *VALOX® Injection Molding, Brochure VAL15C*, General Electric Co., Pittsfield, Mass., pp. 23–25.

82. *RYNITE® General Guide to Products and Properties, Brochure E-36602*, E. I. du Pont de Nemours & Co., Inc., Wilmington, Del., pp. 3 and 4.

83. *Chem. Week* **13**(4), 12, 13 (1985).

84. F. M. O'Neill, *Mod. Plast.* **62**, 12 (Feb. 1985).

85. *Br. Polym. Rubber,* 25 (July/Aug. 1985).

86. A. Bettelheim, *Plast. Technol.* **31**, 20, 22 (1985).

87. D. Bellius, *Adv. Photochem.* **8**, 109 (1971).

88. W. F. Christopher and D. W. Fox, *Polycarbonates,* Reinhold Publishing Corp., New York, 1962, pp. 161–173.

89. *Plast. Brief* **10**(51), 1 (1981).

90. D. C. Clagett and S. J. Shafer, *Polycarbonate Resins,* SPE-RETEC, Mississauga, Ontario, Canada, Oct. 16, 1984, pp. 2 and 3.

91. *Plast. World, 1984 Plast. Directory Plast. Prop. Guide* **42**(4), 54 (1984).

92. *LEXAN® Resin Competitive Data, Brochure CDC 672*, General Electric Co., Pittsfield, Mass., p. 8.

93. R. L. Maffly, J. Bakker, and K. Tsuchiya in Ref. 44, 1983, p. 580.1121 C.

94. R. L. Maffly, J. Bakker, and K. Tsuchiya in Ref. 44, p. 580.1121 D.

95. Ref. 4, p. 127.

96. R. L. Maffly, J. Bakker, and K. Tsuchiya in Ref. 44, 1983, pp. 580.0951 D–I.

97. *DELRIN® General Guide to Products and Properties, Brochure E-43409*, E. I. du Pont de Nemours & Co., Inc., Wilmington, Del., pp. 3 and 4.

98. *CELCON® Acetal Copolymer Properties, Brochure 10M*, Celanese Corp., New York, June 1983.

99. H. E. Grey, R. L. Maffly, E. Helenes, J. Bakker, and J. Shimosato in Ref. 44, 1983, pp. 580.0821 C.

100. R. M. Bonner, M. I. Kohan, E. M. Lacey, P. N. Richardson, T. M. Roder, and L. T. Sherwood, Jr. in Ref. 9, pp. 399 and 400.

101. L. T. Sherwood in Ref. 12, pp. 28–34.

102. H. Doffin, W. Pungs, and R. Gabler, *Kunststoffe* **56**, 542 (1966).

103. H. E. Grey, R. L. Maffly, E. Helenes, J. Bakker, and J. Shimosato in Ref. 44, 1983, pp. 580.0822 D–E, Q–S, V.

104. Ref. 22, pp. 103–106.

105. J. March, *Advanced Organic Chemistry,* 2nd ed., McGraw-Hill, Inc., New York, 1977, p. 1171.

106. P. Kovacic and F. W. Koch, *J. Org. Chem.* **28**, 1964 (1963).

107. Ref. 22, pp. 106 and 107.

108. G. M. Van der Want and C. A. Kruissink, *J. Polym. Sci.* **35**, 119 (1959).

109. *Designing with Plastics, Brochure E-32986*, E. I. du Pont de Nemours & Co., Inc., Wilmington, Del., pp. 18 and 19.

110. D. Bremer, *Plast. Des. Forum* **8**(1), 58 (Jan.–Feb. 1983).

111. *Ibid.,* pp. 57–64.

112. *Mod. Plast.* **60**, 69, 70 (Apr. 1983).

113. H. E. Grey, R. L. Maffly, E. Helenes, J. Bakker, and J. Shimosato in Ref. 44, 1983, pp. 580.0821 E, G. H.
114. Ref. 110, p. 57.
115. R. I. Warren, *Polyphenylene Ethers and the Alloys,* SPE-RETEC, Mississauga, Ontario, Canada, Oct. 16, 1984, pp. 1–7.
116. Ref. 4, p. 130.
117. *NORYL® Design, Brochure CDX-83,* General Electric Co., Pittsfield, Mass., 1983, pp. 6 and 7.
118. *PREVEX™ Polymers, Technical Publications PR-102A,* Borg-Warner Chemicals, Inc., Parkersburg, W. Va.
119. U.S. Pat. 4,315,086 (Feb. 9, 1982), H. K. Uemo and T. T. Marugama (to Sumitomo Chemical Co.).
120. R. L. Maffly, O. Kamatari, and J. Bakker in Ref. 44, 1982, pp. 580.0600 G, I, J.
121. M. Rappaport, *Engineering Blends,* SPE-RETEC, Mississauga, Ontario, Canada, Oct. 16, 1984, p. 3.
122. T. R. Nasson, D. R. Paul, and J. W. Barlow, *J. Appl. Polym. Sci.* **23,** 85 (1979).
123. P. M. Subramunian, *Permeability Barriers by Controlled Morphology of Polymer Blends,* SPE-RETEC, Mississauga, Ontario, Canada, Oct. 16, 1984, pp. 1–9.
124. Ref. 121, p. 12.
125. A. S. Wood, *Mod. Plast. Int.* **14**(10), 43 (Oct. 1984).
126. R. Wehrenberg, *Plast. World* **42,** 39 (Sept. 1984).
127. *Plast. World* **42,** 32 (Oct. 1984).
128. D. McCall and Briefing Panel, *Report of the Research Briefing Panel on High Performance Polymers and Polymeric Composites,* National Research Council, Washington, D.C., Aug. 10, 1984, pp. 11–12, 16–19.
129. R. Wehrenberg, *Plast. World* **42,** 62 (Oct. 1984).
130. A. Sternfield, *Mod. Plast. Int.* **12**(4), 48 (July 1982).
131. *Plast. Rubber Week.,* 6 (July 21, 1984).
132. T. Yamazaki, *Chem. Econ. Eng. Rev.* **14**(7–8), 30 (1982).

DONALD C. CLAGETT
General Electric Company

FATIGUE

Recent estimates by the National Bureau of Standards place the total annual loss of engineering components due to fracture at more than $110 billion dollars. What causes a component to fail? Fracture may occur if the applied stress exceeds the strength limits of the material. This overload results from the failure of a designer to select the proper material or from the application of an excessively high load. Alternatively, failure may be attributed to material property deficiencies resulting from improper thermal mechanical processing. Other fractures include those associated with preexistent defects in the polymeric component, eg, gas or shrinkage pores, inclusions, and weld- and flow-line defects. Component defects can also result from mechanical processing; these include scratches, gouges, and other markings that introduce stress concentrations on the surface. Finally, cracks can be introduced into the material during service as a result of repeated loading at stresses below the nominal yield strength, even though the material has satisfactory properties and contains no adventitious defects. Once formed, such fatigue cracks grow a considerable distance prior to catastrophic failure. The initiation of such cracks and their subsequent propagation form the subject of this article. Earlier reviews on polymer fatigue have appeared (1–7).

Lifetime Curves

Numerous studies have shown that the cyclic lifetime of a test bar or component varies inversely with the magnitude of the applied cyclic stress, strain or deflection. Various test procedures have been developed to obtain mechanical property information useful in fatigue design analysis. Long ago, Wöhler developed a procedure wherein the specimen, notched or unnotched, is subjected to a cyclic stress (S) and the number of loading cycles (N) to cause failure is noted (8). Examples of S–N fatigue curves are shown in Figure 1 (9). Such tests routinely employ tapered specimens (Fig. 2) that are subjected to reciprocating bending loads (see ASTM Standard D 671-71) (10). The cyclic lifetime of a given polymer decreases with increasing applied stress consistent with greater damage accumulation per cycle at higher stress. Some of the plots in Figure 1 exhibit a relatively well-defined plateau, suggesting an endurance limit for a particular polymer below which fatigue damage is not expected. By contrast, other engineering plastics reveal a steadily decreasing allowable cyclic stress with increasing number of loading cycles. Stress cyclic lifetime curves are used to define the permissible operating stress (or strain), so as to preclude fatigue failure in less than a certain number of cyclic excursions. For example, if it is desired that a component withstand at least N_1 loading cycles in a given lifetime, the part should be designed so that the applied cyclic stress is less than S_1. Since fatigue damage accumulation involves stochastic processes, S_1 should represent the lower bound stress level, plus an appropriate safety factor to insure safe life behavior for at least N_1 cycles. Sources of scatter in fatigue data include variation in material properties and internal structure, nature and degree of surface irreg-

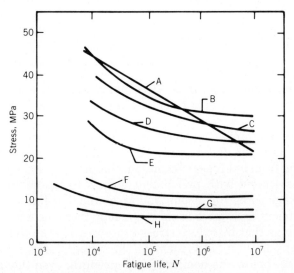

Fig. 1. Stress vs cyclic life plots for typical plastics (9). Cyclic lifetime N_1 corresponds to stress S_1. A, polyurea; B, diallyl phthalate resin; C, alkyd resin; D, cellulose acetate; E, acrylic resin; F, polypropylene; G, polyethylene; H, polytetrafluoroethylene. To convert MPa to psi, multiply by 145.

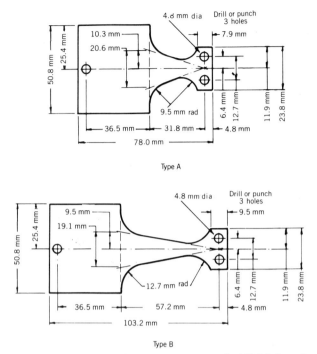

Type A

Type B

Fig. 2. Tapered bend bars used in association with ASTM Standard D 671-71 test method (10).

ularities, presence of adventitious defects, differences in applied stress due to variations in component geometry and alignment, and variations in environment from one component to another.

Hysteretic Heating. The fatigue ranking of the plastics in Figure 1 may not measure the relative intrinsic fatigue resistance of these materials. These test results can be applied in design only when all design factors including magnitude and mode of the stress, part size and shape, temperature, heat transfer conditions, cyclic frequency, and environmental conditions are comparable to the test conditions (10).

Specimen "failure" takes place by either of two modes. Crack nucleation and subsequent growth across the gauge section of a specimen may produce failure. The fatigue life of such a sample would then be defined simply as the number of loading cycles necessary to fracture the specimen. Tests may be conducted on prenotched samples to determine fatigue lifetimes as a function of the local stress at the notch root (11–15). Alternatively, some materials may become overheated as a result of the accumulation of hysteretic energy generated during the loading cycles. Such energy dissipation produces a significant temperature rise within the specimen which, in turn, causes the elastic modulus of the material to decrease. As a result, large specimen deflections are required for continuation of the test at the initial stress level. These larger deflections contribute to even greater hysteretic energy losses with further cycling. In the limit, an autoaccelerating tendency for rapid specimen heating and increased compliance is experienced. Such thermal failure corresponds to the number of loading cycles at a given applied stress range that bring about an apparent modulus decay to 70% of the original modulus of the specimen (10). If one polymer exhibits a greater tendency than another for hysteretic heating under cyclic loading conditions, the relative ranking of the two polymers may differ greatly when the tests are conducted at both high and low cyclic frequencies.

Numerous studies (eg, Refs. 16–18) have shown that the energy dissipation rate under cyclic loading conditions varies directly with the test frequency, the loss compliance of the material, and the square of the applied stress:

$$\dot{E} = \pi f J''(f,T)\sigma^2 \tag{1}$$

where $\dot{E}$ is the energy dissipation rate per unit time, f, the frequency, J'', the loss compliance, and σ, the peak stress. Neglecting heat losses to the surrounding environment, equation 1 may be reduced to show the temperature rise per unit time as

$$\Delta\dot{T} = \pi f J''(f,T)\sigma^2/\rho\, c_p \tag{2}$$

where $\Delta\dot{T}$ = temperature change/unit time, ρ = density, and c_p = specific heat. In Figure 3 (19), a typical curve of stress vs number of cycles to failure for polytetrafluoroethylene (PTFE) is shown with the superposition of temperature rise curves corresponding to various stress levels. For all stress levels above the endurance limit, the polymer heats to the point of melting; ie, heat is generated faster than it can be dissipated to the surroundings. Curve F corresponds to a stress level below the endurance limit. The temperature rise is stabilized at a maximum below the point of thermal failure. This specimen did not fail after 10^7 cycles. For conditions associated with thermal failure, cyclic lives can be enhanced by intermittent rest periods during the cyclic history of the sample (19–21). Temperature rise resulting from adiabatic heating can thus be dissipated pe-

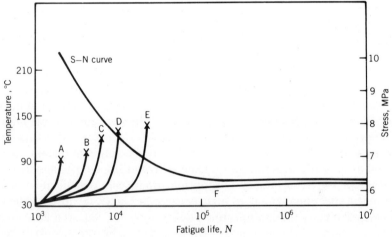

Fig. 3. S–N and temperature cyclic lifetime plots for polytetrafluoroethylene. Thermal failure occurs in specimens A–E (19). Stress values are A = 10.3 MPa; B = 9.0 MPa; C = 8.3 MPa; D = 7.6 MPa; E = 6.9 MPa; F = 6.3 MPa. To convert MPa to psi, multiply by 145.

riodically. Therefore, component life can be enhanced by specimen cooling to develop isothermal test conditions (20,22). For example, fatigue life of plastic gears is improved when the tooth temperature is decreased and the gears are lubricated (Fig. 4).

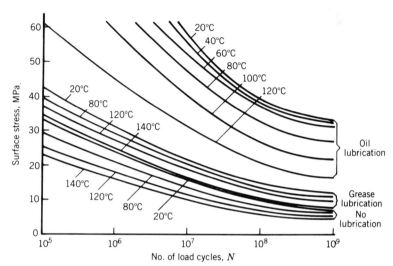

Fig. 4.   Influence of lubrication and tooth temperature on fatigue life in gears (23). To convert MPa to psi, multiply by 145.

Whereas intermittent rest periods during load cycling enhance overall cyclic lifetime at a given stress, damage accumulation takes place during each loading increment (24). Rotating beam bars of polyoxymethylene (acetal resin) were cycled at a stress of 63 MPa until the surface temperature (measured with an infrared microscope) reached 65°C, whereupon the sample was allowed to cool to ambient. Failure would have occurred with repeated cycling by melting at a temperature of ca 110°C. The sample was then immediately load-cycled at the same stress level and the time was recorded for the specimen temperature to again reach 65°C. This time decreased dramatically with increasing test sequence (Fig. 5). To ensure that the interior of the sample was cooled to ambient prior to reloading, three other load-rest sequences were conducted, each following additional 24 h rest periods, and confirmed the trend toward greater susceptibility to hysteretic heating with repeated cycling. From equation 2, it appears that the rapid increase in $\Delta\dot{T}$ with repeated cycling is related to a permanent increase in J''.

The transition from thermal to mechanical failure has been described in terms of a changeover stress level, which depends on test frequency, mean stress, cyclic wave form, and specimen surface area-to-volume ratio (25). An empirical relationship between the uniaxial changeover stress and test frequency is shown in Figure 6 (25) and is given by

$$\sigma_c = (A/V)^{1/2}(C_1 - C_2 \log_{10} f) \tag{3}$$

where σ_c is the changeover stress, A, the specimen surface area, V, the specimen volume, and f, the cyclic frequency.

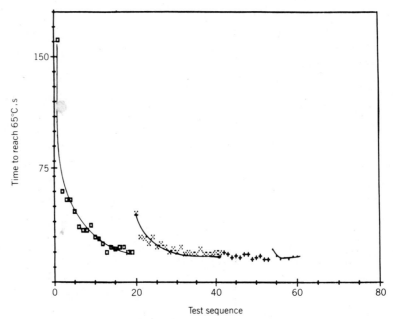

Fig. 5. Time to reach specimen temperature of 65°C vs test sequence at 50 Hz for Delrin acetal resin. Different data symbols refer to sequences following 24 h at ambient (24).

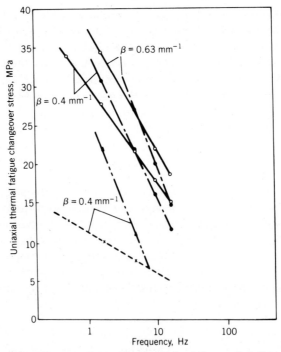

Fig. 6. Changeover stress from mechanical to thermal failure as a function of specimen surface/volume ratio β and test frequency (25). To convert MPa to psi, multiply by 145.

When mechanical, rather than thermal failure is encountered, the S–N curve often reveals several different regimes associated with different slopes (Fig. 7) (26,27). At high stress, fatigue failure is dominated by shear yielding or crazing deformation; at intermediate stress levels, fatigue lifetimes are dominated by crack initiation and propagation processes. The development of a third regime at the lowest stress levels reflects the existence of an endurance limit for that particular material.

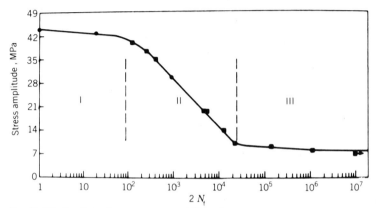

Fig. 7. S–N plot in polystyrene revealing three regimes of material behavior (26). To convert MPa to psi, multiply by 145.

Comparable cyclic tests of unnotched samples have been conducted with deflection or strain as the test variables. A test method involving constant amplitude of deflection has been suggested (28). The machine for this test is a fixed-cantilever, repeated constant-deflection type. As a result of plastic deformation and hysteretic heat-induced softening of the material, the initial stress range induced by the applied deflection limits decay with repeated cycling (Fig. 8). It

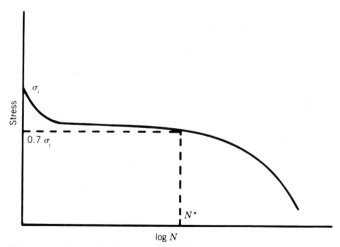

Fig. 8. Deflection-controlled test revealing decay of stress with increasing cyclic life. Fatigue life N^* corresponds to cyclic stress level = to 70% of initial test value σ_i.

has been suggested that the fatigue life be defined when the instantaneous stress level drops to 70% of its initial value. Note the similarity between this fatigue-life criterion and that used to describe fatigue failure under load-controlled test conditions associated with thermal heating. The heat generation rate H_x under fixed displacement conditions has been described (17) as

$$H_x \propto DE_d x^2 \tag{4}$$

where D is the damping capacity, E_d, the dynamic modulus, and x, the displacement amplitude. Initial cycling would be expected to generate hysteretic heating which reduces the dynamic modulus (E_d). Thereafter, specimen heating is lower and a steady-state temperature is quickly reached. Autoaccelerating heat buildup is not experienced during a constant deflection fatigue test.

Cyclic Strain Fatigue. Some experiments have been conducted to evaluate the fatigue response of engineering plastics under constant strain amplitude test conditions involving fully reversed tension–compression loading conditions (26,29–31). During the course of such testing, the height of the hysteresis loops may either increase or decrease, reflecting conditions of cyclic strain hardening and softening, respectively. Both changes in material properties have been identified for a large number of metallic alloy systems. In crystalline, amorphous, and composite engineering polymers, however, only cyclic strain softening has been reported (26). The change in cyclic stress amplitude associated with constant strain amplitude test conditions is shown in Figure 9 where evidence for strain

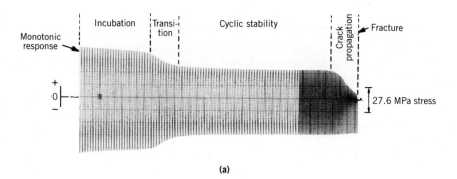

(a)

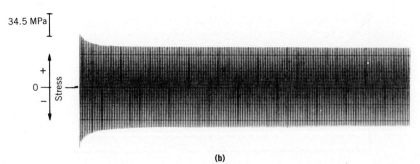

(b)

Fig. 9. Cyclic strain softening corresponding to constant strain amplitude testing in (**a**) polycarbonate and (**b**) nylon at 298 K (26). To convert MPa to psi, multiply by 145.

softening associated with a decrease in the height of the hysteresis loop is shown for both materials. The asymmetry in the loop shape near the end of the test in polycarbonate is associated with the onset of crazing (qv) in the tensile portion of the load cycle.

Cyclically strained specimens exhibit a different stress–strain response compared with the behavior of samples that experience no prior mechanical history. Examples of the cyclic vs monotonic stress–strain curves for nylon-6,6 and ABS are shown in Figure 10. In both cases the cyclically strained material undergoes softening and ductility is diminished. Cyclic stress–strain curves characterize the mechanical response of cyclically deformed material such as occurs in the vicinity of a stress concentration, eg, hole or notch.

Crack Initiation

So far cyclic fatigue lifetimes have been discussed as the number of cycles both to initiate and propagate a defect to some critical length. Hence,

$$N_T = N_i + N_p \tag{5}$$

where N_T = total fatigue life, N_i = number of cycles to initiate a crack, and N_p = number of cycles to propagate a crack to a critical length necessary for fracture. If a defect is present in a component at the outset of its service life, N_T corresponds largely to the propagation stage of crack extension. Depending on the size and shape of the defect, N_i represents either a small or large fraction of N_T. Without knowing the relative importance of initiation and propagation stages of fatigue damage accumulation, it is difficult to identify the role of material characteristics such as molecular weight and impact modifier content on these two stages of the fatigue damage process. Some preliminary experiments have been reported to identify the extent of the fatigue crack-initiation process as influenced by molecular weight ($\overline{M}_w$), impact modifier level (14,15), and stress level (6,32). Figure 11 is a three-dimensional plot showing the interactive effects of $\overline{M}_w$ and % rubber content on fatigue crack initiation (FCI) lifetime (N_i) in impact-modified poly(vinyl chloride) (PVC). The rubber phase is methacrylate–butadiene–styrene (MBS) polymer. N_i is defined as the number of loading cycles necessary to nucleate a crack 0.25 mm in length from a semicircular surface notch with a radius of 1.59 mm. FCI lifetimes increase with increasing molecular weight for all rubbery phase contents. Conversely, in high $\overline{M}_w$ PVC blends, rubber modification lowers FCI resistance although the addition of MBS in low $\overline{M}_w$ PVC slightly improves FCI resistance. The addition of a rubbery phase to epoxy resin and nylon-6,6 also reduces FCI lifetimes and hastens thermal fatigue, respectively.

Crack Propagation

Engineering plastic components may contain material defects, eg, porosity and inclusions, and may also suffer damage as a result of fabrication and service. These defects may not cause immediate fracture under the intended design stress, but these cracks may grow slowly to critical dimensions under the influence of cyclic loading. For this reason, the kinetics of fatigue crack propagation (FCP) must be adequately identified. Fatigue crack propagation data for amorphous,

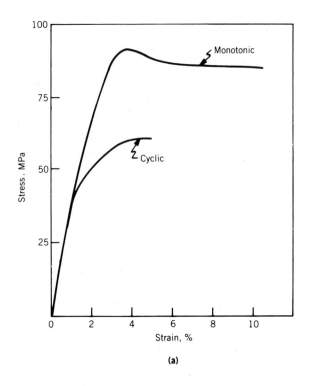

(a)

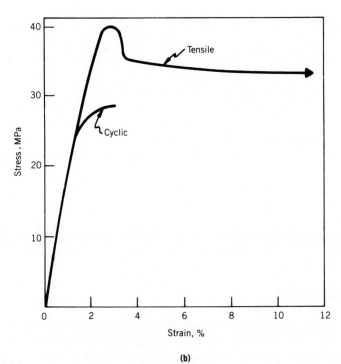

(b)

Fig. 10. Stress–strain curves for monotonic and cyclic loading at 298 K in (**a**) nylon-6,6 and (**b**) ABS (26). To convert MPa to psi, multiply by 145.

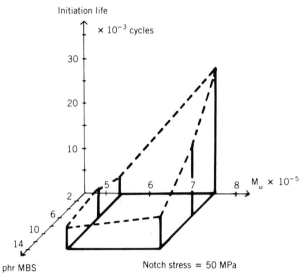

Initiation life

Fig. 11. Three-dimensional plot revealing interactive effects of $\overline{M}_w$ and % rubber content (MBS) on fatigue crack initiation lifetime in PVC (14).

semicrystalline, and multiphase polymeric solids are available in the literature (1,2,33) and may be obtained from various test specimens (34). Beginning with a mechanically sharpened notch, cyclic loads are applied and the resulting change in crack length is recorded as a function of the number of load cycles. Such crack-length monitoring may involve either direct visual examination or indirect compliance methods (34,35). The growth rate increases with increasing crack length and stress (Fig. 12) as described by a relationship of the form

$$da/dN \propto f(\sigma,a) \tag{6}$$

where a is crack length. It has been postulated (36) that the stress intensity factor, itself a function of stress and crack length, is the overall controlling factor in the FCP process (see FRACTURE). Here, the stress intensity factor (K) is defined as a measure of the severity of stress conditions at the advancing crack tip. From the theory of elasticity for cracked bodies, the stress intensity factor is found to be a function of stress and crack length (37):

$$K = Y\sigma\sqrt{a} \tag{7}$$

where Y is a geometrical correction factor representing a function of the ratio of crack length, a, to specimen width, W; σ is the nominal stress applied to the specimen. The relationship between K and the strain energy release rate $\mathcal{G}$ has been identified, and the correspondence between the energy approach to fracture, based on the Griffith theory of failure (38), and the approach based on the stress analysis of cracks has been demonstrated (39):

$$K = \sqrt{E\mathcal{G}} \qquad \text{(plane stress)}$$

$$K = \sqrt{\frac{E\mathcal{G}}{(1 - \nu^2)}} \qquad \text{(plane strain)} \tag{8}$$

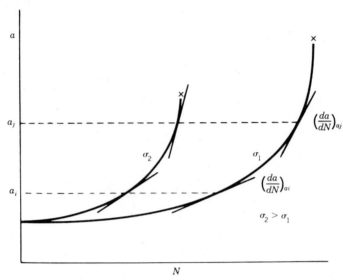

Fig. 12. Crack length (a) increases with increasing number of loading cycles (N). Crack growth rate increases with increasing crack length and stress level. × signifies failure.

The stress field at a crack tip has been described (37,40) using the notation shown in Figure 13, such that

$$\sigma_y = \frac{K}{(2\pi r)^{1/2}} \cos\left(\frac{\theta}{2}\right) \left(1 + \sin\frac{\theta}{2} \sin\frac{3\theta}{2}\right),$$

$$\sigma_x = \frac{K}{(2\pi r)^{1/2}} \cos\frac{\theta}{2}\left(1 - \sin\frac{\theta}{2} \sin\frac{3\theta}{2}\right),$$

$$\tau_{xy} = \frac{K}{(2\pi r)^{1/2}} \cos\frac{\theta}{2} \cos\frac{3\theta}{2} \sin\frac{\theta}{2}, \tag{9}$$

$$\tau_{xz} = \tau_{yz} = 0,$$

$$\sigma_z \cong \nu(\sigma_y + \sigma_x) \qquad \text{(plane strain)},$$

$$\sigma_z \cong 0 \qquad \text{(plane stress)},$$

where ν is Poisson's ratio, r, the distance to the crack tip, θ, the angle above the crack plane, and K, the stress intensity factor. From equation 9 it is apparent that these local stresses rise to extremely high levels as r approaches zero. However, plastic deformation develops over a distance r, where the elastic stresses exceed the yield strength of the material. This distance may be estimated (41) by setting the local stress equal to the yield strength so that

$$r_y \cong \frac{1}{2\pi}\left(\frac{K^2}{\sigma_{ys}^2}\right) \tag{10}$$

where σ_{ys} is the material yield strength and r_y the crack-tip plastic zone radius. As the crack lengthens under cyclic loading conditions, the stress intensity value

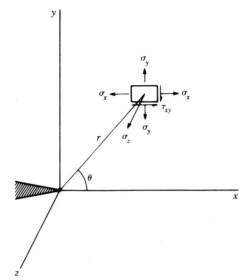

Fig. 13.   Stress distribution around a crack tip.

increases in parallel fashion. At some critical point, the stress intensity level reaches the limit for that particular material and fracture results. This critical value of stress intensity factor, K_c, is referred to as the material's fracture toughness. Similarly, $\mathcal{G}_c$ describes a critical strain energy release rate at fracture.

The growth rates, $(da/dN)_{a_{i,j,\ldots n}}$, at crack lengths, $a_{i,j,\ldots n}$, correspond to respective values of the stress intensity factor range $\Delta K_{i,j,\ldots n}$ (36). Crack growth-rate data (Fig. 14) suggest a relationship of the form

$$da/dN = A\Delta K^m \qquad (11)$$

where da/dN is the fatigue-crack-propagation rate, ΔK, the stress-intensity-factor range ($\Delta K = K_{max} - K_{min}$), and A, $m = f$ (material variables, environment, frequency, temperature, and stress ratio). Many studies have verified the log–log linear relationship between da/dN and ΔK, but others have shown sigmoidally shaped FCP plots. Crack growth rates sometimes decrease to vanishingly low values as ΔK approaches a limiting threshold value ΔK_{th} and increase markedly at very high ΔK values as K_{max} approaches K_c. The asymptote at low ΔK conditions describes FCP rates that diminish rapidly. As such, a limiting stress intensity factor range ΔK_{th} is defined and represents a service operating limit below which fatigue damage is highly unlikely. This threshold stress intensity factor is analogous to the endurance limit described earlier for S–N plots. Several studies have determined the threshold value of ΔK for engineering plastics, eg, polycarbonate (5,43). The da/dN values are in excellent agreement for growth rates $>10^{-5}$ mm/cycle but differ significantly for lower values. Additional threshold data have been reported for poly(methyl methacrylate) (PMMA) as a function of stress ratio R, where $R = K_{min}/K_{max}$ (44,45). Similar to the case of metallic alloys, ΔK_{th} values decrease with increasing R ratio, though differences remain in absolute ΔK_{th} values.

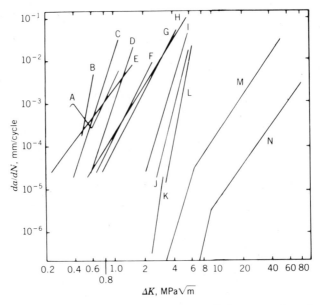

Fig. 14. Fatigue-crack-propagation rates vs ΔK for several engineering plastics and metal alloys (42). A, LDPE; B, epoxy; C, poly(methyl methacrylate) (PMMA); D, polysulfone; E, polystyrene; F, PVC; G, poly(phenylene oxide) (PPO); H, polycarbonate; I, nylon-6,6; J, HI-nylon-6,6; K, poly(vinylidene fluoride); L, acetal resins; M, 2219-T851 aluminum alloy; N, 300 M steel alloy.

Unusual crack growth behavior has been reported in some semicrystalline polymers with FCP rates decreasing with increasing ΔK at intermediate levels (46–48) (Fig. 15). This transient response has been related in LDPE to a change in the scale of the fracture micromechanism (47), whereas others attribute this kinetic transient to localized crack-tip heating, reasoning that such heating would reduce the yield strength and result in a large increase in the depth of the crack-tip damage layer (48). According to another model (49–51), an increase in the volume of the crack-tip damage layer should result in enhanced FCP resistance and associated lower crack growth rates. Localized crack-tip heating should not only increase crack-tip plastic zone size, but also FCP resistance by increasing the material's strain to break (52).

Another unusual form of FCP response seen in certain highly fatigue-resistant semicrystalline polymers involves an oscillatory or "stick-slip" character to the crack growth process. Crack growth-rate values for poly(butylene terephthalate) (PBT) reveal a considerable degree of scatter even when tests are conducted under constant ΔK conditions (Fig. 16) (53). Duplicate tests assure that the oscillatory nature of crack growth in this material is not an artifact, but rather reflects complex crack growth processes. Preliminary examination of the resulting fracture surfaces, however, does not provide any clear indication of the micromechanism responsible for this behavior. The oscillatory nature of crack growth in polyoxymethylene (POM) is attributed to discontinuous crack growth associated with periodic breakdown of the crack-tip plastic zone (54).

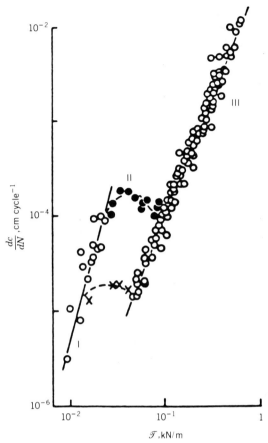

Fig. 15.   Fatigue crack growth vs tearing energy in low density polyethylene. Note transition in growth rates at intermediate tearing energy levels (46).

When rubbery solids are subjected to either tensile or compressive loading conditions, the resulting FCP rates may be correlated with the tearing energy parameter (55). By relating the tearing energy parameter T, a function of the strain energy density and crack length, to the fatigue crack growth rate in rubber, it is possible to characterize the kinetics of fatigue crack extension in materials that do not obey the classical theory of elasticity. An earlier study (56) showed that the FCP rate in natural rubber subjected to cyclic tension could be expressed in terms of the energy parameter, T, by an expression of the form

$$da/dN = A\Delta T^n \tag{12}$$

where da/dN is the FCP rate, A, the material constant, ΔT, the tearing energy range related to $\Delta\mathcal{G}$, the strain energy release rate (eq. 8); the exponent n is approximately 2. More recently, good agreement was demonstrated between fatigue crack growth rates in rubber under both conditions of cyclic tension and compression (Fig. 17) (57,58). No crack growth was found when $T < T_0$, T_0 is a threshold level, similar to ΔK_{th}, below which no mechanically induced crack growth takes place. Instead, crack growth is believed to be driven by chemical

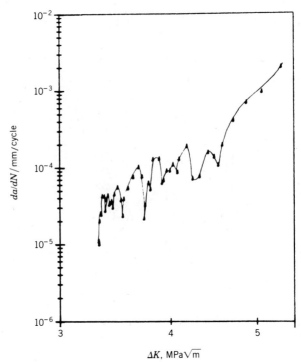

Fig. 16. Oscillatory character of FCP in PBT (53).

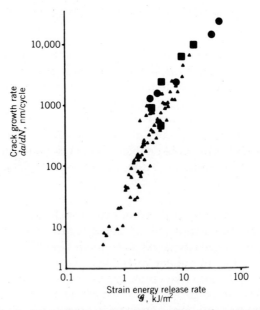

Fig. 17. FCP rates in rubber under tension (▲) and compression (●, ■) loading (57).

degradation due to ozone attack, and is independent of T (59). Recent discussions of fatigue in rubber include a review and a study of several rubbers and blends (60–62).

Cyclic Frequency.   Although hysteretic heating explains a decrease in fatigue resistance with increasing cyclic frequency in unnotched polymer test samples (eq. 2), the fatigue resistance of notched polymers is exceedingly complex. FCP rates can either decrease, remain unchanged, or increase with increasing test frequency (63–65) (Fig. 18). A correlation has been found between the FCP frequency sensitivity factor (FSF) in polymers and the frequency of movement of the main chain segments responsible for generating the β-transition peak at room temperature (64) (Fig. 19).

The highest frequency sensitivity corresponds to a condition wherein the test frequency range straddles the value of segmental jump frequency associated with the β-peak. This resonance condition suggests the possibility that localized crack-tip heating may be responsible for polymer FCP frequency sensitivity. Other materials that are not frequency-sensitive at room temperature might be made so at other temperatures, if the necessary segmental-motion jump frequency is comparable to the mechanical test frequency. Conversely, it might be expected that materials such as PMMA that are frequency sensitive at room temperature, may be made frequency insensitive at other temperatures. These speculations have been confirmed by experiments conducted at temperatures below ambient (66).

The overall frequency sensitivity for several engineering plastics tested has been shown to be dependent on $T - T_\beta$, the difference between the test temperature and the temperature of the β-damping peak within the appropriate test-frequency range (Fig. 20) (66). Studies have verified that the resonance condition brings about localized heating at the crack tip (67). This heating is believed to enhance plastic flow which blunts the crack tip, results in a decrease in the effective stress intensity factor, and brings about a reduction in the corresponding FCP rate. When the heating is not localized near the crack tip, but spread across the entire specimen gauge section, the overall compliance of the specimen increases, resulting in higher growth rates with increasing test frequency in those specimens that exhibit generalized specimen heating under cyclic loading conditions. Figure 18c shows the adverse influence of increased test frequency on material fatigue resistance. Other studies confirm the influence of test frequency on hysteretic heating (68–70), and also show that part of the frequency sensitivity of engineering plastics is due to frequency-induced strain-rate influences relating to these viscoelastic materials (68). For example, increasing strain rate together with increasing test frequency, leads to a reduction in the plastic zone size and critical crack-opening displacement, both detrimental influences on FCP resistance. By contrast, increasing strain rate results in a beneficial increase in dynamic modulus. At least for the case of substantial crack-tip temperature elevations, hysteretic heating appears to represent the dominant mechanism responsible for frequency sensitivity in a given polymer.

The frequency sensitivity factor (FSF) is not a material property since the FSF term varies with the stress ratio, R, and the slope, m, of the da/dN (ΔK) relationship (71).

Test Temperature.   Based on the viscoelastic response of engineering plastics, the FCP response of these materials is expected to depend on test temper-

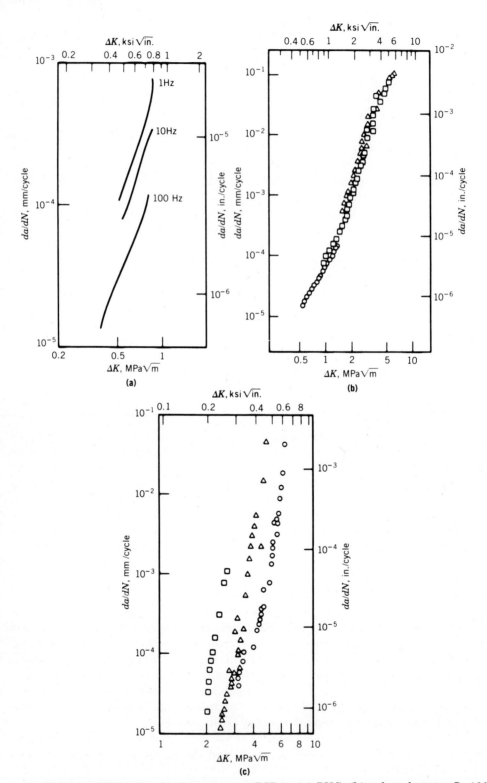

Fig. 18. Effect of cyclic frequency on FCP in (**a**) PVC; (**b**) polycarbonate: $\bigcirc$, 100 Hz; $\square$, 10 Hz; $\triangle$, 1 Hz; (**c**) Zytel ST 801: $\bigcirc$, 1 Hz; $\triangle$, 10 Hz; $\square$, 30 Hz (63–65).

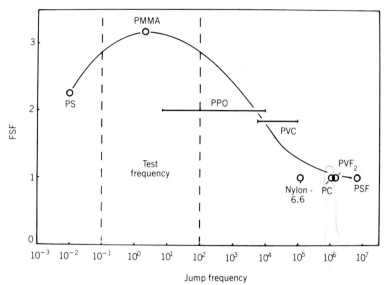

Fig. 19.   Relationship between FCP frequency sensitivity factor (FSF) and RT jump frequency for several polymers (64). PS = polystyrene; PMMA = poly(methyl methacrylate); PPO = poly(propylene oxide); PVC = poly(vinyl chloride); PVF$_2$ = poly(vinylidene fluoride); PC = polycarbonate; PSF = polysulfone.

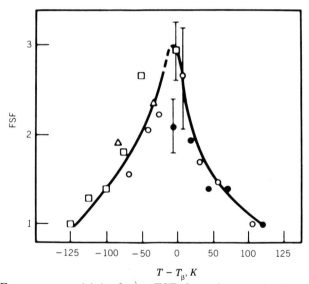

Fig. 20.   Frequency sensitivity factor (FSF) dependence on test temperature relative to β-transition temperature, $T - T_\beta$ (66). ●, PSF; ○, PC; □, PMMA; △, PS.

ature. Indeed, studies of polystyrene and ABS have shown that FCP rates for a given ΔK level generally decrease with decreasing test temperature (72–74). By contrast, a minimum FCP resistance was noted in polycarbonate and polysulfone at intermediate test temperatures (75,76) and confirmed at a test frequency of 1 Hz but not at 100 Hz (63). A complex test-temperature response was also noted in studies on the influence of test temperature and absorbed water on FCP in nylon-6,6 (77). Of particular interest was the possibility to normalize the data (73,77) through empirical use of time–temperature superposition principles (18).

Effects of Chemistry in Homogeneous Polymers

Molecular Weight and Distribution.   Molecular weight and distribution profoundly affect the thermal and mechanical properties of polymers (78–81). Thus the tensile strength of many polymers is very low or negligible when the average M is low ($<10^4 - 10^5$), but increases rapidly above M_0, a characteristic value of molecular weight, and then levels off (80–85). When $M > M_0$, the relationship between strength and molecular weight may often be expressed by an empirical equation of the form

$$\sigma_B = A - B/\overline{M}_x \tag{13}$$

where σ_B is the tensile strength, A and B are constants, and $\overline{M}_x$ is an average molecular weight. Often the number-average molecular weight $\overline{M}_n$, which strongly reflects the presence of chain ends as defects, is appropriate (82,83). $\mathscr{G}_c$ tends to vary in a similar fashion (86–90); a sigmoidal shape over the whole range of M may sometimes be discerned on a log–log plot. The initiation and propagation of fatigue damage also depends strongly on molecular weight. Indeed both static and fatigue fracture are controlled by similar viscoelastic processes, though the balance of processes and the intensity of expression in failure rates differ significantly (2). Energies and volumes of activation tend to be similar for molecular flow and fatigue processes (91,92), and the fatigue resistance of many polymers correlates well with fracture toughness (2). Thus an earlier proposal (85) that the nature of entanglement networks controls M_0 and fracture generally has been confirmed and amplified. Much evidence suggests that only molecules longer than the average length spanning a craze can form a ductile network that can effectively resist crack propagation (86,87,93–97). Equation 13 has been derived on this basis, with $A = \sigma_B$ for infinite M (97), and the role of l_c, the contour length between entanglements, has been shown to control the maximum elongation possible for a polymer (98,99). In addition, high and low values of the ratio of l_c to the rms end-to-end distance between entanglements tend to be associated with crazing and shear yielding, respectively. Experimental evidence clearly shows that dense, stable crazes can develop only when $M > M_0$ (96,100). Although molecular weight may or may not affect the crazing stress (100–102), a direct correlation exists between higher M values and greater ability to resist propagation of a crack through a craze.

However, in spite of the association of high fatigue resistance with high values of M, the sensitivity to molecular weight is much greater in fatigue than in monotonic loading. Thus generalities about fatigue resistance of generic polymers must be viewed cautiously, for rankings depend strongly on molecular weight as well as on structure.

Unnotched Specimens.   As mentioned, fatigue resistance depends strongly, and often dramatically, on molecular weight and its distribution (2,4,79). As $\overline{M}_n$ increases from 1.6×10^5 to 2×10^6 in three polystyrenes with narrow molecular-weight distributions, the S–N curves, (obtained under conditions yielding negligible hysteretic heating), shift upwards and to the right, and the fatigue endurance limit increases from ~¼ to ½ the breaking stress (4,79,103) (Fig. 21). Similar improvements in fatigue life with increasing molecular weight have also been noted in pulsed compression tests on polystyrene (104). Tensile strength increases by 25% over the range of molecular weight, whereas fatigue strength increases by 100% (4,83,103–105). Equation 13 holds for tensile strength, but a different dependence is seen for fatigue strength (83); the ratio of the former to the latter varies from 0.1–0.3. The S–N curve for a polydisperse polystyrene ($\overline{M}_n = 2.7 \times 10^5$) was not parallel to the others in Figure 21, and yielded an endurance limit higher than predicted on the basis of the other specimens (4,103,105). This finding is similar to that in which the addition of only 1% of high molecular weight species ($\overline{M}_w = 7 \times 10^5$) was able to induce craze stability in a brittle, low molecular weight polystyrene ($\overline{M}_w = 10^4$) (96). These results have been attributed to increased entanglement density and hence, orientation hardening of craze fibrils, and decreased concentration of defects in the form of chain ends in long molecules (4,83). At the same time, increases in fatigue life are not noted when $\overline{M}_w$ is increased by irradiation, so that $\overline{M}_n$ is essentially constant (105). Under such conditions, even below the gel point, the benefits of increased molecular weight are clearly overwhelmed by the flaws introduced by chain ends.

Other examples of the beneficial effects of high molecular weight on fatigue life in amorphous or nearly amorphous polymers include poly(vinyl chloride) (PVC) (14,106) and styrene–acrylonitrile copolymer (SAN) (102). Increasing the $\overline{M}_w$ of PVC films increases the fatigue life by orders of magnitude, not only in an inert atmosphere but also in the presence of aggressive alcohols (106). More

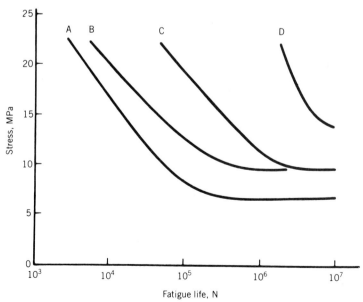

Fig. 21.   Effect of molecular weight on fatigue life in polystyrene. A, $\overline{M}_w = 1.6 \times 10^5$; B, $\overline{M}_w = 3.3 \times 10^5$; C, $\overline{M}_w = 8.6 \times 10^5$; D, $\overline{M}_w = 2.0 \times 10^6$. Specimen B has a broader distribution of molecular weight than the others (4). To convert MPa to psi, multiply by 145.

recently, the number of cycles necessary (in tension–tension fatigue) to initiate a crack in PVC increased from 5.3×10^4 to 7.5×10^4 (Fig. 11) (14); since the crack propagation rate correspondingly increased, so did the overall fatigue life. In the case of SAN tested at 21 Hz in tension–compression, an increase in the number of cycles to failure from 2.7×10^3 to 7.6×10^3 for $\overline{M}_w$ values ranging from 1.2×10^5 to 1.9×10^5, respectively (102), corresponded to a similar increase in the number of cycles for crack initiation. In contrast, the corresponding breaking stress varied only from 59–74 MPa (8,555 to 10,700 psi). Although hysteretic heating resulted in a temperature rise from 6–7°C (higher for higher $\overline{M}_w$), such values are not likely to induce significant softening.

Significant differences in the micromechanism of failure have been noted as a function of molecular weight (102). At low mean stresses, a single surface craze tends to develop; a crack then forms within the craze and propagates (4). At a high mean stress and a low alternating stress, crazes may develop at various points along the specimen. Macroscopically, a fairly smooth mirrorlike region can be seen prior to a rough area corresponding to rapid crack growth. In SAN, the size of the smooth area increases with increasing molecular weight, whereas the tendency towards discontinuous growth bands decreases (102).

Molecular weight also affects the fatigue life of semicrystalline polymers, presumably because long molecules can provide entangled chain sequences (tie molecules) that link crystallites together in a mechanically effective network. For example, at a stress amplitude of 21 MPa (3000 psi) in tension–compression, a polyethylene having $\overline{M}_w = 5 \times 10^4$ failed after 5×10^4 cycles, whereas higher molecular weight samples ($\overline{M}_w \geqq 6.9 \times 10^5$) exhibited no failure after 10^7 cycles (4,103).

Notched Specimens. Striking effects of molecular weight on fatigue crack propagation rates in notched poly(methyl methacrylate) (107) and PVC (108) have been noted (109,110) and similar strong effects have been reported for polycarbonate (111), amorphous PET (112), and polystyrene (113,114). In general, the Paris law is followed over a wide range of ΔK. Empirically, the behavior typified in Figure 22 obeys the following relationship (Fig. 23) (113–116):

$$\frac{da}{dN} = A' \, e^{B'/M} \, \Delta K^n \tag{14}$$

where A', B', and n are constants depending on the material and test conditions. Thus, even though values of static toughness may change relatively little over a given range of molecular weight, small increases in $\overline{M}_w$ may result in up to several orders-of-magnitude decreases in crack growth rates; the effect is greater, the lower the molecular weight.

Effects of molecular-weight distribution were also found to be consistent with those affecting craze stability (96). The addition of small amounts of high or medium molecular weight tails to PMMA results in greater resistance to both fatigue and static loading, whereas the addition of low molecular weight tails has a deleterious effect (110). Indeed the addition of a high M tail to a low M polymer too brittle to test yields enough stability to permit testing.

Because the FCP response can be correlated semiquantitatively with the proportion of high M species estimated by gpc (107), it has been suggested that resistance to cyclic disentanglement is favored by high M species. Whereas the

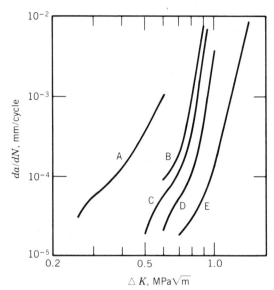

Fig. 22.   Enhancement of resistance to fatigue crack growth in PMMA by increasing molecular weight ($\overline{M}_w$) (107). $\overline{M}_w$ values are A = 1.1 × 10^5; B = 1.9 × 10^5; C = 3.5 × 10^5; D = 2.3 × 10^6; E = 3.6 × 10^6. To convert MPa to psi, multiply by 145.

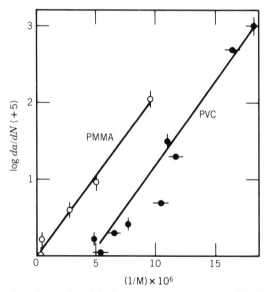

Fig. 23.   Functional relationship between FCP rates in PMMA and PVC and reciprocal of molecular weight at 10 Hz for ΔK = 0.6 MPa$\sqrt{m}$ (107). Data for PVC displaced to the right by a factor of 2.

short time scale of static fracture tests permits relatively little disentanglement, the longer time scale involved in fatigue allows the fracture process to reflect more strongly the energy dissipation concerned in breakdown of the entanglement network (2). In this sense, the fatigue process resembles the longtime relaxation processes in time-to-failure tests (117,118) in which strong effects of molecular weight are also seen. Hence it is likely that during cycling, the entanglements in craze fibrils involving low M species are progressively broken down, but the longer molecules permit considerable strain hardening to occur in the fibrils (2,4,119). This conclusion is supported by fractographic evidence obtained for specimens of PMMA containing high and low molecular weight tails (120).

As with unnotched specimens, FCP resistance in semicrystalline polymers is typically higher at higher molecular weight. Examples include high density and ultrahigh molecular weight polyethylene (121,122), acetal resin (123), nylon-6,6 (123), polypropylene (122,124), and poly(ethylene terephthalate) (112) (Fig. 24). In general, however, the sensitivity of da/dN to molecular weight is less than with an amorphous polymer, perhaps because effects of M primarily reflect the less crystalline regions. The fatigue behavior is determined at least in part by the stability of the entanglement network in the amorphous region, ie, by the tie molecules linking the crystallites.

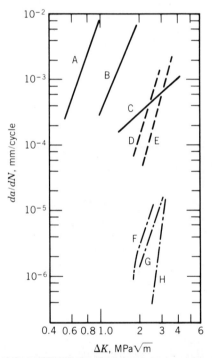

Fig. 24. Effect of molecular weight on fatigue crack growth rates in high density polyethylene: ——— $\overline{M}_w$ = A, 45,000; B, 70,000; C, 200,000 (121); nylon-6,6: ---- $\overline{M}_n$ = D, 17,000; E, 34,000 (123); and acetal resins: ----- $\overline{M}_n$ = F, 30,000; G, 40,000; H, 70,000 (54).

Attempts to develop quantitative theories to rationalize the effects of molecular weight have met with some success, at least with PMMA, PVC, and polystyrene (113–115,125). Thus it has been proposed that the FCP rate can be expressed as the product of two functions, one involving ΔK, and the other characteristic of the relaxation process in the plastic zone (115):

$$\frac{da}{dN} = f(\Delta K) \cdot \frac{1}{\tau_f} \tag{15}$$

where $f(\Delta K)$ reflects the effect of applied stress and τ_f, relaxation time, reflects the resistance to chain disentanglement. Increased FCP resistance has been correlated with increased characteristic stress–relaxation times in PVC (126) and a rubber–polystyrene interpenetrating polymer network (127). By considering τ_f in terms of a Zhurkhov-type rate process (for constant energy of activation)

$$\tau_f = A' \exp(-V'\sigma/RT) \tag{16}$$

where A' is a constant and V' is the volume of activation. Assuming that the stress involved is a localized average mean stress in the plastic zone, the following equation can be derived:

$$\frac{da}{dN} = f(\Delta K) \exp[\sigma_{m,y} V'_\infty/RT \; \psi(M)] \tag{17}$$

Here $\sigma_{m,y} = \sigma_y$ (the yield stress) $\times (1 + R)/2$, where R is the load ratio, V'_∞ is the activation volume for infinite M, and $\psi(M)$ reflects the fraction of molecules that can form a mechanically effective entanglement network. In terms of molecular weight, $\psi(M)$ is defined as $\psi(M) = (V'_\infty/V' = (W - W^*)/(1 - W^*)$, where W is the weight fraction of molecules when $M > M_0$; M_0 is the minimum value of M required for a stable network; and W^* is the minimum weight fraction of those molecules required to form a stable craze. The value of $(1 - W^*)$ is consistent with the void volume of mature crazes in PMMA and PVC (115). Alternatively, to account for dangling chain ends, $\psi(M)$ can be defined as (114,116):

$$\psi(M) = \int_{\overline{M}_c}^{\infty} W(1 - \overline{M}_c)^{2/3} \, dM \tag{18}$$

The model applies to PMMA and PVC and gives values of V' that agree fairly well with values for static fracture. Thus a combination of rate process theory with fracture mechanics can usefully relate molecular and continuum ideas in rationalizing fatigue fracture.

A more recent and more general version of the model was proposed to describe the whole range of fatigue crack growth from $\Delta K'_{th}$, the threshold value of ΔK, to K'_c, the critical value corresponding to catastrophic fracture (113,116):

$$da/dN = \tau_0^{-1} \frac{C}{\nu} \exp[(-E'_a + V'\sigma_{y,m})/RT] \cdot [(\Delta K - \Delta K'_{th})^4/\sigma_b(K'^2_c - K^2_{max})]^m \tag{19a}$$

$$= DK^{*m} \tag{19b}$$

Here E'_a is the activation energy, K_{max}, the maximum value of K attained, ν, the frequency, σ_b, the tensile strength (or 0.2% offset yield stress), and τ_0, C, and m, constants. In equation 19b, the constant term D collects the left-hand terms of equation 19a; ΔK^* collects the right-hand, stress-related terms. Equation 19b describes the FCP behavior of a wide range of polystyrenes and PS blends with both narrow and broad distributions of molecular weight, and structures ranging from linear to branched and cross-linked (113,114,116). The value found for E'_a, 130 kJ/mol (31 kcal/mol), was independent of test conditions and consistent with literature values for plastic flow of polystyrene, and hence the hypothesis of disentanglement.

Vertical shifting of FCP curves gives master curves relative to a very high molecular weight specimen ($\psi \cong 1$) as long as ΔK_{th} is independent of M, ie, when

M is greater than some limiting value (Fig. 25). The shift factor, a_v, is related to V' and $\psi(M)$:

$$a_v = \exp(V'\sigma_{y,m})/\exp(V'_\infty\sigma_{y,m}) \tag{20}$$

where again $V' = V'_\infty/\psi(M)$. Equations 19a and 19b are also useful in correlating effects of waveform, load ratio R, and frequency; extension to crystalline or filled polymers may be possible.

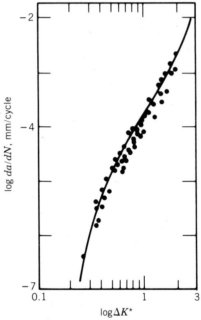

Fig. 25. Master curve for fatigue crack propagation in polystyrene (10 Hz, sine wave) as a function of stress intensity factor (113). Curve includes data for 12 specimens varying greatly in molecular weight and its distribution, and encompasses both the threshold of crack growth and the approach to catastrophic fracture.

An alternative, simpler approach has been proposed in which crack growth essentially is considered the reverse of crack healing (125). In effect, disentanglement is taken to be the reverse of interdiffusion of molecules at the surfaces of two planes in close contact. Thus, assuming scaling laws for characterizing molecular reptation hold, it has been suggested that da/dN is proportional to the quotient of the distance of interpenetration x to t_∞, the time required for complete molecular diffusion and interpenetration. Since $x \propto M^{1/2}$ and $t_\infty \propto M^3$, then

$$\frac{da}{dN} \propto M^{-2.5} \tag{21}$$

Equations 17 and 21 give good agreement when applied to data from Ref. 115 (128).

Generalizations about the generic FCP behavior of typical polymers must be tempered with the realization that molecular weights must be taken into account. For example, although a typical poly(methyl methacrylate) is less resistant to FCP than a typical polycarbonate, a PMMA of very high molecular weight may be about as resistant (or more so at low ΔK) than the latter (compare Figs. 22 and 14). Rational prediction of engineering behavior of both notched and unnotched specimens requires specification of molecular weight and its distribution, and laboratory tests conducted under realistic conditions of cyclic loading.

Cross-linking. Few controlled studies on the effects of cross-linking on fatigue exist. When irradiation was used to increase $\overline{M}_w$ in polystyrene, keeping $\overline{M}_n$ constant, the fatigue life was little affected by the total dose applied, both up to the gel point and beyond. The FCP resistance of cross-linked polystyrene is lower than that of polystyrene or rubber-modified polystyrene (129). A typical epoxy resin is much less resistant to FCP than thermoplastic polymers (2,130). This observation is generally consistent with the ability of cross-linking to inhibit segmental mobility, crazing, and consequent energy dissipation.

Most systematic studies of cross-linked polymers confirm the above hypothesis, but one study reveals an interesting exception. As the percent of a multifunctional acrylate cross-linking agent in PMMA increased from 0–11, the FCP rates increased by over two orders of magnitude (at $\Delta K = 0.5$ MPa$\sqrt{m}$); the slopes of the da/dN curves also increased (131). Similar deleterious effects of cross-linking are described in a study of static and fatigue performance in bisphenol-A epoxies cured with methylenedianiline (132) (Fig. 26). As the amine:epoxy ratio decreased from 2:1–1:1, ie, from a nearly linear to a densely cross-linked

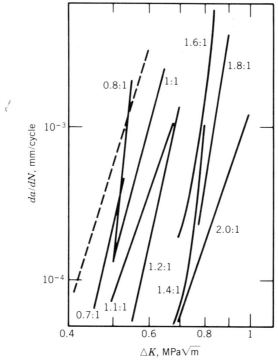

Fig. 26. Fatigue-crack-propagation behavior in Epon 828/MDA epoxy as a function of amine/epoxy ratio (numbers labeling curves) at 10 Hz (132). Dotted line represents other data (130).

resin, K_c decreased and FCP rates increased by over an order of magnitude at a given ΔK. Whereas an excess of amine increased FCP resistance, an excess of epoxy had the opposite effect. In contrast, photopolymerized cross-linked polystyrene exhibited higher values of K_c and lower values of FCP rate, apparently due to energy-dissipating processes associated with entangled microgel particles (116).

Unless mechanisms other than energy dissipation deriving from deformation of segments between cross-links exist, increasing constraints due to network junctions tend to decrease fatigue resistance. This tendency may, however, be offset to some extent by the presence of more deformable molecular species.

Effect of Diluents.   Diluents, eg, plasticizers, residual solvents or monomers, or absorbed environmental agents, greatly affect the viscoelastic response of a polymer and related properties such as damping, modulus, T_g, and yield strength. The resultant changes in properties can therefore affect fatigue behavior whether the diluents are dispersed uniformly in the bulk or only at the surface. The effects of lower stiffness differ depending on whether the testing is conducted under constant stress or constant deflection conditions. In the first case, fatigue behavior may deteriorate due to the increase in strain experienced by a more compliant, ie, less stiff, specimen. In the latter case, the more compliant specimens, which experience lower stresses, may exhibit superior performance.

The S–N behavior in tension–compression at 7 Hz of plasticized PVC containing 10 phr of dioctyl adipate exhibits several features characteristic of other polymers (133). Under conditions of high strain amplitude, high ambient temperature, and specimen cooling by natural convection, thermal failures are observed: a steady increase in damping (tan δ) and a decrease in modulus was observed as failure was approached. In contrast, brittle failure along with crack propagation through crazes occurs under low strain amplitudes or low ambient temperatures, and with an increase in storage modulus and a decrease in tan δ prior to failure. However, when thermal failure conditions are coupled with forced-air cooling, brittle fracture is also typical. Also, the criterion (134) for thermal failure is valid (133).

In PVC plasticized with dioctyl phthalate (DOP), FCP rates reportedly increased as the DOP concentration increased from 0 to 10 to 20% (135) suggesting that lower yield stress and modulus, and resultant tendency towards creep, may play an important role. Quite different results were obtained in a later study at a higher frequency, which may have hindered such creep (108). In this case, FCP rates were relatively unchanged by the presence of DOP, although K_c values were reduced somewhat, especially at the lowest concentration, 6%. Evidently the antiplasticizing embrittling effect of DOP at low concentrations was manifested more in the static than in the fatigue response. It is believed that PVC can form an effective entanglement network and stable crazes in the presence of much DOP, and the long-time-scale loading characteristic of fatigue may permit extensive cyclic disentanglement.

With internally plasticized poly(methyl methacrylate), ie, a copolymer with butyl acrylate, still other effects have been seen (107). As the concentration of butyl acrylate increases to mol fractions of 0.1 and then 0.2, FCP rates increase dramatically and K_c values decrease, presumably because increased ductility is associated with lower strength and modulus. However, with a mol fraction of 0.3

butyl acrylate, localized crack-tip heating and blunting develop from an increase in damping, and FCP rates decrease, so that the overall behavior, with respect to FCP only, resembles that of a commercial PMMA. Hence it appears that a plasticizing comonomer in a chain significantly facilitates the detailed local viscoelastic response and weakens craze fibrils, whereas a semimiscible external plasticizer may not necessarily inhibit the ability to form strong craze fibrils.

The plasticizing influence of absorbed water on the FCP response of polyamides has also been studied (136–142). Earlier data were difficult to interpret because various experiments employed different test frequencies, mean stress levels, amounts of imbibed water (0–8.5%), and different specimen configurations. According to the principles of linear elastic fracture mechanics, specimen configuration should have no bearing on material fatigue response, since the propagation of a fatigue crack is assumed to be controlled entirely by the stress field near the crack tip and therefore to be a function of ΔK only. However, for the case of engineering plastics that exhibit significant viscoelastic heating at the prevailing test conditions, variations in specimen configuration have been shown to alter FCP resistance as a result of changes in the overall stress range ($\Delta \sigma$) acting on the specimen (68,142,143). For example, crack growth rates determined with center cracked (CCT) specimens exceed those using wedge-opening load (WOL) specimens (68,143). In addition, greater hysteretic heating occurs in CCT samples compared with those measured in WOL samples. These results are readily understood in terms of the expression for the stress intensity factor range (eq. 7)

$$\Delta K = Y \Delta \sigma \sqrt{a} \qquad (22)$$

Since Y is much smaller for the CCT specimen than for the WOL specimen (34), it follows that $\Delta \sigma_{CCT}$ is much larger than $\Delta \sigma_{WOL}$; the latter difference in far field stress distribution alters the hysteretic heat generation capability and the relative stiffness of these samples.

Nylon–water interactions and their associated influence on mechanical response are of considerable engineering importance since absorption of moisture occurs readily when the nylon component is in service. Roughly 2.5 wt% water is imbibed after equilibration at 50% rh. Depending on the property and the amount of water absorbed, effects may be beneficial or deleterious. For example, RT modulus increases marginally with the addition of 2.2% water in nylon-6,6 but is significantly reduced with 8.5% water (144) (Fig. 27). Such differences in elastic modulus have been traced to the manner by which the water molecules are incorporated in the nylon. At low absorbed moisture, the polar H_2O molecules interrupt the hydrogen bonds between amide groups in neighboring chain segments and create water bridges between these groups. This tightly bound state is believed to persist until all of the hydrogen bonds in the amorphous fraction of the nylon have been replaced by water bridges, corresponding to a water/amide group ratio of 1/2 (145,146). At higher concentrations, weakly bound water is imbibed and essentially serves as a diluent.

Fatigue tests on unnotched samples have demonstrated that the fatigue strength (the stress corresponding to failure at a given number of cycles) is reduced by as much as 30% when nylon-6,6 is equilibrated at 50% rh (144) (Fig. 28). In marked contrast, it has been found that fatigue crack growth rates in

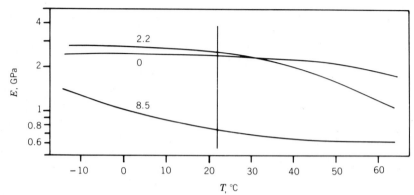

Fig. 27. Modulus as a function of temperature for various moisture contents (% H_2O noted on curves) in nylon-6,6, $\overline{M}_n = 17,000$ (141).

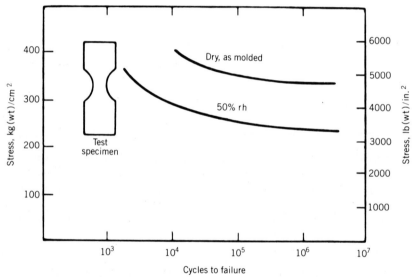

Fig. 28. Unnotched fatigue life in nylon-6,6 as molded and equilibrated to 50% rh (147). The test specimen is 0.775 cm thick.

nylon-6,6 exhibit a pronounced minimum at an absorbed moisture content of 2.6 wt% water, and crack growth rates are higher in saturated material (8.5 wt% water) than in the dry polymer (Fig. 29) (141,148). The decrease in crack growth rates with increasing water content at low moisture can be attributed to enhanced chain mobility and an attendant increase in crack-tip blunting associated with localized crack-tip heating. In addition, fatigue resistance is believed to increase with the slight increase in elastic modulus that accompanies the addition of up to 2.5 wt% water. Conversely, the marked increase in crack growth rates at moisture >2.5 wt% is believed to result from the pronounced reduction of elastic modulus which accompanies high amounts of loosely bound absorbed water (144,146). (Fig. 27).

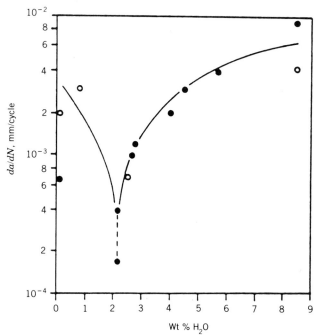

Fig. 29. Fatigue-crack-propagation response of nylon-6,6 as a function of moisture content ($\Delta K = 3$ MPa$\sqrt{m}$) (136); $\bigcirc$ = 8.5 mm, $\bullet$ = 6.4 mm.

The complex influence of absorbed water on FCP resistance in nylon-6,6 and in nylon-6 (141) reflects a competition between the beneficial influence of crack-tip blunting and the deleterious effect of modulus degradation. This competitive interaction is reflected in the fatigue-mechanism transition discussed below. The blunting mechanism dominates the fatigue response at low moisture levels and the decrease in modulus overshadows crack blunting at higher water contents. The same deformation processes occur in fatigue tests on unnotched samples; however, overall global stress levels in these samples are much larger than for the case of precracked specimens. As a result, the greater amount of hysteretic heating in nylon-6,6 samples equilibrated at 50% rh reduces specimen stiffness resulting in more damage per cycle than that experienced by the dry specimens; consequently the wet samples are expected to exhibit inferior fatigue properties in the S–N test.

The influence of absorbed water on FCP resistance is expected to be more significant for high impact nylon-6,6 (HI-N66) since this blend exhibits greater amounts of hysteretic heating than neat nylon-6,6 (69,70). In general, the effect of water content on the relative ranking in FCP resistance for various nylon-6,6 blends is related to specimen temperature. At low water contents and low values of ΔK, heating is beneficial, and blends with more rubber show the best FCP resistance. For these conditions, heating is localized and contributes to crack-tip blunting which lowers the effective ΔK. With increasing water content and/or ΔK, greater specimen heating occurs which is detrimental to fatigue resistance. Since rubbery additions and extensive hysteretic heating bring about a substantial reduction in E', the principal cause for reduced FCP resistance in the rich blends containing water, it is interesting to note that when ΔK is normalized with respect to the elastic modulus, the richer blends exhibit greater resistance to cyclic strain (69). In general, variations in FCP behavior of HI-N66 depend strongly on changes in the dynamic storage and loss moduli resulting from hysteretic heating. Heat-induced changes in modulus are more important than the absolute temperature increase of the sample.

The absorption of water by poly(methyl methacrylate) also influences fatigue resistance in this amorphous polymer (149). There is significant reduction in fatigue lifetime with up to 1% water content for extruded PMMA (molecular weight = 79,300) tested at 27.6 MPa (4000 psi) and 2 Hz (Fig. 30) which is related to plasticization and reduction in resistance to craze initiation and subsequent breakdown. At higher water content, no further deterioration takes place even though the water molecules tend to cluster and aggregate.

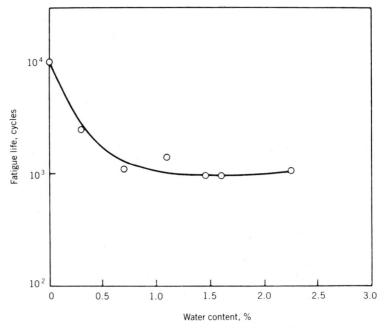

Fig. 30. Fatigue life vs water content in extruded PMMA (molecular weight = 79,300) tested at 27.6 MPa (4000 psi) and 2 Hz (149).

Surface Modification. Previous examples involve essentially homogeneous polymer–diluent systems. Interesting effects with practical consequences also occur when unnotched samples are exposed to various environmental media (4,150). Because fatigue failure in such cases usually involves initiation at the surface, such effects are to be expected.

Deleterious effects of atmospheric exposure on mechanical properties are well known. Surface oxidation in rubbers results in embrittlement, surface cracking, enhanced fatigue crack propagation, and reduced fatigue life. Even in a ductile plastic, the formation of brittle surface layers and cracks invariably facilitates the onset of crack propagation under static (151,152) or fatigue loading.

For example, when exposed to natural weathering (qv), the fatigue strengths of polystyrene, ABS and acetal resin decrease both in an absolute sense and in a relative sense with respect to tensile and bending strength (152).

The role of surface oxidation in the fatigue of polystyrene has been clearly demonstrated (150). Whereas control samples exhibit an average fatigue life of 25,000 cycles in air, lifetimes increase ~90% by coating with a ductile metal (aluminum or gold, 200 nm thick), and ~100% by testing in nitrogen. Coating with flexible, rubber polymers, or with a low molecular weight oligomer of polystyrene also increases fatigue life in polystyrene (150,153). The relative effect is greater for unpolished specimens; interestingly, the coatings themselves are more effective than polishing. It seems likely that these coatings reduce stress concentrations at surface flaws. Even though the natural rubbers used are susceptible to oxidation, the polishing effect is dominant. On the other hand, coating with a high molecular weight brittle polystyrene layer results in a decrease in fatigue life (4); the propensity for facile initiation of craze cracks evidently overcomes any protective effect. Of course, oxidation is favored by stress.

Although liquid environments often induce environmental-stress crazing or crazing in glassy plastics, not all liquids are deleterious to fatigue performance, and some are beneficial. Typical liquids, eg, heptanes, esters, and alcohols, decrease the fatigue life of polystyrene (154); the decrease is greater when the solubility parameter is closer to that of polystyrene. The increased solubility and plasticization lowers the crazing stress and hence facilitates crack development. In contrast, with liquids known to increase crazing stresses in polystyrene, eg, water, ethylene glycol, or glycerol, fatigue lifetimes increase $\sim 3 \times$, $4 \times$, and $9 \times$, respectively. This effect is attributed to enhanced hydrogen bonding, to the filling of surface voids and flaws, or to surface-tension changes.

Increases in fatigue life for nylon-6,6 exposed to petroleum ether, water, and alcohols have also been reported (155), and the effects of various liquids on fatigue of polycarbonate have been described (156). Various vapors can be beneficial to fatigue life in polystyrene (4,157).

Thermal History. This is an important parameter in determining viscoelastic and mechanical behavior (158). For example, in amorphous polymers, quenching from above T_g increases the free volume and hence facilitates relaxation, whereas annealing (qv) below T_g decreases free volume and enthalpy, increases the yield stress, and decreases fracture toughness (see also AGING). With crystalline polymers, the morphology (qv) depends very much on the balance between nucleation and growth corresponding to the rate and pattern of cooling from the melt (see KINETICS OF CRYSTALLIZATION).

Whereas fatigue behavior of semicrystalline polymers can be significantly altered by varying the thermal history, effects of such history on fatigue in amorphous polymers are not well-documented. Even though sub-T_g annealing of polycarbonate reduces the static fracture toughness, relatively little effect on FCP response is noted (159). Similarly, only modest and difficult-to-reproduce effects of quenching and annealing on FCP behavior have been reported for PVC (160,161). Although repetitive cycling can condition the state of a polymer, at least under some conditions (162), the fact that yielding occurs within the plastic zone implies the attainment of a state of mobility equivalent to a temperature high enough to erase some or all of the effects of the applied history.

Crystallinity and Morphology.   As a class, semicrystalline polymers tend to exhibit superior fatigue resistance, relative to that of amorphous polymers, both in terms of S–N behavior and fatigue crack propagation rates (2,4,163), as long as tests are conducted at frequencies low enough to minimize excessive hysteretic heating. This is so even allowing for variations due to molecular weight. Indeed even though performance in a standard S–N test at 30 Hz may be relatively poor, many such polymers behave very well under service conditions involving lower frequencies and strains. Thus polypropylene is an excellent material for integral hinges.

Such inherently superior fatigue resistance is not fortuitous (2). Semicrystalline polymers having adequate intercrystalline linkages can not only dissipate considerable energy when the crystallites are deformed but also form a new and strong crystalline morphology (2,164,165). Crystallization developed during straining of amorphous polymers is responsible for the superior fatigue resistance of strain-crystallizing rubbers (166,167) and of amorphous poly(ethylene terephthalate) (112).

Confirmation of the general fatigue superiority of semicrystalline polymers is implicit in studies of S–N behavior at frequencies low enough to minimize temperature rises (163). Whereas the ratios of the endurance limit to tensile strength for typical amorphous polymers (polystyrene, PMMA, and cellulose acetate) are ca 0.2, values for nylon-6,6 acetal resin, and polytetrafluoroethylene are 0.3, 0.5, and 0.5, respectively. At a low frequency, polyethylene ($M > 50,000$) exhibits no failure after 5×10^6 cycles, even at the relatively high alternating stress of 21 MPa (3000 psi) (163). The general pattern shown in Figure 14 also indicates the importance of crystallinity in determining overall resistance to fatigue crack propagation. Curves for polymers such as nylon-6,6, poly(vinylidene fluoride), and acetal resin appear to the right of the curve; data for nylon-12 (168) and properly heat-treated poly(ethylene terephthalate) (112) also fall in this region. In fact, when FCP data for a wide range of crystalline polymers are normalized with respect to modulus (Fig. 31), the envelope reveals much better relative performance than for typical metals (54). Although the low modulus of LDPE lowers resistance to fatigue crack propagation, HDPE is quite resistant, especially when quenched, and when the molecular weight is high (Fig. 24) (121,123). If the data for LDPE are normalized with respect to modulus (by plotting $\Delta K/E$), the comparative ranking is greatly increased (2).

Relatively few systematic studies have been made on the effects of morphology, eg, % crystallinity or crystallite size, on fatigue behavior. Even when the crystalline texture is demonstrably changed, effects of such changes may be confounded with effects of other factors such as the addition of antioxidants (169–172) or toughening agents (70). Nevertheless, the dominant role of several antioxidants in increasing fatigue life has been attributed to their ability to nucleate smaller and more uniform spherulites. In addition to specific effects of the fine-structured morphology, the associated reduction in hysteretic heating in stress-controlled tests undoubtedly also helps to increase fatigue life (170). Presumably enhanced nucleation rates also increase the proportion and effectiveness of tie molecules (121). In contrast, with rubber-modified nylon-6,6, in which the rubber reduces both spherulite size and FCP rate, the principal contributor to toughening is thought to be the rubber (70).

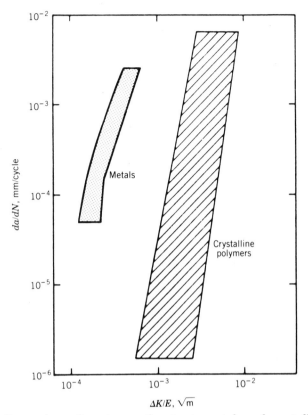

Fig. 31.   Comparison of crack growth rates in metals and crystalline polymers as a function of normalized stress intensity factor $\Delta K/E$ (54).

Annealing and quenching treatments of polymers without additives certainly do affect fatigue behavior, although even here, differences in the amount of crystallinity may be confounded with microstructural differences. By annealing nylon-6 at 180°C in oil, the fatigue life increases manyfold at a given stress, and the endurance limit increases from ~23–33 MPa (3340–4790 psi) (173); the improvement is attributed to observed changes from a nonuniform morphology comprising hexagonal crystal form to a more uniform morphology comprising monoclinic crystal packing. With polytetrafluoroethylene quenched from the melt, the endurance limit increases with increasing quenching rate, which results in decreased crystallinity (19). Processing conditions are also important with poly(vinyl chloride), in which fatigue life is higher, the higher the degree of fusion (174).

So far, effects of stereoisomerism on fatigue are not documented, except for the case of polybutadiene in which an increase in cis-1,4 content (and concomitant crystallinity) has been shown to improve fatigue response (175).

Studies of the effect of thermal treatment on FCP behavior has also revealed significant variations in response. As shown in Figure 32, as the cooling time from molding to 100°C decreases from 8–0.5 min, the slopes of the da/dN curves decrease markedly, and the FCP rates decrease by up to an order of magnitude, depending on ΔK (121). The increased fatigue resistance of the rapidly quenched

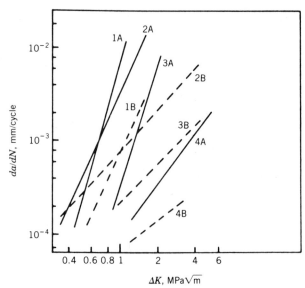

Fig. 32. The effect of morphology on fatigue crack propagation in high density polyethylene of varying density (g/cm³) and $\bar{M}_w$ (121). 1, d = 0.96, $\bar{M}_w$ = 45,000; 2, d = 0.935, $\bar{M}_w$ = 72,000; 3, d = 0.96, $\bar{M}_w$ = 70,000; 4, d = 0.945, $\bar{M}_w$ = 200,000. A, indicates slow cooling; B, indicates fast cooling.

polymers can be explained in terms of the increased propensity towards a greater concentration of tie molecules linking the spherulites together. This reasonable suggestion is also consistent with the fact that high rates of quenching also yield larger damage zones.

Both expected and unexpected effects of thermal history have been reported for polypropylene (176) which was treated to yield three different morphologies characterized by (A) a fine spherulite size, (B) a mixture of fine and large sizes, and (C) a large spherulite size (ca 25, 25 + 250, and 100 μm in diameter, respectively). Effects of temperature were small for morphologies (B) and (C). However, the behavior was strongly dependent on morphology; the ΔK required to drive the crack at 10^{-2} mm/cycle increased in the sequence (B) to (C) to (A) from ~4 to 5 to 5.8, respectively. No value of ΔK exists at which FCP rates of all three types can be compared. Clearly the increasing content of large spherulites increases da/dN by orders of magnitude or infinitely, depending on ΔK.

Effects of crystalline content on FCP have been successfully differentiated from effects of crystallite size and perfection in poly(ethylene terephthalate), in which the percent and nature of the crystallinity can be varied independently (177). At a given crystallization temperature, and hence at a given crystallite size and perfection, FCP rates first increase slightly, then decrease as crystallization proceeds, and then increase catastrophically. Although the reason for the initial increase is not established, the decrease in FCP rate with increasing crystallinity surely reflects the corresponding increase in the ability to dissipate energy through the deformation of crystallites. However, at an even higher crystallinity, the concentration and effectiveness of tie molecules decrease to the point where an effective entanglement network no longer links the profuse crystallites, and fracture readily occurs. This hypothesis is also consistent with the fact that at constant crystalline content, FCP rates increase manyfold as crystallite size and perfection increase. Thus, as with static toughness, fatigue resistance requires a suitable balance between crystalline and amorphous regions, with adequate molecular interconnections.

Anomalous decelerations in FCP rates of semicrystalline polymers are sometimes observed (48,53,178) (see Fig. 16).

Design Concepts

The fatigue life of an engineering component depends markedly on the size of the existing defect, the magnitude of the stresses, and the characteristics of the component material. Some studies of characteristic defects in components have been reported. For example, in polyethylene pipe, a number of preexisting defects at pipe-weld discontinuities caused by pipe-wall misalignment and weld mismatch have been identified (179). Also, three types of defects were found in three different pipe-grade polyethylenes: possible gels, calcium stearate particles which are used in pipe manufacture, and iron-based particles, presumably reflecting wear debris from the plastic processing equipment (180).

Crack Propagation Calculations. Fatigue-life calculations are required to design a new part with a safe operating service life under cyclic loading conditions. Such computation can be performed by integrating equation 11 (where $\Delta K = Y\Delta\sigma\sqrt{a}$) between the starting and final flaw sizes as the limits of the integration. When the geometrical correction factor Y does not change within the limits of integration (34), eg, for the case of a circular flaw where $K = 2/\pi \, \sigma\sqrt{\pi a}$, the cyclic life is given by

$$N_f = \frac{2}{(m-2)AY^m\Delta\sigma^m}\left(\frac{1}{a_0(m-2)/2} - \frac{1}{a_f(m-2)/2}\right) \text{ for } m \neq 2 \quad (23)$$

where N_f = number of cycles to failure, a_0 = initial crack size, a_f = final crack size at failure, $\Delta\sigma$ = stress range, A,m = material constants, and W = specimen width.

Fatigue life predictions in rubbers based on integration of equation 23 give good agreement with razor cuts of known length or are consistent with the size of observable flaws in actual components (181). Usually, however, this integration cannot be performed directly since Y varies with the crack length. Consequently, cyclic life may be estimated by numerical integration procedures by using different values of Y held constant over a number of small crack-length increments. Usually, $a_0 \ll a_f$, so that the computed fatigue life depends predominantly on estimations of the starting crack size a_0. Computations based on an equation of the form given by equation 23 are shown in Figure 33 for PVC pipe (5 cm dia) pressure cycled to the point where leakage was detected (182). Increased lives are associated with decreased pressure and initial crack length.

The theoretical calculations shown in Figure 33 correspond to a leak-before-break condition which represents an important design philosophy in the fabrication of pressure vessels and piping systems (34). From examining the stress-intensity-factor distribution along the surface of a semielliptical surface flaw, it

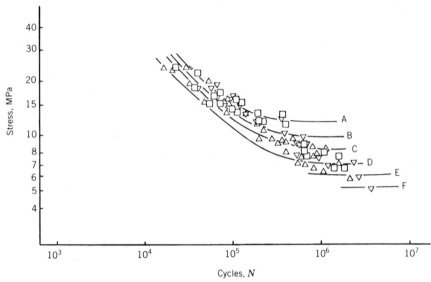

Fig. 33. Fatigue life calculations for PVC pipe. Experimental-data points agree well with theoretical values (182). Starting flaw sizes a_0 in mm are A, 0.35; B, 0.5; C, 0.7; D, 1.0; E, 1.4; F, 1.75.

has been learned that the crack tends to grow more rapidly in a direction parallel to the minor axis of the ellipse until the flaw approaches a semicircular configuration. The crack then continues to grow as an ever-expanding semicircle until it breeches the vessel's outer wall, thereby allowing fluid to escape. At this point, the crack breaks through the remaining unbroken plastic ligament and assumes the configuration of a through-thickness flaw. Assuming that the crack remained semicircular to the point where breakthrough occurred (at which time a equals the wall thickness), the dimension of the through-thickness flaw would be $2a$ or equivalent to twice the vessel-wall thickness, $2t$. Hence the stress intensity factor for this crack becomes

$$K = \sigma\sqrt{\pi a} = \sigma\sqrt{\pi t} \qquad (24)$$

If K is $< K_c$, then fracture would not take place even though leaking had commenced. In general, then, the leak-before-break condition would exist when a crack of length equal to at least twice the vessel-wall thickness could be tolerated under the prevailing stresses.

Two additional design problems illustrate cyclic-life prediction procedures (183). In the first problem, a 10-mm thick large diameter liquid storage cylinder is to be fabricated with an engineering plastic. The cylinder contains a semicircular surface crack 2-mm deep that is oriented normal to the fluctuating stress of 15 MPa. The number of loading cycles necessary for the crack to grow completely through the cylinder wall has been computed for several engineering plastics (Fig. 34). These computations show an almost 4 order of magnitude difference in the number of cycles necessary to cause leakage in the tank. The relative ranking of the plastics parallels their relative fatigue-crack-propagation resistance. Tanks made from acrylic, polystyrene, and epoxy would never leak

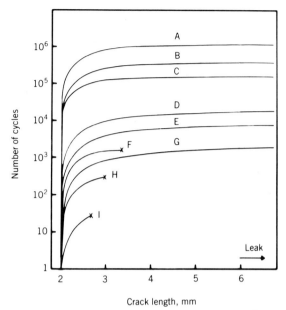

Fig. 34. Theoretical life calculations for fatigue of liquid storage cylinder made from various materials (183). × = fracture. A, poly(vinylidene fluoride) (PVDF); B, nylon-6,6; C, HI-nylon-6,6; D, PC; E, PVC; F, PMMA; G, ABS; H, PS; I, epoxy.

prior to their catastrophic failure. Rather, conditions for unstable crack growth would have been met in these two materials prior to the time when the crack would have penetrated the cylinder wall. That is, abrupt fracture would occur before the warning stage of tank leakage could develop.

In another design problem, a large flat plate is to be made from an engineering plastic. If the plate experiences a cyclic stress of 3 MPa and contains an initial through-thickness crack 2.5 cm in length, the computed cyclic lives of plates manufactured from different engineering plastics would correspond to the data shown in Table 1. Final crack lengths at fracture and the associated fatigue lives for materials differ dramatically. Table 2 shows the effect of molecular

Table 1. Computed Fatigue Lives of Plastic Plates[a]

Material	Molecular weight	Final crack length, cm	Fatigue life (No. of cycles)
polystyrene	2.7×10^5	7.25	20,200
ABS		35.5	95,900
acrylic	4.8×10^6	8.2	266,700
PVC	2.25×10^5	24.6	402,500
polycarbonate	4.8×10^4	39.8	1,031,000
nylon-6,6		38.4	2,902,000
poly(vinylidene fluoride)		44.8	9,826,000

[a] Ref. 183.

Table 2. Computed Fatigue Lives of Acrylic Plates

Molecular weight	Final crack length, cm	Fatigue life (No. of cycles)
1.1×10^5	2.5^a	0
1.9×10^5	3.45	2,300
3.5×10^5	4.16	6,400
2.3×10^6	4.16	39,800
4.8×10^6	8.2	268,300

^{a} Specimen fractured upon initial loading.

weight on fatigue life for acrylic plates. To carry out the necessary integration, the final flaw size based on the condition where unstable fracture would occur must first be computed. This can be done by determining the critical or final stress-intensity-factor value (fracture toughness) from either laboratory fatigue or fracture tests.

From a reassessment of equation 23, the service lifetime of a component can be extended in several ways. Clearly, by reducing the initial size of a defect, the overall service life of a component can be extended greatly. For example, the use of filters in the polymer melt to reduce the size of inclusions in the fabrication of plate sections has been suggested (180). Increased care should also be taken in the handling of components during service and installation. Stress concentrations should be reduced or eliminated wherever possible and the nominal stresses lowered. Equation 23 shows that the fatigue life of a cracked component varies inversely with the mth power of cyclic stress. For example, if $m = 4$ (typical for some engineering plastics), the fatigue life of a component is increased by a factor of 16 if the cyclic stress range is reduced by 50%.

Nonlinear viscoelastic response of engineering plastics, resulting from changes in the magnitude of stresses across stress gradients and/or hysteretic heating-induced temperature variations across the unbroken ligament of a component, introduces variations in elastic modulus throughout the specimen width. The associated breakdown of the linear relationship between stress and strain, thereby violates basic assumptions regarding the applicability of linear elastic fracture mechanics principles and the use of ΔK and relationships to estimate component lifetime such as that given in equation 23, must be viewed with caution. This is particularly true in cases where cracks are physically short and applied stress levels are high. Further studies are needed to determine whether or not alternative crack-tip field parameters such as ΔT (eq. 11) or cyclic J-integral values (184) may be more appropriate for modeling subcritical flaw growth in nonlinear viscoelastic materials.

Fracture Surface Morphology

Macrofractography. Macroscopic examination of numerous service failures generated by repeated loading reveals distinct fracture surface markings. Typically, the fracture surface is flat, indicating the absence of appreciable gross plastic deformation during service life. The absence of observable plastic flow belies the accumulation of irreversible damage through fatigue crack formation. These flat fracture surfaces generally contain parallel sets of lines, referred to as "clamshell markings," "arrest lines," and/or "beach markings," attributed to different periods of crack extension, eg, during the intermittent operation of a pump in a liquid transport system. The bands reflect periods of growth and do not correspond to individual load excursions. Intermittent rest periods where no cyclic damage is accumulated also contribute to the formation of these macro-scopic fracture bands. Beach markings often are curved, with the origin located at the center of curvature; to this extent, they direct the investigator to the fracture initiation site.

When fatigue cracks traverse a metallic or polymeric cross section, the orientation of the crack plane may be either normal to or inclined at an angle of $\pm 45°$ to the applied stress direction. Sometimes fracture surfaces possess both flat and slanted fracture zones. The relative degree of flat and slant fracture depends on the crack-tip stress state. When plane-stress conditions prevail (where the through-thickness stress $\sigma_z \cong 0$ and the plastic zone dimension $\gg$ specimen thickness), the fracture plane often assumes a $\pm 45°$ orientation with respect to the load axis and panel thickness probably because failure occurs on those planes containing the maximum resolved shear stress. The planes of maximum shear lie along $\pm 45°$ lines in the $y - z$ plane.

In plane strain, where $\sigma_z \cong \nu(\sigma_y + \sigma_x)$ and $r_y \ll t$, the planes of maximum shear are located in the $x - y$ plane at $\pm 60°$ from the horizontal plane with crack propagating along the $x - z$ plane. Apparently, the fracture plane under plane strain conditions lies midway between the two maximum shear planes, thereby reflecting the tendency for the crack to remain in a plane containing the maximum net-section stress. Studies of shear lip formation have been reported (111,185,186).

Different fatigue processes in engineering plastics give rise to various colors and textures on the fatigue fracture surfaces. For example, polycarbonate, polysulfone, polystyrene, and low molecular weight poly(methyl methacrylate) (PMMA) develop a mirror-smooth, transparent fatigue fracture surface at low ΔK levels that transforms to a rougher texture along with a misty appearance at intermediate and high ΔK values. Beginning at the mirror-mist transition region and extending for some distance beyond, one often observes a brilliant array of packets of color fringes on these fracture surfaces. By contrast, only one color is revealed for each material within the mirror region. Such color fringes reflect the presence of craze material (187,188). The relatively uniform color found in the mirror region probably reflects the existence of a single craze of relatively uniform thickness; the color fringe packets in the mist region reflect craze bun-dling, ie, multiple parallel crazes.

Stress whitening has been observed on the fatigue fracture surfaces of a number of engineering plastics, particularly semicrystalline polymers such as polyamides, polyoxymethylene, chlorinated polyether, and poly(vinylidene fluo-ride). Such whitening is believed to result from refractive index changes occurring as a result of molecular-chain reorientation and spherulite breakdown in the slowly advancing fatigue crack-tip plastic zone. The size of this damaged zone increases with increasing crack length in accord with an associated increase in the crack-tip plastic zone size (eq. 10).

Rubber-modified amorphous polymers also reveal stress whitening in the fatigue-crack-propagation regime which is believed to be caused by multiple crazing and/or shear banding during the cyclic deformation process. In many instances, the amount of whitening decreases abruptly at the point of fast fracture. Color fringes and stress whitening found on fatigue fracture surfaces may also occur under monotonic loading conditions. However, the relative differences in color-fringed and stress-whitened regions from the rest of the fracture surface in combination with the presence of beach markings, generally identifies the existence of a fatigue crack zone on the overall fracture surface.

Microfractography.   The fatigue-fracture-surface micromorphology of engineering plastics is complex. At relatively high ΔK levels, many polymers reveal the presence of fatigue striations that correspond to the successive location of the advancing crack front after each loading cycle (Fig. 35); the width of each striation, therefore, represents the amount of crack-front incremental advance resulting from the prior load excursion. There is a clear distinction between macroscopically observed clamshell markings, which represent periods of growth during which hundreds or thousands of loading cycles may have occurred, and fatigue striations, which represent the extent of crack-front advance during one load excursion. Each fatigue clamshell marking may, therefore, represent the extent of crack advance corresponding to thousands or tens of thousands of individual load or strain excursions. The microscopic measurement of the fatigue-crack-propagation rate, ie, striation width, is in excellent agreement with macroscopically determined crack growth rates (119,189) (Fig. 36). This strong correlation between macroscopic and microscopic growth rates in polymeric solids reflects the fact that 100% of the fracture surface in this ΔK regime is striated—an observation not typically found in metal alloy systems. In the latter instance, other fracture mechanisms such as cleavage, microvoid coalescence, and intergranular failure are observed in various locations on the fracture surface.

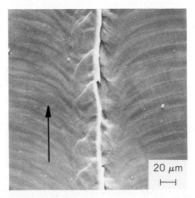

20 μm

Fig. 35.   Fatigue striations in polysulfone corresponding to the incremental advance of a crack as a result of a single loading cycle (119). Arrow indicates crack growth direction.

More quantitative information has been obtained from the measurement of fatigue striations than from any other fracture surface detail. Since the striation represents the position of the crack front after each loading cycle, its width can be used to measure the fatigue-crack-propagation rate at any given location on

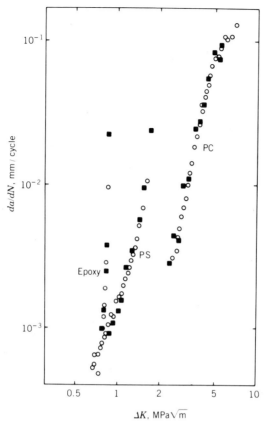

Fig. 36. Macroscopic (○) vs microscopic (■) fatigue crack growth in epoxy, polystyrene, and polycarbonate (119).

the fracture surface. This information is extremely valuable in postfailure analyses of service fractures. For metallic alloys, the striation spacing width varies with the stress-intensity-factor range according to the relationship (190):

$$\text{striation spacing} \sim 6 \left(\frac{\Delta K}{E}\right)^2 \tag{25}$$

where ΔK = stress-intensity-factor range and E = modulus of elasticity. Though most metallic alloys reveal this second-power dependence, no such correlation has been clearly identified for engineering plastics. Since there is a close match between striation spacings and the associated macroscopic growth rates in polymers, the ΔK dependence on polymer striation width is not constant but varies greatly with material. For example, fatigue crack propagation rates have been found to vary with ΔK from the third to twentieth power of the stress intensity range. Whereas striation width-crack-opening displacement correlations have been proposed for the case of metallic systems, (based on the second power dependence of K on the crack-opening displacement), such correlations are suspect for the case of polymer fatigue fracture (189,190). Results for PMMA show that neither the craze length nor craze width agree with the macroscopic-growth increment per unit cycle (191). Since striation spacings should correlate with the macroscopic growth rate in the appropriate ΔK regime (Fig. 36), a correlation between craze width and fatigue striation spacing also cannot be established.

Other fatigue fracture surface markings have been identified. Depending on the chemistry and molecular weight of a particular polymer, as well as the prevailing ΔK level and test frequency, bands oriented parallel to the advancing crack front have been observed (Fig. 37). The spacing between these fracture bands does not correspond to the prevailing macroscopic growth rate in this ΔK regime and, as such, does not correspond to fatigue striations discussed above. Instead, these fracture bands reflect discrete crack advance increments that develop following several hundred to several thousand loading cycles of total crack arrest. Calculations (192) show that these individual band widths vary with the square of the prevailing stress intensity factor and correspond to the dimension of the crack-tip plastic zone (l_D) as formulated by the Dugdale model (193) given by

$$ l_D = \frac{\pi}{8} \left(\frac{K}{\sigma_{ys}} \right)^2 \tag{26} $$

These large bands develop in two distinct stages (119,194). First, a craze forms at the leading edge of the crack which then lengthens and thickens continuously with each successive loading cycle. When the craze reaches some critical dimension, the crack extends abruptly through the craze during the next loading cycle and terminates at or near the craze tip. Other results have been interpreted as showing that the crack jumped through approximately two-thirds of the craze before arresting (195,196). Crack growth occurs again in discontinuous fashion only after the development of a new craze at the crack tip.

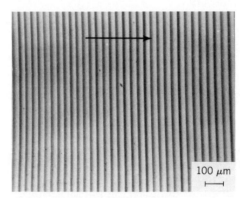

Fig. 37. Discontinuous growth bands in poly(vinyl chloride). Arrow indicates crack growth direction.

It has been suggested that the growth of the craze to the Dugdale dimension, l_D, constitutes a necessary, but not sufficient condition, for fracture of these fracture bands. Instead, the key to the sudden breakdown of the band is believed to correspond to a cyclic strain-induced stretching of the craze fibrils to their ultimate length, t_{max}, which is less than the maximum crack-opening displacement (COD_{max}) for the given ΔK level (197,198). If $t_{max} > COD_{max}$, then continuous crack growth should occur and no discontinuous growth bands (DGB) would be expected (Fig. 38a). The latter condition exists under low test-frequency conditions (199) and when the material possesses either a high molecular weight or high molecular weight tail in a lower average molecular weight polymer (120,194). The model for discontinuous-growth-band formation pictures craze thickening as occurring by fibril stretching which contributes to orientation hardening and fibril strengthening (189,191,197,198,200). At some point, it is expected that the most highly stressed fibrils will disentangle and fracture. With the breakage of each additional fibril, the remaining unbroken ligaments assume an ever-increasing load and associated strain. These fibrils then become stronger as a result of further orientation hardening. It is envisioned that the reduction in fibril-volume fraction is counterbalanced by an increase in the average strength of the remaining fibrils due to orientation hardening. The net effect of this competition (weakening from fibril fracture vs strengthening from remaining fibril-orientation hardening) establishes a relatively constant stress across the craze zone. This constant stress condition is consistent with the underlying boundary conditions associated with the Dugdale plastic-zone formulation and with other findings (190). Sudden crack advance through the craze is then expected to take place when $t_{max} < COD_{max}$ and when the nth fibril fractures and the applied load exceeds the collective load-bearing capacity of the remaining craze-spanning fibrils (Fig. 38b) (196).

Recalling that a discontinuous growth band is developed over many loading cycles, the cyclic stability or life of each band, N^*, may be determined by dividing the band width by the corresponding fatigue crack growth rate such that (2)

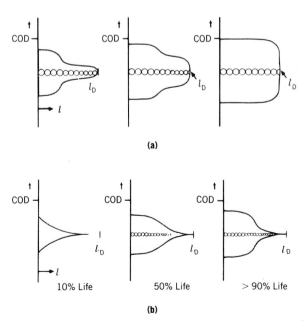

(a)

(b)

Fig. 38.   Schematic representation of (**a**) nondiscontinuous growth band formation and (**b**) DG-band formation where $t_{max} < COD_{max}$ (197).

$$N^* = \frac{\text{band width}}{da/dN} \tag{27}$$

N^* decreases markedly with increasing stress-intensity-factor range (Fig. 39) (54) because of the greater amount of cyclic damage introduced within the craze zone at the higher ΔK levels. The overall craze stability improves dramatically with increasing molecular weight and is consistent with the markedly greater degree of fatigue-crack-propagation resistance observed with high molecular weight specimens (201) (Fig. 23).

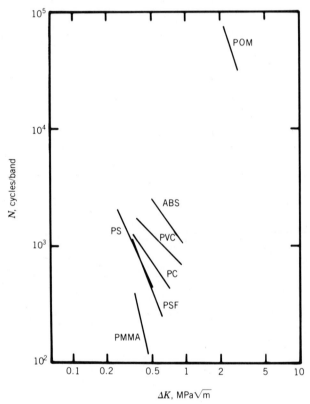

Fig. 39. Cyclic stability of discontinuous growth bands in several polymers decreases with increasing ΔK level (54).

An electron microscope examination of these discontinuous growth bands reveals a clearly defined void gradient with large voids located near the crack tip and small voids found near the craze tip (119,194). These microvoids represent small regions of empty space surrounded by broken fibrils (Fig. 40) (195) (see MICROMECHANICAL MEASUREMENTS). Separating adjacent bands is a narrow stretch zone consisting of elongated microvoids that point back towards the crack origin. These fracture surface markings suggest that the craze breakdown process takes place by void coalescence and fibril rupture with the void-size distribution reflecting the internal structure of the craze just prior to crack extension. The void-

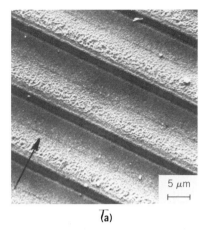

(a)

(b)

Fig. 40.   (a) Discontinuous growth bands in PVC. Arrow indicates growth direction;
(b) note ruptured fibrils (194,195).

size gradient parallels the craze-opening displacement distribution across the craze.

Different craze lengths for a given ΔK level have been reported, depending on the manner in which the sample was viewed (202,203). When crazes are viewed normal to their plane, a series of interference fringes are observed corresponding to the length of the craze ahead of the crack tip. When viewed with the craze on edge, the measured craze length is typically smaller because the craze thickness near the craze tip is believed to be less than the resolution limit in the light microscope. Based on several specimens of PVC, corresponding to different ΔK levels and different molecular weights, the relative craze-growth kinetics are independent of the prevailing stress intensity factor and molecular weight (203). That is, when the instantaneous craze length and associated number of cycles is normalized by the final craze length l_D (defined by the Dugdale dimension) and the cyclic stability N^*, respectively, the data may be described by a relationship of the form

$$(l/l_D) = D(N/N^*)^{0.3} \tag{28}$$

These results point out that, regardless of the PVC material parameters or stress intensity level, the kinetics of craze development and final breakdown during cyclic loading are the same.

Additional optical interference measurements pertaining to discontinuous-growth-band development and breakdown have been reported (191,195,200, 204,205). Of particular note, the craze thickens with continued cycling by fibril stretching and associated orientation hardening (200). This finding is consistent with the DGB model described above; other investigators found that craze thickening in polycarbonate and polystyrene under quasistatic conditions takes place by drawing fresh material into the craze from the surrounding bulk (206,207). It is believed that surface drawing of new material into the craze takes place under static loading conditions and should also occur under cyclic loading at low cyclic frequencies. Since the fibrils are not stretched significantly, but rather are lengthened by the addition of new material, fibril length can exceed COD_{max} before failure; DGB formation would then be precluded (Fig. 38). On the other hand, when fibril stretching occurs under high cycle fatigue conditions, DGB formation occurs more readily since the fibrils are strained to failure at lengths less than COD_{max}. The development of discontinuous growth bands, therefore, reflects a competition between two craze-thickening processes: fibril stretching and surface drawing of new material into the body of the craze.

A modified form of the discontinuous crack growth process has been identified (5,208–210) as a pair of shear bands that form above and below the crack plane associated with discontinuous crack-growth formation in polycarbonate and polysulfone. The combined craze band and double shear band is referred to as an ϵ-DG band since these deformation markings resemble the Greek letter epsilon (209) (Fig. 41). Such shear bands were not found in other studies (194). Subsequent studies traced the existence of these shear bands to test conditions involving unnotched specimens that were loaded to stress levels >25% of the material's yield strength (197). The formation of ϵ-DG bands at high σ_{max}/σ_{ys} ratios in engineering plastics provides proof that the incidence of crazing and shear banding can change markedly when the nominal stress/yield strength ratio is altered.

Another type of fracture surface morphology has been identified in polystyrene and polycarbonate (4,43). A patch-type morphology is seen at low ΔK levels suggesting the passage of the fatigue crack along alternate boundaries of the craze–matrix interface (211). Further studies are needed to clarify the stress–time–temperature parameters pertaining to the development of these various fracture surface markings.

Fatigue striations and DGBs have also been observed in semicrystalline polymers, eg, nylon-6,6 and polyoxymethylene (54,212), but, in general, the fracture surface reveals different micromorphological features. When plastic deformation is limited and fatigue fracture occurs before significant disruption of the spherulitic structure, the fracture surface morphology reflects the original microstructure of the bulk polymer. On the other hand, when plastic deformation

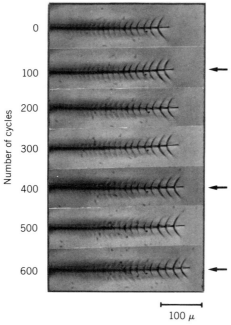

Number of cycles

0
100
200
300
400
500
600

100 μ

Fig. 41. Shear bands associated with discontinuous growth bands to form ε-discontinuous growth bands (209).

occurs readily, compressive yielding within the reverse plastic zone at the crack tip crushes and elongates the spherulites in the direction of crack growth. For nylon-6,6 containing various amounts of imbibed water, the transition from a transspherulitic to highly drawn fracture surface morphology depends primarily on the viscoelastic state of the polymer (Fig. 42) (213).

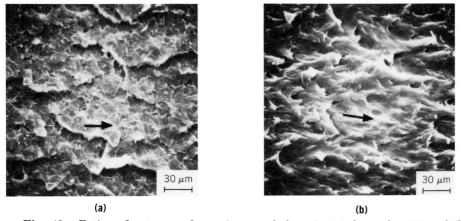

(a)   (b)

Fig. 42. Fatigue-fracture-surface micromorphology in (a) dry nylon-6,6 and (b) nylon-6,6 + 1% H_2O (213). Arrow indicates crack growth direction.

Polymer Blends

Effects of rubbery phases on the fatigue behavior of a rigid polymer can be very complex. As with unmodified polymers, high values of $\Delta\sigma$ or frequency may generate enough hysteretic heating to induce failure by softening (Fig. 43); lower values may lead to failure by the initiation and propagation of a crack in a nominally brittle manner (2,214). Even with the latter, thermal effects at and beyond the crack tip can greatly modify FCP behavior (2,67). Whereas softening due to cyclic damage, with or without hysteretic heating, is typical of most polymers (6,25), a rubbery phase can cause several specific effects, eg, significant decreases in yield stress and modulus and a greater susceptibility to creep. Moreover, the reduction in ultimate elongation associated with cyclic softening may be greater in rubber-modified polymers (26,215) (Fig. 10b). Thermal effects, which may be beneficial or deleterious, are likely to be more intense, at least with cycling between fixed stress limits, for the loss compliance (D'') may be increased by the presence of rubber (2). Also, rubber modification may permit cycling to higher values of load and ΔK, thus increasing the propensity for hysteretic heating as cracks grow further before fracture (67).

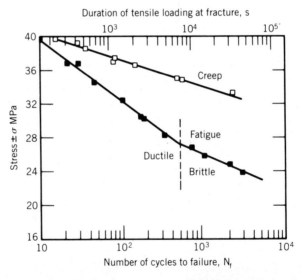

Fig. 43.   Comparison of S–N curve under square-wave loading with creep–rupture data for typical ABS (214). Note that the creep–rupture time scale is equivalent to that of the fatigue cycle scale in terms of time under tensile load. To convert MPa to psi, multiply by 145.

Thus it is not surprising to find exceptional diversity of fatigue behavior in rubber-modified systems. Although results can often be explained in terms of a plausible balance of mechanistic factors, unequivocal quantitative conclusions about the role of rubber are usually difficult because the behavior of the polymer blend is seldom compared with the identical matrix polymer, and loading conditions used have varied widely. Because the rubbery phase may affect crack initiation and propagation differently, it is convenient to consider the behavior of unnotched and notched specimens separately. (For reviews see Refs. 2, 6, and 79).

Unnotched Specimens. Examples of materials using fixed load limits include: acrylonitrile–butadiene–styrene terpolymers (ABS) (6,102,214–217), rubber-modified polystyrene (HIPS) (6,218,219), toughened poly(methyl methacrylate) (PMMA) (220,221), poly(vinyl chloride) (14,15), and epoxy (14,15). Unfortunately, except for the materials of Refs. 14, 15, and 222, comparisons of behavior of matrix materials known to have the same molecular weight as in the blend are unavailable, although some comparisons with matrix materials of the same general type have been made (6,102). All these examples involve sinusoidal wave forms and square-wave loading has been examined with ABS (214); tests involved both tension–tension and tension–compression conditions. Constant-strain-range studies have also been made with ABS and HIPS (26,222), and rubber-modified PVC and PMMA (222); repeated impact tests have been reported for rubber-modified polycarbonate (223).

As with monotonic loading, both crazing and shear mechanisms have been identified during repetitive tensile loading of HIPS (93,224), during cyclic fatigue of ABS and HIPS (6,219), and during impact fatigue of toughened polycarbonate (223). Crazing appears to dominate the deformation process in HIPS (6,214,219), whereas shear is more important in ABS (6,214) and a combination of shear and crazing is present in rubber-toughened PC (223). The relative importance of shear and crazing is clearly evident in Figure 44 (214) which shows a typical distortion of the hysteresis loops in HIPS from repeated opening of crazes in tension and closing in compression (6,214). In contrast, with ABS the loops are essentially symmetrical, as expected for a response dominated by shearing. Figure 44 also illustrates the occurrence of significant progressive softening in both polymers; rubber-modified PMMA behaves like ABS (220,221). Although hysteretic heating is undoubtedly a major cause of softening (102,220), in toughened PMMA compliance tends to rise due to the generation of multiple shear bands even after a steady-state temperature has been attained (Fig. 45) (220). At least with HIPS, crazing in fatigue begins at lower stresses than is the case with monotonic loading (219).

High-load-amplitude failure in ABS, HIPS, and toughened PMMA involves gross yielding accompanied by multiple internal cracking. Whereas crazes are typical in HIPS (6), shear bands have been reported in ABS (102) and toughened PMMA (220); some crazes may also appear in ABS, as well as microvoiding in the rubbery phase (102). Impact failure in toughened PC involves a combination of cracking and shear yielding (223). At low stress amplitudes, on the other hand, brittle craze cracks develop through crazes from the surface in both ABS and HIPS and grow inward (6,102,214). Thus, even though ABS exhibits high impact toughness, brittle fatigue crack propagation is characteristic at low stresses; evidently when plane-strain conditions are present with low load, constraint inhibits the shear banding that otherwise would confer high ductility (214). Interestingly, unnotched toughened nylon-6,6 does not readily fail under repetitive impact loading.

Consideration of overall fatigue life and the relative effects of rubber on

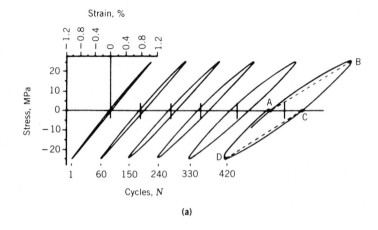

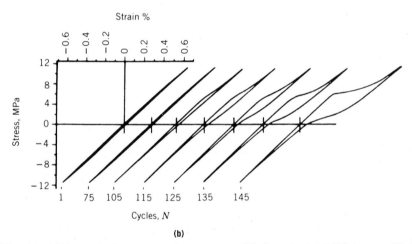

Fig. 44. Hysteresis loops at various stages of fatigue in (**a**) ABS ($\sigma = \pm 25.4$ MPa) and (**b**) HIPS ($\sigma = \pm 11.6$ MPa) (214). To convert MPa to psi, multiply by 145.

crack initiation and propagation reveals interesting and perhaps unexpected behavior. For example, whereas the fatigue performance of a typical ABS resin has been reported to be superior to that of typical PS samples (217), detailed examinations of several rubber-toughened polymers have shown that the high level of impact strength is not necessarily carried over into fatigue resistance, at least under some loading conditions. Indeed, although superior to HIPS in this respect, ABS can exhibit premature fatigue failure (225) and has been compared unfavorably with, for example, rigid PVC (226); other rubber-modified plastics may suffer from a similar problem (214).

At least with HIPS and ABS tested in tension–compression, the rubbery phase decreases both initiation and propagation lifetimes (N_i and N_p) compared to typical polystyrenes or styrene–acrylonitrile copolymers (SAN) (6,102,216,218). The fatigue life is significantly reduced in HIPS compared to a typical PS; evidently the rubber reduces the resistance to both crack initiation and propagation, at least in tension–compression loading with high stress amplitudes. Such stresses

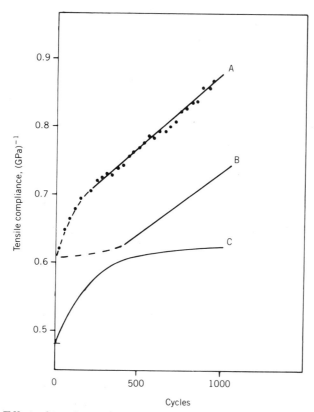

Fig. 45. Effect of tension–compression cycling at 0.5 Hz on compliance in rubber-modified PMMA (220). Curves A, B, and C refer to the overall compliance change, the difference due to structural changes, and the contribution of hysteretic heating. To convert GPa to psi, multiply by 145,000.

are higher than would be anticipated in conventional design. Although ABS exhibits better performance than HIPS or PS, its behavior is still inferior to that of SAN matrix materials. Evidently the lower yield stress in the toughened resins greatly facilitates initiation, and compressive loading severely damages craze fibrils during the crack closure part of the load cycle (Fig. 44). In fact, if the S–N curves for HIPS and PS are normalized by plotting $\Delta\sigma/\Delta\sigma_y$ instead of $\Delta\sigma$, HIPS appears to be relatively superior to PS; a similar superiority for ABS relative to SAN is seen if comparison is made by normalizing $\Delta\sigma$ with respect to tensile strength (102).

The especially deleterious effect of compressive stresses is consistent with the observation that fatigue life of HIPS and ABS is increased significantly by switching to tension–tension loading (6,102). Indeed, when the minimum stress is raised to 8.8 MPa, no failure is observed in HIPS even after 10^7 cycles. With ABS, hysteretic heating dominates fatigue failure; cooling of specimens by interruption of the test yields lifetimes equivalent to those of SAN (102). An increase in frequency (at least up to 10 Hz) causes an increase in fatigue life (6), presumably due to the associated increases in modulus, or to localized crack blunting.

Deleterious effects of rubber modification have also been reported with toughened PVC (14,15,222), epoxy (14,15), PMMA (222), polystyrene (222), and polycarbonate (223). Whereas the MBS rubber in PVC increases the resistance to FCP, the resistance to crack initiation is decreased by the presence of rubber, especially for a high molecular weight matrix (Fig. 11). Crack initiation is also facilitated in epoxy and polycarbonate.

Thus, to avoid unpleasant surprises, caution should be used in subjecting rubber-modified plastics to fatigue loading, especially if compressive stresses or high stress amplitudes are involved. Before application in such situations, careful tests should be run to simulate anticipated loading conditions.

Notched Specimens. Most studies of notched specimens of rubber-toughened polymers have involved tension–tension rather than the tension–compression loading typical of investigations with unnotched materials. Beneficial effects of rubber have been reported for several systems. For example, in early FCP research, typical commercial ABS and HIPS resins were shown to be more resistant to FCP than a typical PS (129); a similar conclusion was reached for modified poly(p-xylyl ether). More recently, extensive studies of polymer blends in comparison with the matrix have revealed significant increases in FCP resistance in PVC modified with a methacrylate–butadiene–styrene terpolymer (227) and rubber-toughened nylon-6,6 (65,70,77,213,228), provided the frequency is low enough to minimize extensive hysteretic heating. A similar beneficial effect has been noted in simultaneous interpenetrating polymer networks (SINs) based on cross-linked polystyrene (127) or PMMA (229) intertwined with a second network consisting of a cross-linked elastomer.

With modified PVC (Fig. 46), the best performance is obtained with the highest molecular weight matrix and the highest rubber content (14 phr rubber). However, the greatest relative improvement is observed with a lower molecular weight matrix. The softening of the high molecular weight PVC due to the rubber probably offsets the benefit of the rubber to some extent. With nylon-6,6, the FCP resistance also increases with rubber content at a frequency of 1 Hz, but at a higher frequency, the curves exhibit a transition above which cracks grow faster, the greater the rubber content (228). Because normalization of FCP rate with respect to Young's modulus (by plotting vs $\Delta K/E$) eliminates the crossover, the behavior is rationalized in terms of a balance between toughening from the rubbery phase and softening from extensive hysteretic heating. Indeed, temperatures as high as 100°C or more have been recorded at the crack tip, depending on the rubber content and frequency.

Toughening appears to be associated with crazing around the rubber particles in HIPS (6,230) and in toughened poly(p-xylyl ether) (231), and with enhanced shear response in modified PVC (232). However, the latter system fails by extension of a craze ahead of the main crack. Thus in these polymers, toughening mechanisms operative in FCP are those seen in impact failure. A similar conclusion is likely with modified PS; in fact the ranking of ABS as superior to HIPS in FCP resistance agrees with the ranking of unnotched specimens, and with the generally higher propensity of ABS for shear deformation. With toughened nylon-6,6, a combination of microcracking with extensive shearing appears to dominate the response (213) (Fig. 47), as it does with impact loading (233).

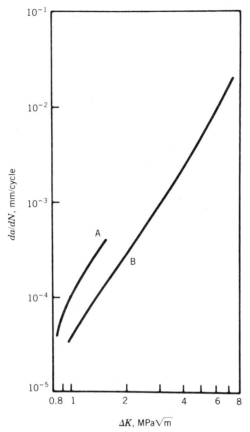

Fig. 46. Effect of MBS rubber (14 phr) on fatigue crack propagation behavior of poly(vinyl chloride) at 10 Hz (mol wt = 9×10^4). A, no MBS; B, with MBS (227).

Fig. 47. Secondary cracks (see arrow) oriented parallel to the advancing crack front in rubber-modified nylon-6,6. Note extensive shear-induced tufting oriented in the crack direction (213).

Little or no significant benefit of rubber modification has been seen in the few epoxies examined, even though the rubber improved the impact strength (234,235). Thus, as the rubber content is increased to 15%, only a slight improvement in FCP resistance is noted (235). The cause of this paradoxical behavior has not been established, but it is speculated that the rubber-matrix interface may fail well ahead of the advancing crack, thus limiting the already low capacity for shear deformation in the matrix.

Effects of frequency on the FCP response of rubber-modified polymers are interesting and complex (2,67); effects of wave form have not been studied. The FCP rate of toughened nylon-6,6 at a constant and fairly high value of ΔK increases with increasing frequency (228). Indeed, FCP tests could not be run at a frequency greater than 30 Hz due to excessive softening. At the same time, increasing frequency (up to 50 Hz) decreases the FCP rate in a typical ABS, but has little effect on the FCP rate of a toughened PVC (67). This diversity of behavior can be rationalized in terms of variations in the scale of hysteretic heating by examining both the loss compliance as a function of temperature and the temperature profile at the crack tip. Thus with the modified PVC, only slight increases in crack-tip temperature are observed at lower frequencies, due to a nearly constant value of D'' around ambient temperatures. At 50 Hz, however, the effect of increased frequency becomes evident and localized crack blunting may occur. With ABS, the higher values of D'' presumably result in such blunting even at low frequency. However, with the toughened nylon-6,6, the higher values of ΔK attained apparently offset the lower value of D'' so that extensive heating occurs. Thus the effect of frequency depends on the balance struck between localized (and beneficial) heating at the crack tip and generalized (and deleterious) heating at the crack tip that causes a significant loss of stiffness in the bulk material.

The improvements in FCP resistance described above are also seen in toughened nylon-6,6 and acetal resin subjected to repeated impact loading (233,236,237). The toughened nylon fails on impact only when notched, in contrast to neat nylon-6,6. With unmodified nylon, little damage accumulates prior to catastrophic fracture. On the other hand, the toughened material progressively undergoes significant energy absorption and damage until it fails; the number of impacts to failure is double that for the control resin. The decreased sensitivity of the toughened resins to repeated impact is evident in Figure 48, which compares the number of impacts to failure for several polymers as a function of impact energy.

In summary, with some fatigue loading conditions, rubber modification of brittle or notch-sensitive polymers may improve fatigue resistance. In such cases, the micromechanisms that result in toughening of monotonic loading may well be operative, though the precise balance has not yet been elucidated. However, with other loading conditions, especially involving a combination of tension and compression, rubber modification may have a deleterious effect on fatigue performance. In view of the current trend towards increased use of polymer blends in load-bearing applications, many of which involve vibrations, caution in selection is advisable. Potential service conditions should be simulated as closely as possible in fatigue testing, and the importance of both crack initiation and propagation stages should be determined.

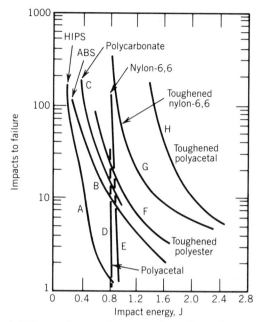

Fig. 48. Impact fatigue of several neat and toughened engineering plastics (237). A, HIPS; B, ABS; C, PC; D, polyacetal; E, nylon-6,6; F, toughened polyester; G, toughened nylon-6,6; H, toughened polyacetal.

Rigid Particle Systems

Fillers (qv), eg, silica, calcium carbonate, and mica, are commonly added to polymeric matrices to increase modulus and dimensional stability, and, some-times, to increase wear resistance and decrease the coefficient of friction (79,238). Preformed rigid polymer particles are also incorporated into casting resins for orthopedic or dental restorations (2,239–245). The effects of filler on ultimate fracture behavior is less predictable and dependent on the filler content, inter-facial adhesion, and temperature and ductility of the matrix (79,238,246). Thus, depending on the system, the value of $\mathscr{G}_c$ may increase, decrease, or exhibit a maximum as filler content is increased. This complex behavior reflects compe-tition between several factors, ie, cracking of the subsurface, increase in surface roughness and modulus, ability of particles to pin the crack (79), and ductility from variations in matrix constraint at the interface (79,238,246,247). In the case of K_c, values may increase or decrease with filler content. Although much remains to be learned about the balance of factors mentioned, typical particulate fillers often are deleterious to strength and toughness.

Fatigue life in several filled nylon and epoxy systems has been described and softening prior to failure attributed to hysteretic heating (248); the study did not compare behavior with that of the unfilled matrix. Several studies of acrylic bone cements have also been reported (239,242–245) as a function of composition, mixing technique, and environment. At frequencies ranging from 20–23 Hz, failure in S–N tests is by crack growth rather than softening (242);

in other studies at 1 Hz, fatigue strength is observed to be lower than for mono-lithic PMMA (244,245), but to be better for the more porous cements cured under typical operating-room conditions. Fatigue life is somewhat improved by testing in bovine serum (242) or Ringer's solution (244,245) to simulate the physiological environment; the presence of a radio-opaque filler ($BaSO_4$) increases fatigue life in air, but not in bovine serum (242). Special specimen designs and test procedures for dental cements have been proposed (241).

The following quantitative expression relating fatigue life to frequency, wave form, load, and static properties has been proposed and shown to hold for talc-filled polypropylene (249):

$$N = kv\sigma^{-n}f(n) \tag{29}$$

where N is the number of cycles to failure, v, frequency (in this case, <1 Hz), $f(n)$, a function of wave form, σ, the maximum load (assuming a minimum load of zero), and k and n, constants derived from creep-rupture tests.

In addition to particulate composites containing $<\sim50$ vol% of filler, systems comprising only enough polymer to serve as a binder (ca 10–20 vol%) are of increasing interest. Such polymer concretes are commonly used in applications such as bridge-deck overlays, patching compounds, and construction materials (238). Although little is known about fatigue performance of polymer concretes (qv), excellent results have been reported for a mineral-based composite (containing up to 88 vol% filler) designed to encapsulate the stator windings of a superconducting turbogenerator, and to undergo compressive and tensile fatigue at 200 Hz (250). After 10^7 cycles, fatigue strengths are $>50\%$ of the static values; prepacking of the filler to give pathways for good thermal conduction reduces hysteretic heating, and gives superior results.

In a study of fatigue crack propagation in typical bone cement, the Paris law was followed, and the response affected by the microstructure of the cured composites. The resistance to fatigue was greater for specimens prepared with microvoids, ie, under typical operating-room conditions (244,245). All specimens exhibited lower slopes of da/dN vs ΔK and higher values of K_c than noted for typical neat PMMA.

Several crack-resisting mechanisms were observed: pinning of the crack front, lower crack growth rate in the matrix (possibly due to plasticization by unreacted monomer), and the nucleation of microcracks at void sites. The importance of crack pinning in increasing FCP resistance, at least up to an optimum filler concentration, has also been reported for several polymers containing hollow or solid glass or fly-ash beads (251,252). At lower filler contents ($<\sim50$ vol%) the Paris law appears to hold, and interpretation of the toughening in terms of interparticle spacing is possible (252). However, at higher filler contents, FCP rates may begin to increase (251) or become irregular (252). A possible analogy has been observed in photochemically polymerized polystyrene (116), in which microgel particles formed *in situ* appear to generate energy-dissipative shear processes that increase the resistance to FCP in comparison to typical linear polystyrenes. Also, FCP rates in PVC can be reduced significantly by the incorporation of high molecular weight particles of emulsion-polymerized PVC (253).

In contrast, the FCP rate of a mineral-filled nylon-6,6 is significantly higher than that of the matrix (254). Although this result is surprising in the light of the ability of the filler to increase strength and toughness markedly under monotonic loading, tests showed the induction of a stronger sensitivity to notch sharpness than in the control, and hence to a fatigue-induced, rather than machined, notch.

One of the most detailed studies of FCP behavior so far has been performed on nylon-6,6 filled with glass beads (avg dia ~100 μm) treated to enhance adhesion (247). Both fracture toughness and fatigue resistance (as measured by a vertical shift in da/dN curves) decreased as the filler content increased to a volume fraction of 0.52. As the filler content increased, the tendencies towards interfacial fracture and matrix yielding increased and decreased, respectively, so that the macroscopic mechanism changed progressively from ductile tearing to brittle, unstable crack growth. Evidently, the increase in yield stress (and consequently smaller plastic zone size and K_c) due to the presence of filler dominates the response in this case.

A simple rule of mixtures reflecting the relative contributions of fracture resistance for the matrix and interface, as well as the tortuosity from the filler has been proposed based on fractographic examination (247) and shown to agree with predictions based on plausible assumptions. For fatigue crack propagation of the composite, the following equation was developed for a given value of ΔK:

$$\frac{da}{dN} = \frac{da}{dN_\mathrm{m}} \cdot (1 - c \cdot f_\mathrm{p}) \cdot b + \frac{da}{dN_\mathrm{i}} \cdot c \cdot f_\mathrm{p} \cdot n^{-1} \tag{30}$$

where the subscripts m and i refer to the matrix and interface, respectively, f_p is the volume fraction of polymer, c, a measure of the contribution of interfacial fracture to the fracture surface ($c \geq 1$), n, a tortuosity factor ($1 < n < 2$), and b, the relative importance of matrix vs interface fracture [$1/(1 - f_\mathrm{p}) \geq b \geq 1$]. In effect, the term b is a measure of the relative magnitude of r_p (the plastic zone size) with respect to the interparticle distance D_p; in low amplitude fatigue, $r_\mathrm{p} < D_\mathrm{p}$, whereas in high amplitude fatigue or static fracture, $r_\mathrm{p} > D_\mathrm{p}$. Reasonable agreement was found between the prediction of equation 29 and the experiment, assuming that the rate of interfacial crack growth rate contributes up to 75% of the total crack growth rate as estimated from the features of the fracture surface.

As with failure under monotonic loading, complete understanding of the role of interfacial adhesion and its effect on the mechanisms of fracture awaits systematic studies involving well-controlled variations in interfacial properties (qv). The use of acid–base concepts may well prove to be advantageous in such fundamental studies, as well as in modifying engineering behavior (246,255).

Fibrous Composites

As the shape of a high modulus phase dispersed in a polymer is elongated from a sphere to a fiber, the ability of the dispersed phase to bear load is progressively increased. At a given concentration of uniformly dispersed reinforcement, several matrix-related properties are usually improved, eg, stiffness, strength, resistance to creep and fatigue, and heat-distortion temperature; fracture energy is also often increased, especially with long fibers. Maximum effects are achieved with long or continuous fibers, but significant improvements are conferred by short fibers as well. In practice, fibers such as glass, boron, carbon, and high modulus organic materials, eg, aromatic polyamides (aramids), are typical, often by themselves, but also in combination with each other, or with particulate fillers (as in sheet-molding compounds). A wide variety of geometrical arrangements are used, including randomly and uniaxially oriented fibers, cross-plied laminates, and woven structures. Compositions range from those in which the matrix is the dominant component to those in which the matrix serves to bond and protect an aggregation of fibers (2,238,256–261) (see also COMPOSITES).

Except for some sheet-molding compounds, the mechanical properties of fibrous composites depend on the concentration, orientation, and nature of the fibers; in the case of laminates (qv), the stacking sequence of the plies is also important. Whether attained by shrinkage of the matrix or by the operation of chemical adhesion, the degree of coupling between the fibers and the matrix strongly affects load transfer across the interface and the micromechanisms of failure (262). With many composite geometries and combinations of loading, the matrix is an important strength-limiting component, eg, when fibers are aligned at angles close to the loading axis, when shear stresses are involved, and with certain polymers.

The overall failure process in composites is complex, both in static and fatigue loading. Complications include the following (2,260): *1*. Although damage is progressive, physical integrity may be maintained for many decades of cycles, and criteria for failure may be based arbitrarily on the degradation of a property such as modulus. *2*. Diverse micromechanisms of failure include fiber deformation and fracture, debonding, delamination of plies, and matrix cracking (262–264). *3*. The balance of micromechanisms depends on such factors as hysteretic heating (modified by the presence of fibers), relative orientation of fiber and stress axes, mode of loading, constraints and coupling of stress fields due to the fibers or laminae, and the presence and nature of flaws. *4*. In principle, linear elastic fracture mechanics is inapplicable to heterogeneous systems, and the concept of a crack requires redefinition in terms of a more diffuse zone of damage. *5*. Under some circumstances, fatigue loading can result in effective crack blunting so that fracture toughness may actually increase, at least during part of the fatigue life (260,265).

Although these complexities hinder fundamental study, their occurrence reflects the considerable advantages of fibrous composites as engineering materials. Indeed many such composites are capable of outstanding performance both in an absolute sense in comparison with metals, and especially when compared relative to density (2,257,259–261,266,267).

Short-fiber Systems.   If uniaxially oriented, short fibers in a composite can be loaded to their maximum possible stress as long as the length $\ell \geqq \ell_c$, where ℓ_c is a critical fiber length, typically in the range of 100–200 μm (268). In terms of length-to-diameter ratio, ℓ/D, the number of cycles to failure (corresponding to a 20% decrease in modulus) in a boron-fiber-reinforced epoxy rises steeply with increasing aspect ratio up to a value of ~200, and then approaches an asymptotic limit (269). Of course, fiber breakage during injection molding must be taken into account. Also, in typical short-fiber systems, eg, sheet-molding compounds containing randomly oriented chopped-glass strands or injection-molded composites, fiber orientation patterns are complex (266,267,270). Thus whereas the fatigue resistance of short-fiber composites is generally very good, and in some cases similar to that of continuous glass-fiber composites and to that of carbon-fiber composites tested off-axis, full optimization of fatigue behavior has not yet been exploited (266). However, improved fatigue performance and fracture toughness relative to that of the matrix is often accompanied by lowered impact strength (266,268). Also, short-fiber composites tend to have lower static strengths than long-fiber composites, and to be more sensitive to notches and flaws. Nevertheless, on balance, the ease of fabrication and the wide spectrum of improved properties of short-fiber-reinforced plastics has led to increased use in automobiles, appliances, and structural components.

Two systems are currently of commercial interest: glass-fiber-reinforced thermosets based on random glass strands, and injection-molded thermoplastics. The former class includes impregnated chopped-strand mats (CSM) and, more recently, sheet molding compounds (qv) (SMC) (266,271,272). Little attention has yet been given to the use of aligned short-fiber composites (269,273–277), though such systems have the potential for providing useful multidirectional laminates in a manner analogous to that used for continuous fibers (266,273).

The S–N behavior of many short-fiber reinforced plastics has been reviewed (266). Examples of matrices include: neat and toughened unsaturated polyesters and epoxies; nylon-6,6; polycarbonate; poly(phenylene sulfide); polyamideimide; polysulfone; polyetheretherketone; polyetherimide; poly(ethylene terephthalate); and poly(vinyl chloride). Although most tests have been conducted in tension, flexure, or shear (266), the effects of impact loading have been studied (277). Typical S–N curves for tensile fatigue are shown in Figure 49 for several injection-molded polysulfone composites (278). The superiority of all composites to the matrix is evident, as is the superiority of carbon relative to glass fibers. Clearly the greater stiffness and thermal conductivity of carbon constitute significant advantages in lowering the strains at a given stress and minimizing hysteretic heating. Other S–N data show similar trends (279,280). However, the conservative prediction of endurance limits from S–N data derived from tests at relatively high frequencies requires caution in view of the possibility of significant hysteretic heating at the frequencies used (2,266). Such heating is known to reduce the fatigue life of at least some short-fiber-reinforced rubbers (281) and plastics (282).

Figure 49 shows two regions: yielding caused by self-heating, and fracture due to the initiation and propagation of a crack. The behavior in the latter region may be approximated by equations 31 or 32 (266):

$$\sigma/\sigma_0 = 1 - b \log N \tag{31}$$

$$\sigma/\sigma_0 = N \exp(-1/m) \tag{32}$$

where σ and σ_0 are the maximum stress and static strength (equivalent to σ at one cycle), respectively, N, the number of cycles, and b and m, constants. For many systems, FCP rates follow the Paris law in this region. Statistical aspects of fatigue failure in carbon-fiber-reinforced nylon-6,6 fabricated by both injection and compression molding have been successfully modeled using Weibull-type expressions (283,284); long life is characterized by high matrix ductility and interfacial adhesion.

The tensile fatigue sensitivity of unnotched chopped-glass-strand composites resembles that of unidirectional glass-fiber laminates tested parallel to the stress

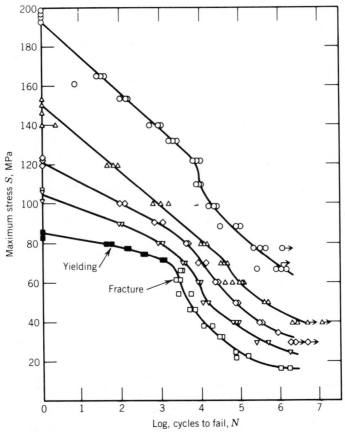

Fig. 49. Tensile fatigue data for injection-molded polysulfone matrix composites (R = 0.1, 5–20 Hz) (278). ■, □, unreinforced; ▽, 10% glass; ◇, 20% glass; △, 40% glass; ○, 40% carbon. To convert MPa to psi, multiply by 145.

axis (266); slopes of the S–N curves correspond to 9–12% decrease in static strength per decade of cycles. Effects of frequency, stress ratio, and small notches are minimal. Debonding of off-axis fibers and matrix cracking at low stresses precede failure, but it is uncertain which micromechanisms dominate failure (282). In-jection-molded glass and carbon-fiber-reinforced thermoplastics exhibit a wider range of behavior, but also tend to approach limiting values for the corresponding continuous-fiber controls, especially with polyetheretherketone and poly(phenylene sulfide) matrices and carbon-fiber laminates with some layers parallel to the stress axis. Failure may involve an accumulation of damage resulting from creep rupture of a ductile, well-bonded matrix, or debonding of a creep-resistant matrix; alternatively, a macroscopic fatigue crack may initiate and grow. Although systems that fail by FCP are usually superior to the matrix, failure by debonding may result in poorer relative performance.

In fatigue crack propagation, the presence of fibers typically results in an increase in both the threshold value of ΔK and K_c (254,267,270,274,275,285–287). Whereas in some cases this effect may be due to energy-dissipating mechanisms of the fiber and interface, the increase in modulus frequently is dominant (2,274).

The Paris law is usually followed at intermediate values of ΔK, often with relatively higher values of m (in the range of 7–9) than in the neat matrix. Such obedience to the Paris law is common in both injection-molded composites and chopped-strand mats (266,288,289) for both uniaxial and biaxial loading. However if damage is diffuse, so that a clear crack cannot be defined or measured, visual observation must be replaced by deduction of an equivalent crack from changes in compliance (288,289) or by the use of Moire techniques (289). Interpretation in terms of the crack-layer theory can also be helpful (50,285). In any case, fatigue life can be predicted by integrating the Paris equation (288), assuming there are no complications from hysteretic heating.

Consistent with the effects of fibers on ΔK_{th} and K_c, fatigue crack growth rates in injection-molded, short-fiber composites tend to be lower in well-bonded systems than in the neat matrix (Fig. 50), at least up to a value ($\sim$50–60 wt%) indicating fiber-bundle development (270). Optimum fiber concentration also occurs when $\Delta \mathcal{G}$ is plotted instead of ΔK (285). Fiber orientation is important in that fibers transverse to the crack direction are more effective than parallel fibers in blunting crack growth (267,275).

Typical failure mechanisms are shown in Figure 51 (185); see also Ref. 290.

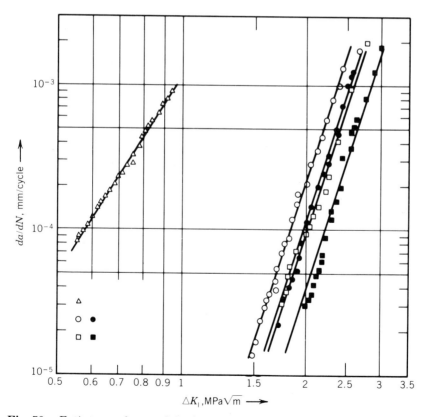

Fig. 50. Fatigue crack growth in short-glass-fiber reinforced polyethersulfone (286). Test conditions: 5 Hz, sine wave; open symbols refer to notch direction parallel, and filled symbols to notch direction transverse to the mold-fill direction. △, matrix; ○, 20 wt % glass fiber; □, 30 wt % glass fiber.

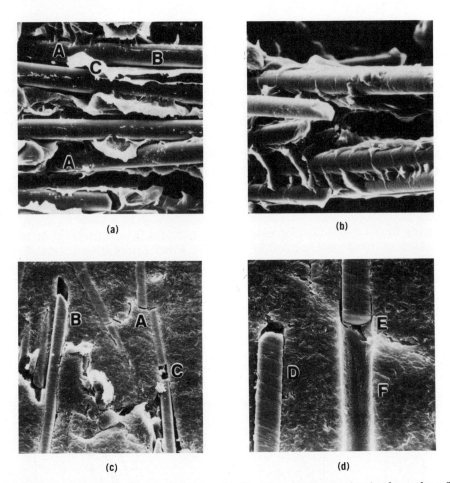

(a)

(b)

(c)

(d)

Fig. 51. Typical micromechanisms of fatigue crack propagation in short glass-fiber reinforced nylon-6,6 (185,254); fibers (~10 μm dia) in (**a**) and (**b**), and (**c**) and (**d**), at low and high angles, respectively, relative to the crack plane. (**a**) Stable FCP region: microvoid formation, A; extensive debonding along fiber, B; and matrix drawing, C. (**b**) Fast fracture regions; (**c**) debonding at fiber ends, A; fiber pullout, B; fiber failure, C; and (**d**) shear-crack formation along fiber interface, D; fiber fracture due to stress concentration associated with adjacent subsurface fiber, E; and fiber bed revealing evidence for a microvoid coalescence mechanism, F.

Debonding is typical in fatigue crack growth even at low values of ΔK in nominally well-bonded systems (254,267). Thus strong interfacial adhesion is advantageous, at least with random fibers and fibers transverse to the crack direction (267,270,275). Indeed composites that exhibit especially poor bonding to the matrix may possess significantly lower fatigue resistance than the matrix (121,287). However, some evidence suggests that weaker adhesion may be preferred with fibers that are parallel (275). Although enhanced matrix ductility is often beneficial to fatigue resistance (2,266,270), such is not the case with short-glass-fiber-reinforced polypropylene (274). A negligible effect of enhanced matrix ductility is seen in glass-fabric polyester composites (291).

Effects of frequency as a function of fiber content are complicated because the high thermal conductivity of typical fibers tends to offset localized crack-tip heating (and consequent crack blunting), so that frequency sensitivity of fatigue crack growth is a function of fiber content (52). In general, water is deleterious to performance because of associated premature debonding and, in systems plasticized by water, a tendency of a ductile matrix to pull away from the fibers (267,274). Water may also induce stress-corrosion cracking in some glass-fiber systems (292). Hydrocarbons can also have adverse effects on nonpolar matrices such as polypropylene (293).

The effects of fibers on strength and fatigue behavior have been expressed quantitatively in terms of a microstructural efficiency factor M^* given by (287)

$$M^* = a + nR^* \tag{33}$$

Here R^* is a function of volume fraction V_f and orientation of the fibers (= $V_f/2$ for random fibers), $a(\geq 1)$, a function of the stress state of the matrix, and $n(\geq 0)$, a function of the energy dissipation from the fibers. For fatigue crack propagation, the Paris law becomes

$$\frac{da}{dN} = A(M^*)\,\Delta K^{m(M^*)} \tag{34}$$

where the constants A and m are functions of M^*. The value of ΔK, which corresponds to an arbitrary value of da/dN, ΔK^*, is given by

$$\log \Delta K^* = C_1 + C_2 \log M = C_1 + C_2 \log (a + nR) \tag{35}$$

With a typical well-bonded system, eg, glass–polyetheretherketone (287), ΔK^* increases with M^* ($n > 0$), but with a poorly bonded system, such as glass–PTFE, ΔK^* decreases with increasing M^* ($n < 0$). The incorporation of carbon fibers in ultrahigh molecular weight polyethylene has a similar deleterious effect (121).

Continuous-fiber Systems. With typical continuous-fiber systems, S–N curves are linear, in accordance with equation 31. Compressive fatigue is much more severe than tensile fatigue (Fig. 52) (294). Compression-induced effects, eg, buckling delamination, and matrix shear (264,295) reduce fatigue resistance; cycling in flexure, torsion, or other shear modes is also especially deleterious (2,258,264,266,282,296–301). In such cases, the dominance of the fibers in determining properties decreases, and the matrix and interface play more important roles. Because such severe modes of loading are often more typical of actual service than is axial tension, more testing should be done under these more rigorous conditions. Similar deleterious effects are seen with off-axis fibers (296,297) and with multiaxial loads (297); in practice, laminates are designed with ply angles selected to optimize resistance to cracking. Carbon fibers (qv) confer greater resistance in constant-stress-range cycling than glass fibers because the much higher modulus of the former implies much lower corresponding strains (300); carbon is also less susceptible to environmental effects. The superiority of carbon fibers, especially with uniaxial orientation, is shown in a different way in Goodman-type diagrams (282), which reveal superior tolerance of high mean stresses and stress amplitudes at a constant life (10^7 cycles). Indeed, when normalized with respect to specific gravity, the ranking of carbon-fiber systems is superior to that of typical steel and aluminum alloys.

As revealed by microscopic evaluation or measurement of, for example,

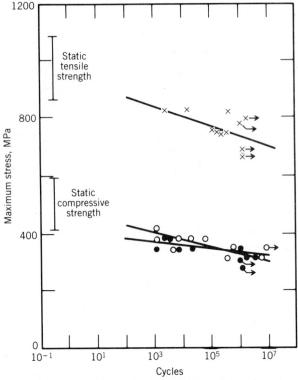

Fig. 52. Axial-load fatigue results for unidirectional, surface-treated, high modulus fiber composites prepared from prepreg tows with a typical epoxy resin system; volume fraction of fiber 0.61; 117 Hz; compressive stress plotted positive (lower curve) (294). × = zero tension; ○ = zero compression; ● = fully reversed.

acoustic emission (302–307) or modulus (266,306–309), localized damage, eg, fiber breakage, debonding, or matrix cracking, usually occurs during the early stages of fatigue. Measurements of damping (307) and radiographic (310) and ultrasonic techniques (305,307–311) are also useful in the monitoring of damage (see also NONDESTRUCTIVE TESTING). At sufficiently high frequencies, significant hysteretic heating may also occur. Although damage may increase progressively, the strength and modulus often attain an essentially steady state after an initial drop, and then drop significantly as small cracks coalesce, delamination becomes extensive, and temperature rises rapidly (266,309). In continuous-fiber systems failure is usually defined in terms of an arbitrary reduction of a service-related property such as modulus (306,311–315).

Although residual strength and toughness tend to decrease during cycling, paradoxical increases may occur as a result of the realignment of fibers (265) or crack blunting associated with a diffuse damage zone (2,316). Fatigue life is also affected by varying the spectrum of loads or frequency (299,317–323); thus sometimes application of high static loads during proof testing, or periodic overloads during cycling, can be beneficial (317,318). The periodic imposition of a high frequency may also increase fatigue life through the retardation of creep (321). However, even if strength is thereby improved, the damage created is still undesirable because of greater susceptibility to environmental degradation. Statistical aspects of failure have been reviewed by several authors, and various models have been proposed based on modified Weibull functions (324–329).

In general, elevated temperatures and exposure to water adversely affect fatigue life (2,330–332). The frequency factor is complex because of concomitant and sometimes conflicting effects of changes in time under load and strain rate, as well as the effects of hysteretic heating (2,321,333); hence the response to frequency changes is often nonlinear.

Until recently, most high-performance continuous fiber composites have been based on epoxy matrices. However, standard epoxies are relatively sensitive to hygrothermal exposure, which tends to affect the strength and toughness of the matrix and interface adversely and thus lower the properties of composites, the behavior of which is not dominated by the fibers (332,334). Although epoxies can be modified to reduce moisture sensitivity, much research has been devoted to the use of thermoplastic matrices that may or may not be subsequently cross-linked (334–338). For example, with good control of process history and with suitable modifications of the resin and interface, excellent fatigue performance has been reported for 0/90 carbon-fiber laminates based on a polyetheretheracetone (aromatic polymer composite, APC) (334–338). The greater ductility and interfacial adhesion of the aromatic matrix make it possible to use relatively compliant fibers to match the performance of a carbon–epoxy composite based on stiffer fibers (334,337). At the same time, fatigue resistance is significantly better than with the epoxy matrix when composites are tested at 45°C to the fiber axis (Fig. 53), and effects of hygrothermal exposure are negligible. Because high T_g aromatic matrices exhibit greater ease of fabrication and repairability, in addition to greater inherent resistance to water, interest may be expected to continue (335).

Although the damage zone in composites containing notches or holes is complex and diffuse, the propagation of damage can often be considered as analogous to the propagation of a crack (2,339). One model suggested for the propagation of damage from a stress raiser is (339):

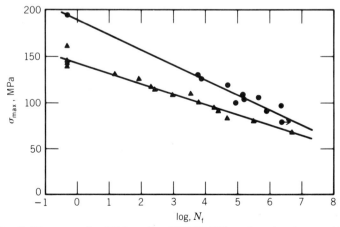

Fig. 53. S–N curves for 0/90 carbon/PEEK (●) and carbon/epoxy (▲) composites tested in the ±45°C orientation (65% rh, 23°C, R ≈ 0.1) (334). To convert MPa to psi, multiply by 145.

$$\frac{dD}{dN} = C\sigma(\ell + pD)D/qN \tag{36}$$

where D is the dimension of the damaged region transverse to the stress direction, ℓ, the original length of the stress raiser, dD/dN, the rate of damage accumulation per cycle, σ, the stress range, and C, p, and q, constants. In some cases, eg, delamination or crack growth in the forward direction in uniaxial tension (340,341), the rate of crack or damage propagation follows the Paris law (with a high m) or a variant thereof. For example, in the latter case, the following equation holds well (341):

$$\frac{da}{dN} = \left(\frac{C}{f}\right)(\mathcal{G}_{max}/\mathcal{G})^m \tag{37}$$

where $\mathcal{G}_{max}$ and $\mathcal{G}$ are the maximum-strain energy release rate and the minimum required for crack extension, and m, a constant. However, sometimes crack growth may follow a discontinuous pattern characterized by repeated decelerations due to crack blunting followed by reaccelerations (332).

Other Composite Systems

Fatigue of adhesive joints is of considerable practical importance. As with other composite systems, a fracture mechanics approach can be useful (342,343); the rate of crack growth may be described by the energy-based equivalent of the Paris law (ie, replacing ΔK with $\Delta \mathcal{G}$) (342). The locus of failure (center of bond vs interface) depends on the material, the frequency, and the environment, although the Paris law variant has been reported to hold in all cases.

The fatigue behavior of Portland cement concretes has been studied to a limited extent (344,345). Significant improvements in fatigue performance, noted with concretes impregnated with PMMA, have been attributed to improved toughness and interfacial bonding between the concrete constituents (see also CEMENT ADDITIVES).

Recently, the incorporation of low concentrations (0.2–2 vol %) of metallic-glass ribbons has been shown to reduce fatigue crack growth rates in an epoxy significantly (346). The reduction is greatest when the width of the ribbon is perpendicular to the crack growth direction; however, when the width is parallel, absolute values of da/dN for the composite are lower by an order of magnitude. A quantitative model for the blunting of crack growth has been proposed. Such composites offer promise of better transverse properties than those found for unidirectional fiber systems.

BIBLIOGRAPHY

1. J. A. Manson and R. W. Hertzberg, *Crit. Rev. Macromol. Sci.* **1**, 433 (1973).
2. R. W. Hertzberg and J. A. Manson, *Fatigue of Engineering Plastics*, Academic Press, Inc., Orlando, Fla., 1980.
3. J. C. Radon, *Int. J. Fract.* **16**, 533 (1980).
4. J. A. Sauer and G. C. Richardson in Ref. 3, p. 499.
5. M. T. Takemori, *Annu. Rev. Mater. Sci.* **14**, 171 (1984).
6. J. A. Sauer and C. C. Chen, *Adv. Polym. Sci.* **52/53**, 169 (1983).

7. J. M. Schultz, "Properties of Solid Polymeric Materials" in J. M. Schultz, ed., *Treatise on Materials Science and Technology, Part B,* Academic Press, Inc., Orlando, Fla., 1977, p. 599.
8. A. Wöhler, *Engl. Abstr. Eng.* **2**, 199 (1871).
9. M. N. Riddell, *Plast. Eng.* **30**(4), 71 (1974).
10. *ASTM D 671, Standard for Metric Practice,* Vol. 08.01, American Society for Testing and Materials, Philadelphia, Pa., 1985, p. 277.
11. K. V. Gotham, *Plast. Polym.* **41**(156), 273 (1973).
12. I. Constable, L. E. Culver, and J. G. Williams, *Int. J. Fract. Mech.* **6**(3), 279 (1970).
13. S. J. Hutchinson and P. P. Benham, *Plast. Polym.* **38**, 102 (1970).
14. J. T. Turkanis (Brennock), R. W. Hertzberg, and J. A. Manson in *Deformation, Yield and Fracture of Polymers,* Plastics and Rubber Institute, London, UK, 1985, paper 54.
15. J. T. Turkanis (Brennock), *The Fatigue Crack Initiation Behavior of Neat and Rubber-modified Poly(vinyl chloride) (PVC) and Other Engineering Polymers,* Master's Thesis, Lehigh University, Bethlehem, Pa., 1985.
16. S. B. Ratner and V. I. Korobov, *Mekh. Polim.* **1**(3), 93 (1965).
17. A. X. Schmidt and C. A. Marlies, *Principles of High Polymer Theory and Practice,* McGraw-Hill Book Co., New York, 1948, p. 578.
18. J. D. Ferry, *Viscoelastic Properties of Polymers,* 3rd ed., John Wiley & Sons, Inc., New York, 1980.
19. M. N. Riddell, G. P. Koo, and J. L. O'Toole, *Polym. Eng. Sci.* **6**, 363 (1966).
20. L. J. Broutman and S. K. Gaggar, *Int. J. Polym. Mater.* **1**, 295 (1972).
21. K. Oberbach, *Kunstoffe* **55**, 356 (1965).
22. L. C. Cessna, J. A. Levens, and J. B. Thomson, *Polym. Eng. Sci.* **9**(5), 339 (1969).
23. H. Hachmann and E. Strickle, *Soviet Plastics and Engineering Annual Technology Conference,* Vol. 26, 1968, p. 512.
24. W. A. Herman, R. W. Hertzberg, and J. A. Manson, unpublished work.
25. R. J. Crawford and P. P. Benham, *Polymer* **16**, 908 (1975).
26. P. Beardmore and S. Rabinowitz in R. J. Arsenault, ed., *Treatise on Materials Science and Technology,* Vol. 6, Academic Press, Inc., Orlando, Fla., 1975, p. 267.
27. J. F. Mandell, K. L. Smith, and D. D. Huang, *Polym. Eng. Sci.* **21**, 1173 (1981).
28. *Annual Book of Standards, Part 27,* American Society for Testing and Materials, Philadelphia, Pa., 1973, p. 916.
29. S. Rabinowitz, A. R. Krause, and P. Beardmore, *J. Mater. Sci.* **8**, 11 (1973).
30. S. Rabinowitz and P. Beardmore, *J. Mater. Sci.* **9**, 81 (1974).
31. B. Tomkins and W. D. Biggs, *J. Mater. Sci.* **4**, 532 (1969).
32. J. F. Mandell and J. P. F. Chevaillier, *Polym. Eng. Sci.* **25**, 170 (1985).
33. *Fatigue of Plastics,* Plastics and Rubber Institute, London, UK, 1983.
34. R. W. Hertzberg, *Deformation and Fracture Mechanics of Engineering Materials,* 2nd ed., John Wiley & Sons, Inc., New York, 1983.
35. C. J. Beevers, ed., *Advances in Crack Length Measurement,* EMAS Publishers, Warley, West Midlands, UK, 1982.
36. P. C. Paris, *Proceedings of the 10th Sagamore Army Materials Research Conference,* Syracuse University Press, 1964, p. 107.
37. G. R. Irwin, *Handb. Physik* **7**, 551 (1958).
38. A. A. Griffith, *Trans. R. Soc. London* **A221**, 163 (1920).
39. G. R. Irwin, *Trans. ASME J. Appl. Mech.* **24**, 361 (1957).
40. H. M. Westergaard, *Trans. ASME J. Appl. Mech.* **61**, 49 (1939).
41. G. R. Irwin, *Trans. ASME Ser. D* **82**(2), 417 (1960).
42. J. A. Manson, R. W. Hertzberg, and P. E. Bretz in S. R. Valuri, D. M. R. Taplin, R. RamaRao, J. F. Knott, and R. Dubey, eds., *Advances in Fracture Research, Proceedings of the Fifth International Conference on Fracture,* Vol. 1, Pergamon Press, Oxford, UK, 1982, p. 443.
43. R. S. Vecchio, private communication.
44. L. Jilken and C. G. Gustafson in J. Backlund, A. Blom, and C. J. Beevers, eds., *Fatigue Thresholds, Fundamental and Engineering Applications,* EMAS, Warley, UK, 1981, p. 715.
45. G. M. Connelly, R. W. Hertzberg, and J. A. Manson, unpublished data.
46. E. H. Andrews and B. J. Walker, *Proc. R. Soc. London* **A325**, 57 (1971).
47. P. E. Bretz, R. W. Hertzberg, and J. A. Manson, *Polymer* **20**, 575 (1981).
48. T. A. Morelli and M. T. Takemori, *J. Mater. Sci.* **19**, 385 (1984).

49. A. Chudnovsky and A. Moet, *Org. Coat. Plast. Chem.* **45,** 607 (1981).
50. A. Chudnovsky and A. Moet, *Polym. Eng. Sci.* **22,** 922 (1982).
51. A. Chudnovsky and A. Moet in Ref. 33, paper 2.
52. R. W. Lang, J. A. Manson, and R. W. Hertzberg in J. C. Seferis and L. Nicolais, eds., *The Role of the Polymeric Matrix in the Processing and Structural Properties of Composite Materials,* Plenum Publishing Corp., New York, 1983, p. 377.
53. G. Attalla, G. M. Connelly, M. R. Morgan, W. M. Cheng, J. A. Manson, and R. W. Hertzberg, Lehigh University, unpublished data.
54. R. W. Hertzberg, M. D. Skibo, and J. A. Manson, *J. Mater. Sci.* **13,** 1038 (1978).
55. R. S. Rivlin and A. G. Thomas, *J. Polym. Sci.* **10,** 291 (1953).
56. A. G. Thomas, *J. Polym. Sci.* **31,** 467 (1958).
57. P. B. Lindley and A. Stevenson in F. Sherratt and J. B. Sturgeon, eds., *Materials, Experimentation and Design in Fatigue,* Westbury House Publishers, Ltd., Surrey, UK, 1981, p. 233.
58. A. Stevenson, *Int. J. Fract.* **23,** 47 (1983).
59. G. J. Lake and P. B. Lindley in A. C. Strickland, ed., *Physical Basis of Yield and Fracture,* Physics Institute, London, UK, 1966, p. 176.
60. G. J. Lake and A. G. Thomas in D. E. Carlson and R. T. Shield, eds., *Proceedings of the IVTAM Symposium on Finite Elasticity,* Nijhoff Publishing, Ltd., The Hague, The Netherlands, 1981.
61. D. G. Young, E. N. Kresge, and A. J. Wallace, *Rubber Chem. Technol.* **55**(2), 428 (1982).
62. D. G. Young, *Rubber Chem. Technol.* **58,** 785 (1985).
63. M. D. Skibo, Ph.D. Dissertation, Lehigh University, 1977.
64. R. W. Hertzberg, J. A. Manson, and M. D. Skibo, *Polym. Eng. Sci.* **15,** 252 (1975).
65. R. W. Hertzberg, M. D. Skibo, and J. A. Manson, *ASTM STP 700, Standard for Metric Practice,* American Society for Testing and Materials, Philadelphia, Pa., 1980, p. 490.
66. R. W. Hertzberg, J. A. Manson, and M. D. Skibo, *Polymer* **19,** 359 (1978).
67. M. T. Hahn, R. W. Hertzberg, R. W. Lang, J. A. Manson, J. C. Michel, A. Ramirez, C. M. Rimnac, and S. M. Webler in *Deformation, Yield and Fracture of Polymers,* Plastics and Rubber Institute, London, UK, 1982, paper 19.
68. R. W. Lang, M. T. Hahn, R. W. Hertzberg, and J. A. Manson, *ASTM STP 833, Standard for Metric Practice,* American Society for Testing and Materials, Philadelphia, Pa., 1984, p. 266.
69. M. T. Hahn, Ph.D. Dissertation, Lehigh University, 1982.
70. M. T. Hahn, R. W. Hertzberg, and J. A. Manson, *J. Mater. Sci.* **21,** 31 (1986).
71. A. M. B. A. Osorio and J. G. Williams in *Fatigue in Polymers,* Plastics and Rubber Institute, London, UK, 1983, paper 7.
72. Y. W. Mai and J. G. Williams, *J. Mater. Sci.* **14**(8), 1933 (1979).
73. Y. W. Mai, K. W. Gock, and B. Cotterell, private communication.
74. Y. W. Mai, K. W. Gock, and B. Cotterell, unpublished work.
75. G. C. Martin and W. W. Gerberich, *J. Mater. Sci.* **11,** 231 (1976).
76. R. J. Wann, G. C. Martin, and W. W. Gerberich, *Polym. Eng. Sci.* **16**(9), 645 (1976).
77. M. T. Hahn, R. W. Hertzberg, J. A. Manson, and L. H. Sperling, *Polymer* **27,** (1986).
78. H. H. Kausch, "Polymer Fracture" in *Polymers Properties and Applications,* Vol. 2, Springer-Verlag, Inc., New York, 1978.
79. A. J. Kinloch and R. J. Young, *Fracture Behavior of Polymers,* Applied Science Publishers, New York, 1983.
80. H. W. McCormick, F. M. Brower, and L. Kim, *J. Polym. Sci.* **39,** 87 (1959).
81. J. R. Martin, J. F. Johnson, and A. R. Cooper, *J. Macromol. Sci. Rev. Macromol. Chem. Part C* **8,** 57 (1972).
82. P. I. Vincent, *Polymer* **1,** 425 (1960).
83. J. A. Sauer in Ref. 66, p. 859.
84. J. H. Golden, B. L. Hammant, and E. A. Hazell, *J. Polym. Sci. Part A* **2,** 4787 (1964).
85. A. N. Gent and A. G. Thomas, *J. Polym. Sci. Part A-2* **10,** 571 (1972).
86. R. P. Kusy and D. T. Turner, *Polymer* **17,** 161 (1976).
87. R. P. Kusy and M. J. Katz in Ref. 75, p. 1475.
88. J. P. Berry in Ref. 84, p. 4069.
89. G. L. Pitman and I. M. Ward, *Polymer* **20,** 895 (1979).
90. R. E. Robertson in R. D. Deanin and A. M. Crugnola, eds., *Toughness and Brittleness of Plastics, Advances in Chemistry Series No. 154,* American Chemical Society, Washington, D.C., 1976, p. 89.

91. J. L. Weaver and C. L. Beatty, *Polym. Eng. Sci.* **18**, 1117 (1978).
92. J. C. Michel, J. A. Manson, and R. W. Hertzberg in Ref. 14, paper 21.
93. R. P. Kambour, *J. Polym. Sci. Macromol. Rev.* **7**, 1 (1973).
94. H. H. Kausch, *Kunstoffe* **65**, 1 (1976).
95. G. W. Weidmann and W. Döll, *Colloid Polym. Sci.* **254**, 205 (1976).
96. S. Wellinghoff and E. Baer, *J. Macromol. Sci. Phys. Part B* **11**, 367 (1975).
97. D. J. Turner, *Polymer* **23**, 626 (1982).
98. A. M. Donald and E. J. Kramer, *J. Polym. Sci. Polym. Phys. Ed.* **20**, 899 (1982).
99. A. M. Donald and E. J. Kramer in Ref. 97, p. 457.
100. D. L. G. Lainchbury and M. Bevis in Ref. 75, p. 2222.
101. J. F. Fellers and B. F. Kee, *J. Appl. Polym. Sci.* **18**, 2355 (1974).
102. J. A. Sauer and C. C. Chen in *Toughening of Plastics,* Plastics and Rubber Institute, London, UK, 1985, paper 26.
103. J. A. Sauer, E. Foden, and D. R. Morrow, *Polym. Eng. Sci.* **17**, 246 (1977).
104. C. L. Beatty and J. L. Weaver, *Bull. Am. Phys. Soc.* **24**, 378 (1979).
105. S. Warty, J. A. Sauer, and A. Charlesby, *Eur. Polym. J.* **15**(5), 445 (1979).
106. J. R. Martin and J. F. Johnson in Ref. 101, p. 3227.
107. S. L. Kim, M. D. Skibo, J. A. Manson, and R. W. Hertzberg in Ref. 103, p. 194.
108. M. D. Skibo, J. A. Manson, and R. W. Hertzberg, *J. Macromol. Sci. Phys. Part B* **14**, 525 (1977).
109. C. M. Rimnac, J. A. Manson, R. W. Hertzberg, S. M. Webler, and M. D. Skibo, *J. Macromol. Sci. Phys. Part B* **19**(3), 351 (1981).
110. S. C. Kim, J. Janiszewski, M. D. Skibo, J. A. Manson, and R. W. Hertzberg in Ref. 91, p. 1093.
111. G. Pitman and I. M. Ward, *J. Mater. Sci.* **15**, 635 (1980).
112. A. Ramirez, J. A. Manson, and R. W. Hertzberg in Ref. 50, p. 975.
113. J. C. Michel, J. A. Manson, and R. W. Hertzberg, *Polym. Prepr. Am. Chem. Soc. Div. Polym. Chem.* **26**(2), 141 (1985).
114. J. C. Michel, J. A. Manson, and R. W. Hertzberg, unpublished work.
115. J. C. Michel, J. A. Manson, and R. W. Hertzberg, *Polymer* **25**, 1657 (1984).
116. J. C. Michel, Ph.D. Dissertation, Lehigh University, 1984.
117. J. F. Rudd, *Polym. Lett.* **1**, 1 (1963).
118. D. M. Bigg, R. T. Leininger, and C. S. Lee, *Polymer* **22**, 539 (1981).
119. R. W. Hertzberg, M. D. Skibo, and J. A. Manson, *ASTM STP 675,* American Society for Testing and Materials, Philadelphia, Pa., 1979, p. 471.
120. J. Janiszewski, R. W. Hertzberg, and J. A. Manson, *ASTM STP 743, Standard for Metric Practice,* American Society for Testing and Materials, Philadelphia, Pa., 1981, p. 125.
121. F. X. de Charentenay, F. Laghouati, and J. Dewas in *Deformation, Yield and Fracture of Polymers,* Plastics and Rubber Institute, London, UK, 1979, p. 6.1; G. M. Connelly, C. M. Rimnac, T. M. Wright, R. W. Hertzberg, and J. A. Manson, *J. Orthoped. Res.* **2**, 119 (1984).
122. C. Sheu and R. D. Goolsby, *Org. Coat. Appl. Polym. Sci. Proc.* **48**, 833 (1983).
123. P. E. Bretz, J. A. Manson, and R. W. Hertzberg, *J. Appl. Polym. Sci.* **27**, 1707 (1982).
124. G. Attalla, M. J. Carling, J. A. Manson, and R. W. Hertzberg, Lehigh University, 1984, unpublished work.
125. R. P. Wool, *Polym. Prepr. Am. Chem. Soc. Div. Polym. Chem.* **23**(2), 62 (1982).
126. C. Su, *M. S. Report,* Lehigh University, 1984.
127. S. Qureshi, J. A. Manson, J. C. Michel, R. W. Hertzberg, and L. H. Sperling in S. S. Labana and R. A. Dickie, eds., *Characterization of Highly Cross-linked Polymers,* ACS Symposium Series No. 243, American Chemical Society, Washington, D.C., 1984, p. 109.
128. J. C. Michel, Lehigh University, 1983, unpublished work.
129. R. W. Hertzberg, J. A. Manson, and W. C. Wu, *ASTM STP 536,* American Society for Testing and Materials, Philadelphia, Pa., 1973, p. 391.
130. S. A. Sutton, *Eng. Fract. Mech.* **6**, 587 (1974).
131. S. Qureshi in Ref. 2, p. 122.
132. S. L. Kim, M. D. Skibo, J. A. Manson, R. W. Hertzberg, and J. Janiszewski in Ref. 91, p. 1093.
133. A. Takahara, K. Yamada, T. Kajiyama, and M. Takayanagi, *J. Appl. Polym. Sci.* **25**, 597 (1980).
134. I. Constable, J. G. Williams, and D. J. Burns, *J. Mech. Eng. Sci.* **12**, 20 (1970).
135. K. Suzuki, S. Yada, N. Mabuchi, K. Seiuchi, and Y. Matsutani, *Kobunshi Kagaku* **28**, 920 (1971).
136. P. E. Bretz, R. W. Hertzberg, J. A. Manson, and A. Ramirez in S. P. Rowland, ed., *Water in*

Polymers, ACS Symposium Series No. 127, American Chemical Society, Washington, D.C., 1980, p. 32.

137. S. Arad, J. C. Radon, and L. E. Culver, *J. Appl. Polym. Sci.* **17,** 1467 (1973).
138. S. Arad, J. C. Radon, and L. E. Culver, *Eng. Fract. Mech.* **4,** 511 (1972).
139. H. A. El-Hakeem and L. E. Culver, *Int. J. Fract.* **1,** 133 (1979).
140. R. W. Hertzberg, M. D. Skibo, J. A. Manson, and J. K. Donald in Ref. 72, p. 1754.
141. P. E. Bretz, R. W. Hertzberg, and J. A. Manson, *J. Mater. Sci.* **16,** 2061 (1981).
142. M. T. Hahn, R. W. Hertzberg, J. A. Manson, R. W. Lang, and P. E. Bretz in Ref. 97, p. 1675.
143. R. W. Lang, M. T. Hahn, R. W. Hertzberg, and J. A. Manson, *J. Mater. Sci. Lett.* **3,** 224 (1984).
144. *Zytel Design Handbook,* E. I. duPont de Nemours & Co., 1972.
145. N. G. McCrum, B. F. Read, and G. Williams, *Anelastic and Dielectric Effects in Polymeric Solids,* John Wiley & Sons, Inc., New York, 1967.
146. R. Puffr and J. Sabenda, *J. Polym. Sci. Part C* **5,** 79 (1967).
147. M. I. Kohan, *Nylon Plastics,* John Wiley & Sons, Inc., New York, 1973.
148. P. E. Bretz, R. W. Hertzberg, and J. A. Manson in Ref. 72, p. 2482.
149. C. C. Chen, J. Shen, and J. A. Sauer, *Polymer* **26,** 89 (1985).
150. J. A. Sauer, S. Warty, and S. Kaka in Ref. 121, paper 26.
151. C. B. Bucknall, *Toughened Plastics,* Applied Science Publishers, London, UK, 1977.
152. S. Suzuki and T. Tsurue, *2nd International Conference on Mechanical Behavior of Materials,* American Society of Metals, Metals Park, Ohio, 1976.
153. C. C. Chen, N. Chheda, and J. A. Sauer in Ref. 109.
154. S. Warty, J. A. Sauer, and D. R. Morrow in Ref. 66, p. 1465.
155. S. B. Ratner and N. L. Barasch, *Mekh. Polim.* **1**(1), 124 (1965).
156. J. Miltz, A. T. DiBenedetto, and S. Petrie in Ref. 54, p. 2037.
157. N. Chheda, C. Chen, S. Kaka, cited in J. A. Sauer in Ref. 4.
158. L. C. E. Struik, *Physical Aging in Amorphous Polymers and Other Materials,* Elsevier Science Publishing Co., New York, 1978.
159. J. A. Manson, R. W. Hertzberg, S. L. Kim, and W. C. Wu in Ref. 90, p. 146.
160. S. Webler, Ph.D. Dissertation, Lehigh University, 1983.
161. C. M. Rimnac, Ph.D. Dissertation, Lehigh University, 1983.
162. T. C. Ho, Ph.D. Dissertation, Rensselaer Polytechnic Institute, Troy, New York, 1973.
163. J. A. Sauer and K. D. Pae in H. J. Kaufman and J. J. Falcetta, eds., *Introduction to Polymer Science,* John Wiley & Sons, Inc., New York, 1977, Chapt. 7.
164. G. P. Koo and L. G. Roldan, *J. Polym. Sci. Polym. Lett. Ed.* **10,** 1145 (1972).
165. G. Meinel and A. Peterlin, *J. Polym. Sci. Polym. Lett. Ed.* **9,** 67 (1971).
166. E. H. Andrews, *Fracture of Polymers,* Elsevier Science Publishing Co., New York, 1968.
167. E. H. Andrews, *J. Appl. Phys.* **32,** 542 (1961).
168. R. Boukhili, F. X. de Charentenay, and J. V. Khanh in Ref. 14, paper 22.
169. I. P. Bareishis and A. V. Stinskas, *Mekh. Polim.* **9**(3), 562 (1973).
170. V. K. Savkin, V. A. Belyi, T. I. Sogolova, and V. A. Kargin, *Mekh. Polim.* **2**(6), 803 (1966).
171. T. I. Sogolova in Ref. 170, No. 5, p. 643.
172. V. K. Kuchinskas and A. N. Machyulis, *Mekh. Polim.* **3**(4), 713 (1967).
173. A. V. Stinskas, Y. P. Basuhis, and I. P. Bareishis, *Mekh. Polim.* **8**(1), 59 (1972).
174. K. V. Gotham and M. J. Hitch, *Br. Polym. J.* **10,** 47 (1978).
175. L. Gargani and M. Bruzzone in Ref. 113, p. 27.
176. K. Friedrich, *Proceedings of the 9th Conference on Scanning Electron Microscopy,* German Society for Testing Materials (DVM), Stuttgart, FRG, 1979, p. 173.
177. A. Ramirez, P. M. Gaultier, J. A. Manson, and R. W. Hertzberg in Ref. 71, paper 3.
178. A. Sandt and E. Hornbogen in Ref. 141, p. 2915.
179. W. E. Cowley and L. E. Wylde, *Chem. Ind.* **3,** 371 (1978).
180. M. B. Barker, J. Bowman, and M. Bevis, *J. Mater. Sci.* **18,** 1095 (1983).
181. A. N. Gent, P. B. Lindley, and A. G. Thomas, *J. Appl. Polym. Sci.* **8,** 455 (1964).
182. J. G. Williams in *Engineering Design with Plastics: Principles and Practice,* Plastics and Rubber Institute, London, UK, 1981, p. 182.
183. R. W. Hertzberg and J. A. Manson, *Plast. World* **35**(5), 50 (1977).
184. J. R. Rice in H. Liebowitz, ed., *Fracture, An Advanced Treatise,* Vol. 2, Academic Press, Inc., New York, p. 192.

185. R. W. Lang, Ph.D. Dissertation, Lehigh University, 1984.
186. M. T. Hahn, R. W. Hertzberg, J. A. Manson, and C. M. Rimnac, *J. Mater. Sci.* **17**, 1533 (1982).
187. J. P. Berry, *Nature (London)* **185**, 91 (1960).
188. R. P. Kambour, *J. Polym. Sci. Part A-2* **4**, 349 (1966).
189. R. W. Hertzberg and J. A. Manson in Ref. 57, p. 181.
190. R. C. Bates and W. G. Clark, Jr., *Trans. Quart. ASM* **62**(2), 380 (1969).
191. W. Döll, L. Könczöl, and M. G. Schinker, *Polymer* **24**, 1213 (1983).
192. J. P. Elinck, J. C. Bauwens, and G. Homès, *Int. J. Fract. Mech.* **7**(3), 227 (1971).
193. D. S. Dugdale, *J. Mech. Phys. Solids* **8**, 100 (1960).
194. M. D. Skibo, R. W. Hertzberg, J. A. Manson, and S. L. Kim, *J. Mater. Sci.* **12**, 531 (1977).
195. M. G. Schinker, L. Könczöl, and W. Döll, *J. Mater. Sci. Lett.* **1**, 475 (1982).
196. M. T. Takemori and D. S. Matsumoto in Ref. 98, p. 2027.
197. C. M. Rimnac, R. W. Hertzberg, and J. A. Manson in Ref. 71, paper 9.
198. R. W. Hertzberg in Ref. 42, p. 163.
199. M. D. Skibo, R. W. Hertzberg, and J. A. Manson in Ref. 75, p. 479.
200. L. Könczöl, M. G. Schinker, and W. Döll in Ref. 71, paper 4.
201. C. M. Rimnac, R. W. Hertzberg, and J. A. Manson, *ASTM STP 733,* American Society for Testing and Materials, Philadelphia, Pa., 1981, p. 291.
202. C. M. Rimnac, R. W. Hertzberg, and J. A. Manson, *J. Mater. Sci. Lett.* **2**, 325 (1983).
203. C. M. Rimnac, R. W. Hertzberg, and J. A. Manson in Ref. 48, p. 1116.
204. W. Döll, *Adv. Polym. Sci.* **52/53**, 105 (1983).
205. R. W. Lang, J. A. Manson, R. W. Hertzberg, and R. Schirrer, *Polym. Eng. Sci.* **24**, 833 (1984).
206. B. D. Lauterwasser and E. J. Kramer, *Phil. Mag.* **A39**(4), 469 (1979).
207. N. Verheulpen-Heymans in Ref. 89, p. 356.
208. D. S. Matsumoto and M. T. Takemori, *GE Technical Information Series No. 82CRD240,* 1982.
209. M. T. Takemori in Ref. 50, No. 15, p. 937.
210. N. J. Mills and N. Walker in Ref. 111, p. 1832.
211. J. Murray and D. Hull, *J. Polym. Sci. Part A-2* **8**, 583 (1970).
212. P. E. Bretz, R. W. Hertzberg, and J. A. Manson in Ref. 118, p. 1272.
213. M. T. Hahn, R. W. Hertzberg, and J. A. Manson in Ref. 70, p. 39.
214. C. B. Bucknall and W. W. Stevens in *Toughening of Plastics,* Plastics and Rubber Institute, London, UK, 1978, paper 24.
215. G. Menges and E. Weigand, *Proceedings of the 21st Annual Technical Conference of the Society of Plastics Engineers,* 1975, p. 465.
216. J. A. Sauer and C. C. Chen in Ref. 205, p. 786.
217. R. L. Thorkildsen in E. Baer, ed., *Engineering Design for Plastics,* Van Nostrand-Reinhold Co., Inc., New York, 1964, p. 279.
218. D. Woan, M. Habibullah, and J. A. Sauer in Ref. 118, p. 699.
219. H. Bartesch and D. R. G. Williams, *J. Appl. Polym. Sci.* **22**, 457 (1978).
220. C. B. Bucknall and A. Marchetti, *J. Appl. Polym. Sci.* **28**, 2689 (1983).
221. C. B. Bucknall and A. Marchetti in Ref. 205, p. 535.
222. R. Murukami, N. Shin, N. Kusomoto, and Y. Motozato, *Kobunshi Ronbunshu* **33**, 107 (1976).
223. M. Takemori and T. A. Morelli in Ref. 182, paper 7.
224. C. B. Bucknall, *Br. Plast.* **40**(11), 118 (1967); **40**(12), 84 (1967).
225. P. J. Clarke cited in Ref. 214.
226. C. B. Bucknall, K. V. Gotham and P. I. Vincent in A. D. Jenkins, ed., *Polymer Science,* North Holland Publishers, Amsterdam, The Netherlands, 1972, p. 621.
227. M. D. Skibo, J. A. Manson, R. W. Hertzberg, S. M. Webler, and E. A. Collins, Jr. in R. K. Eby, ed., *Durability of Micromolecular Materials,* ACS Symposium Series No. 95, The American Chemical Society, Washington, D.C., 1979, p. 311.
228. M. D. Skibo, R. W. Hertzberg, and J. A. Manson in Ref. 121, p. 4.1.
229. T. Hur, Lehigh University, 1986, unpublished work.
230. J. A. Manson and R. W. Hertzberg, *J. Polym. Sci. Polym. Phys. Ed.* **11**, 2483, 1973.
231. T. A. Morelli and M. Takemori, *J. Mater. Sci.* **18**, 1836 (1983).
232. C. M. Rimnac, Lehigh University, 1979, unpublished work; cited in Ref. 2, p. 196.
233. E. A. Flexman, Jr., *Polym. Eng. Sci.* **19**, 564 (1979).
234. D. N. Shah, J. A. Manson, G. M. Connelly, G. Attalla, and R. W. Hertzberg in C. K. Riew and

J. K. Gillham, eds., *Rubber Modified Thermoset Resins, ACS Advances in Chemistry Series No. 208*, American Chemical Society, Washington, D.C., 1984, p. 117.

235. J. A. Manson, R. W. Hertzberg, G. M. Connelly, and J. Hwang in L. H. Sperling and D. R. Paul, eds., *Multicomponent Polymer Materials, ACS Advances in Chemistry Series No. 211*, American Chemical Society, Washington, D.C., 1986, p. 291.

236. G. C. Adams and T. K. Wu in Ref. 71, paper 15.

237. E. A. Flexman, Jr., *Mod. Plast.* **62**(2), 72 (1985).

238. J. A. Manson and L. H. Sperling, *Polymer Blends and Composites*, Plenum Publishing Corp., New York, 1976.

239. J. Lankford, W. J. Astleford, and M. A. Asher in Ref. 75, p. 1624.

240. P. W. R. Beaumont and R. J. Young, *J. Mater. Sci.* **10**, 1334 (1975).

241. F. F. Koblitz, V. R. Luna, J. F. Glenn, K. L. DeVries, and R. A. Draughn, *Org. Coat. Plast. Chem.* **38**(1), 322 (1978).

242. T. A. Freitag and S. L. Cannon, *J. Biomed. Mater. Res.* **13**, 343 (1979).

243. C. F. Stark in Ref. 242, p. 343.

244. R. M. Pilliar, R. Blackwell, I. Macnab, and H. U. Cameron, *J. Biomed. Mater. Res.* **10**, 893 (1976).

245. R. M. Pilliar, W. J. Bratina, and R. A. Blackwell, *ASTM STP 636*, American Society for Testing and Materials, Philadelphia, Pa., 1977, p. 206.

246. J. A. Manson, *Pure Appl. Chem.* **57**, 1667 (1985).

247. K. Friedrich and U. A. Karsch, *Polym. Compos.* **3**, 65 (1982).

248. L. E. Nielsen, *J. Compos. Mater.* **9**, 149 (1975).

249. G. G. Trantina in Ref. 205, p. 1181.

250. B. W. Staynes in Ref. 71, paper 14.

251. M. Breslauer, Lehigh University, 1985, unpublished research.

252. K. P. Gadkaree and G. Salee, *Polym. Compos.* **4**, 19 (1983); **5**, 231 (1984).

253. J. C. Michel, J. A. Manson, and R. W. Hertzberg in *Polymer Blends*, Plastics and Rubber Institute, London, UK, 1981, paper 15.

254. R. W. Lang, J. A. Manson, and R. W. Hertzberg in Ref. 50, p. 982.

255. F. M. Fowkes, D. C. McCarthy, and D. O. Tischler, in H. Ishida and G. Kumar, eds., *Molecular Characterization of Composite Interfaces*, Plenum Press, New York, 1985.

256. L. J. Broutman and R. H. Krock, eds., *Composite Materials*, 8 Vols., Academic Press, Inc., New York, 1974.

257. L. J. Broutman in *Fracture and Fatigue* in L. J. Broutman and R. H. Krock, eds., *Composite Materials*, Vol. 5, Academic Press, Inc., New York, 1974.

258. B. D. Agarwal and L. J. Broutman, *Analysis and Performance of Fiber Composites*, John Wiley & Sons, Inc., New York, 1980.

259. D. Hull, *An Introduction to Composite Materials*, Cambridge University Press, New York, 1981.

260. K. L. Reifsnider in Ref. 3, p. 563.

261. K. L. Reifsnider and R. B. Pipes, eds., *The Fatigue Behavior of Composite Materials*, Elsevier Science Publishing Co., New York, in press.

262. J. K. Wells and P. W. R. Beaumont in Ref. 186, p. 397.

263. M. P. Pigott in Ref. 30, p. 494.

264. C. C. Chamis in Ref. 257, p. 94.

265. E. M. Wu in Ref. 257, Chapt. 3.

266. J. F. Mandell in Ref. 261.

267. R. W. Lang, J. A. Manson, and R. W. Hertzberg in C. D. Han, ed., *Polymer Blends and Composites in Multiphase Systems, ACS Advances in Chemistry Series No. 206*, 1984, p. 261.

268. M. J. Folkes, *Short Fiber Reinforced Thermoplastics*, Research Studies Press, Chichester, UK, 1982.

269. R. E. Lavengood and L. E. Gulbransen in Ref. 22, p. 365.

270. K. L. Friedrich, *Kunstoffe* **72**(5), 290 (1982).

271. S. V. Hoa, A. D. Ngo, and J. S. Sankar in G. C. Sih and V. P. Tamuzs, eds., *Fracture of Composite Materials*, Nijhoff Publishers, Ltd., The Hague, The Netherlands, 1982, p. 477; Ref. 247, p. 44.

272. P. Kundrat, S. Joneja, and L. J. Broutman in Ref. 247, p. 105.

273. B. B. Moore in H. Lilholt and R. Talreja, eds., *Fatigue and Creep of Composite Materials*, Riso National Laboratory, Roskilde, Denmark, 1982, p. 245.

274. M. J. Carling, J. A. Manson, and R. W. Hertzberg, *Polym. Prepr. Am. Chem. Soc. Div. Polym. Chem.* **26**(1), 284 (1985).

275. M. J. Carling, J. A. Manson, and R. W. Hertzberg, *Proceedings of the 43rd Annual Technical Conference of the Society of Plastics Engineers*, 1985, p. 396.

276. K. Friedrich, K. Schulte, K. Horstenkamp, and T. W. Chou, *J. Mater. Sci.* **20**, 3353 (1985).

277. C. Lhymn, *J. Mater. Sci. Lett.* **4**, 1429 (1985).

278. J. F. Mandell, F. J. McGarry, D. D. Huang, and C. G. Li in Ref. 252, p. 32.

279. J. Theberge, B. Arkles, and R. Robinson, *Ind. Eng. Chem. Prod. Res. Dev.* **15**, 100 (1976).

280. P. S. Pao, J. E. O'Neal, and C. J. Wolf, *Polym. Mater. Sci. Eng.* **53**, 677 (1985).

281. S. K. De and V. M. Murty, *Polym. Eng. Rev.* **4**, 313 (1984).

282. M. J. Owen in Ref. 257, Chapts. 7 and 8.

283. A. T. DiBenedetto, G. Salee, and R. Hlavacek in Ref. 64, p. 242.

284. A. T. DiBenedetto and G. Salee in Ref. 91, p. 634; Ref. 233, p. 512.

285. S. Nguyen and S. Moet, in press.

286. H. Voss and R. Walter, *J. Mater. Sci. Lett.* **4**, 1174 (1985).

287. H. Voss and K. Friedrich in Ref. 286, p. 569.

288. M. J. Owen and P. T. Bishop, *J. Phys. D* **7**, 1214 (1974).

289. C. R. Wachniki and J. C. Radon, *Composites* **15**(3), 211 (1984).

290. C. Lhymn and J. M. Schultz in Ref. 180, p. 2029.

291. M. J. Owen and R. G. Rose, *Plast. Polym.* **40**, 325 (1972).

292. J. F. Mandell, F. J. McGarry, A.J.-Y. Hsieh, and C. G. Li, *Polym. Compos.* **6**, 168 (1985).

293. K. Iisaka and Y. Namai, *J. Appl. Polym. Sci.* **30**, 3397 (1985).

294. S. Morris, Ph.D. Thesis, University of Nottingham, UK, 1970, quoted in Ref. 282, Chapt. 8, p. 281.

295. D. Schuetz in Ref. 289, p. 217.

296. B. Harris, *Composites* **8**, 214 (1977).

297. S. S. Wang, E. S. M. Chim, and D. F. Socie, *ASTM STP 787*, American Society for Testing and Materials, Philadelphia, Pa., 1982, p. 287.

298. D. C. Phillips and J. M. Scott in Ref. 296, p. 233.

299. D. Schuetz and J. J. Gerharz in Ref. 296, p. 245.

300. C. K. H. Dharan, *Proceedings of the 1975 International Conference on Composite Materials*, Vol. 1, The Metallurgical Society of the American Institute of Mechanical Engineers, New York, 1976, p. 830.

301. P. W. R. Beaumont and B. Harris, *International Conference on Carbon Fibers, Their Composite Applications*, Plastics and Rubber Institute, London, UK, 1971, paper 49.

302. J. H. Williams and S. S. Lee, *J. Compos. Mater.* **12**, 348 (1978).

303. M. Fuwa, B. Harris, and A. R. Bunsell, *J. Phys. D.* **8**, 1460 (1975).

304. R. F. Dickson, C. J. Jones, B. Harris, T. Adam, and H. Reiter, *Proceedings of the First International Symposium on Acoustic Emission from Fiber-reinforced Composite Materials*, San Francisco, Calif., 1983 (SPI, New York, 1983), paper A–4, pp. 1–14.

305. J. J. Nevadunsky, J. J. Lucas, and M. J. Salkind in Ref. 248, p. 394.

306. G. Maier, H. Ott, A. Protzner, and B. Protz, *Composites* **17**, 111 (1986).

307. M. N. Gibbins and W. W. Stinchcomb in Ref. 297, p. 305.

308. S. S. Wang, D. P. Goetz, and H. T. Corten, *J. Compos. Mater.* **18**, 2 (1984).

309. H. J. Kim and L. J. Ebert, *J. Compos. Mater.* **10**, 139 (1978).

310. G. L. Roderick and J. D. Whitcomb in Ref. 245, p. 73.

311. D. T. Hayford and E. G. Henneke, II, *ASTM STP 674*, American Society for Testing and Materials, Philadelphia, Pa., 1979, p. 184.

312. K. L. Reifsnider, K. Schulte, and J. C. Duke, *ASTM STP 813*, 1983, p. 136.

313. R. L. Ramkumar in Ref. 312, p. 116.

314. J. D. Whitcomb, *ASTM STP 723*, American Society for Testing and Materials, Philadelphia, Pa., 1981, p. 48.

315. K. L. Reifsnider, ed. in Ref. 312, p. 116.

316. R. B. Pipes, S. V. Kulkarni, and P. V. McLaughlin, *Mater. Sci. Eng.* **30,** 113 (1977).
317. H. T. Hahn and R. Y. Kim in Ref. 248, p. 297.
318. J. N. Yang and M. D. Liu, *J. Compos. Mater.* **11,** 176 (1977).
319. E. P. Phillips in Ref. 314, p. 197.
320. R. Badaliance, H. D. Dill, and J. M. Potter in Ref. 297, p. 274.
321. C. T. San and E. S. Chim in Ref. 314, p. 233.
322. J. N. Yang and D. L. Jones in Ref. 314, p. 213.
323. J. N. Yang and S. Du, *J. Compos. Mater.* **17,** 511 (1983).
324. H. T. Hahn in Ref. 311, p. 383.
325. S. L. Phoenix in Ref. 311, p. 455.
326. J. N. Yang and D. L. Jones, *Proceedings of the 20th AIAA/ASME/ASCE/ASH Structures, Structural Dynamics, and Materials Conference,* 1979, p. 232.
327. D. G. Harlow, *ASTM STP 674,* American Society for Testing and Materials, Philadelphia, Pa., 1978, p. 484.
328. P. C. Chou and R. Croman, *ASTM STP 674,* American Society for Testing and Materials, Philadelphia, Pa., 1977, p. 431.
329. W. Hwang and K. S. Hahn, *J. Compos. Mater.* **20,** 125, 154 (1986).
330. A. Rotem and H. G. Nelson in Ref. 314, p. 152.
331. J. V. Gauchel, I. Steg, and J. E. Cowling, *ASTM STP 569,* American Society for Testing and Materials, Philadelphia, Pa., 1975, p. 45.
332. S. C. Kunz and P. W. R. Beaumont in Ref. 331, p. 71.
333. C. T. Sun and W. S. Chan in Ref. 311, p. 418.
334. R. F. Dickson, C. J. Jones, B. Harris, D. C. Leach, and D. R. Moore in Ref. 276, p. 60.
335. F. N. Cogswell and D. C. Leach, *Plast. Rubber Process Appl.* **4**(3), 271 (1984).
336. F. N. Cogswell and M. Hopprich, *Composites* **14,** 251 (1983).
337. J. B. Sturgeon, *RAE Technical Report MAT 75135,* Royal Aircraft Establishment, Farnborough, UK, 1975.
338. J. T. Hartness and R. Y. Kim, *Proceedings of the 28th National SAMPE Symposium,* Society for the Advancement of Material and Process Engineering, Azusa, Calif., April 1983, p. 335.
339. W. S. Carswell in Ref. 296, p. 251.
340. J. F. Mandell and U. Meier in Ref. 331, p. 45.
341. M. D. Campbell and B. W. Cherry in G. C. Sih and C. L. Chow, eds., *Fracture Mechanics and Technology,* Vol. 1, Sijthoff and Noordhof, Alphenaan den Rijn, The Netherlands, 1977, p. 297.
342. S. Mostovoy and E. J. Ripling, *Polym. Sci. Technol.* **9B,** 513 (1975).
343. J. A. Marceau, J. C. McMillan, and W. M. Scardino, *Scientific Advances in Material and Process Engineering Series No. 22,* 1977, p. 64.
344. G. Gebauer and J. A. Manson, Lehigh University, 1974, unpublished research.
345. L. E. Kuckacka, J. Fontana, and A. Romano, *Concrete-Polymer Materials Development Progress Report for the Federal Highways Administration,* Brookhaven National Laboratory for the quarter July–September 1973.
346. K. Friedrich, A. Fels, and E. Hornbogen, *Compos. Sci. Tech.* **23**(2), 79 (1985).

RICHARD W. HERTZBERG
JOHN A. MANSON
Lehigh University

FIBERS, ENGINEERING

Fibers have a long history as a material of construction. Their use in apparel dates back to Adam and Eve. There were also early engineering uses of fibers, eg, the use of straw to reinforce bricks. Also, Mongolian nomads discovered that sheep's wool, kept under the saddle, in time formed a firm, strong structure. Placed on wooden frames, the fabric thus formed turned out to be an excellent exterior wall for yurts. Today, it would be said that the wool had been compacted by mechanical action, the movement of horse and rider, in the presence of heat and moisture supplied largely by the sweating horse, to a nonwoven felt, a fibrous engineering structure evidently useful in the construction of mobile homes.

Since these early applications, fibers and fibrous structures have been investigated by many branches of science and engineering in order to design efficient processes of fiber manufacture and to improve fiber characterization and fiber properties. This development has been accompanied by improved methods for handling fibers and forming new structures from them.

Fibers are now produced to exacting quality standards with reproducible properties and thus can truly be described as engineering materials. Fibers with well-known chemical, physical, mechanical, electrical, and optical properties can be assembled, alone or in combination with other materials, into well-defined structures with engineered properties. Woven or knit fabrics, nonwoven fabrics (qv), and composites (qv) are examples.

Engineering applications of fibers can be divided into five areas: earth sciences, human interaction with nature and the environment; biomedical applications; composites; industrial applications; and the frontiers of science and technology, where fibers are making possible entirely new products.

Earth Sciences

Agriculture. Fibrous structures play an important role in modern agricultural practice. Originally limited to use as shade cloth and tobacco seed-bed covers, lightweight fibrous structures are now used to protect, promote, or control plant growth in greenhouse thermal screens, row-crop covers, turf covers, weed barriers, and hydroponics. In 1980 the European Economic Community (EEC) used more than 45,000 acres of fibrous row covers (1). This segment and others can expect annual growth rates of approximately 10 to 20% as growers become more aware of the benefits (1–3).

Product development of fibrous structures for the agricultural market involves mostly sheet structures. Key properties are light, air, and moisture permeability; the fiber plays a limited role in determining these characteristics. Fiber properties are more important when cost effectiveness is considered in terms of weight, strength, and dimensional stability. Also, fibers are critical in determining the uv stability, moisture pickup, and thermal-use limits of the sheet. Fibers in current use and suppliers of sheet structures are listed in Table 1 by agricultural applications (qv) and fiber type.

Table 1. Fibers in Agricultural Applications

Use	Fiber	Manufacturer
row covers	polyester	DuPont, Unitika, Freudenberg
	polypropylene	Kimberly Clark, Sodoca, Freudenberg
	nylon	Chicopee
	polyamide	Beghin-sey
tobacco seed	polyester	DuPont
bed covers	polypropylene	Kimberly Clark
	nylon	Chicopee
	cotton	Chicopee
turf covers	polyester	DuPont, Hoechst
weed barrier	polyester	Hoechst
	polypropylene	DuPont, Phillips, Dewitt

Soil Engineering.   In soil engineering, fibrous structures are used for separation, drainage, reinforcement, and filtration. When so used, the structures are called geotextiles (4). Separation applications include barriers between subsoil and stony surfaces, as in road building, and structures used over plastic sheeting for puncture protection, as in landfills. In drainage uses, the geotextile keeps soil out of drains. In reinforcement, multiple layers of geotextiles containing granular fill form a wall, and in filtration, a geotextile filters and stills run-off water to remove small soil particles (Fig. 1).

Nonsoil materials have been used to correct soil deficiencies for millenia. Logs and timber were used in corduroy-road construction as early as 2500 BC (5), and bundles of willow and bamboo, called fascines, were used in embankments and levees construction. In 1935 a woven cotton fabric was used in a South Carolina road, but biodegradation prevented widespread use. A vinyl chloride automobile upholstery fabric was used under concrete-block revetments to solve an erosion problem in Florida in the 1950s. This led to the use of geotextiles in erosion control throughout the world.

The modern era of geotextiles began in Europe in the 1960s and 1970s with Rhone-Poulenc's Bidim needle-punched polyester and ICI's Terram (Celanese's Mirafi), a nonwoven fabric consisting of bonded sheath-core fibers. U.S. producers entered the geotextile market, led by DuPont with Typar spun-bonded polypropylene and Phillips Petroleum with Petromat, a needle-punched polypropylene. Tables 2 and 3 list applications and properties of fibers used in geotextiles. Geotextile selection for a project is based on improved design performance such as more efficient installation, reduced maintenance, increased life or a substantial cost savings compared to alternatives (6–9).

Water Purification, Separation.   Reverse osmosis, the production of pure water from saline water utilizing pressure as the driving mechanism, is one of the major markets for hollow-fiber membranes (10,11). Other areas of application of hollow fibers are ultrafiltration, the separation of dissolved species based on size differences, and the separation of gases. The use of membranes (qv) for these separations is the most energy-efficient method known. The greatest advantage of fibrous membranes over other membrane forms is their extremely high surface area per device unit volume. A wide variety of polymers with differing physical

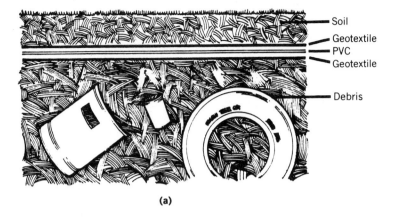

(a)

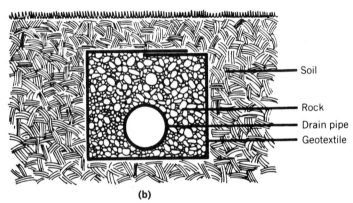

(b)

Fig. 1. Soil engineering applications. (**a**) Separation: Geotextile on both sides of PVC film protects it from puncture in landfill. (**b**) Drainage: The geotextile acts as a matrix to keep the soil out of the drain. (**c**) Reinforcement: Multiple layers of geotextile contain granular fill to form a wall. (**d**) Filtration: A geotextile silt fence filters runoff water to remove small particles of soil.

and chemical properties can be spun as hollow fibers. Selection of the proper polymer and the ability to modify the hollow-fiber membrane structure in the spinning and postspinning manufacturing processes provide true engineering flexibility to tailor products with optimum properties for specific applications.

Reverse Osmosis (RO). Reverse osmosis, the preferred method of producing pure water, uses one-tenth the energy per liter of pure water made from seawater, as evaporation processes such as multistage flash distillation (10), because no phase change occurs during separation. In osmosis, pure water, separated from a solution by a semipermeable membrane, flows into the solution to equalize the concentration on both sides of the membrane. The pressure required to stop this flow is the osmotic pressure. If the pressure on the solution side is raised above the osmotic pressure, the flow is reversed, and pure water can be produced from solutions such as seawater.

RO purification of water falls into two categories, depending on the concen-

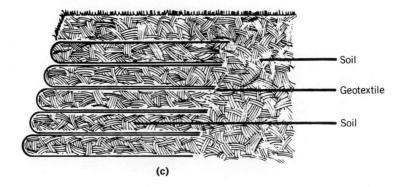

Soil

Geotextile

Soil

(c)

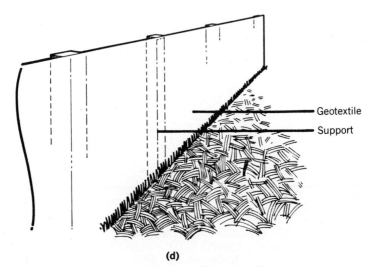

Geotextile

Support

(d)

Fig. 1. (*Continued*)

tration of dissolved solids in the feedwater. Typical pressure requirements and types of hollow fibers used are shown in Tables 4 and 5.

Aramid hollow-fiber permeators designated B-9 were introduced by DuPont in 1970 (Fig. 2). Cellulose acetate fibers in permeators became available from Dow at about the same time and from Toyoba in 1978. The B-9 aramid hollow

Table 2. Uses for Geotextile Fabrics

Application	% of all uses	Function of fabric
stabilization	28	separation and reinforcement
primary road	26	reinforcement and water-proofing
secondary road	13	separation and reinforcement
drainage	13	drainage and separation
erosion control	10	drainage, filtration, and separation
silt fence/other	7	filtration
railroad tracks	4	reinforcement and separation

Table 3. Fibers Used in Geotextiles

Fiber	Strength, N/tex[a]
aramid	2.2
polyamide	0.53–0.88
polyester	0.62–0.88
polypropylene	0.31–0.62
polyethylene	0.31–0.62

[a] To convert N/tex to g-f/denier, multiply by 11.3.

Table 4. Pressure Required for Reverse Osmosis

Type of feed	Total dissolved solids, ppm	Osmotic pressure, MPa	Required back pressure, MPa
brackish water	<10,000	<0.7	<2.8
seawater	30,000–40,000	2–3.4	>5.5

Table 5. Types of Fibers Used in Purification

Polymer	Hydrolytic stability	Oxidative stability	Biological stability	Acceptable pH range
aramid	excellent	poor	excellent	4.0–11.0
cellulose acetate	poor	fair	poor	4.5–7.5

fiber has an outside diameter of 90 μm and an inside diameter of 40 μm. Water flows through the fiber walls into the base of the fibers, and the dissolved solids are rejected at the outer surface. A 20-cm diameter B-9 permeator, approximately 1.2 m long, contains 2.5×10^6 hollow fibers, has an active surface area of 465 m^2 and produces 60,000 L/day of pure water when operated at 2.8 MPa (400 psig) and 25°C.

The B-10 hollow-fiber permeator can produce potable water from seawater with <500 ppm total dissolved solids (TDS). This permeator operates at pressures from 5.5–8 MPa (800–1200 psig) at temperatures to 40°C. A 20-cm diameter unit produces 19,000 L/day of pure water when operated at 5.5 MPa (800 psig), 25°C, while rejecting more than 98.5% of the dissolved solids in the feedwater. Both the product flow and the salt rejection increase as the operating pressure is increased.

The high water-permeation rates and high salt-rejection rates of the hollow fibers result from the development of an asymmetric structure in the fiber walls. This structure consists of a thin, dense salt-rejecting layer at the fiber surface, with a porous substrate providing mechanical support with minimal resistance to water flow. The asymmetric structure is produced by a combination of spinning and postspinning techniques. Hollow fibers are spun from a polymer solution and quenched in a nonsolvent for the polymer. This produces a porous wall fiber with a dense incipient skin at the outer fiber surface. Typically, the rejection of this surface layer is improved by one or more postspinning operations such as thermal annealing or various chemical treatments.

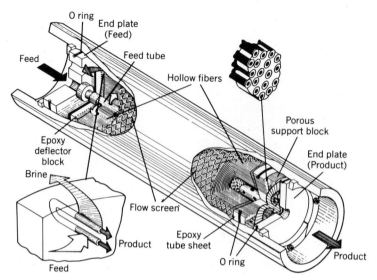

Fig. 2. Hollow fine fiber Permasep permeator.

The water permeability of fibrous permeators is irreversibly reduced with time owing to morphological changes in fiber structure (Fig. 3). After several years of operation, the flow of a permeator may be reduced to 50–75% of its initial flow. Acceptable compromises between initial flow rates and flux-decline rates are reached by controlling the density of the fiber walls. This is achieved by careful control of the spinning operation, including the composition of the spinning solution and by thermal posttreatment of the fibers.

Ultrafiltration and Gas Separation. In ultrafiltration (qv), the separation is based on the molecular size of the dissolved species. Smaller species pass through the fiber wall, but larger species are rejected. The lower limit is typically a molecular weight of 1000, but it can be as high as 10,000. This separation technique is applied in the food, pharmaceutical, and electronic industries.

In ultrafiltration, the rejection layer is on the inner wall of the fiber rather than on the outside as it is in RO. Ultrafiltration fibers also are much larger

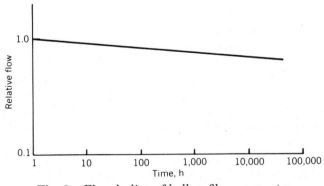

Fig. 3. Flux decline of hollow fiber permeator.

than RO fibers; the smallest inside diameter is 500 μm. In operation, the feed stream is pumped down the core of the fibers using feed pressures >0.7 MPa (100 psig), compared to 1.4–8 MPa (200–1200 psig) for RO systems. Polymers used in ultrafiltration hollow fibers include cellulose, cellulose acetate, polyacrylonitrile, polysulfone, and poly(vinyl acetate).

Hollow-fiber membranes for gas separation have been commercially available for several years. Typical applications include the separation of hydrogen and methane, carbon dioxide and methane, carbon monoxide and carbon dioxide, and oxygen and nitrogen. Advantages over competitive processes such as cryogenic separations are the compactness of the plants and the ability to handle relatively small gas streams more economically.

Monsanto's Prism hollow fiber-gas separators are used on a variety of gas streams, but their primary use is the separation of hydrogen from other gases. The hollow fibers are composed of polysulfone. Dow has two commercially available hollow fiber-gas separation products. Their Cynara subsidiary produces a cellulose-based hollow-fiber permeator for the separation of carbon dioxide from hydrocarbon gas streams. Generon permeators use hollow fibers composed of poly(4-methyl-1-pentene) and are designed for the separation of atmospheric oxygen and nitrogen.

Biomedical Applications

Fibers have been used for many years in medical applications (qv) such as dressings, sanitary pads, tapes, and operating-room apparel. Applications of engineering fibers include sutures, surgical implants, prosthetic devices, and medical equipment. Engineering fibers are also used for separations and purification such as kidney dialysis, and they will play important roles in emerging biotechnology applications such as substrates for large-scale cell culture (12). These fibers aid in maintaining and improving the quality of life.

Fiber Properties. Fibers possess certain advantages applicable to biomedical uses. The surface-to-volume ratio is very large, which is important in surface-dominated applications such as separations. Engineering fibers can also be processed into knitted, woven, or hollow structures that provide the varying degrees of porosity and strength needed in biomedical applications such as synthetic blood vessels, artificial ligaments, and controlled drug release.

Many engineering fibers are used within the body because of their good mechanical properties, long-term stability, and nontoxic character. These fibers also are biocompatible; that is, they do not produce any significant changes in the surrounding tissue over long periods of time. Biocompatibility of the fibers depends on the type of polymer from which they are made, on processing additives, and on the surface properties of the fiber. Woven or knitted fibers may approximate the stress–strain behavior of natural tissue more closely than other structures and thus offer greater compatibility (13). This is of particular importance in applications such as the replacement of natural blood vessels (see also BIOLOGICAL ACTIVITY; BIOMATERIALS).

Fiber Types. Both natural and synthetic engineering fibers are used in the biomedical field (Table 6). Natural fibers such as silk and catgut (collagen)

Table 6. Engineering Fibers Used in Biomedical Applications

Type	Examples
natural fibers	collagen, silk, cellulose, chitin
synthetic organic fibers	polyester, polyamide, polypropylene, polyurethane, polytetrafluoroethylene, polyethylene, poly(vinyl alcohol), polyacrylonitrile, poly(glycolic acid), poly(lactic acid), regenerated cellulose, polydimethylsiloxane, aramid
inorganic fibers	carbon, graphite, alumina, glass

have been used by physicians for many years. However, the introduction of synthetic polymers has brought about the increased use of fibers in biomedical applications (14). Polyester and nylon fibers are used in a large number of medical applications. Recently, higher performance fibers such as carbon have been evaluated, particularly in applications requiring high strength and improved tissue compatibility.

Sutures.   One of the oldest and largest biomedical uses for engineering fibers is surgical sutures (qv). Fiber properties of particular importance in suture applications are tensile strength, strength retention in the body's environment, handling characteristics of the fiber, and knot strength and security (15). Furthermore, the fibers must be biologically compatible with the surrounding body tissue.

There is no universal suture material, but a number of natural and synthetic fibers satisfy the above requirements to varying degrees, including natural fibers such as silk and cotton and synthetic fibers such as polyester, polyamide, and polypropylene. These fibers are nonabsorbable or not broken down by the body. Fibers that are ultimately absorbed by the body are also used, especially to close deep tissue. Examples are catgut and poly(glycolic acid) along with its lactide copolymers.

Silk fibers have been used as sutures for many years. The primary advantage of silk is ease in handling. However, its tensile strength is inferior to that of most synthetic fibers, and it can produce significant tissue reaction. Polyamide, polyester, and polypropylene are the most common nonabsorbable synthetic fibers used as sutures. Polyamide sutures are made from either nylon-6 or nylon-6,6 fibers. Nylon is stronger than silk and produces less tissue reaction; it is, however, more difficult to handle.

Polyester and polypropylene sutures have high strength and excellent strength-retention properties. Polyester fibers based on poly(ethylene terephthalate) make one of the strongest and longest-lasting sutures. Polypropylene sutures offer high tensile strength, strength retention, and low tissue reactivity, but have poorer handling characteristics than nylon.

Fibers based on poly(glycolic acid) and its lactide copolymers are used as absorbable sutures. These fibers are considered superior to natural absorbable fibers because of better tensile strength, more uniform and reproducible properties, and lower tissue reactivity.

Implants.   Engineering fibers are also used in surgical implants. One application is in synthetic replacements for diseased or nonfunctioning blood vessels, such as segments of the aorta or other large-diameter arteries. Polyester fiber and expanded polytetrafluoroethylene (PTFE) are currently the synthetic materials of choice for these applications. Polyester fiber produced from poly(ethylene terephthalate) is knitted or woven into a porous tube with a smooth, lightly napped surface. The PTFE is produced as a continuous microporous tube with a smooth, noncrimped flow surface.

Tensile strength, long-term durability, biocompatibility, and a controlled degree of porosity are requirements of the woven or knitted fiber tubing used in arterial replacements (16). The structure must have adequate strength to resist longitudinal and transverse tension, to withstand the intraluminal blood pressure, and to be suturable. These properties must be maintained sufficiently to prevent rupture or aneurysm formation for many years after implantation. The fibrous structure must also have sufficient porosity to allow tissue in-growth and the formation of a thin fibrin-based thromboresistant neointima on the inner surface of the tubing. On the other hand, the pores must be small enough to prevent the outflow of blood. Finally, the fiber must not elicit an adverse tissue response that would lead to rejection.

Whereas structures based on polyester fibers and PTFE have been acceptable in large-diameter, high-blood-flow applications, synthetic substitutes for smaller-diameter arteries require new elastomeric fibrous structures that are more thromboresistant (17,18).

Damaged ligaments and tendons may be replaced by orthopedic implants produced from engineering fibers. Natural ligaments and tendons are made up primarily of collagen fibers; ligaments connect two or more bones whereas tendons connect muscle to bone. Engineering fibers that are candidate replacements for these structures must approximate their mechanical and biological properties. Carbon, polyester, polypropylene, and polytetrafluoroethylene fibers are currently being investigated.

Carbon fibers (qv) have attracted interest for both ligament (19) and tendon (20) replacements. They can serve as scaffolds upon which new, oriented collagenous fibrous tissue, which ultimately functions as the replacement, grows. The fibers have been coated with poly(lactic acid) to contain migration of carbon fragments produced by premature mechanical degradation from the implant site (21). In addition to the excellent tissue-compatibility properties, other important properties of carbon fiber are its high ultimate tensile strength (2.6 GPa) and high modulus of elasticity (260 GPa). Carbon fibers, however, are brittle and therefore require particular care in surgical installation.

Polyester fibers have also been used in tendon implants (22,23). Composites of carbon fiber with polysulfone resin (24) and carbon fiber-reinforced carbon (25) are being evaluated in orthopedic devices such as artificial hips. The potential advantage is a closer match to the modulus of bone.

Others. High modulus carbon fibers in epoxy matrices are relatively transparent to x-rays and are used in medical equipment such as x-ray tables and film cassettes. Because of the combination of strength and light weight, engineering fibers such as graphite can be used for external prosthetic devices such as artificial limbs.

Hollow fibers are particularly useful for biomedical separations because of large-surface/volume ratios. Hollow fibers of regenerated cellulose are used in kidney dialysis. Other uses for engineered hollow fibers are in plasma-exchange therapy, controlled release of drugs, blood oxygenation, and cell separations (see also DRUG APPLICATIONS).

Fibers are also used as supports for the immobilizaton of enzymes. Enzymatic reactions are of interest because of high specificity and reaction rates compared to conventional catalysts. Extremely fine fibers, such as those produced from poly(vinyl alcohol), have been investigated because of their greater surface area for absorbing particular enzymes (26) (see also ENZYMES; ENZYMES, IMMOBILIZED). The combination of fiber technology and cell-culture techniques to develop bioartificial organs such as the pancreas and liver is expected (27).

Advanced Composite Materials

In order to meet the requirements of advanced military systems, a number of low density fibers have been produced that exhibit both high specific stiffness and high specific strength. In general, these properties are shown by materials based on relatively light elements such as beryllium, boron, carbon, nitrogen, oxygen, magnesium, aluminum, silicon, and titanium which form covalent bonds inherently stiffer and stronger than compounds formed through metallic or ionic bonds. However, many of these materials are brittle in bulk form because of the presence of internal or surface flaws. The probability of the presence of such flaws is minimized by using the material in fibrous form. By binding the fibers so they act in concert, it is possible to obtain composite materials that exhibit physical and structural properties not attainable with conventional engineering materials.

High Performance Fibers

Of the elements mentioned above, carbon and silicon are the principal constituents in commercially available high performance fibers. These fibers can be grouped into three broad categories: carbon, organic resins, and inorganic compounds. Within each group, several classes of high performance fiber materials have been developed that satisfy the basic criteria of low density, high strength, and high stiffness to varying degrees.

Carbon Fibers. Carbon fibers (qv) are currently the predominant high strength, high modulus fiber used in the manufacture of advanced composite materials. These fibers are made by the pyrolytic degradation of a fibrous organic precursor. By heating an organic polymeric fiber under tension to high temperatures in an inert atmosphere, the volatile constituents are driven off. The residual carbon atoms tend to orient themselves along the fiber axis into graphitic crystallites and thus form a high strength, high modulus fiber. The properties of this fiber product are a function of the composition of the precursor and of the time–temperature profile of the pyrolytic process. In general, the higher the maximum processing temperature, the greater the extent of crystallite orientation parallel to the fiber axis, and thus the higher the fiber modulus. Because of increasing sensitivity to flaws, this increase in modulus is usually accompanied by a decrease in strength.

Most of the commercially available carbon fibers are obtained by the pyrolysis of polyacrylonitrile (PAN). Worldwide, there are currently more than a dozen producers of PAN-based carbon fibers. Most manufacturers offer a high strength product (Type I carbon fiber) and a high modulus product (Type II carbon fiber). Some manufacturers also offer an ultra-high modulus product (Type III carbon fiber). Representative properties of these materials are given in Table 7. There is close correlation between carbon contents and physical properties. Carbon fibers differ from other reinforcing fibers in that they are good conductors of electricity and exhibit unusual thermal-expansion characteristics. The thermal coefficient of expansion is negative along the fiber axis, but positive in the transverse direction.

Table 7. Representative Properties of Standard PAN-based Carbon Fibers

Designation properties	Type I high strength	Type II high modulus	Type III ultrahigh modulus
carbon content, wt %	92–94	>99	>99.9
specific gravity	1.7–1.8	1.8–1.9	1.9–2.1
filament diameter, μm	7–8	7–8	8–9
tensile modulus, GPa	220–250	340–380	520–550
tensile strength, GPa	2.5–3.5	2.2–2.4	1.8–1.9
tensile elongation, %	1.2–1.4	0.6–0.7	0.3–0.4
toughness, MPa	20	7.5	3
electrical resistivity, $\mu\Omega/m$	15–18	9–10	6–7
longitudinal coefficient of thermal expansion, 10 m/(m·K)	−0.5	−0.7	−0.9 (est)
range of available tow counts, thousand fibers/bundle	1–320	1–12	<1
price range, 1985 $/kg	40–300	90–1000	>300

The low strain-to-failure characteristics have been the principal limitation of standard PAN-derived carbon fibers, but high strain carbon fibers are now available that exhibit tensile elongations of the order of 1.7 to 1.8%, as shown in Table 8. These fibers not only are unusually strong but also are much easier to handle, which facilitates the fabrication of composite parts.

Table 8. Representative Properties of High Strain Type I PAN-based Carbon Fibers

Properties	Values
carbon content, wt %	92–94
specific gravity	1.7–1.8
filament diameter, μm	5–8
tensile modulus, GPa	240–270
tensile strength, GPa	4.0–4.7
tensile elongation, %	1.7–1.8
toughness, MPa	34–42
longitudinal coefficient of thermal expansion, 10 m/(m·K)	−0.5
range of available tow counts, thousand fibers/bundle	6–12
price range, 1985 $/kg	90–100

The price of commercial PAN-based carbon fibers currently ranges from about \$40/kg to \$1000/kg. The prices are mainly a function of fiber modulus and end count. These high prices reflect, in part, the relatively high cost of using PAN as a precursor. Because petroleum pitch is a readily available, low cost material with a high carbon content, there has been growing interest in its use as a precursor material in order to reduce carbon-fiber manufacturing costs. The pitch-based carbon fibers currently available exhibit a much lower strain to failure at equal modulus than do PAN-based carbon fibers and are not significantly less expensive.

Organic Fibers.   The measured tensile strength and modulus of most natural and synthetic organic fibers are typically less than 1 and 10 GPa, respectively. These values are significantly lower than the theoretical values of the maximum tensile strength (>20 GPa) and modulus (>100 GPa) based on the force constants of bonds in the polymer chain and crystal extensions observed by x-rays on stressed fibers (28). A number of polymeric fibers are now commercially available whose strength and modulus more closely approach these theoretical values and are high enough for these materials to be considered as reinforcements for advanced composites. The materials include primarily aromatic polyamides, but also ultrahigh molecular weight polyolefins and certain classes of polyesters. Compared to textile fibers, these high strength, high modulus materials all exhibit a high degree of molecular orientation and chain alignment (see also HIGH MODULUS POLYMERS; POLYAMIDES, AROMATIC; POLYESTERS, AROMATIC).

Aromatic polyamides, or aramids, were introduced in the early 1960s initially to meet the need for fibers with improved heat and flammability resistance. The first example of this class was poly(m-phenyleneisophthalamide) (MPD-I), which is sold commercially by E.I. du Pont de Nemours & Company, Inc. under the trademark Nomex. In the early 1970s, DuPont introduced Kevlar aramid fibers, which are based on para-substituted aromatic polyamides, most notably poly(p-phenyleneterephthalamide) (PPD-T). Related aramid fibers have been introduced by Enka (Netherlands) and Teijin (Japan).

Because of their rigid chain structure, PPD-T and related p-aramids exhibit liquid-crystalline behavior in solution. Even at low concentrations, the rodlike molecules of these materials aggregate in nematic, ordered domains. When solutions of these materials are subjected to shear, these ordered domains tend to orient in the direction of flow. When passed through a spinneret, liquid-crystalline solutions retain the high degree of orientation imparted by the spinning operation, leading to as-spun fibers with extraordinary degrees of crystallinity and orientation. As-spun fibers obtained by the spinning of a 20% solution of PPD-T in 100% sulfuric acid exhibit a crystalline orientation angle of about 12° (as determined from wide-angle x-ray diffraction) and a modulus of about 72 GPa (28). Heat treatment increases the degree of crystalline alignment. Heat-treated fibers have an orientation of about 9° and a modulus of about 120 GPa.

The several types of Kevlar fibers offer different combinations of properties to satisfy the requirements of a variety of uses. Selected properties of Kevlar 49 and Kevlar 29, two products used in plastics reinforcements, are presented in Table 9.

Kevlar 49, because of its higher modulus, is normally used in advanced composites. Whereas Kevlar 49 behaves elastically in tension, it exhibits non-

Table 9. Representative Properties of Kevlar p-Aramid Fibers

Properties	Kevlar 29, standard	Kevlar 49, standard	Kevlar 49, improved
specific gravity	1.44	1.44	1.44
filament diameter,			
tex[a]	0.17	0.17–0.25	0.13
μm	13	13–16	11
tensile modulus, GPa	62[b]	131	124
tensile strength, GPa			
unimpregnated yarn[c]	2.8	2.7	
impregnated strand[d]	3.8	3.8	4.1
tensile elongation, %	4.0	2.8	3.3
toughness, MPa	87	55	68
electrical resistivity, $\mu\Omega$/m			
longitudinal coefficient of thermal expansion, 10^{-6} m/(m·K)		−2	
range of available tow counts, thousand filaments/strand	0.13–10	0.13–2	4

[a] To convert tex to denier, multiply by 9.
[b] Initial modulus.
[c] As per ASTM Test D 885.
[d] As per ASTM Test D 2343.

linear characteristics under compressive stress. The application of compressive stress results in the formation of structural defects called kink bands and eventual ductile failure (29). The onset of this behavior occurs at compressive yield strains of about 0.3 to 0.5%. As a result of this unusual behavior under compression, the use of p-aramid fibers in applications subject to high strain compressive or flexural loads is limited. At the same time, however, because of their ductile response under compression, p-aramid fibers are inherently tough materials that exhibit good damage-tolerance and energy-absorption characteristics.

Inorganic Fibers. Even though the densities of inorganic fibers are significantly higher than those of carbon and polymer fibers, these materials are of interest in the preparation of advanced composite materials because of their resistance to high temperatures and their compatability with metal and ceramic matrices. Inorganic fibers of commercial interest are based on silica, alumina, and silicon carbide (see also PRECERAMIC POLYMERS).

Silica-based Fibers. A number of quartz (>99.95% silica), high silica (>95% silica), and silicate glass (>50% silica) fibers have been developed as reinforcing materials. Glass fibers, also called fiber glass, have become an important product since their commercial introduction by Owens-Corning Fiberglass Corporation in 1939. With an estimated worldwide consumption of about 1.5×10^6 tons per year, glass fibers are the primary reinforcing material currently being used by the plastics industry.

Glass and other high silica fibers are prepared by melting a mixture of raw materials (silica sand, limestone, boric acid, alumina, etc), extruding the melt through a platinum spinneret, and then cooling the streams to form a multitude of individual fibers, which are lubricated and mechanically gathered into strands.

The lubricant is a sizing that increases the abrasion resistance of the fibers and promotes resin adhesion.

The nominal compositions of E-glass (electrical-glass) and S-glass (strength-glass), the principal grades of glass fibers used in composite applications, are presented in Table 10. Representative properties are in Table 11. Glass fibers generally have a high strength-to-weight ratio, but their moduli, which are in the range of aluminum alloys, are significantly lower than those of most of the other high performance fibers used in advanced composite applications. E-glass is the most common grade of fibrous glass used by the plastics industry. Originally introduced for electrical applications because of its dielectric and loss tangent properties, it is used in many nonelectrical applications because of its combination of mechanical properties and low cost. E-glass is the only reinforcing fibrous material available at a price of about $1/kg.

S-glass is a high strength glass initially developed for military applications. Its modulus is about 20% greater than that of E-glass, it is about one-third stronger, its creep rupture resistance is significantly better, and it is less susceptible to acid corrosion. Because of its high failure energy, S-glass is also a very tough material. A related product, S2-glass, which has the same fiber composition as S-glass but different sizing, has been developed for commercial applications. The two materials have similar properties, but the cost of S2-glass fibers is significantly lower because they are not subject to the stringent and expensive quality-control procedures imposed by military-procurement requirements.

A number of high silica and quartz fibers have been developed for extreme-temperature applications. At lower temperatures these fibers differ from glass fibers in their higher melt temperatures and superior electrical properties. The dielectric-constant and power-dissipation factor are even lower than those of E-glass. The high temperature properties of silica fibers are of little value in resin-matrix composites, where temperature limitations are imposed by the thermal resistance of the matrix material. Because they are significantly more expensive than glass fibers, high silica and quartz fibers are rarely used as composite reinforcing fibers except for specialized electronic applications where an extremely low dissipation factor is required.

Alumina-based Fibers.   Fibers with a high alumina content (>60 wt% alumina) have been developed for both high temperature applications, where they compete with silica and quartz fibers, and as reinforcements for metal- and ceramic-matrix composites. The nominal composition and representative properties of commercially available materials are listed in Tables 10 and 11. These fibers are generally prepared by extruding an aqueous or organic precursor gel through spinnerets, drying the resulting fibers to remove most of the liquid phase, then subjecting the resulting filaments to high temperature (1200°C) heat treatment to form a continuous refractory yarn. DuPont's FP fibers are polycrystalline α-alumina whereas ICI's Saffil fibers have a δ-alumina crystal phase. The Sumitomo and 3M Nextel fibers are alumina-rich mixtures of alumina and silica. These materials are available in continuous form, except for Saffil alumina, which is available only as fibers 1- to 5-cm long (in mat form), or less (in bulk and milled form).

The high alumina fibers exhibit superior high temperature properties and

Table 10. Chemical Composition of Inorganic Oxide Fibers

Material	Trade name	Supplier	SiO_2	Al_2O_3	Fe_2O_3	CaO	MgO	$Na_2O + K_2O$	B_2O_3
						Components			
E-glass		various	53.2	14.8		21.1	0.3	1.3	9.1
S-glass		Owens-Corning	64.3	24.8	0.2	<8.81	18.3	0.27	<8.81
quartz			>99.95						
ceramic	Nextel 312	3M	24	62					14
ceramic	Nextel 448	3M	28	78					2
alumina	Saffil	Sumitomo/Avco	15	85					
alumina		ICI	3	97					
alumina	Fiber FP	DuPont		>99					

Table 11. Representative Properties of Inorganic Oxide Fibers[a]

Properties	E-glass	S-glass S2-glass	Quartz	Ceramic	Ceramic	Alumina	Alumina	Alumina
specific gravity	2.60	2.49	2.2	2.7	3.1	3.25	3.3	3.95
filament diameter, μm	9(*)	9(*)	10	8–12	8–12	17	3	20
range of available tow counts, thousands filaments/strand	0.2–2	0.2	0.3	0.39	0.39	1		0.21
price range, 1985 $/kg	1	6–30	100–400	180	440	760	440	440
Properties at ambient								
tensile modulus, GPa	72	87	69	152	220	200	300	377
tensile strength, GPa	3.45	4.6	0.9	1.5	1.7	1.5	2	1.4
tensile elongation, %	4.8	5.4	1.3	1	0.8	0.8	0.67	0.36
toughness, MPa	83	124	5.9	7.7	6.6	5.6	13.3	2.5
dielectric constant 10e10 Hz	6.1	5.2	4.5	4.7	6.6			9.5
longitudinal coefficient of thermal expansion, 10^{-6} m/(m·K)	5	5.6						6.8
High temperature properties								
melt temperature, °C	1260		1650	1800			>2000	2045
max use temperature °C	600	760	900	1200			1600	
90% tensile strength retention temperature °C	220	275		1100	1400	1250		1100

[a] For types of fibers and suppliers, see Table 10.

can sustain exposure temperatures ranging from 1200 to 1600°C. At 1100°C, the fibers are wetted by, and stable in, light metal alloys such as those of aluminum and magnesium. Their tensile strength at 1100°C is only 10% lower than at ambient temperature. These fibers are thus useful for reinforcing materials for metal-matrix composites.

Silicon Carbide Fibers. Filaments of carbide, nitride, and boride fibers are of interest because of their thermal stability, oxidation resistance, superior mechanical properties at elevated temperatures, and compatibility with most molten metals and many ceramic materials. Silicon carbide fibers offer the most promise and are the furthest developed. They are available as two different types of continuous filaments as well as in whisker form. These fibers, as well as isotropic silicon carbide powder, are being actively examined as reinforcements for metal- and ceramic-matrix applications.

Because of its refractory nature, silicon carbide cannot be produced in fiber form by consolidation from the melt or by sintering. Rather, the silicon carbide fibers currently available are produced by chemical reaction, using very different process routes. Avco Corporation produces large-diameter (140 μm) silicon carbide filaments by a vapor chemical-deposition process in which a blend of methylchlorosilanes are reduced with hydrogen in the presence of an electrically heated, continuously moving carbon substrate filament. Nippon Carbon Corporation produces finer-diameter (10–20 μm) filaments by the pyrolysis of polycarbosilane polymers (30). The physical properties of these continuous fibers are given in Table 12 (see also POLYSILANES and POLYCARBOSILANES).

Table 12. Representative Properties of SiC Fibers

Properties	Chemical vapor deposition[a]	Pyrolysis[b]
specific gravity	3	2.55
filament diameter, μm	140	10–20
range of available tow counts, thousands filaments/strand	not applicable	0.5
price range, 1985 $/kg	1100	500
Properties at ambient		
tensile modulus, GPa	430	190
tensile strength, GPa	2.4	2
tensile elongation, %	0.56	1.1
toughness, MPa	13	11
electrical resistivity, $\mu\Omega$/m		0.1
longitudinal coefficient of thermal expansion, 10^{-6} m/(m·K)		3.1
High temperature properties		
melt temperature, °C	2700[c]	2700[c]
max use temperature, °C	1150	
90% tensile strength retention temperature, °C	800	300

[a] Fiber supplied by Avco.
[b] Fiber supplied by Nippon Carbon Co./Dow-Corning.
[c] Sublime.

Matrix Materials

The matrix of an advanced composite material is the continuous phase that binds the fibers together so they can act in concert. The main characteristics required of a matrix material are chemical compatibility with the reinforcing fibers, compatibility with the manufacturing methods used to fabricate the desired advanced composite component, and environmental stability under conditions of use of the advanced composite component.

The matrix material has the primary influence on mechanical properties of interlaminar shear strength, and compression and flexural strength, especially at high temperatures and also dictates the processability of the composite and its environmental resistance. No single ideal matrix material exists that satisfies all requirements imposed by different applications. Advanced composites have thus been prepared with a diversity of matrix materials that include organic polymers, carbon, metals, and ceramics.

Resin-matrix materials dominate the technology and the commercial use of advanced composite materials for a number of reasons. One is that the relatively low fabrication temperatures of resin-matrix materials, compared to those of carbon-, ceramic-, or metal-matrix materials, reduce the problems of internal stresses induced by differences in the coefficients of thermal expansion of the components, and/or reaction products between components. Another advantage is that the fabrication technology developed for fiber-glass-reinforced plastics is adaptable to the fabrication of advanced composite materials systems.

Carbon-, ceramic-, or metal-matrix composites are being developed principally for use in high temperature environments hostile to resin-matrix materials or in applications that require characteristics such as electrical conductivity or the toughness of metals.

Epoxy resins (qv) are by far the most important polymeric-matrix materials because of the good balance of properties obtained with the relative ease of handling and processing. Their performance limitations are the significant loss of mechanical properties at high temperatures (>177°C), particularly in the presence of moisture, and low toughness. Long processing cycles also are required to develop material properties fully. Thermosetting resins that result in composites with improved environmental resistance, (eg, better properties under hot/wet conditions) are now available, including polyimides (qv) and ethynyl-terminated resins (see ACETYLENE-TERMINATED PREPOLYMERS).

Thermoplastic resins have not been used as matrix materials for continuous-fiber reinforced composites, in spite of their greater ease of processing, because of their lower creep resistance at elevated temperatures. Furthermore, thermoplastics as a class tend to be more susceptible to solvent attack than thermosets. However, thermoplastic materials that exhibit good high temperature properties and reasonable solvent resistance have become available, eg, aromatic polyesters, poly(phenylene ether), poly(phenylene sulfide), polysulfone, aromatic polyetheretherketone (PEEK), and polyetherimides, polyesterimides, or polyamideimides. These materials all contain an aromatic backbone, but differ in the way the aryl groups are linked.

Processing and Fabrication Technology

Most fibers used to reinforce advanced composite materials have a diameter of 5 to 20 μm. Fibers of this diameter are flexible enough to be processed by many standard textile operations, in spite of their high modulus. At the same time, fibers of this diameter are too large to be aspirated into the respiratory tract, and thus do not present the environmental health hazards associated with finer fibrous materials, such as asbestos. Fibers of this size are too fragile to be handled individually and are available only as multifilament strands. The reinforcing materials used in most advanced composite applications have a tow count of 200 to 12,000 filaments per strand. Higher end count bundles are available, but low end count strands are preferred as reinforcements for critical applications because of higher product uniformity.

The manufacture of advanced composite components entails surrounding the fibers with the matrix material in a nonrigid state, arranging the reinforcing fibers in selected patterns and orientations, solidifying the matrix material under constrained conditions that prevent fibers from moving during solidification, and integrating the advanced composite into the finished component or system.

The order of the first three steps may be changed, depending on the matrix material and process technology selected. For example, most advanced composites are made by laying-up sheets of precoated collimated fibers (eg, prepreg material) which are then cured by the application of heat and pressure. However, strands of reinforcing fibers could first be woven into flat fabrics, into three-dimensional or knitted fabrics, or even into braided shapes, on standard textile machinery by slight process modifications. The choice of production methods of advanced composite materials is a function of the matrix material rather than the high performance fiber (see COMPOSITES, FABRICATION).

Properties of Composite Structures

A unique feature of composite materials is the directional dependence or anisotropy of the various properties. This feature is intuitively clear for a unidirectional lamina: properties along the direction of the fiber reflect the higher performance characteristics of the fiber whereas properties perpendicular to the fiber tend toward the behavior of the matrix material. This directional dependence introduces the opportunity to tailor the material to specific applications. The alignment of stiff and strong fibers along anticipated load paths can yield highly efficient structures. The effects of fiber orientation, developed through different stacking sequences, are illustrated in Table 13. Properties of aluminum and cold rolled steel are included for comparison.

In order to compensate for the low transverse properties of unidirectional material (0°), laminates of unidirectional fiber may be cross-plied at right angles to one another (0°, 90°). The resulting structure has improved transverse properties, as compared to the unidirectional structure, but poorer longitudinal properties. Furthermore, the in-plane shear strength is not significantly improved over that of the unidirectional structure. Shear strength can be improved, but at

Table 13. Effect of Fiber Orientation on Mechanical Properties of Fiber Reinforced Composites[a]

Material	Fiber layup geometry	Density, g/cm^3	Elastic moduli, GPa			Ultimate strength, MPa			Fatigue strength, % ultimate strength
			Longitudinal E_x	transverse E_y	Shear G_{xy}	Longitudinal x	transverse τ_y	Shear τ_{xy}	
A-type graphite/epoxy	unidirectional (0°)	1.57	138	6.9	4.5	1517	41	97	70
	crossply (0°, 90°)	1.57	74	74	4.5	838	838	97	59
	crossply (±45°)	1.57	17.2	17.2	31.0	138	138	345	
	isotropic (0°, 90°, ±45°)	1.57	48	48	17.8	604	604	221	78
	harness satin weave cloth (warp)	1.57	62	62		462	476		
IIM-type graphite/epoxy	unidirectional (0°)	1.60	221	6.9	4.8	1206	34	69	69
	crossply (0°, 90°)	1.60			4.8			69	
	crossply (±45°)	1.60	17.2	17.2	44.8	124	124	290	
	isotropic (0°, 90°, ±45°)	1.60	73	73	24.8	345	345	179	
UHM-type graphite/epoxy	unidirectional (0°)	1.68	303	6.9	6.6	758	28	48	
	crossply (0°, 90°)	1.68	159	159	6.6	402	402	48	
	crossply (±45°)	1.68	20.7	20.7	79.3	96.5	96.5	207	
	isotropic (0°, 90°, ±45°)	1.68	103	103	42.8	242	242	128	
Kevlar 49/epoxy	unidirectional (0°)	1.38	86(tens) 41(comp)	5.5	2.1	1517(tens) 276(comp)	28	41	70

crossply (±90°)	1.38	7.6	7.6	20.7	207	207	221	
crossply (±45°)								
isotropic (0°, 90°, ±45°)								
181 fabric (warp) (50v/o)	1.33	31	31	2.0	517(tens) 172(comp)	517(tens) 172(comp)	110	
S-glass/epoxy								
unidirectional (0°)	1.88	48	6.9	3.4	1730	40	10	
crossply (0°, 90°)	1.88	31	31		980	900		
crossply (±45°)	1.88	16	16		170	170		
isotropic (0°, 90°, ±45°)	1.88	25	25		730	730		
E-glass/epoxy								
unidirectional (0°)	1.80	39	9.6	2.1	1104	20		23
crossply (0°, 90°)	1.80	25	25		518	518(32)[b]		22
crossply (±45°)	1.80	11	11		152	152		
isotropic (0°, 90°, ±45°)	1.80	18	18		330	330(42)[b]		
aluminum 6061-T-6	2.70	72	72	26.2	310	310(88)[b]	207	31
steel, cold-rolled (0.2% carbon)	7.85	207	207	83	552	552(75)[b,c] (60)[b,d]	414	40–45

[a] Ref. 31.
[b] Yield strength as percent ultimate strength.
[c] Tension.
[d] Shear.

339

Table 14. Major Applications of Advanced Composite Materials

Application	Advantages[a]							
	Vibration control	Toughness	Controlled thermal expansion	Electro-magnetic properties	Corrosion resistance	Parts integration	Lower life cycle costs	Lower first costs
Aerospace								
satellite structures	X		X	X			X	
rocket component		X	X		X		X	
military aircraft		X		X	X			
rotorcraft	X	X				X	X	
business aircraft	X					X	X	
commercial transports secondary structures	X					X	X	
Marine								
spars for racing boats	X				X			
hulls for small craft	X	X			X	X		
propeller shafts	X				X		X	
Ground transportation equipment								
racing-vehicle components	X							
limited-production automobiles	X	X			X	X		
body and chassis parts								
heavy-duty trucks	X					X	X	X
springs and drive shafts								
engine components	X						X	X
Industrial and scientific equipment								
process piping	X	X			X		X	X(installed)
oil-field structures					X		X	X(installed)
robotic components	X	X						
high speed machinery	X	X						
safety equipment including ballistic protection		X						
communications antennae	X		X		X			
electronic-circuit boards	X	X		X	X	X		
medical devices	X		X			X		
Consumer products								
sporting goods	X	X			X	X		
musical instruments	X					X		

[a] All materials provide weight reduction compared to traditional materials.

the expense of longitudinal and transverse properties, by a stacking sequence of ±45°. A stacking sequence of 0, 45, and 90° plies yields a quasi-isotropic structure in which the in-plane properties are independent of direction.

Woven fabric structures are similar to cross-plied laminate structures; however, they are usually less stiff and strong due to the various weaving patterns. Fabric geometries offer the advantage of drapability and handling characteristics that facilitate the fabrication of geometrically complex shapes.

The structures described in Table 13 consist of a single fiber component. However, the material properties may be further altered through hybrid composites consisting of two or more types of fibers in a common matrix. By judicious positioning of different fibers, it is possible to greatly expand the range of properties that can be achieved with composite materials systems (see also COMPOSITES; COMPOSITES, TESTING).

Table 14 indicates the principal applications for advanced fiber reinforced composites and summarizes the advantages of these materials in the varied uses. Advanced composite materials offer many advantages including weight reduction, cost reduction through parts consolidation, lower life cycle cost and improved systems productivity.

Industrial Applications

Rubber Goods.   Reinforcement of rubber was an early engineering application of fibers in tires (qv), V-belts, conveyor belts, and many other articles. Fiber-reinforced rubber was actually one of the earliest modern composites.

Early tires consisted of cotton fabrics and natural rubber (32). Since then, the development of engineering fibers and improvements in rubber technology, as well as an understanding of the stresses under which tires operate, have led to three methods of tire-design improvement: the development of increasingly stronger reinforcing fibers, from cotton to rayon to nylon, then to polyester and steel, and now to Kevlar aramid; and the development of sophisticated engineering calculations to determine the optimum configuration of the fiber ("cord") and specific design of the cord/rubber composite. Much of this technology is also being applied to other mechanical rubber goods, such as V-belts and conveyor belts. Table 15 shows the present use of various fibers in mechanical rubber goods.

Table 15. Use of Fibers in Mechanical Rubber Goods in 1985, 10³ t

Fiber	Tires and chafers	Belts	Hose	Total
nylon	66.4	4.2	2.7	73.3
polyester	74.5	7.3	4.1	85.9
glass	5.9	very small	very small	5.9
steel	144.5	4.5	20.5	169.5
rayon	3.6	1.4	3.6	8.6
cotton	1.9	4.3	1.3	7.5
Kevlar	3.5[a]	na[b]	na[b]	

[a] 1983 information from Textile Economics Bureau data.
[b] na = not available.

Ropes and Cables.   The art of ropemaking was developed many thousands of years ago, and ropes have been important structures in the growth of our civilization, especially in shipping and the conquest of the sea (33). The advent of synthetic fibers such as Kevlar aramid with its extremely high strength and modulus and low density has introduced new dimensions to the engineering of ropes and cables. Comparative properties of typical fibers used in ropes and cables are shown in Table 16.

Table 16. Properties of Fibers Used in Ropes and Cables

Fiber	Production capacity, 10^3 t	Strength, N/tex[a]	Modulus, N/tex[a]	Elongation, %	Density, g/cm^3
jute (sisal)	na[b]	0.28	18	1.4	1.49
nylon	16	0.53–0.88	2.6	20	1.14
steel	182	0.31		2	7.68
Kevlar	0.11	1.9	44	4	1.44

[a] To convert N/tex to g-f/den, multiply by 11.3.
[b] na = not available.

Other properties that may become critical in many uses are excellent tension-tension fatigue resistance, corrosion resistance, low creep, nonconductivity, and excellent cycling-over-pulley life. The high strength-to-weight ratio of *p*-aramid compared to steel is shown in Figure 4, where the "breaking length" is the length of strength member that will break under its own weight. It is cal-

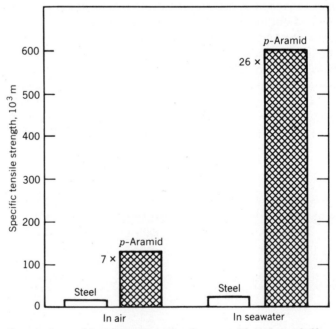

Fig. 4.   Comparison of breaking length of *p*-aramid and steel fibers. (Length at which strength member breaks of its own weight = tensile strength/density.) Data from Naval Undersea Center, Kailua, Hawaii.

culated by dividing the strength by weight per unit length. The comparison is particularly noteworthy in seawater where specific strength of Kevlar may be more than 20 times greater than that of steel. Smaller, lighter, more easily handled lines can do the same application. In long lengths where the self-weight of steel becomes a factor, aramid can offer more payload, higher safety factors, or the possibility of performing missions previously not feasible.

A wide variety of rope and cable structures have been made from *p*-aramid fibers including parallel-lay, single- and double-braid, three-strand, eight-strand-plaited, and stranded-wire, rope-type constructions. In properly made ropes of comparable constructions, those made from aramid are about twice the strength of nylon or polyester ropes at equal weight and about the same strength as steel wire rope, but at 20–25% the weight.

For each use, technology must be developed to capitalize on the unique features of engineering fibers and to minimize such negative features as poor abrasion resistance and damage tolerance, which are lower than steel's. An example of an engineered rope structure is that of a riser-tensioner line used on an offshore floating-oil drilling platform to keep the riser pipe in a constant position and under uniform tension while the drilling platform surges with the waves (Fig. 5). These lines experience considerable cycling over pulleys, called sheaves. A typical incumbent steel wire rope has a 4.4-cm diameter, 103-t strength, and 8.6 kg/m weight.

Through laboratory studies, technology for ropes of Kevlar with dramatically improved performance over steel has been developed. Design changes include optimizing the rope construction to minimize relative strand motion and jacketing each strand with a nylon or polyester fiber lubricated with Teflon and

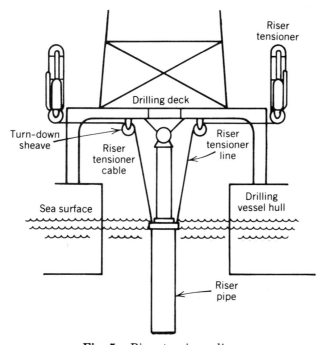

Fig. 5.   Riser-tensioner line.

silicone oil to reduce friction and heating when strands move by each other. The result is a 4.4-cm dia rope of p-aramid with 150-t break strength, weighing 1.97 kg/m and lasting more than three times the life of steel in severe laboratory tests and up to five times longer in actual service.

Mechanical ropes of p-aramid with varying breaking strengths are used for antenna guys, yacht lines, trawl lines, racket strings, buoy moorings, ship moorings, towing, riser tensioners, and pendent (anchor retrieval) lines; they also are under development as oil-rig mooring lines. Electromechanical cables and fiber-optic cables reinforced with p-aramid strength members have been successfully developed for use in fixed and towed sensor arrays and as primary umbilicals and tether cables for unmanned undersea work vehicles for both military and commercial applications. Cables reinforced with p-aramid are also being used for diving-bell life-support systems, submarine tow cables, telecommunication cables, and submarine surveillance and mobile tracking systems.

Future opportunities for engineering-fiber ropes, where inherently high weight limits steel, include mooring lines for deep-water oil-drilling platforms (Fig. 6). Steel-chain mooring systems have a practical water-depth limitation of about 450 m because high weight reduces the line restoring force required to position the drilling platform. Restoring force is proportional to the horizontal component of line tension and is reduced by rope weight. At 720-m water depth, the restoring force with a p-aramid mooring line is twice that of a steel chain, and the same rope weight of steel in water could perform at water depths at least 5–10 times greater.

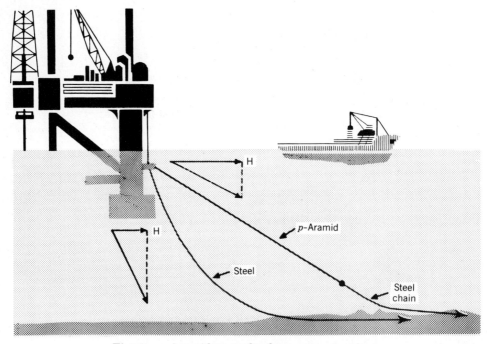

Fig. 6. p-Aramid ropes for deep water mooring.

Friction Products and Gaskets.   Asbestos is a mineral fiber that has been used in friction products and gaskets because of many excellent characteristics (34): hairlike fibers, high tensile strength, flexibility, and heat and chemical resistance. Asbestos also has a stable coefficient of friction when exposed to increased temperature. However, asbestos causes serious health problems and its use is very restricted (35). The favorable cost/performance ratio of asbestos has been hard to achieve with other fibers. For example, no one substitution material could match all the useful properties of asbestos, particularly in friction products. Also, the relatively low cost of asbestos permitted its use as a filler as well as for reinforcement in friction products and gaskets. Evaluation of other low cost fillers was necessary in designing nonasbestos structures. Although asbestos is still used to some extent, many nonasbestos friction and gasket products have been developed.

Friction Products.   Friction products are used to impede relative motion between mating surfaces, specifically, to drive, control the speed of, and stop moving equipment. They usually are a combination of reinforcing fibers, resin binders, and friction-control particles, and fillers. Friction products are used in many forms of brakes and clutches for industrial and transportation applications.

Reinforcing fibers are essential in most friction products for several reasons. First, the thermoset resins used, such as phenolics, are brittle with little tensile strength when cured, and friction products have a low resin content, 15–20% by weight, compared with 50–60% for composites. Second, friction-product structures can experience a variety of operational stresses: flexural, shear, compressive, and rotational. Localized stresses from attachment bolts or rivets can also occur. Finally, friction-product structures expand and contract many thousands of times during their life as kinetic energy is converted into a considerable amount of heat. Therefore, reinforcing fibers are required to knit together a combination of materials so that a noncracking, stable structure that resists service conditions results despite modest resin content.

The mechanism of reinforcement is via adhesion of the reinforcing fiber surface to the cured resin. The degree of adhesion determines how well the physical properties of the fiber are transferred to the matrix structure. Table 17 presents an overview of reinforcing-fiber requirements, and Tables 18 and 19 compare metallic, mineral, and organic fibers that are used commercially for the application.

Table 17. Reinforcing Fiber Requirements for Friction Products

Property	Requirement
thermal stability	stability at or above 370°C
	decomposition without forming slippery surface
adhesion	good adhesion to resin for maximum property transfer
aspect ratio (L/D)	consistent with good processing
tensile strength	high
modulus	high
wear	low
moisture	essentially unaffected by moisture
chemical resistance	compatible with oils and brake fluids
density	low, to reduce weight

Table 18. Comparison of Reinforcing Fibers for Friction Materials[a]

Fiber type and name	Advantages	Potential problems
Metallic		
steel	high strength and modulus	high density
		corrosion
	thermal stability	low cold friction coefficient
Mineral		
asbestos	high strength and modulus	linked to health problems
	thermal stability	
	infusible	
	good wear	
	also acts as filler	
glass, mineral	high strength and modulus	melts at very high temperatures (above 540°C) causing fade
		breaks up in high shear mixing
		expands when molding pressure released
		low wear resistance
		aggressive wear characteristics
Organic		
aramid pulp	high strength and modulus	requires special fiber opening techniques during mixing
	thermal stability	
	nonaggressive wear characteristics	
carbon	high strength, very high modulus	breaks up in mixing
	infusible	
	thermal stability	
cellulosic	infusible	low strength and modulus
		low char temperature
thermoplastic synthetic acrylics	high strength	melts, causing brake fade

[a] Ref. 36.

Improved utilization of most costly asbestos substitute fibers can be expected by further improvements in mixing and resin bonding. Also, engineered fiber shapes, such as the highly fibrillated aramid pulp, may lead to cost reduction by reducing the required amount of fiber, because of increased effectiveness.

Small cars with increasingly smaller metallic brake components for heat sinks increase thermal demands on friction products and their reinforcing fibers. Anticipated brake temperatures of 540°C will require new reinforcing fibers of higher thermal capabilities than present-day asbestos substitutes. Continuously variable transmissions for cars will reduce the number of clutch applications but will require high performance reinforcing fibers in the belt.

Gaskets. Since no joint is perfectly flat or parallel, gaskets are used to seal joints to contain a fluid under pressure or vacuum. A gasket must be comformable to fill voids, resilient to accommodate expansion and/or vibration, resistant to fluids, resistant to creep and flow, and useful over a wide range of temperatures.

Table 19. Reinforcing Fibers for Friction Products[a]

Name	Manufacturer	Relative cost vs asbestos	Specific gravity	Upper useful temp, °C	Shape	Length, mm	Diameter, mm	Aspect ratio	Tensile strength, MPa[b]	Hardness, Mohs
Metallic										
steel	American Steel Wool	2.7	1.7–2.2	540	triangular cross-section fiber	800–2,000	80–100	10+		3.0–4.0
Mineral										
asbestos[c] (chrysotile)	Johns-Manville	1.0	2.4–2.6	650	fiber	250	0.025	10,000	2,100	
fiberfrax	Carborundum	5.9	2.73	1,260	fiber	12–700	2–3	4–7	1,000–2,100	5
franklin fiber	U.S. Gypsum	0.8	3.0	540	whiskers	50+	2	25+		3.5
glass ("E")[c]	Owens-Corning	3.0	2.54	700	fiber	3,175	14	240	3,500	0.5
mineral fiber	U.S. Pipe and Foundry	1.9	2.7	760	fiber	28	1–10	40–60	70	
Wellastonite (WYAD-G)	Interface	1.0	2.9	650	acicular	40–45	2.5	15		4.5
Organic										
aramid fiber	DuPont	20	1.44	430	fiber	6,350	12	530	2,800	
pulp[c]		19	1.44	370	fibrillated fiber	2,000	0.1–12	200–20,000		
carbon WFA	Union Carbide	30	1.0	540	fiber	635+	11	575	210	1.0
graphite WFA	Union Carbide	100	1.4	650	fiber	6,360+	8.4	750	800	1.0
cellulose[c] Solka-Floc	Brown Co.	1.2	1.4	200	fiber	76	18	4		1.0
phenolic Kyrol(F)[d]	Carborundum Co.	15	1.25	1,370	fiber	38–100	14	2,720	200	

[a] Ref. 37.
[b] To convert MPa to psi, multiply by 145.
[c] Also, used in gaskets.
[d] Gasket.

About two thirds of gaskets are used in industrial and appliance applications and the rest in automotive and truck uses.

There are two types of asbestos sheet gaskets. Compressed gaskets are calendered from a solvent-based mix, and beater-addition (BA) gaskets are produced on a papermaking machine (38). The latter type is generally lower in price and performance. Compressed asbestos gaskets usually perform better than beater-addition asbestos gaskets because of longer fibers and higher sheet density. This is not necessarily the case with nonasbestos reinforcing fibers.

In general, nonasbestos gaskets contain lesser amounts of reinforcing fiber (10–20% by weight) than asbestos types. The binder is usually an elastomer, including styrene–butadiene (SBR), nitrile (NBR), neoprene or ethylene–propylene–diene (EPDM) types. Bulk is provided by low cost talc, clay, mica, or mineral fibers, which replace that portion of asbestos used for filler in asbestos gaskets.

Asbestos gaskets are typically 85% (by weight) asbestos and 15% binder. Asbestos functions both as reinforcement and filler. Reinforcement occurs when surface adhesion of the fiber to the rubber binder permits transfer of some degree of fiber properties to the final structure. Reinforcement fibers must have high modulus and adequate creep resistance. Reinforcing fibers for gasket sheeting are compared in Table 20.

Table 20. Reinforcing Fibers for Gasket Sheeting[a]

Mineral	Advantages	Disadvantages
asbestos	high strength and modulus thermal stability cheap enough to use as a filler	linked to health problems high chloride content

glass	high strength and modulus
	thermal stability
		fiber breaks up in high shear
		  mixing, reducing reinforcing
		  efficiency
		irritating to handle

Organic		

aramid
pulp
	excellent creep resistance and sealability
	high strength
	high modulus
	thermal stability
	
	acceptable tensile strength at relatively
	  high temperatures
	low chloride content
	nonbrittle (won't break up during
	  mixing)
	excellent radiation resistance
		mixing procedures required are
		  different from those used for
		  asbestos
		requires care to prevent static
		  build-up

cellulosic	sufficient strength and modulus for ambient temperature service ease of mixing low cost	degrades at 200°C poor dimensional stability
phenolic	thermal stability high aspect ratio	low tensile low modulus

[a] Ref. 38.

The principal gasket-reinforcing fiber today is still asbestos. Nonasbestos gaskets are only about 5% of the annual worldwide sheet-gasket market. Continued replacement of asbestos reinforcement fibers can be expected.

The largest competitive nonasbestos fiber is aramid pulp. Cellulose is used in less demanding and lower temperature beater-addition nonasbestos gaskets, but a variety of mineral and organic reinforcing fibers can be used in both beater-addition and compressed-gasket processes (Table 20). Sometimes more than one kind of fiber is used in a blend.

Besides eliminating health hazards, nonasbestos gaskets offer other performance advantages over asbestos gaskets. For example, a major reduction in fiber content has improved sealability (Fig. 7). Also, the lower chloride content of nonasbestos reinforcing fibers and proper filler selection can produce gaskets with a lower chloride content and reduce corrosion of stainless-steel joints. This is important to the nuclear-power industry.

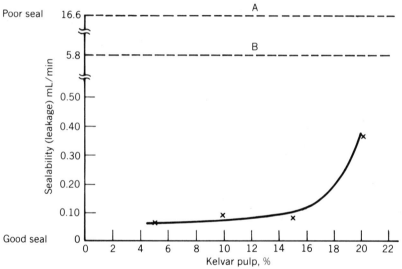

Fig. 7. Sealability (leakage) vs fiber content of gaskets. (A), beater-addition asbestos gaskets; (B), high performance NBR compressed asbestos gaskets.

Ballistics Applications. The use of armor to protect personnel and equipment has a long history. In the days of the Crusades, knights wore head-to-foot armor made of plate or chain mail (tiny rings linked together) (39). The introduction of steel for helmets resulted in a significant savings from death and injury during the World Wars.

Body armor made from leather, wood, bronze, brass, and steel is stiff and heavy, inhibits mobility, and is extremely uncomfortable. Engineering armor with textile fibers in the form of woven fabric has led to the design of lightweight, flexible products for the protection of both personnel and strategic equipment. Nylon antifragmentation vests, commonly known as "flak jackets," were issued throughout the Korean and Vietnam conflicts. The term flak jacket derives from its first use late in World War II by bomber crews to whom the main threat was ground-to-air "flak" (40). Heavy nylon canopies have also served protective purposes.

The introduction of Kevlar aramid fiber has extended armor applications. This fiber is two and one-half times as strong as nylon with a modulus of elasticity an order of magnitude higher (Table 21). Ballistically, aramid has the same ability to stop projectiles as nylon, but at one-half the weight. The U.S. Army initiated development of aramid fabrics for flak jackets to replace nylon ballistic cloth and began to use the material in reinforced plastic helmets replacing steel (41) (see also MILITARY APPLICATIONS).

Table 21. Comparative Yarn Properties

	Kevlar 29 aramid	DuPont nylon type 728	S-glass	E-HTS glass	Type HT graphite
tensile strength, MPa[a]	2760[b] 3620	985[b]	4580[c]	3445[c]	2760[c]
modulus, GPa	90	5.5	87	72	221
elongation to break, %	4.0	18.3	5.4	4.8	1.25
density, g/cm^3	1.44	1.14	2.46	2.57	2.75

[a] To convert MPa to psi, multiply by 145.
[b] Unimpregnated twisted yarn test, ASTM D 2256.
[c] Impregnated strand test, ASTM D 2343.

Although initial developments were concerned with fragmentation resistance the Army has determined that aramid fabrics can arrest an assortment of handgun projectiles at significantly less weight than other known ballistic materials including nylon. The Law Enforcement Assistance Administration (LEAA) of the U.S. Department of Justice has funded development of bullet-resistant vests for law enforcement officers (42). Soft body armor weighing only 0.7–2.3 kg, depending on the degree of handgun protection desired, is now available. These vests are concealable, and they are comfortable enough to wear on a full-time basis. More than 600 police officers have been saved from death or serious injury by wearing protective apparel made from aramid (43).

The strength of aramid is an important factor in its ability to defeat bullets and fragments. However, strength alone does not determine the fiber-projectile interaction. Fiber glass and graphite fibers are strong (Table 21), but in fabric form they cannot defeat a bullet or fragment as efficiently as nylon or aramid. Neither does toughness, the area beneath the stress-strain curve, determine the effect of ballistic impact, since S-fiber glass is "tougher" than aramid (Fig. 8). The key to stopping a projectile from penetrating armor is the unique ability to spread energy.

When woven fabric encounters a high-speed projectile, such as a bullet or a fragment, fibers stretch and transmit energy along their length. As energy is transmitted by the fibers directly hit by the projectile, this energy is transferred and shared by other fibers in the woven fabric. Transfer occurs at "crossover points," where warp and weft fibers are interwoven. Additional energy is absorbed by the fabric parallel to the path of the projectile.

In its simplest form, the impact of a projectile on a single filament or yarn (Fig. 9) elicits the following dynamic response. A longitudinal wave propagates outward from, or perpendicular to, the path of the projectile, at the speed of sound, which is equal to the square root of fiber modulus divided by fiber density. The

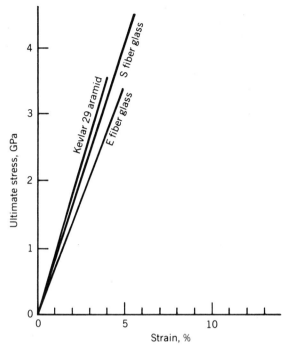

Fig. 8. Stress-strain behavior of yarns. To convert GPa to psi, multiply by 145,000.

higher this wave velocity, the greater the volume of fiber capable of interacting with the projectile. Behind this longitudinal wave, material flows inward and experiences a tensile strain dictated by the magnitude of impact velocity. A second, or transverse, wave propagates along the fiber, behind which material begins to move transverse to the fiber axis, parallel to and at the speed of the projectile. The speed of the transverse wave front is also related to the magnitude of the projectile-impact velocity, but is slower than that of the longitudinal wave. Wave reflections from the edges of the target intensify the tensile strain. Total energy absorbed equals strain energy of fiber behind the longitudinal wave front.

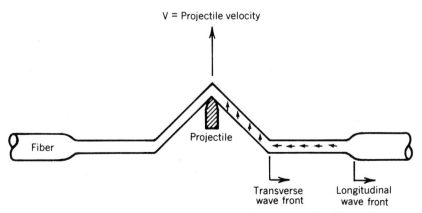

Fig. 9. Ballistic behavior of a fiber when hit by a projectile. Wave fronts at t_1.

With woven fabrics, the longitudinal and transverse stress waves occur in the yarns directly impacted (primary yarns). Partial reflections of the longitudinal wave occur at yarn crossovers, causing a desirable diversion of stress to the secondary yarns, but an undesirable strain gradient in the primary yarns. However, the most important single factor is that the lateral displacement behind the transverse wave front leads to loading of secondary yarns through crossover interactions, thereby unloading the primary yarns. Experimental evidence shows that up to 40–50% of the total energy is absorbed by the secondary yarns, suggesting that tightly woven, plain-weave fabrics are most efficient.

Although fiber glass does not perform well ballistically in fabric form, it can be combined with various resin systems to form composite, or reinforced plastic, armor. Aramid also can be used in these applications, and these two fibers dominate the composite-armor industry. Aramid, the more expensive product, is preferred over fiber glass when weight savings is the prime design consideration as in military helicopters and aircraft or in fast and highly manueverable land and sea vehicles. Other typical applications for composite armor include riot shields, clipboards, helmets, and other personnel protective devices, as well as armor for naval ships, automobiles, military vehicles, tactical shelters, and transportable containers.

The addition of resin to woven fabrics to make composite armor has a strong influence on the ballistic behavior of the structure. Armor with a resin matrix generally has a lower ballistic performance than all-fabric armor. Longitudinal wave velocity is slowed, the rigid crossovers give extensive reflections, resulting in a high strain gradient, and the rigid structure restricts transverse deflections. However, the bonded layers permit load transfer to nonprimary yarns through the continuum of the matrix, allowing wave propagation in the thickness direction and spreading the load over a larger area on the back layers. This is particularly important for thick armor designed to stop very high velocity fragments.

To minimize the effect of rigidity in restricting transverse deflection, some delamination between fiber and resin on ballistic impact is desirable. This ability to delaminate locally is believed to be the reason polyester, phenolic, and vinylester resins perform better than epoxies. Flexible resin matrices can also be used to make a more easily deformed laminate, with good performance. Also, transverse deflection is necessary for maximum efficiency. Thus, armor should not be placed in front of (or sandwiched between) rigid components. Energy absorption capability is a strong function of impact velocity; however, increasing velocity induces higher fiber strain, allows less time for interaction, and promotes steeper strain gradients. Therefore, slower-moving, large fragments can be stopped more easily than smaller, higher velocity fragments, and the relative weight advantage of engineered textile fabrics versus metals (which exhibit a different mechanism) will decrease as projectile velocity increases. As armor thickness increases, the ballistic performance of composite armor improves, relative to all-fabric armor. Front plies become less effective in both cases, since their primary function is to spread the impact over a large area. Composites spread this impact more efficiently. For high-energy fragments, material combinations (for example, ceramic backed with composite armor) may provide the best route to optimum performance. The front face spreads the load, breaks up, and slows the projectile.

Aramid has the best ratio of ballistic resistance to weight. The disadvantage of aramid is its relatively high cost when compared to nylon or fiber glass. Other high strength fibers are emerging as candidates for textile antiballistic armor, eg, gel-spun polyethylene and polypropylene.

Gel-spun Polyethylene.   All high performance fibers are candidates for industrial applications where superior properties and reliability are required. In addition to the polyamides, polyesters, and aramids which have a long history in such applications, Allied-Signal has recently introduced a new fiber, "Spectra 900," based on gel-spun polyethylene. Its high strength (2.6 N/tex (30 g-f/den)) and high modulus (106 N/tex (1200 g-f/den)) suggest applications where temperatures are relatively mild, eg, in ropes and cables and ballistics.

Electro-optical Uses

Engineering fibers are making contributions to the transmission of information, the miniaturization of computer components, and the efficient handling of electrical power.

Optics

The largest use of optical fibers is the transmission of information by light, as a direct substitute for transmission of electronic signals over metal wires. But optical fibers have many other uses based on their unique ability to transmit light. Optical-fiber technology, as well as the large diversity of its applications, is thoroughly described in recent textbooks (44–49) and review articles (50–52) (see also FIBERS, OPTICAL).

Mechanism of Light Transmission in Optical Fibers.   When light moving through a transparent material encounters an interface with a material of lower refractive index, one of three things can happen. At some angles of incidence, θ, the light will be refracted and will move on through the low index material, but at a different angle (Fig. 10**a**). At a critical angle θ_c, the refracted light is parallel to the interface (Fig. 10**b**). But when the angle of incidence is less than the critical angle, the light is reflected back into the material of higher refractive index at an angle equal to the angle of incidence (Fig. 10**c**). This is known as total internal reflection, since none of the light is lost upon reflection at the interface.

Step-index Fibers.   Optical fibers are formed when a transparent material of higher refractive index is made into a fiber and that fiber is coated with a material of lower refractive index. Light injected into the end of such a fiber at angles less than the critical angle undergoes multiple total internal reflections, as shown in Figure 11**a**, and is transmitted along the fiber. Optical fibers can be made of a single transparent material, without a coating of a second material. In this case, when air is in contact with the fiber, it becomes the material of lower refractive index. Such fibers transmit light and are well known to glass blowers. Paintings from ancient Egypt show that such optical fibers made by glass blowers have been known for thousands of years.

However, optical fibers used for engineering purposes are coated to protect the interface. Scratches at the interface will scatter light, thereby decreasing the amount of light transmitted along the fiber. Contact of the interface with ab-

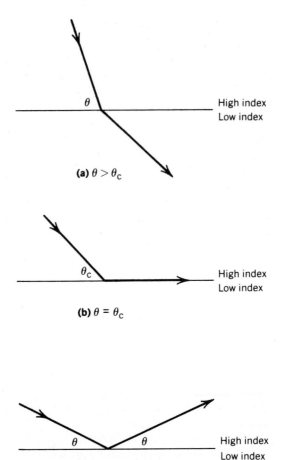

Fig. 10. Refraction and reflection of light at an interface between two materials. (a), At an angle greater than the critical angle; (b), at critical angle; (c) at less than the critical angle.

sorbing materials will also result in a loss of light transmission. For most uses, the interface must be protected by a solid material.

When optical fibers are made by coating a material of low refractive index onto a fiber of high refractive index, there is a step change in the refractive index at the interface. Such fibers are known as step-index fibers. Actually, in most processes, both the core fiber and the coating are formed simultaneously, in a single step.

Graded-Index Fibers.   Optical fibers can also be made with a gradual change in the refractive index from the center to the outside. Such fibers are known as graded-index fibers. There is no sharp interface for internal reflections. In such fibers, the light rays curve or bend along the fiber as illustrated in Figure 11b. The net effect is to keep the light inside the fiber as it moves along the fiber.

Single-Mode Fibers.   Graded-index optical fibers are made with diameters near the wavelength of the light being transmitted; the mechanism can be de-

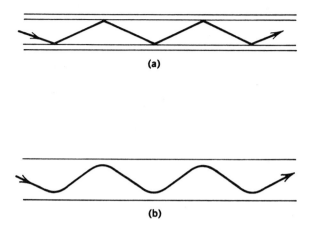

(a)

(b)

Fig. 11.   Light transmission in optical fibers. (**a**), step-index optical fiber; (**b**), graded-index optical fiber.

scribed only by complex mathematical expressions for transmission of electromagnetic waves. These equations have several solutions, or modes, of propagation along the fiber. When the combination of fiber geometry, refractive index, and the light wavelength is just right, light is propagated along the fiber in a "single mode." Such fibers are especially useful at long distances.

Attenuation.   Although no light is lost upon total internal reflection or in ray "bending" in graded-index fibers, light is lost by absorption and scattering. The useful length of an optical fiber is limited by these light losses. No solid material is perfectly transparent. Most materials absorb and/or scatter too much light to be useful in optical fibers. Those which are used are highly purified to minimize absorption and scattering, but even with the best of materials, losses vary with the wavelength of light. Thus, light sources must be matched with the light transmitting characteristics of the fiber. Optical fibers are used both with visible and with near infrared light sources.

Light transmission decreases exponentially with the length of the fiber and is described by the equation:

$$I = I_o \exp\left(-\alpha \times \text{length}\right)$$

where I is the light intensity at a given length from the source, I_o is the light intensity injected into the fiber at less than the critical angle, and α is the attenuation factor. Although α could be expressed in units of reciprocal length, it is usually expressed in decibels per kilometer (dB/km).

Optical fibers have been made and used with attenuation factors of several thousand decibels per kilometer. Such fibers are useful for short run lengths. Other fibers have been made with attenuation factors below 1 dB/km at useful wavelengths and have been used over distances of tens of kilometers.

The useful length of an optical fiber depends not only on the attenuation factor, but also on the light source (amount of light injected into the fiber) and the sensitivity of the detector and amplifier on the receiving end.

Materials of Fiber Construction.   *Plastic Optical Fibers.*   In principle, any two transparent polymers having different refractive indexes can be made into optical fibers by coextrusion. In practice, the choice of materials is severely limited. Commercial plastic optical fibers made by DuPont and by Mitsubishi use a core of poly(methyl methacrylate) ($n_D = 1.50$) and a coating or sheath of a specialty fluorine-containing polymer ($n_D < 1.40$). Commercial plastic optical fibers have also been made with a polystyrene core ($n_D = 1.60$) and a sheath of poly-(methyl methacrylate) ($n_D = 1.50$).

These plastics transmit light in the visible wavelength region but absorb strongly in the near infrared. Plastic optical fibers have also been made with a core of perdeuterated polymethacrylate (53). These fibers transmit light in the near infrared as well as the visible region and are especially useful with certain light-emitting diodes that produce light at near-infrared wavelengths.

Plastic optical fibers are especially useful for run lengths of a few feet. Their relatively large size (200 μm or larger) makes it possible to use simpler connectors, or coupling devices, at the two ends. Plastic optical fibers can be made so that they are very tough and immune to breaking by bending; they can literally be tied into a knot. Therefore, they are easiest to install and handle. Using highly purified materials and special production techniques, plastic optical fibers have been made with sufficiently low attenuation that run lengths of over 100 m have been attained.

Plastic-Coated Silica Optical Fibers.   A pure silica rod ($n_D = 1.45$) can be heated and drawn down to a fiber and coated with a polymer ($n_D < 1.40$) to form an optical fiber. Silica can be purified sufficiently so that the attenuation of such fibers can be made lower than that of plastic fibers. Such fibers can be made with diameters up to 400 μm and still retain a flexibility almost as good as the all-plastic optical fibers. Run lengths of several hundred meters have been achieved. Both the plastic-coated silica and the all-plastic fibers are step-index fibers.

Glass Optical Fibers.   Step-index, graded-index, and multimode optical fibers have been made using various kinds of glass. The refractive index is controlled by the composition of the glass. Typically, a rod is first formed having the desired refractive-index profile. It is heated and drawn into a fiber (typical diameter < 100 μm) that retains the index profile. Such fibers are often coated with plastic to provide mechanical protection of the glass; this outside plastic coating does not affect the light transmission of the optical fiber.

By careful purification of the materials and control of all manufacturing steps, very low attenuations can be achieved in glass optical fibers, and especially in single-mode fibers. Run lengths of tens of kilometers have been demonstrated. The small size of glass optical fibers makes them more difficult to connect to light sources, detectors, and one another. Special techniques and devices have been developed to make these connections possible. Glass optical fibers are the only suitable type for long-distance telecommunications.

Optical fibers of all types are usually protected by some kind of external plastic jacket. A single optical fiber, or several fibers grouped together, inside a single jacket form a "cable." Other types of fibers may be added to the cable to provide a mechanical strength far greater than that of the optical fibers alone. Kevlar is often used to provide this mechanical strength in optical-fiber cables.

Glass optical fibers and cables are manufactured by several companies in the United States, Japan, and Europe.

Applications. Optical fibers are used in decorative objects, such as lamps containing dozens of illuminated optical fibers. In signs, a single light source can supply light for a large number of fibers; light is applied at one end of the fibers; the other ends of the fibers can be arranged in any shape, such as numbers and letters. Illuminated signs for airport taxi strips and runways have been made of optical fibers.

As remote light sources, a bundle of optical fibers can provide "cold" light for flexible illumination for microscopic studies while keeping the hot lamp at a safe distance. Optical fibers are also used to provide a remote indication that a lamp is "on." For example, in automobiles, one end of a small fiber bundle is installed in the housing for each headlight and taillight. The other end is located so that it can be easily seen from the driver's seat and the driver can verify that these lights are functioning.

A bundle of optical fibers can be arranged coherently so that the bundle can transmit an image while still retaining a high degree of flexibility in the bundle. Objects being examined can be illuminated by light transmitted to the object through other optical fibers in the bundle. This flexible bundle of optical fibers can be inserted through a small hole to examine the interior of objects that would otherwise be difficult to reach. Such instruments are used industrially for inspection of interior cavities and by physicians for examining various interior portions of the human body.

Optical fibers are used to conduct laser light to the precise point desired. For example, optical fibers have been used to destroy arteriosclerotic deposits in arteries. One end of an optical fiber is inserted into an artery and "pointed" at the deposit. Laser light is injected into the other end of the fiber and transmitted through the fiber to the deposit, which is then destroyed by the absorption of the laser light.

Telecommunications. By far the largest use for optical fibers is in telecommunications (54). Almost all of these uses rely on digital communication of information. Electronic signals (voice, video, and data) are used to modulate the output of a light source, usually a light-emitting diode or laser. This modulated light is transmitted by the optical fiber to a detector, which may be located at a distance ranging from a few centimeters to many kilometers. The flexible optical fiber can be routed by any convenient path. After detection, the signal is amplified and used.

In this use, optical fibers serve the same function as metallic wires, but with some distinct advantages: high transmission capacity including several channels or circuits per fiber, large repeater spacings, immunity from electromagnetic interference (qv), and the small size and weight of optical-fiber cables. The commercial deployment of optical-fiber systems is proceeding at a rapid pace, and large systems are under construction in the industrial countries of the world (54). A system is now being planned for a transatlantic cable, about 6,650 km long, between the United States and England and France, to be placed in service June 1988. The system includes about 130 optoelectronic repeaters, which are spaced approximately 50 km apart (54).

In 1984, 8×10^7 circuit kilometers of optical-fiber telecommunication capacity had already been installed. Estimates as high as 1.1×10^{10} circuit kilometers have been projected for 1988 (55).

Electronics

Fibers are making substantial contributions to the miniaturization of electronic computer components. A printed-circuit or -wiring board (PCB or PWB) functions primarily as an electrically insulating substrate on which various electrical or electronic components (resistors, capacitors, integrated circuits, etc) are mounted, connected by relatively thin copper lines. PCBs eliminate bulky wiring, increase reliability of circuitry and interconnections, and provide miniaturization advantages. PCBs, mostly composites, are found in computers, telecommunications equipment, and electronics for military, automotive, industrial, and consumer uses. The size of the PCB market is substantial. In 1981, the value of U.S. production of PCBs was \$3.2 billion (56).

PCB Types and Structures.   Printed-circuit-board substrates may be flexible or rigid. Flexible types generally are reinforced polyester or polyimide films. Rigid PCBs are made primarily from thermosetting resins reinforced by dielectric fibers in fabric or nonwoven form.

There are three classes of rigid PCBs, depending on the number of conducting layers in the final circuit board. The oldest and simplest is single-sided, with circuitry etched only on one side for undemanding, low cost application. Double-sided PCBs often have plated-through holes electrically connecting both sides. The third class is multilayer boards with more than four conducting layers of circuitry interconnected with plated-through holes. The monolithic ceramic substrate, though an important rigid subclass, is not discussed here since it does not involve fibers.

Fiber Requirements in Circuit Boards.   The need for very high performance and reliability makes stringent demands on reinforcing fibers since the fiber content is 35–70% by weight of the composite. Reinforcing fibers must have good dielectric properties. A high degree of electrical insulation must be maintained between the copper circuitry traces in the final product.

Low moisture regain is necessary to minimize degradation of electrical performance. Stiffness and strength are important not only for product performance but also for handleability in processing from laminating impregnated B-staged materials, through etching, plating and soldering operations. Chemical resistance to acids, bases, fluxes, and a variety of organic solvents is essential.

Thermal stability is required to withstand soldering and application temperatures, which range up to 260°C. High dimensional stability during temperature excursions is needed for stable performance. Dimensional stability is particularly important in multilayer circuit boards where highly reproducible fabrication techniques are used.

Good adhesion to the resin is important for high board mechanical strength and durability and to eliminate delamination during hole drilling. Drillable reinforcing fibers are required to yield a clean, plateable surface of the drilled hole. Table 22 compares properties of the four reinforcing fibers that are generally recognized as meeting the above requirements for use in rigid printed-circuit boards.

Table 22. Reinforcing Fibers for Printed-Circuit Boards

Properties	E-glass fiber	p-Aramid fiber	Quartz fiber
Bulk			
density, g/cm^3	2.54	1.44	2.2
Electrical			
dielectric construction, 1 MHz	6.32	3.6	3.7
dissipation factor, 1 MHz	0.005		0.0001
volume resistivity, Ω-cm	1×10^{15}	1×10^{15}	1×10^{15}
moisture regain, %			
50% rh	neg	3.2	neg
75% rh	neg	4.8	neg
90% rh	neg	5.7	neg
Mechanical			
tensile stiffness, GPa[a], at 20°C	72	16.	10
tensile strength, GPa[a]	3.5	3.6	0.9
elongation, %	4.6	2.9	1.3
Thermal			
coefficient of thermal expansion	5.0	−5.0	0.5
softening point, °C	841		1670
decomposition temperature, °C		500	
thermal conductivity, W/(m·K)[b]	10.4	NA	
Chemical Resistance			
acids	excellent	fair	excellent
bases	excellent	fair	excellent
organics	excellent	excellent	excellent
Other			
drillable	excellent		
clean hole	excellent	clean when plasma etched	excellent
drill bit wear	acceptable	acceptable	poor
punchable	unacceptable	unacceptable	unacceptable
PWB relative cost[c], $/15 m	1	20	20

[a] To convert GPa to psi, multiply by 145,000.
[b] To convert W/(m·K) to (Btu·in)/(h·ft^2·°F), multiply by 6.94.
[c] Cellulose paper relative cost $0.5/15 m.

Ever-increasing circuit densities, especially in multilayer boards, will require more consistent, high quality reinforcements to meet the dimensional-stability uniformity targets that play such a large role in manufacturing yields. Increased surface mounting of chip carriers for integrated-circuit chips will require circuit-board substrates with tailored thermal-strain capabilities. Faster circuitry will favor board reinforcements having low and controlled dielectric constants and dissipation factors.

Electrical

Development of new electrical insulation materials has been stimulated by technical and environmental factors such as availability of synthetic polymers and fibers, greater power density in electrical equipment requiring thinner insulation materials capable of operating at higher temperatures, automated manufacturing techniques requiring tougher insulation materials, miniaturization of electronic equipment, and health and environmental problems associated with asbestos fibers and askarel liquid insulation. New materials in solid, liquid, and gas form have displaced many traditional insulations such as mica, cellulose, asbestos, and mineral oil. Synthetic varnishes and resins have also contributed to improved insulation systems. Flexible laminates combining two or more sheet insulation materials have been introduced to meet special needs.

Basic Electrical Insulation Requirements. Electrical insulation materials essentially prevent current flow from one current source to another at lower electrical potential. Requirements for electrical insulation are in general high dielectric strength (kV/mm) and low power loss within the insulation. In a-c systems, the needs are low dielectric loss (W/cm^3), low dissipation factor, and low power factor. In low loss dielectrics, the dissipation and power factors are numerically equal because tan δ approaches zero.

In d-c systems, high resistivity to impede flow of leakage current through or across the surface of the insulation is needed, and it is expressed as volume resistivity ($\Omega \cdot$cm) and surface resistivity (ohms/square). Also necessary are good mechanical properties to withstand physical abuse during the manufacture of the electrical equipment and mechanical stresses during use, good thermal stability to withstand operating temperatures for the intended application lifetime, good chemical resistance for compatibility with other materials in the insulation system, and high corona resistance to withstand destructive attack by charged particles (see also ELECTRICAL APPLICATIONS; ELECTRICAL PROPERTIES).

Fibers. The inorganic fibers used for electrical insulation are asbestos and glass. Cellulosic fibers historically represent the dominant organic material, but in the last 30–40 years, petrochemically based synthetic fibers have found important applications. Predominant synthetic fibers for electrical insulation are polyesters and aromatic polyamides (aramids).

In most cases, fibers are used in fabricated form such as a woven or nonwoven fabric or a paper with or without resin impregnation. There are endless variations and combinations of fibrous, film, and resin systems. Table 23 compares the important properties of selected fibrous electrical insulation materials in their most widely used form.

Asbestos. Of the six minerals of fibrous crystalline structure known as asbestos, only one, chrysotile, is useful as electrical insulation. Chrysotile is a hydrated magnesium silicate of empirical formula, $Mg_6(OH)_8Si_4O_{10}$. Fibers can be processed on textile equipment to produce yarns and woven fabrics or on conventional wet-lay equipment to form papers. Papers from 100% asbestos are physically weak and the fiber is often processed with small amounts of binders or reinforcing fibers, or pure asbestos papers are used with saturants or impregnants to improve physical properties.

Asbestos has been used at high temperature for centuries, but has relatively poor electrical properties. Chrysotile fibers lose water of crystallization at about 450–475°C and decomposition begins. The basic sheet will retain its dielectric strength up to 300°C. Actual use temperature will depend on the type and extent of reinforcement or binder. Asbestos is generally used in low voltage applications where temperature and/or flame resistance are important. Health hazards associated with asbestos have greatly limited its use in recent years.

Table 23. Typical Properties[a] of Fibrous Electrical Insulation Materials[b]

Material	Specific gravity	Thickness, mm	Elongation, %	Tensile strength, MPa[c]	Tear strength, kN/m[d]	Thermal class, °C	Dielectric strength, kV/mm	Dielectric constant	Dissipation factor	Volume resistivity, Ω·cm
Inorganic										
asbestos										
paper (with no binder)	0.72	0.088		0.90		300	8–10			
paper (with organic binders)	0.82	0.25	<1			130	9	8.9	0.35	3.5 × 10⁹
board (with organic binders)	0.96	0.8					12	8.3	0.18	
glass										
cloth (with epoxy resin)	1.3–1.45	0.12–0.25	2.3 / 2.4	MD: 110–250 / CD: 66–131		155	36–48			
Organic										
cellulosic										
standard kraft paper	0.68–1.35	0.25	3.5 / 7.5	MD: 90–145 / CD: 19–55		90	8–22			
capacitor kraft paper	0.75–1.15	0.015					28–68	1.5–2.8		
kraft board	0.9–1.35	3	5–6 / 8–10	MD: 83–152 / CD: 28–61			7–17	5.8		
aramid paper (calender, no binder)	1.01	0.25	21 / 18		MD: 32 / CD: 16	220	35	2.7	0.006	1.5 × 10¹⁶
polyester										
paper (with epoxy resin)	1.5	0.38	24 / 5–24	MD: 46 / CD: 30		155	40			
laminates (mat/film/mat)	0.97	0.25	42 / 46		MD: 19 / CD: 14	155	38	2.2	0.0053	

[a] Not for establishing specifications.
[b] Source: published product data and DuPont test results.
[c] To convert MPa to psi, multiply by 145.
[d] To convert kN/m to ppi, multiply by 5.7.

Glass.   Alkali-free alumino-borosilicate glasses are used in drawn-fiber form, in both continuous filament and chopped form to make a range of woven fabrics, nonwoven mats, and papers. Since glass mats and papers cannot be densified with pressure without fiber damage, resin impregnations are normally used. Advantages of glass include high temperature stability (softening point of 538–593°C), high thermal conductivity, low moisture absorption, and good chemical resistance. In actual use, the thermal class depends on the type of resin used.

Cellulose.   Cellulose is the least expensive of fibrous materials and cellulosic papers are available in a wide range of thicknesses and grades. Papers have moderate mechanical strength, are flexible and easy to handle, and are readily impregnable with resins or oils. The dielectric constant of cellulosic papers generally decreases with increasing frequency, whereas dielectric losses increase with frequency and temperature. However, at power frequencies (eg, 60 Hz) and normal temperatures, these variations are of little consequence. Cellulosic papers degrade rapidly at temperatures above 100°C; hence, their use is limited to lower thermal classes. Untreated papers are classified as Class 90°C; treated papers may be advanced to Class 105°C. Cellulose is hygroscopic and absorbed moisture can affect electrical properties. Replacement of the hydroxyl groups can improve thermal resistance and moisture sensitivity; acetylation and cyanoethylation are commonly used for this purpose.

Polyester.   Poly(ethylene terephthalate) is the polyester polymer most used for electrical insulation. The stiffness of the polymer due to the rigid *p*-phenylene groups, crystallinity, and strong interchain forces results in strong and flexible fibers and films. In 100% form it is suitable for electrical insulation exposed for long times to temperatures up to 120°C; with appropriate resins it can fall into the 155°C class. Polyester is attacked by strong acids and is embrittled by phenols and cresols, but exhibits good stability in the presence of most insulating oils. Fibers are used in the form of papers and nonwoven mats. Polyester mat is often sandwiched with polyester film to form tough, flexible laminate suitable for 155°C applications.

Aramid.   The aromatic polyamide used for electrical insulation is poly(*m*-phenylene isophthalamide) (MPD-I), usually in paper form. Aramid paper is made on conventional wet-lay paper equipment from two physical forms: short fibers (floc) and smaller platelike particles (fibrids) that act as binder. The wet-lay sheets are permanently bonded under heat and pressure to form a paper structure with excellent dielectric, tensile, and tear properties, requiring no other binders or resins. Papers are available in a range of thicknesses.

Aramid polymer is nonmelting, does not support combustion, and degrades at temperatures above 375°C. The useful service temperature of aramid papers is 220°C. Dielectric constant and dissipation factor are nearly constant to 220°C, and dielectric strength is not affected by moisture. Aramid papers are compatible with resins (although difficult to impregnate), wire enamels, and varnishes; they also have good resistance to acids and organic solvents. The outstanding resistance to beta and gamma radiation makes them particularly useful in nuclear power and other high energy applications. An intimate blend of aramid fibers/fibrids and mica is also available to improve high voltage endurance and arc and track resistance. Aramid fibers are the most expensive of those discussed, and their application is generally limited to those areas where continuous service

Table 24. Availability of Key Engineering Fibers in 1984

Fiber	Price $/kg	Production capacity, 10^3 t	No. of producers	U.S. consumption including imports, 10^3 t
polyamide	3.2	200	21	180
polyester	1.8–1.9	155	20	470
olefin	1.4–3.6[a]	450	68	190
Kevlar		20	1	10
type 29	15–88			
type 49	29–110			
carbon	22–88	1	8	2
glass		790	12	530
E-glass	1.6			
S-glass	8.8			

[a] 1982 price.

temperatures are high, flame resistance is necessary, or reliability of equipment at intermediate temperatures with potential for hot spots is particularly important.

Today's synthetic polymers offer fiber-engineering capabilities over an extremely wide range of requirements.

New engineering applications for fibers will come in areas where light weight, corrosion resistance, and versatility promise to make possible more cost effective, longer lasting structures (57,58). This will be especially true as fiber properties are brought closer to their ultimate theoretical limits.

Capacity and price information for currently available engineering fibers appears in Table 24.

Acknowledgment

This article was made possible by an interdepartmental effort involving Textile Fibers, Polymer Products, Finishes and Fabricated Products, and Central Research and Development departments in the DuPont Company. Many people in the DuPont Company contributed by providing data, information and stimulating discussions on engineering aspects of fibers. We would particularly like to acknowledge the contributions of the following colleagues who provided papers and data in their areas of expertise: James A. Barron, John K. Beasley, Joyce S. Bloom, Charles J. Brown Jr., Thomas P. Doherty, John A. Hall, David E. Hoiness, Richard D. Hutchins, Paul R. Langston, John D. Matheny, Paul G. Riewald, James C. Romine, John J. Schmieg, Edward W. Tokarsky.

BIBLIOGRAPHY

1. J. C. Garnaud, *Plasticulture*, **49**, 37 (1981).
2. B. J. Bailey, *Materials for Greenhouse Thermal Screens,* National Farmers Union/British Agricultural and Horticultural Plastics Associations, Harrowgate, UK, 1985.
3. J. C. Garnaud, *Perforated Films for Crop Protection,* National Farmers Union/British Agricultural and Horticultural Plastics Associations, Harrowgate, UK, 1985.
4. *Geotextile Engineering Manual,* Federal Highway Administration/National Highway Institute, 1984.
5. P. R. Rankilor, *Membranes in Ground Engineering,* John Wiley & Sons, Ltd., Chichester, 1981.
6. *Proceedings of the First International Conference on Geotextiles,* Paris, 1977.
7. *Proceedings of the Second International Conference on Geotextiles,* Las Vegas, Nev., 1982.
8. R. M. Koerner and J. P. Walsh, *Geotextiles and Geomembranes,* Wiley-Interscience, New York, 1980.
9. *Geotechnical Fabrics Report,* Industrial Fabrics Association International.
10. L. E. Applegate, *Chem. Eng.*, 65 (June 11, 1984).

11. C. H. Gooding, *Chem. Eng.*, 56 (Jan. 7, 1985).
12. H. Strathmann, *Trends in Biotechnology* **3**(5), 112 (1985).
13. T. Hayashi and A. Nakojima, *Sen'j Gakkaishi* **41**(1), 427, (1985).
14. C. Artandi, *Chemtech* **8**, 476 (1981).
15. C. C. Chu in Szycher, ed., *Biocompatible Polymers, Metals and Composites*, Technomics, 1983, p. 417.
16. J. D. Martensen, *Vascular Replacements, TR5-533-011*, University of Utah Research Institute, 1981.
17. A. D. Callow, *Surg. Clin. of North Am.* **62**(3), 501 (1982).
18. D. Annis and co-workers in E. Chiellini and P. Glusit, eds., *Polymers in Medicine*, Plenum Publishing Corp., New York, 1983, p. 323.
19. J. R. Parsons and co-workers in Ref. 15, p. 873.
20. D. H. R. Jenkins and co-workers, *J. Bone Jt. Surg.* **59B**, 53 (1977).
21. U.S. Pat. 4,512,835 (1985), H. Alexander.
22. J. Kolarik and co-workers, *Biomed. Mater. Res.* **18**, 115 (1984).
23. J. M. Hunter and co-workers, *Orthopedic Clinics of North America* **8**(2), 473 (1977).
24. K. St. John in Ref. 15, p. 861.
25. G. Jenkins and co-workers, *Carbon* **15**, 33 (1977).
26. H. Ichijo and co-workers, *J. Appl. Polym. Sci.* **27**, 1665 (1982).
27. M. Curtin, *Biotechnol.*, 131 (Feb. 1984).
28. E. E. Magat, *Philos. Trans. R. Soc. London, Ser. A* **294**, 463 (1980).
29. D. Tanner, A. K. Dhingra, and J. J. Pigliacampi, *J. Met.* **38**(3), 21 (1986).
30. S. Yajima in *Philos. Trans. R. Soc. London, Ser. A* **294**, 419 (1980).
31. R. Kaiser, "Automotive Applications of Composite Materials," Report DOT-HS-804745, prepared for U.S. Dept. of Transportation, NHTSA, July 1978.
32. *Mechanics of the Pneumatic Tire, NBS Monograph 122*, U.S. Dept. of Commerce, Nov. 1971.
33. D. Himmelfarb, *The Technology of Fibers and Rope*, Interscience-Publishers, New York, 1957.
34. S. W. Scott, *Des. News*, 44 (March 26, 1984).
35. *Revised Recommended Asbestos Standard*, U.S. Dept. of Health, Education, and Welfare, Public Health Service Center for Disease Control, National Institute for Occupational Safety and Health, DHEW (NIOSH) Publication No. 77-169, Dec. 1976.
36. H. Y. Loken, "Asbestos Free Brakes and Dry Clutches Reinforced With Kevlar® Aramid Fiber," SAE Earthmoving Industry Conference, Peoria, Ill., Apr. 14–16, 1980, SAE Tech. Paper series 800667.
37. A. Frances, "Nonasbestos Beater Addition Gaskets With Kevlar® Aramid Pulp," 1984 Nonwovens Symposium, TAPPI, p. 125.
38. *Manufacturing Guide for Gasket Sheeting Reinforced With Kevlar® Aramid Pulp, Publication E-69541*, E.I. du Pont de Nemours & Co., Inc., Wilmington, Del.
39. R. T. Masters, "History and Use of Body Armor", paper presented at the International Security Conference, Chicago, Ill., July 1983.
40. S. Dunstan, *Flak Jackets*, Osprey Publishing, London, 1984.
41. *New Fiber Helmet Saves Lives*, Associated Press release, Dec. 8, 1983.
42. "Soft Body Armor—An Executive Summary," prepared by the Aerospace Corporation for the U.S. Dept. of Justice, July–August, 1977.
43. "Soft Body Armor Study," Research and Development Division, International Association of Chiefs of Police, July 1984.
44. S. E. Miller and A. G. Chynoweth, *Optical Fiber Telecommunications*, Academic Press, New York, 1979.
45. J. E. Midwinter, *Optical Fibers for Transmission*, John Wiley & Sons, Inc., New York, 1979.
46. S. D. Personick, *Optical Fiber Transmission Systems*, Plenum Publishing Corporation, New York, 1981.
47. Y. Suematsu, ed., *Optical Devices and Fibers*, North-Holland, Amsterdam, 1982.
48. C. K. Kao, *Optical Fiber Systems*, McGraw-Hill Inc., New York, 1982.
49. T. Okoshi, *Optical Fibers*, Academic Press, New York, 1982.
50. S. E. Miller, *Science* **195**, 1211 (1977).
51. Y. Suematsu, *Proc IEEE* **71**, 692 (1983).
52. T. Li, *IEEEJ. Selected Areas Commun.* **1**, 356 (1983).
53. U.S. Patent 4,138,194 (1979), J. K. Beasley, R. Beckerbauer, H. M. Schleinitz, and F. C. Wilson.
54. H. Kogelnik, *Science* **228**, 1043 (May 31, 1985).

55. *Business Week* (2907), 29 (Aug. 12, 1985).
56. *Analysis of the Market for PWB's and Related Materials for the Year 1982*, Institute for Interconnecting and Packaging of Electronic Circuits, 1982 Program of the Technology/Marketing Research Council, 1982.
57. R. C. Forney, *J. Met.* **38**(3), 18 (1986).
58. A. Dhingra, *J. Met.* **38**(3), 17 (1986).

ASHOK K. DHINGRA
HERBERT G. LAUTERBACH
E.I. du Pont de Nemours & Co., Inc.

FRACTURE

Fracture is defined as stress-biased material disintegration through the formation of new surfaces within a body. The brittle fracture of a plastic knife and the rupture of an inflated elastomeric balloon are typical examples. Fracture is synonymous with rupture and breakage but not with failure. The latter term is more general and also encompasses nonmechanical breakdown through heat (thermal failure) or environmental degradation (chemical attack, irradiation).

For fracture to occur, it is generally necessary that a specimen be subjected to mechanical loads, that it deforms (viscoelastically), and that some damage develops which leads to material disintegration.

Fracture Phenomena

Brittle Fracture at Low Temperatures, High Loading Rates, or High Strain Energy. Brittle fractures, the most conspicuous fractures, occur in highly loaded samples, often in an unstable manner, ie, rapid crack propagation, and at very small additional strains (Figs. 1–6). Different polymers exhibit different behavior. At low temperature, thermosetting resins and thermoplastic polymers show an almost linear increase of stress with strain, and fracture occurs at strains of a few percent without noticeable plastic deformation, resulting in macroscopically smooth surfaces. Under the electron microscope, however, small ridges or markings become apparent (Figs. 2, 3, and 6). Elastomeric networks, on the other hand, need to be strained by several hundred percent before the strain energy becomes sufficient for unstable rapid crack propagation to occur. Such a rupture is also classified as brittle fracture because the sample is not subjected to any noticeable additional deformation between the moment of fracture initiation, usually at some flaw or defect, and termination (Fig. 7).

Dry Crazing, Stress-whitening, and Environmental Stress Cracking. Many glassy polymers, stressed well below breaking, develop large numbers of straight silvery zones which are traversed by fibrillar matter called crazes (6,7) (see CRAZING).

The characteristic features of crazing of polycarbonate (PC), for example, are presented as follows. During a tensile test, below the glass transition ($T_g = 149°C$) conventional crazes are observed well below the yield point (Figs. 8 and 9). These crazes, called extrinsic crazes or crazes I, are separated from each other (Figs. 9 and 10). Surface defects may frequently be identified as craze initiators.

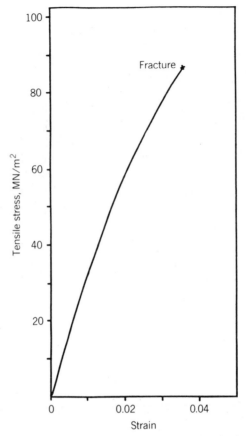

Fig. 1. Stress–strain curve of an unfilled, highly cross-linked epoxy resin. To convert MN/m² to psi, multiply by 145.

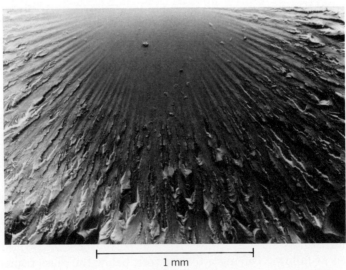

1 mm

Fig. 2. Electron micrograph of a typical brittle fracture surface (same specimen as in Fig. 1).

Fig. 3. Electron micrograph of a poly(methyl methacrylate) (PMMA) specimen bent at 20°C to brittle rupture (1). Magnification: 75×. Courtesy of S. Hashemi.

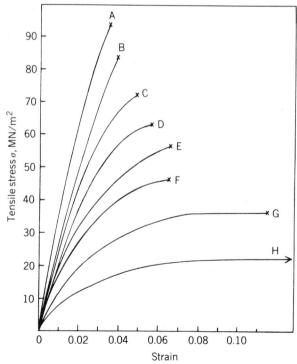

Fig. 4. Stress–strain curves of polyethylene (PE) at different temperatures (2): A, 93 K; B, 111 K; C, 149 K; D, 181 K; E, 216 K; F, 246 K; G, 273 K; H, 296 K.

Fig. 5. Brittle fracture of a polyethylene pipe subjected to a burst test at 500 kPa (5 bar) (3). Courtesy of *3R International*.

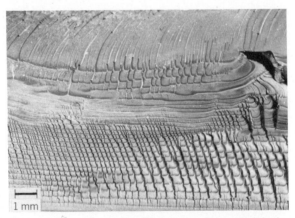

Fig. 6. Electron micrograph of the fracture surface of high density polyethylene (HDPE) bent at liquid-nitrogen temperature to brittle fracture (4).

An example of crazes I initiated at the surface and not grown sufficiently to traverse the whole specimen is shown in Figure 10. Crazes of this type also appear under comparable circumstances in other glassy polymers, such as polystyrene (PS) or poly(methyl methacrylate) (PMMA), and in semicrystalline polymers such as polyethylene (PE) or polypropylene (PP) (7).

At higher strains, many polymers, eg, PVC, PE, PP, tend to deform by opening up small voids throughout the whole volume. The deformed regions

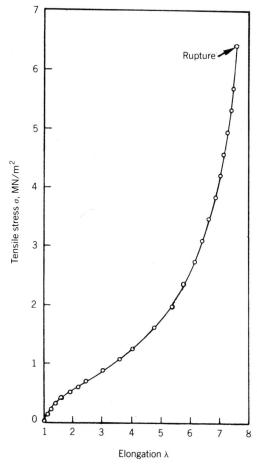

Fig. 7. Stress–strain curve of natural rubber; the deformation up to $\lambda = 7$ is an elastic deformation and completely reversible (5). To convert MN/m² to psi, multiply by 145.

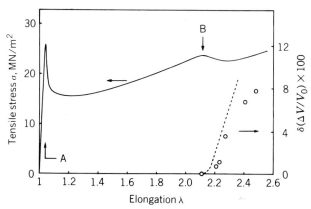

Fig. 8. Engineering stress–strain curve and incremental volume increase, $\delta(\Delta V/V_0)$, for polycarbonate ($T = 129°C$; strain rate $\dot{\epsilon} = 3.5\%/min$): A, crazes I; B, crazes II.

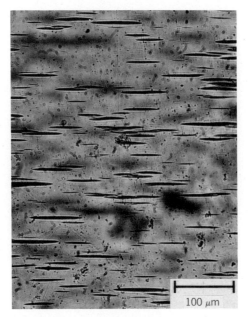

Fig. 9. Optical micrograph of conventional crazes (8).

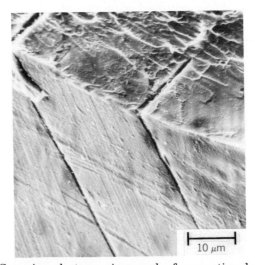

Fig. 10. Scanning-electron micrograph of conventional crazes (8).

appear to be white (stress-whitening) (Fig. 11). In PC, stress-whitening is equivalent to a new crazing mechanism (8). Well above the yield point of the stress–strain curve, at the point marked in Figure 8, this type, called intrinsic craze or craze II, is observed. Its initiation is preceded by some strain hardening and followed by a strain-softening mode, resulting in a flattening of the stress–strain curve.

The marked influence of the initiation and growth of crazes II on the stress–strain curve can easily be understood in view of their large number, as revealed by optical and scanning electron microscopy (Figs. 12 and 13). In Figure

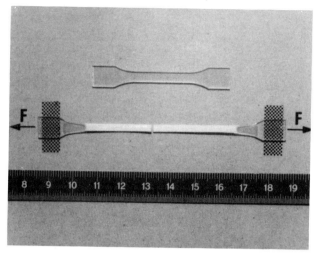

Fig. 11.   Stress-whitened polycarbonate (PC).

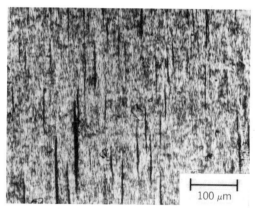

Fig. 12.   Optical micrograph of crazes I and II in oriented PC (8).

12, both types of crazes can clearly be identified, the large and isolated crazes I and the dense pattern of very fine crazes II.

Crazing is greatly enhanced by liquids, greases, and even gases. The crazing agent must be able to diffuse into a polymer to effect initiation. It thus acts through its presence within the polymer matrix. In increasing the chain mobility (lowering T_g), it facilitates craze initiation but also craze breakdown (4,7,9).

Creep Crazing (Fissurelike Brittle Rupture).   Many thermoplastics exposed to low stresses for extended periods of time fail because of an apparently plane fracture zone, usually initiating from a defect at a boundary (Figs. 14 and 15). Brittle fracture at low loading rates and creep crazing (at constant load) lead to similar fracture-surface patterns.

Fatigue.   Polymers are sensitive to fatigue loading, ie, to the repeated application of loads. A specimen may rupture through fatigue crack development at a load smaller than the quasi-static strength (Figs. 16 and 17) (see also FATIGUE).

Yielding, Necking, Creeping, and Flowing.   As opposed to brittle fracture,

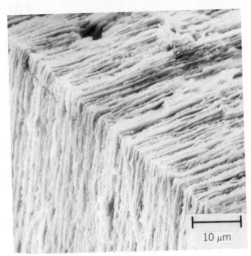

Fig. 13. Scanning-electron micrograph of crazes II in PC (8).

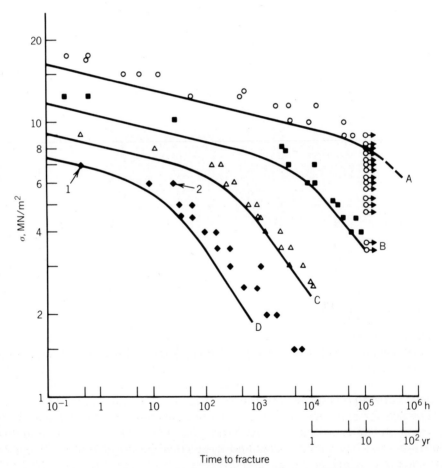

Time to fracture

Fig. 14. Times to failure of high density polyethylene (HDPE) water pipes under internal pressure p at different stresses and temperatures (4): A, 20°C; B, 40°C; C, 60°C; D, 80°C. 1 = ductile failure; 2 = creep crazing. Circumferential stress $\sigma = d_m/2s$, where d_m = average diameter and s = wall thickness. Specimens are shown in Figures 15 and 19. To convert MN/m² to psi, multiply by 145.

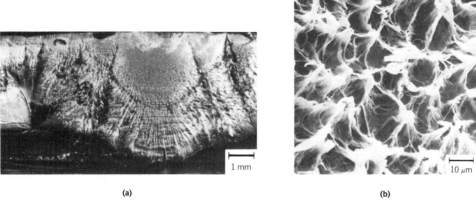

(a)   1 mm   (b)   10 μm

Fig. 15. (**a**) Surface of a creep craze formed in HDPE under conditions shown as point 2 in Figure 14: $\sigma_v = 6$ MN/m²; $T = 80°C$ (4). (**b**) Detail of the fracture surface close to the upper center of the zone which was apparently the point of craze initiation (4).

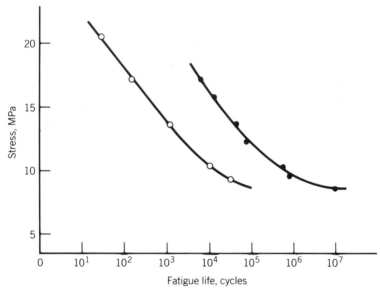

Fig. 16. Stress amplitude vs cycles to fracture of high impact PS tested at 0.2 Hz (○) and 21 Hz (●) (10). To convert MPa to psi, multiply by 145.

a localized event, rupture through ductile deformation generally involves a more extended region, and occasionally the whole specimen (Figs. 18 and 19). Ductile deformation phenomena leading to polymer fracture include yielding with and without neck formation, creep, and flow. The latter two phenomena certainly do not comply with the definition of fracture given above, since neither creep nor flow would give rise to the formation of new surfaces within the body. However, all of these phenomena must be discussed in the context of polymer fracture because in some cases the homogeneous ductile deformation only precedes and modifies a subsequent brittle fracture event; in other cases, slight modifications

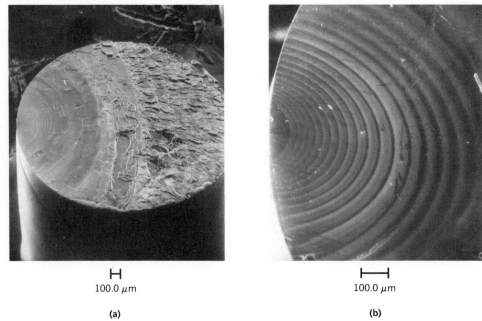

H
100.0 µm

H
100.0 µm

(a)   (b)

Fig. 17.   Fatigue fracture surface of PS tested at 17.2 MPa and 2 Hz (11): (**a**) low magnification; (**b**) higher magnification near source (10). Courtesy of *Advances in Polym. Science*.

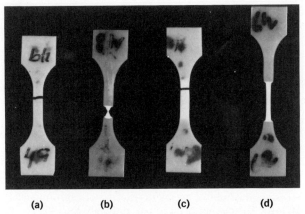

(a)   (b)   (c)   (d)

Fig. 18.   Tensile rupture of polyethylene (2): (**a**) brittle fracture at low temperatures ($T < 95$ K); (**b**) necking and rupture ($T = 100–240$ K); (**c**) non-necking ductile fracture of slowly cooled low molecular weight material ($T = 90–300$ K); (**d**) necking and drawing ($T > 0°C$). Courtesy of I. M. Ward.

of internal (molecular or microscopic structure) or external parameters (temperature, rate of loading) cause the change from brittle to ductile behavior (Fig. 18) (see also POLYMER CREEP).

Abrasion and Wear.   Specimens in sliding or rolling contact with each other mutually exert forces on their surface regions which may easily go beyond the

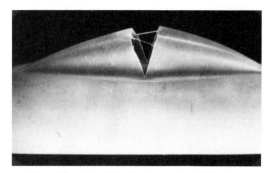

Fig. 19.   Ductile fracture of a PE pipe.

elastic limits and lead to damage and local rupture. Small particles can be torn off the surface, which is thus abraded and worn (see also ABRASION AND WEAR).

Fracture Processes.   A fracture process consists of at least two phases. In the first, defect initiation or activation, there is no visible crack growth. It is even difficult to separate this phase of locally irreversible deformation clearly from the mechanically reversible deformation of a sample.

In the second phase, a deformation zone or a crack starts growing, and crack-growth rates very often increase considerably. The final breakdown of a sample occurs in an unstable manner by rapid crack propagation (Figs. 2, 3, 5, and 6).

Multiaxial States of Stress

In most engineering applications, the load-bearing capability of polymeric materials is utilized. Fracture is highly undesirable and must be avoided. The classical approach to predicting safe operating conditions for metallic specimens is based on the assumption that there is a strategic property, such as uniaxial tensile strength, shear strength, elastic or total extensibility, or energy-storage capability, that determines the failure of the stressed sample. If similar considerations are applied to polymers under multiaxial states of stress, it must be recognized that polymers are viscoelastic materials and that the above properties depend on environmental parameters such as temperature T, strain rate $\dot{\epsilon}$, or the surrounding medium. If these parameters are taken to be constant, failure must be expected whenever the components of multiaxial stress (usually the three principal stresses σ_1, σ_2, and σ_3) combine in such a way that the strategic quantity reaches a critical value C. For different T or $\dot{\epsilon}$, C may assume different values. The condition $f(\sigma_1, \sigma_2, \sigma_3) = C(T, \dot{\epsilon})$, derived from a failure criterion, represents a two-dimensional failure surface in three-dimensional stress space (4,12–14).

As indicated previously, failure in polymeric materials may occur in a number of phenomenologically different ways, eg, as brittle fracture from propagating cracks, as ductile fracture by means of plastic deformation following shear yielding, or as quasi-brittle fracture after normal stress yielding (crazing). These phenomena may respond differently to the magnitude and state of stress. This means that for different fracture phenomena separate failure surfaces exist which may overlap and penetrate each other (4,12–14).

The most common failure criteria not based on fracture mechanics (4) describe yielding, ie, the ductile failure of isotropic, elastic-plastic solids in multiaxial states of stress. Although ductile failure is not considered as fracture in the narrow sense, these criteria must be discussed in view of the possibility that (brittle) fracture could be initiated, preceded, or accompanied by plastic deformation. The physical significance of the principal criteria is described below.

The maximum-principal-stress theory (Rankine's theory) states that the largest principal stress component σ in the material determines failure regardless of the value of normal or shearing stresses. The criterion is formulated as

$$\sigma < \sigma^* \tag{1}$$

where σ^* is a basic material property which can be determined from a tension test. In stress space, the surface of failure, according to the maximum-stress theory, is a cube. A modified form of this theory allows for a larger critical value σ^* if one of the stresses is compressive. Then the surface of failure is a cube again, but with the center displaced from the origin.

The maximum-elastic-strain theory (St. Venant's theory) states that inception of failure occurs if the largest local strain ε_3 within the material exceeds a critical value ε^*. The failure criterion, therefore, is derived as

$$\varepsilon = \frac{1}{E}[\sigma_1 - \nu(\sigma_2 + \sigma_3)] < \varepsilon^* \tag{2}$$

where E is the isotropic Young's modulus and ν Poisson's ratio. The term in square brackets thus must be smaller than an equivalent stress σ^*.

In the theory of total elastic energy of deformation, the criterion of total elastic energy seems to have no significance, since under high hydrostatic pressure large amounts of elastic energy may be stored without causing fracture or permanent deformation.

In the theory of constant elastic strain energy of distortion, consideration of the poor significance of the elastically stored energy in hydrostatic compression or tension led to the idea of subtracting the hydrostatic part of the energy from the total amount of elastically stored energy. Thus, it is assumed that only the energy of distortion W determines the criticality of a state of stress. For small deformations,

$$\frac{6E}{1 + \nu}W = [(\sigma_1 - \sigma_2) + (\sigma_2 - \sigma_3)^2 + (\sigma_3 - \sigma_1)^2] < 2(\sigma^*)^2 \tag{3}$$

If the octahedral shearing stress τ_{oct} is postulated to be smaller than a critical value τ^*, a mathematically similar expression is obtained:

$$(\tau_{oct})^2 = \frac{1}{9}[(\sigma_1 - \sigma_2)^2 + (\sigma_2 - \sigma_3)^2 + (\sigma_3 - \sigma_1)^2] < (\tau^*)^2 \tag{4}$$

In this case, the limiting condition of small deformations does not pertain.

The Coulomb yield criterion states that the critical stress τ^* for shear deformation to occur in any plane decreases linearly with the stress σ_n applied normally to that plane:

$$\sigma_3 - \sigma_1 < \tau^* = \tau_0 - \mu\sigma_n \tag{5}$$

The material constant τ_0 is given by the cohesion of the material and μ is a coefficient of friction. If friction is neglected, the *Tresca yield criterion* is obtained.

An equivalent expression for the case of triaxial stress relates τ_{oct} and the hydrostatic tension $\sigma_{oct} = (\sigma_1 + \sigma_2 + \sigma_3)/3$:

$$\tau_{oct} < \tau_0 - \mu\sigma_{oct} \tag{6}$$

The failure surface in principal stress space corresponding to this criterion is a cone whose axis coincides with the space diagonal (12).

The shape of the failure surfaces in three-dimensional stress space has been analyzed for different isotropic polymers (Fig. 20) (14). It is surprising that there is no difference in the elliptic shape between the envelopes describing brittle fracture (Fig. 20a) and ductile failure (Fig. 20b). The best fit of the brittle-fracture data is obtained by using a ratio m of uniaxial compressive strength σ_{cb} to uniaxial tensile strength σ_{tb} of between 1 and 1.45. In the case of ductile failure, m is between 0.92 and 1.30, thus only slightly lower.

Although the effect of multiaxial states of stress on the brittle and ductile failure of isotropic polymers is sufficiently well represented by the above classical failure envelope, this is not the case for crazing or the failure of anisotropic polymers, ie, fibers, single crystals, etc. As an example, the envelope for craze initiation in PMMA (11) is compared with the pure-shear and shear-yielding envelopes. The stress-bias criterion has been formulated (11) as

$$\sigma_{craze} = |\sigma_1 - \sigma_2| \geqslant A(T) + \frac{B(T)}{\sigma_1 + \sigma_2 + \sigma_3} \tag{7}$$

where A and B are functions of the temperature T.

The experimental data represented in Figure 20 show a certain scatter. In the light of the classical failure criteria, this can be interpreted in two ways: either material differences influence the critical quantities σ^* or τ^* or defects or frozen-in tensions locally modify the principal stresses.

Statistical Aspects

Experimentally determined quantities, such as strength or time to fracture, show a wide scatter of values and may be regarded as stochastic variables. In Figure 21, for example, the service lives t of 500 HDPE tubes under identical test conditions are plotted in the form of a histogram (4). The expectation value of the time to fracture of a specimen subject to the test condition employed is the time that corresponds to the mean value of the (logarithmic) distribution (15), here 247.6 h. The actually determined values of t evidently scatter widely around

Fig. 20. Biaxial stressing of 13 different polymers leading to brittle or ductile fracture (14); parameter is the ratio $m = \sigma_{cb}/\sigma_{tb}$. (**a**) Brittle fracture: A, $m = 1$; B, $m = 1.45$. ●, PMMA; ○, PMMA; ◑, PMMA; ×, cellulose acetate butyrate (CAB); +, cellulose acetate (CA); ★, poly(vinyl chloride acetate) (PVCA); ■, ethylene propylene (EP-Harz 1); □, ethylene propylene (EP-Harz 2). (**b**) Ductile fracture: A, $m = 1$; B, $m = 0.92$; C, $m = 1.30$. +, PS; ◑, PVC; ●, PVC; ○, PVC; ■, PMMA; □, PMMA; ▲, PC; △, PC; ×, CAB; ★, PE; ▽, PP; ▼, polyamide (PA); ◣, ABS. Courtesy of Carl Hanser Verlag.

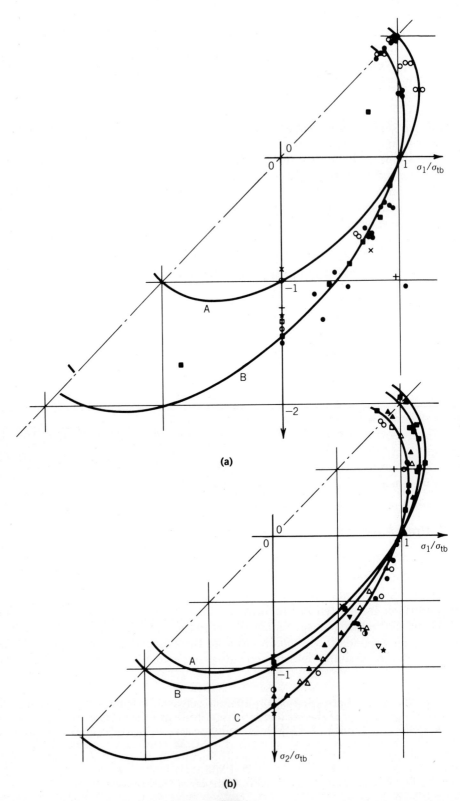

(a)

(b)

378

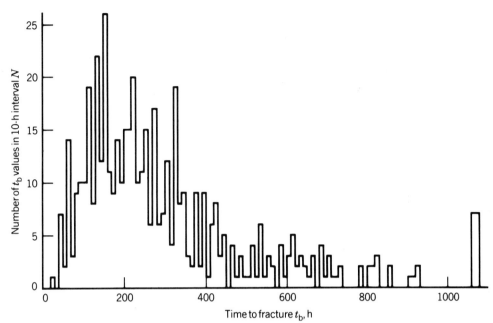

Fig. 21. Scatter of times to fracture of 500 PE tubes under identical loading conditions: 80°C, σ_v = 4 MN/m² (4).

this value. Nevertheless, even such a distribution can be characterized by testing a few randomly selected specimens. For a normally distributed variable, any three random values cover, on average, a range of 1.69 s (where s is the variance). The mean value of the (infinite) distribution is found in 86% of all cases within the range indicated by the three randomly determined values (15). This extreme example was chosen to emphasize the nature of stochastic variables. It is also intended to show the significance of a few samples, ie, 3–10, which are generally used in practice to establish correlations between stochastic variables and environmental or material parameters (16).

The variability of a material property, such as fracture strength, can be explained as follows (4): The beginning of crack propagation is a statistical event, whose occurrence is described by probability laws. Apparently identical specimens of a set are inherently different. They contain, for example, a large number of flaws of different size or severity; the most severe of these determines the fracture strength. The load to which a sample is subjected until it fractures activates a large number of molecular processes. The uncertainty of the cumulative event, eg, macroscopic failure, results from the uncertainty of the molecular events and the mode of interaction of these events.

Based on this analysis, a purely stochastic theory seems to hold for the creep rupture of some rubber vulcanizates (17). Statistical considerations also play an important role in the stressing of bundles (4,18), where a bundle is defined as an ensemble of n load-bearing units subjected to a given total load that is equally divided among the n units. Two statistical effects should be noted: the lower strength of a bundle as compared with that of a monofilament, caused by the accelerating increase in failure probability after rupture of the first unit within a bundle, and the increase of strength with loading rate, arising from the reduced time each filament spends at each load level.

Undoubtedly, the principal source of the statistical variation of strength and lifetime are inherent differences such as nature and location of defects (Figs. 2, 3, and 15) and the presence of internal tensions. As a quantitative example of the individual contributions of these parameters, a statistical analysis of the variance of the time to fracture t_b of PE pipes may be cited. From the slope of the integral distribution shown in Figure 22, the variance of $\log(t_b/h)$ is determined as $s = 0.300$, which corresponds to a variation of t_b by a factor of 2 around the average. A distinct correlation of the location of a defect at the surface or in the interior with the time to fracture was found in a series of low density polyethylene (LDPE) pipes (19). If the influence of flaw location is described by a stochastic variable y and the statistical variation due to flaw growth by x, the total time to fracture t_b can be represented as the product of the variables x and y and of t_m, the logarithmic mean value of t_b:

$$t_b = xyt_m \tag{8}$$

The mean values $\langle x \rangle$ and $\langle y \rangle$ are unity; thus, $\langle \log(t_b/h) \rangle$ is identical to $\log(t_m/h)$.

Statistical analysis of these data (19) shows that $s_x = 0.29$ and $s_y = 0.50$ (20). In this case, the variation of the defect location contributes much more (by a factor of 3) to the variation of t_b than the flaw-growth uncertainty (a factor of 2, as before).

The above observation of surface-related fracture is made frequently and is easily understood, since the surfaces are the first to be exposed to the environment, eg, scratches, liquids, oxygen, and are very often also sites of processing effects which may lead to frozen-in tensions or molecular orientations.

It follows that the scatter of fracture data is more pronounced if the stress–temperature situation favors fracture of the loaded specimen from flaw

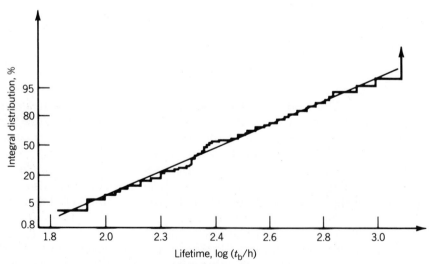

Fig. 22. Cumulative normal distribution of the lifetime logarithms plotted in Figure 21 (4).

growth; this has been observed many times. It is clearly visible in the scatter of the strength of PVC as a function of temperature plotted in Figure 23. With the onset of plastic deformation, some local strain hardening becomes possible, which counterbalances the detrimental effect of a defect. Elimination of defects not only increases the service life of a structure but also reduces the observed scatter (4,21).

Another important conclusion must be drawn with respect to the effect of sample size on sample strength. According to the Weibull theory (22), one and the same flaw population should be responsible for the scatter of strength values and for the size effect. However, the Weibull approach is generally not valid for unfilled polymeric materials (4,14,18,23).

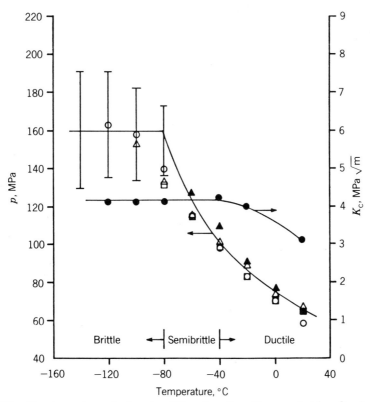

Fig. 23. Fracture stress in tension (T) or bending (three-point bending) of unplasticized PVC (1). The vertical bars indicate the maximum scatter band.

		specimen width, mm	specimen depth, mm
tension	●	6	10
	△	3	12.5
	□	6	15
3-point bending	▲	12.5	15
	○	6	50

Courtesy of S. Hashemi.

Fracture Mechanics

Fracture mechanics describes the behavior of cracks or other defects when a body is loaded. A crack constitutes a discontinuity within a body. Fracture mechanics describes the conditions under which it grows or propagates, creating new surface area. At the boundary of the discontinuity, very high local stresses and strains are usually observed, and in the limit of a sharp crack in an elastic body, these may theoretically be infinite or singular. There are two approaches to describing fracture. In the first, the balance of energy is discussed, some of which is released from the elastic deformation by the creation of the new crack area and some of which is absorbed by the area created. In the second, the local stress field around the crack tip is considered and characterized by some parameter at fracture. The energy approach is taken here, and characterizing parameters are used when convenient.

In general, fracture criteria are material properties, whereas stability criteria are functions of the body concerned. Fracture mechanics is a macro theory, being concerned with the overall behavior of a body. The particular material behavior is considered only in connection with fracture criteria or deformation processes involved. Much of the theory is thus not material specific and applies equally to all materials. It is based on linear elastic fracture mechanics (LEFM), which assumes linear elastic deformation and very local energy absorption at the crack tip. The theory has been extended to plastically deforming materials, which undergo rate-independent, irreversible deformation. Neither of these assumptions is strictly true for polymers, since all are somewhat viscoelastic and exhibit time dependence, giving behavior intermediate between these limiting states. However, much of the behavior observed in the fracture of polymers can be described in terms of LEFM, and plasticity modifications are of value in dealing with ductile polymer fracture. It is possible to modify the analysis to encompass time effects; this is necessary for some phenomena, eg, slow crack growth. Reference 24 covers the topics in much more detail and provides an extensive bibliography.

Toughness.   The toughness of a material may be defined in terms of its resistance or R curve, as shown in Figure 24, where R is the energy per unit area necessary to produce the new fracture surface. For a crack of length a_0 and

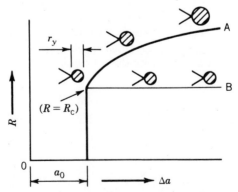

Fig. 24.   The resistance or R curve. See text for significance of curves A and B.

in a plate of uniform thickness B, the R curve may be expressed in terms of the crack growth Δa. Up to some value of R, there is no growth and the line is vertical. However, there is crack-growth initiation at $R = R_c$. Subsequent growth may be described, as in line A, with R increasing as the crack extends. This usually involves the increase of the local plastic flow in a zone of size r_y around the crack tip. If the extent of this zone, which represents the energy absorption, remains the same, R remains at R_c, a constant as shown in line B. This is so for the more brittle fractures, where the energy absorption is very local and the zone is small and very much less than a_0; a constant R_c results. For tougher materials with small a_0, an increasing r_y is more likely. It is possible to describe the rising R curve using LEFM, but the constant R_c case is more important. The plastic zone size r_y is a very important parameter, since it provides a length factor which controls the various size effects.

Energy-release Rate. The driving force for fracture comes from the energy which is released by the growth of the crack (25). This growth violates continuity within the body and results in the final energy state being lower than the initial; the difference is released and available to provide the crack resistance R. A loading line $0A$ is shown in Figure 25 for a linearly elastic body in which crack growth initiates at A, resulting in a reduction in the load P and an increase in the deformation to the point A'. The initial stored energy is the area $0AF$ and that after the crack growth is $0A'F'$. Part of this is generated by external work, given by $AA'F'F$. The shaded area $0AA'$ is the energy dU which is released by the system and given by

$$dU = \tfrac{1}{2}(P\,du - u\,dP) \tag{9}$$

where u is deformation in the direction of applied load.

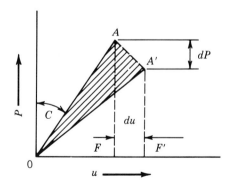

Fig. 25. Energy release shown by the shaded area.

The energy-release rate with respect to crack growth is given the symbol G:

$$G = \frac{1}{B}\frac{dU}{da} = \frac{1}{2}\left(P\frac{du}{da} - u\frac{dP}{da}\right) \tag{10}$$

where a is the crack length. This is conveniently expressed in terms of the compliance of the body:

$$C = \frac{u}{P} \tag{11}$$

which is a function of the crack length and therefore

$$G = \frac{P^2}{2B}\frac{dC}{da} = \frac{u^2}{2B}\frac{1}{C^2}\frac{dC}{da} = \frac{U}{B}\cdot\frac{1}{C}\frac{dC}{da} \tag{12}$$

Thus, G can be determined from the load, the deflection, or the energy at fracture, providing the compliance $C(a)$ is known. For any load, lines of G vs a can be plotted for a cracked body, as shown in Figure 26 together with an R curve. For the load of line A, there is no growth when $G = R$, but for line B there is. Just after growth, it can be seen that $R > G$ and the load would need to be increased for the crack to grow and result in a stable fracture. This condition may be characterized for stable fracture by

$$G = \frac{dU}{Bda} = R \quad \text{and} \quad \frac{1}{B}\frac{dG}{da} = \frac{1}{B^2}\frac{d^2U}{da^2} < \frac{dR}{Bda} \tag{13}$$

For line C, the tangency condition, instability prevails after initiation; only that part of the R curve up to the tangency point gives stable growth. For a constant R_c, there is no growth until $R = R_c$, and unstable growth thereafter; it is usually written as $G = G_c$, where $G_c = R_c$ at initiation of crack growth.

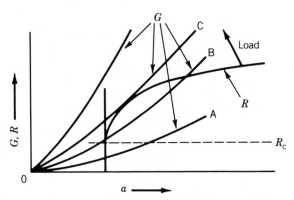

Fig. 26. Energy-release rate G and material resistance R for various loads.

This condition pertains in the most important example in fracture, which has been described (25) for an infinitely wide sheet containing a crack of length $2a$ and loaded with a constant stress σ, as shown in Figure 27. The expression for G is

$$G = \frac{\pi\sigma^2 a}{E} \tag{14}$$

where E is the Young's modulus. In a perfectly brittle material undergoing no plastic work in which $G_c = 2\gamma$, where γ is the true surface work of the solid, at fracture

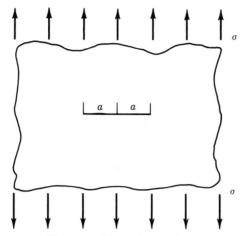

Fig. 27.   The infinite sheet.

$$G = G_c \quad \text{and} \quad \frac{\pi\sigma^2 a}{E} = 2\gamma \tag{15}$$

Thus, the basic relationship of fracture mechanics is that $\sigma^2 a \approx$ constant at fracture. In this case $dR/da = 0$ and $dG/da > 0$ and the fracture is always unstable. However, this does not hold for a fixed displacement.

Another case is the double-cantilever beam (DCB) shown in Figure 28. From beam theory:

$$C = \frac{8a^3}{EBD^3} \tag{16}$$

where B is specimen thickness and D specimen width. In practice $C \propto an$, $n < 3$, because of rotation at the beam ends and this empirical result is frequently used in practice. Using equations 12 and 16;

$$G = \frac{12P^2 a^2}{EB^2 D^3} = \frac{3}{16} \cdot \frac{ED^3 u^2}{a^4} \tag{17}$$

Again, $G = G_c = 2\gamma$ at fracture giving load-crack length and displacement-crack length relationships at fracture, and also $dR/da = 0$. Fixed load gives $dG/da > 0$ as before, resulting in unstable growth, but for constant displacement, $dG/da < 0$ and stable growth occurs.

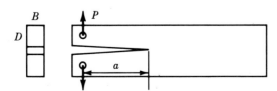

Fig. 28.   The double-cantilever beam (DCB).

In principle, any body can be characterized by finding $C(a)$ by measurement or computation. Thus, by detecting crack-growth initiation, the G value at fracture, termed G_c, can be found. Subsequent stability can be described in terms of dG/da and dR/da. Some more general aspects to the analysis are also important and, in particular, the fact that G may be described for any elastic body, not necessarily linear, as

$$G = \int_s W \cdot dy \qquad \text{displacement constant} \qquad (18)$$

where W is the strain-energy-density function, y the coordinate normal to the crack direction, and s some closed contour taken around the crack tip. This form is particularly useful for the analysis of nonlinear elastic systems as encountered in rubber elasticity (26). An example, shown in Figure 29, is a very wide parallel strip where a contour may be taken, as indicated by the broken line. The horizontal portions give no contribution since $dy = 0$, and if the vertical lines are remote from the crack tip, the part behind it has no stored energy, whereas that in front has the energy per unit volume of the uncracked strip, eg, W_0. Equation 18 then gives

$$G = W_0 D \qquad (19)$$

The result is obvious, since a strip of width dx has an initial stored energy of $W_0 B D dx$. After fracture, this is reduced to zero by the creation of area Bdx, giving the above results. Of course, W_0 may be found for any type of elastic behavior simply by loading the strip initially. This result is also true in this case even when the crack speed is high and there are no kinetic energy effects, since these must all eventually go to zero as in the static case, though it is debatable if kinetic energy is truly reversible in real systems.

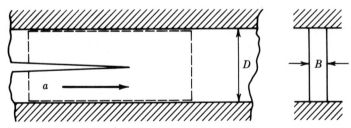

Fig. 29.   The parallel strip.

Stress-intensity Factor.   The stresses around the crack tip, as shown in Figure 30, may be expressed in the form of a series:

$$\sigma = \frac{Kf(\theta)}{\sqrt{2\pi r}} + A(\theta) + B(\theta) \cdot r^{\frac{1}{2}} + \dots \qquad (20)$$

where f, A, B, etc, are functions of the angle θ and r is the distance from the crack tip. As the crack tip is approached ($r \to 0$), all terms other than the first two tend to zero and the first term is dominant; $A(\theta)$ represents nonsingular stresses which can have effects, but, in general, the first, singular term (ie, tending

Fig. 30.   Local stresses around a crack tip.

to ∞ as $r \to 0$), dominates. The form of the stress field is the same for all remote loading states and determined by $f(\theta)/\sqrt{2\pi r}$. The magnitude of the local stresses is defined by K, the stress-intensity factor, which is a function of the remote loading. If a characterizing parameter in the local region of the crack tip is sought, clearly σ (or strain) is not useful since it tends to infinity as $r \to 0$. However, the product $\sigma\sqrt{r}$ is finite in the local zone and for this reason K (and not σ) is taken as the local characterizing parameter; K is defined as

$$\sigma\sqrt{2\pi r} \text{ as } r \to 0 \text{ at } \theta = 0 \tag{21}$$

$K = K_c$ is postulated at fracture. It is convenient to express the loading on a crack in terms of three orthogonal components (Fig. 31), which may be superimposed to give any loading state: Mode I, the opening mode; Mode II, the shear mode; and Mode III, the out-of-plane shear mode.

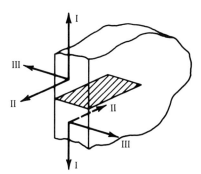

Fig. 31.   Modes of loading.

An applied loading may give rise to a mixture of all three, which can be expressed in terms of K_I, K_{II}, and K_{III}. Mode I is usually the most important, since combinations of loadings often result in a local mode-I fracture.

In practice, the local stress fields for modes I and II are the most important and may be written as (24):

$$\sigma_\theta = \frac{K_I}{\sqrt{2\pi r}} \cdot \frac{1}{2} \cos \frac{\theta}{2} (1 + \cos \theta) - \frac{K_{II}}{\sqrt{2\pi r}} \cdot \frac{3}{2} \sin \frac{\theta}{2} (1 + \cos \theta) \tag{22}$$

$$\sigma_r = \frac{K_I}{\sqrt{2\pi r}} \cdot \frac{1}{2} \cos \frac{\theta}{2} (3 - \cos \theta) - \frac{K_{II}}{\sqrt{2\pi r}} \cdot \frac{1}{2} \sin \frac{\theta}{2} (1 - 3 \cos \theta) \tag{23}$$

$$\sigma_{r\theta} = \frac{K_{\mathrm{I}}}{\sqrt{2\pi r}} \cdot \frac{1}{2} \sin\frac{\theta}{2} (1 + \cos\theta) - \frac{K_{\mathrm{II}}}{\sqrt{2\pi r}} \cdot \frac{1}{2} \cos\frac{\theta}{2} (1 - 3\cos\theta) \tag{24}$$

$$K_{\mathrm{I}} = \sigma_\theta \sqrt{2\pi r} \text{ at } \theta = 0 \tag{25}$$

$$K_{\mathrm{II}} = \sigma_{r\theta} \sqrt{2\pi r} \text{ at } \theta = 0 \tag{26}$$

In mode-I loading, fracture may occur when K_{I} attains K_{Ic}, a critical value. In combined-loading modes, crack propagation is not colinear and some complications result. If it is assumed to occur in a local mode I, from equation 22 a condition is obtained for the local fracture angle θ_c (24):

$$K_{\mathrm{Ic}} = K_{\mathrm{I}} \cdot \frac{1}{2} \cos\frac{\theta_c}{2} (1 + \cos\theta_c) - K_{\mathrm{II}} \cdot \frac{3}{2} \cdot \sin\frac{\theta_c}{2} (1 + \cos\theta_c) \tag{27}$$

and an additional condition that

$$\frac{d\sigma_\theta}{d\theta} = 0(\sigma_{r\theta} = 0) \tag{28}$$

giving

$$0 = K_{\mathrm{I}} \cdot \frac{1}{2} \sin\frac{\theta_c}{2} (1 + \cos\theta_c) - K_{\mathrm{II}} \cdot \frac{1}{2} \cdot \cos\frac{\theta_c}{2} (1 - 3\cos\theta_c) \tag{29}$$

In pure mode I ($K_{\mathrm{II}} = 0$), θ_c is zero as expected, whereas in pure mode II, ($K_{\mathrm{I}} = 0$), $\theta_c = \cos^{-1} \frac{1}{3} = \pm 70.3°$.

The local singular stress field is described by K_{I} and it is from this discontinuity that the energy is released by way of G. It is to be expected, therefore, that G and K_{I} would be related and this can be demonstrated by deriving G around a local contour (24):

$$K_{\mathrm{I}}^2 = EG \tag{30}$$

The π factor in equation 20 is introduced in the definition of K_{I} to remove any numerical constant in equation 30. There is no similar result for mode II alone or for mixed modes, since the propagation is not colinear ($\theta_c \neq 0$) and the result is not valid.

Returning to the case of the infinite plate,

$$K_{\mathrm{I}}^2 = EG = \pi\sigma^2 a \tag{31}$$

If σ is measured at fracture for a given a, K_{Ic} is defined and the fracture behavior is characterized without the need to involve E, which is needed when G is used. This is of considerable benefit in the polymer field, where E is time-dependent and often uncertain. Load measurements are the most commonly used in fracture tests, and K_{Ic} may be defined from these by means of a more general form of equation 31:

$$K_{\mathrm{I}}^2 = Y^2 \left(\frac{a}{W}\right) \cdot \sigma^2 \cdot a \tag{32}$$

where $Y^2(a/W)$ is a calibration function for any geometry and is available for most practical cases (27,28). For the infinite plate $Y^2\pi$, all cases tend to this result for $a/W \to 0$. Of course, Y^2 is another way of expressing the compliance calibrations given in equation 12; for a three-point bend test, as shown in Figure 32, for example,

$$\frac{dC}{da} = \frac{9}{2} \cdot \frac{L^2}{BW^4E} \cdot Y^2 a \qquad (33)$$

where L is the length of the beam.

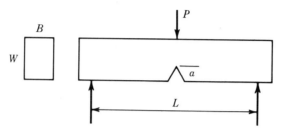

Fig. 32. The three-point bend test.

An important alternative form is the energy version of equation 12 in which

$$G = \frac{U}{BW} \cdot \frac{1}{\phi} \qquad (34)$$

and the calibration factor

$$\phi = \frac{C}{dC/d(a/W)} = \frac{\int Y^2 \cdot x \cdot dx}{Y^2 \cdot x} + \frac{L}{18W} \frac{1}{Y^2 \cdot x} \qquad (35)$$

where $x = a/W$.

Values of ϕ are also available for bending (24), and G_c may be found by measuring the energy to crack initiation and via equation 34. In any test, the three parameters at fracture can be measured and various combinations of E and G determined. Thus, from load in a bending experiment $EG(K^2)$ is obtained, from energy G, and from displacement G/E by

$$\frac{G}{E} = \frac{u^2W^2}{9L^2a} \frac{1}{(\phi \cdot Y)^2} \qquad (36)$$

$G/E = \pi/4 \, \rho$, where ρ is the elastic tip radius of the crack under load. If all three are used, consistency can be checked, and, in addition, E is found.

Plastic Zone and Size Effects. The elastic analysis predicts infinite stresses at the crack tip, but local yielding prevents this from happening. The use of LEFM requires small scale yielding, and two sizes may be considered, as shown in Figure 33. The outer, larger zone is characterizing the elastic field, whereas the inner zone is the plastic zone which must form because of the limitation of the yield stress σ_y. From the stress fields in equation 20, the extent of the zone at fracture is given approximately by

$$r_y \sim \frac{1}{2\pi} \cdot \left(\frac{K_{Ic}}{\sigma_y}\right)^2 \qquad (37)$$

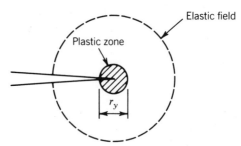

Fig. 33.   Elastic field and plastic zone; $\sigma = \dfrac{K}{\sqrt{2\pi r}} \cdot f(\theta)$.

The imposed stress state around the zone (in mode I) is given when $\theta = 0$ and is $\sigma_\theta = \sigma_r$ and $\sigma_{r\theta} = 0$. The out-of-plane component σ_z is determined by the strain e_z in that direction since

$$e_z = \frac{1}{E} [\sigma_z - \nu(\sigma_r + \sigma_\theta)] \tag{38}$$

If the crack-front length or thickness B is large compared with r_y, contraction in the z direction is constrained and the state of plane strain is approached in which $e_z = 0$ and $\sigma_z = \nu(\sigma_r + \sigma_\theta)$. In the zone $\nu \to \frac{1}{2}$, so the stress state tends to equitriaxial tension $\sigma_r = \sigma_\theta = \sigma_z$, the most severe in fracture terms. For thin sections $\sigma_z \to 0$, a state of plane stress which is much less severe. In testing it is desirable to employ the most severe conditions to explore the worst case. To achieve this, the specimen sizes must be considerably greater than the plastic zone. This is defined by the ASTM size criterion (29) for bend testing which gives a minimum thickness:

$$\hat{B} > 2.5 \cdot \left(\frac{K_{Ic}}{\sigma_y}\right)^2 \tag{39}$$

approximately equivalent to $\hat{B} > 16 r_y$. If $\hat{B} < B$, the stress state tends to plane stress and measured toughness values are unrealistically high.

In many polymers, crazes form at the crack tips (30) and these can be more realistically modeled by a line zone, as shown in Figure 34. Here, the yielding is modeled elastically by a line of tractions and the zone length is

$$r_y = \frac{\pi}{8} \left(\frac{K_{Ic}}{\sigma_y}\right)^2 \tag{40}$$

Another useful parameter arises, since the model predicts the displacement at the crack tip δ_c as

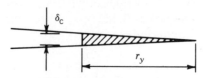

Fig. 34.   The line zone.

$$\delta_c = \frac{K_{Ic}^2}{E\sigma_y} = \frac{G_c}{\sigma_y} \tag{41}$$

This can be used as a fracture criterion to extend LEFM, since a constant δ_c is equivalent to constant K_{Ic} and G_c for LEFM. If δ_c remains constant outside the LEFM range, it can be used to make predictions for viscoelastic and large-scale plasticity behavior.

Viscoelastic Effects. Most polymers show some degree of viscoelasticity, which can have an important influence on fracture behavior. This is most clearly demonstrated by crack propagation in an elastic material. All parameters are independent of time and G_c (or K_{Ic}) would be independent of crack speed, as shown by the horizontal line in Figure 35. Some simplifying assumptions can give some insight into viscoelastic behavior. A constant yield strain is often a reasonable assumption and a power law in time can represent modulus changes:

$$\frac{\sigma_y}{E} = e_y, \quad \text{and} \quad \frac{E}{E_0} = \left(\frac{t}{t_0}\right)^{-n} \tag{42}$$

where e_y, E_0, t_0, and n are constants. If a constant δ_c criterion in the line zone model is assumed, from equation 41:

$$K = (\delta_c e_y)^{1/2} E_0 \left(\frac{t}{t_0}\right)^{-n} \tag{43}$$

This equation predicts an initiation time t_i given by

$$\frac{t_i}{t_0} = \left[\frac{K}{(\delta_c e_y)^{1/2} E_0}\right]^{-1/n} \tag{44}$$

but when the crack moves the time scale becomes

$$t = \lambda \cdot \frac{r_y}{\mathring{a}} \tag{45}$$

where $\mathring{a}$ is a derivative of crack length with respect to time and where λ is a constant which depends on the displacements in the zone and is usually about $1/3$ (24). Substitution in equation 43 gives a relationship between K and $\mathring{a}$:

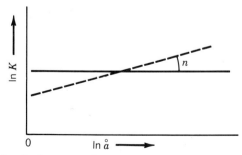

In K

0   In $\mathring{a}$

Fig. 35. Viscoelastic slow growth: – – –, viscoelastic behavior; ———, elastic behavior ($n = 0$).

$$K = (\delta_c\, e_y)^{1/2} E_0 \left(\frac{\mathring{a}}{\mathring{a}_0}\right)^n, \quad \mathring{a}_0 = \frac{\pi}{24} \cdot \frac{\delta_c}{e_y t_0} \tag{46}$$

This form is also shown in Figure 35 as the dashed line which is a rising line of K vs $\mathring{a}$. Thus, for an applied K value, the situation is not the "go–no-go" situation of the elastic case but a continuous relationship between K and $\mathring{a}$. Such curves can be most easily measured using specimens in which K is constant with respect to crack length so that constant loading or loading rates give a constant crack speed.

Such data are of particular importance because they provide a method by which long-term failure may be predicted from much shorter-term results. The tests measure crack-growth rate, which may be integrated to predict failure times. For example, the time for a flaw of original length a_0 to grow to a length much more than a_0 is given by

$$t = t_i \left[1 + \frac{5n}{1-n}\left(\frac{a_0}{r_0}\right)\right], \quad n \ll 1 \tag{47}$$

where $r_0 = \dfrac{\pi}{8}\dfrac{\delta_c}{e_y}$, the zone size, and t_i is the initiation time. For very small flaws, $t \sim t_i$ and very little of the life is spent in crack propagation. Such a result predicts $\sigma \propto t^{-n}$ for a constant stress test, ie, the stress rupture log–log line has the reverse slope of the log K–log $\mathring{a}$ crack-growth line.

The method may be extended to include environmental effects, since the cracks may be grown in their presence. Usually, this results in a lowering of K for a given speed which arises from the local plasticization of the crack-tip zone, reducing the zone stress. Some complications may arise because the effects depend on the environment being able to keep up with the crack. The flow into the crack tip depends on local crack-tip sizes, such as pores, and the characteristics of the fluid, such as viscosity and surface tension. These effects result in a rapid transition from the environmental line to that without environment (Fig. 36). Stage I is the environmental curve, stage II is the transition region where flow rates and crack speeds are similar, and stage III is unaffected by the environment because the crack is growing too fast. There is some evidence to suggest that the

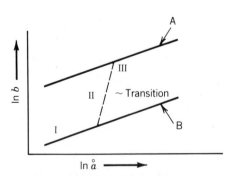

Fig. 36. Environmental crack growth: A, without environment; B, with environment.

speed at which the transition occurs is inversely proportional to the fluid viscosity, as expected from a flow-controlled process.

Fatigue-crack growth may be characterized in a similar fashion as K vs da/dN, the crack growth per cycle. Fatigue is usually a consequence of an additional, separate-damage process, but after the data have been obtained, the lines can be integrated with respect to number of cycles instead of time. Viscoelastic and fatigue effects interact frequently, particularly in heat generation during cycling.

Plasticity Effects. In LEFM, it is assumed that the plastic-zone size r_y is very small, but it is often difficult to take measurements under conditions where this is true because of size limitations. When it is not true, it is possible to find the fracture resistance R, but it is not G_c in this case. A correction can be made for the plastic zone, where the computed toughness is termed J_c. In effect, the crack length is increased by r_y to compensate for the modification of the elastic stress field by the plasticity:

$$J = \frac{\pi\sigma^2}{E}(a + r_y) = \frac{\pi\sigma^2}{E}\left(a + \frac{1}{2\pi} \cdot \frac{EJ}{\sigma_y^2}\right) \tag{48}$$

$$\frac{J}{G} = \left[1 - \frac{1}{2}\left(\frac{\sigma}{\sigma_y}\right)^2\right]^{-1} \tag{49}$$

where $G = \pi\sigma^2 a/E$. This correction is useful for $\sigma/\sigma_y < 0.8$, and it can be seen that $J \to G$ for small stresses. The line zone (Fig. 34) provides a somewhat more precise form:

$$\frac{J}{G} = 2\left(\frac{2}{\pi}\frac{\sigma_y}{\sigma}\right)^2 \ln \sec\left(\frac{\pi}{2} \cdot \frac{\sigma}{\sigma_y}\right) \tag{50}$$

There is very little difference between these two results up to $\sigma/\sigma_y = 0.8$, but as $\sigma \to \sigma_y$ the former tends to 2 and the latter to infinity, a more reasonable result. In fact, as the fully plastic condition is approached, such corrections are not satisfactory because they are very sensitive to the stress and the analysis is better couched in terms of displacement.

A simple case where this can be done is that of three-point bending, shown in Figure 37. If the ligament is fully plastic with a stress distribution as shown, the collapse load is given by

$$P = \sigma_y \frac{B}{L}(W - a)^2 \tag{51}$$

The plastic-zone work is $U_p = Pu_p$, where u_p is the displacement. The general definition of fracture toughness R, analogous to G for stable growth, may be used:

$$R = J = \frac{1}{B}\frac{dU_p}{da} \qquad u_p \text{ constant} \tag{52}$$

therefore,

$$J = \frac{2U_p}{B(W - a)} \tag{53}$$

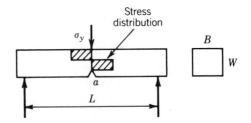

Fig. 37. Fully plastic three-point bending.

This assumes a fully plastic ligament and ignores elastic effects but does allow J to be found in the fully plastic case. In fact, this form is true when elastic energy is included, providing that $L/W = 4$. This analysis has been widely employed for measuring the toughness of very tough materials since the stress state in the ligament is very close to plane strain for the three-point bend test. Thus, the J_c measured at crack initiation should be the same as G_c in an LEFM test. The specimens may be much smaller and the size criterion is

$$B \text{ and } (W - a) > 25 \left(\frac{J_c}{\sigma_y} \right) \tag{54}$$

which is generally about a factor of 3 less than the LEFM value (29,31).

Crack initiation in these fully plastic cases is usually followed by slow, stable growth, and it is often difficult to determine when growth has been initiated. This problem is solved by a procedure (31) in which several identical specimens with $a/W \sim 0.5$ are tested, but each is loaded to a different load-point displacement. The energy under each load-deflection curve is measured and J is found from equation 53. Each specimen is broken open after cooling in liquid nitrogen to reveal the growth reached when the test was stopped. Growth is then plotted as a function of J and, by extrapolation to zero, J_c at initiation is obtained. However, the crack tip is often blunted considerably owing to displacement. This can be modeled by assuming a semicircular crack tip; the apparent blunting growth is then $\delta/2$ and this growth is given by

$$\Delta a_b = \frac{\delta}{2} = \frac{J}{2\sigma_y} \tag{55}$$

The extrapolation of the measured Δa values, which include Δa_b, should then be made to the blunting line of slope $2\sigma_y$. This is shown in Figure 38 in which A is the blunting line and B the initiation condition $J = J_c$.

High Rate Effects. The behavior of polymers at high testing rates is of considerable importance since their viscoelastic nature can result in substantial changes. For fracture, the changes in yield stress are of particular importance since increases result in a tendency to brittle fracture. Such concerns are reflected in the large number of impact tests employed in polymer testing and it is useful to employ fracture mechanics in such evaluations. In principle, it is simply necessary to measure loads, deflections, or energy at fracture and thus find G_c, K_c, or J_c as in the static case, but complications arise because at high rates the

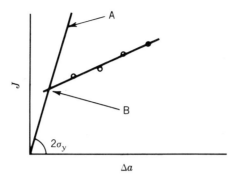

Fig. 38. Determination of J_c in a fully plastic three-point bend test: A, blunting line; B, $J = J_c$ at initiation. $\Delta a = \Delta ab$.

dynamic behavior of the specimen and machine can interfere with measurement and analysis.

In the most common test, a specimen is loaded at high rate, usually by impact, and the crack-initiation conditions are determined. Loads measured in the striker can undergo large cyclic variations because of stress-wave and vibration effects. These loads are related in a complicated manner to those in the specimen. Indeed, over part of the loading cycle contact between specimen and striker can be lost owing to bouncing, and failure can occur. The natural frequency of the specimen, including its contact stiffness, is important since it determines the period of the load oscillations, which are typically in the 1–10 kHz range. For the signal not to be significantly affected by these oscillations, the loading time must be sufficient to encompass at least 10 of these, ie, the loading times must be longer than ca 1 ms. With less time, it is necessary to correct loads for the dynamic effects, and this can be difficult. Such effects are often encountered in testing materials of low toughness, particularly those with long cracks.

Energy measurements circumvent some of these problems, since the readings are essentially an integral over the load-deflection behavior which has the effect of smoothing out many of the oscillatory effects. It is, of course, easy to measure the total energy absorbed, as in the conventional pendulum Charpy or Izod test. Depending on the behavior, G_c may be found. If the failure is unstable, the area under the triangle is measured, as shown in Figure 39a, which is required for the energy form of equation 12. This may be used with the calibration factor ϕ discussed previously (32):

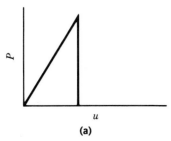
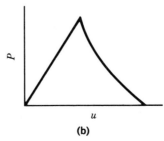

Fig. 39. Impact-load-deflection diagrams: (a) unstable; (b) stable.

$$U = G_c \cdot BW \cdot \phi \left(\frac{a}{W} \right) \tag{56}$$

Thus, a set of specimens of various values of a may be tested, ϕ found from tables, and the energy U plotted vs $BW \cdot \phi$ to give a line of slope G_c. The measured energy, however, includes that imparted in the form of kinetic energy. The plot shows a positive intercept on the energy axis U_k, which is given approximately by $2\ mv^2$, where m is the specimen mass and v the striker velocity; U must be several times larger than U_k in order to obtain sensible results. This is an equivalent condition to that of the number of long cycles mentioned earlier (Fig. 40a).

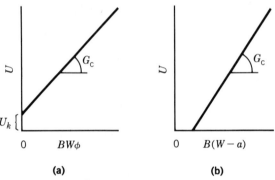

Fig. 40. Determination of G_c: (a) unstable ϕ analysis; (b) stable ϕ analysis (ligament-area basis).

When the fracture is stable, the load-deflection diagram changes to that shown in Figure 39b, and the diagram develops a "tail." The energy is that required to propagate the crack across the ligament and is actually the G_c of propagation and not initiation as in the previous case; these values may or may not be equal, and G_c may be found from

$$G_c = \frac{U}{B(W - a)} \tag{57}$$

It is usually better to plot U vs $B(W - a)$, as shown in Figure 40b, since plasticity end effects can give significant apparent negative intercepts in this case. If only the total energy is recorded, it is not known which type of failure has occurred. If, however, both plots are made, the linearity of one often, but not always, determines this. An exception is the occurrence of unstable fracture in the presence of significant plastic zones. Such data are linear on neither basis and best dealt with by correcting ϕ by a plastic-zone size. Load-deflection diagrams, even with substantial oscillations, can be used to define the various energies, and both initiation and propagation G_c values can thus be found.

The necessity of including the kinetic energy in the unstable fracture case illustrates the problem in performing the calculation of G for a high speed deformation process. Several practical problems, eg, the safety of plastic pipes, make it useful to compare the performance of materials with cracks propagating at high speeds. The crack is almost always unstable and the effective G is now

$$G = -\frac{1}{B}\frac{d}{da}(U_s + U_k) \qquad \text{at constant displacement} \tag{58}$$

where U_s is the strain energy and U_k kinetic energy. In many cases, this results in an expression of the form (33):

$$G = G_{\text{static}}\left[1 - k^2\left(\frac{\mathring{a}}{C}\right)^2\right] \tag{59}$$

where $C = \sqrt{E/\rho}$ is the wave speed (ρ is density) and k^2 some dimensionless parameter of the system. As with the static G values, there are many forms of this relationship describing a wide variety of behavior. The kinetic energy usually absorbs some of the released stored energy, and $G < G_{\text{static}}$, but not always. For an infinite plate, there can be no dimensionless length parameters, since only one length a exists, and k^2 must be a constant. This form illustrates some important aspects of dynamic behavior. For a constant G_c criterion (ie, independent of $\mathring{a}$), for example, if $G_{\text{static}} > G_c$, the crack speed increases until $G = G_c$. A limiting crack speed (for which $G = 0$) is given by C/k. In most polymers, $C \sim$ 1700 m/s and observed maximum speeds are about 600 m/s; thus, $k \sim 3$. Few data exist, but it seems that the G_c value for high speed running cracks is somewhat less than the initiation values and that it rises substantially at higher speeds.

Molecular and Microscopic Theories

Kinetics. In fracture-mechanics applications, the material is frequently treated as an isotropic, homogeneous elastic–plastic solid. Taking the proper precautions as described above, such an approach may also be valid for polymers. The principal polymer parameters are the viscoelastic behavior and nature and extent of plastic deformation at the crack tip. However, on a molecular level, ie, 1–10 nm, all polymers are strongly anisotropic and both chain slippage and chain scission can occur. The stress at which local failure occurs varies from a few MPa (for the slip of adjacent segments in an elastomer) to 10–20 GPa (for stress-induced chain scission). Inhomogeneity must also be taken into account. Crystal lamellae (5–50 nm), elastomeric or mineral fillers, and defects (0.2–20 μm) influence stress distribution, crack initiation, propagation, and ultimate behavior.

As opposed to continuum-mechanical theories, molecular theories of fracture recognize the presence of discrete particles or elements forming the material. These theories relate breakage, displacement, and reformation of these elements with deformation, defect development, and fracture of the structured material (4). Some of these theories share the assumption that macroscopic failure is a rate process, that the basic fracture events are controlled by thermally activated breakages of secondary or primary bonds, and that the accumulation of these events leads to crack formation or breakdown of the loaded sample. These events are sometimes defined in a general manner as damage and are not experimentally related to particular morphological changes or to the existence of microcracks. In contrast, morphological theories are based on the physical evidence of molecular damage obtained from spectroscopic or x-ray scattering methods.

From the original concept of flow, ie, the thermally activated movement of molecules across an energy barrier, various fracture theories of solids have emerged (4), eg, considering the reduction of secondary bonds (34). A more general approach, the rule of cumulative damage (35–38), does not explicitly specify the nature of the damage incurred during loading, but attempts to account for the influence of load history on sample strength. The importance of primary bonds for static strength had been deduced very early from the dependences of sample strength on molecular weight and volume concentration of primary bonds (39–42). Later, this idea was developed into kinetic theories (43–48) based on the breakage of primary bonds. The time to fracture under uniaxial stress σ_0 of, for example, PMMA and PS below their glass-transition temperatures could be expressed by an exponential relationship involving three kinetic parameters:

$$t_b = t_0 \exp (U_0 - \gamma\sigma_0)/RT \tag{60}$$

The three parameters, based on the available data, were interpreted as energy for activation of bond scission U_0, inverse of molecular oscillation frequency t_0, and structure-sensitive parameter γ. Experimental data and theoretical curves according to equation 60 are given in Figure 41.

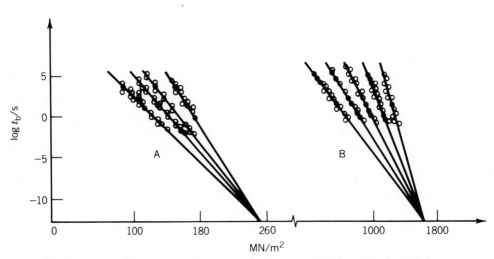

Fig. 41. Times to fracture under constant uniaxial load (4,49). A, Cellulose nitrate (CN) at temperatures of 70, 30, -10, and $-50°$C, reading from left to right. B, Nylon-6 at temperatures of 80, 30, 18, -60, and $-110°$C, reading from left to right. Courtesy of Savitsky and co-workers. To convert MN/m^2 to psi, multiply by 145.

Equation 60 can be considered as a general expression of the kinetic nature of material disintegration. It was proposed that the "rate of local material disintegration" K is proportional to $1/t_b$ and that fracture occurred after a critical concentration of damage had been attained (43–48). Extensive spectroscopical investigations of molecular damage (43–48,50–54) seem to corroborate this theory. However, a more thorough discussion of the meaning of the molecular oscillation frequency, the structure-sensitive parameter γ, and the activation energy term is necessary (4) in view of the given molecular and network structure and sequence of molecular processes. Furthermore, in many cases of creep rupture, more than one mechanism is active. Thus, the energy term U_0 is not *a priori* identical with the thermal dissociation energy.

The above Arrhenius concept is used if the effect of temperature on the time to fracture or on fracture stress is being studied. It gives reasonable representations of the fracture data if there is just one rate-determining mechanism to control crack-growth initiation or slow crack growth effectively.

Chain Scission.   The stress-induced breakdown of practically all polymeric materials is accompanied by chain scission. This phenomenon has been studied in great detail with esr, ir spectroscopy, and molecular weight analysis (4,50–59).

The undisputed fact of the rupture of chains in large numbers during mechanical action is in itself neither proof nor even indication that macroscopic stress relaxation, deformation, and fracture are consequences of chain breakage. The observed total number of broken chains is much too small to account by virtue of their load-carrying capability for the measured reductions in macroscopic stress (4). The most detailed information on the kinetics of stress-induced chain scission has been obtained with highly oriented polyamide fibers. In this material, chain scission leads to the formation of stable, secondary free radicals which can be traced by esr (4,45,55). The first free radicals are observed at a macroscopic strain of about 8–10%; at higher strains, free radicals are produced in increasing numbers. Such strain free-radical histograms reflect the conformation distribution of the broken chains. The primary information to be obtained from them includes the number N_f of free radicals at fracture and the width of the distribution. The volume concentrations of free radicals formed in uniaxially stressed fibers range from $<10^{20}/m^3$ (PMMA, PS), $0.01–2 \times 10^{22}/m^3$ (PE, PET) to about $0.5 \times 10^{24}/m^3$ (nylon-6, natural silk). The surface concentrations of free radicals in ground or milled polymers are on the order of $1–20 \times 10^{16}/m^2$ (4,50,55). If the number of free radicals is taken to correspond to the number of newly formed chain ends, it is much too small to control the deformation process of the fiber and even less its fracture. However, the occurrence of chain scissions is intimately related to the fibril microstructure and it is through the microstructure that fiber strength and radical production are correlated. The more highly oriented nylon-6 fibers are stronger and have a higher modulus of elasticity in amorphous regions and a higher number of extended tie molecules. Because of the higher fiber rigidity, chain scission begins at higher stresses. This should be taken into consideration whenever a failure criterion is to be based on esr data. Since the formation of free radicals in these fibers is a consequence and not an independent cause of fiber deformation, it does not seem advisable to base fracture criteria on the free-radical distribution curves.

This is different, though, in ultrahighly oriented fibers where chain orientation is nearly perfect and the elastic modulus approaches the theoretical value (49,60,61). In PE drawn 150 times, the highest strength values ever determined were measured (49), ie, 6–10 GPa, which is very close to the theoretical strength of a PE chain (4,49,61,62). The same applies for the strength of single-crystal fibers of, for example, polydiacetylene (62).

Other forms of chain scission, not caused by mechanical action, but, for instance, by thermal degradation or chemical attack, evidently influence mechanical properties. A synergistic effect between mechanical stress and rate of chemical (oxidation) reactions has been proved (63).

Structural Implications.   The dominant characteristic of macromolecules is the chain length, which greatly influences polymeric strength and toughness (Fig. 42).

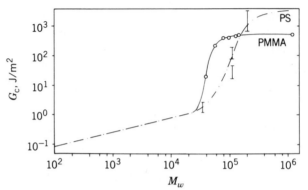

Fig. 42.   Energy-release rate G_c at inception of crack propagation as a function of molecular weight (4). To convert J/m² to ft·lbf/in.², multiply by 476.

Molecular Mobility.   The influence of molecular mobility on fracture strength σ_b is directly shown by temperature dependence (Figs. 4, 14, 23). Obviously, slip and disentanglement of chains and sample ductile deformation are enhanced by chain mobility and usually lead to a reduction of σ_b and K_c. At a given temperature, chemically different molecular chains can have widely different mobilities. If, however, the respective glass-transition temperatures are chosen as reference temperatures, some common characteristics can be established: brittle behavior at $T \ll T_g$ if $E > 4.5$ GPa, semiductile at $T < T_g$ with E between 2 and 4 GPa, tough at $T < T_g$ with $E < 1.5$ GPa, and rubber elastic behavior at $T > T_g$. The transition from brittle to semiductile impact fracture of thermoplastics occurs at the temperature of the β-relaxation peak (64).

Chain Branching, Tacticity, and Crystallinity.   The molecular architecture of a chain influences its mobility and the nature and extent of the crystalline superstructure. As a rule, the degree of crystallinity and the rigidity of a semi-crystalline polymer decrease with chain branching and increase with chain regularity (see also BRANCHED POLYMERS; MORPHOLOGY).

Chain length, architecture, and mobility, but also thermal history, eg, molding conditions, influence the degree of crystallinity (Fig. 43).

Morphology.   Morphology is defined as the supermolecular structure of crystallizable polymers. Such a structure can be formed by crystalline lamellae containing more or less defects, folds, and emanating molecules, spherulites (Fig. 44), shish-kebab or row structures, and extended chain crystallites with interspersed amorphous regions. A semicrystalline polymer is, therefore, heterogeneous on a microscopic level, which leads to large local variations in stress and strain, depending on the size and perfection of crystal lamellae and spherulites and the number and orientation of tie molecules.

Possible failure mechanisms are the ductile or semibrittle failure of the sample as a whole, fracture along spherulite boundaries, and intercrystallite cracking. These mechanisms occur in PE and PP as a function of temperature

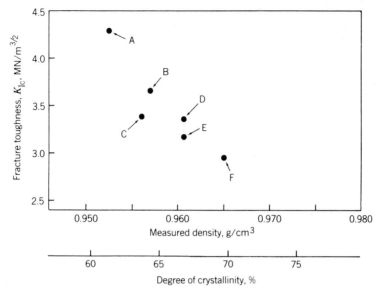

Fig. 43. Plane strain fracture toughness of Hostalen GM5010 Type 2 PE as a function of degree of crystallinity (65): A, quenched plate, $-91°C$, test temperature; B, pipe, $-60°C$; C, pipe, -53 to $-150°C$; D, slowly cooled plate, $-91°C$; E, slowly cooled plate, $-59°C$; F, annealed plate, $-60°C$. Courtesy of *Polymer Engineering and Science*.

Fig. 44. Spherulitic structure of a melt-crystallized PP.

and strain rate (66). The general failure map shown in Figure 45 indicates the influence of molecular and experimental parameters on the three principal failure mechanisms. The separation of crystal lamellae (intercrystallite cracking) occurs preferentially at low temperatures and higher strain rates. The temperature of transition to spherulite boundary cracking is shifted upward with spherulite size and annealing, and downward with parameters reducing crystal perfection, such as higher molecular weight and defective chains. Except for large spherulite size, all these parameters favor ductile fracture behavior at a given temperature.

Cross-link Density. In thermosetting resins and elastomers, the extent of molecular deformation is strongly influenced by cross-linking. Some fundamental relations between chemical structure and physical properties of cross-linked materials have been analyzed (67,68) using epoxy polyaddition. The morphology of

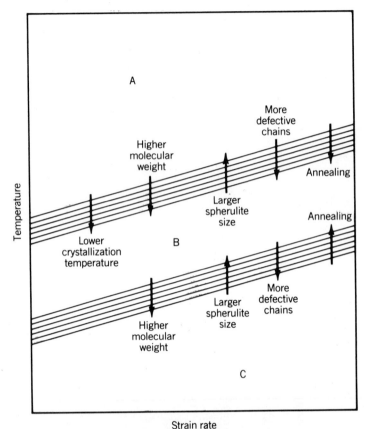

Fig. 45.   Failure map of semicrystalline polymers (66): A, ductile or semibrittle failure; B, spherulite boundary cracking; C, intercrystallite cracking. Courtesy of *Polymer Engineering and Science.*

the polymers is characteristically influenced by the length and nature of the segments between the cross-links and by the cross-linking fragments formed during the polyaddition. Amine-cured epoxy systems show distinct β-relaxations, resulting in greater toughness and flexibility as well as higher glass-transition temperatures than anhydride-cured polymers. The influence of the network structure on the physical properties of the cured specimen is demonstrated by Figure 46, where impact bending strength has been plotted as a function of cross-linking density. The low deformability and high stiffness of densely cross-linked systems results in comparatively low strength, as predicted by fracture mechanics.

In stressed elastomers, two breakdown mechanisms are conceivable: initiation and growth of a cavity in a moderately strained matrix and accelerating, cooperative rupture of interconnected, highly loaded network chains. The second mechanism is more important under experimental conditions, which permit the largest breaking elongation $\lambda_{b\,max}$ to be attained (69). In that case, the quantity $\lambda_{b\,max}$ was expected to be proportional to the inverse square root of the cross-link density ν_e; in fact, an increase of $\lambda_{b\,max}$ with $\nu_e^{-0.5}$ to $\nu_e^{-0.78}$ was found experimentally for a variety of networks. The maximum elongation $\lambda_{b\,max}$ is the only

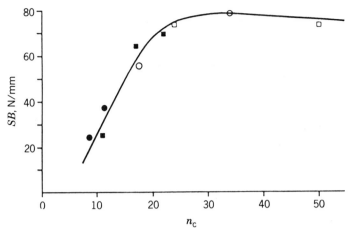

Fig. 46. Dependency of impact-bending resistance (*SB*) of epoxy resins on the mean number n_c of atomic distances between cross-linking points (68). Bisphenol A diglycidyl ether: ○, amine; □, phenol. Epoxy novolac: ●, amine; ■, phenol. To convert N/mm to ppi, divide by 0.175.

feature of large deformation behavior that clearly depends only on network topology (69).

Defects. The influence of defects on the brittle strength and time to failure of load-bearing materials is generally considered as the main source of variation of experimental data (Figs. 3 and 21–23). If it is assumed that the stress concentration caused by a flaw is directly responsible for premature brittle fracture, the size a_m of the largest intrinsic flaw can be determined from the breaking stress σ_b in short-time loading according to

$$a_m = K_{Ic}^2/\sigma_b^2\pi \tag{61}$$

Intrinsic flaw sizes of 50 μm for PMMA and 260 μm for PS were thus obtained (70); values of 62–95 μm were reported for PMMA (71). These values seem very high, since no direct optical observations of such flaws have been reported. It must be concluded, therefore, that the flaw sizes inferred from a tensile experiment or some other mechanism are by no means inherent but the result of crack growth during the experiment.

Starting from this consideration, the fracture surfaces of a large number of well-tested PMMA dumbbell specimens were studied (72). The usual pattern was observed: a semicircular mirror zone (slow crack growth) and a subsequent region covered with parabolae (rapid crack propagation). The center of the mirror regions exhibited semicircular markings with radii of 79 ± 31 μm at 23°C and 108 ± 55 μm at 50°C, which could correspond to former craze zones. The development of an existing defect into an unstable crack in PMMA would then occur in the following manner. A preexisting defect of a few μm or even less develops within Δt_1 into a craze zone fairly rapidly up to the temperature-dependent size of 70–160 μm. This is followed by period Δt_2, the growth-of-the-mirror zone (slow growth up to several 100 μm, depending on stress).

The fracture-mechanics concept that decreasing the flaw size should increase stress-rupture lifetime was confirmed (21). In a controlled experiment, the influence of flaw size was ascertained by purposely adding flaws or removing them by fine melt filtration. For the PE pipe resin investigated, it has been clearly shown that stress-rupture lifetime depends on flaw size and that fine filtration improves pipe performance (21). A similar mechanism was concluded to be in operation in several other polymers (73).

Intrinsic Crazing and Stress-whitening.   The initiation of ordinary crazing is generally controlled by foreign particles and surface grooves, which both act as stress concentrators. Under these conditions, a small number of extrinsic crazes grow in amorphous single-phase polymers and usually cause premature rupture.

A large number of crazes distributed throughout the sample volume are observed in multiphase systems such as rubber-modified polymers where numerous rubber particles act as stress concentrators. The intensive light scattering from the multitude of crazes gives rise to the phenomenon of stress-whitening. This phenomenon is also observed in semicrystalline polymers where the transformation of spherulitic into fibrillar microstructure is usually accompanied by intensive void formation.

Stress-whitening caused by intrinsic crazing has been reported in a few cases of amorphous single-phase polymers (74–78). Strong evidence indicates that the initiation of intrinsic crazes is governed by a temperature- and stress-activated instability of the entanglement network frozen-in during the glass transition (79).

Environmental Parameters.   In a wider sense, all parameters acting on a specimen from outside are environmental. In view of their evident importance, however, stress, strain, and temperature are usually considered independently. The remaining parameters can be classified as physical (electromagnetic radiation, particle irradiation), physicochemical (physical action of liquids and gases), and chemical (oxidation, other forms of chemical attack (see also DEGRADATION; RADIATION-RESISTANT POLYMERS; WEATHERING). In general, chemical degradation and mechanical action do not influence each other. Some notable exceptions include ozone degradation of elastomers and oxidative degradation of some polyolefins.

The action of mechanical forces of many thermoplastic polymers in contact with some liquids or gases has a strong synergistic effect, namely, the formation of environmental stress cracks (ESC). Only the simultaneous action of stresses and physicochemical environment leads to this phenomenon. The origin of ESC is a stress-enhanced sorption of the environmental agent by the polymer, which in turn greatly facilitates crazing and eventually cracking. Extensive investigations (80,81) have shown that this sorption leads to swelling and plasticization of the contacted (surface) zone of the polymer and a notable reduction of the critical strains ε_c of craze initiation as compared with those in air or an inert medium (dry crazing) (4,7,9,80,81). For amorphous polymers, a strong correlation between ε_c and the difference $|\delta_s - \delta_p|$ of the solubility parameters of environment (s) and polymer (p) has been found (9,80,81). The smallest values of ε_c are usually observed for those polymer–solvent pairs where the equilibrium solubility S_v shows a maximum (7,9,80,81).

Environmental agents also influence the later stages of stress cracking, ie, craze growth and breakdown, resulting in crack formation. The fracture-mechanics concept has proved to be useful to explain quantitatively the kinetics of crack growth in a liquid environment if the wetting, spreading, flow, and diffusion behavior of the liquid at the crack tip and within the capillaries, opened through the crazes, are taken into account (9,81–83).

An example of ESC phenomenon in semicrystalline polymers is the appearance of three different zones of behavior in the static loading of LDPE in an active environment (Jaixa Triclean) (84,85) (Fig. 47). Possible mechanisms for the drastic acceleration of sample breakdown at intermediate K values include an increase in the rate of rearrangements due to diffusional interpenetration and plasticization of the polymer matrix by the liquid, and a reduction in critical stress for craze initiation and of the resistance of an initiated craze due to a reduced extensibility of craze material.

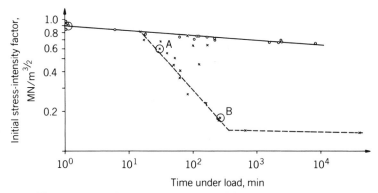

Fig. 47. Environmental stress cracking (ESC) of LDPE: ——, air; ----, ESC; ○, microphoto. Microphoto A refers to Figure 48; B refers to Figure 49.

Experiments have shown (84,85) that the time-to-fracture t_b in air increases exponentially with decreasing K_{10}, as also observed for glassy polymers (PMMA, PC) and that fracture in air occurs in the direction of highest tensile stresses in a brittle manner. In the presence of the stress-cracking agent (SCA), three zones of behavior are distinguished: an incubation period t_i of $\simeq$ up to 15 min at high values of K during which the SCA has no apparent effect; at times $t_b > t_i$ and moderately high values of K, considerable ductile deformation and crack branching occur (Fig. 48); and after long times and at accordingly low K values, fracture is brittle again (Fig. 49). Presoaking of specimens in SCA has no effect and a delayed application to a loaded specimen leads to an increase in bending rate after immersion in SCA.

The morphological features shown in Figure 49 agree to some extent with those of LDPE (double-edge notch) specimens (86). High stress failure in air occurs by crack propagation involving the formation of striations by local cold drawing and necking. High stress ESC by nucleation and growth of holes is followed by viscoelastic deformation, and low stress ESC by interlamellar failure without cold drawing (86).

The three-stage fatigue behavior of LDPE exhibits very similar characteristics (87): brittle trans-spherulitic failure at low K, microductile failure at high K, and a transition zone at intermediate K values. The ESC experiments described

Fig. 48. Ductile deformation and crack branching in high stress ESC of LDPE.

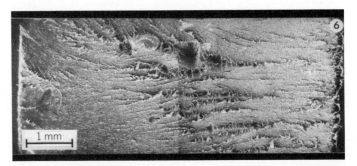

Fig. 49. Interlamellar failure at low stress ESC of LDPE (85).

by Figures 47–49 can be understood on the basis of a stress-activated diffusion of stress-cracking agent into interlamellar regions. The mode of ESC failure of semicrystalline polymers is determined by the state of stress and material resistance. It occurs by local drawing of the practically unplasticized sample ($t_b < t_i$) in a mixed mode involving large-scale plastic deformation including void formation and multiple cracking at high stress ESC ($t_b > t_i$) and by interlamellar failure (chain disentanglement or rupture) at low stress ESC ($t_b \gg t_i$).

The environmental problems of different classes of polymers are given in Table 1 (88,97–99).

Characterization and Test Methods

Fracture as a phenomenon is of great importance and its occurrence must be controlled. Fracture mechanics has been utilized mainly as an analytical method to characterize materials and as a design method to ensure component integrity via the relation

$$K_{Ic}^2 = EG_c = Y^2 \cdot \sigma^2 a \tag{62}$$

Table 1. Problematic Environments of Polymers[a]

Environment	Polymer class	Ref.
organic liquids and vapors	amorphous polymers	89
hydrocarbons, tensioactive agents (detergents)	polyolefins	90
alcohols, hydrocarbons	poly(methyl methacrylate)	91
humidity, acid- or base-catalyzed	polymers with heteroatoms in backbone, ie, polysaccharides, polyethers, polyesters, polyamides, polyurethanes	92
Lewis acids, zinc, copper	polyamides	93
ozone	unsaturated rubbers	94
sunlight	polymers with photosensitive groups, ie, polyacetals, polysulfone, polyphenylene	95
biodegradation	short-chain polymers, nitrogen-containing polymers, polyesters	96

[a] Ref. 88–96.

Toughness and related properties are characterized by K_{Ic}, G_c, or J_c. The design method consists of defining either σ or a when the other is known for some particular material property.

The characterization of toughness or strength is important for many polymers, since they are produced for a particular performance, and although it is important to know absolute values, it is frequently difficult to compare two similar materials. Performances are often very different in, for example, impact testing and environmental crack growth, and the apparent suitability of a material for some application can be strongly influenced by the imposed criteria. For example, a criterion for a material may be suitable for pipes on the basis of resistance to slow crack growth, since it is an important factor in determining long-term life. However, good impact resistance is also required to withstand accidental damage. Usually, a compromise is the result, and balancing the various factors constitutes skill in material development and selection.

Crack Initiation.   In the basic test of fracture mechanics, a cracked specimen is loaded slowly and the onset of crack growth is observed. All fracture-mechanics tests use prenotched specimens, since the analysis is concerned with the behavior of notches or cracks. The nature of the initial notch is of great importance since the measured toughness can be sensitive to its sharpness. Ideally, a natural crack should be grown in a prenotching test, often at low temperatures, but these tests are frequently not easy to perform. The notch must be carefully formed with a very sharp cutter to avoid crazing, crack branching, or blunting; fatiguing is time-consuming.

Convenient test geometries are the single-edge notched plate in tension and bending and the compact tension shown in Figure 50. The specimens are loaded slowly in a testing machine and crack initiation is observed. This can be rate-dependent because of viscoelastic effects, and instability does not always follow. The load-deflection diagram continues to rise after initiation. Close observation of the crack tip is desirable and initiation can be marked on the load record. The two-tension geometries usually give unstable behavior after some stable growth, but bending frequently gives extensive stable crack growth.

A series of specimens of various crack lengths can be tested to determine

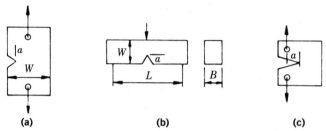

(a) **(b)** **(c)**

Fig. 50. Testing geometries: (**a**) single-edge notch tension (SENT); (**b**) single-edge notch three-point bending (SENB); (**c**) compact tension (CT).

the gross stress σ at crack initiation. A graph of $\sigma^2 Y^2$ versus $1/a$ (or similar form) may be plotted to obtain K_c^2 from the slope. Such data, given in Figure 51 for a polyacetal, demonstrate the high quality that can be achieved with good notching. It is important to measure the yield stress of the material under the same conditions in order to check the size criteria:

$$B > 2.5 \left(\frac{K_{\text{Ic}}}{\sigma_y} \right)^2 , \quad W > 2B \tag{63}$$

This is valid only for three-point bend and compact tension; for tension, extrapolation versus B^{-1} may be used. If these conditions are not met, sizes must be increased or the temperature lowered (to increase σ_y) to achieve a valid plane-strain value. Nonlinearity in the $\sigma^2 Y^2$ versus a^{-1} graph sometimes occurs because of the plastic-zone effect; the correction given in equations 48 and 49 may be incorporated.

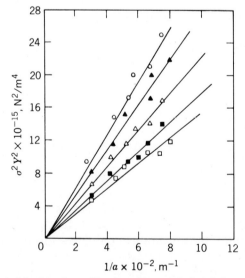

Fig. 51. Crack-initiation data. Single-edge notch bend data for polyacetal at temperatures of $+20$ to $-60°C$. $T = \square$, $+20°C$; ■, $0°C$; $\triangle$, $20°C$; ▲, $-40°C$; $\bigcirc$, $-60°C$.

When the size criteria for LEFM cannot be met, it may be possible to use a J test (eqs. 48 and 49). Initiation is defined by the multiple-specimen method. Data of J vs Δa are shown in Figure 52 for polypropylene. Typical values of K_{Ic} and G_c (or J_c) are given in Table 2 for several polymers at 20°C.

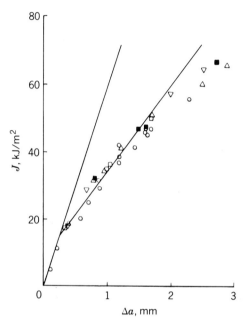

Fig. 52.   Crack-initiation data for polypropylene. J versus Δa at 20°C with various a/W values, $\sigma_y = 30$ MPa (300 bar), $W = 40$ mm. a/W: ○, 0.56; △, 0.50; □, 0.43; ■, 0.38; ▽, 0.31. To convert kJ/m² to ft·lbf/in.², divide by 2.10.

Table 2. Toughness Values at 20°C

Material	K_{Ic}, MPa$\sqrt{m}$[a]	G_c or J_c, kJ/m²[b]
PMMA	1.7	0.5
PS	1.0	0.3
high impact polystyrene	2.0	2.0
epoxy and polyester	0.6	0.1
polyethylene		
high density	2.1	2
medium density	5.0	28
PP	4.6	15.5
PC	2.2	1.5
ABS	2.8	3.8
PVC	2.5	1.7
reinforced PVC	3.2	3.8
nylon	2.8	2.6
reinforced nylon	4.9	10.0

[a] To convert MPa to bar, multiply by 10.
[b] To convert kJ/m² to ft·lbf/in.², divide by 2.10.

Slow Crack Growth. In these tests, K_c–$\mathring{a}$ curves are determined as described above. It is usually more convenient to measure the crack speeds in specimens in which K does not vary with a for a given load or loading rate. The double-torsion and tapered-cantilever beam tests are illustrated in Figure 53. In the first,

$$K^2 = \frac{3(1 + \nu)}{2} \cdot \frac{P^2 l^2}{B^2 D^3} \tag{64}$$

where l is the distance from the center of the specimen to the point of load support.

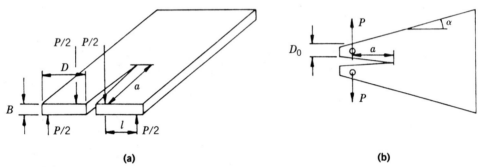

(a)   **(b)**

Fig. 53. Crack-growth testing geometries, constant K tests: (a) double tension; (b) tapered cantilever.

For the tapered beam,

$$K^2 \sim \frac{45P^2}{B^2 D_0^3} \text{ for a tape angle } \alpha \sim 11° \tag{65}$$

Under dead loads (constant P), the constant K values give fixed crack speeds which can be measured over a range of crack growth; the K_c–$\mathring{a}$ curve is determined by changing the load. Environmental data are obtained by conducting the test with the specimen immersed in the environment. Side grooves are often used in these tests to guide the crack; appropriate modifications are made to the formula for K. Typical data for PE in water are given in Figure 54.

Impact Tests. Impact is most conveniently tested in the three-point bend. Load measurement presents problems because of dynamic effects; a typical curve taken from a falling weight test is shown in Figure 55. The oscillations in the signal are apparent, but by careful analysis it is possible to obtain useful K_c values. Data taken from such a test are plotted in the usual $\sigma^2 Y^2$ vs a^{-1} basis in Figure 56. Energy measurements are easier; some data for crack initiation are shown in Figure 57 (see also IMPACT RESISTANCE).

Design Methods. Applications of fracture mechanics in design is based on the assumption that all bodies contain imperfections. The conditions are defined as those under which such flaws make the component unsafe or unserviceable.

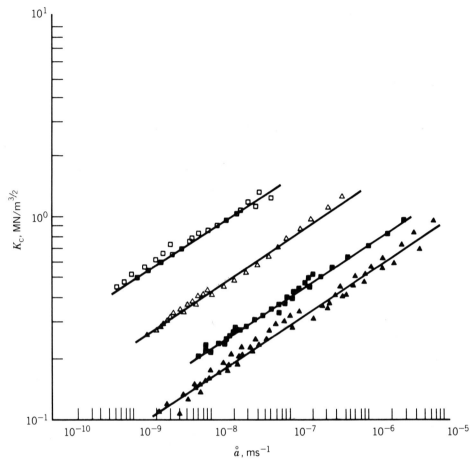

Fig. 54. K_c–$\mathring{a}$ curves for PE in distilled water at different temperatures. Slopes = 0.25. $T = \square$, $+19°C$; $\triangle$, $+40°C$; $\blacksquare$, $+60°C$; $\blacktriangle$, $+75°C$.

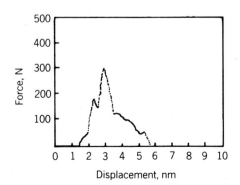

Fig. 55. Load–time curve in impact test. To convert N to dyn, multiply by 10^5.

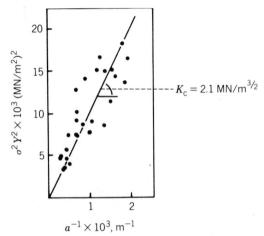

Fig. 56. Crack-initiation data in impact-load analysis.

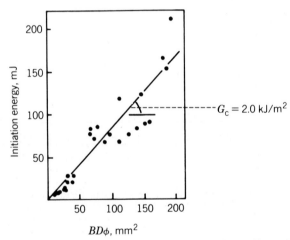

Fig. 57. Crack-initiation data in impact energy analysis. To convert J to cal, divide by 4.184; to convert kJ/m² to ft·lbf/in.², divide by 2.10.

The nature of the original flaw is important, since, if it is very small, most of the life will be spent in an initiation phase and often this is controlled by yielding or crazing mechanisms. Fracture mechanics usually controls the behavior of flaws greater than the plastic-zone size $\frac{1}{2\pi}\left(\frac{K_c}{\sigma_y}\right)^2$. Such flaws arise from accidental damage, substantial inclusions, or voids. If the loading on a component is known and slow crack-growth data are available, it is possible to predict the tolerable flaw size to avoid leakage or instability. An inspection procedure may be instituted to avoid flaws greater than this value. The reverse procedure may also be carried out, since, if a flaw size is known, a safe working stress may be computed. Such calculations are of importance in defining the residual life of a component.

Impact-toughness values may also be used, for example, to define the magnitude of impact loads which can be withstood.

Much of the utility of these methods depends on the criteria adopted for successful operation. In the gas industry, this method has been used with success.

BIBLIOGRAPHY

"Fracture" in *EPST* 1st ed., Vol. 7, pp. 261–361; "Long-Term Phenomena," pp. 261–291, by John B. Howard, Bell Telephone Laboratories; "Short-Term Phenomena," pp. 292–361, by P. I. Vincent, Imperial Chemical Laboratories.

1. S. Hashemi, Ph.D. thesis, Imperial College of Science and Technology, London, 1984.
2. N. Brown and I. M. Ward, *J. Mater. Sci.* **18,** 1405 (1983).
3. E. Gaube and W. Müller, *3R Internat.* **23,** 236 (1984).
4. H. H. Kausch, *Polymer Fracture,* 2nd ed., Springer-Verlag, Berlin, 1986.
5. L. R. G. Treloar, *The Physics of Rubber Elasticity,* 2nd ed., Clarendon Press, Oxford, UK, 1958.
6. C. C. Hsiao and J. A. Sauer, *J. Appl. Phys.* **21,** 1071 (1950).
7. H. H. Kausch, ed., *Advances in Polymer Science,* Vol. 52/53: *Crazing in Polymers,* Springer-Verlag, Berlin, 1983.
8. M. Dettenmaier in Ref. 7, p. 52.
9. E. J. Kramer in E. H. Andrews, ed., *Developments in Polymer Fracture,* Vol. 1, Applied Science Publishers, Ltd., London, 1979, p. 55.
10. J. A. Sauer in Ref. 7, p. 169.
11. S. S. Sternstein and L. Ongchin, *Polym. Prepr. Am. Chem. Soc. Div. Polym. Chem.* **10,** 1117 (1969).
12. N. W. Tschoegl, *J. Polym. Sci. Part C* **32,** 239 (1971).
13. I. M. Ward, *Mechanical Properties of Solid Polymers,* 2nd ed., John Wiley & Sons, Inc., New York, 1983.
14. R. Bardenheier, *Mechanisches Versagen von Polymerwerkstoffen,* Carl Hanser Verlag, Munich, 1982.
15. H. H. Kausch, *Materialprüf.* **6,** 246 (1964).
16. W. Feller, *An Introduction to Probability Theory and Its Applications,* John Wiley & Sons, Inc., New York, 1971; L. Sachs, *Angewandte Statistik,* Springer-Verlag, Berlin, 1984.
17. S. Kawabata and P. J. Blatz, *Rubber Chem. Technol.* **39,** 39, 923 (1966).
18. B. D. Coleman and D. W. Marquardt, *J. Appl. Phys.* **28,** 1065 (1957); S. L. Phoenix, *Text. Res. J.* **49,** 407 (1979).
19. P. Stockmayer and S. Wintergerst, *3R Internat.* **20,** 274 (1981).
20. H. H. Kausch, *Materialprüf.* **20,** 22 (1978).
21. G. Sandilands, P. Kalman, J. Bowman, and M. Bevis, *Polym. Commun.* **24,** 273 (1983).
22. R. A. Hunt and L. N. McCartney, *Int. J. Fract.* **15,** 365 (1979).
23. P. I. Vincent, *J. Appl. Phys.* **13,** 578 (1962).
24. J. G. Williams, *Fracture Mechanics of Polymers,* Ellis Horwood, Ltd., Chichester, UK, 1984.
25. C. Gurney and J. Hunt, *Proc. R. Soc. London, Ser. A* **299** (1967).
26. R. S. Rivlin and A. G. Thomas, *J. Polym. Sci.* **10,** (1953).
27. W. F. Brown and J. E. Srawley, *ASTM STP 410,* The American Society for Testing and Materials, Philadelphia, Pa., 1966.
28. D. P. Rooke and D. J. Cartwright, *Compendium of Stress Intensity Factors,* Her Majesty's Stationery Office (HMSO), London, 1976.
29. *ASTM Standards 31* (1969) and *ASTM STP 463* (1970), The American Society for Testing and Materials, Philadelphia, Pa.
30. Ref. 7, Chapt. 1.
31. ASTM Committee E-24, Task Group E24.01.09, *Recommended Procedure for J_c Determination,* Task Group Meeting in Norfolk, Va., Mar. 1977, The American Society for Testing and Materials, Philadelphia, Pa., 1977.

32. E. Plati and J. G. Williams, *Polym. Eng. Sci.* **15**, 470 (1975).

33. P. K. Roberts and A. A. Wells, *Engineering* **178**, 820 (1954).

34. A. Tobolsky and H. Eyring, *J. Chem. Phys.* **11**, 125 (1943).

35. J. Bailey, *Glass Ind.* **20**, 95 (1939).

36. M. A. Miner, *J. Appl. Mech.* **12**, A-159 (1945).

37. D. C. Prevorsek, G. E. R. Lamb, and M. L. Brooks, *Polym. Eng. Sci.* **10**, 1 (1967).

38. G. B. McKenna and R. W. Penn, *Polymer* **21**, 213 (1980).

39. P. J. Flory, *J. Am. Chem. Soc.* **67**, 2048 (1945).

40. R. N. Haward, *The Strength of Plastics and Glass,* Cleaver-Hume Press, London, 1949, Chapt. II.

41. L. Holliday and W. A. Holmes-Walker, *J. Appl. Polym. Sci.* **16**, 139 (1972).

42. P. I. Vincent, *Polymer* **13**, 557 (1972).

43. S. N. Zhurkov and B. N. Narzullayev, *Zh. Tekh. Fiz.* **23**, 1677 (1953).

44. S. N. Zhurkov and E. Ye. Tomashevskii, *Zh. Tekh. Phys.* **25**, 66 (1955).

45. S. N. Zhurkov, V. A. Zakrevskii, V. E. Korsukov, and V. S. Kuksenko, *Fiz. Tverd. Tela* **13**, 2004 (1971); *Sov. Phys. Solid State* **13**, 1680 (1972); *J. Polym. Sci. Part A-2* **10**, 1509 (1972).

46. F. Bueche, *J. Appl. Phys.* **26**, 1133 (1955).

47. *Ibid.,* **28**, 784 (1957).

48. *Ibid.,* **29**, 1231 (1958).

49. A. V. Savitsky, I. A. Gorshkova, I. L. Frolova, G. N. Shmikk, and A. F. Ioffe, *Polym. Bull.* **12**, 195 (1984).

50. V. R. Regel, A. I. Slutsker, and E. E. Tomashevskii, *Kinetic Nature of Durability of Solids* (in Russian), Isdat. Nauka, Moscow, 1974.

51. K. J. Friedland, V. A. Marichin, L. P. Mjasnikova, and V. I. Vettegren, *J. Polym. Sci. Polym. Symp.* **58**, 185 (1977).

52. V. A. Zakrevskii and V. A. Pakhotin, *Sov. Phys. Solid State* **20**, 214 (1978).

53. A. E. Tshmel, V. I. Vettegren, and V. M. Zolotarev, *J. Macromol. Sci. Phys.* **21**, 243 (1982).

54. S. N. Zhurkov, V. S. Kuksenko, and V. A. Petrov, *Theor. Appl. Fract. Mech.* **1**, 271 (1984).

55. K. L. DeVries and D. K. Roylance, *Prog. Solid State Chem.* **8** (1973).

56. J. Sohma and M. Sakaguchi, *Adv. Polym. Sci.* **20**, 109 (1976).

57. E. H. Andrews and P. Reed in Ref. 9, p. 17.

58. V. S. Kuksenko and V. P. Tamusz, *Fracture Micromechanics of Polymer Materials,* Martinus Nijhoff Publishers, The Hague, 1981.

59. O. Frank, Ph.D. Thesis, Technische Hochschule, Darmstadt, FRG, 1984.

60. P. Smith, P. J. Lemstra, and H. C. Booij, *J. Polym. Sci. Polym. Phys. Ed.* **19**, 877 (1981).

61. T. Kanamoto, A. Tsuruta, K. Tanaka, M. Takeda, and R. S. Porter, *Polym. J.* **15**, 327 (1983).

62. C. Galiotis, R. T. Read, P. H. J. Yeung, R. J. Young, I. F. Chalmers, and D. Bloor, *J. Polym. Sci. Polym. Phys. Ed.* **22**, 1589 (1984).

63. A. A. Popov and G. E. Zaikov, *Rev. Macromol. Chem. Phys.* **23**(1), 1 (1983).

64. F. Ramsteiner, *Kunststoffe* **73**, 148 (1983).

65. J. F. Mandell, D. R. Roberts, and F. J. McGarry, *Polym. Eng. Sci.* **23**, 404 (1983).

66. J. M. Schultz, *Polym. Eng. Sci.* **24**, 770 (1984).

67. H. Batzer, F. Lohse, and R. Schmid, *Angew. Makromol. Chem.* **29/30**, 349 (1973).

68. M. Fischer, F. Lohse, and R. Schmid, *Makromol. Chem.* **181**, 1251 (1980).

69. T. L. Smith and W. H. Chu, *J. Polym. Sci. Part A-2* **10**, 133 (1972).

70. L. J. Broutman and F. J. McGarry, *J. Appl. Polym. Sci.* **9**, 589 (1965).

71. J. P. Berry in B. Rosen, ed., *Fracture Processes in Polymeric Solids,* Wiley-Interscience, New York, 1964, p. 157.

72. W. Döll and L. Könczöl, *Kunststoffe* **70**, 563 (1980).

73. S. Hashemi and J. G. Williams, *J. Mater. Sci.* **20**, 4202 (1985).

74. G. Rehage and G. Goldbach, *Angew. Makromol. Chem.* **1**, 125 (1967).

75. J. Murray and D. Hull, *J. Polym. Sci. Part A-2* **8**, 1521 (1970).

76. D. L. G. Lainchbury and M. Bevis, *J. Mater. Sci.* **11**, 2235 (1976).

77. A. S. Argon and J. G. Hannoosh, *Philos. Mag.* **36**, 1195 (1977).

78. M. Dettenmaier and H. H. Kausch, *Polymer* **21**, 1232 (1980).

79. Ref. 7, p. 83.

80. R. P. Kambour, *Polymer Commun.* **24**, 292 (1983); R. P. Kambour and E. A. Farraye, *Polymer Commun.* **25**, 357 (1984).
81. E. H. Andrews and L. Bevan, *Polymer* **13**, 337 (1972).
82. G. W. Weidmann and J. G. Williams, *Polymer* **16**, 921 (1975).
83. L. Morbitzer, *Colloid Polym. Sci.* **259**, 832 (1981).
84. M. E. R. Shanahan and J. Schultz, *J. Polym. Sci. Polym. Phys. Ed.* **16**, 803 (1978); **18**, 1747 (1980).
85. H. H. Kausch and K. Jud, *International Conference on Fracture*, Cannes, Apr. 1981.
86. S. Bandyopadhyay and H. R. Brown, *Polymer* **19**, 589 (1978); **22**, 245 (1981).
87. E. H. Andrews and B. J. Walker, *Proc. R. Soc. London Series A* **325**, 57 (1971).
88. W. Brostow and R. D. Corneliussen, eds., *Failure of Plastics*, Carl Hanser Verlag, Munich, 1986, Chapts. 15–17.
89. J. Verdu, *Vieillissement des Plastiques*, AFNOR Technique ed., Association Française de Normalisation (AFNOR), Paris, 1984, p. 77.
90. *Ibid.*, p. 96.
91. *Ibid.*, p. 67.
92. *Ibid.*, p. 243.
93. *Ibid.*, p. 242.
94. B. Ranby and J. F. Rabek, *Photodegradation, Photooxidation, and Photostabilization of Polymers*, John Wiley & Sons, Inc., New York, p. 253.
95. *Ibid.*, p. 120.
96. Ref. 89, p. 250.
97. D. V. Rosato and R. T. Schwartz, *Environmental Effects on Polymeric Materials*, Wiley-Interscience, New York, 1968; N. Grassie, *Developments in Polymer Degradation*, Vol. 1, Applied Science Publishers, Ltd., London, 1977.
98. J. Verdu, *Vieillissement des Plastiques*, AFNOR Technique ed., Association Française de Normalisation (AFNOR), Paris, 1984.
99. B. Ranby and J. F. Rabek, *Photodegradation, Photooxidation and Photostabilization of Polymers*, John Wiley & Sons, Inc., New York, 1975.

H. H. KAUSCH
École Polytechnique Fédérale de Lausanne

J. G. WILLIAMS
Imperial College of Science and Technology

HARDNESS

Hardness, or the resistance to local deformation, defines the ease with which a material can be indented, drilled, sawed, or abraded (see also ABRASION AND WEAR). It involves a complex combination of properties (elastic modulus, yield strength, strain-hardening capacity) (1,2). In rubberlike materials, the elastic properties are important (3). With most elastomers, the indentation disappears when the force is removed. Therefore, hardness must be measured with the force applied. On the other hand, although the elastic moduli of polymers are relatively large, the range over which polymers deform elastically is relatively small. Consequently, the deformation of polymers takes place predominantly outside the elastic range, with considerable plastic or permanent deformation. For this reason the hardness of polymers is bound up primarily with plastic properties and only secondarily with elastic properties. In some cases, particularly in measurements of dynamic hardness, the elastic deformation may be as important as the plastic properties (3). Furthermore, polymeric materials differ in their susceptibility to plastic and viscoelastic deformation. Thus indentation occurs at a decreasing rate during application of the force, leading to a local creep effect. The material consequently recovers upon the removal of force at a decreasing rate. Hence the duration of the application of the force must be specified.

Hardness is measured by static penetration of the material with a standard indenter exerting a known force, by dynamic rebound of a standard indenter of known mass dropped from a standard height, or by scratching with a standard pointed tool exerting a known force.

Hardness can be important for assessing product quality and uniformity and is used in identification, classification, and quality control. Hardness tests provide a rapid evaluation of variations in mechanical properties affected by changes in chemical or processing conditions, heat treatment, microstructure, and aging. Morphological and textural changes in crystalline polymers can be detected by hardness tests (4,5). Inasmuch as small loads (<0.1 N or 10^4 dyn) are often used, hardness (H) is often referred to as microhardness (MH), and this designation is seen for some of the data presented in this article.

Mechanics of Indentation

Static indentation, the method most widely used in determining polymer hardness, involves the formation of a permanent surface impression. Hardness is conveniently computed by dividing the peak contact load P by the projected area of impression. Hardness so defined can be considered as indicative of a characteristic irreversible deformation. The indentation stresses, although concentrated in the contact area, may extend into the elastic matrix. The material under the indenter thus contains a zone of severe plastic deformation (ca 4–5 times the penetration h of the indenter into the surface) surrounded by a larger zone of elastic deformation (6). Hardness is, therefore, an elastic-plastic property. To determine the reversible and irreversible contributions to deformation, the elastic recovery after unloading is measured (7). The extremes of depth recovery are exhibited by soft metals in which it is negligible, and elastic rubber in which it is complete. The well-known viscoelastic behavior of polymers is intermediate. Pyramidal indenters exert a pressure which is nearly independent of depth of penetration. They are less affected by elastic release than other indenters (see Fig. 1). The following effects can be distinguished during an indentation cycle:

An *elastic deformation* leads to an instant recovery (Fig. 1, from B to C) upon removal of the stress. This deformation alters the bond angles in the polymer chains.

A *permanent plastic deformation, C,* is governed by the arrangement and structure of microcrystals and their connections by tie molecules and entanglements. Permanent indentation with irreversible rearrangement of molecules is also possible in amorphous polymers below the glass-transition temperature.

A *time-dependent, t, contribution* to plastic deformation during loading can be described by H $= At^{-k}$, where A is a coefficient depending on temperature

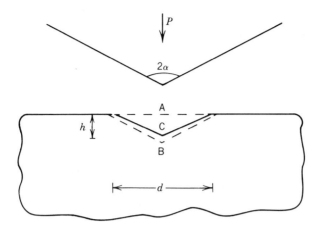

Fig. 1.   Penetration of an indenter with a semiangle α of 74° below the specimen surface at A, zero load; B, maximum load; and C, complete unload. The characteristic contact diameter is given by d, and the residual penetration depth is given by h.

and loading stress and k is a constant which measures the rate of local creep and depends on the microstructure.

A *long delayed recovery of indentation* after load removal is produced by the viscoelastic relaxation of the material.

Correlation of Hardness with Other Mechanical Properties

The hardness value for a pyramid indenter producing plastic flow is approximately three times the yield stress, H $\sim$ 3Y (1). This is the basic relation between indentation hardness and bulk properties. It is, however, only applicable to an ideally plastic solid showing no elastic strains. The correlation between H and Y is given in Figure 2 for linear polyethylene samples with different morphologies. The lower hardness values of 30–45 MPa (3.06–4.6 kgf/mm²) obtained for melt-crystallized material fall below the H/Y ca 3 value, which may be related to a lower stiff–compliant ratio for these lamellar structures (8). Polyethylene annealed at ca 130°C gives an H/Y ratio closer to that predicted for an ideal plastic solid. Thus plastics showing a high stiff–compliant ratio (high crystallinity) approach Tabor's relation, whereas those with a low stiff–compliant ratio (low crystallinity) deviate from classical plasticity theory.

An empirical correlation of the type H = aE^b (where a and b are constants) has been found for linear polyethylene with different crystallinities (8). It is important to realize that H, the plastic deformation of crystals at high strains, depends primarily on crystal thickness and perfection, whereas in the case of the modulus, the principal role is played by the amorphous layer reinforced by tie molecules, elastically deformed at low strains. The hardness increase modulated by the chain extension of the crystals usually parallels the increase in stiffness. The correlation seems, however, to fail for very stiff chain-extended materials.

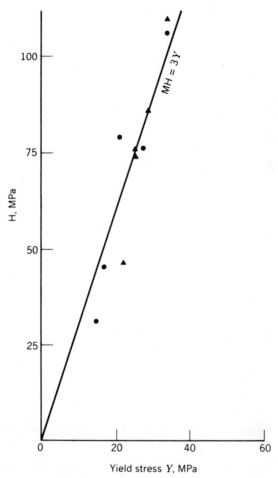

Fig. 2. Correlation between hardness at 6 s (loading time) and yield stress of polyethylene samples with different crystallinities. ●, $\overline{M}_n$ = 170,000; ▲, $\overline{M}_n$ = 2 × 10⁶ (8). The straight line follows Tabor's relation (1). To convert MPa to kgf/mm², divide by 9.81.

Commercial Polymers

Hardness should be intimately related to the critical stresses required for plastic deformation of the polymer crystal. Thus applied compressive and shear stress must overcome the cohesive force between the molecules. Calculated and experimental values of crystal hardness for various polymers are listed in Table 1. The discrepancies are due to the complex microstructure of the polymer solids.

Hardness is a reliable index of water diffusion in certain polymers (9). The striking influence of relative humidity on hardness is shown in Figure 3. Dry samples were stored at 100°C under a pressure of 2 kPa (15 mm Hg) for 12 h, quenched, and tested. Samples with 100% humidity were immersed in distilled

Table 1. Hardness Values[a] Derived from the Ideal Shear Strength of a van der Waals Solid and Experimental Vickers Hardness Data

Polymer	Calculated H, MPa[b]	Experimental H, MPa[b]
polyethylene	150–170	50–125
nylon-6	113	90
polystyrene[c]	94	129–170
polyoxyethylene	159	170
polyoxymethylene	93	70
selenium	37	
poly(trans-1,4-isoprene)	87	

[a] Refs. 8 and 9.
[b] 9.807 MPa = 9.807 N/mm² = kgf/mm².
[c] Isotactic.

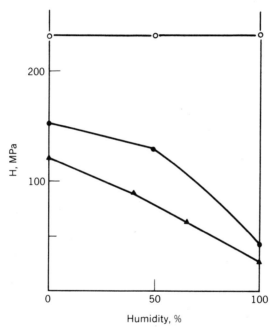

Fig. 3. Vickers hardness as a function of relative humidity for ○, hardened epoxy resin (amide); ●, partially cross-linked amide epoxy resin; and ▲, cellulose acetate (9). To convert MPa to kgf/mm², divide by 9.81.

water for 12 h and tested after drying. Polyamides, epoxy resins, and cellulose acetate are softened rapidly by water. Temperature also influences hardness as shown in Figure 4 (9). It is evident that hardness must be evaluated at controlled temperatures. Shear modulus is similarly affected (9,10). Thermal expansion in crystalline regions may enlarge the chain cross-section and reduce the cohesion energy, leading to a gradual softening with an increase in temperature. An exponential law has been proposed for polyethylene hardness (8).

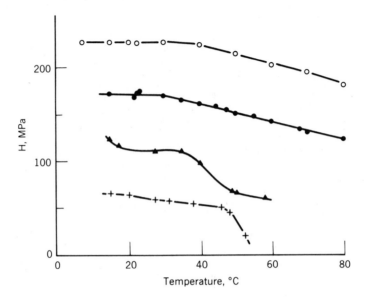

Fig. 4.   Influence of temperature on the Vickers hardness of ○, hardened amide epoxy resin; ●, polystyrene; ▲, acetal resin, and +, polyamide (9).

Microstructure and Hardness

Macroscopic unoriented polymers show a lamellar morphology (qv), with crystalline lamellae intercalated by amorphous regions. Semicrystalline plastics such as TPX poly(4-methyl-1-pentene) and polyethylene can be regarded as composite materials containing distinct compliant (disordered) and hard crystalline elements (5,11). In polyolefins, the two phases H_c (crystalline) and H_a (amorphous) differ in hardness by at least two orders of magnitude. The irreversible deformation modes of crystalline lamellae involve chain slip and tilt leading to lamellar shear, with crystal destruction at large strains (12). The predicted hardness in linear polyethylene is shown in Figure 5 as a function of crystallinity α for the ratio $H_c/H_a = 120$ of the hardness values of the components. The plots correspond to the parallel (eq. 1) and series (eq. 2) models, respectively:

$$H = \alpha H_c + (1 - \alpha)H_a \tag{1}$$

$$\frac{1}{H} = \frac{\alpha}{H_c} + \frac{1 - \alpha}{H_a} \tag{2}$$

For crystallinities above 75% the experimental results follow the parallel model, which also applies to blends of linear and branched polyethylenes (13). However, for decreasing values of α in Figure 5, there is a deviation from the parallel model due to the fact that the upper bound, H_c, is not constant but depends on crystal thickness. With $H_c \gg H_a$, equation 1 reduces to $H \simeq \alpha H_c$. Calculations for ideal polyethylene crystals (Table 1) give a limiting value of $H_c^\infty \sim 150$–170 MPa (15.3–17.3 kgf/mm²). In lamellar polyethylene, the finite crystal thickness produces a hardness below this value. It is further known that the crystal thickness decrease is related to the expansion of the unit cell. In case of low density

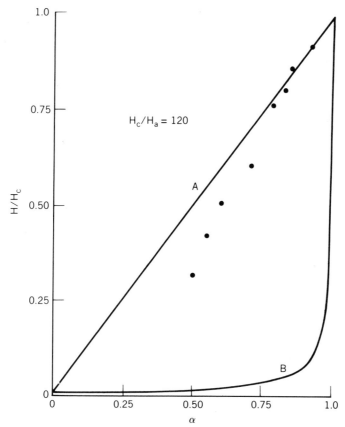

Fig. 5. Parallel (A) and series (B) model for the hardness of a two-component (crystalline–amorphous) composite as a function of crystalline volume fraction. The data correspond to linear polyethylene with different crystallinities (8). Hardness is expressed as H/H_c for a ratio $H_c/H_a = 120$.

polyethylene, the decrease of crystal hardness has thus been correlated with the increasing expansion of the lattice unit cell, caused by the inclusion of a fraction of chain defects within the crystals (11) (Fig. 6).

Surface Hardening of Polymers by Chemical Attack

The controlled exposure of polymers to etching atmospheres greatly strengthens the mechanical properties of the surface, yielding hardness approaching that of some metals, eg, lead–antimony. The increase of H is shown in Figure 7 as a function of chlorosulfonation of two polyethylene samples with different crystallinities (14). The concurrent hardening and weight increase suggest interlamellar attachment of electron-dense groups. With lower crystallinity, hardening is faster. Heavy groups reduce crystal slippage and creep. The final hardening is a function of the volume of disordered regions and of treatment temperature. Similar results are obtained by etching polyethylene with fuming nitric acid (4) or sulfuric acid (15).

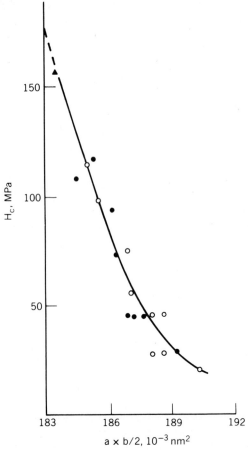

Fig. 6. Crystal hardness $H_c = MH/\alpha$ (for $t = 6$ s) vs cross section ($a \times b/2$) of projected unit cell for melt-crystallized polyethylene samples with different branching ratios (11). To convert MPa to kgf/mm^2, divide by 9.81.

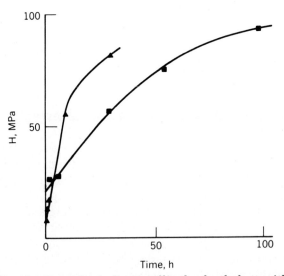

Fig. 7. Vickers hardness for melt-crystallized polyethylene with crystallinities of ▲, $\alpha = 0.35$ and ■, $\alpha = 0.45$, as a function of chlorosulfonation at 20°C (14).

Microindentation Anisotropy of Oriented Polymers

The correlation between molecular orientation and hardness can be applied to the characterization of drawn materials (16). The influence of orientation (qv) on hardness has been demonstrated for poly(ethylene terephthalate) films by using a steel ball (r = 1.5 mm; P = 0.15 N or 0.015 kgf) (9). The difference between the main axes of the elliptical impression increases with draw ratio. The influence of draw ratio on the anisotropy of indentation after load removal has been investigated in polystyrene, polypropylene, poly(vinyl chloride), and polyamides (9). The dependence of H on draw ratio is shown in Figure 8 for polyethylene plastically deformed up to draw ratios of λ ca 24. Before the neck (λ = 1), H of the isotropic microspherulitic sample is equal to ca 34 MPa (3.46 kgf/mm^2) (loading time 1.2 × 10^3 s). At the end of the neck for λ = 5, H = 46 MPa (4.7 kgf/mm^2). The change in H from the isotropic to the fibrous structure (λ > 5) is rather discontinuous in accordance with other structural changes in the neck region such as the sudden change in molecular orientation (17,18). The anisotropic shape of the indentation is caused by the increasing anisotropy of the fiber structure; two hardness values appear for draw ratios λ > 8. The higher

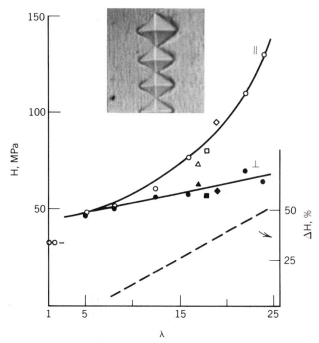

Fig. 8. Vickers microhardness values parallel (‖) (open symbols) and perpendicular (⊥) (solid symbols) to the orientation axis as a function of local–draw ratio, λ = $l\ drawn/l_0$ for polyethylene drawn at ○, 60°C; ◇, 100°C; △, 120°C; and ☐, 130°C (8). The microindentation anisotropy ΔMH = 1 − (d ‖/d ⊥)^2 increases linearly with λ; d ‖ and d ⊥ are the indentation diagonals parallel and perpendicular, respectively, to the orientation direction (dashed line) λ. Vickers indentations for oriented polyethylene show typical anisotropic impression (fiber axis vertical). To avoid interaction, H data are derived from indentations spaced well apart.

value is derived from the indentation diagonal parallel to the fiber axis. The low value is calculated from the diagonal perpendicular to it. The first corresponds to instant elastic recovery in the draw direction, the second to the plasticity of the oriented material. The principal deformation modes of the fibrous structure include a sliding motion of fibrils, which are sheared normally to the fiber axis, and a buckling of fibrils parallel to the axis. The anisotropy arises upon load removal because elastic recovery is greater along the fiber axis. This linear increase of anisotropy ΔH with λ has been correlated to the relative increase of tie-taut (-strained) molecules in oriented PE (4). The linear dependence of ΔH on λ has been verified for ultraoriented PE fibers prepared by solid-state extrusion (19) and ultraoriented polymers drawn through a heated conical die (20). The elastic modulus and ΔH of these fibers are correlated (20). The anisotropy behavior of very thin (50–100 μm) filaments with a shish-kebab structure has been reported (21). These filaments show the largest anisotropy so far achieved for PE (ΔH ca 65%), which is consistent with the large elastic modulus ($\sim$100 GPa) (1.45×10^7 psi) measured.

Test Methods

Brinell. In this test a steel ball is forced against the flat surface of the specimen. The standard method (22) uses a 10-mm ball and a force of 29.42 kN (3000 kgf). The Brinell hardness value is equal to the applied force divided by the area of the indentation:

$$\text{HB} = \frac{2P}{\pi D^2[1 - \sqrt{(1 - d/D)^2}]} \tag{3}$$

in which P is the force, N (kgf); D is the diameter of the ball, mm; and d is the diameter of the impression, mm. A 20-power microscope with a micrometer eye-piece can measure d to 0.05 mm. The minimum radius of a curved specimen surface is 2.5 D. The results of the test on polypropylene, polyoxyethylene, and nylon-6,6 have been interpreted in terms of stress–strain behavior (23).

Vickers. The test uses a square pyramid of diamond with included angles α between nonadjacent faces of the pyramid of 136°. The hardness is given by

$$\text{HV} = \frac{2P \sin (\alpha/2)}{d^2} = 1.854\frac{P}{d^2} \tag{4}$$

where P is the force in N (kgf) and d is the mean diagonal length of the impression in mm. The value of HV is expressed in MPa (kgf/mm^2). The force is usually applied at a controlled rate, held for 10–30 s, and removed. The length of the impression is measured to 1 μm with a microscope equipped with a filar eyepiece (24). Cylindrical surfaces require corrections up to 15% (22).

Knoop. Another and more commonly used hardness test uses a rhombic-based pyramidal diamond with included angles of 172° and 130° between opposite edges. The hardness is given by

$$\text{HK} = C\frac{P}{d^2} \tag{5}$$

where P is the force in N (kgf), d is the principal diagonal length of indentation in mm, and C is equal to 14.23 (22). The Vickers test gives a smaller indentation than the Knoop test for a given force. The latter is very sensitive to material anisotropy because of the twofold symmetry of the indentation (25).

Rockwell. In this test the depth of indentation is read from a dial (22); no microscope is required.

In the procedure used most often, the Rockwell hardness number does not measure total indentation but only nonrecoverable indentation after a heavy load is applied for 15 s and reduced to a minor load of 98 N (10 kgf) for 15 s. The scales, indenter dimensions, and loads used on plastics are given in Table 2. The Rockwell hardness tester has a range of 250 scale divisions. The red scale, used for all except the α test, has a range of -120 to $+130$ divisions. Each division represents 0.002 mm of indentation. The 130 point indicates zero indentation. Results given by testers in different ranges are not readily interconverted. Rockwell hardness data for a variety of polymers are given in Refs. 3 and 26.

Table 2. Rockwell Hardness Scales,[a] (ASTM D 785)

	Load, N[b]		Indenter[c]
Hardness value	Minor	Major	diameter, mm
R	98	588	12.5
L	98	588	6.25
M	98	980	6.25
E	98	980	3.13
K	98	1470	3.13
α^d	98	588	12.5

[a] Red scale, except where otherwise indicated; given in increasing order.
[b] To convert N to kgf, divide by 9.807.
[c] Steel sphere.
[d] Black scale.

Scleroscopy. In this test the rebound of a diamond-tipped weight dropped from a fixed height is measured (3,22). The Model C (HSc) uses a small hammer (ca 2.3 g) and a fall of about 251 mm; model D (HSd) uses a hammer of about ca 36 g and a fall of about 18 mm.

Scratch–hardness Test. This test measures resistance to scratching by a standardized tool (22). A corner of a diamond cube is drawn across the sample surface under a force of 29.4 mN (3 gf) applied to the body diagonal of the cube, creating a V-shaped groove; its width Λ, in μm, is measured microscopically. The hardness is given by:

$$HS = 10,000/\Lambda \tag{6}$$

The constant 10,000 is arbitrary.

Applicability of Tests. The Vickers and Brinell hardness scales are almost identical up to a Brinell hardness of about 5 GPa (510 kgf/mm^2). This range covers all polymeric materials. The Brinell test is preferred for measuring the macrohardness of large pieces in which large indentation (2.5–6 mm diameter) is acceptable. The Vickers macrohardness test ($P > 30$ N or 3.06 kgf) is used mainly where the indentation is limited in size. The Vickers microhardness test

($P < 1$ N or 0.1 kgf) is used mainly with small and inhomogeneous specimens. Forces down to 10 mN (1 gf) are suitable for most commercial instruments. A tester recently developed can operate down to 10 μN (1000 μgf) in conjunction with a scanning electron microscope (27). The Knoop microhardness test is more rapid than the Vickers test and more sensitive to material anisotropy. The Rockwell instrument is used in production and quality control where absolute hardness is unimportant. The scleroscope is used for specimens that cannot be removed or cannot tolerate large indentations. Values given by the scleroscope (dynamic hardness) and the static-ball indenter correspond directly. The scratch test is used where indentation microhardness tests cannot be made close enough to determine local variations. Typical values for polymers are found in Ref. 28.

BIBLIOGRAPHY

"Hardness" in *EPST* 1st ed., Vol. 7, pp. 470–478, by Paul I. Donnelly, Hercules Incorporated.

1. D. Tabor, *The Hardness of Metals,* Oxford University Press, New York, 1951.
2. H. O'Neill, *Hardness Measurement of Metals and Alloys,* Chapman Hall, London, 1976.
3. B. Maxwell, *Mod. Plast.* **32**(5), 125 (1955).
4. F. J. Baltá-Calleja, *Colloid Polym. Sci.* **254**, 258 (1976).
5. J. Bowmann and M. Bevis, *Colloid Polym. Sci.* **255**, 954 (1977).
6. D. M. Marsch, *Proc. R. Soc. London Ser.* **A279**, 420 (1964).
7. B. R. Lawn and V. R. Howes, *J. Mater. Sci.* **16**, 2745 (1982).
8. F. J. Baltá-Calleja, *Characterization of Polymers in the Solid State, Adv. in Polym. Sci.* **66**, 117 (1985).
9. P. Eyerer and G. Lang, *Kunststoffe* **62**, 222 (1972).
10. K. Ito, *Mod. Plast.* **34**(2), 167 (1957).
11. F. J. Baltá-Calleja, J. Martinez-Salazar, H. Cacković, and J. Loboda-Cacković, *J. Mater. Sci.* **16**, 739 (1981).
12. A. Peterlin, *J. Mater. Sci.* **6**, 490 (1971).
13. J. Martinez-Salazar and F. J. Baltá-Calleja, *J. Mater. Sci. Lett. Ed.* **4**, 324 (1985).
14. J. Martinez-Salazar, D. R. Rueda, M. E. Cagiao, E. López Cabarcos, and F. J. Baltá-Calleja, *Polym. Bull.* **10**, 553 (1983).
15. F. J. Baltá-Calleja, C. Fonseca, J. M. Pereña, and J. G. Fatou, *J. Mater. Sci. Lett. Ed.* **3**, 509 (1984).
16. J. Bednarz, *Z. Werkstofftech.* **1**, 195 (1970).
17. G. Meinel, N. Morossoff, and A. Peterlin, *J. Polym. Sci.* **A2**(8), 1723 (1970).
18. N. Morossoff and A. Peterlin, *J. Polym. Sci.* **A2**(10), 1237 (1970).
19. F. J. Baltá-Calleja, D. R. Rueda, R. S. Porter, and W. T. Mead, *J. Mater. Sci.* **15**, 765 (1980).
20. D. R. Rueda, J. Garcia, F. J. Baltá-Calleja, I. M. Ward, and A. Richardson, *J. Mater. Sci.* **19**, 2615 (1984).
21. D. R. Rueda, F. J. Baltá-Calleja, and P. F. van Hutten, *J. Mater. Sci. Lett. Ed.* **1**, 496 (1982).
22. *Annual Book of ASTM Standard Part 10,* American Society for Testing and Materials, Philadelphia, 1978.
23. E. Baer, R. E. Maier, and R. N. Peterson, *SPE. J.* **17**, 1203 (1961).
24. K. Müller, *Kunststoffe* **60**, 265 (1965).
25. F. J. Baltá-Calleja and D. C. Bassett, *J. Polym. Sci.* **58C**, 157 (1977).
26. I. E. Nielsen, *Mechanical Properties of Polymers,* Reinhold Publishing Corporation, New York, 1963, p. 220.
27. H. Bangert, A. Wagendritzed, and H. Aschinger, *Philips Electron Optics Bull.* **119**, 17 (1983).
28. L. Boor, J. Rijan, M. Marks, and W. Bartre *ASTM Bull.* **145**, 68 (Mar. 1947).

F. J. Baltá-Calleja
J. Martinez-Salazar
D. R. Rueda
Institute for Structure of Matter CSIC, Madrid, Spain

HEAT-RESISTANT POLYMERS

Heat-resistant polymers are used in many applications, such as insulators for microelectronic components, sealants for fuel tanks in high speed aircraft, coatings on cookware, binders in brake systems, and structural components in space vehicles. Research was initiated in the late 1950s to meet the requirements of advanced aircraft, weapon systems, and electronics. Many different heat-resistant polymer systems have been reported, of which several are now commercially available.

Heat-resistant resins are defined as materials that retain mechanical properties for thousands of hours at 230°C, hundreds of hours at 300°C, minutes at 540°C, or seconds up to 760°C. These temperatures may be encountered during the manufacture or use of the product.

Requirements for Heat Resistance

Heat resistance is the capacity of a material to retain useful properties for a stated period of time at elevated temperatures ($\geq$230°C) under defined conditions, such as pressure or vacuum, mechanical load, radiation, and chemical or electrical influences at temperatures ranging from cryogenic to above 500°C (see also CRYOGENIC PROPERTIES). Both reversible and irreversible changes can occur. In a reversible change, for example, as a polymer under load approaches the glass-transition temperature T_g, deformation occurs without change in chemical structure. Reversible changes occur primarily as a function of T_g, which for the purposes of this article, ie, for high temperature structural polymers, must be above 230°C. The maximum-use temperature for an amorphous or semicrystalline structural resin usually depends on T_g rather than the crystalline melt temperature T_m. A semicrystalline polymer can exhibit substantial loss of mechanical properties near the T_g, depending upon the degree of crystallinity. The T_m is usually so high that in its vicinity chemical degradation (qv) occurs. Irreversible changes alter the chemical structure. For example, exceeding the thermal stability results in bond breaking.

The chemical factors which influence heat resistance include primary bond strength, secondary or van der Waals bonding forces hydrogen bonding, resonance stabilization, mechanism of bond cleavage, molecular symmetry (structure regularity), rigid intrachain structure, and cross-linking and branching. The physical factors include molecular weight and molecular weight distribution, close packing (crystallinity), molecular (dipolar) interactions, and purity.

The primary bond strength is the single most important influence contributing to heat resistance. The bond dissociation energy (1) of a carbon–carbon single bond is ~350 kJ/mol (83.6 kcal/mol), and that of a carbon–carbon double bond is ~610 kJ/mol (145.8 kcal/mol). In aromatic systems, the latter is even higher. Known as resonance stabilization, this phenomenon adds 164–287 kJ/mol (39.2–68.6 kcal/mol). As a result, aromatic and heterocyclic rings are widely used in thermally stable polymers. The dissociation energy of a carbon-to-fluorine bond of ~430 kJ/mol (102.8 kcal/mol) is increased to ~504 kJ/mol (120.4 kcal/mol) by a second fluorine atom attached to the same carbon atom. Therefore, perfluorinated materials are thermally stable. However, if a fluorine atom is next to a hydrogen, hydrogen fluoride is given off at elevated temperatures. The bond dissociation energy of a nitrogen–nitrogen bond is relatively low (~160 kJ/mol or 38.2 kcal/mol) and would not be expected to be incorporated in thermally stable polymers. However, because of resonance stabilization, the N–N bond in heterocyclic rings, such as 1,3,4-oxadiazole, exhibits reasonably high thermal stability. Although the bond strengths of most inorganic elements are high, eg, B–N, 386 kJ/mol (92.2 kcal/mol), Si–N, 437 kJ/mol (104.4 kcal/mol), Ti–O, 672 kJ/mol (160.6 kcal/mol), these bonds are attacked by oxygen and water. Therefore, few useful high temperature inorganic polymers (qv) have been reported.

Secondary or van der Waals bonding forces provide additional strength and thermal stability. Dipole–dipole interaction and H bonding contribute 25–41 kJ/mol (6.0–9.8 kcal/mol) toward molecular stability and affect the cohesion energy density, which influences the stiffness, T_g, melting point, and solubility.

Thus, heat-resistant polymers often contain polar groups, eg, $-\overset{\overset{\displaystyle O}{\|}}{C}-$, $-SO_2-$, that participate in strong intermolecular association. Polymers containing electron-withdrawing groups, eg, $-\overset{\overset{\displaystyle O}{\|}}{C}-$, as connecting groups are generally more stable than those containing electron-donating groups, eg, $-O-$.

The mechanism of bond cleavage also influences thermal stability. In polysiloxanes, for example, the energy of the silicon–oxygen single bond is ~445 kJ/mol (106.4 kcal/mol), and that of the silicon–carbon single bond ~328 kJ/mol (78.4 kcal/mol). Although the Si–C bond would be expected to cleave at high temperatures more readily than the Si–O bond, the latter breaks at high temperatures to form low molecular-weight cyclic siloxanes because this degradation route is energetically favored (2). The thermal decomposition of heat-resistant organic polymers is very complex. Certain fluorinated aliphatic materials decompose by an unzipping process (a stepwise breakdown), which starts at the chain ends to yield monomer. Most aromatic heterocyclic polymers or polymers formed by step-growth polymerization undergo degradation via random scission along the chain, eliminating fragments that eventually break down further. Even a small degree of chain cleavage serves to destroy the structural integrity.

Molecular symmetry or regularity of the chemical structure arises when moieties are joined in the same position in each repeat unit. The presence of isomers lowers T_g. Rigid intrachain structure refers to substitution of the aromatic or heterocyclic ring. Although para-oriented polymers have the highest thermal stability and T_g, they have the lowest solubility and processability.

Cross-linking improves heat resistance of a polymer primarily because more bonds must be cleaved in the same vicinity for the polymer to exhibit a weight loss or reduction in mechanical properties. Crystalline regions in a polymer serve as cross-links. Branching in a polymer tends to lower the thermal stability.

Molecular weight and molecular-weight distribution influence mechanical and physical properties. Higher molecular-weight polymers are more heat resistant than lower molecular-weight materials because of more entanglements and the ability to accommodate more chain cleavage without significant property reduction. Low molecular-weight polymers exhibit lower T_gs.

Close packing refers to short-range structural order in amorphous regions or long-range order in crystalline regions. The heat resistance of ordered polymers, particularly crystalline polymers, is higher than that of unordered polymers. Strong intermolecular association contributes to high thermal stability and high T_g.

Purity is of prime importance. Even trace amounts of metals can catalyze thermooxidative degradation. Impurities, often present in the monomers or introduced from an outside source, accelerate thermooxidative degradation.

Determination of Heat Resistance

Heat resistance is assessed by thermogravimetric analysis (tga), isothermogravimetric analysis (itga), differential scanning calorimetry (dsc), dynamic mechanical analysis (rheovibron or torsional-braid analyzer), thermal mechanical analysis (expansion, penetration), and dielectrometry. Heating rate, sample form, atmosphere, frequency, load, and thermal history affect the test results. For example, the faster the heating rate in tga, the higher the apparent thermal stability of the polymer. Powders with higher surface area have higher weight losses by itga than films. In addition, itga and tga are influenced by the atmosphere. Weight losses in static air are lower than weight losses in circulating air; weight losses in vacuum or inert atmosphere are even lower.

The thermal history of the polymer affects heat resistance. For example, the polyimide from 3,3',4,4'-benzophenonetetracarboxylic dianhydride (BTDA) and 3,3'-methylenedianiline has a T_g of 211°C after 1 h at 200°C in air, a T_g of 272°C after an additional hour at 300°C, and a T_g of 234°C after an hour each at 200°C and 300°C (3). In addition, different methods, eg, dsc vs tma, often give significantly different T_gs.

The tga of several aromatic heterocyclic polymers and pyrolytic graphite are shown in Figure 1. Polybenzothiazole exhibits the highest decomposition temperature. However, by itga as shown in Figure 2, the polyimide has the lowest weight loss after 200 h at 371°C in air. A different rating may be obtained by measuring the retention of a mechanical property, eg, tensile strength or fracture

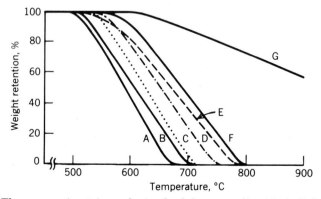

Fig. 1. Thermogravimetric analysis of polyheterocyclics (4). A, Polybenzimidazole; B, polyquinoxaline; C, polyimide D, polybenzoxazole; E, poly(N-phenylbenzimidazole); F, polybenzothiazole; G, pyrolytic graphite. Atmosphere: static air; heating rate: 6.67°C/min.

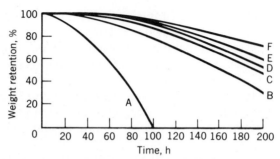

Fig. 2. Isothermal weight loss of polyheterocyclics in air at 371°C (4). A, Polybenzimidazole; B, poly(*N*-phenylbenzimidazole); C, polybenzothiazole; D, polyquinoxaline; E, polybenzoxazole; F, polyimide. Atmosphere: circulating air; sample form: >0.1 mm (<140 mesh) powder.

toughness, after exposure to elevated temperature. A material can undergo substantial degradation and loss of mechanical strength without a similar loss in weight. Some materials gain weight in an oxidizing environment. Since many factors, especially processability, enter into the performance of a material at elevated temperature, the performance of material is evaluated best in an applied form (eg, adhesive, composite, or film).

Polymers with High Heat Resistance

During the early work on heat-resistant polymers in the late 1950s and early 1960s, attention focused on polymers with high heat resistance, as measured primarily by tga. Little or no regard was devoted to processability or the heat resistance of an applied form, such as an adhesive bond. However, processability is of prime importance. The characteristics that contribute to the thermal stability also make it more difficult to process, namely, poor solubility and high melt viscosity. In fact, the melt viscosity of heat-resistant high molecular-weight linear amorphous polymers above T_g is greater than 1 MPa·s (10^7 P) at a low shear rate such as encountered under compression-molding conditions. To obtain polymers processable by compression and injection molding (qv), T_g and thermal stability are often compromised. Processability can be improved by incorporation of the more flexible 4,4'-diphenyl ether, 4,4'-diphenyl ketone or 4,4'-diphenyl sulfone, or 3,3'-diphenyl ketone and 3,3'-diphenyl sulfone in place of 1,4-phenylene or 4,4'-diphenylene and limitation of molecular weight. Copolymerization, reactive oligomers, polymer blends, and plasticizers have also been employed to improve the processability.

The evolution to melt-processable forms is illustrated by polyimides and to a lesser extent by the polyquinoxalines, polyquinolines, and polybenzimidazoles. Processability requirements for polymers potentially useful as fibers or ribbons having all-para structures and commonly referred to as extended chain and rigid rodlike polymers are different. Several of these polymers are processed from solution (generally in strong acids such as methane sulfonic acid) to form fibers and ribbons; other can be molded under high temperature and pressures to form complex shapes.

Polyimides.   Early synthesis (5) involved the reaction of an aromatic tetracarboxylic acid, such as pyromellitic acid or its half-ester, with a diamine (aromatic or aliphatic) to form a salt. Heating gave the polyimide (PI), with substantial evolution of condensation volatiles. With aromatic monomers, the polymeric salt and the polyimide were intractable and of low molecular weight. Polyimides derived from long-chain aliphatic diamines, which remain in a melt under the polymerization conditions, were usually tractable and of high molecular weight. An improved route (eq. 1) was reported later (6) where an aromatic dianhydride (pyromellitic dianhydride) and an aromatic diamine (4,4'-oxydianiline) in a polar solvent (N,N-dimethylacetamide, DMA) formed a high molecular-weight poly(amide–acid), also called polyamic acid. The poly(amide–acid) can be cyclodehydrated by chemical (7) (acetic anhydride and pyridine) or thermal means (8,9) to the polyimide (see also POLYIMIDES).

poly(amide–acid)

$$(1)$$

The poly(amide–acid) is prone to hydrolytic cleavage, and moisture must be excluded by maintaining the reaction temperature low enough (<30°C) to avoid imide formation, by blanketing with an inert atmosphere and by keeping cold and dry during storage. Excess anhydride limits the molecular weight of the poly(amide–acid) (9). Reduction of the molecular weight of the poly(amide–acid) during thermal conversion to polyimide at 150–200°C has been reported (10,11). Upon increasing the temperature to 200–300°C, polymerization increases the molecular weight of the polyimide. A wet film of a high molecular-weight poly(amide–acid) (inherent viscosity η_{inh} = 0.98 dL/g) from pyromellitic dianhydride and 3,3'-diaminobenzophenone was dried and converted to the polyamide. The film became very brittle after 5 h at 175°C, brittle after 2 h longer at 225°C, and flexible after 1 h more at 300°C (3). The increase in flexibility is presumably due to an increase in the molecular weight.

The evolution of solvent and condensation volatiles during the thermal conversion causes severe processing problems, which has led to the development of other synthetic routes.

The properties of polyimides are controlled by chemical structure, molecular weight, and process conditions (thermal history). When assessing structure–property relationships, the method and conditions of determining T_g and the thermal history should be considered. The glass-transition temperatures given in Tables 1–4 demonstrate that increasing molecular flexibility lowers the T_g.

Table 1. Glass-transition Temperature of Polyimides[a]

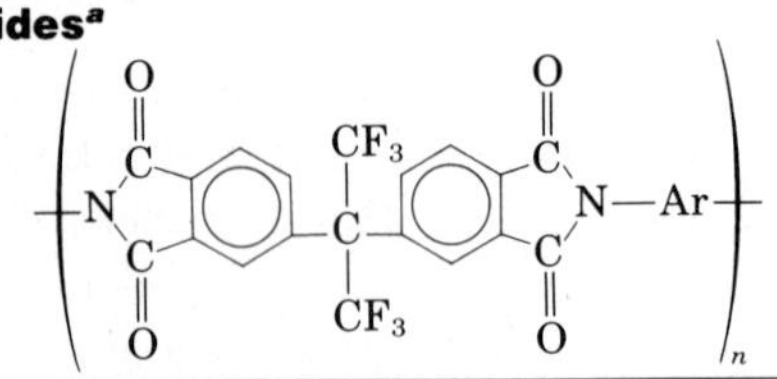

Ar	$\eta_{inh}{}^{b}$, dL/g	T_g, °C[c]
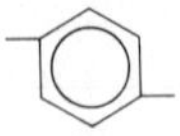	0.35	326
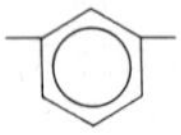	0.41	297
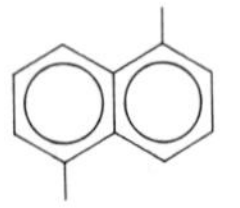	0.64	365
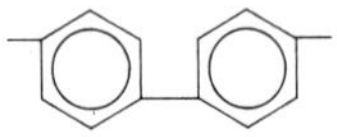	0.40	337
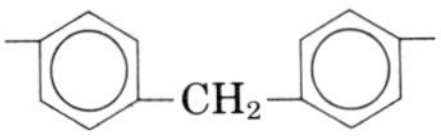 $-CH_2-$	0.38	291
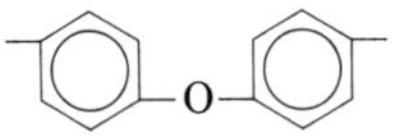 $-O-$	0.46	285
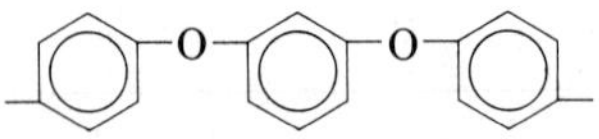 $-O-\ \ -O-$	0.35	229
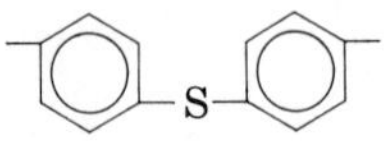 $-S-$	0.35	283
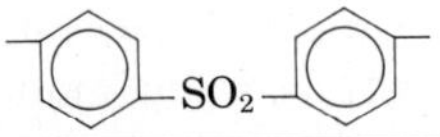 $-SO_2-$	0.31	336

[a] Ref. 12.
[b] 0.5% solution in concentrated sulfuric acid at 25°C.
[c] Determined by dsc at a heating rate of 10°C/min.

Process conditions affect properties as shown in Table 5; tensile strength and tensile modulus increase dramatically upon hot drawing. The PI of Table 5 designated LARC-2 and LARC-TPI (15) shows excellent high temperature adhesive properties for joining titanium to titanium (16) or Kapton to Kapton (17) or other substrates, eg, copper foil. For example, Ti/Ti lap-shear specimens exhibited 24.8 MPa (3600 psi) when tested at 232°C after aging for 37,000 h at 232°C in air (18).

Several materials derived from a poly(amide–acid), such as Kapton film or Vespel moldings, are commercially available and are thought to have the chemical structure shown in equation 1. Kapton has excellent mechanical properties (see Table 6). The time required for a 70 to 1% reduction of elongation (measured at 25°C) is 1 year at 275°C and 3 months at 300°C (20). Kapton is used as an insulator, as substrate and packaging material in the electronic industry, and as bagging material for high temperature components. Various poly(amide–acid)

Table 2. Glass-transition Temperature of Polyimides[a]

Ar	η_{inh} of poly(amide–acid)[b], dL/g	T_g, °C[c]
	0.41	167
	0.87	187
	0.80	202
	1.20	206
	0.44	221

[a] Ref. 13.
[b] 0.5% solution of DMA at 35°C.
[c] Determined by tma at a heating rate of 5°C/min.

Table 3. Polyimide Structure and Glass Transition[a]

Dianhydride[b]	Diamine[c]	η_{inh} of poly(amide–acid)[d], dL/g	T_g of polyimide, °C[e]
BTDA	4,4'-DABP	0.73	295
BTDA	3,4'-DABP	0.64	283
BTDA	3,3'-DABP	0.55	264
PMDA	4,4'-DABP	0.98	380
PMDA	3,4'-DABP	0.84	339
PMDA	3,3'-DABP	0.83	321

[a] Ref. 3.
[b] BTDA = 3,3',4,4'-benzophenonetetracarboxylic dianhydride; PMDA = pyromellitic dianhydride.
[c] DABP = diaminobenzophenone.
[d] 0.5% DMA solution at 35°C.
[e] Determined by tma (heating rate of 5°C/min) on film dried 1 h each at 200°C and 300°C in vacuum.

forms are available from DuPont, such as Pyre-ML for magnetic wire, Pyralin solutions for other coating use, and low metal-content forms such as PI 2545 and 2555 for microelectronic applications, eg, interlevel dielectric passivants.

Other polyimides made via the poly(amide–acid) route are given in Table 7. Although the DuPont NR-150B2 exhibited high thermooxidative and dimensional stability ($T_g \sim 360$°C), serious processing problems were encountered using the poly(amide–acid) or a mixture of monomers in the fabrication of composites and large-area bonds due to a combination of factors such as evolution of con-

Table 4. Glass-transition Temperatures of Polyetherimides Made via Poly(amide–acid) Route[a]

Ar	Isomer[b]	Ar'	$[\eta]$[c], dL/g	T_g, °C[d]
(phenyl)	3		2.25	263
	4		1.71	242
	3		0.45	259
	4		0.96	255
	3		0.94	214
	4		insoluble	199
–SO₂–	3		0.76	267
	4		1.13	260
	3		0.34	266
	4		0.70	265
	3		0.61	230
	4		0.88	219

[a] Ref. 14.
[b] Position of attachment of —OArO—.
[c] Intrinsic viscosity measured in *m*-cresol.
[d] Determined by dsc at a heating rate of 40°C/min.

Table 5. Thin-film Properties of LARC Polyimide at 25°C[a]

Process conditions	Tensile strength[b], MPa[c]	Tensile modulus[b], GPa[d]	Elongation[b], %
dried 1 h at 100, 200, and 300°C in air	111.0	3.25	6.6
same as above but drawn 50% at 260°C before testing	169.6	4.65	6.7

[a] Ref. 3.
[b] ASTM D 882-647.
[c] To convert MPa to psi, multiply by 145.
[d] To convert GPa to psi, multiply by 145,000.

Table 6. Properties of Kapton-type H Film[a]

	Test temperature, °C				
Property[b]	−195	23	200	300[c]	450[c]
ultimate tensile strength, MPa[d]	241	172	117	96.5	41.4
yield point at 3%, MPa[d]		69	41		
tensile modulus, GPa[e]	3.5	3.0	1.86		
ultimate elongation, %	2	70	90	110[c]	110
density[f], g/cm³		1.42			

[a] Ref. 15 except where otherwise stated.
[b] ASTM D 822-64T except where otherwise stated.
[c] Ref. 19.
[d] To convert MPa to psi, multiply by 145.
[e] To convert GPa to psi, multiply by 145,000.
[f] ASTM D 1505-63T.

densation volatiles, low flow, and solvent retention. In spite of these problems, NR-150B2 is still of interest as a high temperature resin for functional and structural applications. The soluble PI, CIBA-GEIGY's XU-218 ($T_g \sim 320°C$) (22,23), was prepared from 5(6)-amino-1(4′-aminophenyl)-1,3,3-trimethylindane isomers and BTDA. This PI and its variations are potentially useful in the microelectronic industry as passivation coatings and interlayer dielectrics. The TRW partially fluorinated PI (PFPI) (24) and NR-150B2 are candidates for use in hot areas of jet engines, eg, stator vane bushings, where the material must be stable at 357°C under an air pressure of 404 kPa (4 atm) for 100 h. PFPI has a T_g of 270°C after a 300°C cure in air (13) and a T_g of ~393°C (probably cross-linked) after a cure at 429°C in air (27). Other potential uses for NR-150B2 and PFPI are fiber sizings for carbon–graphite filament reinforcement for high temperature composites and protective coatings. The American Cyanamid XPI-182 (25), a potentially low cost polymer (dianhydride made from styrene and maleic anhydride), has a T_g of ~265°C and could be compression molded at ~310°C. Processing problems discouraged commercialization. Phenylated PI (26), although expensive and commercially unavailable, has a T_g of 413°C and is readily soluble in chloroform. Flexible transparent yellow films can be cast from chloroform solutions.

Polyimides containing pendent ethynyl and phenylethynyl groups have been prepared by the reaction of aromatic dianhydrides and aromatic diamines containing an ethynyl group, eg, 1,3-diamino-5-phenylethynylbenzene (28). The ethynyl group undergoes a thermally induced cross-link reaction at a lower temperature than that required for the phenylethynyl group. Cross-linked PI with a T_g above 400°C can be obtained but at reduced toughness.

Other synthetic routes include the reaction of aromatic dianhydrides with aromatic diisocyanates with the elimination of carbon dioxide (29,30); the Upjohn 2080 material is made by this route. A nonstoichiometric amount of caprolactam has been reported (31) to react with BTDA to yield a product that is boiled in ethanol and subsequently treated with 4,4′-methylenedianiline. An exchange reaction was proposed where the aromatic amine displaced the aliphatic amine fragment from caprolactam to yield a PI. Although the precise chemistry is unknown, a low density flame-resistant resilient foam can be made by microwave curing of a spray-dried powder of a formulated intermediate made by this route.

Table 7. Polyimides Prepared via Poly(amide–acid) Route

Dianhydride	Diamine	Designation	Manufacturer	Ref.
(structure)	(structure) H_2N—NH_2 [a]	NR-150B2	DuPont	21
(structure)	(structure) CH$_3$ / H$_3$C CH$_3$ [b]	XU-218	CIBA-GEIGY	22,23
(structure)	(structure) CF$_3$—C—CF$_3$	PFPI	TRW	24
(structure)	(structure) H_2N—CH$_2$—NH_2	XPI 182	American Cyanamid	25
(structure) C$_6$H$_5$ ...	(structure) H_2N—O—NH_2	phenylated PI		26

[a] 95% 1,4; 5% 1,3 substitution.
[b] —NH$_2$ substitution at ring position 5 or 6.

The displacement polymerization of activated nitro, chloro, or fluoro groups on bisimides by the dianion of bisphenols, as shown below, is an unusual synthetic route (32,33).

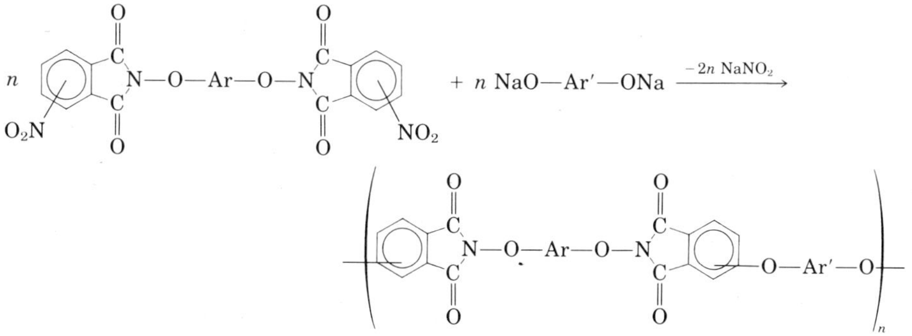

$$(2)$$

where Ar = p-phenyl, m-phenyl, 4,4'-biphenyl,

Ar' = p-phenyl, m-phenyl, 4,4'-biphenyl,

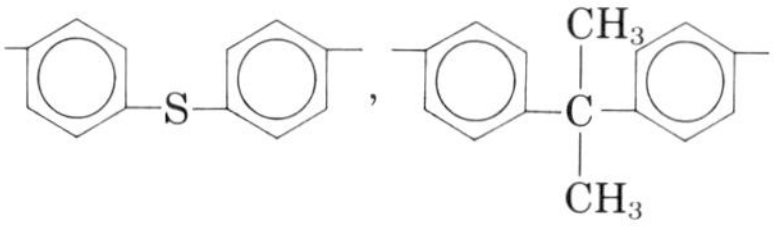

The T_gs of several polyetherimides made by the nitro displacement route are given in Table 8. Ultem 1000, a General Electric polyetherimide, has a $T_g \sim 220°C$ and a number-average molecular weight of ~19,000.

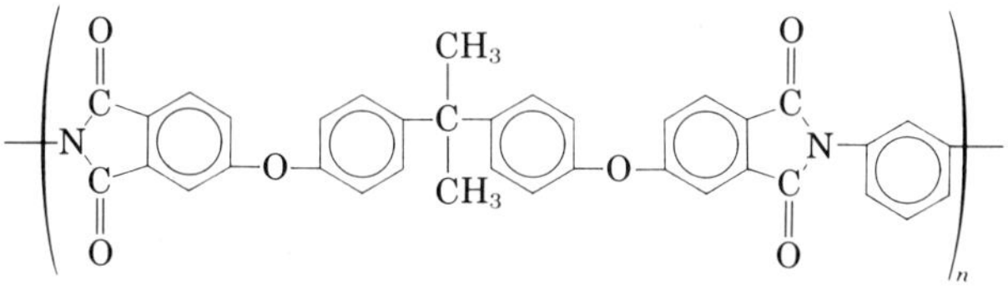

Ultem 1000

The processability of PI was improved by eliminating the hydrolytically unstable, volatile-producing poly(amide–acid). In the early 1970s, imide oligomers terminated with norbornene (34,35) and acetylenic (36,37) groups were developed for structural applications. The first norbornene (nadic) terminated imide oligomers were known as P13N; P means polyimide, 13 stands for oligomeric M_n of ~1300, and N denotes norbornene termination. The same technology gave LARC-160 (38) and PMR-15; PMR means *in situ* polymerization of monomeric reactants yielding an oligomer with M_n ~1500 (39).

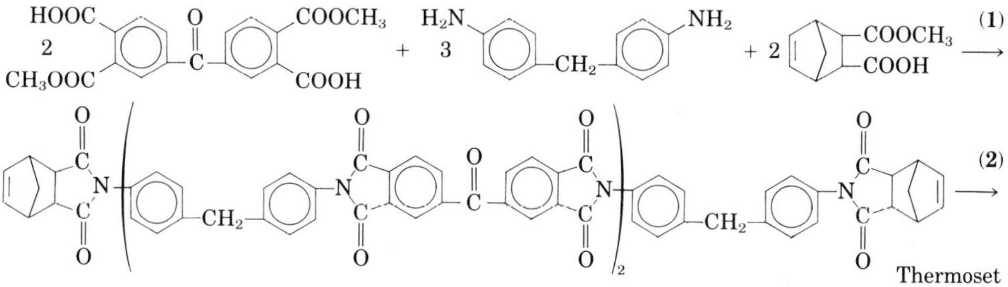

Table 8. Glass-transition Temperatures of Polyetherimides Made via Nitro Displacement Reaction[a]

X	Ar	Isomer	$[\eta]$, dL/g[b]	T_g, °C
CH$_2$	(structure with CH$_3$/C/CH$_3$)	3,3'	0.28[c]	230
CH$_2$	(structure with CH$_3$/C/CH$_3$)	3,4'	0.31	224
—O—	(structure with CH$_3$/C/CH$_3$)	3,3'	0.33	226
—O—	(structure)	3,3'	0.16	226
—O—	(structure)	3,3'	0.45	237
—O—	(structure with O)	3,3'	0.80[d]	239
—O—	(structure)	3,3'	0.60	277
—O—	(structure with O)	4,4'	0.44[e]	215

[a] Ref. 33.
[b] In N,N-dimethylformamide unless otherwise indicated.
[c] In methylene chloride.
[d] In chloroform.
[e] In sulfuric acid.

In the condensation step (1) water and methanol are lost. Heating to ~275–300°C in step (2) yields a thermosetting material. The pyrolytic polymerization of the terminal norbornene groups involves a complex reaction of norbornene with male-imide and cyclopentadiene formed from a reverse Diels–Alder reaction to yield a thermoset. A small amount (~1–2%) of cyclopentadiene is evolved, which may lead to porosity in a composite. PMR-15 and LARC-160 provide an avenue to a heat-resistant polyimide with excellent processability. For example, in practice with PMR-15, a methanol solution of the three monomers is used to impregnate a reinforcement of carbon–graphite fibers. The prepreg is dried to a solvent con-tent of ~14% to retain tack and drape. The prepreg is stacked in the required order (laid-up) and heated to the temperature where most of the volatiles could be removed (B-staged); the monomers are converted to the norbornene-terminated imide oligomeric form. These oligomers have sufficient flow for further compaction at elevated temperatures under pressure and are cured as the temperature is increased to yield composites with a low void content. PMR-15 carbon–graphite filament-reinforced composites give high mechanical performance at 316°C after hundreds of hours at 316°C in air. These composites may find use in high temperature applications, such as jet engines (40). However, they undergo microcracking after thermal cycling, especially while under load, with a reduction in mechanical properties.

Acetylene-terminated imide oligomers (ATI) known as HR-600 and Thermid-600 (see ACETYLENE-TERMINATED PREPOLYMERS) present processing problems.

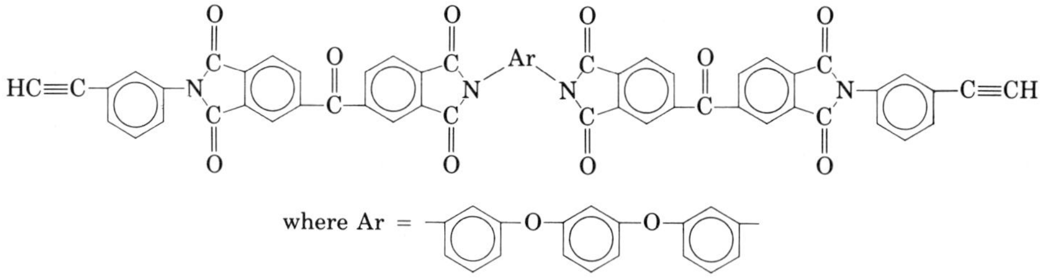

where Ar =

The temperatures required for softening and flow are sufficiently high to initiate reaction of the terminal acetylenic groups, thereby further reducing the flow. Cured neat resin properties of an ATI with the above structure are given in Table 9. In an attempt to improve the processability, acetylene-terminated isoimides have been prepared (42). Although the isoimide is reported to be more soluble and to exhibit a lower softening temperature, processability is still a serious problem.

Table 9. Properties of Cured Acetylene-terminated Imide[a,b]

Property	Value
tensile strength, MPa[c]	96.5
tensile modulus, GPa[d]	3.79
elongation, %	2.6
flexural strength, MPa[c]	124
flexural modulus, GPa[d]	4.48
compressive strength, MPa[c]	213.7
Burcol hardness	45
T_g^e, °C	350

[a] Ref. 41.
[b] Neat resin cured 2 h at 250°C under 137.8 MPa (20,000 psi) and 16 h at 316°C in air.
[c] To convert MPa to psi, multiply by 145.
[d] To convert GPa to psi, multiply by 145,000.
[e] After 48 h at 370°C.

Several imide copolymers, eg, esters (43), phenylquinoxalines (44), and amides (45), have been reported, but the Torlon polyamideimides (Amoco Chemical Corp.) (46) have received the greatest attention and are currently used in many applications. Polyimides in general are used where heat resistance is required. Various forms for use as adhesives, composite matrices, coatings, films, foams, and moldings are commercially available (19,47–50).

Polyquinoxalines. The polyquinoxalines (qv) (PQ) are prepared by the reaction of aromatic bis(o-diamines), such as 3,3′,4,4′-tetraaminobiphenyl (TAB), and aromatic bisglyoxal hydrates, such as 4,4′-oxydiphenylenediglyoxal dihydrate, in m-cresol at 80–100°C (51–55). Soluble, high molecular-weight thermoplastic PQ, governed by the chemical structure, is readily formed. Materials evaluated as heat-resistant adhesives and composite matrices require processing temperatures of ~400°C, but provide outstanding high temperature properties. For example, boron-filled PQ–stainless steel tensile-shear specimens fabricated under 1.38 MPa for 1 h each at 344, 426, and 455°C gave strengths in air of 15.7 MPa (2280 psi) at 316°C after 200 h at 316°C, 17.5 MPa (2540 psi) at 371°C after 50 h at 371°C, and 9.1 MPa (1325 psi) at 538°C after 10 min at 538°C (56). The 371°C strengths are the highest reported for an organic adhesive. The amorphous boron filler (particle size <10 μm) apparently interacts in some manner with the PQ at the high processing temperatures to eliminate the thermoplasticity.

Polyphenylquinoxalines (PPQ), first reported in 1967 (57), are easier to make than PQ and offer superior solubility, processability, and thermooxidative stability (58). High molecular-weight PPQs are readily synthesized from aromatic bis(o-diamines), such as TAB, and bis(phenyl-α-diketones), such as 4,4′-oxydibenzil (ODB), in m-cresol at ambient temperature.

$$\tag{4}$$

Both PPQ and PQ are configurationally unordered with three possible isomers distributed randomly along the chain. This dissymmetry contributes to the amorphous character of the polymers, their excellent solubility in m-cresol, and their thermoplastic nature. The effects of chemical structure variation within the PPQ backbone on T_g is shown in Table 10; the effects of substituents on the pendent phenyl group are shown in Table 11.

The PPQs have excellent high temperature adhesive, film, and composite properties. A PPQ film in air exhibits fingernail creaseability after 631 h at 316°C and a weight loss of only 1.24% after 1100 h at 316°C (60). The Ti–Ti tensile-shear strength in air of specimens cured up to 371°C exhibit strength of 33.8 MPa (4900 psi) at 25°C, 24.8 MPa (3600 psi) at 232°C after 8000 h at 232°C, and 4.8 MPa (700 psi) at 316°C (thermoplastic failure). Additional curing under pressure up to 454°C lowers the strength at 25°C to 24.1 MPa (3500 psi) (more brittle) but increases the strength at 316°C to 13.8 MPa (2000 psi) because of cross-linking (61). Although curing at temperatures of ~400°C can induce cross-linking and thereby improve high temperature mechanical performance, it usually initiates degradation, reducing the long-term thermooxidative stability.

In an attempt to increase the use temperature of PPQ, acetylenic groups were placed pendent along the polymer chain and subsequently thermally cured. First, 2,2′-bis(phenylethynyl)biphenyl moieties were incorporated in a PPQ. Upon curing at elevated temperatures, an intramolecular cyclization was expected to yield 9-phenyldibenz[a,c]anthracene moieties (62). However, both intra- and intermolecular reactions occurred and a brittle polymer was obtained with a T_g of

Table 10. Glass-transition Temperatures of PPQ

Y	Ar	η_{inh}, dL/g[a]	T_g, °C[b]
[c]		2.2	370
[c]		1.3	320
[c]		1.2	290
—O—		1.9	279
O‖—C—		2.3	288
—SO$_2$—		1.4	324

[a] 0.5% solution in m-cresol at 25°C.
[b] Determined by dsc at a heating rate of 20°C/min.
[c] No group intervenes between the rings at this position.

Table 11. Glass-transition Temperatures of PPQ[a]

Ar	X	η_{inh}, dL/g[b]	T_g, °C[c]
	H	0.73	319
	—C$_6$H$_5$	0.81	326
	—O—C$_6$H$_5$	0.45	245
	—OH	1.10	375
	—OCH$_3$	0.64	305
	—C$_4$H$_9$	0.49	231
	—Br	1.10	313
	—CN	1.62	340

[a] Ref. 59.
[b] 0.5% solution in m-cresol at 25°C.
[c] Determined by dsc at a heating rate of 20°C/min.

365°C (62). Pendent ethynyl and phenylethynyl groups were incorporated along the PPQ chain (63,64) followed by thermal cure to raise the T_g. The thermooxidative stability of cured PPQ containing pendent ethynyl or phenylethynyl groups was lower than that of a comparable PPQ without ethynyl groups. Incorporation of phenylethynyl groups improves the mechanical properties at 232°C, as demonstrated on adhesives. The Ti–Ti lap-shear strength at 232°C of a PPQ without phenylethynyl groups is 16.3 MPa (2370 psi); incorporation of 10 mol % of phenylethynyl groups increases it to 21.4 MPa (3100 psi) (65).

Acetylene-terminated phenylquinoxaline polymers have been reported (66–68). Films of an acetylene-terminated phenylquinoxaline polymer fabricated by compression molding at 316°C for 24 h and at 371°C for 5 h have a tensile strength at 25°C of 103.4 MPa (15,000 psi), tensile modulus of 2.61 GPa (380,000 psi), and break elongation of 5% (69). Specimens fabricated at 371°C under 1.38 MPa (200 psi) for 1 h from the acetylene-terminated phenylquinoxaline polymer shown below give Ti–Ti tensile shear strengths of 35.4 MPa (5130 psi) at 25°C; 18.8 MPa (2720 psi) at 260°C after 10 min at 260°C; and 14.5 MPa (2100 psi) at 260°C after 500 h at 260°C (70).

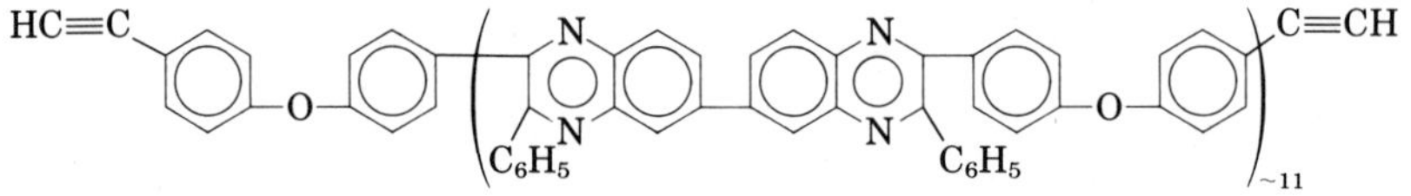

The cured acetylene-terminated phenylquinoxaline polymers are less stable under thermooxidative conditions than the parent polymers.

Polyquinolines. Polyquinolines (qv) and polyphenylquinolines are synthesized by routes similar to those for polyanthrazolines (71,72) and polyisoanthrazolines (72), where monomers such as 4,6-diaminoisophthaldehyde and 4,6-dibenzoyl-1,3-phenylenediamine react with aromatic diacetyl monomers under acid- or base-catalyzed conditions (Friedländer reaction). In the synthesis of polyquinolines (73) or polyphenylquinolines (74), a bis(o-aminoketone) (eg, 4,4′-bis(2-aminobenzoyl)diphenyl ether) reacts at 130°C with aromatic diacetyl (eg, 4,4′-diacetyldiphenyl ether) and di(phenylacetyl) [eg, 4,4′-di(phenylacetyl)diphenyl ether] monomers, respectively, in a 3:1 mixture of m-cresol and polyphosphoric acid (PPA).

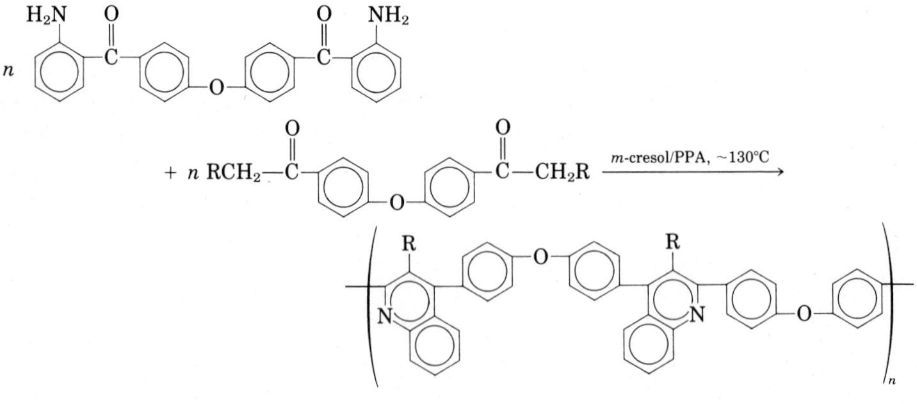

$$\tag{5}$$

where R = H or C_6H_5

Quinoline polymers with $\overline{M}_n$ as high as 78,900 have been reported by this route (73). Upon tga in air at a heating rate of 5°C/min, a polyquinoline started to lose weight at ~420°C; weight loss at ~500°C was 10%. A polyquinoline film exhibited no weight loss in air at 316°C after 100 h and a 10% weight loss at 360°C after 21 h (73). Quinoline polymers exhibit various degrees of crystallinity, depending primarily upon chemical structure and thermal history. Investigation of the thermal transitions of the polyquinoline in equation 5 reveals a T_g of 251°C and T_m of 483°C (~10% crystallinity) (75). The series of quinoline polymers in Table 12 shows that T_g increases with increasing phenyl substitution, presumably because of the bulkiness of the phenyl group, which sterically inhibits molecular motion.

Quinoline polymers have been end-capped with biphenylene groups with the expectation that thermal curing would induce chain extension and cross-linking (77); the curing reaction was catalyzed by bis(triphenylphosphine)dicarbonyl nickel(0). Surprisingly, the catalyst does not affect the thermooxidative stability of the cured polymer as measured by weight loss at 300°C in air. The mechanical properties of these polymers in composite form at 316°C are poor, probably because of insufficient cross-linking (78).

Polybenzimidazoles. Polybenzimidazoles (qv) (PBI) were initially synthesized in 1961 (79) by melt polycondensation of TAB or 1,2,4,5-tetraaminobenzene with diphenyl esters of aromatic dicarboxylic acids.

$$(6)$$

Other aromatic bis(o-diamines) have also been employed (80). A mixture of the monomers is heated under nitrogen from 250 to 280°C with evolution of large quantities of phenol and water. The oligomeric mixture is pulverized and further heated *in vacuo* to 400°C. A PBI prepared in this manner from stoichiometric quantities of TAB and diphenyl isophthalate had tga and itga curves as shown in Figures 1 and 2, respectively, and a T_g of ~408°C. PBIs made by this route are intractable. The oligomeric material formed after heating at ~250°C is the processable form, although it contains large amounts of condensation volatiles, which cause severe processing problems. Composites and bonded parts with useful mechanical properties for short periods of time at elevated temperatures have been reported (80–82). Typical tensile shear strengths on molybdenum stainless steel adherends are 27.6 MPa (4000 psi) at 25°C; 17.2 and 7.58 MPa (2500 and 1100 psi) at 300°C after 100 and 200 h, respectively, at 300°C in air; and 7.58 MPa (1100 psi) at 538°C after 10 min.

Polybenzimidazoles are processed into fibers, films, membranes, and low-to-medium-density foams. Depending upon the density, the foams have excellent strength retention after 100 h at 316°C. The foams, like other forms of PBI, are resistant to thermal shock from cryogenic to very high temperatures. Char yields are very high and, accordingly, ablative characteristics are excellent. Nonflammable fibrous material has a limited oxygen index of 38–43 (83). Sulfonated PBI

Table 12. Thermal Transitions of Polyquinolines[a]

Polymer	η_{inh}, dL/g	T_g, °C	T_m, °C
	0.36	268	none
	0.26	273	none
	0.38	326	475
	0.57	345	500
	1.80	266	448
	2.00	305	476
	3.10	308	480
	4.10	351	480
	1.10	300	455
	3.40	345	475

[a] Ref. 76.

fibers (textile grade) are commercially available (Celanese Corp., Charlotte, N.C.) (84). Sulfonated PBI films exhibit at 25°C a tensile strength of 110 MPa (16,000 psi), tensile modulus of 2.56 GPa (372,000 psi), and break elongation of 11% (85).

Phenyl-substituted PBIs have been prepared from aromatic bis(o-anilino-amino) compounds such as 3,3′-dianilinobenzidine and aromatic dicarboxylic acid derivatives such as diphenyl isophthalate (80). Phenyl-substituted PBIs are more thermoplastic with significantly lower T_g than unsubstituted PBIs primarily because of the absence of hydrogen bonding; in addition, thermooxidative stability is improved as measured by itga (see Fig. 2).

In poly-N-arylenebenzimidazoles, the chain is extended by an arylene moiety on the substituted N of the imidazole ring (86). These polymers are prepared in high molecular weights by melt polycondensation of monomers such as N,N'-bis(o-aminophenyl)benzidine and the diphenyl ester of 4,4′-diphenyl ether dicarboxylic acid (eq. 7).

$$(7)$$

Diphenyl isophthalate does not yield high molecular-weight polymer, because of the formation of a macrocyclic compound. Poly-N-arylenebenzimidazoles are more thermoplastic than the polyphenylbenzimidazoles; thermooxidative stability is superior.

Melt polymerization of 2,2′-bis(o-aminophenyl)bibenzimidazole with diphenyl isophthalate or other aromatic diphenyl esters yields polybenzimidazo-quinazolines (PIQ) (87,88).

$$(8)$$

Although problems were encountered in composite fabrication due to volatile evolution and poor flow, excellent composite properties were reported at 371°C after 200 h at 371°C in air (89).

Rigid Rodlike Polymers. High temperature, rigid rodlike polymers, as exemplified by polybenzothiazoles (qv) (PBT), polybenzoxazoles (PBO), benzimidazobenzophenanthroline-type ladder polymer (BBL), and poly(4-hydroxybenzoic acid) (poly-4-benzoate), are being developed primarily as high temperature reinforcements, eg, fibers, and ribbons. Poly-4-benzoate is also considered as molding material. The PBTs and PBOs are lyotropic liquid crystalline polymers (qv) with high thermooxidative stability due to their morphology and high molecular weight. PBTs with intrinsic viscosities of 30.3 dL/g were prepared in polyphosphoric acid (PPA) at 190–200°C by the polycondensation of 1,4-dimercapto-2,5-diaminobenzene dihydrochloride with terephthalic acid (90).

$$(9)$$

The PBT exhibited weight losses of 2% at 316°C after 200 h and 47% at 371°C after 200 h. PBT fibers, spun from anisotropic acid solutions using coagulation baths and subsequent heat treatment, had moduli of 250 GPa (2100 g/den), tenacity of 2.4 GPa (20 g/den), and elongation of 1.5% (91). These high modulus strength properties can be attributed to a high degree of molecular orientation along the fiber axis and a well-developed two-dimensional order (91) (see also HIGH MODULUS POLYMERS). PBT fibers exhibit an 82 and 73% retention of tensile modulus and strength, respectively, when tested at 200°C (91). Like Kevlar composites, the compressive strength of composites from PBT fibers is poor because of fiber buckling and interlaminar shear failure.

Similar to PBTs, PBOs are prepared in PPA by the polycondensation of 4,6-diaminoresorcinol dihydrochloride with terephthalic acid (92) or terephthaloyl chloride (93); PBO fiber properties include a modulus of 84.6 GPa (711 g/den), tenacity of 570 MPa (4.8 g/den), and elongation of 0.7% after a 425°C heat treatment. A PBO ribbon (0.32 cm wide × 0.05 mm thick) exhibits a tensile strength of 103 MPa (15,000 psi), tensile modulus of 7.6 GPa (1.1×10^6 psi), and elongation of 0.8% (93). More recent work indicates that PBO can be made in higher molecular weight and with thermooxidative stability and fiber properties better than those of PBT (94).

From the standpoint of thermal stability, the concept of ladder polymers is intriguing; two bonds must be broken to cleave the polymer chain. However, the synthesis and processing of true ladder polymers are extremely difficult. BBL is generally recognized as the thermally stable ladder-type polymer with the fewest defects. It is synthesized by the polycondensation of 1,2,4,5-tetraaminobenzene and 1,4,5,8-naphthalenetetracarboxylic acid in PPA at ~180°C (95).

$$(10)$$

+ isomers

A weight loss of ~10% at 600°C in air was determined by tga (heating rate 3°C/min) and an isothermal weight loss of ~15% was observed after 500 h at 371°C in air (95). BBL film cast from methanesulfonic acid solution followed by drying gave a tensile strength of 114.4 MPa (16,600 psi) at 25°C, tensile modulus of 7.6 GPa (1.1 × 10⁶ psi), and break elongation of 2.9% (96). The elongation is higher than expected and may be a result of plasticization by residual solvent (see also SPIRO AND LADDER POLYMERS).

Liquid crystalline polyesters such as a poly(4-benzoate) (Ekonol) exhibit a higher heat resistance than normally expected for a polyester. This is attributed to the morphology. Poly(4-benzoate) is prepared in high molecular-weight form from the thermal polycondensation of an AB monomer, phenyl-4-hydroxyben-zoate or 4-acetoxybenzoic acid (97). The polymer has a double-helix structure and exhibits a phase change in the 325–360°C range. The high temperature transition is attributed to the disruption of intermolecular ester–ester dipole interaction and not a crystalline melt (97). No distinct second-order transition (T_g) can be observed. The weight loss of a powder sample of poly(4-benzoate) is only 1.5% at 260°C in air after 1500 h (98). Moldings fabricated by compression molding at 438°C and 41.4–82.7 MPa (6,000–12,000 psi) exhibit a flexural strength of 75.8 MPa (11,000 psi) and flexural modulus of 6.9 GPa (1,000,000 psi) (98).

The Ekonol technology was used to develop the thermally stable liquid crystalline polyester XYDAR (Dartco Manufacturing Inc., Augusta, Ga.) based on 4-hydroxybenzoic acid, terephthalic acid, and 4,4'-diphenol (99,100); its properties are given in Table 13. A distortion temperature under load (ASTM D 648) of 337°C and 5% weight-loss temperature at 546°C (by tga) were reported (99). The injection-moldable polymer should find use in a variety of applications such as electronics (connectors and printed circuit boards), automobiles (molded components for use in hot areas), aerospace (structural components), and household articles (microwave cookware and appliances) (see also POLYESTERS, AROMATIC).

Table 13. Mechanical Properties of XYDAR (SRT-500)[a]

Mechanical property	ASTM test	Test temperature, °C		
		23	232	302
tensile strength, MPa[b]	D 1708	125.5	25.5	8.9
tensile modulus, GPa[c]	D 1708	8.3	3.3	2.6
flexural strength, MPa[b]	D 790	135.8	36.5	24.8
flexural modulus, GPa[c]	D 790	13.1	5.7	4.9
elongation	D 1708	4.8		

[a] Ref. 101.
[b] To convert MPa to psi, multiply by 145.
[c] To convert GPa to psi, multiply by 145,000.

Applications

The potential uses for heat-resistant polymers, given below, place stringent demands upon materials with heat resistance.

Electronic and microelectronic components (circuit boards, moldings, flexible cables, insulators, passivants, wire wraps, coatings)

Gaskets, sealants, and tubing

Binding system in brake shoes, abrasive wheels, and cutting disks

Casings, nozzles, and other parts for geothermal energy conversion systems

Structural resins (adhesives, composites, foams, and moldings) for advanced aircraft, space vehicles, and missiles

Engine components (fan blades, flaps, ducting, bushings, cowling, races, rods)

Nuclear reactor components (coolants, insulation, structural parts)

Conveyor belts for treating and drying materials

Filters (exhaust stacks)

Pipes for chemical processing and energy generators

Fire-resistant materials (protective clothing, parachutes)

Reinforcements (high modulus and high strength structural fibers and ribbons)

Ablators (thermal protection systems)

Automotive components (connecting rods, wrist pins, pistons, electrical components)

Coating on cookware (nonstick interior and decorative exterior surfaces)

Electronic applications (qv) probably constitute the largest potential market for heat-resistant polymers. There are many different applications where low moisture absorption, low dielectric constant, long life in an electrical field, high resistivity, and resistance to embrittlement are required. A recent application in multilevel circuitry requires a polymer that can form a planarizing coating to conform to sharp contours. After drying, the 0.5-μm uniform film should have a T_g of ~400°C and be able to withstand a temperature of 400°C for several hours in an inert atmosphere without dimensional change or degradation. In addition, the polymer should be soluble in a processable form in a nontoxic solvent such as diglyme (diethyleneglycol dimethyl ether) and have a very low metal content. Materials of this type are specialty polymers, sold at a low volume but at a high price.

In 1984, DuPont announced a joint venture with Toray, providing a world-wide capacity for Kapton of more than 1132.5 t/yr (102). The largest markets are in flexible printed circuitry, traction motors, aerospace, and magnetic wires and cables. Each space shuttle contains 906 kg of Kapton; new commercial airplanes, such as the Boeing 767, use several hundred kilograms. Lockheed used Kapton to seal the interior-wall fiber glass-insulation blankets for fire protection in the L-1011 Tristar jet (103). Epoxies, acrylics, and FEP (fluorinated ethylene propylene) teflon are used to bond Kapton to itself as well as to other substrates. There is a need for high temperature adhesives to bond Kapton to itself and other substrates; the polyimide LARC-TPI performs well in this application (17) and is currently under commercialization by Mitsui Toatsu Chemical, Inc.

Better high temperature sealants are needed for gaskets, sealants, and tubing in advanced high speed aircraft and other applications. The most serious material problem associated with the U.S. SST program in the late 1960s and early 1970s was a fuel-tank sealant. The leading edge of the wings, which serve as fuel tanks, were projected in the flight profile to experience temperatures from −54 to 248°C. There are still no sealants that perform under these conditions for the life of the aircraft, estimated to be ~50,000 service hours.

The potential market for a high temperature polymer to serve as a binding system in brake shoes, abrasive (grinding) wheels, and cutting disks is significant. These polymers should be inexpensive to compete with phenolics, but must exhibit good processability, compatibility with fillers, and good wear resistance over a wide temperature range.

High temperature polymers, by virtue of their chemical inertness, are candidates for applications in geothermal energy systems. The polymer must be stable upon exposure to the combination of superheated steam, brine, and sulfur compounds. Polyphenylquinoxaline casings and nozzles perform well in this environment, whereas linear polyimides, which are more prone to hydrolysis, do not. In a similar application, a cured acetylene-terminated imide functioned well as a binder in composite components used in deep-oil-well drilling operations.

The largest effort for heat-resistant polymers has been in the area of structural resins for advanced aircraft, space vehicles, and missiles, where processability is of utmost importance. In the fabrication of composites or large area adhesive bonds, no or very little volatile evolution can be tolerated. Even in the complex bleeder systems used in composite fabrication, 1% volatiles can lead to high porosity in the final part and accordingly lower thermooxidative stability and mechanical performance. Many high temperature polymers have been discarded because of processing problems arising from condensation volatiles or retained solvent that evolves at high temperatures. High molecular-weight linear heat-resistant polymers generally have limited solubility, and their solutions are viscous. As a result, it is difficult to obtain quality prepreg and adhesive tape, because filaments are not thoroughly wetted. With multiple-coating operations, resin can be absorbed, but prepreg made by this route often contains solvent. High molecular-weight linear heat-resistant polymers have poor flow and as a result, the escape of residual solvent or trapped air is difficult. In spite of these problems, certain linear polyimides and polyphenylquinoxalines exhibit outstanding high temperature adhesive (104) and composite (105) properties.

Processability is improved by end-capping with norbornenyl and acetylenic groups capable of addition polymerization at elevated temperatures. These materials form low viscosity, high solids content solutions, which thoroughly wet the reinforcement. In some cases, solutions of the monomers as previously discussed for PMR-15 polyimides are used for impregnation. After solvent and condensation volatiles are eliminated during a B-stage cycle, the resin has sufficient flow under pressure at elevated temperatures for further compaction. Polymerization occurs through end-group chain extension and cross-linking. Large complex composite structures are fabricated with this technique with good mechanical performance.

Other applications have special requirements. For example, certain components in nuclear reactors may have to be resistant to molten metal. Corrosion resistance under harsh conditions is needed for pipes in the chemical industry. Flame resistance and wear comfort are required for clothing exposed to high heat. For this application, aromatic polybenzimidazoles (79) are fabricated in fiber form. Sulfonated polybenzimidazoles provide a favorable combination of flame resistance and fiber properties (84).

Thermally stable polymers are used in coatings on cookware. For example, the nonstick coating on the interior of cookware is primarily polyphenylene sulfide or a polyimide blended with ~15 wt % of Teflon.

Outlook

Virtually every thermally stable heterocyclic polymer that can be prepared through the formation of the ring has been reported. It is unlikely any significant improvement in the thermal stability of organic polymers will be obtained. Polymers with outstanding heat resistance include:

Kapton, with excellent heat resistance and wide industrial use (19,103,106)

PI adhesive with no strength loss at 232°C after 37,000 h at 232°C in air (18)

PMR-15 graphite filament reinforced composites with good retention of mechanical properties at 316°C after hundreds of hours at 316°C in air (40)

PQ adhesive with high strength at 371°C after 50 h at 371°C in air (56)

PPQ film with high thermooxidative stability and weight loss of ~1% at 316°C in air after 1,100 h (60)

PBI composites with useful mechanical properties for short time at temperatures >538°C (60,82)

PIQ composites with excellent retention of mechanical properties at 371°C (89)

Advances will be in improved processability, lower costs, and fabrication methods. New synthetic routes via addition-type polymerization will be developed. More stringent process conditions used to fabricate components from heat-resistant polymers will be accepted.

Of the many heat-resistant polymers reported, only a few are commercially available because of high cost, difficult processability, and a questionable market. As the processing characteristics are improved, larger markets will be developed, and the cost of neat resins and finished products will become more attractive. A variety of heat-resistant polymers with a favorable combination of price, processability, and performance will be offered in different product forms in the future.

BIBLIOGRAPHY

"Heat-resistant Polymers" in *EPST* 1st ed., Vol. 7, pp. 478–506, J. E. Mulvaney, University of Arizona.

1. T. L. Cottrell, *The Strength of Chemical Bonds,* 2nd ed., Butterworths, London, 1958.
2. T. H. Thomas and T. C. Kendrick, *J. Polym. Sci. Part A-1* **7**, 537 (1969).
3. V. L. Bell, B. L. Stump, and H. Gager, *J. Polym. Sci. Polym. Chem. Ed.* **14**, 2275 (1976).
4. P. M. Hergenrother, *SAMPE Q.* **3**, 1 (1971).
5. U.S. Pat. 2,710,853 (June 1955), W. M. Edwards and I. M. Robinson (to E. I. du Pont de Nemours & Co., Inc.).
6. U.S. Pat. 3,179,614 and 3,179,634 (April 20, 1965), W. M. Edwards (to E. I. du Pont de Nemours & Co., Inc.).
7. U.S. Pat. 3,179,630 (April 20, 1965), A. L. Endrey (to E. I. du Pont de Nemours & Co., Inc.).
8. G. M. Bower and L. W. Frost, *J. Polym. Sci. Part A* **1**, 3135 (1963).
9. L. W. Frost and I. Kesse, *J. Appl. Polym. Sci.* **8**, 1039 (1969).
10. R. A. Dine-Hart and W. W. Wright, *J. Appl. Polym. Sci.* **11**, 609 (1967).

11. P. P. Nechayev, Ya. S. Vygodskii, G. YeZaikov, and S. V. Virogradova, *Polym. Sci. USSR* **18**, 1903 (1978).
12. H. H. Gibbs and C. V. Breder, *Polym. Prepr.* **15**(1), 775 (1974).
13. A. K. St. Clair, T. L. St. Clair, and K. T. Shevket, *Polym. Matl. Sci. Eng. Prepr.* **51**, 62 (1984).
14. T. Takekoshi, W. B. Hillig, G. A. Mellinger, J. E. Kochanowski, J. S. Manello, M. J. Webber, R. W. Bulsen, and J. W. Nehrich, *NASA CR-145007*, National Aeronautics and Space Administration, Hampton, Va., July 1975.
15. U.S. Pat. 4,094,862 (June 1978), V. L. Bell (to the National Aeronautics and Space Administration).
16. A. K. St. Clair and T. L. St. Clair, *Sci. Adv. Matl. Proc. Eng. Sci.* **26**, 165 (1981).
17. A. K. St. Clair, W. S. Slemp, and T. L. St. Clair, *Sci. Adv. Matl. Proc. Eng. Ser.* **23**, 113 (1978).
18. C. Hendricks, Paper presented at *The Second Symposium on Welding, Bonding, and Fastening,* October 1984, NASA Langley Research Center, Hampton, Va.
19. C. E. Sroog in N. M. Bikales, ed., *Encyclopedia of Polymer Science and Technology,* 1st ed., Vol. 11, John Wiley & Sons, Inc., New York, 1969, p. 247.
20. N. W. Todd, F. A. Wolff, R. S. Mallouk, and J. R. Courtwright, *Machine Design Plastics Ref. Issue,* June 16, 1966.
21. H. H. Gibbs, *Sci. Adv. Matl. Proc. Eng. Sci.* **19**, 154 (1974).
22. J. H. Bateman, W. Geresy Jr., and D. S. Neiditch, *Coat. Plast. Prepr.* **35**(2), 77 (1975).
23. U.S. Pat. 3,856,752 (Dec. 24, 1974), J. H. Bateman and D. Gordon (to CIBA-GEIGY Corp.).
24. U.S. Pat. 4,111,906 (Sept. 5, 1978), R. J. Jones, M. K. O'Rell, and J. M. Hom (to TRW Systems).
25. U.S. Pat. 3,501,443 (Mar. 17, 1970), R. R. Dileone (to American Cyanamid Co.).
26. F. W. Harris, W. A. Feld, and L. H. Lanier, *J. Polym. Sci. Part B* **13**, 283 (1975).
27. R. J. Jones, TRW Systems, Redondo Beach, Calif., private communication, Nov. 1984.
28. F. W. Harris, S. M. Padaki, and S. Varaprath, *Polym. Prepr.* **21**(1), 3 (1980).
29. W. J. Farrissey, J. S. Rose, and P. S. Carleton, *Polym. Prepr.* **9**(2), 1581 (1968).
30. U.S. Pat. 3,708,458 (Jan. 2, 1973), L. M. Alberino, W. J. Farrissey, and J. S. Rose (to Upjohn Co.).
31. U.S. Pat. 4,394,464 (July 19, 1983), J. Gagliani and J. V. Long (to Chem-Tronics).
32. U.S. Pat. 3,730,940 (May 1, 1973), J. G. Wirth and D. R. Heath (to General Electric Co.).
33. D. M. White, T. Takekoshi, F. J. Williams, H. M. Relles, P. E. Donahue, H. J. Klopfer, G. R. Loucks, J. S. Manello, R. O. Matthews, and R. W. Schluenz, *J. Polym. Sci. Polym. Chem. Ed.* **19**, 1635 (1981).
34. U.S. Pat. 3,528,950 (Sept. 15, 1970), H. R. Lubowitz (to TRW Systems).
35. H. R. Lubowitz, *Polym. Prepr.* **12**, 561 (Oct. 29, 1971).
36. U.S. Pat. 3,845,018 (Oct. 29, 1974), N. Bilow, A. L. Landis, and L. J. Miller (to Hughes Aircraft Co.).
37. A. L. Landis, N. Bilow, R. H. Boschan, R. E. Lawrence, and T. J. Aponyi, *Polym. Prepr.* **15**(2), 533, 537 (1974).
38. T. L. St. Clair and R. A. Jewell, *Soc. Adv. Matl. Proc. Eng. Ser.* **23**, 520 (1978).
39. T. T. Serafini, P. Delvigs, and G. R. Lightsey, *J. Appl. Polym. Sci.* **16**, 905 (1972); U.S. Pat. 3,745,149 (July 10, 1973), (to NASA).
40. T. T. Serafini, P. Delvigs, and W. B. Alston, *Soc. Adv. Matl. Proc. Eng. Ser.* **27**, 320 (1982).
41. N. Bilow and A. L. Landis, *Natl. SAMPE Tech. Conf. Ser.* **8**, 94 (1976).
42. A. L. Landis and A. B. Naselow, *Natl. SAMPE Tech. Conf. Ser.* **14**, 236 (1982).
43. D. F. Loncrini, *J. Polym. Sci. Part A-1* **4**, 1531 (1966).
44. J. A. Augl, *J. Polym. Sci. Part A-1* **8**, 3145 (1970).
45. U.S. Pat. 3,049,518 (1962), C. W. Stephens (to E. I. du Pont de Nemours & Co., Inc.).
46. *Torlon Engineering Resins, Technical Data Sheets,* Amoco Chemical Corp., Naperville, Ill.
47. N. A. Adrova, M. I. Bessonov, L. A. Lauis, and A. P. Rudakov, *Polyimides—A New Class of Thermostable Polymers,* Technomic Publishing Co., Stamford, Conn., 1970, p. 151.
48. J. P. Critchley, G. J. Knight, and W. W. Wright, *Heat-Resistant Polymers,* Plenum Press, New York, 1983, p. 200.
49. M. I. Bessonov, *Polyimides—Class of Thermally Stable Polymers,* Nauka, Lenings. Otd., Leningrad, USSR, 1983.
50. K. L. Mittal, *Polyimides: Synthesis, Characterization and Applications,* Vols. 1 and 2, Plenum Press, New York, 1984.

51. G. P. deGaudemaris and B. J. Sillion, *Polym. Lett.* **2**, 203 (1964).
52. J. K. Stille and J. R. Williamson, *Polym. Lett.* **2**, 209 (1964).
53. G. P. deGaudermaris, B. Sillion, and J. Preve, *Bull. Soc. Chem. France* **1964**, 1793 (1964).
54. U.K. Pat. 1,021,005 (Feb. 1966) B. Sillion and G. deGaudemaris (to Institut Francais du Petrole).
55. U.S. Pat. 3,661,850 (May 9, 1972), J. K. Stille.
56. P. M. Hergenrother and H. H. Levine, *J. Appl. Polym. Sci.* **14**, 1037 (1970).
57. P. M. Hergenrother and H. H. Levine, *J. Polym. Sci. Part A-1* **5**, 1453 (1967).
58. P. M. Hergenrother, *J. Macromol. Sci. Revs. Macromol. Chem.* **6**(1), 1 (1971).
59. P. M. Hergenrother, *Macromolecules* **7**, 575 (1974).
60. P. M. Hergenrother, *Appl. Polym. Sym.* **22**, 57 (1973).
61. P. M. Hergenrother, *Polym. Eng. Sci.* **16**(5), 303 (1976).
62. F. L. Hedberg and F. E. Arnold, *J. Polym. Sci. Polym. Chem. Ed.* **14**, 2607 (1976).
63. P. M. Hergenrother, *Macromolecules* **14**, 891 (1981).
64. *Ibid.*, 898 (1981).
65. P. M. Hergenrother, *J. Appl. Polym. Sci.* **28**, 355 (1983).
66. R. F. Kovar, G. F. L. Ehlers, and F. E. Arnold, *J. Polym. Sci. Polym. Chem. Ed.* **15**, 1081 (1977).
67. P. M. Hergenrother in S. S. Labana, ed., *Chemistry and Properties of Crosslinked Polymers*, Academic Press, Inc., New York, 1977, p. 107.
68. F. L. Hedberg and F. E. Arnold, *J. Appl. Polym. Sci.* **24**, 763 (1979).
69. C. E. Browning, A. Wereta, Jr., J. T. Hartness, and R. F. Kovar, *Sci. Adv. Matl. Proc. Eng. Ser.* **21**, 94 (1976).
70. P. M. Hergenrother, *Polym. Eng. Sci.* **21**, 1072 (1981).
71. W. Bracke, *Macromolecules* **2**, 286 (1969).
72. Y. Imai, E. F. Johnson, T. Katto, M. Kurihara, and J. K. Stille, *J. Polym. Sci. Polym. Chem. Ed.* **13**, 2233 (1975).
73. J. F. Wolfe and J. K. Stille, *Macromolecules* **9**, 489 (1976).
74. S. O. Norris and J. K. Stille, *Macromolecules* **9**, 496 (1976).
75. W. Wrasidlo and J. K. Stille, *Macromolecules* **9**, 505 (1976).
76. W. Wrasidlo, S. O. Norris, J. F. Wolfe, T. Katto, and J. K. Stille, *Macromolecules* **9**, 512 (1976).
77. J. P. Droske and J. K. Stille, *Macromolecules* **17**, 1 (1984).
78. J. P. Droske, J. K. Stille, and W. B. Alston, *Macromolecules* **17**, 14 (1984).
79. H. Vogel and C. S. Marvel, *J. Polym. Sci.* **50**, 511 (1961).
80. H. H. Levine in N. M. Bikales, ed., *Encyclopedia of Polymer Science and Technology*, 1st ed., Vol. 11, Wiley–Interscience, New York, 1969, p. 188.
81. H. H. Levine and co-workers, *AFML-TR-64-365*, Vol. 1, U.S. Air Force report, Air Force Wright Aeronautical Laboratories, Wright-Patterson AFB, Ohio, Pt. 1, 1963.
82. J. F. Jones, J. C. Waldorf, and R. Fountain *Soc. Adv. Matl. Proc. Eng. Ser.* **29**, 777 (1984).
83. J. H. Ross, *Soc. Adv. Matl. Proc. Eng. Ser.* **19**, 166 (1974).
84. G. M. Moelter, R. F. Tetreault, and M. J. Hefferson, *Polym. News* **9**(5), 134 (1983).
85. E. J. Powers, Celanese Corporation, Charlotte, N.C., private communication, Sept. 1984.
86. U.S. Pat. 3,708,439 (Jan. 2, 1973), A. R. Sayjigh, B. W. Tucker, and H. Ulrich (to Upjohn Co.).
87. H. H. Levine and co-workers, *AFML-TR-67-23, Air Force report, 1967; AFML-TR-67-429,* Air Force report, Air Force Wright Aeronautical Laboratories, Wright-Patterson AFB, Ohio, 1968.
88. U.S. Pat. 3,503,929 (Mar. 31, 1970), B. L. Loudas (to Minnesota, Mining and Manufacturing Co.).
89. T. J. Aponyi and C. B. Delano, *Soc. Adv. Matl. Proc. Eng. Ser.* **19**, 178 (1974).
90. J. F. Wolfe, B. H. Loo, and F. E. Arnold, *Macromolecules* **14**, 915 (1981).
91. S. R. Allen, A. G. Filippov, R. J. Farris, E. L. Thomas, C. P. Wong, G. C. Berry, and E. C. Chenevey, *Macromolecules* **14**, 1136 (1981).
92. J. F. Wolfe and F. E. Arnold, *Macromolecules* **14**, 909 (1981).
93. E. W. Choe and S. N. Kim, *Macromolecules* **14**, 920 (1981).
94. J. F. Wolfe, SRI International, Menlo Park, Calif., private communication, Dec. 1984.
95. F. E. Arnold and R. L. VanDeusen, *Macromolecules* **2**, 497 (1969).
96. F. E. Arnold, *Polym. Prepr.* **12**(2), 179 (1971).
97. J. Economy, R. S. Storm, V. I. Matkovich, S. E. Cottis, and B. E. Nowak, *J. Polym. Sci. Polym. Chem. Ed.* **14**, 2207 (1976).

98. J. Economy, B. E. Nowak, and S. G. Cottis, *Polym. Prepr.* **11**(1), 332 (1970).
99. *Mod. Plast.* **61**(12), 16 (1984).
100. *Plast. World* **42**(13), 8 (1984).
101. *XYDAR® SRT-500,* technical data sheet, Dartco Mfg., (Dart and Kraft Co.), Augusta, Ga., Aug. 1985.
102. *SAMPE J.* **20,** 71 (1984).
103. *Polym. News* **7**(3), 125 (1981).
104. P. M. Hergenrother, *Chemtech.,* 496 (Aug. 1984).
105. P. M. Hergenrother and N. J. Johnston, *ACS Symp. Ser.* **132,** 1 (1980).
106. *Kapton Polyimide Product Bulletin E-50533,* E. I. du Pont de Nemours & Co., Inc., Wilmington, Del., May 1983.

General References

Refs. 19 and 47–50 are good general references for polyimides.
Ref. 48 is also good for other heat-resistant polymers.
P. E. Cassidy, *Thermally Stable Polymers,* Marcel Dekker, Inc., New York, 1980.
A. H. Frazer, *High Temperature Resistant Polymers,* Wiley–Interscience, New York, 1968.
V. V. Korshak, *The Chemical Structure and Thermal Characteristics of Polymers,* Israel Translation, Keter Press, Jerusalem, 1971.

P. M. HERGENROTHER
NASA Langley Research Center

HIGH MODULUS POLYMERS

High modulus polymers are potential materials of choice in applications requiring high mechanical properties and light weight. In Figure 1 the display of specific modulus and strength of high performance materials, demonstrates that high modulus polymers show a distinct advantage over metals and ceramics on a property–weight basis. However, the penetration of polymers into these applications has been relatively modest for several reasons: only the aramid fibers, Kevlar (DuPont), and Arenka (Akzo), have a significant history of commercial availability. The thermotropic copolyesters, Vectra (Celanese) and Xydar (Dart), and high modulus polyethylene fibers, Spectra 900 (Allied), were not introduced commercially until 1985. Because they are new materials, a design criteria and performance expectation experience base is lacking. With the exception of the thermotropic copolyesters, high modulus polymers are available only as fibers and hence must be used in composite form to produce complex three-dimensional parts. The materials are relatively high cost, ~$4.5/kg, and more.

These polymers are highly anisotropic, and the high modulus is achieved only in the direction of molecular chain orientation. Properties normal to the molecular axis tend to be an order of magnitude or more lower than the longi-

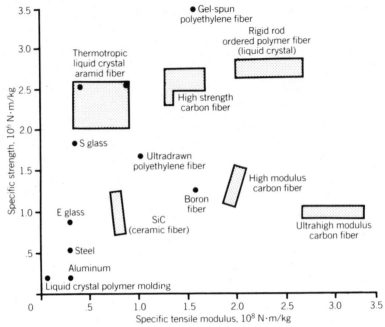

Fig. 1. Comparison of specific strength vs specific modulus for a variety of high performance materials (1). Specific properties are normalized by material density (Pa or N/m² divided by kg/m³).

tudinal values resulting in low compressive and/or shear properties. Compensation for the low shear and compressive properties is accomplished through composite design concepts.

In this article, modulus is taken to mean the tensile modulus as defined by the slope of the initial linear portion of the load extension response (stress–strain curve) of a specimen deformed at room temperature. Some typical stress–strain curves are shown in Figure 2. The term polymer will include materials whose properties are determined by their polymeric nature alone: copolymers, polymer blends, and molecular composites. Highly filled systems or reinforced composites with polymeric matrices are not discussed (see COMPOSITES; REINFORCED PLASTICS).

The tensile modulus of a polymer depends on several factors: the chemical moieties and bonding comprising the molecular chain, the conformation of the molecular chain, the orientation of the molecular chains with respect to the axis of measurement, and the packing density of the molecular chains in the solid state. For a given polymer the crystal structure defines the highest packing density.

For a given molecular structure, the tensile modulus increases as the straightness or rodlike character of the molecular chain increases; the orientation of the molecular chain approaches perfection (in the direction of measurement); and the packing density of the chains, ie, the number of chains per cross-sectional area, is maximized. The modulus is a strong function of the solid-state structure

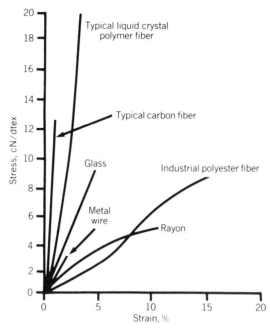

Fig. 2. Comparison of typical stress–strain curves of high performance polymers and related materials. To convert cN/dtex to gf/denier, multiply by 1.13 (2).

which, in turn, is determined by the total process history of the specimen. In the absence of high molecular orientation, no polymeric structure exhibits high modulus. In the case of conventional polymers, high modulus properties can only be produced by careful processing to highly perfected structures. The theoretical limit of tensile modulus is that of the perfect polymeric single crystal, measured parallel to the molecular chain axis. Some representative calculated modulus values as of 1985 are given in Table 1. A definition of high modulus polymer may be one that exhibits a tensile modulus > ca 25% of theory. But caution must be used in comparing the theoretical modulus of a straight chain polymer, eg, polyethylene, to that of a helical chain such as polyoxymethylene, because a high modulus variant of the helical polymer may be significantly lower in absolute modulus vs a low modulus variant of a straight chain polymer.

Typical articles produced from glassy or semicrystalline polymers exhibit tensile modulus values from ca 1–3 GPa (145,000–435,000 psi) for unoriented structures, eg, moldings to ~15 GPa (2×10^6 psi) for highly oriented structures such as tire cords or strapping tapes. The definition of high modulus polymer as a structure possessing a tensile modulus > ca 25 GPa distinguishes these materials from more typical polymeric articles. Table 2 summarizes the modulus range of some common polymers. All highly oriented polymer structures are anisotropic; as modulus is increased by processing to high molecular chain orientation in one direction, it decreases commensurately in all other directions.

Table 1. Tensile Moduli, Maximum Values[a]

Material	Modulus, GPa[b]
polyethylene	240
polypropylene[c]	42
polystyrene	12
polyoxymethylene[c]	53
poly(vinyl alcohol)	250
poly(ethylene terephthalate)	137
poly(p-phenyleneterephthalamide)	200
poly(p-phenylene benzobisthiazole)	>300
thermotropic copolyester (typical)	200
aluminum	69
steel	208
glass	69–183
carbon fiber	285–980

[a] Ref. 3–5.
[b] To convert GPa to psi, multiply by 145,000.
[c] Molecule has helical conformation.

Table 2. Moduli and Glass-Transition Temperatures of Some Common Polymers[a,b]

Polymer	Modulus, GPa[c]	Order	T_g, °C
polyethylene	0.6–170	semicrystalline	−85
polypropylene	0.5–25	semicrystalline	−18
polystyrene	2.8	amorphous	100
polyoxymethylene	3–30	semicrystalline	−82
poly(ethylene terephthalate)	6–18	semicrystalline	75
nylon-6	6–12	semicrystalline	50
poly(p-phenyleneterephthalamide)	62–124	crystalline	360
poly(p-phenylene benzobisthiazole)	47–282	ordered	
thermotropic copolyester	14–124	ordered	150–250

[a] Ref. 6.
[b] At room temperature.
[c] To convert GPa to psi, multiply by 145,000.

In addition to the molecular and structural criteria, the temperature of modulus measurement and the relation of that temperature to the melting temperature, glass temperature or other transition temperatures of the polymer, is important in determining the modulus value of a given polymer. Any process leading to relaxation of molecular orientation or lowering of chain-packing density lowers the modulus. Sample restraint or the lack of it during heating also plays a role. The degree of crystallinity of the polymer determines the extent of modulus drop-off in the vicinity of T_g. A typical amorphous polymer can experience a modulus drop of an order of magnitude or more in the domain of T_g. Thus, a polymer with a high transition temperature or a highly crystalline polymer with an unimpressive modulus at room temperature, may become a relatively high modulus polymer at elevated temperature when compared to less thermally stable molecules. Table 2 also shows degrees of order and glass-transition temperatures of polymers.

There is no single, all encompassing consistent definition of high modulus polymer. In this article the practical but imprecise higher than typical criteria will be employed. Polymeric products with moduli of $70+$ GPa ($>10 \times 10^6$ psi) can be produced through a variety of technologies. In general, the tensile strength of polymers tends to follow the same molecular and structural principles as the modulus. Strength development is complicated, however, by the effects of molecular weight and the statistical probabilities of finding flaws or stress concentrators leading to failure of the specimen (see also MECHANICAL PROPERTIES).

Production

Two fundamentally different approaches to the production of high modulus polymers have emerged. One is the transformation of conventional, flexible-chain, random-coil polymers into almost perfectly oriented, extended-chain structures in the solid state. The most successful example of this approach is gel spinning of high molecular weight polyethylene. Other examples include solid-state extrusion (qv), die drawing, and superdrawing. These methods are extensions of fiber processing (3,7).

The second approach is the molecular design of stiff, planar molecules that exhibit liquid-crystal behavior in solution (lyotropic) or in the melt (thermotropic) and form highly oriented, extended-chain structures in the solid state. Three important examples are poly(p-phenyleneterephthalamide), the lyotropic aramid commercialized first by DuPont as Kevlar; ordered polymers, a group of ultrastiff lyotropic molecules that exhibit the highest modulus values yet observed in organic polymers; and thermotropic, aromatic copolyesters.

To a first approximation, whether the approach is chemical design or morphological manipulation, the end result is a solid state structure which may be represented as a close-packed array of rods. Although crystallinity to hold the rods together is desirable, evidence suggests that long-range, three-dimensional order is not necessary for high modulus properties. The highly crystalline aramid fibers and the much lower crystallinity thermotropic copolyester fibers show overlapping modulus values (Fig. 1). Similarly, many of the very high modulus rigid-rod polymers exhibit only one- or two-dimensional degrees of order in the solid state. True, three-dimensional crystallinity in these highly oriented structures influences the temperature range of tensile modulus stability and transverse properties more than the absolute value of tensile modulus achieved. That is, the crystal unit cell defines the closest molecular chain packing; crystallinity mitigates the lowering effects of secondary transitions and molecular motions on the tensile modulus as a function of temperature; and crystallinity maximizes interchain interactions which, in turn, should maximize shear and compressive properties at the molecular level.

However, much of the mechanical behavior typical of high modulus polymers may be deduced from a simple close-packed rod model. The high mechanical property anisotropy is a direct consequence of the molecular bond anisotropy inherent in these structures. Weakness in shear or compression is also a simple reflection of the structural anisotropy. The ease with which molecular chains or units comprised of a number of molecular chains, can slip past each other, is a key parameter in defining mechanical values in a given high modulus system. In the absence of a uniform and infinite molecular weight, this overlap interaction is the stress–transfer mechanism of these structures and controls shear strength, tensile strength, etc.

Highly oriented polymers form microfibrillar morphologies in the solid state (8). Recently, it has been shown that the solid state of the thermotropic co-polyesters can best be described with a fibrillar hierarchical model, as illustrated in Figure 3. Similar structures have been noted in the aramids (10), the ordered polymers, and in high modulus polyethylene (3). It is not yet clear what role the fibrillar morphology plays in determining tensile modulus in a given polymer system. Most effects can be explained from purely molecular considerations.

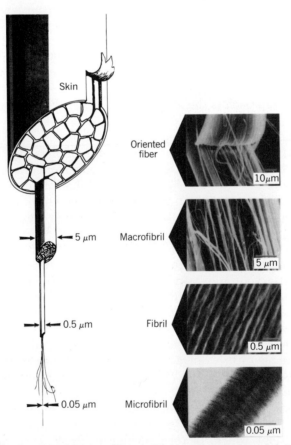

Fig. 3.   Liquid-crystalline polymer structural model (9).

Uniform production of a close-packed array of rods from flexible macro-molecules requires careful lining up of the molecular chains, which, in turn, necessitates the production of one- or two-dimensional structures, ie, fibers and tapes. At least one very thin dimension is needed in the specimen to preclude the formation of thermal or stress gradients that would severely limit the ultimate orientation and hence, properties. One exception is the solid state extrusion of conventional polymers that can be applied to reasonably large cross sections. The most important, and only general exception to the fiber limitation is the thermotropic copolyesters, which may be processed by any thermoplastic processing method, ie, extrusion, injection molding, etc, to produce relatively high modulus articles (4).

High Modulus Structures From Conventional Polymers

The production of high modulus articles from conventional polymers involves fiber-spinning technology in which the control of molecular orientation allows control of mechanical properties of the final fiber. Molecular orientation may be imparted to a fiber during the extrusion (spinning) process or after spinning in a solid state orientation (drawing) process, and will be permanent if conditions include a faster orientation time than polymer relaxation time (see also ORIENTATION PROCESSES). Typical processes are shown diagrammmatically in Figure 4. Detailed process history is essential in determining ultimate fiber properties. Reviews of spinning technology are available (12,13).

Critical insight into the structure of oriented polymers and the nature of the drawing process, especially when applied to polyethylene, has been provided (7,14–21). Axial properties are a function of the molecular orientation which is a function of draw ratio. In semicrystalline polymers the relative importance of the orientation of the crystalline and noncrystalline phases and how these parameters build with process conditions has been assessed. To a first approximation, the average molecular orientation is independent of phase details and increases monotonically with increasing draw ratio.

Draw ratio is a function of molecular weight, and the density, nature and permanence of molecular chain entanglements. To produce high modulus structures from conventional polymers, the objectives are to maximize molecular orientation, minimize structural heterogeneities in the precursor materials to prevent specimen failure before high orientation is achieved, ie, eliminate stress concentrators in the precursor material, and maximize molecular weight.

Composite concepts such as the aggregate and Takayanagi models are used (20). The perfection of molecular orientation and the critical role of stress transfer within the highly oriented microstructure are of foremost importance in determining the ultimate properties of the sample. Application of these concepts to polymer extrusion has led to three routes of production; most of the reported research discusses polyethylene. The routes are solid state extrusion (qv), superdrawing, and gel spinning; the last method is the most successful and commercially advanced.

Solid State Extrusion. In the solid state extrusion process, a polymer billet is oriented by forcing it through a die in the solid state (Fig. 5). The draw-ratio relationships are as follows:

maximum extrusion draw ratio

$$DR_M = \left(\frac{R_a}{r}\right)^2$$

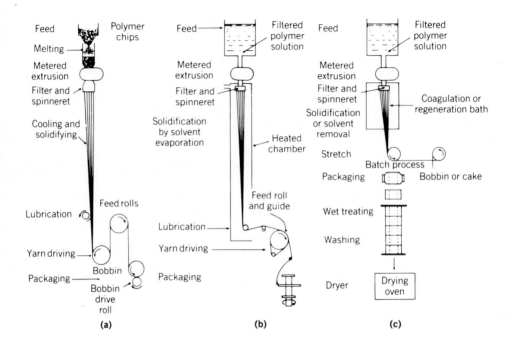

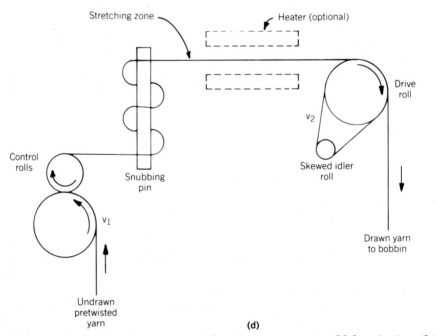

Fig. 4. Typical fiber-spinning and drawing processes. (**a**) Melt spinning; (**b**) dry spinning; (**c**) wet spinning; (**d**) fiber drawing (11).

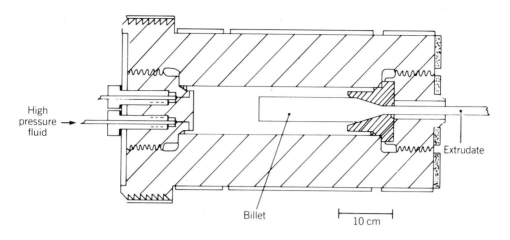

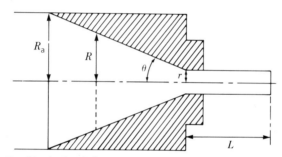

Fig. 5. Typical solid state extrusion processes (7).

conical die, DR(R) < DR$_M$

$$DR(R) = \left(\frac{R_a}{R}\right)^2 = \left(1 - \frac{3r^2L \tan \theta}{R_a^3}\right)^{-2/3}$$

$$= \left[1 - \frac{(DR(r)^{3/2} - 1)}{DR_M^{3/2}}\right]^{-2/3}$$

fiber, DR(r) < DR$_M$

$$DR(r) = \left(\frac{R}{r}\right)^2 = \left(1 + \frac{3L \tan \theta}{r}\right)^{2/3}$$

where R_a, R, r, and θ are shown in Figure 5.

Cross sections of up to several centimeters with moduli on the order of 60 GPa (9×10^6 psi) have been produced from polyethylene (3,7). Acetal resins have also been successfully extruded in the solid state (22).

Variations include hydrostatic extrusion in which a pressure fluid is used to force the extrudate through the die from a uniform pressure environment (7). In die drawing, simultaneous pushing and pulling of the billet through the die is done (7). In split billet extrusion, a billet sandwich of an easy-to-extrude polymer (outside) and a difficult-to-extrude polymer (inside) is prepared (3).

Superdrawing.   By carefully drawing a polymer sample, usually fiber or tape, much higher draw ratios than normal are achieved. Because superdrawing is similar in principle to conventional drawing, the process has been applied to polyethylene (7,23) and polyoxymethylene (24). The success of this approach requires melt production of low orientation precursors with a minimum of chain entanglements, necessitating the use of low-to-intermediate molecular weights. Typical tensile properties for a superdrawn polyethylene fiber include a modulus of 70 GPa (10 × 10^6 psi) at a tenacity of ca 1 GPa (145,000 psi). A marked improvement in the creep performance and modulus retention at elevated temperature can be achieved with high modulus polyethylene through radiation cross-linking after processing (25). This was noted for superdrawn, relatively low molecular weight polyethylene; the effect on gel-spun high modulus polyethylene fibers with ultrahigh molecular weights is unknown.

Polyethylene single crystal mats are an excellent precursor for the production of high modulus polyethylene; both superdrawing and solid-state extrusion processes are effective orienting methods (26). Using molecular weights of 2 million, draw ratios up to 250× and tensile moduli of up to 220 GPa (32 × 10^6 psi) have been reported.

Gel Spinning.   Most reports of gel spinning employ high molecular weight polyethylene and other polyolefins (27–33). Gel spinning of poly(vinyl alcohol) (34) and polyacrylonitrile (35) has also been described.

In gel spinning high density polyethylene (HDPE) there are three unit operations: formation of a high molecular weight HDPE solution capable of gelation, extrusion of the solution to form gel fiber, and hot stretching of the fiber. A typical gel spinning apparatus is shown in Figure 6. The extruded solution passes through an air gap into a cooling bath, precipitating the unoriented gel fiber. The fiber is then hot-drawn in the conventional way to produce the final

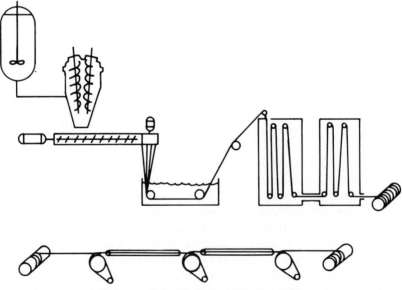

Fig. 6.   Gel-spinning apparatus (33).

structure. The method is similar to the dry-jet, wet spinning of aramids. Spin orientation as a means of developing properties in the gap has not been reported.

HDPE with M_w 10^6 and low polydispersity is dissolved in a common solvent, eg, decalin, at a concentration of ca 2–8 wt % at elevated temperatures with stirring. This solution is then spun at 130–140°C into a room-temperature bath, often water, to form a gel fiber with up to 98% solvent remaining in the fiber. The solvent may be removed prior to hot drawing or allowed to evaporate during the drawing process. Processing speeds are very slow, ie, cm to m/min. The draw temperature, typically 120°C, is below the polymer melting point, but above the gel temperature and the crystal dispersion temperature. Draw ratios vary from 30 to <100.

In the gel-spinning process, a high molecular weight gel precursor structure capable of being hot drawn to very high draw ratios must be formed. As shown in Figure 7, the mechanical properties of HDPE change monotonically as a function of draw ratio; the modulus increases in an essentially linear fashion. Gel spinning allows the formation of ultrahigh molecular weights not possible in conventional processing. The effects of concentration, molecular weight, solvent content, draw ratio, draw temperature, and draw rate on the structure and properties of the resulting fiber have been studied (18,19,27–33,36–38). Mitsui Petrochemical Industries has announced a variation of the gel spinning manufacturing process for their Tekumilon polyethylene fiber. Rather than using the usual low concentration solution, high molecular weight polyethylene is plasti-

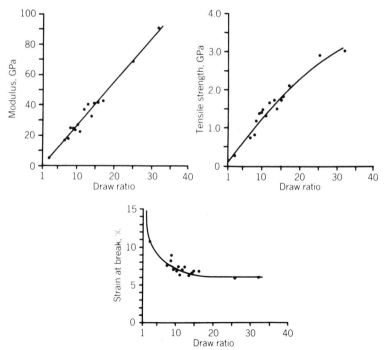

Fig. 7.   Tensile properties of high density polyethylene as a function of draw ratio (30–32).

cized with paraffins and the solution concentration raised to ca 15%. Reported properties are similar to those of other gel-spun products.

The gel structure of polyethylene is that of a molecular network with crystalline entanglements; low concentrations minimize the number of entanglements. At or above the crystal dispersion temperature of ~90°C, the molecular chains can be pulled out of the crystal and straightened if the operation is performed gently enough to prevent fracture (qv). Deformation of this network by hot-drawing allows the production of a very highly oriented, essentially extended chain-microfibrillar structure similar to that observed in high pressure crystallized HDPE (39) and all other highly oriented polymer structures (9,17).

Mechanical properties, especially strength, tend to increase with increasing molecular weight. For HDPE with a given molecular weight distribution, the strength and modulus obtained for a given orientation are independent of the method of formation (Fig. 8). The strength–molecular weight relationship is not well understood. Empirically, the strength increases as the absolute molecular weight increases and the molecular weight distribution narrows (38).

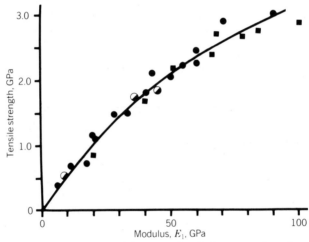

Fig. 8. Mechanical properties of drawn polyethylene as a function of the production technique. ●, surface grown crystals; ◓, solution spun/drawn wet; ■, solution spun/drawn dry (37).

It is also found operationally in melt and solid-state processes that maximum draw-ratio decreases with increasing molecular weight, in contrast to single-chain calculations which suggest the opposite. The reason for this contradiction is that as molecular weight increases, the density of trapped entanglements in the precursor structure, whether melt or solid, increases, leading to draw stresses sufficiently high to fracture the material at relatively low draw ratios. Hence, for a given process the improved properties available through higher molecular weight must be balanced with the reduced extensibility caused by increased entanglement density. Gel spinning allows the attainment of very high draw ratios with high molecular weights because the process fixes a low entanglement density in the precursor fiber. The combination of high draw ratio and high molecular weight, in turn, gives rise to the very high mechanical properties observed. Typical tensile properties claimed for commercial gel-spun polyethylene fiber are a modulus of ca 100 GPa (15 $\times$ 10^6 psi) and strength of ~1.4 GPa (200,000 psi). Problems are the creep performance (25) and the inherently low melting point (see DIMENSIONAL STABILITY, FIBERS). Expected markets are ballistic protection and composite reinforcement (see FIBERS, ENGINEERING).

Liquid Crystalline Polymers

The liquid crystalline nature of stiff polymer molecules in solution was predicted by Flory (1956) (40) and Onsager (1947) (41), and the concept was verified by the synthesis of aromatic polyamides (aramids) (42). Flory suggested that as the molecular chain becomes more rodlike, a critical aspect ratio is reached, above which the molecules necessarily line up to pack in three dimensions efficiently. The packing typical of different classes of liquid crystals is shown in Figure 9. These concepts have been extended to encompass a vast number of homopolymer and copolymer compositions which exhibit mesophase behavior in solution (lyotropic) or in the melt (thermotropic) (4,43,44).

In addition to the main-chain nematics, which are inherently high modulus in the solid state, mesogenic polymers exhibiting smectic and cholesteric phases have been synthesized.

Four categories can be recognized:

Rodlike
  aromatic polyamides, esters, azomethines, benzbisoxazoles
Helical
  polypeptides, nucleotides, cellulosics
Side-chain mesogenic (comb polymers)

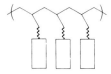

Block copolymers with alternating rigid and flexible units

Polymers exhibiting nematic behavior are highly aromatic, utilize linear linkages of restricted rotation, eg, esters, amides, or azomethines, and exhibit a high aspect ratio and planar conformations in the melt or solution, rather than the random coils typically associated with polymer molecules.

The advantages of nematic liquid-crystalline polymers include both processing ease and solid-state properties. The nematic state can be viewed analogously to "pencils in a box" (Fig. 9). This view shows that the viscosity along the axis of the molecules during elongational flow is low, and hence there is a tendency to form highly uniaxial structures. The nematic liquid–crystal structure is close to the extended-chain crystalline structure desired in the solid state to maximize properties (Fig. 10). Tractability is a problem in liquid-crystal polymers; the

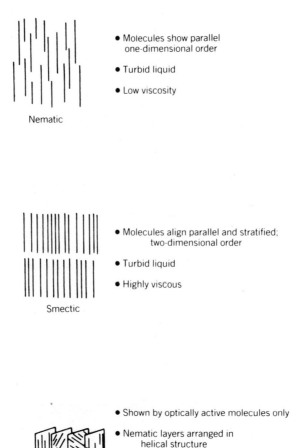

• Molecules show parallel
  one-dimensional order

• Turbid liquid

• Low viscosity

Nematic

• Molecules align parallel and stratified;
  two-dimensional order

• Turbid liquid

• Highly viscous

Smectic

• Shown by optically active molecules only

• Nematic layers arranged in
  helical structure

• Iridescent liquid with
  optical rotatory power

Cholesteric   • Highly viscous

Fig. 9.   Characteristic packing of liquid crystals.

molecular structures are so perfect, that crystals which form in the solid state, decompose before melting. The two predominant approaches to achieving tractability in liquid-crystal polymers are solution processing (lyotropic systems) and copolymerization to lower the melting point to a usable range (thermotropic polymers).

Thermotropic Polymers

Thermotropic copolyesters are perhaps the most versatile class of high modulus, mesogenic polymers amenable to fabrication (4). Since the 1970s, the laboratories of Carborundum, Celanese, DuPont, and Eastman Kodak have produced hundreds of compositions and processing patents. The critical factor in molecular design of tractable, thermotropic polymers is the balance between molecular perfection to preserve the liquid crystal nature of the chain and the molecular im-

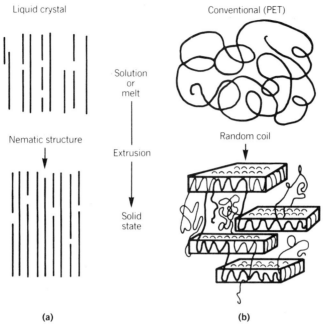

Fig. 10. Structure formation model contrasting conventional and liquid crystalline polymers. (**a**) Extended chain structure with high chain continuity and high mechanical properties; (**b**) lamellar structure with low chain continuity and low mechanical properties.

perfection necessary to allow melting at temperatures far enough below decomposition to allow melt fabrication into useful structures. Copolymerization breaks up the inherent rigidity of these molecular chains; comonomers are chosen to retain the linearity, and hence the mesogenicity, of the resulting copolymer. The nature of the comonomers investigated is summarized in Figure 11. The effect of comonomer concentration on polymer melting is illustrated in Figure 12.

As in the case of the lyotropic nematic polymers, thermotropic copolyesters orient easily in an extensional flow field, producing highly oriented uniaxial structures with moduli as high as 140 GPa (20×10^6 psi) in melt spinning, film extrusion, and similar processes. Drawdown is the critical extrusion variable and low drawdowns are sufficient to achieve almost perfect molecular orientation (Fig. 13). The extrusion processes employed are identical to those used for conventional thermoplastics, with the exception that full molecular orientation is achieved during the spinning process and no further orientation through drawing in the solid state is possible. Property improvements in extrudates, most notably tensile strength increases up to 5–6× the as-spun values, may be achieved by annealing (qv) close to the polymer melting point. This process combines structural perfection with solid-state polymerization (Fig. 14). Heat treatment improves tenacity, elongation, melting point, chemical stability, and thermal retention of tensile properties. Typical fiber properties for thermotropic copolyesters are a tensile modulus of 65–124 GPa (9–18 $\times$ 10^6 psi) at a strength of ~3 GPa (435,000 psi). The broad modulus range reflects differences in molecular architecture within the known thermotropic copolyesters.

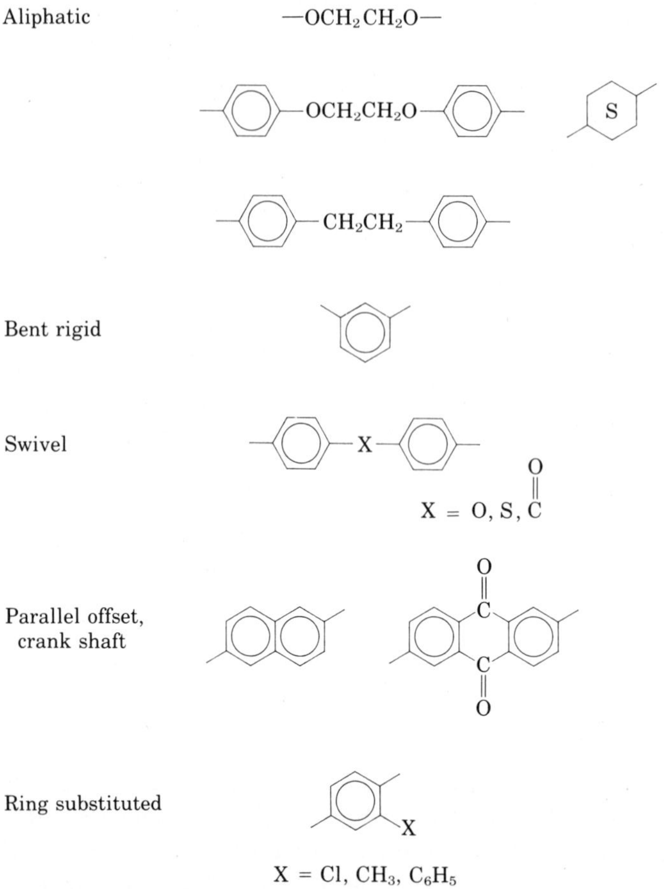

Aliphatic   —OCH$_2$CH$_2$O—

Bent rigid

Swivel

X = O, S, $\overset{\overset{\displaystyle O}{\|}}{C}$

Parallel offset, crank shaft

Ring substituted

X = Cl, CH$_3$, C$_6$H$_5$

Fig. 11. Structures incorporated into thermotropic copolyesters to impart tractability (4).

High modulus thermotropic copolyesters can be thermally fabricated into three-dimensional articles by all methods of conventional melt processing. They offer a balance of properties unobtainable with any other material including moderate density, excellent mechanical properties, controllable range of melting points, low melt viscosity, easy orientation, low shrinkage, controllable thermal expansion, high use temperatures, little property reduction under cryogenic conditions, and excellent solvent resistance.

In 1985, both the Dart division of Kraft, Inc. and the Celanese Corporation, announced decisions to proceed toward commercialization of their thermotropic copolyester compositions, primarily as plastics designed for injection-molding (qv) and shaped-extrusion applications. The Dart material is similar to the Excel polymer developed by Carborundum in the 1970s. The Celanese polymer is a proprietary composition developed internally. Both companies offer a broad range of compounded resins, designed for a variety of specific uses. The properties of these resins are summarized and contrasted in Table 3. The room- and moderate-temperature performance of resins are quite similar; the Celanese material offers improved processability at the expense of property stability at elevated temper-

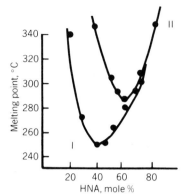

Fig. 12.   Effect of comonomer concentration on thermotropic copolyester melting (4). I is a two-component system, 6-hydroxy-2-naphthoic acid (HNA) and *p*-hydroxybenzoic acid (HBA)

II is a three-component system, HNA/terephthalic acid (TA)/hydroquinone (HQ)

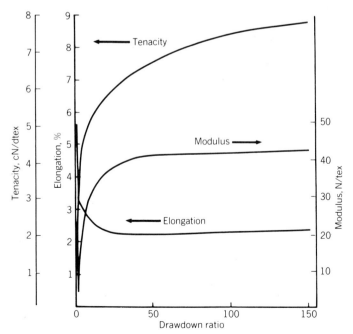

Fig. 13.   Property development in a typical thermotropic copolymer fiber as a function of drawdown. To convert N/tex to gf/den, multiply by 11.3 (4).

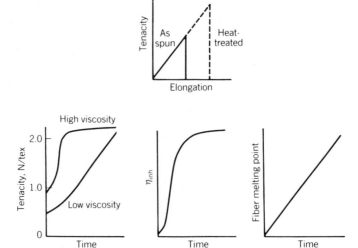

Fig. 14. Fiber-property changes as a function of annealing near T_m for a typical thermotropic copolyester (4). To convert N/tex to gf/den, multiply by 11.3.

Table 3. Injection-molded Properties of Thermotropic Copolyesters

Property	Vectra[a]	Zydar[b]
tensile strength, MPa[c]	140–240	120
tensile modulus, GPa[d]	9.6–38	9.6
break elongation, %	1.2–6.9	
flexural strength, MPa[c]	150–30	130
flexural modulus, GPa[d]	9–34	
notched Izod, J/m[e]	53–534	240–300
dielectric constant	2.6–3.3	2.9–3.7
dissipation factor	0.03–0.004	0.04–0.006
arc resistance, s	63–180	138
dielectric strength, 1.6 mm, kV/mm	31–43	31
melting point, °C	275–330	421
HDT, 1.8 MPa[c], °C	180–240	337–355

[a] Celanese.
[b] Dart.
[c] To convert MPa to psi, multiply by 145.
[d] To convert GPa to psi, multiply by 145,000.
[e] To convert J/m to ft·lb/in., divide by 53.4.

ature. The thermal characteristics of these polymers are controllable through molecular design (Fig. 12), and it is expected that families of products will emerge as these products mature.

Lyotropic Polymers

Stiff, essentially linear polymers, which decompose before melting, must necessarily be processed into useful products from solution. Considerations of mass transfer during solution processing require that the extrudates possess at least one very thin dimension; hence, these processing methods are applicable to fiber, or perhaps film, formation only. A typical polymer of this type is the aramid, poly(p-phenyleneterephthalamide) (PPT), commercialized by DuPont as Kevlar (see POLYAMIDES, AROMATIC). Typical polymerization and processing conditions are shown in Figure 15; typical fiber properties are tenacity of 1.8–2.7 N/tex (20–30 gf/den), modulus of 44–97 N/tex (500–1100 gf/den), and elongation of 2–5%. In solution, PPT is lyotropic; with the proper choice of solvent, (commonly 100% sulfuric acid concentration), and molecular weight, these polymers form nematic liquid crystals. Such solutions possess properties highly propitious for spinning fibers: the nematic structures orient easily in the presence of an orienting field such as the elongational flow field of the spinline. As concentration and molecular weight increase, the viscosity of the solutions pass through a minimum, allowing higher concentration dopes to be spun than with conventional polymer of similar molecular weight (Fig. 16). Detailed reviews of PPT technology are available (2,45,46) (also see LIQUID CRYSTALLINE POLYMERS).

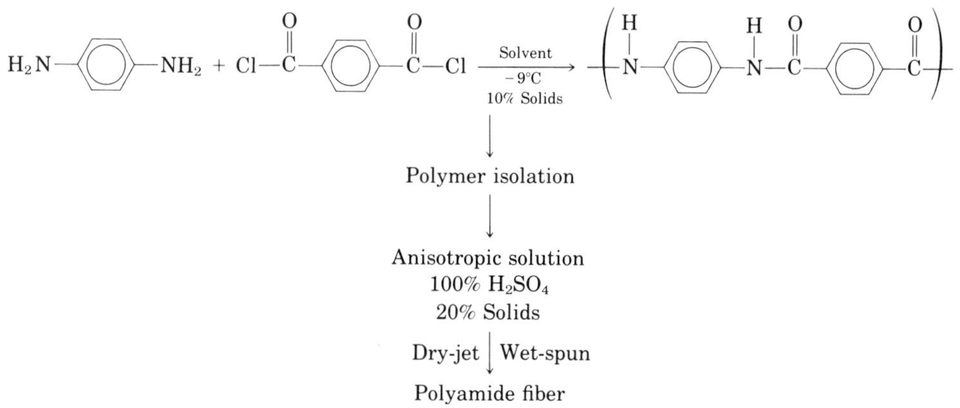

Fig. 15.   Poly(p-phenyleneterephthalamide) processing.

At spinning concentrations and reasonable molecular weights, these solutions are crystalline solids at room temperature and must be melted before spinning. However, to coagulate the spun fibers with a minimum of voidness, coagulation temperatures must be kept low. To satisfy these requirements, the dry jet-wet spinning method is preferentially employed (Fig. 17). For PPT the process is analogous to melt spinning of the nematic phase; the fiber orients and precipitates in the air gap and the solvent is removed during coagulation. The process achieves very high mechanical properties as shown diagrammatically in Figure 10. The development of PPT properties with processing conditions has been detailed in a patent (49) and has been reviewed (2,45,50). For modulus development, fiber drawdown in the airgap has been shown to be the principal variable (45). The range of moduli found in commercial Kevlar products can be produced directly in spinning. An alternative route to the higher modulus product (Kevlar 49) is to heat-treat the spun fiber at temperatures > ca 200°C under tension (51). This is a structure-perfecting process, rather than a true draw process as practiced with conventional polymers. The wet spinning of PPT is more analogous to or-

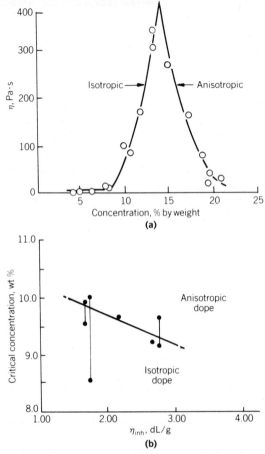

Fig. 16. Viscosity properties of aramid dopes. (**a**) Poly(*p*-benzamide) in HF at 0°C, $\eta_{inh} = 3.9$ dL/g; (**b**) poly(*p*-benzamide) (42). To convert for Pa·s to poise, multiply by 10.

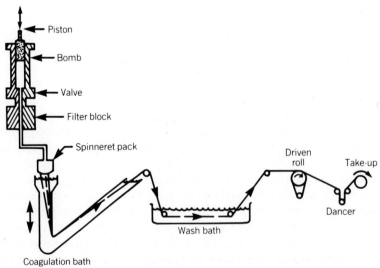

Fig. 17. Dry-jet, wet-spinning apparatus. Piston is attached to reversible drive. Bomb is 50 mL full charge. Filter block contains 2 sintered metal cups (47,48).

dinary fiber processing. Commercial PPT fibers such as Kevlar 29 and 49, are available as filament, staple, and pulp products (2). These materials possess an excellent balance of tensile properties, plus thermal and chemical stability. The tenacity of both Kevlars is 2.7 GPa (390,000 psi); their respective moduli and elongations are Kevlar 29, 62 GPa (9 × 10^6 psi) and 4%; Kevlar 49, 124 GPa (18 × 10^6 psi) and 2.5%. The weakness of these fibers is their poor compressive properties, which probably result directly from the fiber morphology (3).

A second class of polymeric molecules which exhibit lyotropic behavior and are dry jet-wet spun are the rodlike polymers, polybenzimidazoles (PBI), poly-benzoxazoles (PBO), and polybenzothiazoles (PBT).

p-PBI

PBO

PBT

These molecules were developed in the Air Force Ordered Polymer Program and have shown the highest tensile properties produced with organic polymers (Table 2, Fig. 1). As in the aramid case, improvements to spun-yarn properties can be reached through heat treatment. These fibers, consistent with other highly oriented polymeric fibers, also suffer from poor compressive properties.

Polymer Blends and Molecular Composites

An interesting approach to high modulus polymers is the blending of stiff and flexible polymers to create *in situ* composites. Blends range from ca 80/20–20/80 stiff/flexible. Properties follow the rule of mixtures; phase separations and poor adhesion between phases are clearly evident. Systems investigated include both lyotropic and thermotropic polymers as the stiff phase. Most descriptions of these materials appear in the patent literature (52,53).

Molecular composites use truly rodlike polymers, eg, poly(p-phenylene benzobisthiazole) to reinforce a flexible polymer such as a nylon in the same way that carbon or glass fiber is used to reinforce engineering plastics. The idea also comes from the Air Force Ordered Polymer Program (54,55) and is based on Flory's treatment of the rodlike polymer-random-coil polymer-solvent phase diagram (56). A concentration of rodlike polymer is chosen that precludes anisotropic phase formation, but still allows molecular interactions. Figure 18 shows the desired molecular composite structure contrasted with dilute solutions and phase-separated blends. Whether or not a true molecular composite structure has been achieved is an ongoing debate. Some researchers believe that molecular composites are, in fact, polymer blends with highly dispersed phases. Moduli as high as 116 GPa (17 × 10^6 psi) have been reported (54,56). Both thermoplastic and

$c \ll c_{cr}$
Dilute solution

$c \lesssim c_{cr}$
Molecular composite

$c > c_{cr}$
Phase separated system

Fig. 18.   Diagrammatic representation of a molecular composite structure. c_{cr} = critical concentration (54).

solution-processible matrices have been tested with poly(p-phenylene benzobisthiazole) reinforcement (54,57). Properties are comparable to those of liquid-crystalline polymers. An unanswered question is whether the structural and property anisotropy of these materials is reduced in relation to that of the mesogenic component alone (see also POLYMER BLENDS).

Applications

Increased understanding of the origins of high modulus performance in polymers has led to hundreds of molecules and a variety of processing routes which yield tensile modulus values between 35–200 GPa (5–29 $\times$ 10^6 psi). The research emphasis is shifting from demonstrations of incremental modulus improvements, to the quantification of mechanisms, improved predictive models, and analyses of the influence of morphology on overall balance of properties. Industrial concerns are raw-material availability, capital-costs design of conversion equipment, potential market size, and finished-part cost-effectiveness.

The total high modulus polymer market in 1984 was ~20,000 t and almost exclusively filled by aramid fiber. The most important market sectors include high performance composites (for aerospace, sporting goods, etc), ballistic protection devices, tire reinforcement, and ropes and cables. They are also used in electronics, telecommunications, hostile environment applications, and other specialty markets where cost is not critical. Commercial evaluations of new high modulus polymer will be in established markets. As the technology and products mature, it is expected that superior cost-effective products will evolve.

BIBLIOGRAPHY

1. J. P. Riggs, "Carbon Fibers" in J. I. Kroschwitz, ed., *Encyclopedia of Polymer Science and Engineering,* Vol. 2, John Wiley & Sons, Inc., New York, 1985, p. 640.
2. J. R. Schaefgen in A. E. Zachariades and R. S. Porter, eds., *The Strength and Stiffness of Polymers,* Marcel Dekker, Inc., New York, 1983, Chapt. 8.
3. A. E. Zachariades and R. S. Porter, eds., *The Strength and Stiffness of Polymers,* Marcel Dekker, Inc., New York, 1983.
4. G. W. Calundann and M. Jaffee, *Proc. Robert A. Welch Found. Conf. Chem. Res.* **26,** 247 (1982).
5. L. Holliday in I. M. Ward, ed., *Structure and Properties of Oriented Polymers,* John Wiley & Sons, Inc., New York, 1975, Chapt. 7.
6. J. Brandrup and E. H. Immergut, *Polymer Handbook,* 2nd ed., Wiley-Interscience, New York, 1975.
7. A. Cifferri and I. M. Ward, eds., *Ultra High Modulus Polymers,* Applied Science Publishers, Barking, UK, 1979.
8. L. H. Sawyer and W. George in R. Malcolm Brown, Jr., ed., *Cellulose and Other Natural Polymer Systems: Biogenesis, Structure, and Degradation,* Plenum Publishing Corp., New York, 1982, p. 429.
9. L. C. Sawyer and M. Jaffe, *J. Mater. Sci.,* to be published, 1986.
10. M. Panar, P. Avakian, R. C. Blume, K. H. Gardner, T. D. Gierke, H. H. Yang, *J. Polym. Sci. Polym. Phys. Ed.* **21,** 1955 (1983).
11. M. Jaffee in E. A. Turi, ed., *Thermal Characterization of Polymeric Materials,* Academic Press, Orlando, Fla., 1982, Chapt. 7.
12. W. E. Morton and J. W. S. Hearle, *Physical Properties of Textile Fibers,* Textile Institute, Heinemaners.
13. A. Ziabicki, *Fundamentals of Fiber Formation,* John Wiley & Sons, Inc., New York, 1976.
14. A. Peterlin, *J. Mater. Sci.* **6,** 490, (1971).
15. A. Peterlin, *Polym. Eng. Sci.* **18,** 277 (1978).
16. A. Peterlin and A. Cifferri and I. M. Ward, eds., *Ultra-High Modulus Polymers,* Applied Science Publishers, Barking, UK, 1979, p. 379.
17. A. Peterlin in Ref. 3.
18. P. J. Barham and R. G. C. Arridge, *J. Polym. Sci. Polym. Phys. Ed.* **15,** 1177 (1977).
19. P. J. Barham, M. J. Hill, and A. Keller, *Colloid Polym. Sci.* **258,** 899 (1980).
20. I. M. Ward, *Structure and Properties of Oriented Polymers,* John Wiley & Sons, Inc., New York, 1975.
21. I. M. Ward, *Developments in Oriented Polymers-1,* Applied Science Publishers, Barking, UK, 1982.
22. P. D. Coates and I. M. Ward, *J. Polym. Sci. Polym. Phys. Ed.,* 1977.
23. A. G. Gibson, G. R. Davies, and I. M. Ward, *Polymer* **19,** 683 (1978).
24. E. S. Clark and L. S. Scott, *Polym. Eng. Sci.* **14,** 682 (1974).
25. D. W. Woods, W. K. Busfield, and I. M. Vilard, *Plast. Rubber Proc. Appl.* **5**(2), 157 (1985).
26. T. Kanamoto, A. Tsurata, K. Tanaka, M. Takeda, and R. S. Porter, *Rep. Prog. Polym. Phys. Jpn.* **26,** 287 (1983).
27. P. Smith and P. J. Lemstra, *Makromol. Chem.,* 180 (1979); "Influence of Solvent on Drawability," *Makromol. Chem.,* (1983).
28. P. Smith, P. J. Lemstra, and H. C. Booij, *J. Polym. Sci. Polym. Phys. Ed.* **19,** 877 (1981).
29. P. J. Lemstra and P. Smith, *Br. Polym. J.,* 212 (1980).
30. P. Smith and P. J. Lemstra, *J. Mater. Sci.* **15,** 505 (1980).
31. P. Smith and P. J. Lemstra, *Colloid Polym. Sci.* **258,** 899 (1980).
32. P. Smith and P. J. Lemstra, *Polymer* **21,** 1341 (1980).
33. U.S. Pat. 3,413,110 (1984), S. Kavesh and D. Prevorsek.
34. Eur. Pat. Appl. EP 114,793 (June 19, 1985), Y. Kwon, S. Kavesh, and D. Prevorsek (to Allied Corp.).
35. Eur. Pat. 0 144 983 A2 (Dec. 10, 1985) R. M. A. M. Schellekens and P. J. Lemstra (to Dutch State Mines).
36. P. Smith, P. J. Lemstra, J. P. L. Pijipers, and A. M. Kiel, *Colloid Polym. Sci.* **259,** 1070 (1981).
37. P. Smith, P. J. Lemstra, and J. P. L. Pijipers, *J. Polym. Sci. Polym. Phys. Ed.* **20,** 2229 (1982).
38. P. Smith and P. J. Lemstra, *J. Polym. Sci. Polym. Phys. Ed.* **19,** 1007 (1981).

39. B. Wunderlich, *Macromolecular Physics,* Vols. 1–3, Academic Press, New York, 1980.
40. P. F. Flory, *Proc. R. Soc. London Ser. Part A* **234,** 73 (1956).
41. L. Onsager, *Ann. N.Y. Acad. Sci.* **51,** 627 (1949, 1956).
42. U.S. Pat. 3,600,350 (1971), S. L. Kwolek.
43. A. Cifferri, W. R. Krigbaum, and R. B. Meyer, eds., *Polymer Liquid Crystals,* Academic Press, Orlando, Fla., 1982.
44. M. G. Dobb and J. E. McIntyre, *Properties of Liquid Crystalline Main-Chain Polymers,* Vols. 60, 61 of *Advances in Polymer Science,* Springer Verlag, Inc., New York, Munich, 1984.
45. M. Jaffee and R. S. Jones in M. Lewin and J. Preston, eds., *High Technology Fibers, Part A,* Marcel Dekker, New York, 1985, Chapt. 9.
46. J. R. Schaefgen, T. I. Bair, W. Ballou, S. L. Kwolek, P. W. Morgan, M. Paner, and J. Zimmerman in A. Cifferri and I. M. Ward, eds., *Ultra-High Modulus Polymers,* Applied Science Publishers, Barking, UK, 1979, Chapt. 6.
47. S. R. Allen, A. G. Filippov, R. J. Farris, E. L. Thomas, C. P. Wong, G. C. Berry, and E. C. Chenevey, *Macromolecules* **14,** 1135, 1981.
48. S. R. Allen, A. G. Filippov, R. J. Farris, and E. L. Thomas in Ref. 3, Chapt. 9.
49. U.S. Pat. 3,767,756 (Oct. 23, 1973), H. Blades (to E. I. du Pont de Nemours & Co., Inc.).
50. J. R. Schaefgen in Ref. 3, Chapt. 8.
51. U.S. Pat. 3,869,430 (March 4, 1975), H. Blades (to E. I. du Pont de Nemours & Co., Inc.).
52. U.S. Pat. 4,408,022 (Oct. 4, 1983), D. E. Cincotta and F. M. Barandinelli (to Celanese).
53. U.S. Pat. 4,267,289 (May 12, 1981), M. F. Froix (to Celanese).
54. W.-F. Hwang, D. R. Wiff, C. L. Benner, and T. E. Helminiak, *J. Macromol. Sci. Phys. Part B* **22**(2), 231 (1983).
55. W.-F. Hwang, D. R. Wiff, C. Verschoore, G. E. Price, T. E. Helminiak, and W. W. Adams *Polym. Sci. Eng.* **23,** 784 (1983).
56. P. Flory, *Macromolecules* **11,** 1138 (1978).
57. W.-F. Hwang, D. R. Wiff, and C. Verschoore, *Polym. Eng. Sci.,* 789 (1982).

Michael Jaffe
Celanese Research Company

IMPACT RESISTANCE

Impact resistance is a measure of the ability of a material or structure therefrom to withstand the application of a sudden load without failure. The impact resistance of a structure is therefore a complex function of geometry, mode of loading, load application rate, environment, material properties, and, quite importantly, the definition of failure.

The issue of stress geometry or mode is all-important: a material highly resistant to failure in one mode of loading can fail catastrophically in another. Therefore, the impact resistance of a structure is relevant only with respect to a particular mode of loading. A soundly designed structure can fail unexpectedly if the geometry of a structural component is altered by the gradual change of shape, material wear, or crack growth, or if the mode of loading is changed, as when a force is applied in an unforeseen manner. Thus, it is often necessary to test the material or structure under conditions more severe than those of actual use. It is also prudent to follow certain well-established design guidelines.

Impact often implies a high rate, but the type of failure usually associated with impact can also occur under apparently mild conditions, even at low rates of load. When the rate of strain energy accumulation is higher than can be dissipated by the material in the vicinity of a crack or a flaw, unstable fracture (qv) can occur, often entailing crack speeds of hundreds of meters per second. Higher speeds are more likely to cause unstable fracture, other conditions being equal, for then even small flaws, less severe geometries, or milder temperatures can become critical. This is partly due to the viscoelastic nature of polymers. Under truly high speed conditions, ie, when inertial effects are involved, the deformation and fracture behavior is complicated by wave propagation and constraint effects (1–4).

In contrast to impact resistance, the fracture toughness of a solid polymer is much better defined for a given set of environmental conditions (see also FRACTURE AND FATIGUE). Toughness is improved by the incorporation of a soft, elastomeric phase into the rigid polymer matrix. The art and science of material modification for the purpose of toughening is extremely complex (5). Enhancement of toughness improves impact resistance.

The performance of a material or a structure depends on local geometrical variations and processing conditions, such as cooling rate, shear stress, and melt-flow paths, which result in orientation and residual stress. A structure produced by a given process may possess significant morphological differences from test specimens. Ultimately, it may be unavoidable to test the impact resistance of components.

Since the impact toughness of a material, as it is commonly called, depends on the technique of measurement, the testing instruments and the technique must be considered first.

Testing Machines and Techniques

Several types of machines are in use for different types of tests, offering advantages and disadvantages as well as different information. Impact strength is measured by many different empirical methods, some of which may not be appropriate for the performance evaluation of finished products. More sophisticated and meaningful testing devices are under development.

Standard Test Methods. Many impact tests are sufficiently standardized to have ASTM designations (see Table 1).

Pendulum-type Instruments. Pendulum-type machines are used for notched or unnotched specimens that may be of different sizes and supported as a cantilever (Izod) or as a bar supported at its ends (Charpy). These tests are also referred to as flexed-beam impact tests.

The Izod impact test is based on an old, established metallurgical test in which a notched-bar specimen is tested in cantilever fashion with an excess-energy pendulum machine. Izod (ASTM D 256), Charpy (ASTM D 256), and the tensile-impact (ASTM D 1822) tests can be performed with pendulum machines.

A mass is attached to the end of an arm that rotates about a pivot point; a striker is placed near this mass. The arm is raised to a predetermined position. When the arm is released, it swings downward and strikes the specimen, which is mounted rigidly in a vise or a fixture. The precise mounting position and the type of fixture depends on the test. The mass, length of the arm, and angle at the raised position determine the amount of available energy. The total energy must be high enough to give a clean break. Typical pendulum machines of the

Table 1. Standard Tests for Impact Resistance[a]

Test	Designation	Description
brittleness temperature	ASTM D 746	The temperature is determined at which plastics and elastomers exhibit brittle failure under impact.
falling weight	ASTM D 3029	Impact resistance indicated by energy to break or crack rigid plastics by means of a falling weight (tup)[b]; constant height and variable weight are recommended.
falling weight	ASTM D 1709	Similar to D 3029 but for measuring impact resistance of polyethylene film by free-falling dart.
falling weight	ASTM D 2444	For impact resistance of thermoplastic pipe and fittings by falling weight (tup)[b].
fracture toughness	ASTM E 399	For plane-strain fracture toughness.
high rate stress/strain (tension)	ASTM D 2289	Area under stress–strain curve corresponds to measure of impact resistance at testing speeds up to 254 m/min.
Izod impact	ASTM D 256[c]	Energy to break a notched (cantilever beam) specimen upon impact by a pendulum. Notch tends to promote brittle failure. Unnotched impact strength is obtained by reversing the notched specimen in the vise. Notch sensitivity can be determined by using Method D.
tensile impact	ASTM D 1822	Recommended for plastic materials too flexible, too thin, or too rigid to be tested by ASTM D 256. Measures energy to break by "shock in tension" imparted by a swinging pendulum.

[a] Ref. 6.
[b] Designation of the weight.
[c] Includes Charpy impact testing.

Izod and Charpy types are depicted in Figures 1 and 2, respectively; specimen geometries are shown in Figure 3. After the striker breaks the specimen at the bottom of its swing, it continues to swing upward until all the kinetic energy is exhausted. The energy absorbed by the specimen E_S is found by

$$E_S = E_I - E_R - E_F - E_W$$

where E_I is the initial energy available, E_R the energy at the maximum angular travel, E_F the energy dissipated by friction, and E_W the energy to remove the broken section. The last is usually small; E_F can be found by performing the test without a specimen in place. Machines of this type are usually equipped with a dial with calibration marks mounted in the plane of rotation in such a way that E_I and E_R can be read directly. An "instrumented" pendulum machine is equipped with a load cell, usually a piezoelectric device for high frequency response, mounted on the striker or on the mounting block to which the specimen vise is attached. The electric signal from the load cell is amplified, digitized, and recorded by electronic memory. The energy is calculated from the load–time relationship by a computer connected to the instrument. The total energy can be displayed by a digital readout, or the entire load–time trace can be displayed on a cathode-ray tube screen. In the latter instance, a microcomputer is attached and the information can be processed as desired.

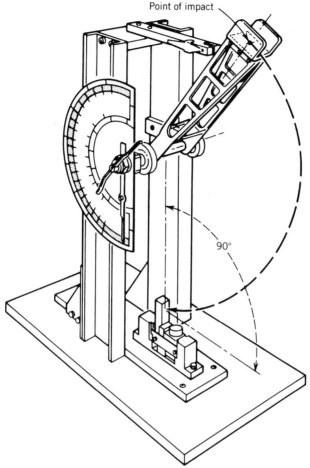

Fig. 1. Izod-type pendulum impact machine, ASTM D 256.

The Izod test is used mostly in the United States and the United Kingdom [British Standard (BS) 2782, Method 306A]. The notched specimen is firmly clamped in a vertical position in a vise fixed in the base of the apparatus (Fig. 1). A pendulum striker of 3.2 mm radius, falling from a height of 0.61 m (ASTM) or 0.31 m (BS), hits the specimen horizontally at a point above the notch. At least ten samples are required to obtain a satisfactory assessment of impact strength.

The U.S. test method has now been accepted as an International Standardization Recommendation ISO/R180, 1961. The material from which the test pieces are cut must be prepared under carefully controlled conditions to ensure development of full strength and freedom from strain caused by uneven heating or rapid cooling. The cutting of the notch requires special care, particularly with notch-sensitive materials. Small variations in the notch can give wide variations in impact values. The notch should be cut with a tool in good condition and shaped smoothly and accurately to the dimension called for in the test specification. Errors are caused by variations in clamping pressure, failure to strike the spec-

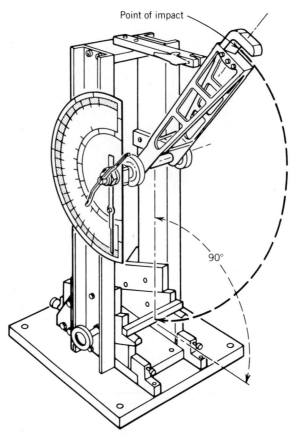

Point of impact

90°

Fig. 2. Charpy-type pendulum impact machine, ASTM D 256.

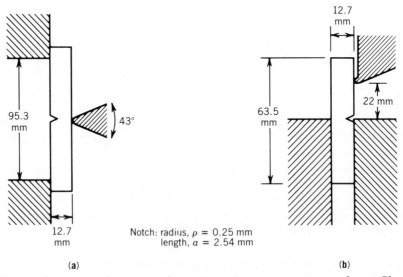

95.3 mm

43°

12.7 mm

Notch: radius, ρ = 0.25 mm
length, a = 2.54 mm

(a)

12.7 mm

63.5 mm

22 mm

(b)

Fig. 3. Geometry of testing and specimen support arrangement for (**a**) Charpy and (**b**) Izod tests specified in ASTM D 256 (7). Courtesy of Applied Science Publishers Ltd.

imens squarely, and the condition of the cutter and the cutting techniques for machining notches (8,9). Impact strength decreases linearly with increasing clamping pressure. A taper of 13 m on the specimen induces errors up to 25% of the expected impact strength of a perfectly parallel specimen. Variations in excess of 10% in expected impact strength were observed from differences in the notch cutter. Torsional stresses caused in the Izod specimen by skewness between the pendulum anvil and the surface of the specimen increase impact strengths even for small angles of skewness (Table 2).

Table 2. Effect of Skew Striking on Impact Strength of Poly(methyl methacrylate) Sheet[a,b]

| Angle of skewness | Impact strength, J/m[c] of notch[d] | |
| | Clamping pressure | |
	16.5 MPa[e]	49.6 MPa[e]
10'	26.05	22.42
50'	27.75	25.78
1°10'	28.02	26.95
4°50'	29.78	29.36
5°10'	29.78	29.52

[a] Ref. 8.
[b] Perspex, Imperial Chemical Industries.
[c] To convert J/m to ft·lbf/in., divide by 53.38.
[d] British standard.
[e] To convert MPa to psi, multiply by 145.

The total breaking energy as measured by the Izod test comprises: energy to initiate fracture; energy to propagate fracture across the specimen; energy for plastic deformation; energy to separate the fragment from the test piece; and energy lost through friction and vibration (9).

The energy to initiate and propagate fracture depends largely on the notch geometry. The energy for plastic deformation is important in materials that, even when notched, break by ductile failure. The energy to separate the fragments is important in materials of low impact strength and high density, eg, mineral-filled phenolic resin. The effect is corrected by placing the fragment on the anvil of the machine, striking it again, and subtracting the energy absorbed from the original apparent impact strength. However, the fragment may adopt different modes of movement, and may spin only under the first strike. For high-impact, low density materials, this source of error is less important. Points of increased stress or of stress concentration can be introduced during machining; their effect is less for the ASTM specimen than for the BS specimen (8).

As can be seen in Table 3, in the ASTM test the energy required for crack propagation of PMMA is more than twice that required for crack initiation; in the BS test they are about equal. Therefore the ASTM test results are much less likely to be affected by the specimen preparation than are BS test results. The BS specimens are more susceptible to imperfections in the notch surface and to the direction of machine marks (ie, stress-producing imperfections). Moreover, since a machined surface is never as smooth as a molded surface, molded (notched)

Table 3. Energy for Crack Initiation and Propagation for Standard Cast PMMA[a,b]

	Notched impact strength, J/m[c]	Energy, %		Energy ratio[d]
		Crack propagation	Crack initiation	
ASTM	17.08	69	31	2.22
BS	24.02	49	51	0.96

[a] Ref. 8.
[b] Perspex, Imperial Chemical Industries.
[c] To convert J/m to ft·lbf/in., divide by 53.38.
[d] Crack propagation to crack initiation.

specimens are expected to give higher values. This has been confirmed with specimens prepared from poly(methyl methacrylate) moldings where molded notch values are higher than machined notch values by a factor of 1.8. In addition, specimens with molded notches, especially those from crystalline polymers, exhibit higher impact strength than those with machined notches because of the difference in morphology or thermal stresses due to rapid cooling near the surface and slow cooling in the interior.

In the Charpy impact tester the specimen is mounted on a span support and struck centrally with a swinging pendulum (see Fig 2). The results are expressed in terms of breaking energy per unit of cross-section or volume.

The Charpy pendulum impact test is most commonly employed on the European continent, especially in the FRG. Both the standard test bar and the small standard test bar are specified in the German Standard DIN 53453. The sample is notched and is supported horizontally against the stops at either end. The pendulum striker hits the sample centrally behind the notch and the excess energy of the pendulum is measured by the angle of the swing. A large number of tests are necessary to give an average value. The Charpy-type test used in the United States is ASTM D 256, Method B; in the UK, it is BS 2782, Methods 306 D and E.

Although the Charpy impact test is accurate and reproducible, significant errors may occur (10). Testing and machine variables tend to cause abnormally high values, since they retard the pendulum swing or otherwise cause excess energy losses. The recorded energy may be in error by as much as 100–200% at low energy and 15–20% at high energy levels. Differences are usually caused by the condition of the test machines; methods of machining and finishing the specimens; and cooling and testing (5). With strict controls and carefully prepared specimens, an average impact value should be accurate to within 5% for values up to 27.1 J (20 ft·lbf).

Pendulum-type machines are inexpensive and simple to use and maintain. Data can be produced from which fracture toughness G_{IC} (more correctly known as the critical strain-energy release rate in Mode I) can be calculated (see FRACTURE and later in this article). However, these machines are limited in the range of impact speeds, with a maximum of 3.4 m/s. Furthermore, the common practice of reporting only the total energy does not allow easy interpretation of the data, which can be overcome by using the instrumentation described previously. The results of these tests usually depend on specimen thickness and notch geometry.

Falling-weight Impact Instruments.   Falling-weight methods are usually employed for sheet. The weight may be gradually increased in mass or dropped from increasing heights on the same specimen or onto a series of specimens. Like the pendulum tester, the machine is driven by gravity. The specimen, ordinarily in the form of sheet, is subjected to a direct blow from a falling weight. The weight is raised to a known height and is allowed to fall almost freely inside a guide tube or along a set of guide rails (see Fig. 4). A striker is mounted beneath the weight or in a guide resting on top of the specimen; the Gardner impact tester, usually preferred, is of the latter type. In recent years drop towers of considerable height have become commercially available, allowing a maximum speed of nearly 8 m/s. However, as the terminal velocity depends on the square root of the drop height, the maximum velocity that can be conveniently obtained is limited. An arrangement similar to that described for the instrumented pendulum impact tester is necessary to determine the load applied to the specimen. Since the maximum velocity achieved with this type machine is considerably higher than that available from the pendulum-type machine, vibration and inertial effects can interfere with the recording of the load signal. Therefore all supporting components must be as rigid as possible, and the piezoelectric load cell assembly must be of low mass and high rigidity to produce a high natural frequency. In

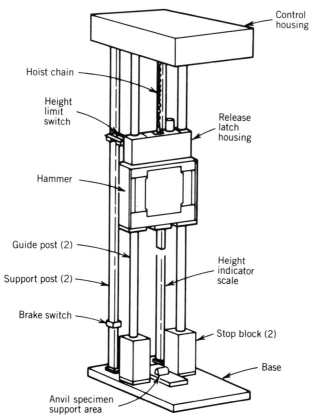

Fig. 4. Typical drop-weight testing machine (11). Courtesy of ASTM.

addition to the load, the velocity of the falling weight must be determined. Frictional effects can result in significant errors when calculating the velocity by assuming free fall. The velocity must therefore be measured directly by a linear array of, for example, photoelectric sensors positioned near the end of the fall, just above where the specimen is struck. An accelerometer attached to the falling weight measures load and velocity separately. The electric signal from the accelerometer is recorded digitally as before. The absorbed energy and the force are calculated from the change in velocity caused by the impact. The results are obtained by processing the acceleration signal in the computer. The accelerometer must be mounted near the tip of the striker to minimize interference from reflected shock waves. An accelerometer is preferred over a load cell at high velocities.

In the falling-weight test (BS 2782, Method 306B) also known as a dart-drop test or Gardner impact test, a spherical ball striker, 12.7 mm in diameter, is fastened to a load-carrying device to which weights can be attached. The striker assembly slides freely in vertical guides and is released from a predetermined height to strike centrally on a specimen, which is supported on the base of the equipment. In the so-called staircase method the load is increased on successive specimens, each struck once, until a specimen fails. The weight is then reduced until a specimen withstands the impact, after which it is increased again by fixed increments until a specimen fails again. This procedure is repeated on at least 20 samples to determine the energy level at which 50% of the specimens break; the result is quoted as the impact strength for 50% failure (F_{50}) (see Fig. 5).

Drop height, m	Impact, J	$F_{50} = 268$ J	Failure	No failure
2.74 m	366			
2.64	352			
2.54	339			
2.44	325			
2.34	312			
2.24	298			
2.13	285	o o o o o o o o o	9	0
2.03	271	x o o x x o o o x x o o o x x o o o o o x x o	14	9
1.93	258	x x o x x x x x x x x o x x x x	2	14
1.83	244	x x	0	2
1.73	230			
1.63	217			

Fig. 5.   Typical results of a dart-drop test using the "staircase" method of analysis (12). To convert J to ft·lbf, divide by 1.35. Courtesy of the Society of Plastics Engineers.

In the Probit method (13), a set of energy levels for the falling weight is selected. A series of specimens is tested at each level, and the failure–impact energy relation determined. With the help of a detailed failure–impact energy curve a simple control test can be carried out by testing a small number of specimens at a single fixed energy level. This procedure is far more reliable and economical than the F_{50} values.

In the common drop-weight test, force and displacement are not recorded by electronic instruments. On the other hand, in fully instrumented procedures a weight–height combination at which the specimen is consistently broken is found by trial and error, or by choosing a height that produces a desired terminal velocity and adjusts the weight to break the specimen consistently. A number of drop tests are then performed under identical conditions. Each test is recorded by an electronic instrument. The exact number of tests required depends on the divergence of the data and the confidence limit desired. This method requires fewer specimens. The velocity can be varied as desired (within the limits of the tester), and an entire set of data can be obtained at the same velocity. The deformation behavior, including the point of crack initiation, can in many cases be deduced from the load–deflection curve, reducing operator errors. Further-more, the ability of a material to decelerate a moving mass can be assessed.

Many materials that give low Izod impact values on notched specimens fail in a ductile manner when tested in sheet form. Notched impact strength and the impact properties of sheet are not correlated because the stress state and the material response differ.

The cause of failure of sheet material in practice is more likely to be revealed by a falling-weight impact test than by an Izod test (14). Falling-weight tests thus give a better guide to practical behavior unless the molded article contains ribs, sharp corners, and similar geometry. Any anisotropy present, such as ex-cessive orientation leading to weakness in a given direction, is revealed in the test. For example, an oriented injection-molded specimen may fail with a single straight crack at a low energy, whereas a substantially unoriented compression-molded specimen of the same material would fail with cracks in several directions after having withstood much higher impact energies.

In falling-dart testers impact is produced by means of a pointed weight or dart. This method assesses the toughness of plastic film under certain specified conditions, such as sample geometry, dimensions of the dart head, and velocity. In one such apparatus, a perfectly flat specimen is clamped by a vacuum and held in uniform tension. The test film is placed over the specimen hole and kept wrinkle-free by the vacuum. The dart is automatically released from an electro-magnet as soon as the correct pressure differential is reached. Darts of various weights are dropped, and the percent failure is plotted against dart weight on probability graph paper. A straight line is drawn through the plot. The inter-section of the line with 50% failure is the weight in grams for F_{50} impact failure. ASTM D 1709 gives a formula for the calculation of impact strength without a graph. In the examination of tubular film the edge fold as well as the face of the film may be tested. The U.S. specification for the falling-dart method (ASTM D 1709-62T) refers specifically to polyethylene film, whereas the UK standard (BS 2782, Method 306F) covers all flexible plastic film. In an equipment modification two electromagnetic counters are provided: one records the number of specimens tested and the other, which is operated by a push switch, the number of breaks. For tests of thick film, a clamping ring is provided in addition to the vacuum clamping to prevent film slippage, which is especially important in this case. In both specifications the dart, which has a 38-mm-dia hemispherical head, falls freely 0.66 m to strike the specimen. The weight of the dart is adjusted with loose weights to determine the limit required to fracture 50% of the specimens. Other more recent versions are commercially available, usually equipped with digital recording devices and computer analysis of the results.

Tensile-impact Instruments.   A standard Izod impact-testing machine, a variation of the swinging pendulum, is easily adapted at little cost to tensile-impact testing. The conventional Izod vise is replaced by one capable of holding the fixed end of a tensile specimen and a metal cross-head is secured to the other end of the specimen (see Fig. 6). The swinging pendulum strikes the cross-head removing both the cross-head and half of the specimen. Development of this instrument led to the specification ASTM D 1822-61 (tensile impact energy to break plastics and electrical insulating material). This method can be used to measure the impact resistance of test pieces cut from actual moldings, and help to assess directional differences in stress that arise from the flow-induced molecular orientation during mold filling. It is also very useful for the evaluation of very small samples. The energy required to produce failure can be estimated by varying the height of fall or the mass of the falling weight or by measuring the excess swing of a pendulum. Transducers and electronic instruments can provide load-deflection curves. The results show reasonable correlation with the falling-ball test of film and with the performance of moldings and sheet. Tensile-impact instruments require less material than the usual falling-weight impact test. At relatively low speeds they give good correlation with falling-weight tests.

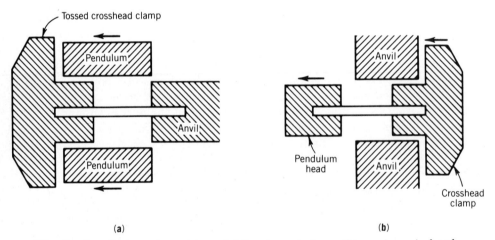

Fig. 6.   Tensile impact machines. (a) Specimen in base; (b) specimen in head.

Special Test Machines.   Hydraulically and pneumatically driven machines have found increasing use in research and quality control laboratories; ASTM standards have not yet been established. These machines require higher operator skills and are more expensive, but offer better control and produce more information.

Hydraulic and pneumatic impact testing machines differ only in the way in which the impact energy is derived, and in the amount of energy and maximum velocity that can be developed. A hydraulic fluid or a gas under high pressure is stored in an accumulator. Upon receiving an electrical signal, a valve is opened, and the fluid or gas is transferred to a cylinder, wherein a low friction piston is connected to the load application device. The pressure gradient on the two sides of the piston causes acceleration to a high velocity. Different designs in the valve and pressure return arrangement allow the velocity to be controlled. The maximum load depends on the pressure of the fluid, piston area, and pressure drop in the valve and upon expansion. The maximum velocity depends on the pressure, the flow rate allowed by the valve, the size of the accumulator, the inertia of the moving mass, and the distance allowed for acceleration. Commercial hydraulically driven impact testers are usually capable of a maximum speed of 22 m/s at a maximum load of 900 kg (see Fig. 7); different designs are available including

(a)

(b)

Fig. 7.   Hydraulic high speed impact tester. (**a**) Specimen enclosed in environmental chamber; (**b**) hydraulic drive unit at the end of the rails for component testing. Courtesy of Instron Corp., Canton, Mass.

computers for control and data acquisition and analysis. Possible tests include tensile impact, puncture (flexed-plate) impact, and flexed-beam impact. A pneumatic machine is shown in Figure 8; anvils for various types of impact tests are shown in Figure 9.

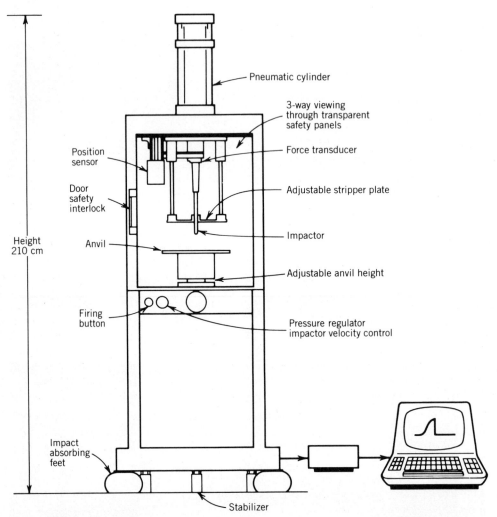

Fig. 8. Pneumatic impact tester. Various types of impactors, anvils, and force transducers are available for different applications. Courtesy of ICI Australia Operations Pty Ltd., Dingley, Victoria, Australia.

In hydraulic and pneumatic machines a much higher velocity is attainable which remains constant during the test; sufficient excess energy is present in the system to prevent deceleration of the moving mass by the specimen.

In pendulum machines the velocity upon striking the specimen is initially constant at 2.4 m/s in the Izod test, and 3.4 m/s in the Charpy test. After the

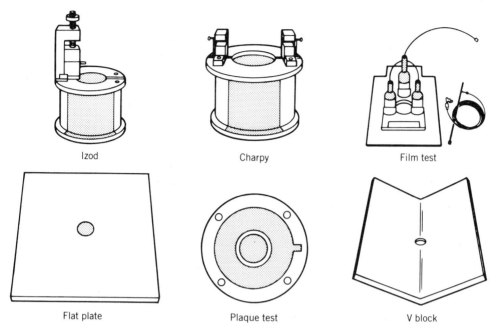

Izod   Charpy   Film test

Flat plate   Plaque test   V block

Fig. 9. Accessory anvils for gravity, hydraulic, and pneumatic testers. Courtesy of ICI Australia Operations Pty Ltd., Dingley, Victoria, Australia.

specimen is struck the velocity decreases, depending on the amount of energy absorbed. This problem is more serious in tough specimens. In drop-weight tests, the velocity upon impact depends on the drop height, and yet the test results often report only the product of weight and drop height, ignoring the impact velocity. Since more polymeric materials are rate sensitive in their mechanical behavior, tests using gravity-driven machines can be very misleading. The use of the special machines described here avoids such problems.

A very useful guide for high-speed tests is given in ASTM D 2289-64T (Tensile Properties of Plastics at High Temperatures), which also explains how to overcome mechanical ringing from the high-frequency load cell, which also may occur.

Product Testing. The specimens are usually specially prepared for falling-weight tests, swinging-pendulum tests, tensile impact tests, and others. However, frequently the fabricated articles perform differently. The conventional forms of impact testing are more suitably applied as control or material tests. This limitation has become more serious in recent years with the introduction of new polymers. The fabricating process can influence impact strength. A typical example is the effect of residual flow orientation in thin-walled injection moldings. The flow properties of the material during the filling of the mold may be more important than the basic impact properties of the unoriented material in governing the impact behavior of the thin-wall moldings. Specimen geometry is also important because at shock loading, the rate of stress buildup is governed by the article shape. Therefore the impact strength of the finished product should be tested in addition to the usual control and quality tests.

Mechanical Behavior

Tensile Impact.   The tensile impact test subjects a specimen to uniaxial tension at a strain rate of approximately $10^2 \ \text{s}^{-1}$. This strain rate is three to six orders of magnitude higher than that encountered in conventional uniaxial tension testing (7,15,16). The effect of strain rate on uniaxial tensile behavior depends on the scale of deformation and the temperature range. At small strains the tensile modulus E increases with strain rate, but the magnitude of change depends on the temperature range. If the relaxation process is significant, eg, glass transition or strong secondary relaxation occurs near the test temperature, the modulus is also rate dependent. The change can be about a factor of two for a secondary relaxation, and an order of magnitude or more for a temperature somewhat below T_g. If no significant relaxation occurs near the test temperature, E can be unaffected by the strain rate. At room temperature, PMMA has a strain-rate sensitive modulus whereas PC has not, although at sufficiently high rates strain-rate sensitivity develops. At strains beyond 1–2%, some materials become brittle even at low strain rates. An example is commercial PS, which typically contains small amounts of foreign material that serves as nucleation centers for craze initiation. Pure PS without surface flaws may be susceptible to shear yielding; increasing strain rate reduces flaw tolerance. If the material is not brittle enough to fracture, it eventually reaches a yield point defined as the maximum in the load-elongation curve. This yield point is not necessarily due to shear yielding. Impact-modified polystyrene, for example, exhibits a yield point independent of shear yielding. If the load maximum is due to shear yielding, further straining results in a neck because of strain softening (16). Some crystalline polymers do not exhibit a maximum in the load-elongation curve but merely a knee because of the strengthening effect of the spherulites. Yield strength usually increases with strain rate (15,16). The strain-rate sensitivity of the shear yield stress behaves in much the same fashion as the modulus; it also depends on a significant relaxation process near the test temperature. After neck formation the strain-rate sensitivity stabilizes and neck extension occurs (16). On the other hand, thinning of the neck results in ductile fracture, which can also be due to high strain rates.

Uniaxial tensile true stress–true strain behavior of polymers deformed at low strain rates is shown in Figure 10. Common engineering stress–strain curves cannot be used to determine the elongational strain because the shape for a given set of testing conditions depends on the specimen geometry. Some polymers show ductile deformation after yield by a process of necking and cold drawing. The apparent neck extension depends on the length of the specimen gauge section if the conventional method of calculating engineering strain $\epsilon = \Delta l / l_0$ is used, where Δl is the elongation and l_0 the original gauge length. The true value of the extensional strain can be ascertained only by direct measurement of the strain in the necked section. If the specimen dimensions are identical and the extents of necking comparable, the apparent engineering strain can be used as a relative ductility index. Ductility usually decreases with strain rate or decreasing temperature. At a given strain rate and temperature, ductility depends on molecular weight and amount of filler or impurity in the material. Stresses applied orthogonal to the principal stress direction can also affect ductility. Tensile impact

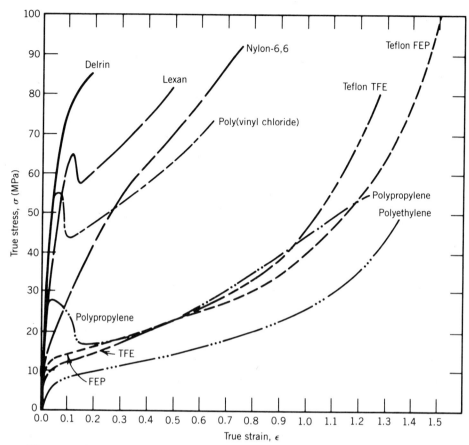

Fig. 10. True stress–true strain behavior of polymers. Delrin is an acetal resin; Lexan, a polycarbonate; Teflon TFE designates the tetrafluoroethylene homopolymer; Teflon FEP is a copolymer of tetrafluoroethylene and hexafluoropropylene (17). To convert MPa to psi, multiply by 145. Courtesy of *Polymer Engineering and Science.*

toughness, a measure of the total energy absorbed by a dog-bone shaped specimen in uniaxial tension, is the integrated value of the force-elongation curve. It is highly sensitive to the ductility of the material at the testing rate. The tensile impact behavior of polypropylene is shown in Figure 11. The curve is typical of a pendulum impact tester equipped with transducers and digital storage instruments. The strain–softening behavior is clearly discernible even at such a high strain rate. The advantage of electronic instrumentation can be seen immediately.

Flexed-plate Impact.   In flexed-plate impact, a ball, dart, or striker is driven by gravity or pressure into a thin plate supported by or clamped over an annular area. The nose of the dart is usually hemispherical, though sometimes square. A flexed-plate impact test corresponds closely to conditions of actual use. However, geometric factors, such as specimen thickness and dart-to-annular opening ratio, cannot be readily standardized, and furthermore, the geometry itself can cause a change in the deformation mode (14). The progression of deformation processes associated with the load-deflection curve can be seen in Figure 12. In

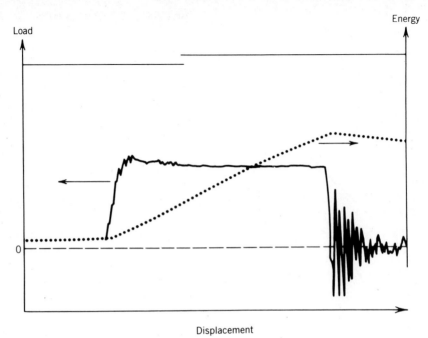

Fig. 11. Typical load and energy data from tensile impact test on polypropylene using an instrumented pendulum tester. The dotted line shows the accumulated energy as the deformation increases. Courtesy CEAST S.p.A., Torino, Italy.

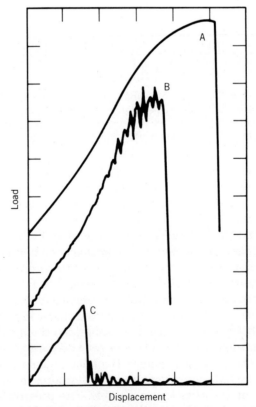

Fig. 12. Typical load-deflection data from puncture impact tests. (A), ductile failure; (B), cracking started at about 50% of the maximum load; (C), brittle failure (18).

Figure 12A the specimen failed in a ductile manner and the plastic deformed extensively around the dart nose. The remaining unsupported material was cold drawn. Partially deformed specimens and finite-elements analysis reveal yielding under the dart early at about 25% of the maximum load. This initial yielding is marked by a slight decrease in the slope. Subsequently the yielded zone in the shape of a hemispherical dome develops around the dart nose. As deflection continues, the net area undergoing yielding also increases, which results in geometric hardening. The processes of material softening and geometric hardening eventually merge which, in this case, result in a slight increase in slope. The unsupported area of the specimen is deforming as a membrane rather than as a plate. The stress in the membrane is essentially uniaxial tension in the radial direction. Finally, excessive thinning takes place somewhere in the specimen, the exact location being dependent on the geometry, and rupture causes the load to drop precipitously (18,19). Thus, the total energy absorbed, which is the area under the load-deflection curve, depends in a very complex manner on both the testing geometry and the yield and deformation behavior of the material. The curve of Figure 12B is obtained when the specimen develops a crack in the membrane-stretching process. The sudden release of elastic energy upon cracking causes the load-cell assembly to vibrate at its natural frequency. The crack is stabilized by the specimen geometry and does not propagate. Consequently the specimen is able to carry an increasing load even though, by some definitions, the specimen would be considered to have failed (18). The curve in Figure 12C is obtained when the specimen fails in a totally brittle manner; no decrease in the slope of the load-deflection curve is discernable. Cracking occurs on the tension side opposite the area where the dart contacts the plate. Little or no plastic deformation occurs anywhere on the specimen. Brittle fracture of this type occurs more readily in thick plates. Ductile-type failure (Fig. 12A) can be transformed to brittle-type failure by decreasing temperature, increasing deformation rate, or environmental aging. With a ball or a hemispherical dart it is commonly assumed that the resultant stress state is equal to the biaxial tension in the center of the plate. This assumption is erroneous because the normal force on the plate can be significant and lead to overoptimistic values on ductile-to-brittle transition conditions, be it temperature or aging time. This behavior is caused by normal stress suppressing brittle failure (15,18,20).

Flexed-beam Impact. Flexed-beam impact tests are sometimes carried out on unnotched specimens, though notched specimens are preferred. The Izod and Charpy testing geometries are different; in the German Charpy test a square notch is used rather than a V-shaped. Regardless of the geometry, the notch creates a severe stress concentration. Many materials that are tough in the tensile and flexed-plate impact tests fail in notched flexed-beam tests because they are brittle. For a given specimen thickness, the impact toughness depends on the notch tip radius (see Fig. 13). Subjecting a thin, notched specimen to bending creates a biaxial tensile stress concentration: the main one along the specimen length, the other parallel to the notch or crack direction. The magnitude of the former depends on the notch sharpness. The minor stress is due to constraints created by notch-free sections of the specimen, which are at a lower stress. Because the specimen is thin, it contracts freely in the thickness direction; a state of plane stress is said to exist in this instance. The biaxial tension state is more favorable than the uniaxial tension state for craze formation because of the larger dila-

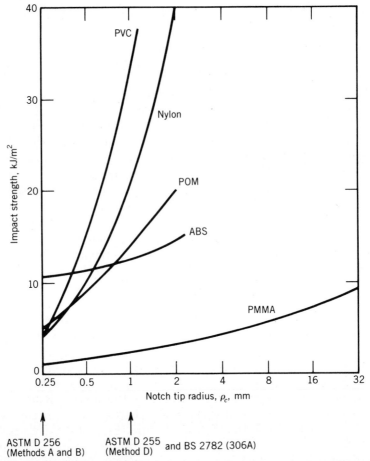

Fig. 13. Impact strength as a function of notch tip radius for different polymers (23). To convert kJ/m^2 to ft·lbf/in.^{2}, divide by 2.10. POM = polyoxymethylene. Courtesy of The Plastics and Rubber Institute.

tational stress component in the former case. Many polymers ductile in uniaxial tension exhibit brittle behavior in this stress state; other polymers are able to shear-yield. Shear yielding creates a large plastic zone at the notch tip, and subsequent failure is due to ductile tearing of the material. This mode of failure usually absorbs a large amount of energy. Other impact-modified materials are able to form a plastic zone ahead of the notch consisting of a high density of crazes. Considerable energy can be absorbed by this mechanism. The load-deflection behavior of tough plastics is shown in Figure 14.

Subjecting a thick notched specimen to bending creates a tensile stress component that is perpendicular to the specimen faces. This third stress is due to constraint in the thickness direction because the material ahead of the notch is unable to contract freely in this direction; the constraint effect is said to be due to plane strain. The triaxial tension state is even more favorable than the biaxial tension state for the formation of crazes, again because of the larger dilatational stress component in the former. With sufficiently large thickness

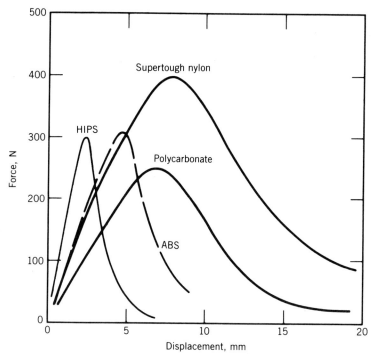

Fig. 14. Load-deflection behavior of a number of tough polymers tested in the notched Charpy geometry at low rates (21). To convert N to kgf, multiply by 0.102. Courtesy of the Society of Plastics Engineers.

and sharp cracks few unmodified polymers are able to deform in a completely ductile manner. The fracture is due to craze formation and rapid crack propagation. At intermediate thicknesses or notch radii the specimen can deform in a mixed plane–strain mode in the specimen center and a plane–stress mode near the surfaces. In the former, plastic deformation zones are often found. Their size determines the amount of energy absorbed by the specimen; the plane–strain zone absorbs little energy.

The notched flexed-beam impact toughness of a material depends on the thickness of the specimen and the radius of the notch, and therefore cannot be considered a material property. Furthermore, the transition from plane–stress to plane–strain is unpredictable when complex geometries are encountered in structural components. Because of this transition, some polymers normally considered to be very tough and ductile need impact modification. The modified versions are insensitive to geometric variations.

The transition from plane–stress to plane–strain can also be brought about by impact rate (22), temperature (Fig. 15), molecular weight, and thermal history (24). The effects of thickness and rate are shown in Figure 16. Notched PC specimens are tested at various bending speeds. The thick (6.4 mm) specimens are uniformly brittle, whereas the thin (3.2 mm) specimens are uniformly ductile. The intermediate thickness (4.4 mm) specimens exhibit a ductile-to-brittle transition at about 0.3 cm/s. In modified PC even the 6.4 mm thick specimens are tough up to about 40 cm/s (see Fig. 17).

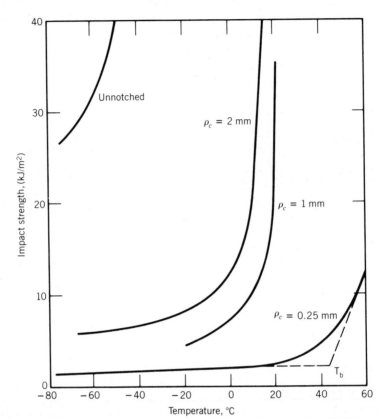

Fig. 15.   Effect of test temperature on the impact strength of PVC specimens containing various notch radii (23). T_b = brittle temperature. To convert kJ/m² to ft·lbf/in.², divide by 2.10. Courtesy of The Plastics and Rubber Institute.

Fracture Toughness from Flexed-beam Impact Tests.   The so-called impact toughness values obtained from Izod and Charpy tests are not material properties because they depend on specimen thickness, notch depth, notch radius, and other factors. Even the mode of failure is affected by a changing specimen geometry. These values are therefore useful only for the comparison of different materials, and are useless for design calculations. Yet, it is desirable to have fracture toughness values that are material properties taken under impact conditions.

The fracture toughness G_{IC} can be determined from the impact energy U if the geometry of the specimen is known (25). It is assumed that linear elastic fracture mechanics (LEFM) applies. When a specimen containing a sharp notch of depth a is subjected to bending with a load P, a deflection x results. The energy U stored is given by

$$U = \frac{1}{2}Px \tag{1}$$

When the crack grows, compliance $C = x/P$ increases as the stored energy is released. Thus, if it is known how C changes with a, the strain-energy release rate G can be calculated from equation 1

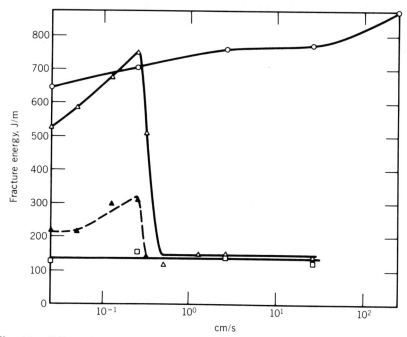

Fig. 16. Effect of test speed on the fracture energy of polycarbonate specimens of various thicknesses tested in the notched Charpy geometry. ○, 3.2 mm; △, 4.4 mm; ▲, energy up to load maxima for 4.3 mm; □, 6.4 mm (22). To convert J/m to ft·lbf/in., divide by 53.38.

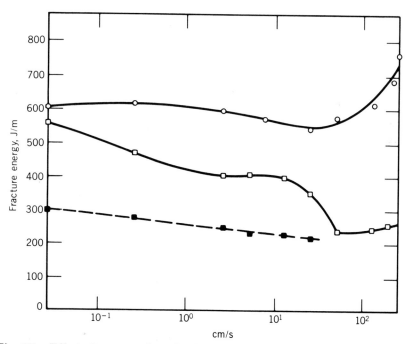

Fig. 17. Effect of test speed on the fracture energy of an impact-modified PC as in Figure 16. ○, 3.2 mm; □, 4.4 mm; ■, energy up to load maxima for 6.4 mm (22). To convert J/m to ft·lbf/in., divide by 53.38.

$$U = \tfrac{1}{2}P^2C \tag{2}$$

For a specimen of uniform thickness B,

$$G = \frac{1}{B}\frac{dU}{da} = \frac{1}{B}\frac{dU}{dC}\frac{dC}{da} \tag{3}$$

When fracture occurs, the critical strain-energy release rate G_c is calculated:

$$G_c = \frac{P^2}{2B}\frac{dC}{da} \tag{4}$$

In an impact test, usually only the energy U is determined. Combining equations 2 and 4 gives

$$U = G_c B\frac{C}{dC/da} \tag{5}$$

In practice, U is determined for a series of specimens with various notch depths a. It is therefore more convenient to express equation 5 in terms of

$$G_c = \frac{U}{BDC}\frac{dC}{d(a/D)} \tag{6}$$

where D is the specimen width. The dimensionless compliance function

$$\phi = \frac{C}{dC/d(a/D)}$$

can be determined experimentally from quasi-static bending tests, and has been calculated analytically for this geometry (26). Equation 6 can then be rewritten as

$$U = G_c BD\phi \tag{7}$$

predicting that a linear relationship should exist between the impact energy U and $BD\phi$, the slope of which should be equal to G_c. This method can be tested by determining the linear relationship for a series of specimens with various notch depths. If the impact tester is instrumented, a further criterion would be a rapid drop in load P after reaching a peak. If the notch is too blunt and a significant plastic flow occurs, a correction factor may be needed, ie, the crack length a is increased by the plastic zone size r_p (25,27):

$$a_f = a + r_p$$

The G_c values for some plastics for both the Izod and Charpy geometries (25) are in Table 4. With the same method, the G_{IC} of a number of polymers was determined (3) at different rates (Fig. 18).

The procedure outlined above can be useful if the specimens are prepared carefully to ensure that LEFM applies. The specimen should be as thick as is practicable, and the notches should be very sharp. If a razor blade is used for cutting, a compressive plastic zone must be avoided at the notch tip. If the fracture surface contains plastic flow, the crack size most likely has to be corrected. The method for determining fracture toughness can be extended to tensile impact (28).

Table 4. Strain-energy Release Rates[a]

Material	G_c, J/m² [b]	
	Charpy	Izod
polystyrene		
general purpose	0.83×10^3	0.83×10^3
HIPS	15.8×10^3[c]	14.0×10^3[c]
PVC		
Darvic 110	1.42×10^3	1.38×10^3
Modified	10.05×10^3	10.00×10^3
PMMA	1.28×10^3	1.38×10^3
nylon-6,6	5.30×10^3	5.00×10^3
polycarbonate[d]	4.85×10^3	4.83×10^3
PE		
medium density[e]	8.10×10^3	8.40×10^3
high density[f]	3.40×10^3	3.10×10^3
low density	34.7×10^3	34.4×10^3
ABS		
Lustran 244	49.0×10^3[c]	47.0×10^3[c]

[a] Ref. 25. Courtesy of *Polymer Science and Engineering*.
[b] To convert J/m² to ft·lbf/in.² divide by 2.10×10^{3}.
[c] J_c as defined in Ref. 1.
[d] Specimens cut in the extrusion direction.
[e] Density = 0.940; MI = 0.2.
[f] Density = 0.960; MI = 7.5.

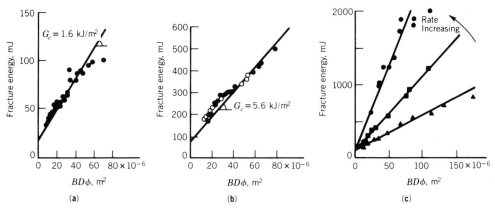

Fig. 18. Fracture energy as a function of $BD\phi$ for (**a**) PMMA at 20°C; (**b**) PTFE at 25°C (○, Charpy; ●, Izod); (**c**) HDPE at 20°C and several rates (▲, 26 s⁻¹; ■, 122 s⁻¹; ●, 620 s⁻¹) (3). To convert J to ft·lbf, divide by 1.355. To convert kJ/m² to ft·lbf/in.², divide by 2.10. Courtesy of The Institute of Physics, Great Britain.

Impact-resistant Materials

Impact resistance is a relative term. It depends on structural parameters as much as it does on intrinsic material toughness. Impact resistance is sometimes gained at the cost of reduced modulus and strength, and perhaps fatigue resistance. The impact toughness of commercial plastics is shown in Table 5.

Table 5. Typical Notched Izod Impact and Drop-weight Impact Toughness[a]

Thermoplastic	Notched Izod, J/m[b]		Falling dart[c], N·m[d]	
	22°C	−40°C[c]	22°C	−40°C[c]
acetal	53–53			
nylon, amorphous	587–1121	213–480[f]		
nylon-6[g]	53–81			
impact modified nylon 6[g]	961	294[f]	65	57[f]
nylon-6,6[g]	27–64			
toughened nylon-6,6[g]	907–1174			
33% glass[g]	219			
nylon/ABS	998	96[f]	73	35[f]
nylon/PPO	213	133[f]	51	40
polyarylate	294	213		
polycarbonate	641–854	133–598		
polycarbonate/ABS	454–641	267–614[f]	38–61	30–46[f]
polycarbonate/PBT[g,h]	694–854	160–641	56	49[f]
polycarbonate/PET	907	694[f]	57	50[f]
polyester, aromatic (LCP)[i]	53–534			
thermoplastic polyester, PBT[h]	43–53			
impact modified PBT[h]	801–854	747–801[f]	43–54	43[f]
impact modified PBT/PET	801–854		54	
30% glass-reinforced PET, thermoplastic polyester	101	96		
toughened 30% glass PET	139–235	123–160		
polyetheretherketone	59–81			
polyetherimide	32–53		34–36	
polyethersulfone	53–81			
poly(phenylene oxide) modified	267–534	133–187	15–34	3–11
poly(phenylene sulfide)	27–75			
poly(phthalate carbonate)	534			
polysulfone	69	64		
polyurethane, engineering thermoplastic	53–213			

[a] Ref. 6.
[b] To convert J/m to ft·lbf/in., divide by 53.38.
[c] Gardner or falling dart.
[d] To convert N·m (J) to in.-lb, divide by 0.113.
[e] Unless otherwise stated.
[f] At −29°C.
[g] Dry as molded.
[h] Poly(butylene terephthalate).
[i] LCP indicates liquid crystalline polymer.

High Impact Polystyrene (HIPS), Acrylonitrile–Butadiene–Styrene Copolymer (ABS), and Modified Poly(xylenol ether) (PXE).   Polystyrene is a transparent, rigid, easily processable, inexpensive material. Its chief disadvantages are extreme brittleness and high susceptibility to crazing by organic liquids. It was also among the first materials to be produced in a toughened form, commonly known as HIPS, in order to allow the material to be used in applications with moderate loading such as electrical appliances.

In the most common production method rubbery polybutadiene (PBD) is precipitated from a styrene–polybutadiene solution during the polymerization of styrene. The precipitated phase itself contains 2–10 μm particles consisting of a rubbery shell surrounding spheroidal polystyrene inclusions that are separated by thin membranes of PBD rubber (see Fig. 19). It is believed that this type of composite particle has an advantage over solid rubber spheres because, for a given weight fraction, it maximizes the volume fraction of the rubbery phase, thus enhancing the impact toughness without causing undue reduction in modulus. The rubbery phase can also be introduced by blending PS with a triblock copolymer such as poly(styrene–butadiene–styrene) (SBS). The toughening is believed to be due to the nucleation and controlled growth of extensive crazing at the rubber–PS interface (29). The appearance of stress-whitening is due to the high density of crazes generated by deformation. The stress–strain behavior of ABS, PS, and a typical HIPS are shown in Figure 20. It is readily apparent that HIPS has much lower modulus and tensile strength, but much higher elongation

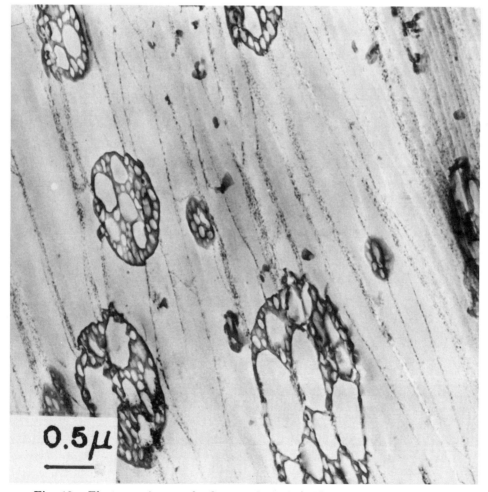

Fig. 19.   Electron micrograph of an osmium tetroxide-stained ultrathin section of HIPS. This micrograph shows the typical morphology of the dispersed rubber particles. The thin lines connecting the particles are crazes formed after the material has undergone deformation.

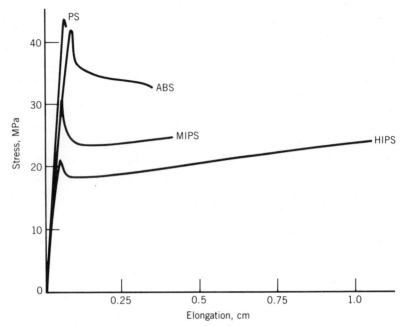

Fig. 20. Typical uniaxial tensile stress–strain behavior of PS, medium-impact PS (MIPS), high-impact PS (HIPS), and ABS (33). To convert MPa to psi, multiply by 145. Courtesy of Springer Verlag.

than PS. At the cost of a decrease in modulus and strength, ductility and impact toughness are greatly enhanced because of the higher volume fractions of rubber particles in the tougher materials.

Acrylonitrile–butadiene–styrene polymers (ABS) (qv), the impact-modified version of SAN (styrene–acrylonitrile copolymer), are produced in the same manner as HIPS. Block copolymers such as SBS may also be used as the second phase. The morphologies of ABS and HIPS are very similar except that for optimal impact toughness the rubber particles in ABS are smaller than those in HIPS, ca, 1–5 μm in diameter. The difference between ABS and HIPS, as far as mechanical behavior is concerned, is due to the fact that SAN is more ductile than PS. Consequently, the impact toughness of ABS is higher than that of HIPS. Because of compositional and morphological differences, these materials exhibit a range of mechanical properties. For a given volume fraction of rubber particles in ABS, higher AN (acrylonitrile) content enhances toughness. The toughening mechanism is believed to consist of the formation of numerous shear bands in addition to massive crazing (29); both are due to the presence of the rubber particles. In certain ABS grades deformation in tension causes the formation of cavities inside the rubber particles, and shear bands are formed from these cavitated particles (30).

The toughest member in this category of plastic materials is a blend of poly(xylenol ether) [PXE, poly(2,6-dimethylphenylene oxide)] and PS and a rubbery impact modifier; PXE is also known as PPO poly(phenylene oxide). However, PPO, a registered trademark, can be confused with poly(propylene oxide), and therefore PXE is the preferred abbreviation. PXE is completely miscible with PS and forms a true solid solution (31); T_g is high and the material is ductile. Melt viscosity is also very high. Processability is improved by blending with PS; the viscosity of the mixture is approximately the weighted average of the two components, whereas T_g varies with the composition (31). Moduli and yield strengths are higher but the fracture toughness is lower than the weighted average (31,32). Commercial grades are usually impact-modified by a dispersion of rubbery particles, which are sometimes obtained by directly blending PXE with HIPS instead of PS. The stress-strain behavior of HIPS/PS/PXE blends is shown in Figure 21. The PXE/HIPS blends exhibit the best overall combination of mechanical properties. The toughening mechanism is believed to be a combination of shear banding and massive crazing (29). The PXE/HIPS blends are much more expensive than HIPS or ABS. Furthermore, they require precise processing control in injection molding because of the relatively high melt viscosity and poor thermal stability of commercial PXE.

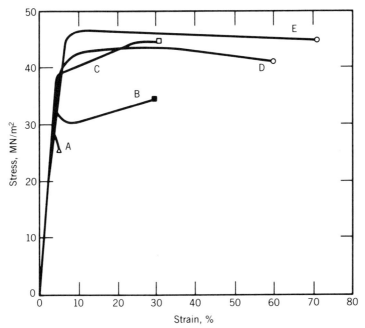

Fig. 21.   Uniaxial tensile stress–strain behavior of HIPS–PS–PXE blends at 20°C showing the effects of matrix composition. Strain rate 4×10^{-4} s^{-1} (29).

HIPS–PS–PXE blends, wt %

Blend	HIPS	PS	PXE
A	50	50	0
B	50	37.5	12.5
C	50	25	25
D	50	12.5	37.5
E	50	0	50

To convert MN/m² to psi, multiply by 145. Courtesy of Applied Science Publishers Ltd.

Neither HIPS, ABS, nor PXE possess unique properties due to the range in compositions obtainable from the manufacturing techniques. This is especially true of PXE blends, which exhibit behaviors resembling HIPS or PXE. Processing and applications must be specific to each grade. Pigments, flame retardants, and especially reinforcing fillers can have significant effects on impact toughness. The relevant properties must be determined for each composition.

Since the toughness of these materials depends on the presence of rubber particles, the destruction of these particles reduces impact toughness. Excessively high temperatures or shear rate during processing and prolonged exposure to oxidative environments destroy the rubber and must be avoided. Prolonged exposure to ultraviolet radiation degrades these materials because of the decomposition of the PBD rubber and sometimes PXE. Saturated elastomers resist oxidative degradation better than unsaturated PBD. In such instances, the low temperature impact toughness of the material might be impaired because the T_g of the saturated rubber is usually higher than that of PBD. Uv stabilizers improve the resistance of PXE to sunlight. The rubber particles act as effective sites for the nucleation of crazes in cyclic fatigue. Therefore fatigue initiation occurs more readily in the impact-modified versions than in the unmodified versions; crack propagation is slower in the former (33).

Polycarbonates, Other Ductile Glassy Polymers, and Their Impact-modified Versions.   A number of glassy polymers possess some degree of ductility at or near room temperature at moderate rates of straining ($<100\%$ s^{-1}). Ductility is defined here as elongation after yield in a uniaxial tension test. These polymers include polycarbonate (PC), polysulfone (PSU), polyethersulfone (PES), polyetherimide (PEI), polyarylates (PAR), and others as well as copolymers derived therefrom. Some low-T_g difunctional epoxies also belong in this category. Poly(vinyl chloride) (PVC), which may possess a low degree of crystallinity, and which loses its ductility more readily than other members in this category, may be included here. The class of materials discussed here, whether impact-modified or not, generally exhibits high values of tensile-impact toughness compared with the materials discussed previously. Drop-weight impact toughness is also high. However, a flexed-beam impact test (ie, Izod or Charpy) performed on notched specimens of these materials gives different results. Using typical notched Izod impact geometries (notch depth of 2.5 mm, notch tip radius of 0.25 mm) and a specimen thickness of 3.2 mm, some of these materials fail in a brittle mode and exhibit impact toughness values that are surprisingly low, given the fact that these same materials are ductile in uniaxial tension. Other materials, PC for example, fail in a ductile mode and absorb large amounts of energy when the specimen thickness is 3.2 mm, but dramatically change to the brittle fracture mode when the specimen thickness is increased or the temperature lowered. The specimen thickness effect is due to the difference in constraints at the surfaces and in the interior of the specimen. The temperature effect is not well understood, but is undoubtedly due to the different temperature dependencies of the shear yield stress and crazing stress. These dependencies may be correlatable to molecular relaxations.

The need for impact modification of this type of materials is not based on brittleness but on notch sensitivity. Impact-modification does not necessarily enhance toughness when tensile impact and drop-weight impact tests are the criteria. Impact modification is achieved by melt blending a second soft, usually elastomeric, phase, which may be in the form of a dispersion of spherical particles or a network-like structure. To aid the dispersion of the second phase, compatibilization is usually required, that is, the elastomer is copolymerized or altered chemically. Inadequate compatibilization can lead to unstable morphologies after melt processing, which can in turn produce weak knit lines in injection-molded articles, or skin and core morphologies that interfere with the proper functioning of the processed article and reduce the impact resistance of the molded articles.

It appears that the toughening occurs as follows: after initial deformation, cavitation occurs within or around the soft phase; localized shear deformations, often referred to as shear bands, grow from these cavities. If the volume fraction of the soft phase is sufficiently large, these shear bands merge or coalesce. Since the localized deformation in shear bands cannot be recovered, it is known as plastic deformation. It absorbs energy and blunts the crack tip (34).

The cavitation in or around the soft phase causes stress-whitening, which is often observed at and near the fracture surface of broken specimens. Stress-whitening is sometimes mistakenly taken to be due to crazing. The cavitation is an essential step in the toughening process, because the close proximity of many cavities relieves the triaxial tension that is built up at the crack tip, firstly because of the plane strain (which, in turn, is caused by the constraint generated by the thickness of the specimen) and secondly because of the initial shear yielding at the specimen surfaces. After the triaxial tension is relieved by cavitation, only biaxial tension remains, which is more favorable for shear yielding. Since these materials are more prone to shear yielding than to crazing, ductile fracture is restored.

Evidence for the deformation process can be seen in the three micrographs of an epoxy toughened with elastomer shown in Figure 22. In Figure 22**a**, cavitated rubber particles can be seen in a scanning electron micrograph of the fracture surface. In Figure 22**b**, the cavitated rubber particles can be clearly discerned beneath the fracture surface. The shear bands between the cavities are revealed by viewing the specimen between crossed polarizers in Figure 22**c**.

Semicrystalline Polymers and Their Toughened Versions. Semicrystalline polymers are different in mechanical behavior from glassy polymers. Conventional melt processing techniques produce semicrystalline morphologies.

The polymers in this category include high density polyethylene (HDPE), polypropylene (PP), polyacetals (POM), polyamides (PA, or nylons), poly(ethylene terephthalate) (PET), poly(butylene terephthalate) (PBT), and more recent materials such as poly(phenylene sulfide) (PPS), and polyetheretherketone (PEEK). Low modulus polymers such as low density polyethylene (LDPE) and certain thermoplastic elastomers can be included. Crystallinity is typically between 30 and 40%, though it can be as low as 10% and as high as 80% in HDPE. Mechanical behavior depends strongly on the degree and distribution of crystallinity and the differential cooling rates between the surface and the interiors of the specimens. In molded articles, the interior of a thick section contains many large spherulites. Since, in the presence of a notch, the interior is in plane strain, resulting in triaxial tension, large spherulites invariably cause brittleness, apparently due to weak interspherulitic boundaries. In very thick sections voiding can also occur during the crystallization process. Sometimes, in the mistaken belief that larger cross-sections provide greater strength and a more gentle radius reduces stress

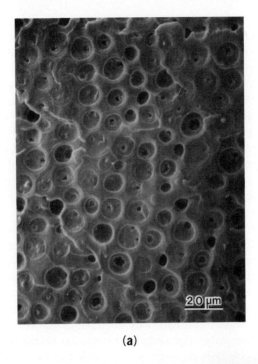

(a)

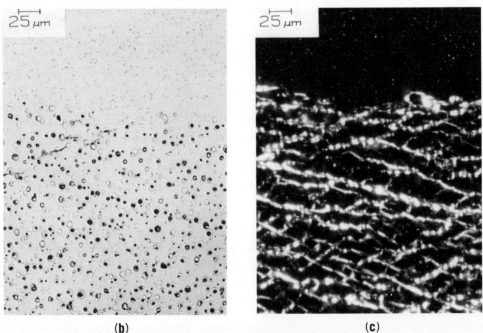

(b)   (c)

Fig. 22. (a) Scanning electron micrograph of the fracture surface of an elastomer-modified epoxy. (b) Optical micrograph of a thin section of the same epoxy. (c) The same optical micrograph viewed between crossed-polarizers.

concentration, a designer specifies very thick sections at the junction between two surfaces. This causes the growth of very large spherulites and voids, resulting in brittleness. Brittleness is avoided by using thin sections. Stiffness can be overcome by the proper choice of cross-sectional geometries.

Enhancement of impact toughness in semicrystalline polymers is similar to techniques used for glassy polymers. Usually an elastomeric phase is melt-blended, resulting in a dispersion of spheroidal elastomeric particles or a network. Compatibilization is obtained by copolymerization or chemical reaction at the interface. The toughening mechanisms are not well understood. Both massive crazing and cavitation–shear banding mechanisms can occur, depending on the polymer blend, morphology, and degree of adhesion. The elastomeric phase may alter the morphology of the semicrystalline phase, for example, by nucleating additional spherulites, perhaps of a different crystalline structure, thus improving the inherent toughness of the crystalline phase by reducing the spherulitic size. If toughening is caused by the formation of shear bands, deformation of spherulites contributes to energy absorption.

Incompatible Blends of Rigid Polymers.   Some blends are composed of rigid polymers, one or both of which may be rubber-modified. Examples of these materials include PC/ABS, PVC/ABS, PC/PBT, and nylon/PXE. Other combinations are also possible. An important difference between these materials and the PS/PXE blend described above is that the latter is thermodynamically compatible, and behaves as a homogeneous material, whereas the former are thermodynamically incompatible. The incompatibility causes the components, though intimately mixed, to form a continuous–dispersed phase morphology, or a cocontinuous morphology; in the latter neither component can be considered a dispersion. The ABS blends are tough, presumably because the ABS component is already toughened. The commercial versions of the other two examples contain an elastomeric impact modifier. The morphologies and mechanical behaviors of these materials have not yet been studied in detail. However, because of the potential for being able to combine the desirable properties of each of the components, this type of material will undoubtedly become increasingly important.

BIBLIOGRAPHY

"Impact Resistance" in *EPST* 1st ed., Vol. 17, pp. 574–629, by C. A. Brighton and D. A. Lannon, British Geon Ltd.

1. J. G. Williams, *Fracture Mechanics of Polymers,* Halsted Press, New York, 1984, p. 237ff.
2. P. E. Read in E. H. Andrews, ed., *Developments in Polymer Fracture-1,* Applied Science Publishers, London, 1979, Chapt. 4.
3. J. M. Hodgkinson and J. G. Williams, *Phys. Technol.* **13,** 152 (1982).
4. S. N. Kukureka and I. M. Hutchings, *Proceedings of the Conference on High Energy Rate Fabrication,* Leeds, UK, 1981, p. 29.
5. C. B. Bucknall, *Toughened Plastics,* Applied Science Publishers Ltd., London, 1977.
6. *Guide to Engineered Materials,* American Society for Metals, Metals Park, Ohio, 1986.
7. A. J. Kinloch and R. J. Young, *Fracture Behavior of Polymers,* Applied Science Publishers, London, 1983, p. 107ff.
8. C. E. Stephenson, *Brit. Plastics* **30,** 99 (1957).
9. C. E. Stephenson, *Brit. Plastics* **34,** 543 (1961).

10. N. H. Fahey, *Mater. Res. Std.* **1,** 871 (1961).
11. J. K. Rieke, "Instrumented Impact Measurements on Some Polymers," in R. E. Evans, ed., *Physical Testing of Plastics, Correlations with End-Use Performance, ASTM STP 736,* American Society For Testing and Materials, Philadelphia, Pa., 1981, pp. 59–76.
12. W. J. Moritz, *Proceedings of the Society of Plastics Engineers, 33rd ANTEC,* 1975, p. 540.
13. S. Turner, *Mechanical Testing of Plastics,* 2nd ed., The Plastics and Rubber Institute, 1983, p. 199ff.
14. *Ibid.,* p. 139ff.
15. I. M. Ward, *Mechanical Properties of Solid Polymers,* 2nd ed., John Wiley & Sons, Inc., Chichester, UK, 1983, p. 329ff.
16. P. B. Bowden, in R. N. Haward, ed., *Physics of Glassy Polymers,* Halsted Press, New York, 1973, Chapt. 5, p. 287ff.
17. S. Bahadur, *Polym. Eng. Sci.* **13,** 266 (1973).
18. L. M. Carapellucci, A. F. Yee, and R. P. Nimmer, *Proceedings of the Society of Plastics Engineers, 44th ANTEC,* 1986, p. 622.
19. R. P. Nimmer, *Polym. Eng. Sci.* **23,** 155 (1983).
20. P. B. Bowden and J. A. Jukes, *J. Mater. Sci.* **3,** 183 (1968).
21. G. C. Adams and T. K. Wu, *Proceedings of the Society of Plastics Engineers, 39th ANTEC,* 1981, p. 185.
22. A. F. Yee, *J. Mater. Sci.* **12,** 757 (1977).
23. P. I. Vincent, *Impact Tests and Service Performance of Thermoplastics,* Plastics Institute, London, 1971.
24. J. T. Ryan, *Polym. Eng. Sci.* **18,** 264 (1978).
25. E. Plati and J. G. Williams, *Polym. Eng. Sci.* **15,** 470 (1975).
26. W. F. Brown, Jr., and J. F. Srawley, *ASTM Special Technical Publication,* ASTM, Philadelphia, Pa., 1966, p. 12.
27. J. G. Williams and M. W. Birch, *Proceedings of the 4th International Conference on Fracture (ICF4),* Vol. 1, Part IV, 1977, p. 501.
28. L. V. Newman and J. G. Williams, *Polym. Eng. Sci.* **20,** 572 (1980).
29. Ref. 5, p. 182ff.
30. H. Breuer, F. Haaf, and J. Stabenow, *J. Macromol. Sci. Phys.* **B14**(3), 387 (1977).
31. A. F. Yee, *Polym. Eng. Sci.* **17,** 213 (1977).
32. A. F. Yee and M. A. Maxwell, *J. Macromol. Sci. Phys.* **B17,** 543 (1980).
33. J. A. Sauer and C. C. Chen, in H. H. Kausch, ed., *Crazing in Polymers,* Vol. 52/53 of *Advances in Polymer Science,* Springer-Verlag, New York, 1983.
34. A. F. Yee, *Proceedings of the International Conference on Toughening of Plastics II,* London, 1985, p. 19.1.

General References

A. J. Kinloch and R. J. Young, *Fracture Behavior of Polymers,* Applied Science Publishers, London, 1983. This is a particularly comprehensive treatise and contains much up-to-date (early 1980s) information.
I. M. Ward, *Mechanical Properties of Solid Polymers,* 2nd ed., John Wiley & Sons, Inc., Chichester, UK, 1983. Provides many of the background concepts, mostly from a polymer physics point of view.
J. G. Williams, *Stress Analysis of Polymers,* 2nd ed., Halsted Press, Chichester, 1980. Presents fundamental mechanics using relatively simple mathematics. The last chapter is particularly relevant.
C. B. Bucknall, *Toughened Plastics,* Applied Science Publishers, London, 1977. Contains mid-1970s state-of-the-art information on the science and technology of impact-modified plastics.
ASTM Standards, Philadelphia, Pa. The current edition of this series contains detailed information on testing techniques and data analysis.

ALBERT F. YEE
University of Michigan

LIQUID CRYSTALLINE POLYMERS

Low molecular weight liquid crystalline (LC) compounds have been known for about 100 years (1). However, LC polymers have attained prominence only in the last 20 years, owing largely to the discovery of the aramids (eg, poly(p-phenyleneterephthalamide) (PPD-T) (1), the polyamide from p-phenylenedi-amine and terephthalic acid) and thermotropic polyesters (eg, copolymers of p-hydroxybenzoic acid (HBA), p,p'-biphenol, and terephthalic acid (2)), and the commercial development of strong fibers and unique plastics from these materials (2–5).

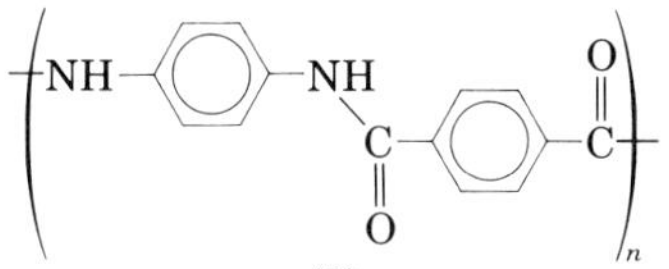

(1)

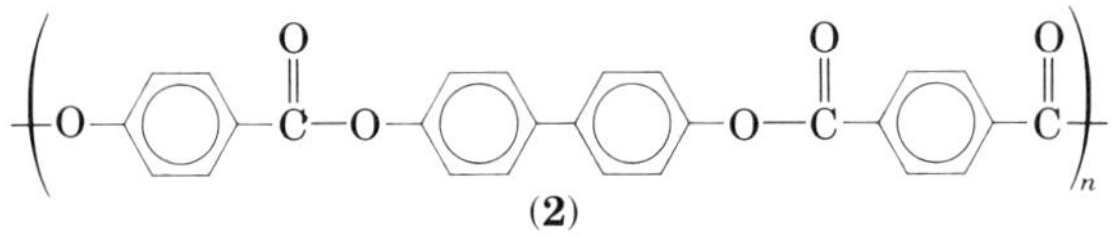

(2)

Such polymers may exhibit LC order either in solution (lyotropic liquid crystals) (1) or in the melt (thermotropic liquid crystals) (2). Specific structural features of LC polymers usually involve a succession of para-oriented ring structures to give a stiff chain with a high axial ratio (ratio of length of molecule to its width, aspect ratio) x. This field has expanded rapidly, as shown by a 100-fold increase in annual publications since 1970.

The common structural feature of low molecular weight LC compounds is asymmetry of molecular shape, manifested either as rods characterized by a uniaxial order with an axial ratio usually greater than three or by thin platelets with biaxial order. Some illustrative structures follow (6):

Nematic:

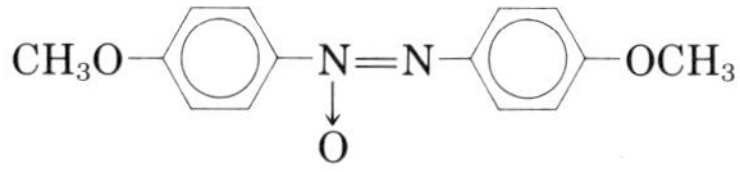

p-azoxyanisole

Smectic:

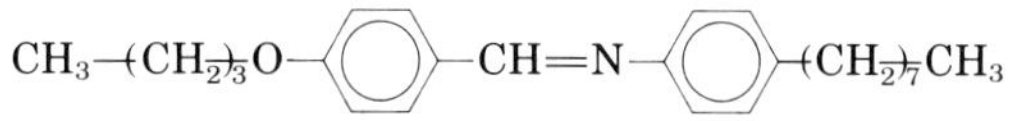

N-(p-butoxybenzylidene)-4-octylaniline

Cholesteric:

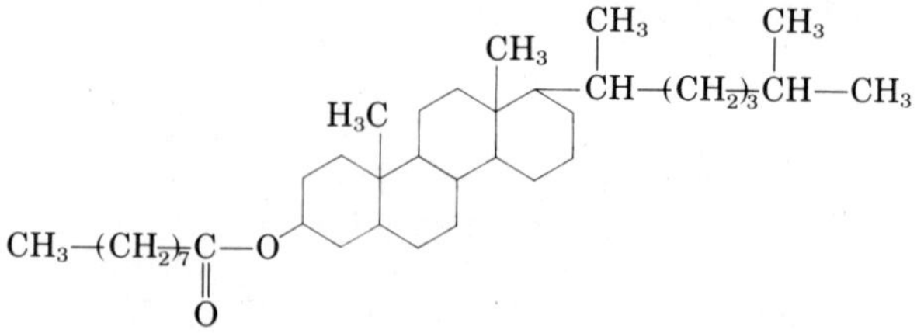

cholesteryl nonanoate

Discotic:

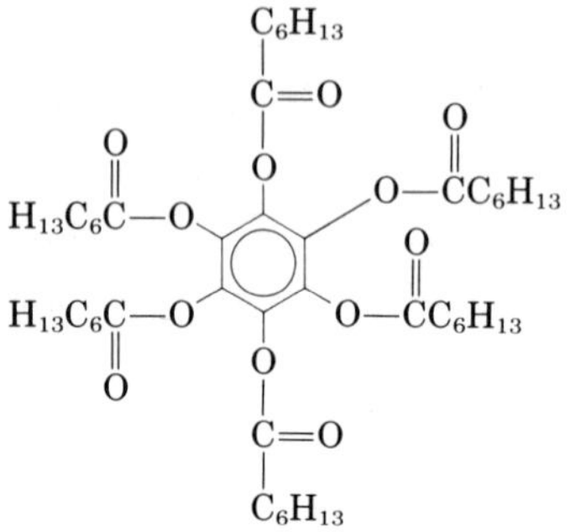

benzene hexa-*n*-heptanoate

A 1923 review (7) of low molecular weight LC compounds posed the questions "What happens to the molecule when it is made ever longer?" and "Does the LC state finally disappear?" However, the attempt to answer these questions with oligomers of *p*-hydroxybenzoic acid was hampered by decomposition before melting and insolubility of the oligomers as the number of oxybenzoyl units increased beyond four.

The types of ordered structures that LC materials form may be represented as shown in Figure 1 (8). These ordered LC structures give birefringent solutions and melts and have been studied by optical methods with polarized light.

Rigid structures that cause high extension of the backbone chain, such as structures (**1**) and (**2**), are conducive to liquid crystal formation in solution or in the melt. Stiff rodlike (or platelet) molecules organize in LC domains in such a way that molecular alignment occurs in a preferred direction along their long axes (for rods). This results in birefringent solutions or melts, visible with a polarizing microscope under crossed polarizers as bright and, often, colored areas with characteristic textures. The rodlike shape is common, the platelet shape unusual. An example of the latter is mesophase pitch (9,10). Many examples of the former exist, including lyotropic solutions of natural products such as tobacco mosaic virus (11), polypeptides existing in solution as helices in nonprotonating solvents such as dioxane, eg, poly(γ-benzyl L-glutamate) (PBLG) (12), and cellulose and its derivatives (13–17). The discoveries of lyotropic synthetic aramids (18), eg, poly(1,4-benzamide) (PBA) (**3**), and of synthetic thermotropic aromatic

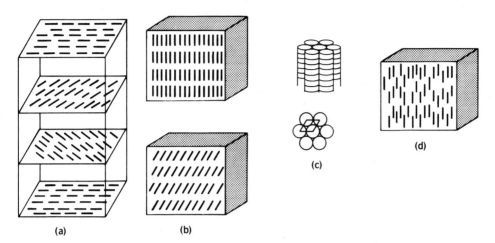

Fig. 1. Mesophasic structures: (**a**) cholesteric; (**b**) smectic A (top) and smectic C (bottom); (**c**) discotic; (**d**) nematic (8). Courtesy of the American Chemical Society and Springer-Verlag.

polyesters (19–23) and polyazomethines (24) set the stage for a great number of inventions and publications.

$$\left(\!\!-\!\!\underset{}{\bigcirc}\!\!-\!\!\overset{\displaystyle\overset{O}{\|}}{\underset{}{C}}\!\!-\!\!\underset{\underset{H}{|}}{N}\!\!-\!\!\right)_n$$

(3)

New synthetic fibers of great strength and stiffness were developed; these included Kevlar aramid fiber (18,25–28) and new plastic products and resins (29) such as Xydar and Vectra (see FIBERS, ENGINEERING; POLYAMIDES, AROMATIC; POLYESTERS, AROMATIC).

It also was discovered (30) that long stretches of rigid units were not required to produce liquid crystallinity in polymer melts. In fact, in some cases where rigid units alternate with flexible spacers, two adjacent para-oriented aromatic rings suffice (31). In another class of LC polymers, stiff side chains attached to a flexible backbone associate to form LC domains. These polymers (3,32–37) closely resemble low molecular weight liquid crystals but differ, for example, by having longer relaxation times because they are polymeric.

Theory

The formation of liquid crystals is solely a consequence of molecular asymmetry. It is not due to intermolecular attractions but is dominated by intermolecular repulsions, ie, the fact that units in two molecules cannot occupy the same space. There is, therefore, a limit to the number of rodlike chains that can be accommodated in random arrangement in solution or in the melt. When this limiting concentration is exceeded, a crystalline phase or liquid crystalline phase separates, depending on a number of factors. Crystallization may take place in two steps (38): cooperative ordering of the chains in a given region into a parallel alignment without change in the intermolecular interactions; and increase in intermolecular interactions made possible by the more efficient packing of the chains in the parallel state. The first step is similar for each event. The second step is separation of an LC phase or a crystalline phase. Solvent–polymer vs polymer–polymer interaction is important in solution and chain regularity plays a role in both cases, especially in melts. The rate of crystallization also could be a factor. With extensive chain alignment and good chain regularity, crystallization should proceed so quickly that the two steps are inseparable, as evidenced from the very rapid crystallization of polyester fiber subjected to elongational flow during high speed spinning.

In the separated LC phase, the molecules align to accommodate the higher concentration. As overall concentration increases further, additional ordered phase is formed at the expense of the random or isotropic phase. This phase separation phenomenon was treated semiquantitatively in an early theory (39). The mathematical complexity of this theory restricts those results to low concentrations. The preferred lattice theory of Flory (38,40–42) is widely applicable for quantitative treatment of phase-separation phenomena in solutions of rodlike chain polymers, including melts. At a concentration V_p^*, ie, volume fraction of polymer, athermal solutions of hard rods (no intermolecular attractions) of axial ratio x show incipiency of metastable order, ie,

$$V_p^* \approx (8/x)[1-(2/x)] \approx 8/x \text{ at high values of } x$$

The first expression is accurate within 2% for $x > 10$. This relation is widely used to represent the threshold volume fraction for appearance of a stable anisotropic phase, an assumption that now appears to be well-justified.

The variation of biphasic composition with axial ratio x is shown in Figure 2 (38,40–42). The difference in composition of the two phases is small and the curves merge at $x = 6.42$, the calculated minimum value of x for stable nematic order in a melt of hard rods. The principal distinguishing feature of these two phases is the order in the anisotropic phase rather than the composition difference.

Theory predicts that V_p and V_p', the volume concentrations of polymer in the isotropic phase and anisotropic phase, respectively, are at equilibrium in monodisperse systems. In real systems, polydispersity is encountered and polymer–solvent interactions can change axial ratio or modify chain stiffness, eg, by changing the equilibrium cis–trans ratio in the amide linkages in aramids or by effecting a helix–coil transition (qv) in PBLG. Despite these difficult-to-quantify variables, the theory fits experimental results quite well. Values of V_p as a function of x for aramids, eg, PPD-T (**1**) and PBA (**3**), are in approximate agreement with theory (solid line in Fig. 3) (38,40–42). For unfractionated polymer systems, x is a weight-average axial ratio calculated from $\overline{M}_w$ to suppress the effect of polydispersity. Values of V_p'/V_p are in the range of 1.3–1.6, also agreeing with theory for monodisperse polymers (qv). For polydisperse systems, one should

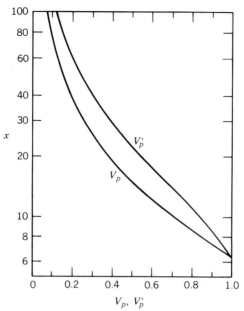

Fig. 2. Compositions of coexisting isotropic and anisotropic phases expressed in volume fractions V_p and V'_p, respectively, as functions of the axial ratio x of hard rods in athermal solutions (42). Courtesy of Springer-Verlag.

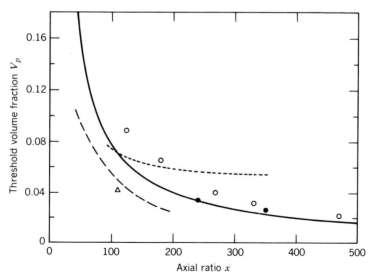

Fig. 3. Threshold volume fractions V_p for incipience of a nematic phase vs axial ratio x (42). The solid line is theoretical. The broken lines and points show experimental results of various authors on aromatic polyamide (structures (1) and (3)) solutions in sulfuric acid and in N,N-dialkylamide solutions. ○, PPD-T–DMAC, LiCl; ●, PPD-T–85% H_2SO_4; △, PBA–DMAC, LiCl. -------, PPD-T–concentrated H_2SO_4; – –, PBA–DMAC, LiCl. DMAC = dimethylacetamide. Courtesy of Springer-Verlag.

expect a higher ratio. Polymers that owe their rigidity to helix formation, eg, PBLG and shear-degraded deoxyribonucleic acid (DNA), are in even better agreement; these are more nearly rodlike structures. Poly-N-alkylisocyanates (**4**)

$$\left(\begin{array}{c} \overset{R}{\underset{|}{}} \ \ \overset{O}{\underset{\|}{}} \\ -\!\!\!\underset{}{N}\!-\!\underset{}{C}\!\!\!- \end{array}\right)_{\!\!n}$$

(**4**)

where R = n-hexyl or n-octyl, are more flexible, and experimental values of V_p are about twice those calculated from theory.

Another consequence of theory is that, in a polydisperse system, the longer species (higher x) preferentially populate the anisotropic phase. This fractionation of species expressed as a concentration ratio between phases can be as high as four (42).

Rodlike and Flexible Chain Mixtures.   Theory predicts that anisotropic phases tolerate only a negligible concentration of flexible random-coil molecules, whereas isotropic phases tolerate a fair concentration of rodlike molecules. This obvious consequence of the lattice treatment method was confirmed experimentally (43) for the system tetrachloroethane–polyisocyanate–polystyrene. Polystyrene does not occur in detectable amounts in the nematic phase. Even for the mixture of poly(1,4-benzamide) with the X-500 copolymer composition —(p-C₆H₄— CONH—NHCO—p-C₆H₄—CONH)— in N,N-dimethylacetamide and 3% LiCl, the similar X-500 composition with only limited flexibility is not detectable in the nematic phase containing PBA. In rejecting the random-coil polymer, the LC phase shows a principal feature of crystalline materials in general, ie, rejection of a foreign material, thus justifying the term liquid crystal (42).

Semiflexible Chains.   Most rodlike polymers are not truly rods. Even aramids, which experimentally verify Flory's theory for rods, depart slightly from linearity and, as $x \to \infty$, can be considered quite flexible. This is apparent from plots of log [η] vs log M, where the slope of the curve (a in the Mark-Houwink relationship [η] = KM^a) is close to that expected, ie, 1.3–1.8 for rodlike molecules at low [η]. At very high M (molecular weight), values of a approach the ⅔ value usually observed for random coils (44). Thus complete rodlike character is not required for LC phase formation. However, as flexibility increases, higher solution concentrations are required to achieve the critical V_p for phase separation. Therefore, for cellulose derivatives (45), V_p (critical) ranges from 0.25 to 0.39, depending on the solvent and molecular weight M. These values are higher than for stiffer aramids of comparable or lower M.

The most favorable conditions for LC phases are in the melt. An axial ratio of 6.42 or about three aromatic rings with connecting linkages should be sufficient to produce LC order. Since longer chains can be built by alternating rigid units and flexible units, liquid crystallinity in such melt systems should be expected, and, indeed, has been reported (30,46).

These systems melt at a relatively low temperature and exhibit both a melting transition and a nematic–isotropic transition. This is in contrast to the aramids (**1,3**) which do not melt or the thermotropic aromatic polyesters (**2**) which melt but usually decompose when heated before reaching the nematic– isotropic transition temperature. In the case of (**5**),

$$\left(\begin{array}{c} O \\ \parallel \\ C-O-\bigcirc-\overset{\overset{\displaystyle CH_3}{|}}{C}=CH-\bigcirc-O-\overset{\overset{\displaystyle O}{\parallel}}{C}-(CH_2)_n \end{array}\right)$$

(5)

the nematic–isotropic transition ($n \rightarrow i$) decreases from 300°C for $n = 6$ to 190°C for $n = 12$. The values of $n \rightarrow i$ oscillate, depending on whether n in (5) is odd or even; it is lower for the odd members. For this to occur, correlations of rigid chain segments must be propagated through as many as 12 bonds. In the absence of such correlation, structures analogous to a mixture of random coils and rigid rods could be considered except that the two are interconnected in the same molecule. There should be a tendency to exclude the flexible unit from the LC phase. In copolymers of p-hydroxybenzoic acid (HBA) and poly(ethylene tere-phthalate) (PET), heterogeneity due to phase separation has been observed (47). According to another interpretation, the flexible segments act as a solvent for the rigid segments corresponding to a single-phase lyotropic solution. A signifi-cant difference is in the promotion of entanglements due to the random-coil nature of the spacer units. This should lead to lower tensile properties as observed, eg, in the fibers from HBA–PET LC copolymers (as opposed to all-ring copolymers (2)). This could be indicative of difficulty in orienting the HBA units because of intertwining of the flexible units. In fibers, however, uniaxial orientation is ob-tained via the small cross section, but even here persisting entanglements prevent higher tensile properties in as-spun fibers. In thick moldings, this condition is enhanced and there is a greater difference between skin and core orientation (21,48).

Intermolecular Interactions. The original theories were developed for ather-mal solutions of rigid rods. Most rigid polymers discussed here, eg, aramids, require strong polymer–solvent interaction to achieve the concentrated solutions necessary for exceeding V_p' in solution to ensure a nearly pure LC phase. Such interaction as measured by χ, the usual polymer–solvent interaction parameter in solution theory, is shown in Figure 4, which indicates stronger interaction as χ decreases. The features of this figure (40) are well-substantiated by data on PBLG and aramids (42). Even the two LC phases, one concentrated and one dilute, coexisting at small positive values of χ, have been observed for PBLG.

Orientation-dependent Interactions and Semiflexible Chains. Polymer–polymer interactions are important in determining the stability of LC phases if these forces are orientation dependent. Groups in polymer chains with polariz-abilities higher along their bond axes than perpendicular thereto can contribute to the stability of the LC state. Examples of such groups are p-phenylene, —N=N—, —C≡C—, and —C≡N. The interactions are calculated in terms of segmental contacts and are secondary and additive to the effect of molecular asymmetry as measured by chain extension or chain stiffness in promoting LC phase formation. The net effect of the interaction is to lower the critical axial ratio needed for LC phase separation (42). These interactions can be the dominant forces in low molecular weight compounds and play an important role in semiflexible long-chain polymers, particularly in the melt.

Monte-Carlo lattice treatment demonstrates that semiflexibility alone does not lead to LC phase separation (49). It is necessary to invoke local orientation-

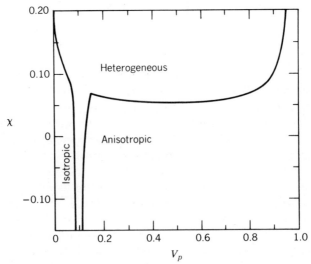

Fig. 4.   Volume fractions V_p of coexisting phases for rods of axial ratio, $x = 100$, as a function of the interaction parameter χ (42). The cusp marks a triple point where three phases coexist. Courtesy of Springer-Verlag.

dependent intermolecular interactions to obtain a nematic transition. A statistical theory (50) takes into account orientation-dependent chain interactions for long semiflexible polymers in the melt, predicting that polytetrafluoroethylene (PTFE) forms a stable nematic melt phase (51). In semiflexible chains, eg, PBLG, that exhibit rigid or coil-like configurations, formation of an anisotropic phase is accompanied by a pronounced induced rigidity in which the ordered phase contains chains substantially more rigid than those populating the coexisting isotropic phase (52). With the help of a simple relationship between chain stiffness and geometry for semiflexible chains, a criterion was arrived at for deciding whether stable nematic order in a homopolymer melt is possible (53).

Lyotropic Solutions

Polypeptides.   The formation of liquid crystalline solutions by high polymers was first noted in natural polypeptides. In 1950, a liquid crystalline solution of a synthetic polymer was discovered (54) in the course of evaporating a chloroform solution of poly(γ-benzyl L-glutamate) (PBLG).

In studies on lyotropic PBLG solutions in chloroform, dichloromethane, dioxane, and m-cresol, the critical concentration point for the first formation of a birefringent, anisotropic phase was noted (12). The point of complete anisotropy and other features such as the high optical rotation and the presence of visible, evenly spaced retardation lines was also observed. The anisotropic phase was concluded to be cholesteric.

In suitable solvents, polypeptides assume the shape of rodlike α-helices as a result of intramolecular hydrogen bonding and association of the solvent with the substituents. Upon increasing concentration, a point is reached where more rodlike molecules can only be accommodated in the medium if some of the rods are arranged in approximately parallel arrays called nematic order. These ordered domains become a separate phase through entropy effects. In the cholesteric state, which is induced by chirality in the polymer or the medium, a superstructure is imposed upon these ordered domains wherein the rods in the anisotropic layers deviate from parallel in a regular way; on a microscopic level, a spiral structure develops in successive layers (Fig. 1a). The pitch of the spiral is twice the spacings of the observable retardation lines; the handedness is affected by the solvent, the nature of the substituents, and the temperature. Racemic mixtures of polymer chains yield nematic solutions. The cholesteric structure can also be overcome by electrical or magnetic fields or by shear (55).

The early observations on PBLG were followed by Flory's theoretical derivation for rodlike molecules of uniform length in 1956 (38,40–42,56).

Experimental support for Flory's theory was provided (57) by data on phase separation and the viscosity at low shear as a function of concentration for PBLG solutions in m-cresol (Fig. 5). This type of plot with its sharp reversal in viscosity slope has become a common characterization test for liquid crystalline solutions. The initial formation of an anisotropic phase was reported to occur at the viscosity peak. The critical concentration point was inversely proportional to molecular weight.

A viscosity–concentration plot of solutions of racemic poly(γ-benzyl glutamate) (PBG) in m-cresol showed a shoulder on the low concentration side (58). Microscopic examination revealed the initial formation of droplets of anisotropic phase in this region (Fig. 6). The occurrence of anisotropy before maximum viscosity was deduced (59) in calculations for solutions of rodlike polymers with uniform chain length and colloidally dispersed anisotropic phase. Regardless of whether or not an anisotropic phase first forms precisely at the maximum apparent viscosity for all lyotropic polymer systems, the biphasic condition begins in this region first as a dispersion of anisotropic phase in isotropic phase, followed with increasing concentration by an inversion to isotropic phase in anisotropic phase, and eventually by an essentially wholly anisotropic phase. Increasing shear rates moved the viscosity peak to lower concentration and eliminated the shoulder (Fig. 6).

The literature on PBLG and other polypeptides is extensive, elaborating on solution properties, conditions for the transition of the chains from flexible coils to helical rodlike structures, and effects of magnetic and electrical fields (60–64).

Aromatic Polyamides.   The para-linked aromatic polyamides are representatives of a class of polymers with a rodlike structure which form lyotropic solutions; they were first prepared at DuPont (65–69). These polymers yield liquid crystalline solutions not because of helix formation, as observed with the polypeptides, but because of an inherently extended rigid chain. This chain is produced

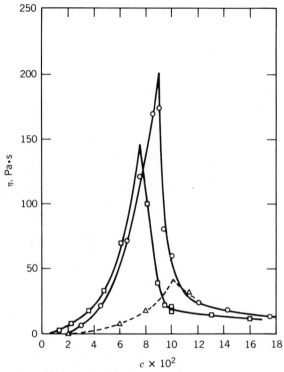

Fig. 5. Viscosity at low shear stress vs concentration for the fractions of PBLG with mol wt 342,000 (□); 270,000 (○); and 220,000 (△) in m-cresol (57). To convert Pa·s to P, multiply by 10. Courtesy of Academic Press, Inc.

by aromatic units with coaxial or parallel and oppositely extending bonds combined with the partial double-bond character of the C—N bond in the predominantly trans amide linkage. The bonds extending from the amide group in solid benzanilide deviate from being exactly parallel by ca 13° (70). Other polyamides that form lyotropic solutions are given in Table 1.

The formation of liquid crystalline solutions of polyamides is affected by the polymer structure (extended chain and trans configuration of the amide group), the molecular weight and axial ratio (which must exceed a minimum value), solvent–polymer interaction, and the solubility must be sufficiently high to exceed the critical concentration. The temperature affects solubility and liquid crystalline range.

Rodlike aromatic polyamides are soluble in tertiary organic amides such as dimethylacetamide, N-methylpyrrolidinone, and tetramethylurea containing lithium or calcium chlorides, or in concentrated sulfuric acid, oleum, substituted sulfuric acids, and hydrogen fluoride. Solubility in the amide–salt solvents results from association of chloride ions with the amide protons in the polymer, as indicated in Figure 7. Solubility in strong acids arises from adducts between acid and amide units and dissolution of this complex with additional acid. The amide–salt solvents are useful media for the synthesis of aromatic polyamides and can be used in the direct preparation of liquid crystalline solutions. The effect of the

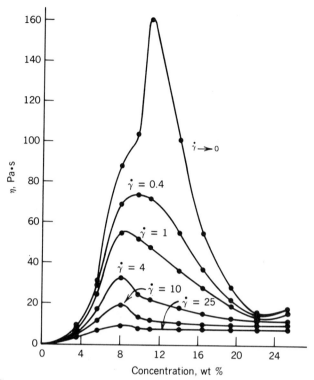

Fig. 6. Concentration dependence of steady-shear viscosity at several shear rates ($\dot{\gamma}$) for PBG in *m*-cresol (58). To convert Pa·s to P, multiply by 10.

salt concentration on solubility and formation of liquid crystalline solutions of PBA is shown in Figure 8.

The liquid crystalline solutions of aromatic polyamides are normally nematic. Cholesteric order has been induced in a solution of poly(1,4-benzamide) in dimethylacetamide–LiCl by addition of (+)-2-methylcyclohexanone (81), and a copolymer of 1,4-benzamide units with 3 mol % of L-valine formed a cholesteric solution at >6.5% in dimethylacetamide–LiCl (82). Some, but not all, polymers show appreciable inverse dependence of the critical concentration on molecular weight. This effect is shown for PBA in Figure 9. The liquid crystalline solutions often appear turbid and exhibit residual schlieren lines at rest. With moderate stirring, the solutions become opalescent. Under the polarizing microscope, samples transmit light in the anisotropic regions, which usually appear highly colored and show mobile boundaries and sometimes threadlike lines (Fig. 10).

The plots at low shear viscosity of aromatic polyamide solutions against increasing concentration show a sharp peak similar to polypeptide curves (Fig. 5). Microscopic observations of solutions of aromatic polyamides in the region of the peak viscosity usually have not produced evidence of anisotropic phase formation before attainment of the viscosity peak. However, one of two samples of PBA in dimethylacetamide–LiCl had an anisotropic phase before the viscosity peak was reached (83). Along the initial part of the down slope, a two-phase system exists. Exactly where the isotropic phase disappears is difficult to deter-

Table 1. Polyamides That Form Liquid Crystalline Solutions

Diamine	Diacid	Solvent[a], %	Concentration[b], wt %	Refs.
1,4-phenylene	fumaric	[c]	27[d]	71
chloro-1,4-phenylene	fumaric	96	31	71,72
4,4'-benzanilide	*trans,trans*-muconic	96	20–30	67,73
1,4-phenylene	mesaconic	100	25[e]	71
1,4-phenylene	*trans*-1,4-cyclohexylene	100.6	14.3	74
chloro-1,4-phenylene	*trans*-1,4-cyclohexylene	100.6	14.4	74
1,4-phenylene	1,4-dimethyl-*trans*-1,4-cyclohexylene	100.6	10	74
1,4-phenylene	2,5-pyridine	96	10	75,76
chloro-1,4-phenylene	2,5-pyridine	100	17	75
3,3'-dimethyl-4,4'-biphenylene	2,5-pyridine	100	14.6	75
1,4-phenylene	4,4'-stilbene	100	12	77
chloro-1,4-phenylene	4,4'-stilbene	99.6	20	77
1,4-phenylene	4,4'-azobenzene	97	12	78
4,4'-azobenzene	4,4'-azobenzene	99.9	18	78
1,4-phenylene	4,4'-azoxybenzene	101.5	12	78
4,4'-azobenzene	4,4'-azoxybenzene	101.2	12	78
1,4-cyclohexylene	4,4'-azobenzene	100	10	78
4,4'-azobenzene	terephthalic	96	12	78
3,8-phenanthridinone	terephthalic	99	>7	79
4,4'-biphenylene	terephthalic	101.0	12.9	18
4,4'-biphenylene	4,4'-bibenzoic	96	insoluble	80
1,4-phenylene	4,4'-bibenzoic	100.3	10.9	18
1,4-phenylene	4,4'-terephenylene	96	insoluble	80
1,4-phenylene	2,6-naphthalic	98.7	10	18
1,5-naphthylene	terephthalic	101.3	10	18
3,3'-dimethyl-4,4'-biphenylene	terephthalic	101	12	18
3,3'-dimethoxy-4,4'-biphenylene	terephthalic	101	12	18
3,3'-dimethoxy-4,4'-biphenylene	4,4'-bibenzoic	101	12	18

[a] Solvent is H_2SO_4 unless otherwise noted.
[b] Concentration is not the critical concentration; temperature is 25–30°C unless otherwise noted.
[c] FSO_3H.
[d] Above 50°C.
[e] Above 80°C.

mine. However, this point may not be reached even at the right-hand minimum of the curve. With increases in solution temperatures, the critical concentration is shifted to higher concentrations (68,84).

In the biphasic region of a lyotropic system, the anisotropic phase has a higher polymer concentration and therefore a different density. The two phases may separate spontaneously upon standing or by centrifugation (66–68). The proportions of phases found in a sample of poly(chloro-1,4-phenyleneterephthalamide) (ClPPD-T) are shown in Figure 11. High polymer is fractionated into the anisotropic phase and residual lower molecular weight polymer in the isotropic phase (Fig. 12).

Substitution of the rings by small groups, especially in unsymmetric posi-

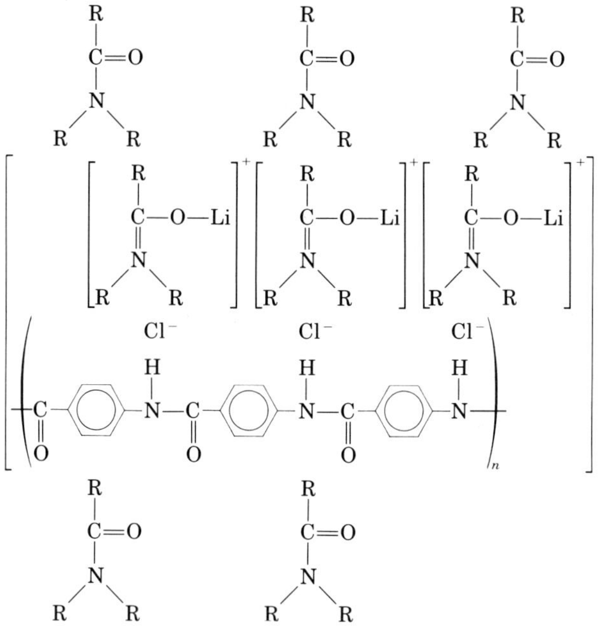

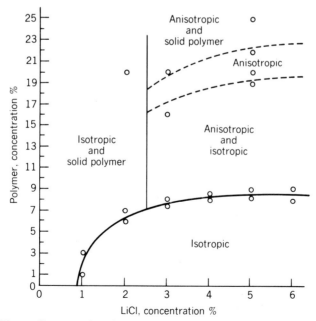

Fig. 7. Postulated structure of solvated aromatic polyamide in *N,N*-dialkylamide–LiCl solvent (81). Courtesy of the American Chemical Society.

Fig. 8. Phase diagram for the system poly(1,4-benzamide)–dimethylacetamide–LiCl; solutions contain 1% water by volume (81). Phase state is determined by polarizing microscope. Courtesy of the American Chemical Society.

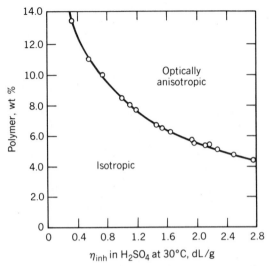

Fig. 9. Effect of inherent viscosity on critical concentration for poly(1,4-benzamide) in dimethylacetamide–4% LiCl; points determined by polarizing microscope (68). Courtesy of the American Chemical Society.

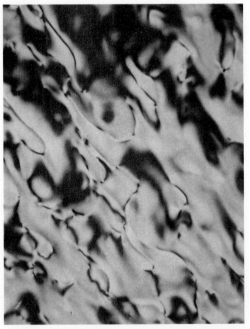

Fig. 10. Nematic solution of poly(chloro-1,4-phenyleneterephthalamide) in dimethylacetamide–LiCl observed between crossed polarizers at 340× (67). Courtesy of the American Chemical Society.

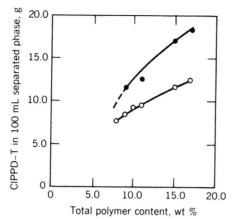

Fig. 11. Poly(chloro-1,4-phenyleneterephthalamide) (ClPPD-T) concentration in equilibrium anisotropic and isotropic phases at room temperature of liquid crystalline solutions prepared with various weight percent polymer (η_{inh} = 1.09 dL/g) in dimethylacetamide–2.18% LiCl (69). ●, ClPPD-T concentration in separated anisotropic phase; ○, ClPPD-T in separated isotropic phase. Courtesy of the American Chemical Society.

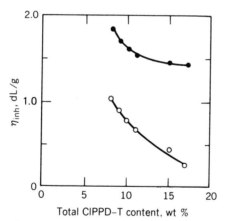

Fig. 12. Inherent viscosity of poly(chloro-1,4-phenyleneterephthalamide) (ClPPD-T) in equilibrium anisotropic and isotropic phases from liquid crystalline solutions prepared with various weight percent of polymer (η_{inh} = 1.09 dL/g) in dimethylacetamide–2.18% LiCl (69). ●, η_{inh} of ClPPD-T in separated anisotropic phase; ○, η_{inh} of ClPPD-T in separated isotropic phase. Courtesy of the American Chemical Society.

tions, enhances polymer solubility and may improve mesogenic capability. Mesogenic copolymers can be formed with any stoichiometrical proportion of units of the type illustrated in Table 1. Copolymers may also contain units that provide flexible joints as with polymethylene, or enforced bends, as with a unit such as 1,3-phenylene. The limiting amount of comonomer(s) of this class varies with the type of unit and the distribution along the polymer chain (85). Strictly alternating polyamides composed of aromatic amide rodlike segments and flexible polymethylene segments have been prepared (73). Anisotropic solutions in concentrated sulfuric acid or dimethylacetamide–5% LiCl were obtained when the aspect ratio of the rigid blocks was above 3.5–4.5, ie, 3–4 1,4-phenylene–amide units. The critical concentration point was lower as the length of the rodlike block increased.

Extensive N-substitution of extended-chain polyamides by small alkyl units increases the flexibility of the polymer chain and destroys the mesogenic character. On the other hand, solutions of N-octadecylated poly(1,4-phenylene-terephthalamide) in tetrahydrofuran yielded results indicative of a lyotropic solution (86). The reaction of 1,4-phenylenediamine and terephthalic acid in liquid sulfur trioxide yields an anisotropic solution of highly sulfonated poly(1,4-phenyleneterephthalamide) (87). The sulfonic acid groups are on the diamine unit. Halogen substitution of the diamine reduces sulfonation, and the polymer from tetrafluoro-1,4-phenylenediamine was unsulfonated (88). All halogen-substituted polyamides yielded anisotropic solutions.

Polyoxamides and Polyhydrazides. Polyoxamides and polyhydrazides may be regarded as polyamides in which two amide groups are linked through carbonyl or imino units (see also POLYHYDRAZIDES AND POLYOXADIAZOLES).

$$-HN-\overset{\overset{\displaystyle O}{\|}}{C}-\overset{\underset{\displaystyle O}{\|}}{C}-NH- \qquad -\overset{\overset{\displaystyle O}{\|}}{C}-NH-NH-\overset{\underset{\displaystyle O}{\|}}{C}- \qquad -\overset{\overset{\displaystyle O}{\|}}{C}-\overset{\underset{\displaystyle O}{\|}}{C}-NH-NH-\overset{\overset{\displaystyle O}{\|}}{C}-\overset{\underset{\displaystyle O}{\|}}{C}-NH-NH-$$

oxamide group   hydrazide group   poly(oxalic hydrazide) units

Such polymers, containing a high proportion of aromatic units with coaxial or parallel and oppositely extending bonds, have highly extended chains and yield liquid crystalline solutions. Similar characteristics are found in oxamide–hydrazide copolymers. However, poly(oxalic hydrazide), composed entirely of amide units, lacks the solubility in the solvents thus far examined to attain the liquid crystalline state.

Extended-chain polyoxamides usually do not melt and have limited solubility. Polyoxamides from 2,2′-dimethyl-4,4′diaminobiphenyl and chloro-1,4-phenylenediamine form liquid crystalline solutions at 15–20 wt % solids above 50°C in fluorosulfonic and chlorosulfonic acids, respectively (71).

Extended-chain polyhydrazides are readily synthesized and are soluble in a variety of polar solvents. Those derived from oxalic, terephthalic, chloroterephthalic, and 2,5-pyridinedicarboxylic acids have been examined for formation of anisotropic solutions in ca 100% sulfuric acid at 10–20% solids (89,90). Of the homopolymers, only the poly(chloroterephthalic hydrazide) was soluble enough to form anisotropic solutions. A mixture of fluorosulfonic acid with 100% sulfuric acid improved the solubility of the hydrazide copolymers and increased the range of formation of liquid crystalline solutions. The critical concentration points vs inherent viscosity are shown in Figure 13 for an alternating copolymer. Polyhydrazides containing appreciable proportions of nonconforming units in the chain, such as isophthalic and alkylene units, give only isotropic solutions.

Many polyhydrazides are readily soluble in various aqueous organic bases (91,92), such as quaternary ammonium hydroxides and certain secondary aliphatic amines. Polymers with the required structures and sufficient solubility and molecular weight form nematic liquid crystalline solutions (Table 2 and Fig. 14). Anisotropic solutions may exhibit nematic–isotropic transitions as the temperature is raised, although some systems may gel before that point.

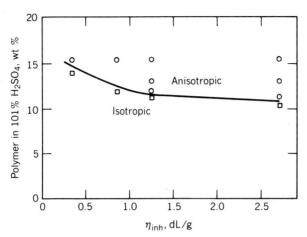

Fig. 13. Critical concentration in sulfuric acid at 0°C vs inherent viscosity (5% aq diethylamine) for poly(terephthalic–chloroterephthalic hydrazide) (89,90).

Table 2. Liquid Crystalline Solutions of Poly(terephthalic hydrazide) in Aqueous Organic Bases

| | Approximate solution range[a], wt % | |
Base	Base	Polymer
tetramethylammonium hydroxide	3–30	6–27
tetraethylammonium hydroxide	7.5–25	7.4–26
diethylamine	10–15	7.5–9
piperidine	10	7.5–9

[a] Determined at 27°C on polymer with inherent viscosities of 3.2–4.5 dL/g.

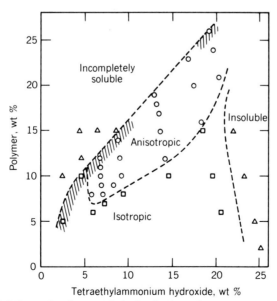

Fig. 14. Solubility of poly(terephthalic hydrazide) with inherent viscosity of 3.2 dL/g (5% aqueous diethylamine) in aqueous tetraethylammonium hydroxide at 27°C (92). Courtesy of Plenum Press.

The four homopolymers and their copolymers, described previously in sulfuric acid solutions, were examined in basic solvents. Both poly(terephthalic hydrazide) and poly(chloroterephthalic hydrazide) yielded anisotropic solutions. Polyhydrazides with ca 20% or more of meta-aromatic units or methylene units gave only isotropic solutions. Similarly, copolymers with more than 15–20% of amide units had greatly reduced solubility and failed to form anisotropic solutions. Polymers with mono-N-methyl substitution gave only isotropic solutions. Addition of 10% of compatible polar solvents, eg, dimethyl sulfoxide and tetrahydrofuran, caused liquid crystalline solutions to become isotropic.

The solutions of extended-chain polyhydrazides in organic bases are a deep yellow and often appear very turbid or opaque in the anisotropic state in bulk form. The uv-visible absorption maximum for poly(terephthalic hydrazide) is 370 nm. The coloration may be due to the association of the organic base with an enol form of the hydrazide unit, as illustrated below. The color is removed upon recovery and washing of the polymer.

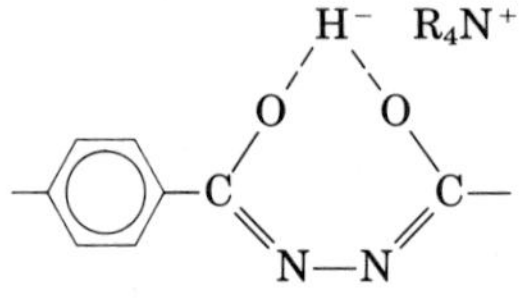

yellow 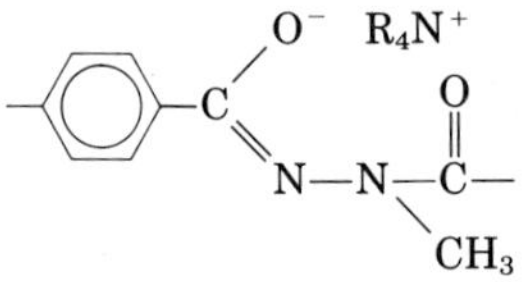   uncolored

Poly(amide–hydrazide)s.   Para-linked poly(amide–hydrazide)s have been prepared from 4-aminobenzhydrazide, oxalic dihydrazide, and terephthalic dihydrazide and para-aromatic diacid chlorides. With copolymers of high molecular weight and solubility, anisotropic solutions were obtained in 101.1% sulfuric acid or a 1:1 mixture of fluorosulfonic acid and sulfuric acid (89,90).

The poly(amide–hydrazide) from 4-aminobenzhydrazide and terephthalic acid (93) gives high modulus, high tenacity fibers by dry-jet wet spinning an isotropic solution in amide–salt solvent followed by drawing. Although the polymer forms anisotropic solutions in 100% sulfuric acid, dimethylacetamide–salt and dimethyl sulfoxide yield only isotropic solutions (94–99). The sulfuric acid presumably increases the persistence length of the polymer chains.

The concentrated dimethyl sulfoxide solutions of poly(terephthaloyl 1,4-aminobenzhydrazide) are close to anisotropy. Under high shear, high birefringence develops and the viscosity–concentration plots exhibit peaks resembling the critical concentration curves of lyotropic polymers at low shear. However, similar peaks were not found for dimethylacetamide–LiCl solutions (99).

Polyesters.   Very few lyotropic polyester systems have been described, partly because of the low solubility of the stiff-chain esters. Copolymers that give lyotropic solutions have been synthesized by interaction of cyclohexylene ester oligomers with aromatic ester oligomers. The compositions include poly(oxy-trans-1,4-cyclohexyleneoxycarbonyl-trans-1,4-cyclohexylenecarbonyl-b-oxy-1,4-phenyleneoxyterephthaloyl) and poly(oxy-cis-1,4-cyclohexyleneoxycarbonyl-trans-1,4-cyclohexylenecarbonyl-b-oxy-1,4-phenyleneoxyterephthaloyl) in methylene chloride–o-cresol (100); poly[oxy-trans-1,4-cyclohexyleneoxycarbonyl-trans-1,4-cyclohexylenecarbonyl-b-oxy-(2-methyl-1,4-phenylene)oxyterephthaloyl)] in 1,1,2,2-tetrachloroethane–o-chlorophenol–phenol (60:25:15 vol/vol/vol) (100); and poly[oxy-

trans-1,4-cyclohexyleneoxycarbonyl-*trans*-1,4-cyclohexylenecarbonyl-*b*-oxy(2-methyl-1,3-phenylene)oxyterephthaloyl] in *o*-chlorophenol (101).

Polyazomethines.   Para-linked aromatic polyazomethines form lyotropic solutions in concentrated sulfuric acid and methanesulfonic acid (24,102–104). Complex formation with the acids gives deeply colored solutions. Regions for the formation of a nematic phase for the polymer from methyl-1,4-phenylenediamine and terephthalaldehyde at molecular weights from 1500 to 7000 ($\overline{M}_w$) in sulfuric acid have been defined (103). Stability in sulfuric acid is limited and the polyazomethines cannot be regenerated as cohesive fibers by wet-spinning into aqueous baths (24). A polyazomethine–amide from 4,4′-diaminobenzanilide and terephthalaldehyde also forms a lyotropic solution in sulfuric acid (105).

Polymers with Heterocyclic Units.   Condensation polymers with heterocyclic units have been studied for use as high temperature-resistant materials (see HEAT-RESISTANT POLYMERS). A number of them are sufficiently rodlike to form liquid crystalline solutions. Although processing is difficult, some give fibers with outstanding strength and modulus values. Structures are given in Table 3 for heterocyclic polymers, not all of which, however, yield lyotropic solutions.

Copolymers of PBT and PBO with up to 25 mol % of 3,3′-biphenylene units remain soluble enough to form lyotropic solutions in polyphosphoric acid or methanesulfonic acid (115). Evidence for liquid crystalline solutions of AB–PBI is inconclusive. The formation of lyotropic solutions from poly(1,4-phenylene-1,3,4-oxadiazole) has been the subject of controversy, but there is clear evidence that such solutions are not formed in 100% sulfuric acid (115) even though the polymer can be processed to high strength fibers.

Polyisocyanides.   Polyisocyanides (qv) are prepared by the catalytic polymerization of monomeric isocyanides (116).

$$\text{---}(\text{C})_n\text{---}$$
$$\underset{\text{NR}}{\overset{\|}{}}$$

Liquid crystalline solutions were first reported (117,118) for poly(α-phenylethyl isocyanide) and poly(β-phenylethyl isocyanide). Poly(*n*-butyl isocyanide) and poly(*n*-octyl isocyanide) were also prepared (119,120). The octyl derivative gave anisotropic solutions in chloroform at ≥10 wt/vol % (23°C). The butyl derivative gave anisotropic solutions in chloroform above 50°C and in tetrachlorethane at ambient temperature.

The Mark-Houwink constant *a* for fractionated poly(α-phenylethyl isocyanide) was 1.3, indicating a rodlike structure (118). On the other hand, a high molecular weight and low intrinsic viscosity were noted for poly(*n*-butyl isocyanide), which is characteristic of high polymer chain flexibility (120).

Polyisocyanates.   Hydrodynamic studies showed that poly(*n*-butyl isocyanate) has a rodlike structure (121). However, it is not soluble enough to form lyotropic solutions. The mesomorphic behavior of polyisocyanates with specific R groups (**6**) in concentrated solution and in pure form has been described (122).

$$\overset{\text{O}}{\overset{\|}{\text{---}(\text{C}\text{---}\text{N})_n\text{---}}}$$
$$\underset{\text{R}}{\overset{\|}{}}$$

(**6**)

Table 3. Lyotropic Solutions of Polyheterocyclic Compounds

Compound	Structure	Lyotropic solution	Refs.
poly(1,4-phenylene-2,6-benzobisimidazole)		methanesulfonic acid	106
poly(1,4-phenylene-2,6-benzobisoxazole) (PBO)		methanesulfonic acid / chlorosulfonic acid / 100% sulfuric acid	107,108 / / 109
poly(1,4-phenylene-2,6-benzobisthiazole) (PBT)		5–10% in polyphosphoric acid / methanesulfonic acid	110 / 108,110,111
poly[2,6-(1,4-phenylene)-4-phenylquinoline]		1.0–1.5% in m-cresol–di-m-cresyl phosphate	112
poly[1,1′-(4,4′-biphenylene)-6,6′-bis(4-phenylquinoline)]		>9% in m-cresol–di-m-cresyl phosphate	112
poly[2,5(6)-benzimidazole] (AB–PBI)		methanesulfonic acid	113
poly(1,4-phenylene-1,3,4-oxadiazole)		none	114

These polyisocyanates are homopolymers with n-alkyl substituents (R) from C_6 to C_{12} or copolymers with mixtures of alkyl substituents or of alkyl substituents with aralkyl substituents; the aryl groups in the latter are spaced two or more methylene units from the main chain. The main chains are rigid, but flexible side groups enhance solubility without distorting the main chain or reducing the length-to-diameter ratio. Longer n-alkyl substituents result in side-chain crystallization (123). Solvents for the lyotropic systems include chloroform, bromoform, tetrachloroethane, and toluene. Phase separation and the consequent polymer fractionation, critical concentration curves, temperature effects, copolymer effects, and compatibility with flexible polymers have been studied. The critical concentration plots for apparent viscosity vs volume fraction of poly(n-hexyl isocyanate) show a shoulder on the low concentration side of the plot (124,125) (Fig. 15). The first appearance of an anisotropic phase observed with a polarizing microscope occurs at this shoulder and not at the viscosity peak. This effect has also been reported for poly(γ-benzyl L-glutamate). The transition to a wholly anisotropic system is on the down viscosity slope and not at the minimum point.

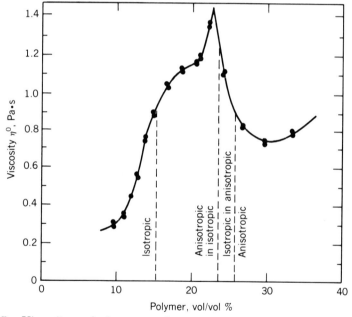

Fig. 15.   Viscosity and phase transformations of poly(n-hexyl isocyanate) (M_w = 73,000) in toluene, as a function of concentration (125). To convert Pa·s to P, multiply by 10.

Polyorganophosphazines.   Polyphosphazines (qv) have very high molecular weights, ie, $>10^6$, and are highly soluble. Many are low melting, but the formation of true liquid crystalline melts is not clearly established (126–128). However, polybisphenoxyphosphazine forms liquid crystalline solutions above 30% in toluene; poly[bis(2,2,2-trifluoroethoxy)phosphazine] yields liquid crystalline solutions at 40% or above in tetrahydrofuran (120,129).

$$-(\text{P}=\text{N})_n-$$

with R above and below P

$$R = \langle\bigcirc\rangle-\text{O}- \text{ or } CF_3CH_2O-$$

Metal Polyynes.　Unusual polymers with rodlike structures, such as (**7**) and (**8**), are formed by condensation of *trans*-bis(tri-*n*-butylphosphine)platinum dichloride with a bisacetylene or *trans*-bis(tri-*n*-butylphosphine)bis(1,4-butadiynyl)platinum and similar combinations in the presence of cuprous iodide and an amide (130) (see also ORGANOMETALLIC POLYMERS).

$$-(\text{Pt}-CH=C=CH)_n- \qquad -(\text{Pt}-CH=CH-\langle\bigcirc\rangle-CH=CH)_n-$$

(**7**)　　　　　　　　　　　　　　(**8**)

$$R = -P(n\text{-}C_4H_9)_3$$

Related polymers have been prepared containing palladium and nickel as well as copolymers with different metals (131). These polymers have high molecular weights and are soluble in toluene, trichloroethylene, and dichloromethane. They form lyotropic, nematic solutions with typical characteristics. The average a exponent in the Mark-Houwink equation was 1.7 for structure (**7**). When lyotropic solutions are placed in a magnetic field, some of the polymers are oriented with the field and others perpendicular to it (131).

Cellulose and Cellulose Derivatives.　The liquid crystalline solutions of cellulose and cellulose derivatives at high concentrations at rest are cholesteric and frequently appear pearlescent and display pale, but marked coloration (Fig. 16). The discovery of liquid crystalline solutions of cellulose derivatives was made independently in several laboratories at about the same time (13,14,133–136).

The formation of liquid crystalline solutions of cellulose derivatives requires appropriate molecular weight, nondegrading solvents, high solubility, and suitable temperature. Solvents vary greatly in nature and range from strong acids to strong bases and from highly polar to nonpolar liquids (Table 4).

These solutions exhibit the typical peak in the bulk viscosity curve as concentration is increased past the point at which the initiation of anisotropy occurs (C^*). Molecular weight seems to be less important than for other polymers. However, the range of concentrations at which anisotropy begins is generally much higher (18–60 wt %) than that required for rigid rod polymers such as the aromatic polyamides (3–10 wt %). The difference is ascribed to the semiflexible nature of the cellulose chain. Elevation of the solution temperature may produce an anisotropic–isotropic transition and a shift of the critical concentration to higher concentrations, provided solubility is sufficient.

In strong to moderately strong acids, the critical concentration C^* increases linearly with pK_a (147); C^* has also been found to be inversely proportional to the molar volume of the solvent (132,133). Other factors affecting solubility and

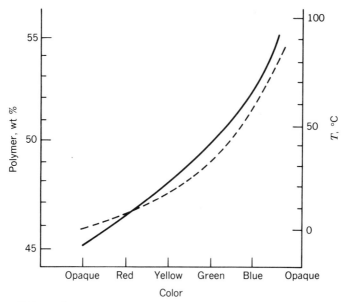

Fig. 16.   Effect of concentration and temperature on the irridescence of ethylcellulose–chloroform solution (132): ——, concentration (25°C); ----, temperature (50% polymer).

C^* are the degree of substitution and the size, polarity, and distribution of substituents.

Hydroxypropylcellulose (15,148), ethylcellulose (148), and propionoxypropylcellulose (149) form liquid crystalline melts.

Liquid crystalline solutions of cellulose derivatives have been studied by nmr and dsc, and for circular dichroism, rheological behavior, and solution viscosity.

Liquid crystalline solutions of unsubstituted cellulose in N-methylmorpholine–water (17,150,151), trifluoroacetic acid–chloroalkanes (138), and aqueous zinc chloride (152) have been described. Dimethylacetamide–LiCl gives solutions containing an anisotropic phase (153,154).

Thermotropic Systems

Main-chain Polymers

Persistence Length.   According to the Flory lattice theory (38,40–42), an axial ratio of only 6.42 is required for a polymer to be stiff enough to form an LC melt. Molecular weight M, however, must be high enough to achieve good mechanical properties. Chain stiffness and high M in a rodlike molecule, eg, para ring structures and linear interlinking groups, result in high T_m. Poly(1,4-benzoate), for example, does not melt but decomposes at $T > 500$°C. The measure of the rodlike character, or straightness, of a chain is the persistence length, ie, the vector sum of the projections of the $(n + i)$th segments where $i = 0$ to ∞ on the nth segment (chosen at random) of a very long chain. The more rodlike a

Table 4. Cellulose Derivatives Forming Lyotropic Solutions

Derivative	DS[a]	Solvents	Refs.
Esters			
triacetate		chlorophenol–dichloromethane, *m*-cresol, dichloroacetic acid, difluoroacetic acid, formic acid–1,2-dichloroethane, trifluoroacetic acid, trifluoroacetic acid–dichloromethane	13,133
		acetic acid, acetone, aniline, *m*-cresol, dimethylacetamide, methyl acetate, nitromethane, 3-picoline, 4-picoline, pyridine, tetrachloroethane	137
		dichloroacetic acid, dichloromethane, trifluoroacetic acid, trifluoroacetic acid–dichloromethane	138
acetate	2.41	acetic acid, chloroform–formic acid, *p*-chlorophenol, dichloroacetic acid, diethylacetamide, dimethylacetamide, dimethyl sulfoxide, dioxane, ethylene sulfite (cyclic), phosphoric acid, trifluoroacetic acid	13,133
acetate–butyrate		acetic acid, dichloromethane–methanol, dimethylacetamide, formic acid, methyl ethyl ketone, phenol	13,133
nitrate		γ-butyrolactone, cyclopentanone, diethyl oxalate, dimethylacetamide, dimethyl carbonate, dimethyl cyanamide, dimethylformamide, dimethyl maleate, dimethyl sulfoxide, *N*-methyl-2-pyrrolidinone, 3-pentanone, pyridine	13,133
		acetone, 2-hexanone, methyl acetate, methyl ethyl ketone, 4-methyl-2-pentanone, methyl propyl ketone, 3-pentanone, *n*-pentyl acetate, *n*-propyl acetate	137
sulfate[b]	1.76	water	13,133
Ethers			
ethyl	2.2–2.4	acetic acid, carbon tetrachloride, chlorophenol–dichloromethane, *m*-cresol, dichloromethane–methanol, trifluoroacetic acid, trifluoroethanol	13,133
		chloroform, 1,1,2-trichloroethane	132
	2.5	acetic acid, benzene–methanol, chlorophenol–dichloromethane, *m*-cresol, dichloromethane, formic acid, phenol, toluene–methanol, trifluoroacetic acid	13,133
		acetic acid, dichloroacetic acid	139
hydroxymethyl		dimethyl sulfoxide	140

Table 4. (*Continued*)

Derivative	DS[a]	Solvents	Refs.
hydroxypropyl		acetic acid, acetone, acetone–water, acetonitrile, dichloromethane, dimethylacetamide, dimethylformamide, dimethyl sulfoxide, dioxane, 1,2-propanediol, pyridine, trifluoroacetic acid, trimethyl phosphate, water	13,14,133–136,141–143
carboxymethyl[b]	1.74	water, water–9% NaOH	13,133
ethyl hydroxyethyl		chloroform, dichloromethane–methanol, trifluoroacetic acid	13,133
cyanoethyl	2.8	dichloromethane–methanol, diethylacetamide, dimethylformamide, dimethyl sulfoxide, trifluoroacetic acid, trifluoroethanol	13,133
Ether–esters[c]			
acetoxyethyl		trifluoroacetic acid	13,133,144
benzoyloxypropyl		acetone, benzene, chloroform, dimethylacetamide, dioxane, ethyl acetate, xylene	145
Urethane			
phenyl		dimethyl sulfoxide, dioxane, N-methyl-2-pyrrolidinone	13,133
		methyl ethyl ketone	146

[a] DS is substituents per glucose unit, ie, degree of substitution (qv).
[b] Sodium salt.
[c] Acyl groups substituting glucose hydroxyls and the hydroxyalkyl units.

chain, the more the $(n + i)$th bond remembers the direction of the nth bond. The persistence length can be calculated from chain-conformation data or derived from measured polymer solution properties; values for some rodlike chains are shown in Table 5.

Useful Structures.   As seen from persistence-length data, polyesters are likely thermotropic melt candidates if the melting point can be depressed without seriously affecting chain stiffness. These polyesters are not soluble as are polyamides (**1**) and (**3**) in Table 5. Polymer (**11**) is also a prime candidate and illustrates the first method of decreasing T_m, that of asymmetric substitution. Like polymers (**1**), (**3**), (**9**), (**11**), unsubstituted polyazomethine does not melt. Polymer (**11**) is a copolymer in the sense that each added diamine unit can add with the methyl group ortho or meta to the reacting amino group. In polymer (**10**), a methyl or chloro substituent is insufficient to depress T_m below the decomposition temperature. Other methods besides asymmetric substitution include copolymerization with saturated 1,4-extended ring structures; units of different length, eg, 1,4-phenylene and 4,4'-biphenylene; units with parallel displacement, eg, 2,6-naphthalene; units with angular displacement, eg, 4,4'-oxydiphenylene; units with pseudo nearly parallel displacement, eg, 3,4'-oxybiphenylene; short flexible units, eg, ethylenedioxy; and rigid kinked units, eg, 1,3-phenylene.

Table 5. Persistence Lengths, nm

Polymer	Structure no.	Calculated[a]	Found	T_m, °C
poly(1,4-benzamide)	(3)	41	40–60[b]	>500
poly(p-phenyleneterephthalamide)	(1)	42.2	15–>45[c]	>500
poly(1,4-benzoate)[d]	(9)	74		>500
poly(p-phenylene terephthalate)[d]	(10)	75.1		>500
	(11)		15[e]	255[f]
	(12)		53[g]	>500
	(13)		150[i]	>500

[a] Ref. 155.
[b] Ref. 44.
[c] Refs. 44, 156, and 157.
[d] Potentially useful thermotropic melt polymers.
[e] Ref. 104.
[f] Ref. 24.
[g] Ref. 158.
[h] Stiff enough to form LC solutions and LC melts but does not because solubility is too low and compound does not melt.
[i] Ref. 107.

Useful monomers for thermotropic melt polymers and copolymers include diols (and dithiols) where X = Y = OH(SH); diamines, X = Y = NH$_2$; dicarboxylic acids, X = Y = COOH; or the functions may be mixed. Combinations are limited by persistence length and T_m considerations.

$$R = -O-, -CO-, -SO_2-, -S-, \overset{\diagdown}{C}=C^{\diagup}, -C\equiv C-, -CH_2CH_2-, -OCH_2CH_2O-, -\overset{\overset{\displaystyle CH_3}{|}}{\underset{\underset{\displaystyle CH_3}{|}}{C}}-$$

$$R' = -O-, -S-, -\overset{\overset{\displaystyle O}{\|}}{C}-$$

Typical Z substituents are Cl, Br, CH$_3$, H, or aryl.

Several naphthalene isomers maintain near linearity of structure; particularly useful is the 2,6-isomer with parallel and oppositely directed bond extensions. Short stiff units (eg, *trans*-vinylene) maintain chain linearity, whereas ethylene and 1,2-ethylenedioxy are flexible. Examples of useful compositions illustrating T_m reduction and stiffness in terms of fiber modulus are shown in Table 6. If an extended conformation can be stabilized in the solid polymer, modulus is maintained despite a flexible or kinked unit. Aminobenzoic acid (ABA) units in polymers with PET (168) result in LC melts at lower mol % ABA than of HBA units in the corresponding copolymers with HBA. Relationships in naphthalene-containing polymers are discussed in Ref. 169. Acetoxypropylcellulose ($\overline{M}_w = 140{,}000$) forms a cholesteric thermotropic melt at 85–125°C (16). Thermotropic behavior has also been observed in polyisocyanates (170). Only a few homopolymers, R = *n*-butyl, *n*-hexyl, and *n*-octyl in poly(*N*-alkyl isocyanates) (4), and several copolymers (to reduce T_m) have been studied because of thermal decomposition at ca 150–180°C.

Synthesis.   Main-chain polymers are prepared by melt or low temperature condensation. A type of the former is best carried out by acidolysis (159,160), eg,

Very high melting compositions are condensed in solution in a high boiling solvent of the Therminol type. The reaction with phenyl esters and diols is slower and less convenient because of phenol generation. During condensation, phase separation at a low molecular weight is observed; as the amount of LC phase increases, the isotropic phase gradually disappears. Vacuum or purging with nitrogen or argon volatilizes the acetic acid, resulting in higher molecular weight. Direct condensation of chlorohydroquinone, quinitol, or *p*-hydroxybenzoic acid is impractical because of thermal decomposition. Direct condensation of dihydric phenols and aromatic dicarboxylic acids catalyzed by tin, titanium, or antimony compounds has been described (171).

Table 6. Polymers Forming Thermotropic Melts

Structural units	Mole ratios	T_m, °C	Fiber modulus, N/tex[a]	Refs.
(trans)	1:1	305	33	159,160
	1:1	340	80	161
	3:1	302	44	162
	5:3.5:1.5		44	163

Structure	Ratio			
—NH, —O—, C=O, C, C=O	2:2:6	280	51	164
—O—, Cl, C=O, C=O, C=O	1:0.7:0.3	302	47	23
—OCH₂CH₂O—, —O—, C=O, C=O, C=O	2:2:6	260	17	165
—O—, C=O, —O—, C=O, C=O [b]	43:28.5:28.5	304	49	166
Cl, —O—, C=O, —O—, C=O	1:1	339	26	167
=N=, N=, CH₃, =CH—CH=	1:1	255	89	24

[a] To convert N/tex to g/den, divide by 0.08826.
[b] If an extended conformation can be stabilized in the solid polymer, modulus is maintained despite a flexible or kinked unit.

537

Condensation of *p*-acetoxybenzoic acid with preformed PET is highly effective. The PET is degraded by acidolysis and then built up as a copolymer (21,165).

Single crystals with a unique morphology have been made by solid-state polymerization (qv) of the simplest AB polyamide (eg, poly(1,4-benzamide) (67)) and polyester (eg, poly(1,4-benzoate) (172)).

Polymer Structure and Morphology. Thermotropic melts generally are nematic. Phases may be identified with the aid of a polarizing microscope, but some polymeric mesophases do not exhibit recognizable textures (173). Mutual miscibility with known mesogens is at times useful for phase identification, as is x-ray diffraction.

The structure of PET–*p*-hydroxybenzoic acid copolymers, expected to be block copolymers, is random, as shown by nmr (21) and x-ray studies (174); more recent work favors a two-phase system (47,175).

Structure can be altered by synthesis. A copolymer normally made from HBA (2 mol), 4,4'-biphenol, formerly called *p*-biphenylene glycol (PPG) (1 mol), and terephthalic acid (T) (1 mol) has a T_m of ca 420°C. An alternating copolymer melts at >500°C; a copolymer made with blocks about 10 units long melts at 350°C; a copolymer made with blocks of ca 5 units has a T_m of 410°C and matches the normal copolymer. Hence rates for HBA and PPG-T polymerizations differ and the copolymer equilibrates only very slowly (176). Equilibration favors a block structure rather than a random one. An initially soluble PET–HBA or 3-chloro-4-hydroxybenzoic acid (ClHBA) copolymer upon heating with an ester interchange catalyst in the melt or preferably in the solid state near the T_m forms an insoluble copolymer with higher T_m. This could be due to a structural reorganization caused by crystallinity to generate longer blocks of HBA or ClHBA in the insoluble portion of the copolymer. A similar crystallinity-induced reorganization took place in a spacer-group-type main-chain LC polymer (177).

In naphthalene copolymer fibers (178), chains consist of random sequences of monomer units oriented parallel to the fiber axis. These chains are in highly extended conformations, but adjacent units are randomly oriented with respect to the chain axes and exhibit a range of torsional angles for rotatable bonds. The chains are arranged in a distorted hexagonal network with random axial stagger in cylinders to give a nematic structure.

X-ray data revealed (179) that PPD-T fiber, like LC polyester fiber, is chain-extended but almost 100% crystalline, highly oriented, and has random tie points between fibrils (Fig. 17). Like LC polyesters, PPD-T fibers have a skin-core structure with parallel arranged fibrils along the fiber axis.

A similar morphology has been found for LC extrusions (48). The structure of the melt is frozen into the crystalline solid on extrusion (naphthalene-type copolymers) (180). The oriented fiber or extrudate consists of a hierarchy of structures from microfibrils (0.05 μm in diameter) to fibrils (0.5 μm) to macrofibrils (0.5 mm) in a highly oriented skin. The difference in organization of these structures within the fiber as contrasted to the molding or bar is one of degree of orientation: organization is much higher in the fiber (48). The fiber threadline can be far better oriented during drawdown, and because of its small diameter, can be rapidly cooled to freeze in the oriented structure as compared with moldings. Thus orientation-dependent properties in the extrusion direction decrease with bar thickness (21).

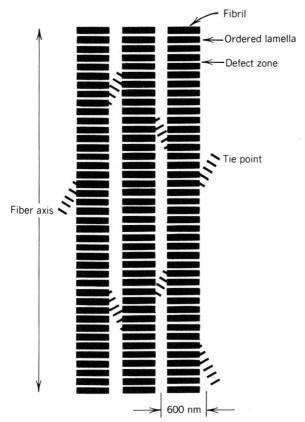

Fig. 17.   Structure of poly(p-phenyleneterephthalamide) (PPD-T) fiber (179).

Rheology and Processing.   It has been suggested that the morphology of extrudates described above develops in the melt and freezes into the solid product during processing; the mechanism of this development is still being studied. The rheology of LC systems is complex (181–184). The liquid-crystal melt shows a yield stress at low stress values (185) and an initial rapid decrease in viscosity with increasing shear rate unlike PET, which is essentially Newtonian at low shear rates. Under normal processing conditions, therefore, the bulk viscosity of LC polyesters can be two orders of magnitude less than that of PET (186). The viscosity is a sensitive function of shear history (183) as well. Because of the long relaxation time of the LC systems (21), a more fluid melt is formed by preshearing or heating to a higher temperature for a short time and cooling to the extrusion temperature (97). In fiber spinning, a die swell near zero is observed, which can be negative at lower temperatures (186). Fiber and molding properties depend markedly on processing conditions. Aramid spinning is described in Ref. 187, LC polyester spinning in Ref. 188, and polyester film formation in Refs. 189 and 190.

Groups of molecular chains associate into small domains and these persist through shearing, which tends to break up larger aggregates and some domains as the yield stress is exceeded (191). The conversion of cholesteric to nematic organization of thermotropic cellulose derivatives in shear is evidence for this behavior (182,192). The rodlike molecular conformation precludes many entanglements, such as exist in random-coil melt systems, and facilitates elongation and reorganization of domains in the shear field. Extensional flow in extrusion results in formation of oriented fibrillar structures. High extensional flow, as in a very thin filament threadline, produces excellent fibrillar orientation since it is subjected to higher forces while being cooled rapidly; orientation in larger extruded moldings is lower but still considerable, as evidenced from properties several times higher than those observed in moldings of conventional resins and exceeding in some cases even glass-reinforced engineering resins (191).

The characterization of thermotropic melts by rheological and optical methods has been the subject of much theoretical and experimental analysis (181–184). Compared with conventional melts, they show a high elastic response to small-amplitude oscillations, a three-zone flow curve, and a strong dependence of rheology on thermomechanical history (184). The rheology of an LC polymer greatly depends on the composition and texture of its mesophases. The three regions (182) in the log–log plot of viscosity as a function of shear rate are an initial shear thinning or rapid decrease in viscosity at very low shear rates, a constant viscosity region at intermediate shear rates, and a second shear-thinning region at higher shear rates. Individual systems can vary widely in conforming to this general picture (181). The initial region suggests a yield stress indicative of a fixed structure due, for example, to the presence of domains in which chains are oriented more or less parallel to one direction; such domains, in turn, form a randomly oriented larger structure. Application of stress, according to one interpretation (182), results in yield and flow at very low shear rates by "plastic flow of piled domains." Restoration of the original larger structure and viscosity is very slow because of long relaxation times. As shear rate increases, the polydomain structure is first dispersed and then a monodomain continuous phase forms at high shear rates as orientation of domains proceeds in the flow field.

The constant viscosity of a thermotropic-melt polyester can be two or more orders of magnitude lower than that of an isotropic melt such as PET (186). A spacer-type thermotropic polyester (183) shows higher bulk viscosity in the isotropic state than in the anisotropic state except at extremely low shear rates, even though the observation temperature is higher. The isotropic state has PET-like rheology, whereas the nematic phase has a very different flow behavior both qualitatively and quantitatively; it is similar to that observed in small-molecule liquid crystals.

The low die swell of LC thermotropic polymers is attributed to the long relaxation times resulting in nonrecoverable energy at the exit of the capillary (186). The variation of viscosity with composition is demonstrated (21) in a series of copolymers of HBA and PET. With increasing HBA content, melt viscosity at 275°C increases to 30 mol % HBA because the chain is stiffened by the HBA component. At this concentration, thermotropic behavior starts and melt viscosity decreases by an order of magnitude to a minimum at 60–70 mol % HBA before increasing again because of a rapid concurrent increase in T_m.

Main-chain LC Polymers Containing Spacer Groups

These polymers (193) were made by use of spacer groups to interlink low molecular weight mesogens (5). They form thermotropic nematic or smectic melts when heated above the melting point. The first report was followed by research on a great variety of mesogenic units, similar to those in low molecular weight compounds, joined by flexible spacer units of varying length and constitution (Table 7).

Characterization.   The polymers discussed here have been characterized by x-ray scattering, dsc, nmr, viscometric techniques, and polarized microscopy. These methods were used to determine T_m, the melting transition to form an LC melt, T_{cl}, the clearing or LC–isotropic melt transition, the values of ΔH and ΔS for these transitions, phase diagrams, and rheological properties. The effect of electrical and magnetic fields to produce orientation has also been studied (46).

Clearing temperatures and melting points are determined by the flexible spacer groups. These polymers differ from low molecular weight compounds because of the coupled interaction of the mesogenic and spacer groups.

Structure–Property Relationships.   The following observations can be made from Table 7. As few as two para-linked aromatic rings (axial ratio about three) suffice for a mesogenic unit. The mesogenic group may consist of a combination of aromatic and cycloaliphatic rings.

The rings in the mesogen may be jointed by stiff polar linkages such as

$$-CH{=}N-, \ -C{=}C-, \ -N{=}N-, \ -COO-, \text{ or } -N{=}N-$$
$$\downarrow$$
$$O$$

The rings may be substituted by alkyl, aryl, or halogen, provided the substituents are not too large. Too large a group interferes with the parallel organization of the mesogenic units in the LC phase. The spacer groups are flexible,

eg, $-(CH_2)_n-$, $-(CH_2CH_2O)_n-$, $-(\underset{\underset{CH_3}{|}}{\overset{\overset{CH_3}{|}}{Si}}O)_n-$.

All types of LC phases have been observed, including cholesteric and discotic melt phases. At longer spacer lengths, the LC phases shift from nematic to smectic. At even longer spacer lengths, liquid crystallinity can disappear. An interesting interpretation of this observation is that the spacers act as solvents for the mesogenic units and that the volume fraction of mesogenic units must exceed V_p^* in the Flory lattice theory to form an LC phase.

Linear and disklike mesogens should be possible as side chains and main chains (199) or as combinations.

Transition Temperatures and Entropies.   *The Effect of Molecular Weight.*   As molecular weight increases, transition temperatures rapidly increase up to a molecular weight $\overline{M}_n$ of ca 10,000, where it levels off (Fig. 18) (200). This is about two to three times the value of $\overline{M}_n$ for the corresponding spacer length in some LC side-chain polymers (201).

The Effect of Mesogens.   The conclusion that the longer the mesogen, the higher the T_m is in accord with theory, since aspect ratio is of primary importance. In a comparison (202) of T_m, T_{cl}, and the mesomorphic range for a number of mesogens with a fixed spacer group, $-(CH_2)_{10}-$, the longer mesogens have a higher T_m ranging from 256°C for p-terphenyldicarbonyloxy (length 1.55 nm) to 225°C for azobenzene-4,4′dioxycarbonyl (length 1.35 nm) to 154°C for biphenyl-4,4′-dicarbonyloxy (length 1.10 nm). Branching lowers T_m and is particularly

Table 7. LC Polymers With Mesogenic and Spacer Groups[a]

Mesogenic unit	Spacer unit	n
–Ph–C(CH₃)=N–N=C(CH₃)–Ph–[b]	$-OCO-(CH_2)_n-OCO-$	6,8,10, 12
–Ph–C(CH₃)=N–N=C(CH₃)–Ph–[b]	$-OC(=O)-(CH_2)_n-CO-$	8,10,12
–Ph–CH=CH–Ph–	$-C(=O)-O-(CH_2)_n-O-C-$	5,6, 10 4–10[c]
–Ph–CH=N–Ph–	$-OC(=O)-(CH_2)_n-CO-$	10
–Ph–N=N–Ph–	$-O-(CH_2CH_2O)_n-$	2–4
–Ph–N=N(→O)–Ph–	$-C(=O)-O-(CH_2CH_2O)_n-C-$	2–4
–Ph(CH₃)–N=N(→O)–Ph(CH₃)–	$-OC(=O)-(CH_2)_n-CO-$	10
–Ph–Ph–	$-OC(=O)-(CH_2)_n-CO-$	5,6,8, 10,12
–Ph–Ph–Ph–	$-C(=O)-O-(CH_2)_n-O-C-$	2,4–6, 10
–Ph–CO–O–Ph–O–OC–Ph–	$-O-(CH_2CH_2O)_n-$	4
–Ph–CO–O–Ph–O–OC–Ph–	$-O-(CH_2)_3-(SiO)_n-Si-(CH_2)_3-O-$ (with CH₃ substituents)	2,3,5
–Ph–CO–O–Ph–O–OC–Ph– (biphenyl substituent)	$-O-(CH_2)_n-O-$	10
–Ph–CO–O–Ph(X,Y)–OC–Ph–[d]	$-O-(CH_2)_n-O-$	10
–Ph–CO–Ph–Ph–OC–Ph–	$-O-(CH_2)_n-O-$	9–11
–Ph–CO–S–S–OC–Ph–	$-O-(CH_2)_3-(SiO)_n-Si-(CH_2)_3-O-$ (with CH₃ substituents)	2,3,5

Table 7. (Continued)

Mesogenic unit	Spacer unit	n
(aromatic diester unit)	$-O-(CH_2)_n-O-$	4–6,8, 9
(aromatic diester unit)	$-CO-(CH_2CH_2O)_n-C-$ (with two C=O)	2–4, 8,13
(triphenylene unit, OR^e, $O-$, OR, RO, OR) or reversed[f]	$-C-(CH_2)_n-C-$ (with two C=O)	10,14[g] 20[h]
(benzene unit with $R-C-O$, $O-C-R^i$, $-O-$, $-O-$, $R-C-O$, $O-C-R$)	$-C-(CH_2)_n-C-$ (with two C=O)	8,20[f] 10,12,14[g]
$-O-$(biphenyl)$-O-$	$-C-(CH_2)_n-C-$ (with two C=O) and $-C-CH_2\overset{*}{C}HCH_2CH_2C-$ with CH_3	7[j]
(aromatic azomethine unit with CH_3 groups) $-N=CH-[()]_k CH=N-$	$-O-(CH_2)_n-O-$	1–10,12, 14,16
(aromatic ester–amide unit) $-C-O-()-NH-C-[()]^l$	$-O-(CH_2)_n-O-$	2–10, 12

[a] Ref. 46 unless otherwise noted.
[b] Ref. 193.
[c] Ref. 194.
[d] X = H, Cl; Y = H, Cl, Br, CH_3.
[e] R = $CH_3-(CH_2)_n-CH_2-$; n = 3.
[f] Ref. 195; ratio of units = as shown:reversed = 86:14.
[g] Discotic.
[h] Isotropic.
[i] R = $CH_3-(CH_2)_n-CH_2-$; n = 5.
[j] Ref. 196; cholesteric.
[k] Ref. 197; the central phenyl may be replaced by biphenyl or the group $-p\text{-}C_6H_4-OCH_2CH_2O-p\text{-}C_6H_4-$.
[l] Ref. 198.

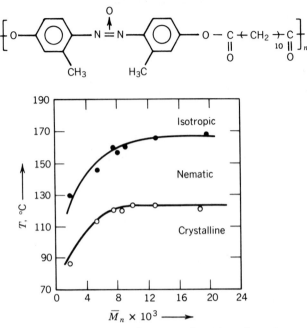

Fig. 18. Transition temperatures shown as a function of number-average molecular weight (200). Courtesy of Springer-Verlag.

effective if there is steric hindrance between methyl substituents, as in

$$T_m = 118°C; \text{length} = 1.35 \text{ nm}$$

The polymer with the mesogenic unit

melts at 196°C and the mesogenic unit has a length of 1.35 nm. The substituent methyl groups in this case cause far less steric hindrance. Lowering T_m is frequently important for molding or spinning LC polymers without decomposition. T_{cl} varies with length of the mesogenic group, and the LC range is quite short for low mesogenic group lengths.

An amide group in the mesogenic group increases the transition temperature and the stability of the smectic and nematic states (Fig. 19) (198). The transition temperatures are higher than in the corresponding ester-containing mesogenic group because of the effect of hydrogen bonding. In the azomethine series with two, three, and four rings in the mesogenic group, T_{cl} is more sensitive to mesogen

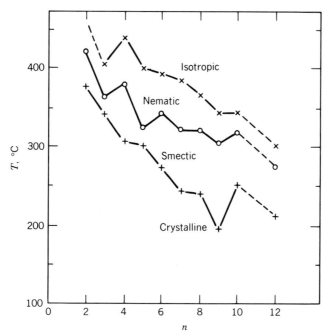

Fig. 19. Variation of transition temperatures with length of spacer group —(CH₂)ₙ— in the last entry (a polyester polyamide) in Table 7 (198). Areas, not curves, are labeled. Courtesy of Butterworth and Co. (Publishers) Ltd.

length than T_m, the mesomorphic range is as high as 100–200°C, and clearing temperatures are very high, in some cases >400°C.

Single substituents increase melt order, as judged by the change in ΔS_{cl} from 7.1 J/(mol·K) (1.7 cal/(mol·K)) for X = Y = H to 19.2 J/(mol·K) (4.6 cal/(mol·K)) for X = H, Y = CH₃ to 21.7 J/(mol·K) (5.2 cal/(mol·K)) for X = H, Y = Br for the polyester in Table 7 designated by footnote d.

The Effect of Spacer Groups. The most frequently used spacer groups are —(CH₂)ₙ—, —(CH₂CH₂O)ₙ—, and —(SiO)ₙ— with CH₃ groups. The transition temperatures of the polymers are about equal for the first two types, but much lower for polymers with siloxane spacers. The transition temperature decreases progressively with increasing spacer-group length. Higher transition temperatures are observed for polymers containing an even number of spacer atoms along the chain, compared with values for the adjacent odd-numbered members (Figs. 19 and 20) (30). Usually, T_{cl} is not as sensitive to the odd–even effect as is T_m. This results in a broader mesomorphic range for the odd-numbered series members. Since odd–even correlations persist through as many as 12 spacer atoms, the effect of the stiff mesogenic unit is felt as far away as 6 chain atoms. The decrease in T_m is greatest for polymers with siloxane spacer groups. Some of these polymers are amorphous with a $T_g < 0$°C and are LC melts at ambient temperatures (46).

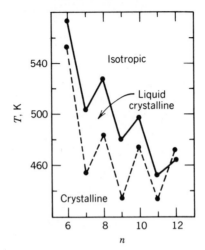

Fig. 20. Variation of T_m and T_{cl} with length of spacer group $-(CH_2)_n-$ in homopolyesters of 4,4′-dihydroxy-α-methylstilbene and aliphatic acids (30,203). Areas, not curves, are labeled. Courtesy of Hüthig & Wepf Verlag.

The ΔS_{cl} for LC isotropic transition can be calculated from dsc measurements. The value of ΔS_{cl} was found to exhibit an odd–even effect for members of a polymethylene series (higher for polymers with even spacer groups). This would imply a lower entropy and increased order in the melt for the polymers with even spacer groups. It also correlates with the observed odd–even alternation in transition temperatures. Entropies of clearing are much higher than for the corresponding low molecular weight mesogens.

The ΔS_{cl} values were lower for a series of polymers containing a poly(ethylene oxide) spacer than for the corresponding polymethylene series, as expected for more flexible chains. Transition temperatures can sometimes be about the same for these series, ie, methylene was found equivalent to oxygen. A two-phase system can be postulated where such flexible groups are excluded from the LC mesogenic phase and constitute a liquid unordered phase (204). Thus in the more flexible polymers containing poly(ethylene oxide) spacer, ΔS_{cl} would be measured for the mesogenic groups.

Thus spacer groups play an important role in determining transition temperatures, mesomorphic range, entropy changes, and nature of the LC phase. They do not completely decouple mesogenic groups, as judged from the odd–even effect in transition temperatures and ΔS_{cl}.

The Effect of Interlinking Groups. The nature and direction of the linkages within the mesogenic group or those joining the mesogenic group with the spacer group can affect transition temperature and the nature of the LC phase. The direction of the ester groups within the mesogenic groups (from one to the next along the chain) can change a polymer from an LC melt to an isotropic melt (205). A specific case where the LC organization is affected is in the smectic polyesters prepared from $HO(CH_2)_nOH$ and 4,4′-bibenzoic acid, compared with those prepared from $HOOC(CH_2)_nCOOH$ and 4,4′-biphenol. The latter exhibit a nematic phase for odd values of n and a broad smectic phase for even values (206).

LC Polymers with Mesogenic Side Chains

Polymers with mesogenic side chains are usually thermotropic. Low molecular weight mesogens are attached to backbone carbon chains or siloxane chains. Although the LC properties of the polymers parallel those of the low molecular weight analogues, the differences are due mainly to (1) high melt viscosity, which impedes structural rearrangements affected by external fields; and (2) the interaction of the polymer backbone chain, which tries to assume a random-coil configuration, and the mesogenic groups, which try to organize in LC phases. This interaction may be interrupted to some extent by inserting spacer groups between the backbone chain and the mesogenic groups (35,37).

Synthesis.   Polymers with mesogenic side chains are prepared by polymerizing a monomer containing an activated vinyl double bond attached by way of a spacer group to a mesogenic group. Methacrylates and acrylates are convenient starting materials. Another method is by addition of a vinyl mesogenic compound to a silyl group on a preformed polymer (207).

$$\begin{array}{ccc}
\text{CH}_3 & & \text{CH}_3 \\
| & & | \\
-\!(\text{Si}\!-\!\text{O})_{\overline{m}} + \text{CH}_2\!=\!\text{CH} \longrightarrow & -\!(\text{Si}\!-\!\text{O})_{\overline{m}} \\
| & | & | \\
\text{H} & (\text{CH}_2)_{n-2} & (\text{CH}_2)_n \\
& | & | \\
& \text{R} & \text{R}
\end{array}$$

where R is a mesogenic group and $-(\text{CH}_2)_n$ a spacer group

Some homopolymers and copolymers exhibit a wide variety of LC structures, including smectic (most often observed), nematic, or cholesteric mesophases. Cross-linking of such systems leads to mesogenic thermosets or elastomers (208); disk-like mesogens may be used (209). Mesogens may be incorporated into both main chains and side chains (199). Examples of smectic LC side-chain polymers are given in Table 8 (210).

The steric effects of bulky mesogenic side groups greatly increase the T_g of a glassy polymer; T_g increases rapidly until a degree of polymerization of ca 10 is reached, and then levels off.

Structure and Properties.   The first work in this area was on mesogens joined directly to the backbone chain, which, in most cases, yields amorphous polymers with high T_g. A mesogenic polymer in the glassy state would lose its LC character when heated above the T_g because of the constraints imposed by the random arrangement of the backbone chain which would prevent LC reorganization upon cooling. It can also be said that the clearing temperature T_{cl} is below the T_g.

The LC range of glassy LC polymers lies between the T_g and the T_{cl}; that of crystallizing polymers lies between the melting point and T_{cl}. These transitions may be monitored by polarized microscopy. The usual effect of the polymer is to broaden the mesomorphic range over that of the corresponding low molecular weight LC compound. The polymeric backbone chain stabilizes the LC phase in such a way that some nonmesogenic monomers give LC polymers, eg,

$$\text{RCOO}-\!\!\!\bigcirc\!\!-\!\!\bigcirc$$

Table 8. Smectic Polymers Obtained by Lengthening the Segments[a]

Polymer	$\Delta \bar{H}_{LC-i}$, J/g[b]	Phase transitions[c] at K

Segment A (end group)

A

$$\cdots -CH_2-\underset{\underset{(CH_2)_2O-\bigcirc-COO-\bigcirc}{|}}{\overset{\overset{CH_3}{|}}{C}}- \cdots \quad -OCH_3$$

| 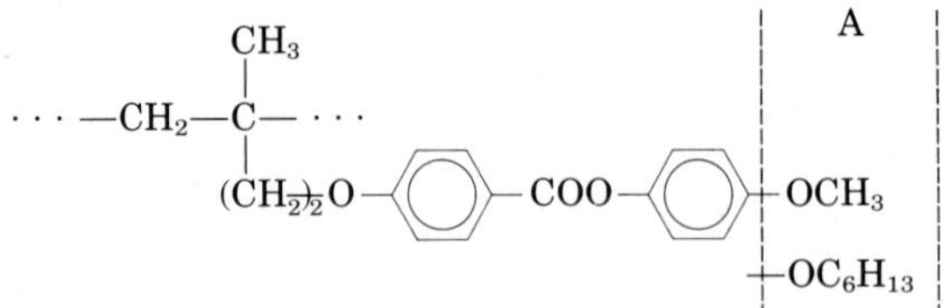 | 2.3 | g 369 n 394 i |
| $-OC_6H_{13}$ | 11.5 | g 410 s 451 i |

$$\cdots -\underset{\underset{(CH_2)_3O-\bigcirc-COO-\bigcirc}{|}}{\overset{\overset{CH_3}{|}}{Si}}-O- \cdots \quad -OCH_3$$

| 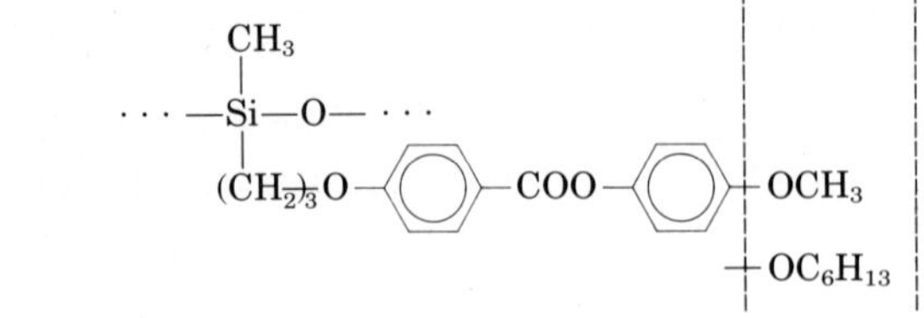 | 2.2 | g 288 n 334 i |
| $-OC_6H_{13}$ | 11.6 | g 288 s 385 i |

Segment B (spacer group)

B

$$\cdots -CH_2-\underset{\underset{COO-(CH_2)_2O-\bigcirc-COO-\bigcirc-CH=N-\bigcirc-CN}{|}}{\overset{\overset{CH_3}{|}}{C}}-$$

| 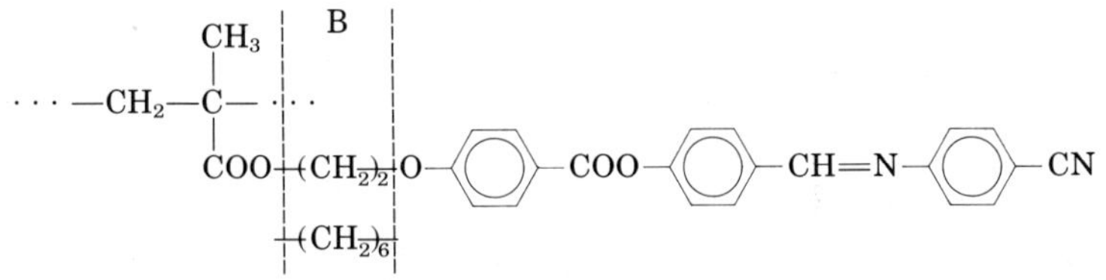 | | g 361 n 580 i |

$-(CH_2)_6-$

| | | g 324 s 607 i |

[a] Ref. 210. Courtesy of Springer-Verlag.
[b] To convert J to cal, divide by 4.184.
[c] g = glassy; n = nematic; i = isotropic; s = smectic.

is not mesogenic, but the following structure is (211):

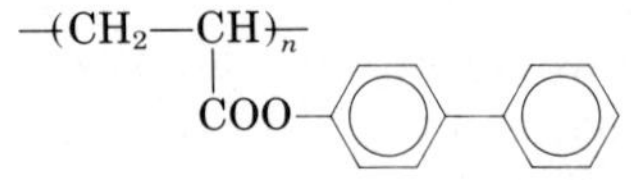

$$-(CH_2-CH)_{\overline{n}}-\underset{\underset{COO-\bigcirc-\bigcirc}{|}}{}$$

The effect of structure on the type of LC organization and transition temperatures is shown in Table 8. Segment A at the end of the mesogen favors nematic organization if short, or smectic if long, just as in low molecular weight mesogens. The spacer group (Segment B) likewise influences the nature of the LC phase, longer groups enabling the side chains to organize into more stable smectic phases.

Since the polymerization of a mesogenic monomer stabilizes the LC state, a nematogenic monomer often yields the higher-ordered smectic polymer, and mesogenic but isotropic monomers yield nematic polymers. Therefore, it is not surprising that efforts to form a cholesteric polymer from a cholesterol derivative invariably give smectic polymers. Cholesteric copolymers, however, can be obtained by copolymerization of a binary mixture of chiral monomers or of a nematogenic monomer with a chiral comonomer. The addition of the chiral monomer converts the nematic guest phase into a cholesteric phase (212). Cholesteric homopolymers can be made (213) from monomers that do not exhibit LC order.

The mesogenic phases exhibited by the polymers shown below vary with length of the methylene spacer group.

$$CH_3$$
$$-(Si-O)_n-$$
$$(CH_2)_m O-\bigcirc-COO-\bigcirc-R$$

At $m = 3$, a cholesteric phase is observed; at $m = 4$, a smectic phase appears at lower temperatures and a cholesteric phase at higher temperatures; and at $m = 5$, only a smectic phase forms because of the longer spacer group. Although the polymer stabilizes the LC phase, it destabilizes the crystalline phase, ie, it is extremely difficult, if not impossible, for the side chains to crystallize in the polymer.

The higher organization of the smectic phase is seen by the presence of discrete equatorial x-ray scattering at low angles, whereas nematic LC polymers show only an amorphous halo, indicating its liquidlike organization. A model of the smectic structure and its effect on ordering of the main or backbone chain are shown in Ref. 214. The greater order of the smectic state is also apparent from the ca five-fold greater ΔH of the smectic–isotropic transition compared with that of the corresponding nematic transition (see Table 8). The order in LC polymers is not as complete as in monomeric LC compounds because of the influence of the backbone-chain structure.

Applications

Fibers

Aramid Fibers. The discovery of LC solutions of aramids (18), high molecular weight aramid compositions (66–69), and novel spinning technology and products (25) led to the commercialization of high tenacity, high modulus Kevlar aramid fibers in the 1970s. The as-spun fiber has over twice the tenacity and nine times the modulus of high strength nylon (28). On a weight basis, it is stronger than steel wire and stiffer than glass. Creep and linear coefficient of thermal expansion are low and thermal stability is high. These properties resemble those of inorganic fibers more than those of organic fibers (Fig. 21) and are attributed to an extended chain morphology (Fig. 22), high molecular weight, and excellent orientation (almost complete in a nearly 100% crystalline fiber) in a thermally stable structure that does not melt. Kevlar aramid fibers owe their utility to a combination of superior properties with the features of organic fibers, eg, low density, easy processibility, and good fatigue and abrasion resistance.

Kevlar 29 aramid fiber with a ratio of tenacity to initial modulus T/M_i of 1.9/43 N/tex (22/490 g/den) is used mainly for rubber reinforcement in tires and belts, for ropes and cables (eg, for anchoring offshore oil rigs), for pressure-wound vessels (eg, rocket-engine cases), and for protective clothing (eg, bullet-resistant vests) (see also FIBERS, ENGINEERING).

A controlled, short (1–5 s) treatment at high temperature increases crystallinity and orientation. Kevlar 49 aramid fiber ($T/M_i = 1.9/86$ N/tex or 22/974

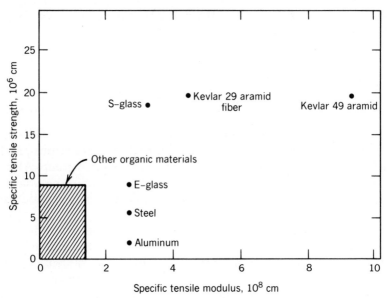

Fig. 21.   Specific tensile strength and specific tensile modulus of fibers and other materials (tensile strength or tensile modulus divided by density) (215). Courtesy of ASTM.

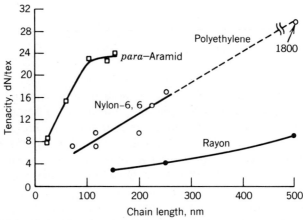

Fig. 22.   Tenacity vs number-average chain length for various fibers (28). To convert dN/tex to g/den, divide by 0.883.

g/den) is used alone or with glass and carbon fibers (qv) in various composite applications. These include aircraft panels, flooring, doors, fairings, seats, race-car bodies, marine applications, and sporting goods. In such uses, composites reinforced with Kevlar aramid fiber combine high tenacity and modulus with low density and resistance to crack propagation (at nail or rivet holes), fatigue, and impact damage. In asbestos replacement, a small amount, ie, <5%, of short fiber or fibrid reinforces an inorganic/organic matrix for friction materials such as brake linings and dry clutches. It increases wear and fade resistance and is nontoxic (28).

In the research that led to Kevlar aramid fiber, a large number of other compositions were made in high molecular weight and spun from LC solutions. Many of these did not yield the high as-spun tenacities observed for Kevlar fiber but did increase in tenacity (and modulus) upon short heat treatment under tension. Since thermal polycondensation of aromatic amines and aromatic acids is at best very slow, the increase in fiber T/M_i upon heat treatment is attributed to increased orientation and crystallization of the polymer chains. Some polymer compositions and fiber properties are given in Table 9; compositions are prepared from the intermediates shown. In the case of acids, acid chlorides are used (219).

Poly(p-phenylenebenzobisthiazole) (PBT) Fibers. The PBT fibers (220) were developed as part of the U.S. Air Force Ordered Polymers Research Program (see also HEAT-RESISTANT POLYMERS; POLYBENZOTHIAZOLES). A 10% LC high molecular weight PBT solution in polyphosphoric acid was dry-jet wet-spun, and the fibers were heat-treated under tension to give the highest modulus organic fiber known to date with a best tenacity and modulus of 22 and 2200 dN/tex (25 and 2500 g/den), respectively. The heat treatment relieves residual stresses and improves lateral order. The tensile properties of the treated fiber exceed those of carbon fiber, but the compressive strength is lower.

Aromatic Polyester Fibers. Thermotropic melts of aromatic polyesters have yielded fibers with tenacity and modulus values equivalent to those of aramid fibers (see POLYESTERS, AROMATIC). These fibers were developed from LC melt polyesters and heat treatment of the relaxed as-spun fiber at temperatures just below the T_m or flow temperature with purging of reaction by-products (160,221). The second step increases the order, orientation, and molecular weight of the fiber, in contrast to ordinary drawn fibers which would shrink and lose orientation. The fibers are formed from polyesters of low molecular weight. They are highly oriented and maintain and increase this orientation during heat treatment and further polymerization.

Since aromatic polyesters can be melt-polymerized from moderately priced intermediates and melt-spun thereafter, they have the potential advantage of lower cost. Addition of 1.5–5% PET as a comonomer (222) reduces cost, and spinning temperatures can be reduced by up to 80°C without affecting tensile properties. Several of the structures are given in Table 6. Additional compositions, fiber properties, and references are shown in Table 10. An excellent candidate is a heat-treated HBA (60%) copolymer (231). The ingredients are reasonable in cost, but the process is complex and probably expensive. It involves two critical long annealing steps (0.5–30 and 50 h) at high temperatures of a finely ground polymer and a melt-spun fiber. Sumitomo Chemical Co. in cooperation with Japan Exlan Industry has developed and announced such polyester fiber with $T/M_i = $ 2.7/78 N/tex (31/880 g/den) (234). Polyazomethines can be heat-treated in minutes instead of hours because the condensation reaction is much faster. Values of T/M_i can be very high (Table 10).

Fibers from Main-chain LC Polymers with Spacers. The use of flexible chains in LC main-chain polymer fibers lowers T and M_i values (Table 11) (197). A definite odd–even effect can be seen and a decreasing tenacity and modulus as n increases. Earlier work (235) with shorter mesogenic groups with $n = 6, 7,$

Table 9. Fibers from Lyotropic Polymers

Constituents	Mol %	Polymer η_{inh}[a], dL/g	Fiber[b]	Tenacity, dN/tex[c]	Elongation, %	Initial modulus, dN/tex[c]	Tex[d]	Orientation angle, deg	Refs.
Polyamides from									
p-aminobenzoic acid		1.38	as spun	7.3	3.1	449	0.34	18	65
			HT 525°C	14.9	1.9	918	0.29	10	65
p-aminobenzoic acid	60	3.77	as spun	9.0	7.8	233	0.24	28	216
p-phenylenediamine	20		HT 586°C	13.8	1.8	740	0.22	10	216
terephthalic acid	20								
p-phenylenediamine	50	6.54	as spun	8.6	8.7	226	0.38	40	216
4,4'-bibenzoic acid	27.5		HT 600°C	19.9	3.2	590	0.31	18	216
terephthalic acid	22.5								
2-chloro-p-phenylenediamine		3.73	as spun	10.2	5	363	80	31	66
terephthalic acid			HT 445°C	12.7	1.4	904	75		66
p-phenylenediamine		6.6	as spun	22.2	3.2	688	21	11.2	25
terephthalic acid			HT 350°C	21.1	2.1	998	20	7.9	25
			as spun	27.4	4.8	627	0.21	11.2	25
			HT 350°C	28.3	3.7	812	0.20	7.9	25
p-phenylenediamine		2.88	as spun	3.8	5.6	178	0.34	29	78
p,p'-azodibenzoic acid			HT 500°C	6.9	2.2	333	0.29	15	78
p-phenylenediamine		5.81	as spun	15.9	5.8	416	0.39	22	75
2,5-pyridinedicarboxylic acid									
1,4-phenylenediamine	50	4.61	as spun	21.5	5.72	597		17	72
fumaric acid	10								
terephthalic acid	40								

Material		Condition						Ref.
3,8-diaminophenan-thridinone terephthalic acid		as spun	11.0	3.3	376	0.33		79
		HT 450°C	22.5	2.3	962	0.28		79
Polybenzoxazole from								
4,6-diamino-1,3-benzenediol terephthalic acid	23.9[e]	as spun	18.9	4.2–6	571	33[f]		217
		HT 500°C	22.0	1.8–2.4	2020	31.4[f]		217
3-amino-4-hydroxybenzoic acid	12.0[e]	as spun	18.0	4.8	389	21[f]		217
		HT 400°C	21.3	3.6	733			217
Other								
ethylcellulose[g]		as spun	3.4	5.6	108	8.8	35	13,133
cellulose triacetate	6.3	as spun	10.4	10.8	133	33	31	218
		HT 255°C	12.6	6.1	199		12	218
polyquinoline from 3,3'-dibenzoylbenzidine 4,4'-diacetylbiphenyl	14.5[e]	as spun	7.5	7.7	141	2.8		112
		HT 380°C	7.9	2.5	300	2.5		112
poly(amide–hydrazide) from p-aminobenzoylhydrazide terephthalic acid	5.3	as spun	4	3.8	215	0.27		90

[a] Unless otherwise stated.
[b] HT = heat treated.
[c] To convert dN/tex to g/den, divide by 0.883.
[d] Unless otherwise stated; to convert tex to den, divide by 0.111.
[e] Intrinsic viscosity.
[f] In μm.
[g] 49.7% ethoxy.

Table 10. Polyester Fibers from Thermotropic Polymers

Constituents	Mol %	Polymer η_{inh}, dL/g	Fiber[a]	Tenacity, dN/tex[b]	Elongation, %	Initial modulus, dN/tex[b]	Tex[c]	Orientation angle, deg	Refs.
Polyesters from									
bis(β-hydroxyethyl) terephthalate	5.6		as spun	2.2	0.7	342	0.40	13	223
terephthalic acid	50.0		HT to 280°C	8.5	2.2	402	0.50		223
methylhydroquinone diacetate	44.4								
2-chlorohydroquinone diacetate	50		as spun	5.8	1.8	483	21	21	23
terephthalic acid	35		HT 260–290°C	26.8	4.7	465	65	18	23
2,6-naphthalenedicarboxylic acid	15								
2-phenylhydroquinone diacetate		1.03	as spun	3.0	0.9	360	44	14	161
terephthalic acid			HT 230–325°C	18.9	3.5	482	41		161
2-chlorohydroquinone diacetate		2.4	as spun	3.9	2.1	179	0.30	13	160,221
1,4-cyclohexanedicarboxylic acid			HT 305°C	17.7	3.3	356	105	11	160,221
3,5-dichloro-4,4'-diacetoxybenzophenone		0.86	as spun	3.7	2.2	256	0.44	28	224
terephthalic acid			HT 205–300°C	15.7	2.3	547			224
hydroquinone	15	1.30	as spun	2.0	0.9	206	1.45		225
isophthalic acid	15		HT	12.0	2.9	370			225
6-hydroxy-2-naphthoic acid	5								
4-hydroxybenzoic acid	65								
chlorohydroquinone	35	1.29	as spun	5.0	1.2	429	1.80		226
3,4'-dihydroxybenzophenone diacetate	15		HT 304°C	24.5	3.7	683	1.3		226
terephthalic acid	50								
hydroquinone diacetate	50	0.80	as spun	5.9	1.9	404	0.52		163
3,4'-dicarboxydiphenyl ether	35		HT 100–288°C	15.2	3.0	442	0.48		163
terephthalic acid	15								
p-acetoxybenzoic acid	75	5.7	as spun	10.7	2.80	478	~1.67		162
6-acetoxy-2-naphthoic acid	25		HT 250°C	17.7	5.0	486	~1.67		162

Composition	%		Condition[a]						Ref
6-acetoxy-2-naphthoic acid	4.9	7.04	as spun	7.3	1.48	535	0.56		227
4-acetoxybenzoic acid	60.4		HT 300°C	22.8	2.6	841	~0.56		227
4,4'-biphenol diacetate	17.4								
terephthalic acid	17.4								
6-acetoxy-2-naphthoic acid	4.8	4.09	as spun	6.7	1.52	477	0.26		228
4-acetoxy-4'-carboxybiphenyl	47.6		HT 300°C	8.9	1.4	672			228
4-acetoxybenzoic acid	47.6								
diphenyl terephthalate	33	3.6	as spun	6.2	2.0	353	4.4		229
phenyl-p-hydroxybenzoate	33								
t-amylhydroquinone	34								
6-acetoxy-2-naphthoic acid	66.6	1.10	as spun	4.2	1.6	330	0.30		230
terephthalic acid	16.7		HT to 280°C	12.9		341			230
3-chloro-4,3'-diacetoxybenzophenone	16.7								
p-hydroxybenzoic acid	60		as spun	4.9	1.4	362			231
terephthalic acid	15		HT to 320°C	27.2	2.7	1254			231
isophthalic acid	5								
4,4'-dihydroxybiphenyl	20								

Polyazomethine from

Composition	%		Condition[a]						Ref
1,2-bis(3-methyl-4-aminophenoxy)ethane		0.69	as spun	4.0	2.0	283	0.84	24	197
terephthalaldehyde			HT 170–280°C	21.6	2.9	741	1.2		197
2-methyl-p-phenylenediamine		6.0	as spun	6.4	1.1	808	2.2		232
terephthalaldehyde			HT to 250°C	24.7	3.2	829	0.50		232
			HT to 234°C	33.6	4.4	894	0.41		232

Other

Composition	%		Condition[a]						Ref
poly(ester–amide) from									
6-acetoxy-2-naphthoic acid	60	4.24	as spun	8.2	2.0	547	0.98		164
terephthalic acid	20		HT 300°C	25.8	6.6	512			164
p-acetoxyacetanilide	20								
poly(ester–carbonate) from									
6-hydroxy-2-naphthoic acid	34.2	1.74	as spun	3.3	2.0	238	3.3		233
hydroquinone	14.6								
diphenyl carbonate	51.2								

[a] HT = heat treated.
[b] To convert dN/tex to g/den, divide by 0.883.
[c] To convert tex to den, divide by 0.111.

Table 11. Heat-treated Fibers from LC Polyazomethines with Flexible Spacers[a]

$$-\text{\Large\bigcirc}-\text{N=CH}-\text{\Large\bigcirc}-\text{CH=N}-\text{\Large\bigcirc}-\text{O}-(\text{CH}_2)_n\text{O}-$$
$$\underset{\text{CH}_3}{\qquad\qquad\qquad} \qquad\qquad \underset{\text{CH}_3}{\qquad\qquad}$$

n	Polymer η_{inh}, dL/g	Tenacity, dN/tex[b]	Initial modulus, dN/tex[b]
1		4.4	41
2	0.69	21.2	741
3	1.05	2.8	50
4	0.46	8.7	217
5	0.73	1.9	112
6	0.23	5.0	149
7	1.94	3.4	34
8	2.02	5.9	147
10	1.04	2.6	126

[a] Ref. 197. Courtesy of the American Chemical Society.
[b] To convert dN/tex to g/den, divide by 0.883.

and 8 gave very weak, brittle fibers. Similar results were found (194,197) for as-spun fiber properties of polymers with 4,4'-stilbene residues as the mesogenic group.

Films

Properties of films from LC polymers are given in Table 12, and those of multiaxially oriented laminates in Table 13.

Since aramids and other lyotropic polymers are spun from high boiling solvents, eg, dialkylamides containing salts, the process does not lend itself easily to film formation. Another difficulty, present even in extruded thermotropic melt compositions, is the ease of uniaxial orientation in the machine direction, resulting in high longitudinal but very weak transverse properties. Films produced from poly(p-phenyleneterephthalamide) (239) by wet spinning with biaxial stretching or by dry-jet spinning with drawdown usually have unsatisfactory properties even after heat treatment.

Thermotropic melt polymers give films with superior properties, especially in the machine direction. Values of T/M_i can be higher than for PET films. More balanced properties are afforded by two-way stretching (240), and cross-lamination of films or sheets (Table 13). Heat treatment is effective (Table 12), but by-product evolution can cause variations in thickness and surface imperfections. Rotation compression-injection-molding techniques produce thin, circumferentially oriented films with an angular orientational component in the thickness direction (190). These films of poly(ethylene terephthalate-co-1,4-benzoate) (20/80) have a tensile modulus along the planar flow line of ca 20 GPa (2.9×10^6 psi), compared with 3.5 GPa (0.51×10^6 psi) for the isotropic molded copolyester, an indication that effective three-dimensional microorientation in domains has occurred. Polymer films are useful in the information industry and as a coating for optical fibers. Toughness and low linear coefficient of thermal expansion are important properties.

Molecular composite films have been made from LC PBT and a more flexible type benzimidazole polymer. Properties are superior to those of LC thermotropic polyesters (Table 12).

Plastics and Resins

Poly(1,4-benzoate) (PBHA) (242) can be compression-sintered (19,243) to a product marketed under the name Ekonol. This product does not melt without decomposition, and hence does not form LC melts. Modification of this material by copolymerization with 4,4'-biphenol and terephthalic acid (244) gives Ekkcel (245), marketed in plain and reinforced compression-molding grades. Ekkcel I-2000 is an injection-moldable material (246). Xydar LC polymer is produced by this technology (247,248); Dartco annual plant capacity is 10,000 metric tons (249). The product is an improved version of Ekkcel I-2000.

Modification of poly(1,4-benzoate) with 20–70% PET (165) produces self-reinforced injection-molded bars with T/M_i four and five times that of conventional polyester (21,165). These LC compositions (21,22) are potentially inexpensive. They are currently being developed in Japan (250).

A large number of LC copolyesters were patented by Celanese Corp. (162,251–256). Naphthalene units were used to modify PHBA to give bars, fibers, and films. Celanese announced Vectra LC polyester polymers (LCPs) in late 1985 (249). The line includes several compositions: PHBA modified with 2-oxynaphthalene-6-carbonyl units, or 2,6-dioxynaphthalene and terephthaloyl units (162,251–256).

Compositions and molded-bar properties appear in Table 14. Properties are superior to those of PET moldings and compare favorably with those of moldings of PET reinforced with 55% glass fibers.

Products. Because of the liquid crystallinity, Xydar and Vectra represent a new dimension in plastics (Table 15). They have superior tensile properties comparable to glass-filled PET (Table 14), and ease of processing. Melt viscosity is low and not sensitive to temperature at processing shear rates. The viscosity is lower than those of some high performance engineering resins. Vectra uses standard injection-molding equipment; Xydar requires some modification because of a higher melting point. The latter has improved high temperature performance with a heat-deflection temperature of 337°C. Both products have 100% regrind capability with more cycles possible for Vectra because of lower processing temperature. The transparency to microwave and high heat-deflection temperature make Xydar highly suitable for microwave ovenware, a large captive market. Because of low creep, low linear coefficient of thermal expansion, and low mold shrinkage, intricate parts as thin as 0.4 mm are possible. In addition, good electrical properties make LC polymers suitable for electrical and electronics applications. Chemical resistance is manifested in molded chemical distillation tower packing where a single Vectra molding replaces several expensive metal parts (261).

Deficiencies include weak weld lines, poor compressive strength, and poor transverse properties caused by anisotropy. Fillers such as talc (50%), glass fiber, and metal powder improve these properties (29,260).

Bars made of injection-molded alternating-spacer mesogen-containing polymers show an alternation in flexural modulus but not in tensile strength in one case (262) and in tensile strength and in modulus in another (263).

Table 12. Films from LC Polymers

Polymer	Mol %[a]	Polymer η_{inh}, dL/g[b]	Strength, MPa[b]	Elongation, %	Modulus, GPa[c]	Thickness, mm[a]	Comments	Ref.
polyester from p-hydroxybenzoic acid 6-hydroxy-2-naphthoic acid	75 25	7.57	437	2.16	25		melt-passed through grid and die-slit at 315°C	236
polyester from phenyl p-hydroxybenzoate hydroquinone diphenyl isophthalate	43 28.5 28.5	0.95	248	4.0	6.9	100[d]	melt extruded through die slit and wound on rotating drum	166
			393	4.0	9.6		heat-treated at 240–280°C at 10°C/h in nitrogen	166
poly(ester–carbonate) from p-hydroxybenzoic acid diphenyl carbonate hydroquinone diphenyl isophthalate	54 8 23 15	1.38	317	2.0	8.9	0.1	formed at 335°C at 10 m/min	237
			413	2.0	11.7		heat-treated at 280°C/6 h	237
molecular composite film from poly(p-phenylenebenzo-bisthiazole) (PBT) poly[2,5(6)-benzimidazole] (AB-PBI)	30[e] 70[e]	31 16	689	5.5	30.6		polymer–methane sulfonic acid solution passed through a die onto a drum; resulting film was wet-drawn between two rotating rolls	238
			917	2.4	88.2		heat-treated at 540°C	238

Biaxially oriented films							
poly(p-phenylene-terephthalamide)		248[f]	8.2[f]	0.026–0.2	unequal biaxial orientation with orientation in MD[f]	239
		15[g]	0.59[g]		biaxially oriented film; equal properties in various directions in the plane of the film	239
		97	2.2			239
polyester from						
phenyl p-hydroxybenzoate	43	1.05	165[f]	4.7[f]	melt-extruded at 360°C and extended five times in MD[f] at 200°C and twice in TD[g] at 250°C; heat-treated at 290°C/30 s	240
diphenyl isophthalate	28.5		344[g]	5.6[g]		
hydroquinone	28.5					

[a] Unless otherwise noted.
[b] To convert MPa to psi, multiply by 145.
[c] To convert GPa to psi, multiply by 145,000.
[d] In μm.
[e] Wt %.
[f] Machine direction.
[g] Transverse direction.

Table 13. Multiaxially Oriented Laminates from Thermotropic Uniaxially Oriented Sheets

Polymer	Mol %	Sheet description	Axis	Tensile strength, MPa[a]	Elongation, %	Initial modulus, GPa[b]	Ref.
Polyester from							
p-hydroxybenzoic acid	75	nonlaminated, as extruded	MD[c]	477	4.42	15.3	241
6-hydroxy-2-naphthoic acid	25	heat-treated	MD[c]	663	7.36	13.5	
		as extruded	TD[d]	55	33.0	1.65	
		heat-treated	TD[d]	48	88.6	1.46	
		cross-laminated		160	6.82	66.5	241
		nonheat-treated laminated	MD/TD	73.7	18.4	2.19	241
		heat-treated laminated	MD/TD	155	33.1	2.19	241

[a] To convert MPa to psi, multiply by 145.
[b] To convert GPa to psi, multiply by 145,000.
[c] Machine direction.
[d] Transverse direction.

Table 14. Anisotropic Polymers for Injection-molded Bars

Polymer	Mol %	Polymer η_{inh}, dL/g	Tensile properties			Flexural properties		Impact, Izod strength, J/m[c]		Ref.
			Break strength, MPa[a]	Break elongation, %	Modulus, GPa[b]	Strength, MPa[a]	Modulus, GPa[b]	Notched	Unnotched	
Polyester from										
2-phenylhydroquinone	50	2.34	274	7		213	8.2	198	737	257
95% *trans*-1,4-cyclohexanedicarboxylic acid	50									
p-acetoxybenzoic acid	60		118	2.67	6.8	140	8.2	397		251
2,6-naphthalenedicarboxylic acid	20									
hydroquinone diacetate	20									
p-acetoxybenzoic acid	33.3	2.9	168	11			19.4	59		258
phenylhydroquinone diacetate	33.3									
terephthalic acid	33.3									
p-acetoxybenzoic acid	60	0.67	232	20			12.5	416	1479	21
PET	40									
6-acetoxy-2-naphthoic acid	20	1.02	219	9	6.1	141	4.7	1036		256
p-acetoxybenzoic acid	30									
terephthalic acid	25									
resorcinol diacetate	25									
Poly(ester–amide) from										
p-aminobenzoic acid	25	0.67	168	10		139	5.2	43	187	168
PET	75									
For comparison										
PET		0.62	49.6	243			2.3	16	507	22
Allied Petra 150[d]			200[e]	1.4[e]		307[e]	17.2	96		259

[a] To convert MPa to psi, multiply by 145.
[b] To convert GPa to psi, multiply by 145,000.
[c] To convert J/m to ft·lbf/in., divide by 53.38.
[d] 55% glass filled.
[e] At yield.

Table 15. Properties of Xydar and Vectra

Property	Xydar[a]	Vectra[b]
tensile strength, MPa[c]	126	138–241
tensile modulus, GPa[d]	8.3	9.6–38
elongation, %	5	1.2–5.5
flexural strength, MPa[c]	131	151–296
flexural modulus, GPa[d]	13.1	8.9–34.4
Izod impact strength, J/m[e]		
notched	208	53–534
unnotched	454	
density, g/cm^3	1.35	
T_m or Vicat softening point, °C	358	275–330
heat-deflection temperature at 1.8 MPa[c], °C	337	180–240
coefficient thermal expansion, ppm, °C		
flow		0–27
transverse		27–54
mold shrinkage		0.006
gamma radiation, 5 MGy		no effect
microwave radiation		transparent
flammability (UL 94)	V-0	V-0
limiting oxygen index, %	42	35–50
dielectric strength, V/μm, 1.6 mm	31	31–43
arc resistance, s	138	63–146
processing, stock temperature, °C	360–398	285–325
mold temperature, °C		50–150
production capacity, t/yr	10,000	230

[a] SRT-500; Refs. 247 and 259.
[b] Range of properties for several compositions; Refs. 29 and 260.
[c] To convert MPa to psi, multiply by 145.
[d] To convert GPa to psi, multiply by 145,000.
[e] To convert J/m to ft·lbf/in., divide by 53.38.

Information-storage Devices

Highly oriented LC polymer films with mesogenic side chains can be used to store information. The LC melt is held between two conductive glass plates and the side chains are oriented homeotropically by an electric field to produce a transparent film. The electric field is turned off and a focused laser beam is used to imprint the film by heating selected areas above T_{cl}. These isotropic areas scatter light when the cooled film is viewed. The "writing" is stored indefinitely at temperatures below the polymer T_g but may be erased by heating or applying an alternating electric field (264). A polysiloxane with highly polar cyanobiphenyl, smectic side chains can be used as an electrooptical storage system (265). Because of polydispersity, these materials exhibit a wide biphasic region above 80°C. An opaque randomly oriented polymer layer in the biphasic region can be converted to homeotropically transparent areas by an alternating electric current. Such textures have been retained at ambient temperatures well above the polymer T_g for at least 18 months.

Glossary

Liquid crystalline indicates a fluid, ordered state existing temperature-wise between the solid state and a wholly disordered melt or an ordered fluid solution.

Thermotropic indicates the capacity to form an ordered melt or the ordered melt itself. *Lyotropic* indicates the capability of forming an ordered solution or the ordered solution.

An *isotropic* solution or melt is one in which the polymer molecules are randomly dispersed. An *anisotropic* solution or melt is one in which the polymer molecules are present in fluid, ordered domains (equivalent to the liquid crystalline state).

Critical concentration C^* is that concentration of a solution at which an anisotropic phase first appears. A special case is the volume concentration of polymer V_p^*.

Mesomorphic denotes the capability of forming, or the presence of, the liquid crystalline state. *Mesogenic* usually indicates a unit or structure expected to impart, or contribute to, the formation of a liquid crystalline state.

Nematic, smectic, and *cholesteric* describe types of order in liquid crystalline states. *Nematic* indicates an approximate parallel array of polymer chains or chain segments in a liquid crystalline domain but with no special order with regard to end groups or chain units. The *smectic* state, of which there are several types, indicates a nematic type of order plus some additional order because of the alignment of end groups or chain units. *Cholesteric* order is nematic-type order with a superimposed spiral arrangement of nematic layers.

T_m is the *melting temperature,* as measured, for example, by dsc. T_{cl} is the *clearing temperature,* eg, the nematic $\rightarrow$ isotropic transition temperature. T/M_i is the ratio of tenacity or tensile strength to initial modulus.

BIBLIOGRAPHY

1. F. Reinitzer, *Monatsh.* **9,** 421 (1888); O. Lehmann, *Flüssige Kristalle,* Engelmann, Leipzig, 1904.
2. M. Gordon, ed., *Liquid Crystal Polymers I,* Vol. 59, and *Liquid Crystal Polymers II/III,* Vol. 60/61 of *Advances in Polymer Science Series,* Springer-Verlag, New York, 1984.
3. A. Blumstein, ed., *Liquid Crystalline Order in Polymers,* Academic Press, Inc., New York, 1978.
4. A. Ciferri, W. R. Krigbaum, and R. B. Meyer, eds., *Polymer Liquid Crystals,* Academic Press, Inc., New York, 1982.
5. *J. Appl. Polym. Sci. Appl. Polym. Symp.* **41** (1985).
6. J. L. White in Ref. 5, p. 3; M. Sorai and H. Suga, *Mol. Cryst. Liq. Cryst.* **73,** 47 (1981).
7. D. Vorländer, *Z. Phys. Chem.* **105,** 211 (1923).
8. G. H. Brown and P. C. Crooker, *Chem. Eng. News* **61,** 25 (Jan. 31, 1983); B. Wunderlich and J. Grebowicz in M. Gordon, ed., *Liquid Crystal Polymers II/III,* Vol. 60/61 of *Advances in Polymer Science Series,* Springer-Verlag, New York, 1984, p. 20.
9. J. D. Brooks and G. H. Taylor, *Nature London* **206,** 697 (1965); *Carbon* **3,** 185 (1965); *Chem. Phys. Carbon* **4** (1968).
10. L. S. Singer in A. Ciferri and I. M. Ward, eds., *Ultra High Modulus Polymers,* Applied Science Publishers, Ltd., London, 1979, Chapt. 9, p. 251.
11. H. Boedtker and N. S. Simmons, *J. Am. Chem. Soc.* **80,** 2550 (1958); F. C. Bawden and N. W. Pirie, *Proc. R. Soc. London Ser. B* **123,** 274 (1937).
12. C. Robinson, *Trans. Faraday Soc.* **52,** 571 (1956); C. Robinson, J. C. Ward, and R. B. Beevers, *Discuss. Faraday Soc.* **25,** 29 (1958).
13. Ger. Pat. Appl. 2,705,382 (Aug. 11, 1977), M. Panar and O. B. Willcox (to E. I. du Pont de Nemours & Co., Inc.).
14. R. S. Werbowyj and D. G. Gray, *Mol. Cryst. Liq. Cryst. Lett.* **34,** 97 (1976).
15. K. Shimamura, J. L. White, and J. F. Fellers, *J. Appl. Polym. Sci.* **26,** 2165 (1981).
16. S. L. Tseng, A. Valente, and D. G. Gray, *Macromolecules* **14,** 715 (1981).
17. H. Chanzy, A. Penguy, S. Channis, and P. Monzie, *J. Polym. Sci. Polym. Phys. Ed.* **18,** 1137 (1980).
18. U.S. Pat. 3,671,542 (June 20, 1972), S. L. Kwolek, reissue of U.S. Pat. 30,352 (July 29, 1980), S. L. Kwolek (to E. I. du Pont de Nemours & Co., Inc.).
19. J. Economy, R. S. Storm, V. I. Matkovich, S. G. Cottis, and B. E. Novak, *J. Polym. Sci. Polym. Chem. Ed.* **14,** 2207 (1976).
20. W. J. Jackson, Jr., *Br. Polym. J.* **12**(4), 154 (1980).
21. W. J. Jackson, Jr. and H. F. Kuhfuss, *J. Polym. Sci. Polym. Chem. Ed.* **14,** 2043 (1976).
22. F. E. McFarlane, V. A. Nicely, and T. G. Davis in E. M. Pearce and J. R. Schaefgen, eds., *Contemporary Topics in Polymer Science,* Vol. 2, Plenum Press, New York, 1977, p. 109.
23. Belg. Pat. 828,935 (Nov. 12, 1975), J. R. Schaefgen, J. J. Kleinschuster, and T. C. Pletcher and U.S. Pat. 4,118,372 (Oct. 3, 1978), J. R. Schaefgen (to E. I. du Pont de Nemours & Co., Inc.).

24. P. W. Morgan, T. C. Pletcher, and S. L. Kwolek, *ACS Symp. Ser.* **260**, 103 (1984).
25. U.S. Pats. 3,869,429 and 3,869,430 (Mar. 4, 1975), H. Blades (to E. I. du Pont de Nemours & Co., Inc.).
26. J. R. Schaefgen, T. I. Bair, J. W. Ballou, S. L. Kwolek, P. W. Morgan, M. Panar, and J. Zimmerman in Ref. 10, Chapt. 6, p. 173.
27. R. E. Wilfong and J. Zimmerman, *J. Appl. Polym. Sci. Appl. Polym. Symp.* **31**, 1 (1977).
28. J. R. Schaefgen in A. E. Zachariades and R. S. Porter, eds., *The Strength and Stiffness of Polymers,* Marcel Dekker, Inc., New York, 1983, Chapt. 8, p. 327.
29. *Mod. Plast.* **62**, 78 (Apr. 1985).
30. A. Roviello and A. Sirigu, *Makromol. Chem.* **183**, 895 (1982).
31. W. R. Krigbaum, J. Asrar, H. Toriumi, A. Ciferri, and J. Preston, *J. Polym. Sci. Polym. Lett. Ed.* **20**, 109 (1982).
32. E. Perplies, H. Ringsdorf, and J. H. Wendorff, *J. Polym. Sci. Polym. Lett. Ed.* **13**, 243 (1976).
33. H. Finkelmann, H. Ringsdorf, and J. H. Wendorff, *Makromol. Chem.* **179**, 273 (1978).
34. H. Finkelmann, W. Helfrich, and G. Heppke, eds., *Liquid Crystals of One- and Two-dimensional Order,* Springer-Verlag, Berlin, 1980, p. 238.
35. H. Finkelmann and G. Rehage in M. Gordon, ed., *Liquid Crystal Polymers II/III,* Vol. 60/61 of *Advances in Polymer Science Series,* Springer-Verlag, New York, 1984, p. 99.
36. V. P. Shibaev, J. S. Freidzon, and N. A. Platé, *Dokl. Akad. Nauk SSSR* **227**, 1412 (1976).
37. V. P. Shibaev and N. A. Platé in Ref. 35, p. 173.
38. P. J. Flory, *Proc. R. Soc. London Ser. A* **234**, 60 (1956).
39. L. Onsager, *Ann. N.Y. Acad. Sci.* **51**, 627 (1949).
40. Ref. 38, p. 73.
41. P. J. Flory and G. Ronca, *Mol. Cryst. Liq. Cryst.* **54**, 289 (1979).
42. P. J. Flory in M. Gordon, ed., *Liquid Crystal Polymers I,* Vol. 59 of *Advances in Polymer Science Series,* Springer-Verlag, New York, 1984, p. 1.
43. S. M. Aharoni, *Polymer* **21**, 21 (1980).
44. M. Arpin and C. Strazielle, *Makromol. Chem.* **177**, 581 (1976).
45. Ref. 42, p. 20.
46. C. K. Ober, J.-I. Jin, and R. W. Lenz in Ref. 42, p. 103.
47. L. C. Sawyer, *J. Polym. Sci. Polym. Lett. Ed.* **22**, 347 (1984).
48. L. C. Sawyer and M. Jaffe, "Morphology of Thermotropic Liquid Crystal Polymers," *paper presented at New Materials Concepts for Emerging Technologies,* Chicago, Sept. 8–13, 1985, American Chemical Society, Washington, D.C.
49. A. Baumgärtner, *J. Chem. Phys.* **84**, 1905 (1986).
50. G. Ronca and D. Y. Yoon, *J. Chem. Phys.* **76**, 3295 (1982).
51. R. Lovell, G. R. Mitchell, and A. H. Windle, *Faraday Discuss. Chem. Soc.* **68**, 46 (1979).
52. P. J. Flory and R. R. Matheson, Jr., *J. Phys. Chem.* **88**, 6606 (1984).
53. R. R. Matheson, Jr., E. I. du Pont de Nemours & Co., Inc., Wilmington, Del., *Macromolecules,* forthcoming communication to the editor.
54. A. Elliot and E. J. Ambrose, *Discuss. Faraday Soc.* **9**, 246 (1950).
55. M. Panar and W. D. Phillips, *J. Am. Chem. Soc.* **90**, 3880 (1968).
56. P. J. Flory and A. Abe, *Macromolecules* **11**, 1119 (1978); P. J. Flory and R. S. Frost, *Macromolecules* **11**, 1126 (1978).
57. J. Hermans, *J. Colloid Sci.* **17**, 638 (1962).
58. G. Kiss and R. S. Porter, *J. Polym. Sci. Polym. Symp.* **65**, 193 (1978).
59. R. Matheson, *Macromolecules* **13**, 643 (1980).
60. E. T. Samulski in Ref. 3, pp. 167–190.
61. D. B. Dupre, *Polym. Eng. Sci.* **21**, 717 (1981); *J. Appl. Polym. Sci. Appl. Polym. Symp.* **41**, 69 (1985).
62. K. Jeremic and F. E. Karasz, *Eur. Polym. J.* **19**, 1037 (1983).
63. R. Sakamoto, *Colloid Polym. Sci.* **262**, 788 (1984).
64. W. G. Miller, P. S. Russo, and S. Chakrabart, *J. Appl. Polym. Sci. Appl. Polym. Symp.* **41**, 49 (1985); W. G. Miller, *Ann. Rev. Phys. Chem.* **29**, 519 (1978).
65. U.S. Pat. 3,600,350 (Aug. 17, 1971) and Fr. Pat. 1,526,745 (1968), S. L. Kwolek (to E. I. du Pont de Nemours & Co., Inc.).
66. U.S. Pat. 3,673,143 (June 27, 1972), T. I. Bair and P. W. Morgan (to E. I. du Pont de Nemours & Co., Inc.).

67. P. W. Morgan, *Macromolecules* **10,** 1381 (1977).
68. S. L. Kwolek, P. W. Morgan, J. R. Schaefgen, and L. W. Gulrich, *Macromolecules* **10,** 1390 (1977).
69. T. I. Bair, P. W. Morgan, and F. L. Killian, *Macromolecules* **10,** 1396 (1977).
70. C. J. Brown, *Acta Crystallogr.* **21,** 442 (1966).
71. T. C. Pletcher and P. W. Morgan, *J. Polym. Sci. Polym. Chem. Ed.* **18,** 643 (1980).
72. U.S. Pat. 3,869,419 (Mar. 4, 1975), P. W. Morgan and T. C. Pletcher (to E. I. du Pont de Nemours & Co., Inc.).
73. S. M. Aharoni, *J. Polym. Sci. Polym. Phys. Ed.* **19,** 281 (1981).
74. U.S. Pat. 3,827,998 (Aug. 6, 1974), P. W. Morgan (to E. I. du Pont de Nemours & Co., Inc.).
75. U.S. Pat. 3,836,498 (Sept. 17, 1974), L. W. Gulrich and P. W. Morgan (to E. I. du Pont de Nemours & Co., Inc.).
76. N. Ogata, K. Sanui, and T. Koyama, *J. Polym. Sci. Polym. Chem. Ed.* **19,** 151 (1981).
77. U.S. Pat. 3,801,528 (Apr. 2, 1974), P. W. Morgan (to E. I. du Pont de Nemours & Co., Inc.).
78. U.S. Pat. 3,804,791 (Apr. 16, 1974), P. W. Morgan (to E. I. du Pont de Nemours & Co., Inc.).
79. T. Kaneda, S. Ishikawa, H. Daimon, T. Katsura, and M. Ueda, *Makromol. Chem.* **183,** 417 (1982).
80. Z. B. Li and M. Ueda, *J. Polym. Sci. Polym. Chem. Ed.* **22,** 3063 (1984).
81. M. Panar and L. F. Beste, *Macromolecules* **10,** 1401 (1977).
82. W. R. Krigbaum, F. Salaris, and A. Ciferri, *J. Polym. Sci. Polym. Lett. Ed.* **17,** 601 (1979).
83. C. Balbi, E. Bianchi, A. Ciferri, A. Tealdi, and W. R. Krigbaum, *J. Polym. Sci. Polym. Phys. Ed.* **18,** 2037 (1980).
84. T. S. Sokolova, S. G. Efimova, A. V. Volokhina, G. I. Kudryavtsev, and S. P. Papkov, *Polym. Sci. USSR* **15,** 2832 (1973).
85. S. M. Aharoni, *J. Appl. Polym. Sci.* **25,** 2891 (1980).
86. M. Takayanagi and T. Katayose, *J. Appl. Polym. Sci.* **29,** 141 (1984).
87. F. M. Silver, *J. Polym. Sci. Polym. Chem. Ed.* **17,** 3519, 3535 (1979).
88. F. M. Silver and W. B. Black, *J. Polym. Sci. Polym. Chem. Ed.* **17,** 3543 (1979).
89. P. W. Morgan, *J. Polym. Sci. Polym. Symp.* **65,** 1 (1978).
90. U.S. Pat. 4,070,326 (Jan. 24, 1978), P. W. Morgan (to E. I. du Pont de Nemours & Co., Inc.).
91. U.S. Pats. 4,039,502 (Aug. 2, 1977), J. D. Hartzler and P. W. Morgan and 3,966,656 (June 29, 1976), J. D. Hartzler (to E. I. du Pont de Nemours & Co., Inc.).
92. J. D. Hartzler and P. W. Morgan in Ref. 22, pp. 19–53.
93. W. B. Black and J. Preston, eds., *High Modulus Wholly Aromatic Fibers,* Marcel Dekker, Inc., New York, 1973.
94. G. Marrucci and A. Ciferri, *J. Polym. Sci. Polym. Lett. Ed.* **15,** 643 (1977).
95. B. Valenti and A. Ciferri, *J. Polym. Sci. Polym. Lett. Ed.* **16,** 657 (1978).
96. L. L. Chapoy and N. F. La Cour, *Rheol. Acta* **19,** 731 (1980).
97. F. N. Cogswell, *Br. Polym. J.* **12**(4), 170 (1980).
98. E. Bianchi, A. Ciferri, J. Preston, and W. R. Krigbaum, *J. Polym. Sci. Polym. Phys. Ed.* **19,** 863 (1981).
99. P. R. Dvornic, *Macromolecules* **17,** 1348 (1984).
100. M. B. Polk, K. B. Bota, E. C. Akubuiro, and M. Phingbodhipakkiya, *Macromolecules* **14,** 1626 (1981).
101. M. B. Polk, K. B. Bota, and E. C. Akubuiro, *Ind. Eng. Chem. Prod. Res. Dev.* **21**(2), 154 (1982); M. B. Polk, K. B. Bota, M. Mandu, M. Phingbodhipakkiya, and C. Edeogu, *Macromolecules* **17**(2), 129 (1984).
102. B. Millaud, A. Thierry, and A. Skoulios, *J. Phys.* **39,** 1109 (1978).
103. B. Millaud, A. Thierry, and A. Skoulios, *J. Phys. Paris Lett.* **40,** 607 (1979).
104. B. Millaud and C. Strazielle, *Polymer* **20**(5), 63 (1979).
105. S. Y. Park, T. W. Son, H. S. Yoon, and S. I. Hong, *Pollimo* **8,** 279 (1984); *Chem. Abstr.* **101,** 192598 (1984).
106. T. E. Helminiak, F. E. Arnold, and C. L. Benner, *Polym. Prepr. Am. Chem. Soc. Div. Polym. Chem.* **16**(2), 659 (1975).
107. G. C. Berry, P. M. Cotts, and S.-G. Chu, *Br. Polym. J.* **13,** 47 (1981); C. P. Wong, H. Ohnuma, and G. C. Berry, *J. Polym. Sci. Polym. Symp.* **65,** 173 (1978).
108. J. F. Wolfe, B. H. Loo, and F. E. Arnold, *Polym. Prepr. Am. Chem. Soc. Div. Polym. Chem.* **20**(1), 82 (1978); *Macromolecules* **14,** 915 (1981).

109. E. W. Choe and S. N. Kim, *Macromolecules* **14**, 920 (1981).
110. J. F. Wolfe, B. H. Loo, and E. R. Sevilla, *Polym. Prepr. Am. Chem. Soc. Div. Polym. Chem.* **22**(1), 60 (1981).
111. S.-G. Chu, S. Venkatraman, G. C. Berry, and Y. Einaga, *Macromolecules* **14**, 939 (1981).
112. P. D. Sybert, W. H. Beever, and J. K. Stille, *Macromolecules* **14**, 493 (1981).
113. A. Wereta, Jr. and M. T. Gehatia, *Polym. Eng. Sci.* **18**, 204 (1978).
114. D. G. Baird and F. M. Silver, *J. Appl. Polym. Sci.* **23**, 941 (1979).
115. R. C. Evers and G. J. Moore, *Polym. Prepr. Am. Chem. Soc. Div. Polym. Chem.* **25**(2), 207 (1985).
116. R. J. M. Nolte and W. Drenth, *Rec. Trav. Chim. Pays-Bas* **92**, 788 (1973).
117. F. Millich, E. W. Hellmuth, and S. Y. Huang, *J. Polym. Sci. Polym. Chem. Ed.* **13**, 2143 (1975).
118. F. Millich, *Adv. Polym. Sci.* **19**, 117 (1975); *Macromol. Rev.* **15**, 207 (1980).
119. S. M. Aharoni, *J. Polym. Sci. Polym. Phys. Ed.* **17**, 683 (1979).
120. S. M. Aharoni, *J. Macromol. Sci. Phys.* **21**(1), 105 (1982).
121. A. J. Bur and L. J. Fetters, *Macromolecules* **6**, 874 (1973); A. J. Bur and D. E. Roberts, *J. Chem. Phys.* **51**, 406 (1969).
122. S. M. Aharoni, *Macromolecules* **12**, 94 (1979).
123. *Ibid.*, **14**, 222 (1981).
124. S. M. Aharoni and E. K. Walsh, *Macromolecules* **12**, 271 (1979).
125. S. M. Aharoni, *J. Polym. Sci. Polym. Phys. Ed.* **18**, 1439 (1980).
126. H. R. Allcock, *Phosphorus-Nitrogen Compounds*, Academic Press, Inc., New York, 1972; *Science* **193**, 1214 (1976).
127. R. E. Singler, N. S. Schneider, and G. L. Hagnauer, *Polym. Eng. Sci.* **15**, 321 (1975).
128. C. R. Desper and N. S. Schneider, *Macromolecules* **9**, 424 (1976).
129. S. M. Aharoni, *Polym. Prepr. Am. Chem. Soc. Div. Polym. Chem.* **22**(1), 116 (1981).
130. S. Takahashi and co-workers, *Macromolecules* **11**, 1063 (1978); **12**, 1016 (1979); *J. Polym. Sci. Polym. Chem. Ed.* **18**, 349 (1980); **20**, 565 (1982); *Mol. Cryst. Liq. Cryst.* **82**, 139 (1982).
131. S. Takahashi, Y. Takai, H. Morimoto, and K. Sonogashira, *J. Chem. Soc. Chem. Commun.* **1984**(1), 3 (1984).
132. R. R. Luise, P. W. Morgan, M. Panar, and O. B. Wilcox, *paper presented at the 53rd Colloid and Surface Science Symposium*, Rolla, Mo., June 1979, Division of Colloid and Surface Chemistry, American Chemical Society, Washington, D.C.
133. Belg. Pat. 656,359 (1976) and Fr. Pat. 2,340,344 (Oct. 7, 1977), M. Panar and O. B. Wilcox (to E. I. du Pont de Nemours & Co., Inc.); *Chem. Abstr.* **87**, 153310 (1977).
134. D. G. Gray, *J. Appl. Polym. Sci. Polym. Symp.* **37**, 179 (1983).
135. R. D. Gilbert and P. A. Patton, *Prog. Polym. Sci.* **9**, 115 (1983).
136. Ger. Pat. Appl. 2,704,776 (Aug. 11, 1977) and U.S. Pat. 4,132,464 (Jan. 2, 1979), J. Maeno (to Ishii, Hideki).
137. S. M. Aharoni, *J. Macromol. Sci. Phys.* **21**, 287 (1982).
138. D. L. Patel and R. D. Gilbert, *J. Polym. Sci. Polym. Phys. Ed.* **19**, 1231, 1449 (1981).
139. J. Bheda, J. F. Fellers, and J. L. White, *Colloid Polym. Sci.* **258**, 1335 (1980); *J. Appl. Polym. Sci.* **26**, 3955 (1981).
140. P. A. Patton and R. D. Gilbert, *J. Polym. Sci. Polym. Phys. Ed.* **21**, 515 (1983).
141. R. S. Werbowyj and D. G. Gray, *Macromolecules* **17**, 1512 (1984).
142. G. Conio, E. Bianchi, A. Ciferri, A. Tealdi, and M. A. Aden, *Macromolecules* **16**, 1264 (1983); **17**, 2010 (1984).
143. R. S. Werbowyj and D. G. Gray, *Macromolecules* **13**, 69 (1980).
144. G. V. Laivins and D. G. Gray in Ref. 134, p. 181.
145. S. N. Bhadani and D. G. Gray, *Makromol. Chem. Rapid Commun.* **3**, 449 (1982).
146. P. Zugenmaier and U. Vogt, *Makromol. Chem.* **184**, 1749 (1983); *Makromol. Chem. Rapid Commun.* **4**, 759 (1983).
147. S. M. Aharoni, *Mol. Cryst. Liq. Cryst. Lett.* **56**, 237 (1980).
148. S. Suto, J. L. White, and J. F. Fellers, *Rheol. Acta* **21**(1), 62 (1982).
149. S.-L. Tseng, G. V. Laivins, and D. G. Gray, *Macromolecules* **15**, 1262 (1982).
150. H. Chanzy, S. Nawrot, A. Peguy, P. Smith, and J. Chevalier, *J. Polym. Sci. Polym. Phys. Ed.* **20**, 1909 (1982).
151. P. Navard and J. M. Haudin, *Br. Polym. J.* **12**(4), 174 (1980).
152. Jpn. Kokai Tokkyo Koho JP 83 151,217 (Sept. 8, 1983) (to Asahi Chemical Industry Co.).
153. G. Conio, P. Corazza, E. Bianchi, A. Tealdi, and A. Ciferri, *J. Polym. Sci. Polym. Lett. Ed.* **22**, 273 (1984).

154. C. L. McCormick, P. A. Callais, and B. H. Hutchinson, Jr., *Polym. Prepr. Am. Chem. Soc. Div. Polym. Chem.* **24**(2), 271 (1983).
155. B. Erman, P. J. Flory, and J. P. Hummel, *Macromolecules* **13**, 484 (1980).
156. M. Arpin, C. Strazielle, G. Weill, and H. Benoit, *Polymer* **18**, 262, 591 (1977).
157. P. Metzger Cotts and G. C. Berry, *J. Polym. Sci. Polym. Phys. Ed.* **21**, 1255 (1983).
158. C. C. Lee, S.-G. Chu, and G. C. Berry, *J. Polym. Sci. Polym. Phys. Ed.* **21**, 1594 (1983).
159. M. Kyotani and H. Kanetsuna, *J. Polym. Sci. Polym. Phys. Ed.* **21**, 379 (1983).
160. S. L. Kwolek and R. R. Luise, *Macromolecules* **19**, 1789 (1986).
161. U.S. Pat. 4,159,365 (June 26, 1979), C. R. Payet (to E. I. du Pont de Nemours & Co., Inc.).
162. U.S. Pat. 4,161,470 (July 17, 1979), G. W. Calundann (to Celanese Corp.).
163. U.S. Pat. 4,487,916 (Dec. 11, 1984), R. S. Irwin (to E. I. du Pont de Nemours & Co., Inc.).
164. U.S. Pat. 4,330,457 (May 18, 1982), A. J. East, L. F. Charbonneau, and G. W. Calundann (to Celanese Corp.).
165. U.S. Pats. 3,778,410 (Dec. 11, 1973) and 3,804,805 (Apr. 16, 1974), H. F. Kuhfuss and W. J. Jackson, Jr. (to The Eastman Kodak Co.).
166. Jpn. Pat. 139,698 (Oct. 30, 1979), T. Urasaki, A. Aoki, and Y. Matsuda (to Teijin, Ltd.).
167. U.S. Pat. 3,991,014 (Nov. 9, 1976), J. J. Kleinschuster (to E. I. du Pont de Nemours & Co., Inc.).
168. W. J. Jackson, Jr. and H. F. Kuhfuss, *J. Appl. Polym. Sci.* **25**, 1685 (1980).
169. W. J. Jackson, Jr., *Macromolecules* **16**, 1027 (1983).
170. Ref. 125, p. 1303.
171. U.S. Pat. 4,093,595 (June 6, 1978), S. P. Elliott (to E. I. du Pont de Nemours & Co., Inc.).
172. J. Economy and W. Volksen in Ref. 28, Chapt. 7, p. 293.
173. W. R. Krigbaum, *J. Appl. Polym. Sci. Appl. Polym. Symp.* **41**, 105 (1985).
174. J. Blackwell, G. Lieser, and G. Gutierrez, *Macromolecules* **16**, 418 (1983).
175. J. Menczel and B. Wunderlich, *J. Polym. Sci. Polym. Phys. Ed.* **18**, 1433 (1980); W. Meesiri, J. Menczel, U. Gaur, and B. Wunderlich, *J. Polym. Sci. Polym. Phys. Ed.* **20**, 719 (1982); A. E. Zachariades, J. Economy, and J. A. Logan, *J. Appl. Polym. Sci.* **27**, 2009 (1982).
176. T. Tsay, W. Volksen, and J. Economy in A. C. Griffin and J. F. Johnson, eds., *Liquid Crystals and Ordered Fluids*, Vol. 4, Plenum Press, New York, 1984, p. 363.
177. R. W. Lenz, J. Jin, and K. A. Feichtinger, *Polymer* **24**, 327 (1983); G. Chen and R. W. Lenz, *Polymer* **26**, 1307 (1985).
178. R. A. Chivers and J. Blackwell, *Polymer* **26**, 997 (1985).
179. M. Panar, P. Avakian, R. C. Blume, K. H. Gardner, T. D. Gierke, and H. H. Yang, *J. Polym. Sci. Polym. Phys. Ed.* **21**, 1955 (1983).
180. S. B. Warner and M. Jaffe, *J. Cryst. Growth* **48**, 184 (1980).
181. K. F. Wissbrun, *J. Rheol.* **25**, 619 (1981); *Br. Polym. J.* **12**, 163 (1980); J. L. White, *J. Appl. Polym. Sci. Appl. Polym. Symp.* **41**, 241 (1985).
182. S. Onogi and T. Asada in G. Astarita, G. Marucci, and L. Nicolais, eds., *Rheology*, Vol. 1, Plenum Press, New York, 1980, pp. 127–147.
183. K. F. Wissbrun and A. C. Griffin, *J. Polym. Sci. Polym. Phys. Ed.* **20**, 1835 (1982).
184. F. N. Cogswell in L. L. Chapoy, ed., *Recent Advances in Crystalline Polymers*, Elsevier Applied Science Publishers, Ltd., Barking, UK, 1984, Chapt. 10, p. 165.
185. M. Prasadarao, E. M. Pearce, and C. D. Han, *J. Appl. Polym. Sci.* **27**, 1343 (1982).
186. R. E. Jerman and D. G. Baird, *J. Rheol.* **25**, 275 (1981).
187. M. Jaffe, *Polym. Prepr. Div. Polym. Chem. Am. Chem. Soc.* **19**(1), 355 (1982); M. Jaffe and R. S. Jones in M. Lewin and J. Preston, eds., *High Technology Fibers*, Vol. 3 of *Handbook of Fiber Science and Technology*, Marcel Dekker, Inc., New York, 1985, Pt. A, Chapt. 9, p. 349.
188. D. Acierno, F. P. La Mantia, G. Polizzotti, A. Ciferri, and B. Valenti, *Macromolecules* **15**, 1455 (1982).
189. Y. Takeuchi, F. Yamamoto, and S. Yamakawa, *Polym. J.* **16**, 579 (1984).
190. A. Zachariades and J. Economy, *Polym. Eng. Sci.* **23**, 266 (1983).
191. Y. Ide and Z. Ophir, *Polym. Eng. Sci.* **23**, 261 (1983).
192. T. Asada, K. Toda, and S. Onogi, *Mol. Cryst. Liq. Cryst.* **68**, 231 (1981).
193. A. Roviello and A. Sirigu, *J. Polym. Sci. Polym. Lett. Ed.* **13**, 445 (1975); *Eur. Polym. J.* **15**, 61, 423 (1979).
194. W. J. Jackson, Jr. and J. C. Morris, *J. Appl. Polym. Sci. Appl. Polym. Symp.* **41**, 307 (1985).
195. W. Kreuder, H. Ringsdorf, and P. Tshirner, *Makromol. Chem. Rapid Commun.* **6**, 367 (1985).
196. J. Watanabe and W. R. Krigbaum, *J. Polym. Sci. Polym. Phys. Ed.* **23**, 565 (1985).
197. P. W. Wojtkowski, *Macromolecules*, in press.

198. A. H. Khan, J. E. McIntyre, and A. H. Milburn, *Polymer* **24,** 1610 (1983).
199. B. Reck and H. Ringsdorf, *Makromol. Chem. Rapid Commun.* **6,** 291 (1985).
200. C. K. Ober, J.-I. Jin, and R. W. Lenz in Ref. 42, p. 111.
201. H. Stevens, G. Rehage, and H. Finkelmann, *Macromolecules* **17,** 851 (1984).
202. C. K. Ober, J.-I. Jin, and R. W. Lenz in Ref. 42, p. 115.
203. Ref. 4, p. 76.
204. Yu. S. Lipatov, V. V. Tsukruk, and V. V. Shilov, *Polym. Commun.* **24,** 75 (1983).
205. C. K. Ober, J.-I. Jin, and R. W. Lenz in Ref. 42, pp. 110 and 112.
206. J. Asrar, H. Toriumi, T. Watanabe, W. R. Krigbaum, A. Ciferri, and J. Preston, *J. Polym. Sci. Polym. Phys. Ed.* **21,** 1119 (1983); W. R. Krigbaum, J. Watanabe, and T. Ishikawa, *Macromolecules* **16,** 1271 (1983).
207. V. P. Shibaev and N. A. Platé in Ref. 35, p. 178.
208. Ref. 35, p. 155.
209. W. Kreuder and H. Ringsdorf, *Makromol. Chem. Rapid Commun.* **4,** 807 (1983).
210. Ref. 35, p. 146.
211. V. P. Shibaev and N. A. Platé in Ref. 35, p. 184.
212. *Ibid.,* p. 136.
213. H. Finkelmann and G. Rehage, *Makromol. Chem. Rapid Commun.* **3,** 859 (1982).
214. H. Finkelmann and G. Rehage in Ref. 35, p. 147.
215. L. H. Miner, R. A. Wolffe, and C. H. Zweben, *ASTM Special Technical Publication 580,* The American Society for Testing and Materials, Philadelphia, 1975, pp. 549–559.
216. U.S. Pat. 3,819,587 (June 25, 1974), S. L. Kwolek (to E. I. du Pont de Nemours & Co., Inc.).
217. U.S. Pat. 4,533,693 (Aug. 6, 1985), J. F. Wolfe, P. D. Sybert, and J. R. Sybert (to SRI International, Inc.).
218. U.S. Pat. 4,501,886 (Feb. 26, 1985), J. P. O'Brien (to E. I. du Pont de Nemours & Co., Inc.).
219. P. W. Morgan, *Condensation Polymers by Interfacial and Solution Methods,* Wiley-Interscience, New York, 1965.
220. S. R. Allen, A. G. Filippov, R. J. Farris, and E. L. Thomas in Ref. 28, Chapt. 9, p. 357.
221. U.S. Pat. 4,183,895 (Jan. 15, 1980), R. R. Luise (to E. I. du Pont de Nemours & Co., Inc.).
222. R. N. Demartino, *J. Appl. Polym. Sci.* **28,** 1805 (1983).
223. U.S. Pat. 4,075,262 (Feb. 21, 1978), J. R. Schaefgen (to E. I. du Pont de Nemours & Co., Inc.).
224. U.S. Pat. 4,399,270 (Aug. 16, 1983), A. H. Frazer (to E. I. du Pont de Nemours & Co., Inc.).
225. U.S. Pat. 4,370,466 (Jan. 25, 1983), R. K. Siemionko (to E. I. du Pont de Nemours & Co., Inc.).
226. U.S. Pat. 4,500,699 (Feb. 19, 1985), R. S. Irwin and F. M. Logullo, Sr. (to E. I. du Pont de Nemours & Co., Inc.).
227. U.S. Pat. 4,473,682 (Sept. 25, 1984), G. W. Calundann, L. F. Charbonneau, and B. C. Benicewicz (to Celanese Corp.).
228. U.S. Pat. 4,431,770 (Feb. 14, 1984), A. J. East and G. W. Calundann (to Celanese Corp.).
229. Eur. Pat. Appl. 72,540 (Feb. 23, 1983), W. Funakoshi and T. Urasaki (to Teijin, Ltd.).
230. U.S. Pat. 4,447,592 (May 8, 1984), J. F. Harris (to E. I. du Pont de Nemours & Co., Inc.).
231. U.S. Pat. 4,503,005 (Mar. 5, 1985), K. Ueno, H. Sugimoto, and K. Hayatsu (to Sumitomo Chemical Co.).
232. U.S. Pat. 4,048,148 (Sept. 13, 1977), P. W. Morgan (to E. I. du Pont de Nemours & Co., Inc.).
233. U.S. Pat. 4,371,660 (Feb. 1, 1983), G. W. Calundann and A. J. East (to Celanese Corp.).
234. Announcement by Sumitomo Chemical Co., Oct. 24, 1985, as reported in *The Daily Industrial News,* Osaka, Japan.
235. W. R. Krigbaum, A. Ciferri, and D. Acierno, *J. Appl. Polym. Sci. Appl. Polym. Symp.* **41,** 293 (1985).
236. Eur. Pat. Appl. 45,575 (July 10, 1982), Y. Ide (to Celanese Corp.).
237. Ger. Pat. 2,704,315 (Aug. 25, 1977), H. Inata, T. Kuratsuji, and S. Kawase (to Teijin, Ltd.).
238. W. F. Hwang, D. R. Wiff, C. Verschoore, G. E. Price, T. E. Helminiak, and W. W. Adams, *Polym. Eng. Sci.* **23,** 784 (1983).
239. J. E. Flood, J. L. White, and J. F. Fellers, *J. Appl. Polym. Sci.* **27,** 2965 (1982).
240. Eur. Pat. Appl. 24,499 (June 24, 1981), U. Takanori, Y. Hirabayashi, M. Ogasawara, and H. Inata (to Teijin, Ltd.).
241. Eur. Pat. Appl. 72,210 (Feb. 16, 1983), Y. Ide (to Celanese Corp.).
242. J. Economy, B. E. Nowak, and S. G. Cottis, *Polym. Prepr. Am. Chem. Soc. Div. Polym. Chem.* **11**(1), 1 (1970).
243. U.S. Pat. 3,759,870 (Sept. 18, 1973), J. Economy and B. E. Nowak (to Carborundum Co.).

244. U.S. Pat. 3,637,595 (Jan. 25, 1972), S. G. Cottis, J. Economy, and B. E. Nowak (to Carborundum Co.).
245. *Mod. Plast.* **50,** 20 (Aug. 1973).
246. *Mod. Plast.* **51,** 20 (Aug. 1974).
247. *Mod. Plast.* **61,** 14 (Dec. 1984).
248. *Plast. Technol.* **30,** 82 (Dec. 1984).
249. *Mod. Plast.* **63,** 18 (Jan. 1986).
250. Chemical correspondence, *Japanese High Technology Monitor,* Osaka, Japan, Nov. 5, 1985.
251. U.S. Pat. 4,067,852 (Jan. 10, 1978), G. W. Calundann (to Celanese Corp.).
252. U.S. Pat. 4,083,829 (Apr. 11, 1978), G. W. Calundann, H. L. Davis, F. J. Gorman, and R. M. Mininni (to Celanese Corp.).
253. U.S. Pat. 4,184,996 (Jan. 22, 1980), G. W. Calundann (to Celanese Corp.).
254. U.S. Pat. 4,219,461 (Aug. 26, 1980), G. W. Calundann (to Celanese Corp.).
255. U.S. Pat. 4,256,624 (Mar. 17, 1981), G. W. Calundann (to Celanese Corp.).
256. U.S. Pat. 4,318,841 (Mar. 9, 1982), A. J. East and G. W. Calundann (to Celanese Corp.).
257. U.S. Pat. 4,342,862 (Aug. 3, 1982), W. J. Jackson, Jr. and W. R. Darnell (to The Eastman Kodak Co.).
258. U.S. Pat. 4,242,496 (Dec. 30, 1980), W. J. Jackson, Jr., G. G. Gebeau, and H. F. Kuhfuss (to The Eastman Kodak Co.).
259. *Plastics Technology Mfg. Handbook and Buyer's Guide,* special issue of *Plast. Technol.* **31,** 518 (1985).
260. *LCP High Performance Resins,* Celanese Specialty Operations, New York; *Plast. Technol.* **31,** 85 (Apr. 1985).
261. *Adv. Mater. Process.* **2**(2), 15 (1986).
262. W. J. Jackson, Jr., *J. Appl. Polym. Sci. Appl. Polym. Symp.* **41,** 25 (1985).
263. Ref. 194, p. 316.
264. V. P. Shibaev, S. G. Kostromin, N. A. Platé, S. A. Ivanov, V. Yu Vetrov, and I. A. Yakovlev, *Polym. Commun.* **24,** 364 (1983).
265. H. J. Coles and R. Simon in Ref. 184, Chapt. 22, p. 323.

S. L. Kwolek
E. I. du Pont de Nemours & Co., Inc.

P. W. Morgan
Consultant

J. R. Schaefgen
Consultant

MECHANICAL PROPERTIES

Applications of the unique properties of materials now known to be polymeric long predated any understanding of their molecular complexity. Humans have made use of natural fibers of both animal and vegetable origin since the dawn of civilization, and the characteristics of natural rubber were investigated and described over one hundred years ago. In more recent years, it generally has been held that the mechanical properties of interest are a result of a material being polymeric, and that they might be improved by modifying the composition or morphology, eg, by changing the temperature at which they soften and become rubberlike, or the overall degree of three-dimensional order. Even so, the incentive for studies of mechanical properties is often the need to correlate the response of different materials under a range of conditions in order to predict the performance of these polymers in practical applications. For a long time, attempts have been made to understand the mechanical behavior of rubberlike materials through the deformation of a network of cross-linked and tangled molecular chains, but efforts to describe the deformation of other polymeric solids in terms of processes operating on a molecular scale are more recent. This article considers the main types of response shown by polymeric solids to different levels of stress: elasticity, viscoelasticity, plastic flow, and fracture.

In rubberlike elasticity (qv), the strains are large and are assumed to occur instantaneously with the addition or removal of stress. The physical mechanism is one of chain uncoiling, which occurs as a consequence of rotation about main-chain bonds. Unhindered rotation occurs only above an adequately high temperature, the glass–rubber transition temperature T_g, whose value depends on the detailed molecular form of the polymer. In most cases, the value of T_g can be related to a free-volume model in which the fraction of the overall macroscopic volume not occupied by the volume of individual molecular chains is considered.

In contrast, viscoelastic materials show, even at small strains, a delayed component of strain both on loading and unloading. The application of a constant strain leads to the corresponding process of stress relaxation. For very small strains, both the instantaneous and delayed components of strain are proportional to the applied stress. The behavior is then called linear viscoelasticity, an approximation most likely to be valid for noncrystalline polymers above T_g. Linear viscoelasticity can be simulated by mathematical models of springs and dashpots which can be used to define the relaxation and retardation times that characterize the time scale of the delayed response. These simple models cannot be related directly to processes, such as chain entanglement, responsible for the delayed response; models that have a greater physical realism must then be considered.

The small-amplitude oscillatory behavior of linear viscoelastic polymers as a function of temperature is characterized by a series of local maxima in energy dissipation observed as an increase in the phase angle between stress and strain. Study of these relaxation transitions, particularly in semicrystalline polymers, gives a revealing insight into physical behavior at the molecular level.

Unfortunately, at useful strains and over long times, viscoelastic behavior is rarely linear, and no general method for the treatment of nonlinear viscoelasticity has yet been developed. Engineers may find it adequate to derive empirical relations from a series of simple tests at different stresses or strains, but such relations have no physical significance. A set of step-loading programs can indicate the relation between strain and a function of stress, but the stress function is found to be dependent on experimental conditions and on small changes at a molecular level in the test polymer. Another approach that has led to some physical understanding considers creep and stress relaxation as thermally activated processes.

Many polymers of commercial importance and scientific interest are anisotropic, having been oriented either along a single-stretch direction (fiber orientation) or along all three axial directions to give a polycrystalline analogue of single-crystal texture. Small-strain elastic and linear viscoelastic measurements on such preoriented materials have been interpreted mainly in terms of two contrasting models: the first seeks to show how anisotropy arises through the orientation of an aggregate of preexisting units, whereas the second considers a semicrystalline polymer as a composite of linked crystalline and amorphous regions.

Under high enough strain a solid polymer shows characteristic necking and cold drawing. In oriented polymers extended in a direction different from that of the initial draw, the deformation may be concentrated in a narrow band. In addition to the development of yield criteria, it is important to gain some understanding at the molecular level, and here a model of the deformation of a molecular network appears to be useful.

At low temperatures polymers fracture in a brittle manner forming a wedge of porous material, termed a craze, at the crack tip (see also CRAZING). The brittle-ductile transition can then be considered a competition between the independently activated processes of crazing and yielding. To date only tentative suggestions have been made concerning the relation between fracture behavior and molecular processes. For instance, impact tests are simple to perform and are useful in characterizing specific engineering materials, but the results can be interpreted in several ways.

The Rubberlike State

A rubberlike polymer is capable of showing complete recovery from large deformations. The recovery usually occurs so rapidly that, to a good approximation, the response may be considered elastic even at large strains. Because the deformations are large, it is necessary to generalize the simple definitions of strain and stress that are adequate for small-strain elasticity. The behavior of polymers in the rubberlike state can be treated by finite-strain elasticity, a branch of continuum mechanics in which features of molecular structure are ignored, or by statistical molecular theories.

Finite-strain Elasticity. The two principal mathematical methods to determine finite-strain elasticity are an extension of the stress–strain relations used for isotropic incompressible bodies at low strains, and the development of a strain–energy function involving the strain invariants.

Hooke's law can be generalized by taking second-order quantities into account. The analysis is simplified by assuming the material to be isotropic in the undeformed state and neglecting volume changes on deformation, a consequence of the bulk modulus of natural rubber being $\sim 10^4$ the shear modulus. For finite strains, expressions are obtained of the form (1)

$$1 + 2e_{ii} = \frac{3}{E}(\sigma_{ii} - p), \; e_{ij} = 0 (i \neq j)$$

where E is Young's modulus and

$$p = \frac{\nu}{1 + \nu}\Sigma\sigma_{ii}$$

is equivalent to a hydrostatic pressure. For an incompressible elastic solid, the Poisson's ratio ν is $\frac{1}{2}$, and e_{ii} and σ_{ii} denote the components of finite strain and stress.

The case of homogeneous pure strain, with extension ratios of λ_1, λ_2, and λ_3, is usually considered in the directions of the coordinate axes. For uniaxial stress, the nominal stress (defined in terms of unstrained cross-section) is derived from

$$f = \frac{E}{3}\left(\lambda_1 - \frac{1}{\lambda_1^2}\right)$$

an expression usually presented as a consequence of molecular theories of elastomeric networks. Materials obeying this relation are called neo-Hookean.

The alternative approach considers the strain–energy function U, which defines the energy stored in the body as the result of the strain; the form of U is dependent on the experimental conditions. For adiabatic changes of state, the relevant functions are for: constant volume U_1 that is identical to the thermodynamic internal energy; and constant pressure U_2 that is identical to the enthalpy. Isothermal changes of state at constant volume define the function U_3, which corresponds to the Helmholtz free energy.

For stress–strain relations at constant pressure and temperature, the appropriate function U_4 is identical with the thermodynamic Gibbs function G. Originally, it was considered that the strain–energy function should be a homogeneous quadratic function of the strain components. For an isotropic solid, the strain energy must be independent of the direction of the coordinate axes, and on this basis would therefore be a function of the strain invariants I_1, I_2, and I_3. For an isotropic incompressible solid undergoing a pure homogeneous deformation with extension ratios λ_i:

$$I_1 = \lambda_1^2 + \lambda_2^2 + \lambda_3^2, \qquad I_2 = \lambda_1^2\lambda_2^2 + \lambda_2^2\lambda_3^2 + \lambda_3^2\lambda_1^2, \qquad I_3 = \lambda_1^2\lambda_2^2\lambda_3^2 = 1$$

For a neo-Hookean material $U = C_1(I_1 - 3) = C_1(\lambda_1^2 + \lambda_2^2 + \lambda_3^2 - 3)$, which corresponds exactly to an expression derived from the statistical theory for a Gaussian network if $C_1 = \frac{1}{2}NkT$.

The most general first-order relation $U = C_1(I_1 - 3) + C_2(I_2 - 3)$ was first derived on the assumption of a linear stress–strain relation in shear (2). The physical significance of C_2 is unclear and has been the subject of discussion.

Solutions to problems in finite-strain elasticity are most readily obtained if strain components are specified and the stress components obtained by using the strain–energy function. For most purposes, it is necessary only to derive results for homogeneous pure strains in which a unit cube deforms to a rectangular parallelepiped λ_1, λ_2, and λ_3. The forces acting on the faces f_1, f_2, and f_3 are defined in terms of the applied force per unit area in the undeformed state. Stress components in the deformed state are then $\sigma_{11} = \lambda_1 f_1$, etc.

By considering the work done per unit volume in an infinitesimal displacement from the deformed state, it can be shown that

$$\sigma_{ii} = 2\left(\lambda_i^2 \frac{\partial U}{\partial I_1} - \frac{1}{\lambda_i^2}\frac{\partial U}{\partial I_2}\right) + p, \qquad \sigma_{ij} = 0$$

Homogeneous pure strain has been examined in thin sheets of vulcanized rubber stressed only along the sheet edges (3,4). Under these conditions,

$$\frac{\partial U}{\partial I_1} = \left[\frac{\lambda_1^2 \, \sigma_{11}}{\left(\lambda_1^2 - \dfrac{1}{\lambda_1^2 \lambda_2^2} \right)} - \frac{\lambda_2^2 \, \sigma_{22}}{\left(\lambda_2^2 - \dfrac{1}{\lambda_1^2 \lambda_2^2} \right)} \right] \bigg/ 2(\lambda_1^2 - \lambda_2^2)$$

$$\frac{\partial U}{\partial I_2} = \left[\frac{\sigma_{11}}{\left(\lambda_1^2 - \dfrac{1}{\lambda_1^2 \lambda_2^2} \right)} - \frac{\sigma_{22}}{\left(\lambda_2^2 - \dfrac{1}{\lambda_1^2 \lambda_2^2} \right)} \right] \bigg/ 2(\lambda_2^2 - \lambda_1^2)$$

By varying λ_1 and λ_2 in such a way that I_1 and I_2 in turn remained constant, the values of $\dfrac{\partial U}{\partial I_1}$ and $\dfrac{\partial U}{\partial I_2}$ shown in Figure 1 were obtained. It can be seen that $\dfrac{\partial U}{\partial I_1}$ is approximately constant and independent both of I_1 and I_2; $\dfrac{\partial U}{\partial I_2}$ is independent of

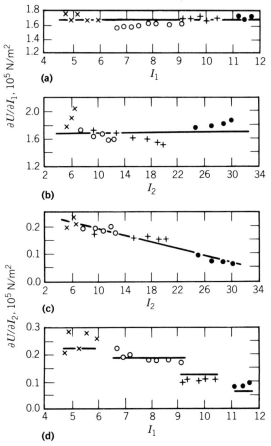

Fig. 1. Dependence of $\partial U/\partial I_1$ and $\partial U/\partial I_2$ on I_1 and I_2. (a) and (d): $\times$, $I_2 = 5$; $\bigcirc$, $I_2 = 10$; $+$, $I_2 = 20$; $\bullet$, $I_2 = 30$. (b) and (c): x, $I_1 = 5$; $\bigcirc$, $I_1 = 7$; $+$, $I_1 = 9$; $\bullet$, $I_1 = 11$ (3). To convert N/m² to psi, multiply by 1.45×10^{-4}.

I_1 but decreases with increasing I_2. The polynomial form for the strain–energy function then becomes

$$U = C_1(I_1 - 3) + f(I_2 - 3)$$

that is, the sum of the first-order neo-Hookean term and a function of I_2. The second term is generally small compared with the first and decreases as I_2 increases. Further experiments in pure shear, with and without a superimposed initial extension, confirm that the assumed form of the strain–energy function provides a good mathematical model of the material.

A further key experiment is simple elongation in which the load f required to extend a specimen of initial cross-section A is

$$f = 2A\left(\lambda - \frac{1}{\lambda^2}\right)\left(\frac{\partial U}{\partial I_1} + \frac{1}{\lambda}\frac{\partial U}{\partial I_2}\right)$$

This is demonstrated by the curve of Figure 2 in which $\left(\dfrac{\partial U}{\partial I_1} + \dfrac{1}{\lambda}\dfrac{\partial U}{\partial I_2}\right)$, ie,

$\dfrac{f}{2A\left(\lambda - \dfrac{1}{\lambda^2}\right)}$, is plotted against $\dfrac{1}{\lambda}$.

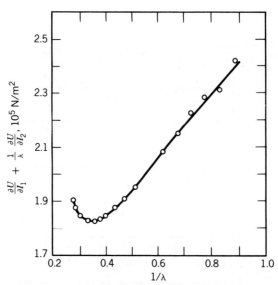

Fig. 2. Plot of $[\partial U/\partial I_1 + (1/\lambda)\,\partial U/\partial I_2]$ against $1/\lambda$ from experiment in simple extension (3). To convert N/m² to psi, multiply by 1.45×10^{-4}.

The Mooney equation $U = C_1(I_1 - 3) + C_2(I_2 - 3)$ can be fitted to this plot for $1/\lambda > 0.45$, but the values of C_1 and C_2 so obtained do not agree with those deduced from two-dimensional extension and pure shear, indicating the inadequacy of the Mooney expression; C_1 and C_2 should each be replaced by linear combinations of at least three constants and other powers of λ should be considered. Subsequent work suggests that the formulation given above for homogeneous strain may not be adequate at low strains (5).

In simple shear, it is necessary to apply normal components of stress (σ_{11}, σ_{22}) in addition to the shear stress σ_{12}, if finite shear is to be produced—a situation equivalent to the Weissenberg effect in viscous flow.

$$\sigma_{12} = 2K\left(\frac{\partial U}{\partial I_1} + \frac{\partial U}{\partial I_2}\right), \qquad \sigma_{13} = \sigma_{23} = \sigma_{33} = 0,$$

$$\sigma_{11} = 2K^2\frac{\partial U}{\partial I_1}, \qquad \sigma_{22} = -2K^2\frac{\partial U}{\partial I_2}$$

where K is a constant defining the magnitude of the shear strain. The normal stresses vanish for small strains as they are second order in K.

If U is a linear function of I_1 and I_2, the shear stress is proportional to the shear displacement, a result that led to the Mooney expression discussed above. For large amounts of shear I_1 and I_2 remain small. If higher order derivatives of U are small compared with $\frac{\partial U}{\partial I_1}$ and $\frac{\partial U}{\partial I_2}$, which is the case for rubber, an approximately linear load-deformation relation should result. Experiments on rubbers involving only simple extension or simple shear may not lead to a satisfactory understanding of the form of the strain–energy function. Subsequently, more general forms have been considered (6,7).

The strain–energy function expressed in terms of extension ratios has the form (8)

$$U(\lambda_1, \lambda_2, \lambda_3) = U(\lambda_1) + U(\lambda_2) + U(\lambda_3)$$

This gives stress–strain relations of the form

$$\sigma_{xx} - \sigma_{yy} = \lambda_1 U'(\lambda_1) - \lambda_2 U'(\lambda_2)$$

where

$$U'(\lambda_i) = \frac{\partial U}{\partial \lambda_i}$$

Experimentally, λ_2 is maintained constant and λ_1 is varied. For $\lambda_1 < 2.5$, it was proposed that $U'(\lambda) = 2\mu \ln \lambda$, where μ is the shear modulus; the excellent agreement of experimental data with these proposals can be seen in Figure 3.

The original formulation of the strain–energy function involved extension ratios squared, because it was considered that negative values of λ are possible mathematically. Assuming that it is not necessary for the strain–energy functions to be restricted to even integral powers of the extension ratios, a more general form for U satisfies the Valanis-Landel representation (9). The strain–energy function is taken as having the form

$$U \sum_n \frac{\mu_n}{\alpha_n} (\lambda_1^{\alpha_n} + \lambda_2^{\alpha_n} + \lambda_3^{\alpha_n} - 3)$$

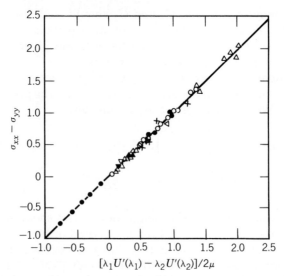

Fig. 3.   Experimental tests for natural rubber using experimental data: $+$, uniaxial; $\triangle$, biaxial, where $\mu = 4.1 \times 10^5$ N/m² (59 psi); $\bigcirc$, $\mu = 3.9 \times 10^5$ N/m² (57 psi); $\bullet$, uniaxial compression, where $\mu = 4 \times 10^5$ N/m² (58 psi) (8).

and it can be shown that

$$U(\lambda) = \sum_n \frac{\mu_n}{\alpha_n} (\lambda^{\alpha_n} - 1),$$

giving

$$\lambda U'(\lambda) = \sum_n \mu_n \lambda^{\alpha_n}$$

This may be written as

$$\lambda U'(\lambda) - c = \sum_n \mu_n (\lambda^{\alpha_n} - 1)$$

where

$$c = [U'(\lambda)]_{\lambda=1} = \sum_n \mu_n$$

A good fit to experimental data can be obtained by suitable adjustment of numerical parameters (Fig. 4). The significant difference from the predictions of the neo-Hookean model are indicated.

Statistical Molecular Theories.   Assuming that rubberlike deformations are reversible in a thermodynamic sense, it is readily shown that the tension in stretched rubber is a consequence of the fact that a stretched molecular network seeks to contract to a state of lower free energy, ie, higher entropy. A statistical model of a network of molecular chains can then be developed in which changes in configurational entropy are related to the force causing the change. These molecular changes are then related to experimentally measured extension ratios.

For a reversible isothermal change at constant volume the work done is

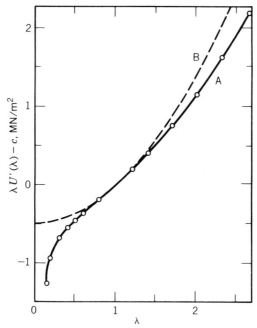

Fig. 4. The function $\lambda U'(\lambda) - c$. A, Experimental; B, statistical theory (7). To convert MN/m² to psi, multiply by 145.

related to the change in the Helmholtz free energy function ($dW = dA = dU - TdS$). Therefore, when an elastic solid of initial length l is extended uniaxially by a tensile force f, the change in internal energy is given by

$$\left(\frac{\partial U}{\partial l}\right)_T = f - T\left(\frac{\partial f}{\partial T}\right)_l$$

For an ideal rubber, the tensile force at constant length is proportional to the absolute temperature, demonstrating that internal energy changes are negligible and that the elasticity arises largely from changes in entropy.

For extensions greater than ca 10%, the force increases with increasing temperature, but at lower strains the trend is reversed (Fig. 5). This thermoelastic inversion effect is a consequence of the thermal expansion of the rubber. Experimentally, it is appropriate to correct for this effect by measuring the tensile force as a function of temperature at constant extension ratio.

Careful experimentation has shown that the changes in internal energy with extension cannot be neglected and, moreover, that all rubbers are slightly compressible; the differences between constant volume and constant pressure conditions must also be taken into account. These small effects are, however, ignored in the simple statistical theory of rubber elasticity.

Long molecular chains can adopt many configurations of equal energy in response to the micro-Brownian motion of their constituent atoms. It is assumed that the chains in a rubber form a cross-linked network in which the number of links is small enough so as not to affect the motion of the chains. External forces constrain the arrangement of the chains and decrease the configurational entropy.

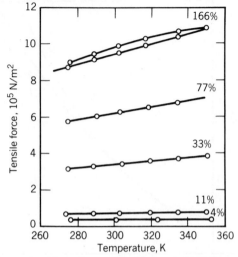

Fig. 5.   Force at constant length as a function of absolute temperature. Elongation as indicated (10). To convert N/m² to psi, multiply by 1.45×10^{-4}.

An evaluation of the stress–strain relation thus involves calculating the entropy of a single chain and then the change in entropy of the complete network of chains as a function of strain.

The problem is approached initially through the assumption of a freely jointed chain of equal links without the restriction that the valence angle between links must remain constant. One end of the chain of n links, each with length l, is considered fixed at the origin, and the probability distribution for the position of the other end is defined in terms of the Gaussian error function, ie, an approximation valid only when the distance between the ends of the chain is less than ~30% of the length of the fully extended chain.

$$p(x,y,z)dxdydz = \frac{b^3}{\pi^{3/2}} \exp\left(-b^2 r^2\right) dxdydz$$

where

$$b^2 = \frac{3}{2nl^2}$$

The probability that the free end of the chain is in an elemental volume between r and $(r + dr)$ is found by multiplying p by the volume of the spherical shell of radius r and thickness dr.

$$P(r)dr = p(r)4\pi r^2 dr = \left(\frac{4b^3}{\pi^{1/2}}\right) r^2 \exp\left(-b^2 r^2\right) dr$$

Furthermore, the mean-square chain length $\overline{r^2} = nl^2$ so r_{rms} is proportional to the square root of the number of links in the chain. From the Boltzmann relation for entropy, $S = k \ln \Omega$, where k is the Boltzmann constant and Ω the number of configurations; the entropy of the chain is derived as $S = C - kb^2(x^2 + y^2 + z^2)$, where C is an arbitrary constant.

The strain–energy function for a molecular network is determined by replacing the actual network by one in which each segment of a molecule between successive cross-links is considered as a Gaussian chain, and the mean-square end-to-end distance for the assembly of chains in the unstrained state is the same as for a corresponding set of free chains. A further assumption is that of affine deformation in which the components of the vector length of each chain are assumed to change in the same ratio as the dimensions of the bulk material.

The change in entropy for a single chain is calculated and a summation is made for all chains with the same value of b. A further summation for all the chains in the network gives the overall change of entropy on deformation as

$$\Delta S = \tfrac{1}{2}Nk(\lambda_1^2 + \lambda_2^2 + \lambda_3^2 - 3)$$

where N is the number of chains per unit volume. Neglecting changes in internal energy and assuming that the strain–energy function U is zero in the undeformed state,

$$U = \Delta A = \tfrac{1}{2}NkT(\lambda_1^2 + \lambda_2^2 + \lambda_3^2 - 3)$$

or

$$U = \tfrac{1}{2}NkT(I_1 - 3)$$

already derived for a neo-Hookean solid through continuum arguments. For uniaxial orientation, where $\lambda_2 = \lambda_3 = \lambda_1^{-1/2}$, as before,

$$f = \frac{\partial U}{\partial \lambda_1} = NkT\left(\lambda_1 - \frac{1}{\lambda_1^2}\right)$$

There have been several developments of this simplified theory:

1. The effective number of cross-links, when entanglements, loops, and loose ends are considered, is obtained through the use of a correction factor $\left(1 - 2\dfrac{M_c}{M}\right)$, where M_c is the mean molecular weight between cross-links and M is the molecular weight before cross-linking.

2. The inclusion of fixed-bond angles, possibly with hindered rotation, also taking into account excluded volume (a chain cannot occupy the same volume more than once) results in an increase in the equilibrium chain length. The concept of an equivalent freely jointed chain may be useful.

3. Attempts have been made to remove the unnecessarily restrictive assumption that junction points remain fixed in the material.

4. The Gaussian assumption, which limits the application of the model to relatively small extensions, can be replaced by the inverse Langevin approximation. The probability distribution for a freely jointed chain is then given by

$$\ln p(r) = \text{const} - n\left[\frac{r}{nl}\beta + \ln\left(\frac{\beta}{\sinh \beta}\right)\right]$$

where

$$\frac{r}{nl} = \coth \beta - \frac{1}{\beta} = \mathscr{L}(\beta), \qquad \beta = \mathscr{L}^{-1}\left(\frac{r}{nl}\right)$$

This method predicts a large increase in tensile stress at high extensions. An increase of this form is observed with natural rubber, but there it is attributed to strain-induced crystallization. The crystal melting point $T_c = \Delta H_f / \Delta S_f$, where H_f is the enthalpy and S_f the entropy of fusion. Stretching reduces the entropy of the noncrystalline state, permitting T_c to rise above the temperature of the experiment.

Simple theory assumes not only that different chain conformations have identical internal energies but also that these are negligible. Additionally, the thermodynamic relations are applicable to experiments at constant volume, whereas most experiments are performed under conditions of constant pressure. Further consideration suggests that there is an anisotropic compressibility in the strained state. Many investigations have suggested that in natural rubber the energy contribution f_e to the total force may be as much as 10–20% at room temperature (6). A number of experiments show f_e to decrease with increasing extension ratio, which is in disagreement with theoretical predictions. The value of f_e/f appears unaffected by the presence of a swelling liquid, confirming that it is determined by the intramolecular energy contributions and not by intermolecular forces.

Linear Viscoelasticity

A simple constitutive relation for the behavior of a linear viscoelastic solid can be obtained by combining Hooke's law for an elastic solid with Newton's law for a viscous liquid. The simplest relation, which assumes that the shear stresses related to strain and strain rate are additive, has the form

$$\sigma_{xy} = Ge_{xy} + \eta \frac{\partial e_{xy}}{\partial t}$$

where G is the shear modulus and η the dynamic viscosity. Initially, it is assumed that strains are small and departures from linearity negligible.

Creep.   Creep is the change in deformation with time under constant stress (see POLYMER CREEP). The deformation under constant stress for a linear viscoelastic solid is compared with that for an elastic solid in Figure 6. The total strain e is the sum of the immediate elastic deformation e_1, the delayed elastic deformation e_2, and the flow e_3. At any given time, the total strain is exactly proportional to the magnitude of the applied stress. The recovery also consists of immediate and delayed components. A simple loading experiment defines a creep compliance $J(t)$, which is a function only of time:

$$J(t) = \frac{e(t)}{\sigma} = J_1 + J_2 + J_3$$

where J_1, J_2, and J_3 correspond with e_1, e_2, and e_3. For cross-linked and highly crystalline polymers and amorphous polymers below their glass transition, e_3 is usually considered to be negligibly small. The division between e_1 and e_2 is arbitrary, and J_1 is the limiting compliance at the shortest experimentally accessible time.

The variation of compliance with time at constant temperature is shown in Figure 7 over a wide time scale for an idealized amorphous polymer with a single

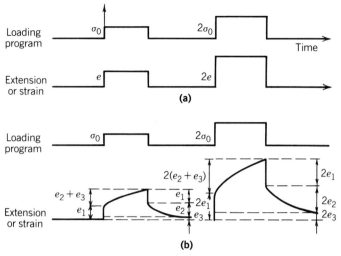

Fig. 6. (a) Deformation of an elastic solid; (b) deformation of a linear viscoelastic solid.

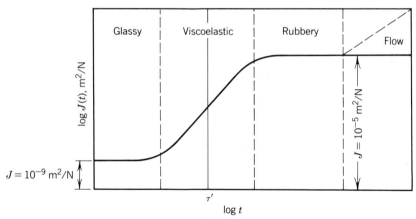

Fig. 7. The creep compliance $J(t)$ as a function of time t; τ' is the retardation time. To convert m^2/N to $in.^2/lb$, multiply by 6897.

relaxation transition. At very short times, the behavior is that of a glassy solid, at very long times that of a rubber, and at intermediate times the compliance lies between these values and is time-dependent. The parameter governing the time scale in creep is known as the retardation time, assumed in Figure 7 to be a single retardation time τ'. The distinction made here between a rubber and a glassy plastic thus depends on the value of τ' at the experimental temperature.

The Boltzmann superposition principle is a useful means of analyzing the creep deformation resulting from several distinct loading or unloading steps. Boltzmann proposed that the creep is a function of the entire loading history, that each loading step makes an independent contribution to the final deformation, and that the final deformation is obtained by the simple addition of each contribution. For the multistage loading program shown in Figure 8, the creep

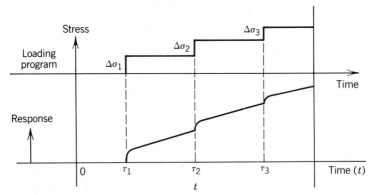

Fig. 8.   The creep behavior of a linear viscoelastic solid.

at time t is given by

$$e(t) = \Delta\sigma_1 J(t - \tau_1) + \Delta\sigma_2 J(t - \tau_2) + \Delta\sigma_3 J(t - \tau_3) \cdots$$

where $J(t - \tau)$ is the creep-compliance function. The expression above can be generalized to give

$$e(t) = \left[\frac{\sigma}{G_u}\right] + \int_{-\infty}^{t} J(t - \tau) \frac{d\sigma(\tau)}{d\tau} d\tau$$

where the immediate elastic contribution is included in terms of the unrelaxed modulus G_u. The form of the integral indicates that the entire loading history must be taken into account. Experimentally, a conditioning procedure may be called for to destroy the long-term memory of the specimen.

The Duhamel integral can be evaluated by treating it as the summation of a number of response terms. For example, two-step loading consists of a stress σ_0 at a time $\tau = 0$, followed by an additional σ_0 at $\tau = t_1$. For this case

$$e(t) = e_1 + e_2 = \sigma_0 J(t) + \sigma_0 J(t - t_1)$$

The additional creep $e_c'(t - t_1)$ produced by the second loading step is

$$e_c'(t - t_1) = \sigma_0 J(t) + \sigma_0 J(t - t_1) - \sigma_0 J(t) = \sigma_0 J(t - t_1)$$

It is identical to the creep that would have occurred had σ_0 been applied at t_1 in the absence of previous loading.

A similar analysis holds when the second load step is of magnitude $a\sigma_0$, where a is a constant. The case of particular interest is $a = -1$, which is equivalent to creep and recovery, ie, the removal of σ_0 at time t_1. The recovery is given by

$$e_r(t - t_1) = \sigma_0 J(t) - [\sigma_0 J(t) - \sigma_0 J(t - t_1)] = \sigma_0 J(t - t_1)$$

and is the difference between the measured response and the creep that would have resulted had unloading not occurred. The recovery is identical to the creep response to σ_0 applied at time t_1, demonstrating that the creep and recovery responses are of identical magnitude.

Stress Relaxation. Stress relaxation, the decay of stress with time at constant strain, is the counterpart of creep. The assumption of linear behavior permits the definition of a stress–relaxation modulus, $G(t) = \sigma(t)/e$. When viscous flow occurs, the stress can decay to zero at sufficiently long times. In the absence of viscous flow, a relaxed modulus G_r is attained at infinite time. The time scale of the viscoelastic behavior may be characterized by a relaxation time τ, which is usually not identical with τ' for the same specimen under creep conditions.

The Boltzmann superposition principle may be applied to stress relaxation to give

$$\sigma(t) = [G_r e] + \int_{-\infty}^{t} G(t - \tau) \frac{de(\tau)}{d\tau} d\tau$$

A formal relation between stress relaxation and creep is given by

$$\int_{0}^{t} G(\tau)J(t - \tau)d\tau = t$$

Viscoelastic Models. A general relation between stress and strain can be expressed in the form $P\sigma = Qe$, where P and Q are linear differential operators with respect to time

$$a_0\sigma + a_1\frac{d\sigma}{dt} + a_2\frac{d^2\sigma}{dt^2} + \cdots = b_0e + b_1\frac{de}{dt} + b_2\frac{d^2e}{dt^2} + \cdots$$

It may be adequate to consider only one or two terms on each side of the equation for representing experimental data obtained over a linear time scale. An equivalent procedure is to construct mechanical models from (massless) Hookean springs and Newtonian dashpots. The simplest models consist of a single spring and a single dashpot. The elements in series form the Maxwell model and, in parallel, the Kelvin (otherwise known as the Voigt) model.

For the Maxwell model (Fig. 9)

$$\frac{de}{dt} = \frac{1}{E_m}\frac{d\sigma}{dt} + \frac{\sigma}{\eta_m}$$

which, when applied to stress relaxation $\left(\dfrac{de}{dt} = 0\right)$, yields

$$\sigma = \sigma_0 \exp\left(-\frac{E_m}{\eta_m}t\right)$$

where σ_0 is the stress at $t = 0$. The model thus indicates an exponential decay of stress with a relaxation time $\tau = \dfrac{\eta_m}{E_m}$. For creep, where $\dfrac{d\sigma}{dt} = 0$, the model predicts that $\dfrac{de}{dt} = \dfrac{\sigma}{\eta_m}$. Neither of these results is applicable to polymers.

For the Kelvin model (Fig. 10)

$$\sigma = E_k e + \eta_k \frac{de}{dt}$$

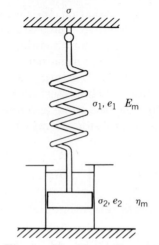

Fig. 9.   The Maxwell model.

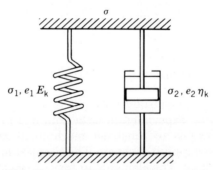

Fig. 10.   The Kelvin or Voigt model.

which reduces to the expression for an elastic solid, $\sigma = E_k e$ when $\dfrac{de}{dt} = 0$. The model is thus incapable of describing stress relaxation.

The model can represent creep, although only in an approximate manner. Under a constant stress σ_0

$$e = \frac{\sigma_0}{E_k}\left[1 - \exp\left(-\frac{E_k}{\eta_k}\right)t\right]$$

with the rate of creep described by the retardation time $\tau' = \dfrac{\eta_k}{E_k}$.

The combination of a Maxwell unit in parallel with a spring (Fig. 11), known as the standard linear solid, yields the relation

$$\sigma + \tau\frac{d\sigma}{dt} = E_a e + (E_m + E_a)\frac{de}{dt}$$

where

$$\tau = \frac{\eta_m}{E_m}$$

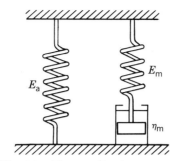

Fig. 11.   The standard linear solid.

Although this model (and an equivalent combination of a Kelvin unit in series with a spring) predicts the main features of both creep and stress relaxation, the exponential response is clearly inadequate. A quantitative description of observed viscoelastic behavior is equivalent either to a large number of Maxwell units in parallel or a large number of Kelvin units in series. Such a model is linked with the previous description in terms of Boltzmann integrals through the concepts of retardation-time and relaxation-time spectra.

For a set of Maxwell units linked in parallel at constant strain e,

$$\sigma(t) = e \sum^{n} E_n \exp\left(\frac{-t}{\tau_n}\right)$$

where E_n and τ_n are the spring constant and relaxation time, respectively, of the nth unit. In an integral representation, the stress–relaxation modulus is given by

$$G(t) = [G_r] + \int_0^\infty f(\tau) \exp\left(\frac{-t}{\tau}\right) d\tau$$

where $f(\tau)$ defines the concentration of Maxwell units with relaxation times between τ and $\tau + d\tau$.

In practice, a logarithmic time scale is more convenient because the rate of relaxation normally decreases with time. Then

$$G(t) = [G_r] + \int_{-\infty}^\infty H(\tau) \exp\left(\frac{-t}{\tau}\right) d(\ln \tau)$$

where $H(\tau)$ is the relaxation-time spectrum, defined in terms of the contributions to stress relaxation associated with relaxation times between $\ln \tau$ and $\ln \tau + d(\ln \tau)$.

Analogously, using Kelvin units in series, a comparable expression is obtained for the creep compliance:

$$J(t) = [J_u] + \int_{-\infty}^\infty L(\tau)(1 - \exp\left(\frac{-t}{\tau}\right) d(\ln \tau)$$

where J_u is the instantaneous (unrelaxed) compliance and $L(\tau)$ is the retardation-time spectrum.

Relaxation- and retardation-time spectra are mathematical descriptions of macroscopic behavior and do not necessarily have a simple interpretation in molecular terms.

Dynamic Mechanical Behavior. When equilibrium conditions have been established in a linear viscoelastic specimen subjected to a sinusoidally varying strain, the stress also varies sinusoidally but out of phase with the strain:

$$e = e_0 \sin \omega t, \qquad \sigma = \sigma_0 \sin (\omega t + \delta)$$

where ω is the angular frequency and δ the phase lag. The stress–strain relation can be defined by a quantity G_1 in phase with, and a quantity G_2, which is 90° out of phase with the strain, where $G_1 = \dfrac{\sigma_0}{e_0} \cos \delta$ and $G_2 = \dfrac{\sigma_0}{e_0} \sin \delta$.

The representation $e = e_0 \exp i\omega t$ and $\sigma = \sigma_0 \exp i(\omega t + \delta)$ leads to the definition of a complex modulus $G^* = G_1 + iG_2$, with $\tan \delta = G_2/G_1$. The storage modulus G_1 defines the energy stored in the specimen by the applied strain; the loss modulus G_2 defines the dissipation of energy. The energy dissipated per cycle is

$$\Delta = \pi G_2 e_0^2$$

Typical values of G_1, G_2, and $\tan \delta$ for a solid polymer are 10^9, 10^7, and 0.01 N/m², respectively (145,000, 1,450, and 1.45×10^{-6} psi).

A similar treatment is used to define a complex compliance $J^* = J_1 - iJ_2$, where $J^* = \dfrac{1}{G^*}$.

The frequency response of complex-modulus behavior for a polymer without flow has the form shown in Figure 12. In terms of the relaxation-time spectrum, it can be shown that

$$G_1(\omega) = [G_r] + \int_{-\infty}^{\infty} \frac{H(\tau)\omega^2\tau^2}{1 + \omega^2\tau^2} \, d(\ln \tau)$$

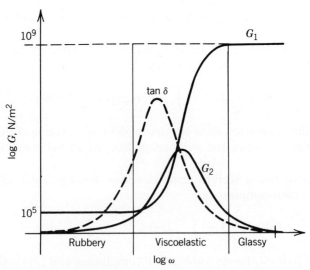

Fig. 12. The complex modulus $(G_1 + iG_2)$ as a function of frequency ω. To convert N/m² to psi, multiply by 1.45×10^{-4}.

$$G_2(\omega) = \int_{-\infty}^{\infty} \frac{H(\tau)\omega\tau}{1 + \omega^2\tau^2} \, d(\ln \tau)$$

similarly, in creep

$$J_1(\omega) = [J_u] + \int_{-\infty}^{\infty} \frac{L(\tau)}{1 + \omega^2\tau^2} \, d(\ln \tau)$$

$$J_2(\omega) = \int_{-\infty}^{\infty} \frac{J(\tau)\omega\tau}{1 + \omega^2\tau^2} \, d(\ln \tau)$$

Approximate methods are often adequate for evaluating the main visco-elastic functions. In the Alfrey approximation the exponential term in the relaxation-spectrum equation is replaced by a step function (11); $\exp\left(\dfrac{-t}{\tau}\right)$ is assumed to be zero up to time $\tau = t$ and 1 at longer times. Then $H(\tau) = -\left[\dfrac{dG(t)}{d(\ln t)}\right]_{t=\tau}$ which is minus the gradient of the stress–relaxation plot. The equivalent approximation in terms of complex modulus is

$$H(\tau) = \left[\frac{dG_1(\omega)}{d(\ln \omega)}\right]_{1/\omega=\tau} = \frac{2}{\pi}[G_2(\omega)]_{1/\omega=\tau}$$

The above relations are indicated in Figure 13, where the longer part of $H(\tau)$ can be found from the stress–relaxation data, and the shorter part of $H(\tau)$

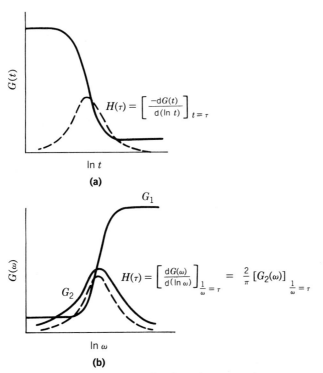

Fig. 13. The Alfrey approximations for the relaxation-time spectrum $H(\tau)$: (a) from the stress-relaxation modulus $G(t)$; (b) from the real and imaginary parts (G_1 and G_2) of the complex modulus $G(\omega)$.

from the dynamic mechanical data. Equivalent relations may be deduced for compliance experiments.

In a complete exposition of the mathematical structure of linear viscoelasticity (12), Laplace and Fourier transforms are used to establish formal connections between the various viscoelastic functions (see also VISCOELASTICITY).

Measurement of Viscoelasticity.   It is desirable to measure viscoelastic behavior over as wide as possible a range of frequency (or time) and temperature, applying a variety of techniques (Fig. 14).

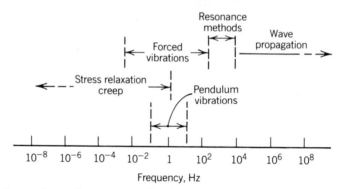

Fig. 14.   Approximate frequency scales for different experimental techniques (13).

Creep and stress-relaxation methods, which cover the very low frequency range, are most effective in revealing the nature of the nonlinear viscoelasticity typical of most polymers other than at very small strains. Specimens used in creep experiments must be mechanically conditioned by subjecting them to several loading and unloading cycles at the highest strains and temperatures. After conditioning, the deformation becomes reproducible since the specimen has lost its long-term memory. It is usually completely recoverable, provided there is an adequate recovery period.

In extensional creep measurements, the change in length is measured between fiduciary marks distant from the ends of the specimen in order to eliminate end effects arising at the clamps. A dead-loading procedure may be used, but at higher strains a device is required for reducing the load in a manner proportional to the decrease in the cross-sectional area. For small strains, a linear differential transformer readout offers advantages over using a traveling microscope for direct measurements of length. For adequately sized specimens of sufficient stiffness, the most accurate technique is to attach an extensometer directly to the specimen. Strain may be converted into a rotation of mirrors or into an electrical signal using a displacement transformer. Creep in torsion can be measured with high accuracy, as the deformation can be used to cause mirror rotation directly and give a large deflection of a light beam.

Stress relaxation can be measured by placing the specimen in series with a spring of stiffness high enough for its deformation to be small compared with that of the polymeric specimen. The spring should incorporate a transducer in order to derive the stress from an electrical measurement. Unlike a conventional tensometer, the strain must be applied as rapidly as possible to minimize the relaxation occurring during loading.

A torsion pendulum gives the most direct measurement of dynamic mechanical properties (qv). It consists of a disk or bar with end weights and is clamped to one end of the test polymer. Following an initial twist, the system undergoes damped sinusoidal oscillations. A viscoelastic solid has a complex torsional modulus, which may be written as $G^* = G_1 + iG_2$. For a cylindrical specimen of mass M, length l, and radius r, the equation of motion can be solved in terms of the observed angular frequency ω and logarithmic decrement Λ to yield for the case of small damping

$$\frac{\pi r^4}{2l} \frac{G_1}{M} = \omega^2, \qquad \frac{\pi r^4}{2l} \frac{G_2}{M} = \omega^2 \frac{\Lambda}{\pi}$$

or

$$\frac{G_2}{G_1} = \frac{\Lambda}{\pi} = \tan \delta$$

The above expressions can be modified for noncircular specimens by using the St. Venant formula. Corrections must be applied if the axial load is large enough to cause extensional creep. A limitation of the method is the small frequency range (0.01–50 Hz); the upper limit is set when the wavelength of the stress waves becomes comparable with the specimen length.

The vibrating-reed method is often used for polymer evaluation in industrial laboratories; a small specimen is vibrated at frequencies of several thousand Hz. One end of a thin, fairly narrow rectangle of the test polymer is clamped in a head driven from a variable frequency oscillator. As the frequency is changed, the specimen passes through a resonance and an optical technique can be used to record the changes in vibrational amplitude.

The modulus ($|E|$) can be obtained from the solution to the elastic problem and tan δ from the resonance curve. For a specimen of length l, thickness h, and density ρ

$$E = 38.24 \frac{l^4 \rho}{h^2} f_r^2$$

where f_r is the resonant frequency; furthermore, tan $\delta = \Delta f/f_r$, where Δf, the band width, is the difference between those frequencies which have an amplitude of vibration $1/\sqrt{2}$ times its maximum value.

Free-vibration and resonance methods are relatively simple to perform and can be highly accurate; however, the main disadvantage of these methods is that the frequency of measurement depends on the stiffness of the specimen which changes with temperature. To determine the frequency and temperature-dependence of viscoelastic behavior, forced-vibration, nonresonant methods are preferable.

For measurements of dynamic extensional modulus, a sinusoidal extensional strain is applied to a specimen in the form of a filament or a thin strip and the stress is measured simultaneously. The viscoelastic behavior is specified from the relative amplitudes of the stress and the strain and from the phase shift between them. It is difficult to control mechanical test equipment operating at lower than 10^{-3} Hz, and an upper frequency limit in the region of 100 Hz (ie, for a typical specimen 10 cm in length) is set in order to keep the length of the specimen small compared to the wavelength of the stress waves. A fixed alternating strain of 0.5% is typical, which is close to the limit at which the behavior approximates linear viscoelasticity. This strain is superimposed on an initial extensional strain sufficient to prevent the tension from dropping to zero. An apparatus that uses a mechanical drive to produce the alternating strain and a Servomex transfer-function analyzer (TFA) to make simultaneous comparison of stress and strain is illustrated schematically in Figure 15. Similar equipment operating over a wide range of temperature is available from several manufacturers.

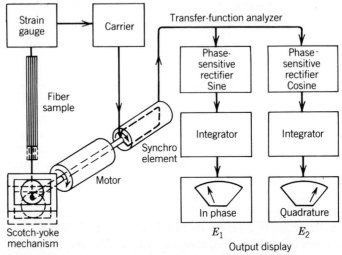

Fig. 15.   A transfer-function analyzer (TFA) dynamic mechanical apparatus (14).

Wave-propagation methods are useful at high frequencies. A movable crystal pickup detects the signal amplitude and phase along the length of a long filament, one end of which is attached to and driven by a vibrating diaphragm. The quantities measured are the ratios of the amplitudes of displacements at the diaphragm and the pickup and the phase difference between these displacements. A typical experiment is illustrated in Figure 16, which shows damped oscillations about the line $\theta = kl$, where k, the propagation constant, is the ratio of the angular frequency ω to the longitudinal-wave velocity c. The attenuation coefficient α can be determined from the amplitude–distance relation. For a specimen of density ρ with a small attenuation coefficient, the in-phase modulus E_1 is given by $E_1 = \dfrac{\omega^2}{k^2} \rho$ and $\alpha = \dfrac{\omega}{2c} \tan \delta$, giving $E_2 = \dfrac{2\alpha c^3}{\omega} \rho$. Other wave-propagation methods are described in Refs. 16 and 17.

Linear Viscoelastic Behavior.   Early studies concerned amorphous polymers because with such materials changes of behavior in terms of frequency and temperature are most pronounced. Particularly noteworthy is a survey of the complex modulus and compliance of polyisobutylene (PIB) over a wide range of frequency in the early 1950s (18). As the frequency is reduced, the material passes

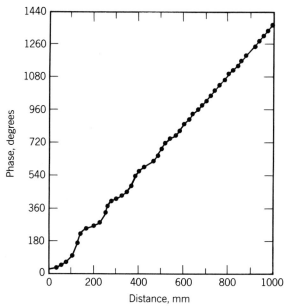

Fig. 16. The variation of phase angle with distance along a polyethylene monofil-ament for transmission of sound waves at 3000 Hz (15).

from a glassy state through viscoelastic and rubberlike ranges to a flow region (Fig. 17). The relaxation- and retardation-time spectra (Fig. 18) can be approx-imated by so-called wedge and box distributions useful for numerical calculations. Cross-linking modifies the viscoelastic behavior by preventing flow. Except at very long times, a similar effect results from an increase in molecular weight sufficient to give rise to a significant degree of molecular-chain entanglement.

Investigations into the temperature-dependence of viscoelasticity have con-siderable practical importance and provide evidence toward a molecular inter-pretation of viscoelastic behavior as the material changes from a glassy to a rubberlike state. The main relaxation is the glass transition, characterized by the glass–rubber transition temperature T_g, during which the stiffness of amor-phous polymers may decrease by a factor of 10^4 as shown for PIB in Figure 17. At lower temperatures, there are usually several subsidiary relaxation transi-tions involving much smaller changes of modulus; they are attributable to fea-tures such as side-group motions. In crystalline polymers, the stiffness at T_g usually changes by a factor of about 10; the relaxation-time spectrum is very broad. At high temperatures the crystalline regions severely curtail molecular mobility, as can be seen from Figure 19, where the behavior of crystalline and amorphous poly(ethylene terephthalate) (PET) is compared.

A single experimental technique is inadequate to investigate the full range of the relaxation-time spectrum at a single temperature, but a change in tem-perature may bring relaxation features of interest within an accessible time scale. Time–temperature equivalence in its simplest form implies that the viscoelastic behavior at two temperatures can be related by a change in time scale only, indicated in Figure 20 by the relative displacement of two creep-compliance plots

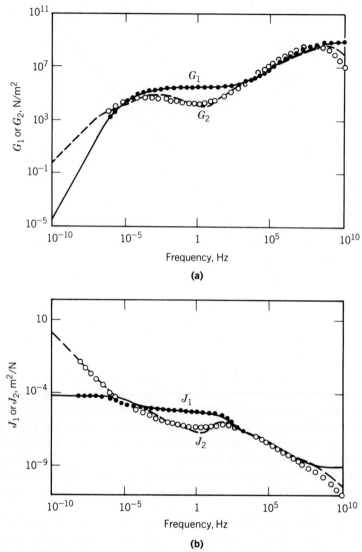

Fig. 17. (a) Complex shear modulus and (b) complex shear compliance for polyisobutylene reduced to 25°C. Points are from averaged experimental measurements; curves are from a theoretical model (19). To convert N/m² to psi, multiply by 1.45×10^{-4}.

by an amount log a_T. In certain cases, where it is necessary to take into account changes in the relaxed and unrelaxed compliances with temperature, a vertical shift in log (creep compliance) is needed in addition to displacement along the log (time) axis (21). For time–temperature superposition to be exact, the spectrum of relaxation times must shift to shorter times as a unit when the temperature is increased. Materials with this characteristic are said to be thermorheologically simple. For such polymers, it is possible to solve complex problems involving viscoelastic deformation under variable temperature conditions.

The simplest treatment of temperature dependence is through transition-

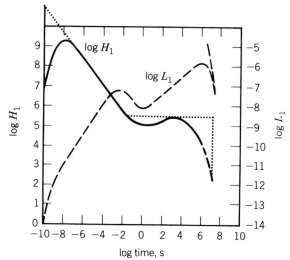

Fig. 18. Approximate distribution functions of relaxation (H_1) and retardation (L_1) times for polysiobutylene (18).

state (or barrier) theories. These theories are developed from the model in which two reacting molecules must first form a transition state which then decomposes into the final products.

In terms of the Arrhenius equation, the frequency of molecular conformational changes depends on the barrier height through

$$\nu = \nu_0 \exp\left(-\frac{\Delta H}{RT}\right)$$

where ΔH is the activation enthalpy (or energy) and R is the molar gas constant. Application to the two frequency curves of Figure 20 predicts that

$$\frac{\nu_1}{\nu_2} = \frac{\exp\left(-\dfrac{\Delta H}{RT_1}\right)}{\exp\left(-\dfrac{\Delta H}{RT_2}\right)}$$

or

$$\log\left(\frac{\nu_1}{\nu_2}\right) = \log a_T = \frac{\Delta H}{R}\left(\frac{1}{T_2} - \frac{1}{T_1}\right)$$

In practice the model is inadequate to describe the glass transition (qv), as only some of the small localized molecular relaxations of polymers can be described in terms of a single activation energy. For these minor relaxations, it may be appropriate to use a site model developed originally to describe the dielectric behavior of crystalline polymers (22); two sites are considered separated by an equilibrium free energy difference $\Delta G_1 - \Delta G_2$ per mole, where ΔG_1 is the height of the barrier of the first side and ΔG_2 that on the second. An applied stress changes the populations in the two sites and this change is assumed to

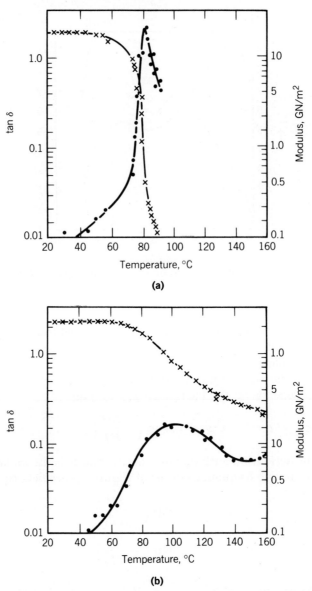

Fig. 19. Tensile modulus and loss factor tan δ for (**a**) unoriented amorphous poly(ethylene terephthalate) and (**b**) unoriented crystalline poly(ethylene terephthalate) as a function of temperature at ~1.2 Hz (20). To convert GN/m² to psi, multiply by 145,000.

relate directly to the strain. On a molecular level, the process may represent an extension of the main chain through internal-bond rotations.

It is assumed that the stress causes a small linear shift in the free energies of the sites in such a way that the changes in the barrier height are

$$\delta G_1 = \lambda_1 \sigma \qquad \text{and} \qquad \delta G_2 = \lambda_2 \sigma$$

where λ_1 and λ_2 are constants with the dimensions of volume. Calculation showing that the model bears a formal resemblance to a Kelvin unit yields a retardation

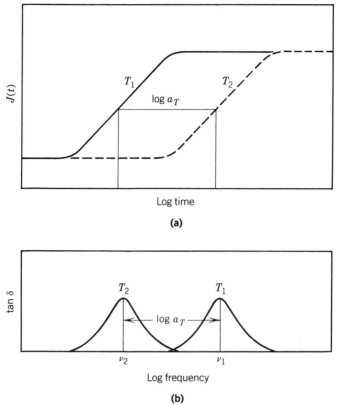

Fig. 20.   The simplest form of time–temperature equivalence for (**a**) compliance $J(t)$ and (**b**) loss factor tan δ (schematic).

time

$$\tau' = \frac{1}{A} \exp\left(\frac{\Delta G_2}{RT}\right)$$

The magnitude of the relaxation is proportional to

$$p\left[\frac{\exp\left[-(\Delta G_1 - \Delta G_2)/RT\right]}{(1 + \exp\left[-(\Delta G_1 - \Delta G_2)/RT\right])^2}\right]\frac{(\lambda_1 - \lambda_2)^2}{RT}$$

where R is the gas constant, T, the temperature, and p, the number of species per unit volume. This expression passes through a maximum when the free-energy difference $(\Delta G_1 - \Delta G_2)$ and RT are of the same order.

Compliance–frequency plots for an amorphous polymer (Fig. 21) may be superimposed by applying a horizontal displacement and, if necessary, a small vertical shift to account for the change in density with temperature. The latter is related to changes in equilibrium with temperature in a manner that is in accord with the theory of rubber elasticity. The master curve obtained (Fig. 22) covers a wide range of frequencies and allows the retardation-time spectrum to be calculated. The variation factor of the horizontal shift factor log a_T with temperature is shown in Figure 23.

Williams, Landel, and Ferry showed that the shift factor–temperature re-

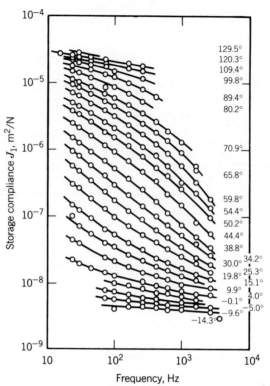

Fig. 21. Storage compliance of poly(n-octyl methacrylate) in the glass-transition region, plotted against frequency at the temperature (°C) indicated (23). To convert m²/N to in.²/lb, multiply by 6897.

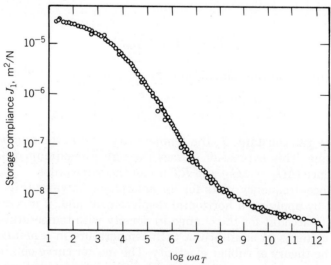

Fig. 22. Composite curve obtained by replotting the data of Figure 21 with appropriate shift factors over an extended frequency scale at a single temperature (23). To convert m²/N to in.²/lb, multiply by 6897.

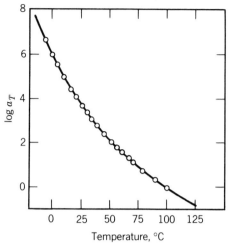

Fig. 23. Temperature dependence of the shift factor a_T used in plotting Figure 22; the curve is the WLF equation with a suitable choice of T_g (23).

lation was approximately identical for all amorphous polymers (24) having the form

$$\log a_T = \frac{C_1(T - T_s)}{C_2 + (T - T_s)}$$

where C_1 and C_2 are constants and T_s is a reference temperature appropriate for a particular polymer. This WLF equation which holds over the range $T = T_s \pm 50°C$ was originally derived empirically, but was later explained in terms of the concept of free volume. In terms of the glass transition T_g, measured from near-equilibrium dilatometric measurements, the WLF equation becomes

$$\log a_T = \frac{C_1^g(T - T_g)}{C_2^g + (T - T_g)}$$

where C_1^g and C_2^g are new constants and $T_g = T_s - 50°C$.

In analogy with low molecular weight liquids, the total volume V of an amorphous polymer is taken as the sum of an occupied volume, which increases linearly with temperature, plus the free volume V_f, which is constant at low temperatures but increases linearly with temperature above T_g. An observed discontinuity of expansion coefficient at T_g corresponds to the onset of expansion of V_f. If the polymer is represented by a series of Maxwell units and the change in modulus with temperature is assumed to be small compared with the change in viscosity, it can be shown that the WLF equation has a form analogous to that of the Doolittle viscosity equation for liquids, $\eta = a \exp (bV/V_f)$, where a and b are constants.

From the WLF equation, it is found that at the glass-transition temperature most amorphous polymers have a fractional free volume of 0.025 ± 0.003 with an average value of the temperature coefficient of free volume being $4.8 \times 10^{-4} \text{ K}^{-1}$.

In a transition-state model, the assumption was made that the required unit of structure can move when the local fractional free volume exceeds some critical value (25). This gives an expression of the form of the WLF equation. The WLF equation may be rewritten in terms of viscosity as

$$\log \eta_T = \log \eta_{T_g} + \frac{C_1^g(T - T_g)}{C_2^g + (T - T_g)}$$

where η_T and η_{T_g} are the viscosities at T and T_g. In this form the equation implies that at a temperature $T_2 = T_g - C_2^g$, where $C_2^g = 51.6$ K, the viscosity is infinite. An attempt has been made to modify the theory, allowing the changes in free volume with temperature to relate to a discontinuity at T_2 rather than T_g; T_2 represents a true thermodynamic-transition temperature. A modified transition-state theory has since been developed in which the frequency of molecular jumps relates to the cooperative movement of a group of segments of the chain (26). The number of segments acting cooperatively is calculated from statistical thermodynamic considerations.

Although there is a considerable amount of support for free-volume theories, it has been shown that the relaxations of poly(methyl acrylate) and poly(propylene oxide) behave rather similarly under both constant pressure and constant volume conditions (27). Since the amount of occupied volume should increase with temperature, the results for the two cases were expected to be very different.

Normal-mode Theories.   The normal-mode theories that attempt to predict the relaxation spectrum for amorphous polymers, as well as time–temperature equivalence, are based on the idea of representing the motion of polymer chains in a viscous liquid by a series of linear differential equations (28–30). Although essentially dilute-solution theories, they can be extended to predict the behavior of the pure polymer.

Each molecular chain is considered an assemblage of submolecules consisting of beads connected by springs whose behavior is that of a freely jointed chain. Only the beads interact with the solvent molecules. The forces acting on a displaced bead are due to the viscous interaction with the solvent molecules together with the tendency of chains to return to a state of maximum entropy. Consideration of the restoring force when a bead is displaced from its equilibrium position leads to the expression

$$\eta \dot{x}_i + \frac{3kT}{zl^2}(2x_i - x_{i-1} - x_{i+1}) = 0$$

where η is the coefficient of friction defining the viscous interaction between the beads and the solvent, l, the length of each link in a chain, z, the number of links in a submolecule, and $\dot{x}_i$, the time differential of the displacement of a bead situated at the point $(x_i y_i z_i)$ between the $(i - 1)$th and $(i + 1)$th submolecules.

There are $3m$ of these equations, where m is the number of submolecules, each of which is equivalent to the equation of a Kelvin unit ($\eta \dot{e} + Ee = 0$). These equations, which define a set of compliances and moduli, can be uncoupled using a normal coordinate transformation, giving, for example, a stress-relaxation modulus

$$G(t) = NkT \sum_{p=1}^{m} \exp\left(\frac{-t}{\tau_p}\right)$$

and a real part of the complex modulus

$$G_1(\omega) = NkT \sum_{p=1}^{m} \left(\frac{\omega^2 \tau_p^2}{1 + \omega^2 \tau_p^2} \right)$$

where there are N molecules per unit volume and τ_p is the relaxation time of the pth mode. The submolecule is considered the shortest length that can undergo relaxation, and the motions of segments within submolecules are ignored; therefore, the theory is applicable only when $m \gg 1$. Under this limitation

$$\tau_p = \frac{n^2 l^2 \eta_0}{6\pi^2 p^2 kT}$$

where $\eta_0 = \eta/z$ is the friction coefficient per unit link. The relaxation times depend on temperature through $1/T$, through nl^2, which defines the mean-square separation of the chain ends, and through changes in η_0. The last of these varies rapidly with temperature and is primarily responsible for changes in τ_p. Each τ_p has the same temperature dependence; thus the theory satisfies the requirements of thermorheological simplicity and gives theoretical justification for time–temperature equivalence.

Later theories give a modified relaxation spectrum by taking into account the fact that the solvent may be affected by movement of the polymer molecules and by considering the hydrodynamic interaction between the moving submolecules. For the longer relaxation times, which involve the movement of submolecules, these theories have been confirmed not only for dilute polymer solutions, but also for solid amorphous polymers at temperatures above the glass transition.

Dynamics of Entangled Polymers. A more recent approach is based on the constraint that, as the molecular chains cannot pass through one another, each chain is effectively confined within a tube (31–33). The center line of the tube (the primitive path) defines the overall path of the tube in space (Fig. 24). There are so many entanglements that even though all the chains are moving at any one time, a tube is well-defined. The motions, named reptation because

Fig. 24. A chain segment in dense rubber. A and B denote the cross-linked points and the dots represent other chains. Because of the entanglements, the chain is confined to the tubelike region defined by the broken lines. The central line shows the primitive path (32).

of their snakelike nature, are of two kinds: short-term wriggling motions corresponding to the migration of a molecular kink along the chain, and larger-scale movement of the polymer chain as a whole. The longest relaxation time associated with the first motion is proportional to molecular weight squared ($\tau_d \propto M^2$), but the characteristic time for the movement of the whole chain is $\tau_R \propto M^3$. For long enough chains, τ_R and τ_d are sufficiently separated to represent distinctly different relaxation times.

An equivalent model is the slip-link network where the primitive path is defined by small rings through which the chains can pass freely (Fig. 25). This model represents long-term stress relaxation by

$$\sigma \propto F(e) \sum_{p \text{ odd}} \frac{8}{p^2 \pi^2} \exp\left[\frac{-tp^2}{\tau_R}\right]$$

where $F(e)$ is a function of strain, p, the mode number, and τ_R, the relaxation time corresponding to the diffusion of the chain out of its original tube or slip-links. The equation shows separability between strain and time variables, which is a feature of the nonlinear BKZ theory discussed later. Theory and experiment

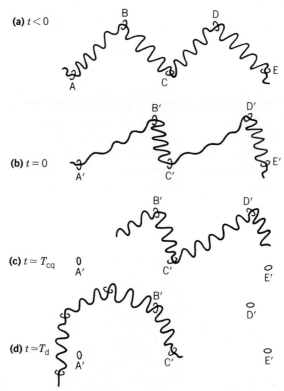

Fig. 25. Relaxations after sudden deformation: (**a**) initial equilibrium state; (**b**) immediately after the deformation each part of the chain is stretched or compressed; (**c**) after the first relaxation process the primitive chain recovers its equilibrium arc length, but the conformation of the primitive chain is still in a nonequilibrium state; (**d**) in the second equilibrium process, the chain disengages from the deformed slip links and returns to the final equilibrium state (32).

appear to be in good agreement except for the effects of molecular weight distribution, where the longest chains are probably modeled better by the Rouse model in which tube constraints are absent.

Relaxation Transitions and Molecular Structure

Amorphous Polymers. For both crystalline and amorphous polymers, relaxation transitions are customarily labeled in alphabetical order, α, β, γ, etc, with decreasing temperature irrespective of their molecular origin. For amorphous polymers, the α-relaxation is the glass transition. It is accompanied by a large change in stiffness and loss modulus as seen in Figure 26 for poly(methyl methacrylate) and some related polymers. Comparative studies on similar polymers using nmr and dielectric techniques have shown the β-relaxation to be associated with side-chain motions of the ester group. The γ- and δ-relaxations involve motion of the methyl groups attached to the main and side chain, re-

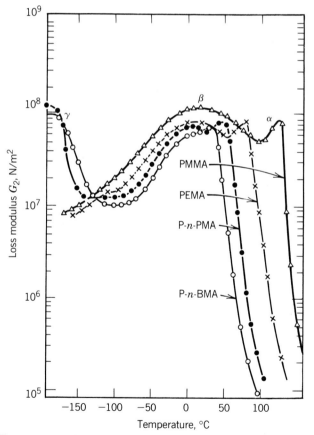

Fig. 26. Temperature dependence of loss modulus G_2 for poly(methyl methacrylate) (PMMA), poly(ethyl methacrylate) (PEMA), poly(n-propyl methacrylate) (P-n-PMA), and poly(n-butyl methacrylate) (P-n-BMA) (34). To convert N/m² to psi, multiply by 1.45 × 10^{-4}.

spectively. The assignment of a relaxation process is, however, frequently less straightforward than in the present example.

The influence of features such as main-chain structure, side groups, and plasticizers on the glass transition can be treated in terms of molecular flexibility or changes in free volume. The transition temperature is increased by an inflexible group in the main chain, eg, a terephthalate residue, or as a side group. Increasing main-chain polarity has a similar effect, presumably because of the increase in intermolecular forces.

Molecular weight does not affect the dynamic mechanical properties of glassy polymers below T_g, although at low molecular weights the temperature of the transition is reduced because of the increase in free volume resulting from a large number of chain ends. Molecular weight has a large effect in the glass-transition range, where the behavior transforms from viscous flow to rubberlike elasticity with increasing chain length, the result of chain entanglements preventing irreversible flow. Increased chemical cross-linking raises T_g and broadens the transition (Fig. 27); in highly cross-linked materials there is no glass transition. This behavior is a consequence of the chains being brought close together reducing the free volume.

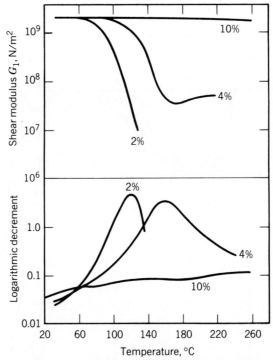

Fig. 27. Shear modulus G_1 and logarithmic decrement of a phenol–formaldehyde resin cross-linked with hexamethylenetetramine at stated concentrations (35). To convert N/m² to psi, multiply by 1.45×10^{-4}.

The mechanical properties of blends and graft copolymers are determined primarily by mutual solubility; if the two polymers are completely soluble in one another, the properties of the mixture resemble those of a random copolymer of the same composition (Fig. 28). A free-volume model of such copolymers gives (37)

$$\frac{1}{T_g} = \frac{1}{(W_1 + BW_2)} \frac{W_1}{T_{g1}} + \frac{BW_2}{T_{g2}}$$

where W_1 and W_2 are the weight fractions of the two monomers whose homo-polymers have transitions at T_{g1} and T_{g2}; B is a constant close to unity. If the two polymers in a mixture are insoluble, they exist as two separate phases and two glass transitions are observed.

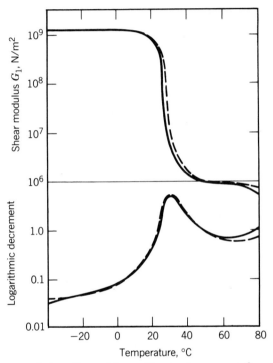

Fig. 28. Shear modulus G_1 and logarithmic decrement for a miscible blend of: ——, poly(vinyl acetate) and poly(methyl acrylate); ------, a copolymer of vinyl acetate and methyl acrylate (36). To convert N/m² to psi, multiply by 1.45×10^{-4}.

Plasticizers, which must be soluble in the polymer, lower T_g by facilitating conformational changes. They also broaden the loss peak with the width increasing as the plasticizer becomes a poorer solvent. Poly(vinyl chloride) (Fig. 29) is a well-known example, where the rigid RT properties of the base polymer contrast with the plasticized materials used for flexible sheeting and footwear. The γ-relaxation in amorphous and some crystalline polymers has been associated with the so-called crankshaft mechanism attributed to a restricted motion of several successive CH_2 groups on a linear part of the main chain.

Crystalline Polymers. Interpretation of the viscoelastic behavior of crystalline polymers is yet at a very speculative stage and is much influenced by the simplified structural model that is selected. Appreciable progress has been made in unraveling the complex relaxation processes in polyethylene, and it is mainly

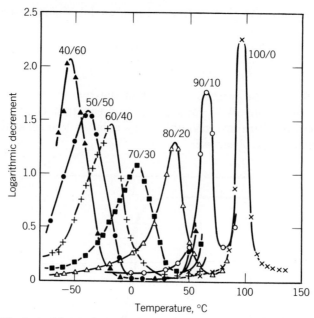

Fig. 29. The logarithmic decrement of poly(vinyl chloride) plasticized with various amounts of di(ethylhexyl) phthalate (38).

data on this polymer that are discussed here. A plausible though unproven assumption is that the phase angle δ, or rather tan δ, is a good measure of the relaxation strength.

In the simplest model, some relaxation transitions are associated with crystalline regions and some with amorphous regions. In polytetrafluoroethylene the γ- and, less convincingly, the α-transitions, which decrease in magnitude with increasing crystallinity, are identified with processes in the amorphous regions (39); the β-relaxation exhibits the opposite behavior. In PET the effect of crystallinity on the β-relaxation is small, but striking changes occur for the α-relaxation (Fig. 30): with increasing crystallinity the loss peak decreases in height, broadens, becomes asymmetrical, and moves to higher temperatures. Although the relaxation occurs in the amorphous regions, the molecular movements associated with the process seem to be constrained by the presence of crystallites. This has been confirmed by nmr measurements of molecular mobility.

The presence of chain folding implies that relaxations within the crystalline phase can involve both cooperative motions of the molecular chains along the length of the crystallite, which are related to the lamellar thickness, and also motions associated with defects, eg, end groups, within the crystal. In addition, there may be relaxations, such as the crankshaft mechanism, which occur in both the crystalline and amorphous phases, and also relaxations, such as the motion of a chain fold, associated with specific morphological features.

Relaxation Processes in Polyethylene. The temperature variation of tan δ at 1 Hz is shown in Figure 31 for low density polyethylene (LDPE) with about three short side branches per 100 carbon atoms and high density polyethylene (HDPE) with no more than five branches per 1000 carbon atoms. Both polymers

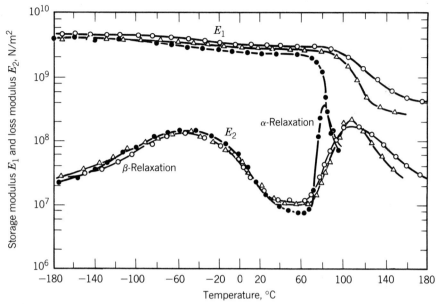

Fig. 30. Storage modulus E_1 and loss modulus E_2 as a function of temperature at 138 Hz for poly(ethylene terephthalate) specimens of differing degrees of crystallinity. ●, 5%; △, 34%; and ○, 50% (40). To convert N/m² to psi, multiply by 1.45×10^{-4}.

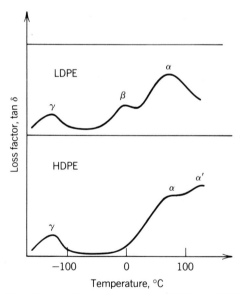

Fig. 31. Relaxation processes in LDPE and HDPE.

exhibit a low temperature γ-relaxation, decreasing as crystallinity increases, which is assigned to an amorphous relaxation. The β-peak, almost absent in the unbranched material, has been associated with the relaxation of side groups or short branch points. The α-peak has been assigned to a crystalline relaxation because on chlorination it disappears at the degree of chlorination required to remove any evidence of crystallinity by x-ray diffraction (41). In HDPE the high temperature relaxation behavior depends to some extent on the experimental technique, but appears to consist of at least two relaxation processes with different activation energies.

Relaxations in HDPE can be understood by comparing the bulk polymer with mats of solution-crystallized polyethylene in which the molecular-chain axes are approximately perpendicular to the plane of the mat (42). The α'-relaxation was absent as the orientation was unfavorable. Annealing and increasing lamellar thickness reduced the α-relaxation, which was associated with chain folds, but increased the γ-relaxation, presumably because more defects were generated.

In a theoretical analysis based on a site model for mechanical relaxations, a chain-relaxation process was considered to involve the rotation of a chain through 180° together with a simultaneous translation along its axis to a new equilibrium position (43). The theory was tested against a series of n-alkanes in which the temperatures of maximum loss for the α- and γ-relaxations increased with increasing numbers of $-CH_2$ groups, but the authors found it difficult to accept the conclusion that the molecular processes were the same in all these materials.

Mechanical relaxations can be dependent on the direction of the applied stress in both low and high density polyethylene, as shown on both uniaxially drawn sheets and single-crystal texture material with orthorhombic symmetry (44–46). In cold-drawn LDPE, the maximum loss occurs in 45°C specimens (Fig. 32), suggesting a relaxation involving shear parallel with the draw direction in a plane containing that direction. This relaxation moves to 70°C in annealed material, where a second relaxation with different anisotropy is seen around 0°C. It was proposed that this second relaxation process is an interlamellar shear process. It corresponds to the β-relaxation in LDPE and occurs also for the α-relaxation in drawn and subsequently annealed HDPE. The mechanical similarity of the two processes does not indicate that they are associated with identical molecular processes. It is likely, as suggested earlier, that the α-relaxation in HDPE requires mobility of the fold surfaces, whereas the β-relaxation in LDPE requires mobility of chains close to branch points. In LDPE, the α-relaxation is

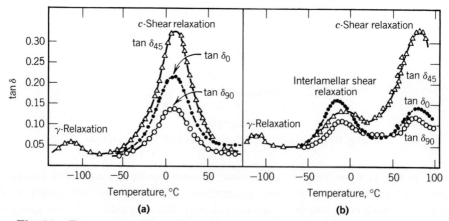

Fig. 32.   Temperature dependence of tan δ in three directions in (**a**) cold-drawn and (**b**) cold-drawn and annealed sheets of LDPE at ~500 Hz (45).

related to the orientation of the crystalline regions, and it seems probable that the chains involved thread these regions, forming interlamellar ties that transmit the stresses through the bulk of the polymer. The nature of relaxations in HDPE and LDPE are thus quite different, which indicates that the conventional labeling of relaxations by α, β, γ, etc, in order of decreasing temperature, may be misleading when comparing closely related polymers.

Nonlinear Viscoelasticity

Even at small strains, the tests of linearity required by the Boltzmann superposition principle may be satisfied only for short times. Beyond these limits the behavior is nonlinear viscoelastic and, at present, there is no adequate comprehensive mathematical representation that also provides physical insight into this form of response.

Engineering Treatment. The requirement here is the ability to predict performance in terms of a limited number of initial experiments. Descriptive empirical relations are adequate, which need not have any physical significance. A three-dimensional stress–strain–time relation can be constructed from a series of creep measurements under constant stress. It is useful to combine two types of measurement at a fixed temperature (47): the isochronous curve relating stress to strain at a given time after loading and at least two creep curves at different stresses (Fig. 33). Although the procedure requires only a modest experimental effort, the data obtained are inadequate to predict behavior in tests, such as recovery, following the removal of load (Fig. 34).

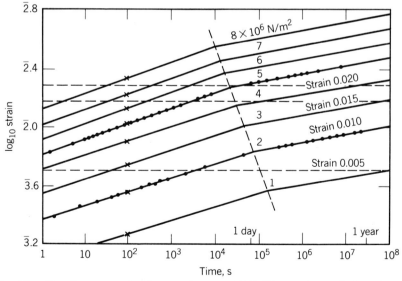

Fig. 33. Tensile creep of polypropylene at 60°C. The stress and time dependence are approximately separable, allowing creep curves at intermediate stresses to be interpolated from a knowledge of: ●, two creep curves and ×, the isochronous stress–strain relationship (47).

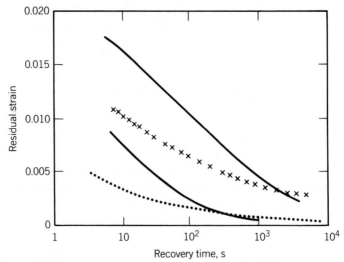

Fig. 34.   Recovery of polypropylene at 20°C after creep under a stress of 20 MN/m² (2900 psi): x, after creep for 1000 s; ●, for creep after 100 s; ——, recovery predicted for linear viscoelastic behavior (47).

Additional parameters can be introduced: the fractional recovery FR and the reduced time t_R.

$$FR = \frac{\text{strain recovered}}{\text{maximum creep strain}}$$

When this is used as the parameter, the recovery curves from creep under different stresses for the same period of time are brought near coincidence.

The reduced time, defined by

$$t_R = \frac{\text{recovery time}}{\text{creep time}}$$

is used in relation to a series of recovery curves from creep under a given stress for variable times. Again, apparently disparate data can be brought into near coincidence (Fig. 35).

Such recovery behavior in terms of FR and t_R is expected if linear superposition can be applied to a material which obeys a power-law relation of the form $e_c(t) = At^n$, where A is a nonlinear function of stress. The essential idea behind these simplifications is that the nonlinearity in stress and time can be separated.

A number of separable stress and time functions have been suggested to describe the creep behavior at strains on the order of a few percent (48–50). However, good agreement with creep data has been obtained only for single-step loading, despite the necessity for introducing a significant number of material-dependent parameters. The Andrade creep law may hold for several polymers and gels over a limited range of time (51). Such empirical methods have severe limitations since they cannot provide a general representation for complicated loading programs, and creep data cannot be related to stress relaxation and dynamic mechanical experiments.

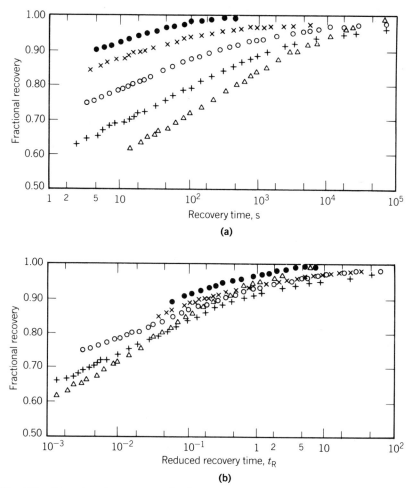

Fig. 35. Fractional recovery of polypropylene after tensile creep under a stress of 10 MN/m² (1450 psi) for various times. Creep times: ●, 50 s; x, 100 s; ○, 1000 s; +, 3000 s; △, 10,000 s. (**a**) Fractional recovery as a function of recovery time; (**b**) fractional recovery as a function of reduced recovery time (47).

Rheological Treatment. Several attempts have been made to extend the formal descriptions of linear viscoelastic behavior. In one approach the separability of stress, strain, and time in the functional representations is preserved, which can lead to single-integral representations consistent with rigorous continuum mechanics. A further development is based on a multiple-integral method.

A semiempirical extension of linear viscoelastic theory describes the large-strain behavior of elastomers (52). A Maxwell model, for which the strain rate is constant, is generalized for a continuous distribution of relaxation times to yield the constant strain-rate modulus

$$F(t) = \frac{1}{t} \int_{-\infty}^{\infty} \tau H(\tau) \left[1 - \exp\left(\frac{-t}{\tau}\right) \right] d(\ln \tau) + E_e$$

where E_e is the equilibrium modulus. For large strains

$$F(t) = \frac{g(e)\sigma(e,t)}{e}, \text{ ie, } \log F(t) = \log \left[\frac{g(e)}{e} \right] + \log \sigma(e,t)$$

where $g(e)$ is a function of strain that approaches unity as the strain approaches zero.

Stress–strain curves at each strain rate are analyzed by selecting the results for each value of strain and constructing plots of $\log \sigma$ vs $\log t$. These plots are parallel straight lines, displaced from one another by the factor $\log [g(e)/e]$. From time–temperature equivalence, $g(e)/e$ would be expected to be independent of temperature also, which is found to be the case over a wide range. For extensions up to 100%, $g(e)/e$ has a simple form, as $\lambda\sigma = Ee$, where λ is the extension ratio.

At higher strains the empirical formula $\sigma = E \dfrac{e}{\lambda^2} \exp A\left(\lambda - \dfrac{1}{\lambda}\right)$ is used.

The results imply that nonlinearity can be attributed to large strains, and that time and strain dependence are separable; similar treatments have been used successfully (53,54).

The creep and recovery of plasticized poly(vinyl chloride) have been analyzed (55); the form of the recovery curve differs from that in creep. If $\dfrac{1}{3}\left(\lambda - \dfrac{1}{\lambda^2}\right)$, a quantity equivalent to the Lagrangian strain measure in the theory of finite elasticity, is used as a measure of the deformation, both creep and recovery and creep curves at different strains can be described by a single time-dependent function. Experiments with textile fibers in the early 1940s indicated that nonlinear effects can occur at small strains (56). With nylon and cellulosic fibers, creep-compliance curves did not coincide for different stresses, although creep and recovery curves for a given stress did. In addition, the compliance at the shortest time of measurement was identical for all stresses. A modified superposition principle was proposed,

$$e(t) = \frac{\sigma}{E} + \int_{-\infty}^{t} \frac{df(\sigma)}{d\tau} (t - \tau) d\tau$$

where $f(\sigma)$ was an empirical material-dependent function of stress. However, such a simple relation is not adequate for all textile fibers and does not describe the behavior in more complicated loading programs.

A single-integral representation has been developed (BKZ theory) in forms suitable for a solid (ie, an elastomer at short times and low temperatures) and for a fluid (an elastomer at long times and high temperatures) (57); in both cases incompressibility is assumed. The first form of BKZ theory derives the stress relaxation behavior with two simplifying assumptions: there exist only single-integral terms of the form

$$\int_{-\infty}^{t} A(t - \tau) f[E(\tau)] d\tau$$

where $E(\tau)$ is a generalized measure of strain; and in equilibrium at long times, the stress–strain relation is consistent with a strain–energy function U, in such a way that $U = MJ_1 + \frac{1}{2}A_1 J_1^2 + A_2 J_2$, where M is a material-dependent constant, A_1 and A_2 are the limiting values of the two time-dependent terms, and $J_1 = \frac{1}{2}(I_1 - 3)$, $J_2 = \frac{1}{4}(I_1 - 3)^2 + (I_1 - 3) - \frac{1}{2}(I_2 - 3)$; I_1 and I_2 are the strain invariants already discussed. The stress relaxation at extension ratio λ is given by

$$\frac{\sigma(t)}{\lambda^2 - \lambda^{-1}} = (\lambda^2 - 1)[\tfrac{1}{2}A_1(t) + A_2(t)] + \frac{1}{\lambda}[A_1(t) + A_2(t)] + A_3 - A_1(t)$$

The second form of BKZ theory describes an incompressible elastic fluid with an elastic potential (58). The stress depends on the deformation history at all previous times, and the elastic potential depends on the strain measured at time τ only through the strain invariants, a form that conforms to finite elasticity theory. The stress–relaxation behavior is then given by

$$\frac{\sigma(t)}{\lambda^2 - \lambda^{-1}} = (\lambda^2 - 1)[\tfrac{1}{2}A_1(t) + A_2(t)] + \frac{1}{\lambda}[A_1(t) + A_2(t)] + A_3(t) - A_1(t)$$

which differs from the previous expression in replacing the constant term A_3 by the single-integral function $A_3(t)$. The second expression provides good agreement with the experimental data on polyisobutylene, vulcanized butyl rubber, and plasticized PVC. The BKZ theory has also been found capable of correlating results from biaxial strain and simple shear, and with polyisobutylene solutions has related torsion pendulum and capillary rheometer experiments.

In an extension of the second form of BKZ theory, stress relaxation is described by 12 functions of the single-integral type and three steady-state coefficients (59). This representation is called finite linear viscoelasticity. A polyurethane specimen satisfies a simple representation where only four independent relaxation functions are required, involving only the strain invariant I_1. An ethylene–propylene rubber is more complicated, involving a linear combination of the strain invariants I_1 and I_2 with five relaxation functions required. In its simplest form, the Lianis theory is equivalent to the BKZ treatment (60). There have been several other related treatments, particularly one in which time and strain again are separated (61).

The fibers studied in Refs. 55 and 56 appear to obey time–temperature equivalence and could be approximated by a model of a standard linear solid with an activated dashpot. Incorporation of the Boltzmann single-integral representation and temperature dependence into a comprehensive scheme has been attempted (62). For creep after single-step loading, the nonlinear compliance is given by

$$J_n(t) = \frac{e(t)}{\sigma} = g_0 J_{\mathrm{u}} + g_1 g_2 \, \Delta J(t/a_\sigma)$$

where J_{u} is the immediate elastic compliance, and a_σ, g_0, g_1, and g_2 are material properties which are functions of stress and related to the thermodynamic functions. A comparable representation is obtained for the stress-relaxation modulus $G_n(t)$. This theory also incorporates time–temperature equivalence, because it has both horizontal shifts in time-scale and vertical shifts. The way in which moisture (in polymers such as nylons and polysulfones) changes relaxation behavior in a manner similar to temperature can be dealt with using this treatment.

Experimentally it is necessary to separate the terms in the product $g_1 g_2$ and in the similar product $h_1 h_2$ for stress relaxation by using multiple-loading programs. The representation provides a good description of the behavior of polymers as different as nitrocellulose, a fiber-reinforced phenolic resin, and polyisobutylene (in the last case, providing a link with BKZ theory).

Multiple-integral Representation.   In the treatments of nonlinearity previously discussed, it has been assumed that the stress (or strain) and time dependence of the response are separable. The inadequacy of such representations was highlighted in a study of the creep of oriented polypropylene fibers (63). Although creep compliance was independent of stress at very low stress, recovery curves never coincided. Furthermore, the instantaneous or short-term recovery was always greater than the instantaneous creep. In two-step loading experiments, the additional creep was always in excess of the initial creep, and the increase in instantaneous strain was of the same magnitude as the instantaneous recovery at an equivalent time. These results were interpreted in terms of a multiple-integral representation for viscoelasticity proposed several years earlier (64–66). In a loading program in which incremental stresses $\Delta\sigma_1(\tau_1),\Delta\sigma_2(\tau_2)$, etc are added at times τ_1,τ_2, etc, for a continuous-load history the deformation can be written in integral form as

$$e(t) = \int_{-\infty}^{t} J_1(t - \tau)\frac{d\sigma(\tau)}{d\tau}\,d\tau + \int_{-\infty}^{t}\int_{-\infty}^{t} J_2(t - \tau_1, t - \tau_2)\frac{d\sigma(\tau_1)}{d\tau_1}\frac{d\sigma(\tau_2)}{d\tau_2}\,d\tau_1\,d\tau_2$$

$$+ \cdots + \int_{-\infty}^{t}\cdots\int_{-\infty}^{t} J_N(t - \tau_1, \ldots, t - \tau_N)\frac{d\sigma(\tau_1)}{d\tau_1}\cdots\frac{d\sigma(\tau_N)}{d\tau_N}\,d\tau_1\cdots d\tau_N$$

The first term is a statement of the Boltzmann superposition principle, but the complete representation is so general that data must be fitted because there is an infinite number of curve-fitting constants. For such a complex representation to be useful the following must apply: data can be fitted to a small number of multiple-integral terms; behavior under complex loading schedules can be predicted from a few simple loading programs; and the representation should assist an empirical understanding of the relationships between viscoelastic behavior and the structure of the polymer.

For creep at constant stress σ_0, the multiple-integral representation gives

$$\frac{e_c(t)}{\sigma_0} = J_1(t) + J_2(t,t)\sigma_0 + \cdots + J_N(t, \cdots t)\sigma_0^{N-1}$$

which suggests the plotting of isochronous curves of compliance against stress for various fixed values of time (Fig. 36). The small region close to the origin approximates to a series of straight lines parallel with the load axis and represents a linear viscoelastic response. The curves are usually parabolic, which suggests that

$$\frac{e_c(t)}{\sigma} = A + B\sigma_0^2$$

where $A = J_1(t)$ and $B = J_3(t,t,t)$ are functions of time only. If this representation were adequate, recovery should take the form

$$\frac{e_r(t - t_1)}{\sigma_0} = A' + B'\sigma_0^2$$

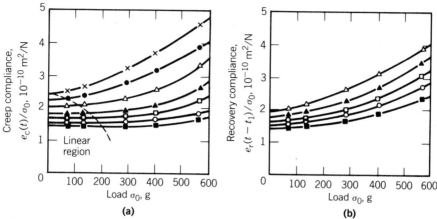

Fig. 36. (a) Creep compliance and (b) recovery compliance as a function of applied load for an oriented polypropylene monofilament. The time of loading for the recovery test is 9.3×10^3 s. Symbols indicate the times in seconds for both creep and recovery: x, 9300; ●, 3000; △, 1000; ▲, 300; □, 100; and ○, 40; ■, 15 (63). To convert m²/N to in.²/lb, multiply by 6897.

where $A' = J_1(t - t_1)$ and $B' = J_3(t - t_1, t - t_1, t - t_1) + 3[J_3(t,t,t - t_1) - J_3(t,t - t_1,t - t_1)]$ are functions of time only. The results of Figure 36b demonstrate the correctness of this assumption. The difference between 15-s recovery and 15-s creep was measured, and, as expected, was also a parabolic function of stress.

For a two-step loading program, in which a stress of σ_0 is applied at $\tau = 0$ and a further σ_0 is applied at $\tau = t_1$, the additional creep response is $e'_c(t - t_1) = e'(t,\sigma_0,t_1,\sigma_0) - e_c(t,\sigma_0), t > t_1$, where $e'(t,\sigma_0,t_1,\sigma_0)$ is the elongation after the second step of loading, and $e_c(t,\sigma_0) = J_1(t)\sigma_0 + J_3(t,t,t)\sigma_0^3$. Evaluating and retaining only the first-order terms $e'_c(t - t_1) = J_1(t - t_1)\sigma_0 + J_3(t - t_1,t - t_1,t - t_1)\sigma_0^3 + 3[J_3(t,t,t - t_1) + J_3(t,t - t_1,t - t_1)]\sigma_0^3$. Subsequent experiments showed that the RT behavior of a range of polypropylene specimens could always be represented by a small number of integral terms, but the particular terms involved changed in a systematic manner with variations in molecular orientation and other features of supermolecular structure (67).

An attempt has also been made to predict the response to complex loading patterns from creep, recovery, and the superposition of identical loads (68). The corrections obtained were small, and the predictions gave only a slight improvement over those of a simple addition superposition rule. Nevertheless, the multiple-integral representation illustrates the complexity of viscoelastic behavior and offers an insight into the failure of the simple superposition laws discussed earlier.

The further development of multiple-integral representations is discussed in Ref. 69, and several experimental studies using different approximation procedures have been made (70).

The effect of temperature changes in the region immediately above room temperature has been studied on oriented polypropylene and PET fibers (71). In both cases, the creep curves for a given stress at different temperatures could be superimposed by a simple horizontal shift along the log (time) axis; the shifts were identical at all stresses. Provided that unloading occurred always at the same strain, the shift factors for recovery were identical with those in creep.

Abbreviation of the multiple-integral representation in an arbitrary manner has unsatisfactory consequences (72). An implicit equation method readily relates creep, stress relaxation, and constant strain-rate data. Study of an oriented PET sheet shows that stress could be represented by a convergent series:

$$\sigma(t) = \dot{e}_0 I_1(t) - p\dot{e}_0^2 I_1^2(t) + p^2 \dot{e}_0^3 I_1^3(t) + \cdots + p^8 \dot{e}^9 I_1^9(t)$$

where

$$\dot{e}_0 = \frac{de}{d\tau} \quad \text{and} \quad I_1(t) = \int_{-\infty}^{t} G_1(t - t_1)\tau_1 \, d\tau_1$$

If it is assumed that the series extends to infinity, its sum yields

$$\sigma(t) = I_1(t)\dot{e}_0 - pI_1(t)\dot{e}_0\sigma(t)$$

This is a simple integral equation with a constant p, of dimensions I^{-1}, which generates the multiple-integral representation. For the specimen studied, p was not quite constant but depended on time, and a slightly more general integral equation was required. It was predicted that the creep strain under a constant stress σ_0 would be given by

$$e(t) = \frac{j(t)\sigma_0}{1 - k\sigma_0}$$

where j is a linear term relating the coefficients of stress and strain. This equation predicts that creep becomes infinite for $\sigma_0 > 1/k$, which was identified with a critical stress for yielding on flow (Fig. 37). The stress–strain curve altered with changes in (constant) strain rate and, consistent with observation, the yield stress

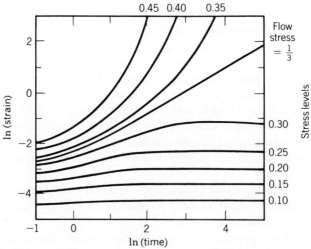

Fig. 37. A plot of ln (strain) against ln (time) as predicted by the implicit representation, parameterized for the stated applied stresses (72).

was predicted to increase almost linearly with log (strain rate) over a wide range of strain rates. The implicit representation was therefore capable of providing a phenomenological base for describing several aspects of nonlinear viscoelasticity.

Multiaxial Deformation: Three-dimensional Nonlinear Viscoelasticity. Although multiaxial deformation adds further complexity to the discussion, it is one of considerable technological importance. The first approach was an octahedral stress theory based on the hypothesis that creep is caused by the shearing of molecules past one another. The principal components of strain (e_i) are related to those of stress (σ_j) by expressions such as (49)

$$e_1 = \frac{1}{E}[\sigma_1 - \nu(\sigma_2 - \sigma_3)] + \tfrac{1}{2}(2\sigma_1 - \sigma_2 - \sigma_3)(J_2)^{(n-1)/2}\{K[1 - \exp(-qt)] + Bt\}$$

where E is Young's modulus, ν is Poisson's ratio, and K, n, q, and B are constants for the materials; $J_2 = \tfrac{1}{2}[(\sigma_1 - \sigma_2)^2 + (\sigma_2 - \sigma_3)^2 + (\sigma_3 - \sigma_1)^2]$ is proportional to the octahedral shear stress. Further assumptions are those of zero-volume change during the time-dependent part of the deformation (Poisson's ratio applied only for the immediate elastic deformation) and a response identical in tension or compression. This final assumption did not hold for poly(methyl methacrylate), but it was satisfactory to equate the mean value of the octahedral shear strains in tension and in compression with the octahedral strain in pure shear. Shear and uniaxial creep have also been related, and a number of tests were performed under hydrostatic compression. There appeared to be two contributions to the total creep response: a linear viscoelastic term from the hydrostatic components of the stress and a nonlinear term dependent on both the deviatoric and hydrostatic stress components. The effect of hydrostatic strain on stress relaxation in poly(methyl methacrylate) undergoing tension and torsion simultaneously has been investigated (73); the governing factor was the magnitude of the mean-normal strain.

Extensional $e_\tau(t)$ and shear $\gamma(t)$ creep strains in a polypropylene tube subjected to combined tension-torsion were interpreted in terms of the stress invariants J_1 and J_2 (74):

$$e_\tau(t) = [\tfrac{1}{3}J(t,J_1,J_2) + \tfrac{1}{9}B(t)]\sigma$$
$$\gamma(t) = J(t,J_1,J_2)\tau$$

where $J_1 = \sigma_1 + \sigma_2 + \sigma_3$, $J_2 = \sigma_1\sigma_2 + \sigma_2\sigma_3 + \sigma_3\sigma_1$, and B was obtained by extrapolation to $\sigma \to 0$, $\tau \to 0$. The values of $J(t,J_1,J_2)$ obtained from the tensile and shear responses were in good agreement. This treatment separates the hydrostatic and deviatoric components of stress and implies that the viscoelastic nonlinearity enters through the shear-deformation behavior. The nonlinearity can be described by allowing the shear-creep compliance to change with stress history.

Despite the success with a single-integral representation, a multiple-integral description is usually needed to account for the response to a combination of applied stresses. Under these circumstances, an investigation of the creep and stress–relaxation responses of an isotropic material, including terms up to the third order, would require seven single stress tests, each at several strain levels, five biaxial stress tests, a triaxial test, and four combinations of normal shear stresses (69). If ten time values are chosen, a total of 463 tests are required, compared with 78 in the one-dimensional situation. However, this form of representation can be used provided that reasonable simplifying assumptions are made and the creep responses parameterized, so that fitting procedures may be applied.

Creep and Stress Relaxation as Thermally Activated Processes. An alternative treatment of nonlinear viscoelasticity is based on the assumption that deformation involves the motion of parts of polymer chains over potential energy barriers, where the process can be either intermolecular, eg, chain sliding, or intramolecular, eg, changes in conformation. In the absence of stress, dynamic equilibrium exists and chain segments move with a frequency ν over the potential barrier, height ΔH, in each direction, with $\nu = \nu_0 \exp(-\Delta H/RT)$. An applied stress σ raises the barrier in one direction by an amount $v\sigma$ and reduces the height by a similar amount in the reverse direction. The net flow in the forward direction is then given by

$$\frac{de}{dt} = \dot{e} = \dot{e}_0 \exp(-\Delta H/RT) \sinh\left(\frac{v\sigma}{RT}\right)$$

where v is termed the activation volume for the molecular event. This equation defines an activated non-Newtonian viscosity and was first developed in terms of a spring and dashpot model. It gives an appropriate algebraic form for creep in general agreement with experimental observations. Activation energy and activation volume may give some indication of the underlying molecular mechanisms, and the activated rate process may provide a common basis for the discussion of creep and yield behavior.

The creep data on poly(methyl methacrylate), obtained over a range of temperatures below its glass transition, were superimposed by assuming that temperature shifts could be interpreted in terms of an activated process where the activation energy fell in a linear fashion with increasing stress (75). This treatment was extended in a study of the compressive creep of polycarbonate resin (76) over a range of stress values. Superposition was obtained by assuming an activation volume of 5.7 nm^3, a value close to that obtained from measurements of the strain-rate dependence of the yield stress (Fig. 38).

The Eyring rate process is useful in modeling the temperature variation of the creep of ultrahigh modulus polyethylene (77,78). For high strains (corresponding to long creep times), the creep rate reaches a constant value at a given temperature. For low molecular weight polymers, this final creep rate could be modeled by a single activated process with an activation volume that is reasonable in molecular terms. For high molecular weight specimens, it was necessary to use two Eyring processes with different preexponential factors and activation volumes coupled in parallel; the second process predominates at low stresses.

Low Strain Elasticity and Anelasticity in Anisotropic Polymers

An oriented polymer is generally an anisotropic nonlinearly viscoelastic material. However, this discussion is confined to strains so small that nonlinear effects can be neglected. In static experiments, compliances and moduli refer to

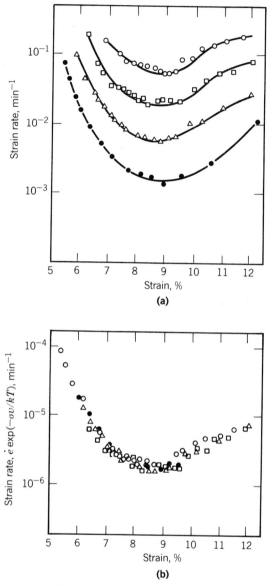

Fig. 38. (a) Creep rates for PMMA in compression at various stresses. ○, 71 MN/m² (10,300 psi); □, 70 MN/m² (10,200 psi); ○, 69 MN/m² (10,000 psi); ●, 68 MN/m² (9860 psi). (b) Superposition of creep data shown in (a) using an activation volume of 5.7 nm³ (76).

instantaneous, unrelaxed values and dynamic experiments are discussed under constant frequency conditions.

Mathematical Description. The mechanical properties of an anisotropic elastic solid at small strains are defined by a generalized Hooke's law:

$$\varepsilon_{ij} = S_{ijkl}\sigma_{kl}, \qquad \sigma_{ij} = C_{ijkl}\varepsilon_{kl}$$

where S_{ijkl} are compliance constants and C_{ijkl} are stiffness constants. In practice, an abbreviated notation is often used in which

$$e_p = S_{pq}\sigma_q, \qquad \sigma_p = C_{pq}e_q$$

It is important to realize that the engineering strains e_p are not the components of a tensor.

Experimentally, it is convenient to apply a simple stress and measure the corresponding strain components. Thus it is usual to work in terms of compliance constants rather than stiffness constants.

$$e_1 = S_{11}\sigma_1 + S_{12}\sigma_2 + S_{13}\sigma_3 + S_{14}\sigma_4 + S_{15}\sigma_5 + S_{16}\sigma_6$$

Drawn fibers and uniaxially drawn films both show isotropy in a plane perpendicular to the direction of drawing. The number of independent elastic constants is reduced to five and, choosing the $z(x_3)$ direction as the axis of symmetry, the compliance matrix S_{pq} reduces to

$$\begin{pmatrix}
S_{11} & S_{12} & S_{13} & 0 & 0 & 0 \\
S_{12} & S_{11} & S_{13} & 0 & 0 & 0 \\
S_{13} & S_{13} & S_{33} & 0 & 0 & 0 \\
0 & 0 & 0 & S_{44} & 0 & 0 \\
0 & 0 & 0 & 0 & S_{44} & 0 \\
0 & 0 & 0 & 0 & 0 & 2(S_{11} - S_{12})
\end{pmatrix}$$

Young's modulus $E_3 = \dfrac{1}{S_{33}}$; Poisson's ratio $\nu_{13} = -\dfrac{S_{13}}{S_{33}}$. Similarly, E_1 (transverse modulus) $= \dfrac{1}{S_{11}}$ and the corresponding Poisson's ratios $\nu_{21} = -\dfrac{S_{12}}{S_{11}}$, $\nu_{31} = -\dfrac{S_{31}}{S_{11}}$. The shear or torsional modulus $(G) = \dfrac{1}{S_{44}}$.

Methods are available for producing oriented films with orthorhombic rather than transversely isotropic symmetry. In this case, the behavior is specified by nine independent elastic constants. The initial rolling or drawing direction is taken as the z axis, the x axis lies in the plane of the film, and the y axis is normal to the film plane. The compliance matrix is

$$\begin{pmatrix}
S_{11} & S_{12} & S_{13} & 0 & 0 & 0 \\
S_{12} & S_{22} & S_{23} & 0 & 0 & 0 \\
S_{13} & S_{23} & S_{33} & 0 & 0 & 0 \\
0 & 0 & 0 & S_{44} & 0 & 0 \\
0 & 0 & 0 & 0 & S_{55} & 0 \\
0 & 0 & 0 & 0 & 0 & S_{66}
\end{pmatrix}$$

There are now three Young's moduli, six Poisson's ratios, and three independent shear moduli.

Measurements on Films or Sheets. The Young's moduli of films or sheets can be determined from long and narrow strips cut in selected directions. A high aspect ratio is needed to minimize the end effects that arise from nonuniform stress conditions near the clamps. Strips cut along and perpendicular to the draw direction yield $1/S_{33}$ and $1/S_{11}$ directly, and those cut at intermediate angles give information on combinations of other compliances, eg, for a strip cut at 45° to the initial draw direction in an orthorhombic sheet

$$\frac{1}{E_{45}} = \frac{1}{4}(S_{11} + S_{33} + 2S_{13} + S_{55})$$

The stiffness normal to the plane of the sheet C_{22} can be measured in a compressional creep apparatus, although precautions need to be taken to ensure adequate accuracy. The lateral compliance S_{13}, which defines the contraction in the x direction for a stress applied along the z direction, has been determined by measuring the distortion of a fine grid evaporated onto the surface of the sheet. For measurement of compliances S_{12} and S_{13} that correspond to changes in thickness, several sensitive mechanical extensometers have been described. A method suitable for thin specimens of high optical clarity measures the fringe shift produced when the specimen is inserted in one arm of a Michelson interferometer.

Torsional measurements can be used to determine the shear compliances S_{44}, S_{55}, and S_{66}. For orthorhombic sheets, a solution can be found only when the specimens are rectangular prisms with surfaces normal to the three axes of symmetry and where the torsion axis coincides with one of these axes. It is necessary to carry out measurements on specimens covering a range of aspect ratios with the effects of axial stresses and the influence of the end clamps taken into consideration. Such complications can be avoided by measuring S_{44} and S_{66} in simple shear.

In the Hall effect simple shear apparatus (79) (Fig. 39), identical specimens are held in position between two plates that are forced together by a calibrated

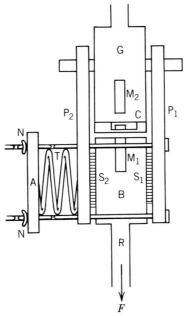

Fig. 39.   The Hall effect simple shear apparatus. C is the Hall plate between magnets M_1 and M_2. Specimens S_1 and S_2 are mounted between plates P_1 and P_2, and the movable block B is pulled by force F. T is a calibrated spring (79).

spring T and a movable block B. Applying a load F to B causes a shear displacement, which is sensed through the Hall voltage arising from the plate C positioned between magnets M_1 and M_2.

Measurements on Fibers.   The torsional modulus $G = 1/S_{44}$ can be measured from free torsional vibrations, where the fiber supports known inertia bars at its free end, although other convenient dynamic methods have been described. The Poisson's ratio $\nu_{13} = -S_{13}/S_{33}$ has been measured with a microscope by observing the radial contraction corresponding to a known change in length. With strains on the order of 0.01–0.02, the uncertainty in measurement was ~10%. The transverse modulus $1/S_{11}$ can be derived from the width of the contact strip when a filament is compressed between two parallel glass plates. Using the same apparatus it is possible to determine the transverse Poisson's ratio $\nu_{12} = S_{12}/S_{11}$ from the change in filament diameter parallel to the plane of contact.

Ultrasonic Measurements.   Mechanical anisotropy in oriented polymers can be determined from measurements of the velocity and attenuation of elastic waves at ultrasonic frequencies. Quartz transducers, to cut disks, bonded at 0, 45, and 90° with the symmetry axis of a uniaxial rod, generate three different linearly polarized elastic waves that propagate along the geometric axes of each disk (80). For each sample, it is possible to measure nine different velocities from which the five independent elastic constants can be determined. A specimen in the form of a rectangular parellelepiped can be immersed in a tank of water containing an ultrasonic transmitter and receiver. The sample can be rotated in various directions and the elastic constants deduced as in the previous method.

Mechanical Anisotropy.   Polymers whose molecules take the form of a planar zigzag or a helix show a large difference between the forces involved when the structure is deformed, first parallel with, and then perpendicular to, the chain axis. For polyethylene, a planar zigzag, the crystal modulus along the chain axis has been calculated as 300 GN/m² (43.5×10^6 psi), the high value arising because the deformation involves the bending and stretching of covalent bonds. In polypropylene, a helix, the corresponding value is reduced to 50 GN/m² (72.5×10^6 psi) as rotation around bonds is involved, as well as the bending of bonds. In each case, the tensile modulus perpendicular to the chain axis and the shear moduli are much lower, 1–50 GN/m² (145,000–72.5×10^6 psi), as they relate to the weak van der Waals or dispersion forces between the chains. The determination of the elastic constants is of considerable importance as the results provide a base line against which practical achievements can be judged. The experimental methods include the changes in the x-ray diffraction pattern with stress, Raman spectroscopy, and inelastic neutron scattering (81,82).

For most commercially available fibers and films the mechanical anisotropy is much less than considerations of the molecular chain would imply, since crystalline polymers are essentially composite materials with alternating crystalline and noncrystalline regions; the latter are not well-oriented during fabrication. It has been hypothesized that Young's modulus of an oriented fiber is determined essentially by the proportion of taut tie molecules or narrow clusters of crystalline chains that form bridges linking the crystalline blocks to provide a degree of overall crystal continuity (83).

In noncrystalline polymers, where few chains achieve a high alignment, the degree of mechanical anisotropy correlates well with molecular orientation. Some polymers with low overall crystallinity, such as poly(ethylene terephthalate), appear to occupy an intermediate position.

Interpretation of mechanical anisotropy requires quantitative models, which take as their starting points either variations in molecular orientation or the composite nature of a crystalline polymer. The aggregate model assumes that the polymer consists of an aggregate of anisotropic units, with increasing orientation corresponding to an alignment of the units (84). The mechanical properties of the units are taken from those of a highly oriented polymer obtained from experimental results. Although this model originated as a single-phase model applicable to amorphous polymers, it has been applied successfully to studies of PET and other crystalline materials (85,86).

The Takayanagi model, which recognizes the two-phase nature of crystalline polymers, is concerned only with tensile moduli parallel with and perpendicular to the draw direction in a highly oriented polymer (87). Fully oriented crystalline and noncrystalline phases are considered being in series along the draw direction and in parallel in the perpendicular direction. The effect of tie molecules or crystal bridges can be taken into account by allowing some proportion of crystalline material to be in parallel with the noncrystalline component. Many applications have been made of these and related models (88).

Experimental Studies and the Aggregate Model.   In amorphous polymers, the degree of mechanical anisotropy is rather low. Dynamic data obtained over a range of frequencies (89,90) are given in Table 1 and Figure 40. It appears that the anisotropy can usually be discussed satisfactorily in terms of the aggregate model.

Table 1. Elastic Compliances of Oriented Amorphous Polymers, 10^{-9} m²/N[a]

Material	Draw ratio	S_{33}	S_{11}	S_{44}
poly(vinyl chloride)	1.0	0.313	0.313	0.820
	1.5	0.276	0.319	0.794
	2.0	0.255	0.328	0.781
	2.5	0.243	0.337	0.769
	2.8	0.238	0.341	0.763
	∞	0.204	0.379	0.730
poly(methyl methacrylate)	1.0	0.214	0.214	0.532
	1.5	0.208	0.215	0.524
	2.0	0.204	0.215	0.518
	2.5	0.200	0.216	0.510
	3.0	0.196	0.217	0.505
polystyrene	1.0	0.303	0.303	0.769
	2.0	0.296	0.304	0.769
	3.0	0.289	0.305	0.769
polycarbonate	1.0	0.376	0.376	1.050
	1.3	0.314	0.408	0.980
	1.6	0.268	0.431	0.926

[a] To convert m²/N to in.²/lb, multiply by 6897.

The first attempt to determine a complete set of elastic constants for drawn crystalline polymers was made on sheets of LDPE (92,93). Within a few years, data had been reported on a range of uniaxially drawn fibers and sheets (86).

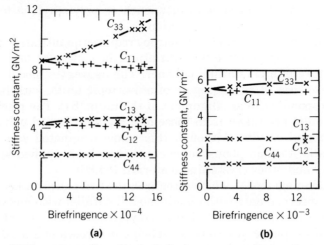

Fig. 40. Stiffness constants of uniaxially drawn amorphous polymers, measured at room temperature, as a function of birefringence: (**a**) poly(methyl methacrylate) and (**b**) polystyrene (91). To convert GN/m² to psi, multiply by 145,000.

With the exception of some experiments on LDPE (94), the measurements were performed at room temperature.

The tensile modulus of uniaxially drawn LDPE, measured in various directions relative to that of drawing, shows a number of unexpected features (Figs. 41 and 42). The minimum at 45° to the draw direction implies [by substituting $\theta = 45°$ in $S_\theta = S_{11} \sin^4\theta + S_{33} \cos^4\theta + (2S_{13} + S_{44})\sin^2\theta \cos^2\theta$] that $(2S_{13} + S_{44})$ is much greater than S_{11} or S_{33}. It was unexpected that at low draw ratios E_0 was less than the isotropic value and also less than E_{90}. Later experiments showed that this anomalous behavior disappears at sufficiently low temperatures (94).

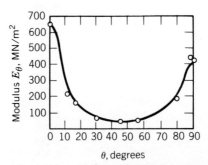

Fig. 41. The experimental variation of modulus at angle θ to the draw direction compared with the theoretical relation (full curve) calculated from E_0, E_{45}, and E_{90} for a sheet of LDPE at a draw ratio of 4.65 (92). To convert MN/m² to psi, multiply by 145.

Results for highly oriented specimens are given in Table 2. The principal effect of increasing orientation is to increase Young's modulus E_3 in the direction of the fiber axis. For nylon and PET there is a corresponding small decrease in the transverse modulus E_1; for polypropylene and HDPE, E_1 remains almost

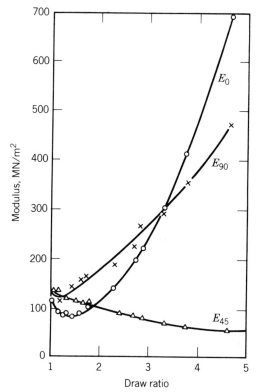

Fig. 42. The variation of E_0, E_{45}, and E_{90} with draw ratio in cold-drawn sheets of LDPE at room temperature (92). To convert MN/m^2 to psi, multiply by 145.

constant, but for LDPE it increases significantly at the higher draw ratios. The anomalous behavior of LDPE is confirmed. An indication of the overall characteristics of LDPE and PET can be obtained from Figures 43 and 44, where the larger change in shear modulus for LDPE is evident. It is the only material in which S_{44} does not have much the same value as S_{11}. The compliance S_{13} varied with draw ratio in a similar manner to S_{33}, giving an extensional Poisson's ratio $\nu_{13} = -S_{13}/S_{33}$ which is insensitive to orientation, and except for HDPE, does not vary significantly from 0.5. Thus it is generally a valid approximation to assume that the fibers are incompressible. Values of the transverse Poisson's ratio are subject to large uncertainties but appear to vary widely for different polymers.

In PET, the molecular orientation, as measured from birefringence, appears to be the primary factor in determining the mechanical anisotropy; the influence of crystallinity is small. To a first approximation, the unoriented polymer can be regarded as an aggregate of anisotropic elastic units whose properties are those of the most highly oriented specimen. The average elastic constants for the aggregate may be obtained by assuming uniform stress throughout the aggregate, which implies a summation of compliance constants, or by assuming uniform strain, implying that stiffness constants are to be totaled. For a random aggregate, the correct value is between these extremes.

The uniform-stress case can be considered a series model of elemental cubes

Table 2. Elastic Compliances of Oriented Fibers, 10^{-10} m²/N[a,b]

Material	Birefringence Δn	S_{11}	S_{12}	S_{33}	S_{13}	S_{44}	$\nu_{13} = -\dfrac{S_{13}}{S_{33}}$	$\nu_{12} = -\dfrac{S_{12}}{S_{11}}$
low density polyethylene film		22	−15	14	−7	680	0.50	0.68
low density polyethylene	0.0361	40 ± 4	−25 ± 4	20 ± 2	−11 ± 2	878 ± 56	0.55 ± 0.08	0.61 ± 0.20
low density polyethylene	0.0438	30 ± 3	−22 ± 3	12 ± 1	−7 ± 1	917 ± 150	0.58 ± 0.08	0.73 ± 0.20
high density polyethylene	0.0464	24 ± 2	−12 ± 1	11 ± 1	−5.1 ± 0.7	34 ± 1	0.46 ± 0.15	0.52 ± 0.08
high density polyethylene	0.0594	15 ± 1	−16 ± 2	2.3 ± 0.3	−0.77 ± 0.3	17 ± 2	0.33 ± 0.12	1.1 ± 0.14
polypropylene	0.0220	19 ± 1	−13 ± 2	6.7 ± 0.3	−2.8 ± 1.0	18 ± 1.5	0.42 ± 0.16	0.68 ± 0.18
polypropylene	0.0352	12 ± 2	−17 ± 2	1.6 ± 0.04	−0.73 ± 0.3	10 ± 2	0.47 ± 0.17	1.5 ± 0.3
poly(ethylene terephthalate)	0.153	8.9 ± 0.8	−3.9 ± 0.7	1.1 ± 0.1	−0.47 ± 0.05	14 ± 0.5	0.43 ± 0.06	0.44 ± 0.09
poly(ethylene terephthalate)	0.187	16 ± 2	−5.8 ± 0.7	0.71 ± 0.04	−0.31 ± 0.03	14 ± 0.2	0.44 ± 0.07	0.37 ± 0.06
nylon-6,6	0.057	7.3 ± 0.7	−1.9 ± 0.4	2.4 ± 0.3	−1.1 ± 0.15	15 ± 1	0.48 ± 0.05	0.26 ± 0.08

[a] To convert m²/N to in.²/lb, multiply by 6897.
[b] Errors quoted are 95% confidence limits.

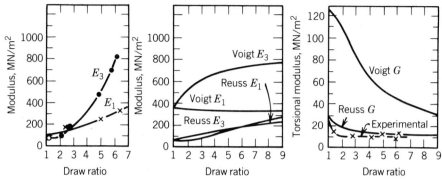

Fig. 43. Moduli of LDPE filaments at room temperature: extensional (E_3), transverse (E_1), and torsional (G). Experimental results are compared to simple aggregate theory. To convert MN/m² to psi, multiply by 145.

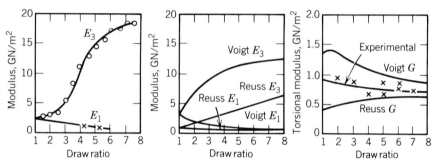

Fig. 44. Moduli of PET filaments at room temperature: extensional (E_3), transverse (E_1), and torsional (G). Experimental results are compared to simple aggregate theory. To convert GN/m² to psi, multiply by 145,000.

arranged end to end, with each cube taken as a transversely isotropic elastic solid. For a random aggregate with zero overall orientation, the average extension compliance

$$\frac{e}{\sigma} = \overline{S}'_{33} = {}^{8}\!/_{15}S_{11} + {}^{1}\!/_{5}S_{33} + {}^{2}\!/_{15}(2S_{13} + S_{44})$$

Similarly, uniform strain can be taken as a parallel packing of elemental cubes to yield

$$\frac{\sigma}{e} = \overline{C}'_{33} = {}^{8}\!/_{15}C_{11} + {}^{1}\!/_{5}C_{33} + {}^{4}\!/_{15}(C_{13} + 2C_{44})$$

For an isotropic elastic solid, the shear constants are

$$\overline{S}'_{44} = {}^{14}\!/_{15}S_{11} - {}^{2}\!/_{3}S_{12} - {}^{8}\!/_{15}S_{13} + {}^{4}\!/_{15}S_{33} + {}^{2}\!/_{5}S_{44}$$

$$\overline{C}'_{44} = {}^{7}\!/_{30}C_{11} - {}^{1}\!/_{6}C_{12} - {}^{2}\!/_{15}C_{13} + {}^{1}\!/_{15}C_{33} + {}^{2}\!/_{5}C_{44}$$

The uniform-stress model gives the so-called Reuss average and the uniform strain model, the Voigt average. In the latter case, it is convenient to invert the matrix to obtain $\overline{S}'_{33}$ and $\overline{S}'_{44}$. Measured compliances for unoriented specimens and those derived from highly oriented material are given in Table 3. For PET and LDPE, the measured isotropic compliances lie between the calculated bounds, suggesting that molecular orientation is the primary factor in determining mechanical anisotropy. In nylon, the measured values lie just outside the bounds suggesting that other structural features, in addition to molecular orientation, play an important role. For HDPE and polypropylene, where other experiments suggest that simultaneous changes occur both in morphology and molecular mobility, the agreement is poor.

Table 3. Extensional and Torsional Compliances for Unoriented Fibers, 10^{-10} m²/N[a]

	Extensional compliance, $\bar{S}'_{11} = \bar{S}'_{33}$			Torsional compliance, $\bar{S}'_{44}$		
	Calculated			Calculated		
	Reuss average	Voigt average	Measured	Reuss average	Voigt average	Measured
LDPE	139	26	81	416	80	238
HDPE	10	2.1	17	30	6	26
PP	7.7	3.8	14	23	11	2.7
PET	10.4	3.0	4.4	25	7.6	11
nylon	6.6	5.2	4.8	17	13	12

[a] To convert m²/N to in.²/lb, multiply by 6897.

The aggregate model has been extended to specimens of overall but intermediate orientation by considering birefringence arising from the rotation of rodlike units toward the direction of drawing. Such a scheme has been called pseudoaffine deformation; it differs from affine-deformation models of rubbers in the units maintaining constant length. Examples in Figures 43 and 44 illustrate the general form of anisotropy predicted by the aggregate model. For LDPE, the Reuss average appears appropriate; it includes the minimum in the extensional modulus, as a consequence of S_{44} being much larger than S_{11} and S_{33}. At low temperatures, a more conventional pattern appears, since S_{44} is no longer much greater than the other compliances. Some differences in detail between prediction and experiment, eg, a small minimum in the transverse modulus of LDPE and in the extensional modulus of HDPE, have been attributed to the influence of mechanical twinning. A more important discrepancy is that mechanical anisotropy increases with draw ratio much faster than is predicted. For LDPE, a much better agreement is obtained when the orientation functions are derived from x-ray diffraction and nmr rather than from birefringence, suggesting that the anisotropy relates to the orientation of the crystalline regions and is then predicted by the Reuss averaging scheme.

In PET, the experimental compliances lie approximately midway between the Reuss and Voigt bounds, but in nylon the Voigt average is closer to experimental data. An aggregate model does not appear to be applicable for HDPE. For polypropylene it is appropriate only at low draw ratios because at higher orientations changes occur both in morphology and molecular mobility. The differences between particular polymers reflect details of stress and strain distributions at a molecular level and are ultimately related to the supermolecular structure.

All nine independent elastic constants have been measured for single-crystal texture PET sheet produced by drawing uniaxially at constant width (95). The high degree of mechanical anisotropy (Table 4) has been related to the high chain axis orientation and the preferential orientation of the terephthalate residues. The low value of extensional compliance S_{33} is thought to relate to the deformation involving bond stretching and bending in extended chain molecules. The much higher transverse compliances S_{11} and S_{22} are related principally to dispersion forces. It follows that if the material is stretched perpendicular to the draw direction, the principal contraction occurs in the other perpendicular direction, rather than in the draw direction. Shear compliances S_{44} and S_{66} are large compared with S_{55}, reflecting easy shear in the 23 and 12 planes, respectively, presumably caused by the planar terephthalate chains sliding over one another constrained only by weak dispersion forces. The compliance S_{55}, which involves distortion of the plane of the polyester molecules, is of the same order as S_{11} and S_{22}. When compared with the aggregate model, experimental values are between the Reuss and Voigt bounds, affording further support for the contention that the mechanical anisotropy of PET can be considered in terms of a simple, single-phase aggregate theory.

Table 4. Full Set of Compliances for the Oriented PET Sheet with Orthorhombic Symmetry

Compliance	10^{-10} m^2/N[a]
S_{11}	3.61 ± 0.12
S_{22}	9.0 ± 1.6
S_{33}	0.66 ± 0.01
S_{12}	−3.8 ± 0.4
S_{13}	−0.18 ± 0.01
S_{23}	−0.37 ± 0.05
S_{44}	97 ± 3
S_{55}	5.64 ± 0.25
S_{66}	141 ± 8

[a] To convert m^2/N to in.^{2}/lb, multiply by 6897.

Composite Models for Oriented Polymers. Crystalline polymers that show a clear lamellar structure when oriented may be considered composite solids. The orientation of the lamellae, as distinct from molecular orientation, plays a dominant role in mechanical anisotropy. This analogy with composite materials was discovered when the temperature variation of dynamic moduli was measured (87). For HDPE (Fig. 45), the modulus in the draw direction (E_0) is greater than E_{90} at low temperatures; with increasing temperature the situation is reversed. Takayanagi proposed that crystalline and amorphous regions alternated in series along the draw axis and so formed a parallel array in the transverse direction. The large drop in E_0 with temperature resulted from the stiffness being determined primarily by the amorphous regions, which are very compliant above the

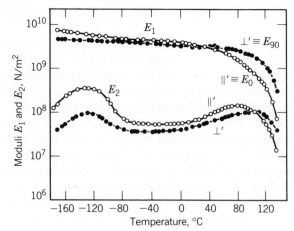

Fig. 45. Temperature dependence of the components of the dynamic modulus in directions parallel with and perpendicular to the direction of draw for annealed specimens of HDPE (87). To convert N/m² to psi, multiply by 1.45×10^{-4}.

relaxation transition. Because the crystalline and amorphous components are in series, each is subjected to the same stress and compliances are being added as in the Reuss averaging scheme. In the transverse direction, the components are in parallel, and each is subject to the same strain and, as in the Voigt scheme, stiffnesses are added. Above the relaxation transition, the stress is supported by the crystalline regions and a comparatively high stiffness is maintained.

The two limiting cases for stress transfer, when one phase is dispersed in the other, are indicated in Figure 46 where (**a**) represents the condition for efficient stress transfer normal to the direction of tensile stress and (**b**) the condition for weak stress transfer across planes containing the tensile stress. Where the continuous phase is considered to represent the crystalline components, there is a clear analogy with the concept of tie molecules extending through the amorphous phase.

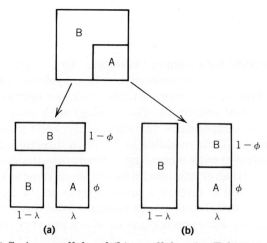

Fig. 46. (**a**) Series-parallel and (**b**) parallel-series Takayanagi models.

A particularly informative application of a composite model was used in an investigation (96–99) of the temperature variation of static modulus and dynamic phase angle for three distinct textures of rolled and drawn sheets of LDPE (Fig. 47). An indication is given of the texture of each type of sheet, showing the molecular orientation within parallel crystalline lamellae separated by amorphous regions. For all sheets the anisotropy of the high temperature α-relaxation was consistent with a process involving shear in the c-axis direction on planes containing the c-axis of the crystallites (the c-shear process). The β-relationship had already been assigned to interlamellar shear, and Takayanagi's concept was extended to include the orientation of the lamellae as dominating mechanical anisotropy. Detailed study of the anisotropy yielded a considerable insight into the deformation processes involved in a particular relaxation for the different textures. Sheets with fiber symmetry, analogous to the specimens measured earlier, were also examined, and the crossover between E_0 and E_{90} at high temperatures was explained in terms of a model in which the lamellar planes form angles of 35–40° with the draw direction and have a conical distribution about that axis.

Studies of HDPE include materials produced through a variety of processes (100,101). In all cases, the mechanical anisotropy could be illustrated by a series-parallel Takayanagi model with a continuous fraction of amorphous material. This model was appropriate even for materials crystallized under strain. It had been anticipated that in specimens crystallized under strain, the parallel lamellae structure would surround a core of extended-chain crystals. In practice this material performed much like conventional hot-drawn HDPE, indicating that the proportion of extended-chain material must be small.

Measurements have been carried out in both the wet and dry states on two different textures of nylon-6 sheets with orthorhombic symmetry (102,103). A high degree of mechanical anisotropy was recorded, suggesting that the noncrystalline regions, in contrast with those in LDPE, retained appreciable molecular orientation. This feature led to significant differences between the shear compliances, and complicated the requirements for an understanding of extensional compliances. For the latter, the plasticizing effect of water was analogous to that produced by increasing temperature, and an explanation was given in terms of a Takayanagi model with a small amount of continuous amorphous phase. Shear measurements also reflected the importance of composite-phase morphology, but additionally showed the significant influence of molecular structure.

Solution-spinning techniques or the application of high draw ratios allow polyethylene to be produced in so highly oriented a state that the Young's modulus approaches the theoretical crystal chain modulus of some 300 GN/m² (43.5×10^6 psi) at low temperatures. Even at room temperature, the 10-s isochronous creep modulus can attain 60 GN/m² (8.7×10^6 psi) at draw ratios in excess of $30\times$. The oriented polymer has been considered as a short-fiber composite (104) consisting of 75% oriented crystal fibers ($E_c \sim 300$ GN/m² or 43.5×10^6 psi) embedded in 25% of a partially oriented amorphous phase with a shear modulus $G_m \sim 1$ GN/m² (145,000 psi) (Fig. 48). The modulus E of the oriented polymer is then

$$E = 0.75\, E_c \left(1 - \frac{\tanh x}{x}\right), \qquad x = \frac{L_c}{r_c}\left(\frac{G_m}{E_c \ln 2\pi/\sqrt{3c}}\right)^{1/2}$$

where c is the crystalline fraction.

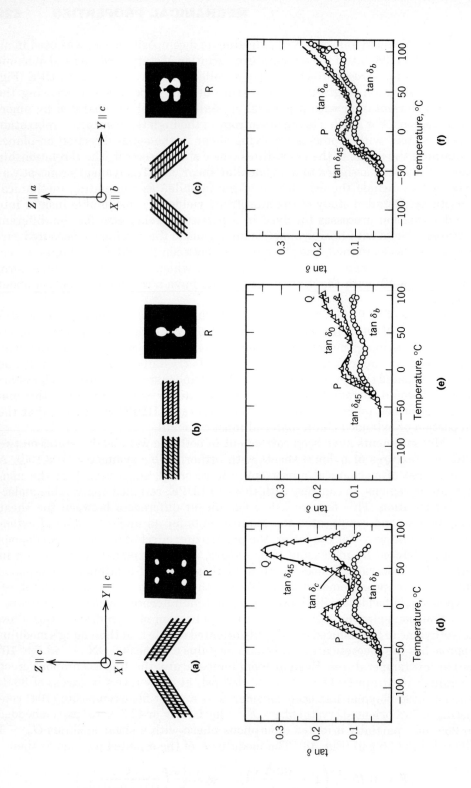

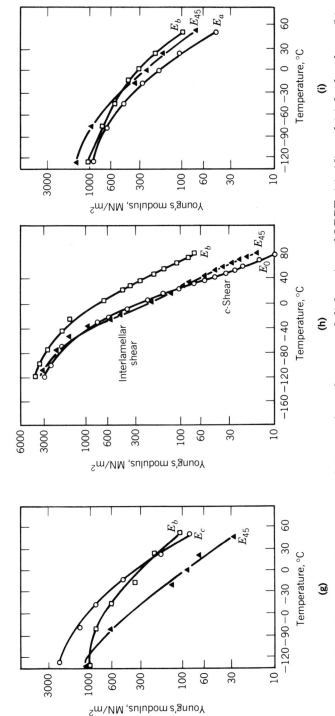

Fig. 47. Mechanical loss spectra and 10-s isochronal creep moduli for oriented LDPE: (**a**), (**d**), and (**g**) for *b-c* sheet; (**b**), (**e**), and (**h**) for parallel lamellae sheet; (**c**), (**f**), and (**i**) for *a-b* sheet. P, Interlamellar shear process; Q, *c*-shear process (*c*-shear process absent in (**f**)); R, small-angle x-ray diagram, beam along *X*. To convert MN/m² to psi, multiply by 145. 183

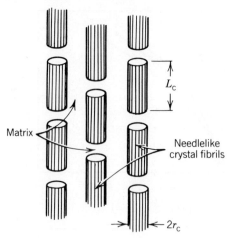

Fig. 48. Model for ultrahigh modulus polyethylene.

These materials are produced by first drawing the fiber through a neck and continuing the extension. It has been postulated that the increase in modulus on postneck drawing is a consequence of an increase in the aspect ratio $\dfrac{L_c}{2r_0}$ of the fibrils, which thus become more efficient reinforcing elements (105). The observed change in modulus with draw ratio is in accord with the assumptions of this fiber-composite model.

An alternative model is based on x-ray measurements, which for material of draw ratio $10\times$ are consistent with a regular stacking of crystal blocks whose length is less than the long period to accommodate the noncrystalline regions (106). Changes on further stretching are consistent with the linking of adjacent crystal sequences by crystalline bridges (Fig. 49), which are assumed to be randomly placed; the probability that adjacent crystal blocks are linked is defined by a single parameter p. Young's modulus is given as a function of the crystallinity χ by

$$E = E_c \chi p(2 - p) + E_a \frac{[1 - \chi + \chi(1 - p^2)]^2}{1 - \chi + \chi(1 - p)^2 E_a/E_c}$$

where the first term is the contribution of the crystalline-bridge sequences and the second, the contribution of the amorphous component and the remaining lamellar material, which are considered to act in series in a Takayanagi-type model. Dynamic mechanical-tensile measurements on drawn polyethylene show a plateau at $-50°C$, marking the region where E_a can be neglected; the Young's modulus becomes $E = \chi p(2 - p)E_c$. The fall in modulus above the $-50°C$ plateau occurs because of a reduction in E_c associated with the α-relaxation.

Both tensile and shear behavior can be brought together by an analogy with a short-fiber composite in which the parallel crystal component is regarded as the fiber phase, with the matrix being the remaining mixture of lamellar and noncrystalline material. The fiber phase is identified here with the sequences associated with the crystal bridges and not with the fibrils discussed previously (104). A best-fit model, in which the radius of the crystalline-bridge sequences was taken as 1.5 nm, predicted the correct magnitudes for the moduli at the

Fig. 49. Structure of the crystalline phase in ultrahigh modulus polyethylene (constructed for $p = 0.4$) (106). Courtesy of IPC Business Press, Ltd.

$-50°C$ plateau, the fall in the in-phase modulus with temperature, and the magnitudes of tan δ.

Yield Behavior

Plasticity theory has been applied to forming, rolling, and drawing processes. Although the yield behavior is temperature- and strain-rate dependent, test conditions can be chosen to provide yield stresses that satisfy conventional yield criteria. Temperature and strain-rate sensitivity, together with the molecular orientation associated with plastic deformation, can provide information at the molecular level to complement phenomenological data.

The most obvious demonstration of yield in tensile tests is the characteristic necking and cold drawing. Textile fibers are useful only when they have been subjected to a stabilized draw process. A typical load-extension curve (Fig. 50) shows a steady increase in load up to point B, where a neck is formed. Further elongation brings a decrease in load, and with continuing extension, the shoulders of the neck travel along the specimen as it thins to the drawn cross-section. The existence of a limiting draw ratio is an important aspect of polymer deformation; where the limit appears to be independent of the stretching conditions, it is referred to as the natural draw ratio. For a ductile material the yield stress defines the practical limit of performance much more than does the ultimate fracture.

A load-extension curve that relates deformation to the original length and cross-section of the specimen should, for greater understanding, be replotted to show the relation between true stress and true strain (Fig. 51), where three distinct regions are evident: first, the stress rises in an approximately linear manner as strain increases; second, there is a drop in true stress at the yield

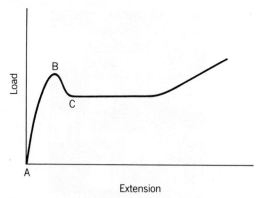

Fig. 50.   Typical load-extension curve for a cold-drawing polymer.

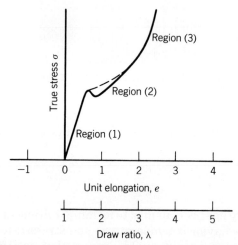

Fig. 51.   The true stress–strain curve for a cold-drawing polymer.

point and a region where the stress rises less steeply with the strain (the dotted line represents the region of strain-softening, a concept adopted before it was realized that a fall in stress might occur); and third, the slope increases again (strain-hardening). Localized deformation at the neck continues until strain hardening sufficiently increases the effective stiffness. The deformation then stabilizes in the highly strained element and, in order to accommodate further overall extension, the neck propagates along the length of the specimen.

The simplest theories of plasticity consider a rigid–plastic solid. Time dependence is neglected, and features occurring below the yield stress are ignored. In practice, these are the very features that throw light on molecular characteristics; indeed, the definition of yield stress itself depends on the shape of the deformation curve. Finally, the situation in standard extension tests differs from that in elasticity: the magnitude of the plastic-strain increments is determined by the movement of externally applied loads, demonstrating that there is no unique relation between stress and strain.

Yield Criteria for Polymers.   The behavior of polystyrene and other polymers was examined in tension, torsion, and compression, with particular attention paid to the effect of the hydrostatic component of stress on yield and on volume changes that occurred before and during yielding (107). There was asymmetry between the uniaxial-tensile and compressive-yield stresses and the results appeared to be compatible with a Coulomb criterion, where the yield stress was assumed to depend linearly on the mean-normal stress. In addition, the direction of yield, ie, the deformation bands, was consistent with the Coulomb criterion. Further studies on poly(methyl methacrylate) indicate that the yield data on this polymer also could be analyzed in terms of the Coulomb criterion, although attempts to adapt the direction of the deformation bands to this criterion were unsatisfactory (108,109).

Subsequently, many studies have shown that both the modulus and the yield stress of polymers increase with increasing hydrostatic pressure. For example, measurements of the torsional yield behavior of poly(methyl methacrylate) were made under superimposed hydrostatic pressures up to 700 MN/m² (101,500 psi) (110). In these experiments, shear yield stress increased substantially up to a pressure of 300 MN/m² (43,509 psi). Above this pressure, brittle failure occurred, which could have been prevented by protecting the specimens from the hydraulic fluid. Although the Coulomb yield criterion could be applied, it was equally reasonable to use the expression $\tau = \tau_0 + \alpha p$, where τ is the yield stress at high pressure, τ_0, that at atmospheric pressure, and α is the coefficient of increase of shear yield stress with hydrostatic pressure. In physical terms, the hydrostatic pressure compresses the polymer significantly, in contrast to metals where the bulk moduli are much larger ($\sim$100 GN/m² or 14.5 $\times$ 10⁶ psi compared with $\sim$5 GN/m² or 725,000 psi).

Further studies on the yield behavior of both glassy and crystalline polymers have been undertaken by several groups of workers (111–117).

Yield in Oriented Polymers.   If an oriented polymer is extended in a direction different from that of the initial draw, the deformation may be concentrated in a narrow deformation band (Fig. 52) that is analogous to the formation of a neck in an unoriented polymer. One type of band is nearly parallel to the initial-draw direction and resembles a slip band in metals; another type is more diffuse, produces a larger angle with the draw direction, and has the appearance of a kink band in metals. The yield criterion is more complex than for unoriented polymers but gives greater scope for an understanding of yield processes in molecular terms.

Slip bands were first recorded when oriented bristles of nylon-6,6 and nylon-6,10 were compressed lengthwise (118). Considerable reorientation occurred in the bands and the results were interpreted in terms of crystal plasticity theory.

For oriented HDPE sheets tested in tension at small angles with the initial-draw direction (119), it was suggested that the deformation process was a combination of (001) slip and twinning. Mechanical twinning in polyethylene had already been proposed as an explanation of crystal reorientation during rolling. At a larger angle to the initial-draw direction, kink bands with gross reorientation were observed, as in nylon.

In PET, despite a crystallinity little more than one-third that of HDPE, the direction of the deformation band differed significantly from the initial drawing

Fig. 52. Deformation band in an oriented PET sheet.

direction (120). The variation of yield stress with the angle θ between initial-draw direction and the tensile axis was consistent with a theory (121) applied to anisotropic sheets of orthorhombic symmetry. A particular feature is that the three orthogonal planes of symmetry, rather than the principal axes of stress, are chosen as the Cartesian axes of reference. The yield criterion then takes the form,

$$F(\sigma_{yy} - \sigma_{zz})^2 + G(\sigma_{zz} - \sigma_{xx})^2 + H(\sigma_{xx} - \sigma_{yy})^2 + 2L\sigma_{yz}^2 + 2M\sigma_{zx}^2 + 2N\sigma_{xy}^2 = 1$$

where F, G, H, L, M, and N are parameters that characterize the anisotropy of the material. The strain increments are derived by analogy with the rules for isotropic materials.

Poly(ethylene terephthalate) sheets have also been measured in simple shear, where the direction of shear displacement produces angles ϕ with the initial-draw direction (122,123). The stress–strain curves were not symmetrical in terms of ϕ, an effect that could be understood by resolving the applied shear stress into compressive and tensile components at right angles. For $\phi = 45°$, the initial-draw direction is parallel with the compressive component, but at $\phi = 135°$ the mean direction of the chains is parallel with the tensile component of stress; at 45° the chains are disoriented by the plastic deformation and a clear yield drop is observed. At 135°, the chains are further oriented by plastic deformation and a yield drop is not evident.

Rods of PET with long axes parallel with the draw direction show a compressive yield stress appreciably lower than that in tension (the so-called Bauschinger effect).

Molecular Interpretation of Yield.   Yield has been considered an activated rate process, with the applied stress inducing molecular flow in a manner comparable with Eyring viscosity theory, where the internal viscosity decreases with increasing stress. The yield stress denotes the point at which the internal viscosity falls and the applied strain rate is identical with the plastic strain rate $\dot{e}$ predicted by the Eyring equation. The measurement of yield stress in a constant strain-rate test is thus analogous to the measurement of the creep rate at constant applied stress.

The Eyring equation is

$$\dot{e} = \dot{e}_0 \exp\left(-\frac{\Delta H}{RT}\right) \sinh \frac{v\sigma}{RT}$$

where ΔH is the activation energy, σ, the applied tensile stress, and v, the activation volume, ie, the volume of the polymer segment that must move as a whole for plastic deformation to occur.

For high values of stress, $\sinh x = \frac{1}{2} \exp x$, which leads to an expression for tensile yield stress (τ)

$$\frac{\tau}{T} = \frac{R}{v}\left[\frac{\Delta H}{RT} + \ln \frac{2\dot{e}}{\dot{e}_0}\right]$$

Experiments on polycarbonate are in accord with the prediction that plots of τ/T against log (strain rate) give a series of parallel straight lines at different temperatures (124). The Eyring activation volumes obtained from yield-stress data are very large in molecular terms (6–18 nm^3), suggesting that yield involves the cooperative movement of a significant number of chain segments (125). The effect of hydrostatic pressure on yield stress can be accounted for by adding a pressure-activation volume term to the Eyring equation, which is equivalent to a linear increase in activation energy with increasing pressure.

Studies on poly(methyl methacrylate) and polycarbonate have shown that at low temperatures and high strain rates, the enhancement in yield stresses with increasing strain rate and decreasing temperature is greater than at high temperatures and low strain rates. Such behavior can be represented by the addition of two activated processes, one of which is dominant at high temperatures and low strain rates.

Attempts have been made to relate yield and creep processes, particularly by extending the linear models of springs and dashpots to embrace Langevin springs and Eyring dashpots. At best, such models provide a semiquantitative fit to experimental yield data and require much smaller activation volumes than those considered earlier (88).

In a modification of the Eyring theory, only two rotational conformations are considered: the trans low energy state and the cis, or flexed, high energy state (126). The energy difference between the two states is reduced in the presence of a shear stress in such a way that the application of a stress changes the fraction of elements in the high energy state. Further modification includes the effect of the hydrostatic component of stress, which increases the energy difference between the trans and cis states (127); this theory brings consistency to yield data in tension and compression for poly(methyl methacrylate) (Fig. 53).

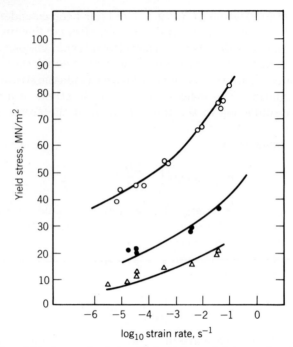

Fig. 53.   The yield stress of poly(methyl methacrylate) as a function of strain rate: ○, compression at 23°C; △, tension at 90°C; ●, tension at 60°C. Curves represent the best theoretical fit. To convert MN/m² to psi, multiply by 145.

Eyring-type theories imply that yield is not concerned with the initiation of the deformation process, but the application of stress changes the rate of deformation until it equals the imposed strain rate. The deformation mechanisms are considered as being present at zero stress and identical with those observed in linear viscoelastic measurements.

In a contrasting viewpoint, yield is assumed to be essentially nucleation controlled arising through the formation of a pair of molecular kinks (128). The resistance to kink formation is due to elastic interactions between a chain molecule and its neighbor, involving intermolecular forces rather than the intramolecular forces considered previously. The associated energy change is calculated by modeling the double kink as two wedge disinclination loops (129), giving the shear yield stress as

$$\tau = \frac{0.102G}{(1 - \nu)} \left[1 - \frac{16(1 - \nu)}{3\pi G\omega^2 a^3} kT \ln \frac{\dot{\gamma}_0}{\dot{\gamma}} \right]^{6/5}$$

where G is the shear modulus, ν, Poisson's ratio, a, the molecular radius, and ω, the angle of rotation of the molecular segment; the shear strain rate $\dot{\gamma} = \gamma_0 \Omega C \nu_a$ exp $\left(-\dfrac{\Delta H^*}{kT} \right)$, where γ_0 is the shear strain in the local volume Ω, C, the total volume density of potentially rotatable segments, ν_a, a frequency factor of the order of the atomic frequency, and ΔH^*, the activation energy for the formation of a kink pair.

Although the fit to experimental data for PET is very good (Fig. 54), the difference from the predictions of an Eyring model cannot be distinguished at the curve-fitting level. It is not surprising, therefore, that the large values of activation volume (>1 nm^3) are comparable with those obtained from fits to Eyring theory. Such values imply that the yield process must require the cooperative change of several adjacent molecular segments, rather than the formation of a single pair of adjacent kinks.

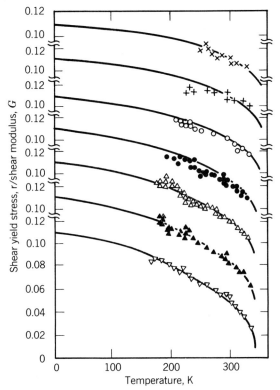

Fig. 54. Ratio of shear yield stress to shear modulus as a function of temperature at different strain rates for amorphous PET. Points are from unpublished data, curves from Ref. 128. Strain rates: x, 1.02×10^2 s^{-1}; +, 21.4 s^{-1}; $\bigcirc$, 1.96 s^{-1}; $\bullet$, 9×10^{-2} s^{-1}; $\triangle$, 9×10^{-3} s^{-1}; $\blacktriangle$, 9×10^{-4} s^{-1}; $\triangledown$, 9×10^{-5} s^{-1}.

Necking and Cold-drawing. Early experiments on the cold-drawing of filaments of PET suggested that stretching produced a local temperature increase, with necking occurring through the strain softening arising from the decrease in stiffness with rising temperature (130). The stability of the drawing process was attributed to an adiabatic process of heat transfer through the shoulders of the neck, with extension taking place at constant tension throughout the neck. A proposed heat-balance equation predicted with reasonable accuracy the length of the shoulder of the neck in cold-drawn PET. In nylon-6,6, there is the possibility of local melting in the neck caused by a combination of hydrostatic tension and temperature (131). Although temperature increases must be appreciable at conventional drawing speeds, necking can occur at very low draw speeds below the glass-transition temperature (132); at low speeds, the temperature rise may be less than 10°C.

Further examination of PET (133) has shown that although the drawing process is affected by adiabatic heat generation, the yield process is not affected at high strain rates (Fig. 55). At a low strain rate, any heat generated is conducted away from the neck at sufficient speed to prevent any rise in temperature. At higher draw rates, the effective temperature at which drawing takes place is increased; heat is conducted into the unyielded material, lowering its yield stress and reducing the force necessary to propagate the neck. Differences between the actual drawing stress and that which would have occurred in the absence of heating (the dashed line A in Fig. 55) allowed the increase in temperature to be calculated, assuming a similar dependence on temperature for drawing and yield stress. The calculated temperature rise approximately agreed with previous calculations derived from the work done in drawing; such elevated temperature appears reasonable for PET, since there is little change of crystallinity on drawing (in contrast with nylon-6,6).

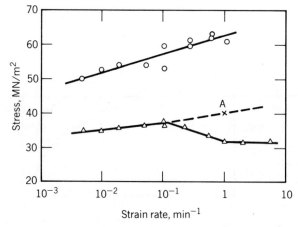

Fig. 55. ○, Yield stress and △, drawing stress as a function of strain rate for PET. To convert MN/m² to psi, multiply by 145.

Drawing and the Deformation of a Molecular Network. In amorphous polymers, yield and subsequent cold-drawing are associated with molecular rearrangements between points of entanglement and cross-linkages, and do not involve long-range molecular flow. Cross-linking certainly does not prevent the required molecular rearrangements. The natural draw ratio of fibers of amorphous polymers is sensitive to the degree of molecular orientation existing before the drawing stage, and experimental data support the theory that a network is formed at the moment the extruded filament first cools (134). The network is subsequently stretched in a rubberlike state before the fiber is collected as a frozen stretched rubber. Cold-drawing then expands the network to its limiting extensibility. Data confirm the theory that the ratio of the extended-to-unextended network is a constant independent of the division of the extension between the various stages of processing. It is possible that strain-hardening requirements directly determine the natural-draw ratio.

In crystalline polymers, plastic deformation causes drastic reorganization at the morphological level with the structure changing from a spherulitic to a fibrillar type. Molecular orientation processes are far from affine and can also involve mechanical twinning. Nevertheless, molecular topology and the deformation of a molecular network predominate in determining strain-hardening behavior and the ultimate-draw ratio (135). Both physical entanglements and crystallites, where more than one molecular chain is incorporated, can provide the network junction points, but the latter may be of a temporary nature since large draw ratios involve the breakdown of crystal structure and the unfolding of molecules.

Measurements of the molecular reorientations that occur in a deformation band can lead to a molecular understanding of the deformation process. For PET, an extension of the aggregate theory gives a reasonable prediction of the refractive indexes in the deformation band, indicating that the material undergoes deformation as a continuum (136); a similar treatment also is appropriate for polypropylene and linear polyethylene. Optical and x-ray studies suggest that the material within a deformation band, whether crystalline or not, becomes aligned about the new direction of maximum elongation, as if controlled by the deformation of an effective molecular network.

Fracture Phenomena

Brittle Fracture.   At low temperatures, fracture in polymers occurs in a brittle manner with the specimen failing at its maximum load at comparatively small strains (see also FRACTURE AND FATIGUE). The Griffith fracture theory, although strictly applicable only to perfectly elastic materials, provides a good framework for the failure of polymers under such conditions. Cracks are believed to spread from preexisting flaws, and, for rupture to occur, the energy required to produce the new surface must be balanced by a decrease in stored elastic energy. A number of recent studies of polymers have followed an equivalent representation (linear elastic fracture mechanics) in which the stress field near an idealized crack is considered (137). Fracture toughness is stated in terms of the critical stress-intensity factor (K_{Ic}), which defines the stress field at fracture for all loadings normal to the crack. Another important concept is the strain-energy release rate G; fracture occurs when G reaches a critical value G_c.

When flat strips of polystyrene and poly(methyl methacrylate) were cleaved lengthwise by propagating a crack down the middle (138), fracture occurred so slowly that the kinetic-energy dissipation associated with the movement of the crack could be neglected. It was shown, and confirmed by later experiments, that the fracture surface energy was independent of sample dimensions, suggesting that it was a basic material property. Measured values of surface energy were $\sim 10^2$ J/m^2, about two orders of magnitude higher than if the energy required to form a new surface arose only through the simultaneous breaking of chemical bonds. This large discrepancy, similar to that found in metals, is attributed to a viscous flow process that deforms the material close to the propagating crack. Work was expended in the alignment of polymer chains ahead of the crack, and the subsequent crack growth left a thin, highly oriented layer on the fracture surface (139). A thin wedge of porous material, termed a craze, forms a crack tip in a glassy polymer (see also CRAZING) (140). A craze forms under plane-strain conditions in such a way that the polymer is not free to contract laterally and consequently there is a reduction in density. The craze profile (Fig. 56) is very similar to the plastic-zone model for metals.

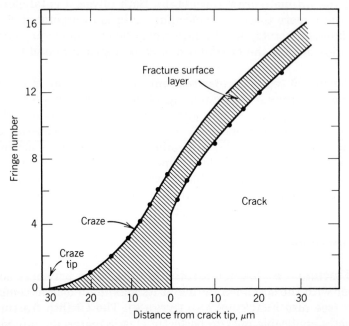

Fig. 56. Craze formation (141). Courtesy of IPC Business Press, Ltd.

The brittle–ductile transition can be considered in terms of competition between crazing and yielding. Both processes are activated, but with different temperature and strain-rate sensitivities; which is favored depends on the particular circumstances. An additional complexity arises from the nature of the stress field, which may favor one process rather than the other.

In some glassy polymers, a raised line of material, called a shear lip, forms on the fracture surface where the polymer has yielded. By analogy with metals, it has been proposed that the overall strain-energy release rate is the sum of contributions from the craze and the shear lip; results for polycarbonate and poly(ether sulfone) agree well with such an assumption (142,143).

More than forty years ago, Flory proposed that brittle strength is related to the number-average molecular weight, and later it was shown that the fracture-surface energy of poly(methyl methacrylate) depends on viscosity-average molecular weight (144) (Fig. 57). Subsequent studies have indicated a dramatic decrease in the size of the craze at molecular weights below 90,000. It has been suggested that the smallest molecule that could contribute to the surface energy would have its ends on the boundaries of the craze region on opposite sides of the fracture plane and be fully extended between these points. This supposition

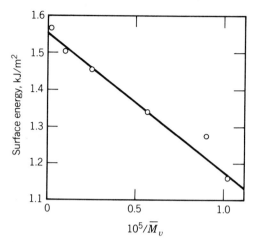

Fig. 57. Dependence of fracture surface energy on reciprocal molecular weight, where $\overline{M}_v$ = viscosity-average molecular weight (144). To convert J to cal, divide by 4.184.

is, however, inconsistent with the observed larger crack-opening displacements, and it would appear that the key factors in craze structure are the presence and average spacing of random entanglements.

Crazes appear also as opaque striations in planes whose normals are the direction of tensile stress when certain polymers, particularly poly(methyl methacrylate) and polystyrene, are subjected to tensile strains with constrained lateral contraction. It has been confirmed that both bulk crazes and surface layers are oriented structures of low density consistent with the deformation of a molecular network. The crazes consist of oriented fibrils with 50% void.

In an attempt to obtain a stress criterion for craze formation, the crazes formed in the vicinity of a small circular hole were examined, and the craze pattern was compared with solutions for the elastic stress field (145). The crazes grew parallel to the minor principal stress vector, ie, the principal stress acted along the craze plane normal and parallel with the molecular orientation axis of the crazed material. Although stress and strain criteria were derived, it was predicted that crazing would require a dilational stress field; this, however, has been refuted by experimental evidence.

It has been proposed that crazing arises from precursor micropores nucleated by stress concentrations of inhomogeneous plastic deformation on the order of 5–10 nm (146). Void formation increases with time by means of a thermally activated rate process. The free energy for the formation of a round pore contains two terms: one relating to the formation of the slip nucleus under the local deviatoric stress in the vicinity of a flaw, the other to the additional plastic deformation required to form a stable round cavity. The time after which the craze nucleus forms can be obtained by combining equations that describe the rate process and the stress field requirements. The growth of a pore to visible size occurs through the elastic unloading of the surroundings of a pore, and the craze front advances through the repeated breakup of the concave air–polymer interface at the crack tip (Fig. 58). The steady-state craze velocity relates to the

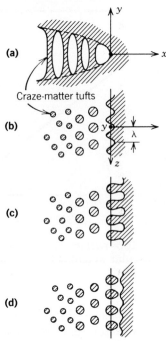

Fig. 58. Production of craze matter by the mechanism of meniscus instability: (**a**) outline of a craze tip; (**b**) cross-section in the craze plane across craze-matter tufts; (**c**), (**d**) advance of the craze front by a completed period of interface convolution (147).

5/6th power of the maximum principal tensile stress; this is supported by experimental results on poly(methyl methacrylate) and polystyrene.

In the presence of fluids, the stress and strain required to initiate crazing can be reduced to an extent determined by the solubility parameters of the solvent and the polymer. The behavior has been modeled after assuming that the work done in producing the craze depends on the expansion of a spherical cavity under a negative hydrostatic pressure (148). This pressure, which decreases with temperature, depends both on the surface tension between the solvent in the void and the surrounding polymer, and on the yield stress. At low enough temperatures, gases produce crazing in almost all linear polymers (149); a key factor is the surface concentration of the absorbed gas.

Molecular Treatment.   To date, there have been only tentative suggestions to explain why fracture behavior may be dependent on molecular processes. It is clear that fiber strength is at least an order of magnitude less than that calculated from bond breaking, and it is tempting to attribute this deficiency to the breaking of the tie molecules considered to determine the overall stiffness. Experiments in which electron-paramagnetic and infrared methods were used to measure the number of broken chains did not bear out this hypothesis (150). The number of breaks was much too small and appeared to relate to inhomogeneous stress distribution rather than strength.

The relation between time and fracture, stress, and temperature can be represented by an Eyring-type relation in which an activation barrier must be surmounted for fracture to occur (151). It was shown experimentally that the barrier height was approximately equal to that observed in thermal breakdown and that free radicals were produced at a rate correlating with the time to break. Microcracks, comparable with those noticed in work on crazing, were observed, but there is disagreement regarding their relevance to chain scission.

Brittle–Ductile Transitions. It has been assumed that brittle fracture and plastic flow in metals are independent processes, with the former occurring when the yield stress exceeds a critical value (Fig. 59). The characteristic curves depend on the strain rate and on the temperature at a constant strain rate, and the operative process (taken as that which occurs at the lower stress) is determined by the experimental conditions. The higher the temperature, the more likely is it that the material is ductile. Although this simple treatment ignores the relevance of fracture mechanics to brittle failure, it can provide a useful insight into material behavior. For common plastics, the yield stress may be ten times more sensitive than the brittle stress to changes in strain rate and temperature. Therefore, the brittle–ductile transition moves to higher temperatures with increasing strain rate. For instance, at room temperature, nylon-6,6 cold-draws at low strain rates, but at higher rate it fractures in a brittle manner. For most polymers, the brittle–ductile transition temperature does not appear to be closely related to a prominent mechanical relaxation. This is not unexpected, as such relaxations are defined by dynamic mechanical experiments performed in the linear, low strain region.

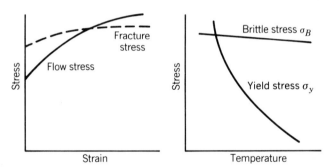

Fig. 59. Brittle–ductile transitions.

A complication at high strain rates occurs when the heat evolved in cold-drawing cannot be conducted away with sufficient rapidity. This isothermal-adiabatic transition does not affect the yield stress but reduces the energy to break. Since strain-hardening is prevented, the specimen fails in a ductile manner.

The brittle–ductile transition may be extremely sensitive to the details of molecular architecture. Increasing molecular weight has little influence on yield stress but appears to reduce the fracture stress. There is no general rule regarding the influence of side groups or branching; cross-linking increases the yield stress but does not have a great effect on the brittle strength so that the temperature of the transition is increased. The addition of a plasticizer has the opposite effect. Molecular orientation introduces mechanical anisotropy, with the brittle strength being more anisotropic than the yield stress. Hence, a uniaxially oriented polymer is more likely to fracture when the stress is applied perpendicular to the symmetry axis. This process of fibrillation forms the basis for the manufacture of synthetic fibers from polymer films.

As in metals, the presence of a sharp notch can change the fracture of a polymer from ductile to brittle. Experiments tend to confirm the suggestion that an ideally sharp and deep notch in an infinite solid should raise the yield stress by a factor of ca 3 (152); that is, a material that was previously ductile fails by brittle fracture when notched, if the brittle stress is less than three times the yield stress.

The above discussion ignores size effects stemming from basic ideas of energy scaling and similarity in which the severity of a test is associated with a characteristic length, eg, the size of the plastic zone or the length of a craze. A theory of fracture transitions was developed in which a lower transition marks the point where platic flow commences and an upper transition corresponds to the size of the plastic zone reaching a maximum dimension characteristic of the test (153,154). Each test is regarded as relating to a particular characteristic length,

$$x_{0_c} = \alpha G_c E / \sigma_y^2$$

where α is a numerical factor determined by the stress field, ie, by the type of test, G_c is the critical strain energy release rate, E is Young's modulus, and σ_y is the yield stress. The fracture transition occurs at the temperature at which $\alpha G_c E / \sigma_y^2$ is equal to the appropriate critical length. In a given test, σ_y decreases with increasing temperature until the equality is satisfied and the transition from brittle to ductile behavior occurs.

Impact Strength.   Impact tests, in which a small bar of polymer, usually notched, is struck by a heavy pendulum are simple to perform but can be interpreted in several alternative ways:

1. The specimen deforms in a linear elastic fashion up to failure, which occurs when the strain-energy release rate reaches a critical value G_c (155). This approach has given values of G_c that are independent of geometry on razor-notched specimens of several glassy polymers and also polyethylene.

2. Deformation occurs in a linear elastic manner up to the point of failure, which occurs when the stress at the root of the notch reaches a critical value and is otherwise independent of specimen geometry (156). The assumption that the fracture criterion is determined by the maximum local stress appears to hold for blunt notched specimens of poly(methyl methacrylate) and several other polymers.

3. The impact test is a measure of the energy required to propagate a crack across the specimen, which is the situation for rubber-toughened ABS polymers (see IMPACT RESISTANCE).

A satisfactory theoretical analysis has been considered only for Charpy and Izod impact tests, and even for these cases there is no accepted interpretation in physical terms. When a craze occurs, there is no convincing link between the craze parameters and fracture toughness, and when a critical-stress criterion appears relevant, the magnitude of the stress may be unexpectedly high. Additionally, attempts to correlate impact performance with dynamic mechanical behavior or the area under the stress–strain curve have met with mixed success, although there is suggestive evidence that in some cases viscoelastic losses during loading need to be taken into account (157).

The impact strength of polystyrene and other thermoplastics can be increased by dispersing rubber throughout the material in the form of small aggregates. Explanations for the improvement have included the absorption of energy by the stretched rubber and the action of the rubber particles as stress-concentration points, giving rise to many small cracks with a high surface area rather than a single large crack. High impact strength in rubber-modified polystyrene has been linked to stress-whitening and craze formation, which occur only when the rubber is able to relax (158). At sufficiently low temperatures, the rubber is unable to relax, there is no craze formation, and brittle fracture occurs.

The difference between stress-whitening and crazing appears to be merely that in the former the craze bands are much smaller in size and greater in quantity. Rubber particles appear to play a complex role; they lower the craze-limitation stress relative to the fracture stress, and though they relax, they are able to sustain part of the load since they are constrained by the surrounding polystyrene matrix to which they are firmly bonded.

The Nature of the Fracture Surface. Fracture surfaces in polymers show similarities to those in metals. Often there is a clearly defined mirror area where the fracture commences surrounded by parabolic markings caused by the interference of the main crack with new cracks nucleating ahead of it. Changes in the rate of loading or the molecular weight produce identical changes in the fracture surface of poly(methyl methacrylate). The mirrorlike area becomes more extensive when the former is decreased (Fig. 60) or the latter increased—an effect which corresponds to a change in the work done during fracture.

Glass has an internal strength much greater than the surface strength because fracture arises from the enlargement of preexisting surface flaws. Rapid extension, stress-pulse, and surface treatment experiments indicate that, in contrast, cracks in brittle polymers originate at flaws throughout the material.

The Ultimate Properties of Rubberlike Polymers. The Griffith fracture theory is too restrictive in relating the reduction in elastic energy caused by crack propagation simply to changes in surface energy (160). Tearing energy T (expended per unit thickness per unit increase in crack length) includes surface energy, energy dissipated in plastic flow processes, and energy dissipated irreversibly in viscoelastic processes. Using a standard trouser tear test on a rubber sheet, two characteristic tearing energies are obtained: for very slow rates of tear $T = 37$ kJ/m^2 (155 kcal/m^2) and for catastrophic crack growth $T = 130$ kJ/m^2 (544 kcal/m^2); both quantities are independent of the shape of the test piece. The tearing energy would not be expected to relate directly to tensile strength, as both the shape of the stress–strain curve and the viscoelasticity of the rubber are involved.

The Griffith equation relating tensile strength to surface energy, modulus, and crack length was applied to rubbers (161). The calculated surface energies implied crack lengths of 10^{-4} mm, two orders of magnitude less than those indicated by direct examination, and it was concluded that most of the energy involved must have been dissipated in viscoelastic and flow processes. A linear rather than the predicted square root dependence of tensile strength on both crack length and modulus was found, which led the data to be reconsidered in terms of the critical stresses developed at breaking points.

Most molecular theories of rubber strength treat rupture as a critical stress

Very slow loading

~3 × 10² N/m² · s

~1.5 × 10³ N/m² · s

1.5 × 10⁶ N/m² · s

Fig. 60. The effect of loading rate on fracture appearance for poly(methyl methacrylate) (159).

phenomenon. A model three-dimensional network of cross-linked chains predicts that the breaking stress σ_B is proportional to $[1 - 2\overline{M}_c/\overline{M}_n]^{2/3}$, where $\overline{M}_c$ and $\overline{M}_n$ are the average molecular weight between cross-links and the number-average molecular weight, respectively (162). Butyl rubber appears to follow such a relationship, although several observers have noted that the tensile strength of both natural and synthetic rubbers drops again at very high degrees of cross-linking. The simple model fails in this region because each chain no longer holds the load at fracture.

The increase in tensile strength of rubbers through the inclusion of reinforcing fillers of carbon black and silicone is of major technological importance. These fillers allow the applied load to be shared among a group of chains, decreasing the chance that a break will propagate.

The effects of strain rate on the tensile strength and ultimate strain were studied for a number of elastomers (163–165). The results for different temperatures could be superimposed by shifts along the strain-rate axis to give master curves for tensile strength and ultimate strain as a function of strain rate (Fig. 61). The shift factors took the form predicted by the WLF equation for the su-

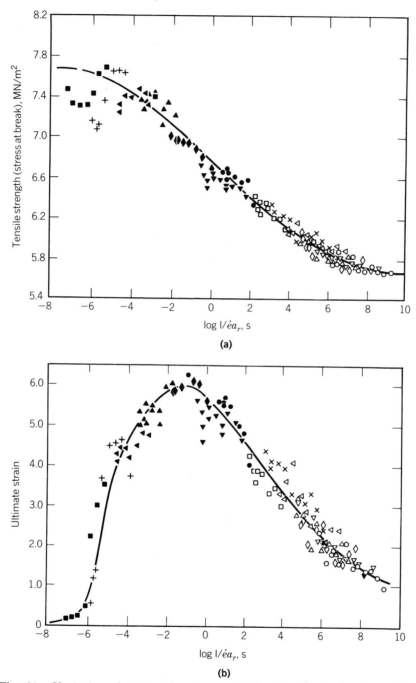

Fig. 61. Variation of (a) tensile stress and (b) ultimate strain of a rubber with reduced strain rate. Values measured at various temperatures and rates were reduced to a temperature of 263 K. Temperatures (°C) for (a) and (b): $\bigcirc$, 93.3; $\triangledown$, 87.8; $\diamondsuit$, 71.0; $\triangle$, 60.0; $\triangleleft$, 43.3; x, 25.0; $\square$, 10.0; $\bullet$, -12.2; $\blacktriangledown$, -23.3; $\blacklozenge$, -34.4; $\blacktriangle$, -42.2; $\blacktriangleleft$, -48.4; $+$, -53.8; and $\blacksquare$, -67.8 (163). To convert N/m^2 to psi, multiply by 1.45×10^{-4}.

perimposition of low strain linear viscoelastic behavior of amorphous polymers, with a glass-transition temperature T_g in good agreement with that obtained from dilatometric measurements. This result suggests that except at very low strain rates and high temperatures, where the molecular chains have complete mobility, the fracture process is dominated by viscoelastic effects.

A plot of σ_B/T against log e yields a unique curve for all strain rates and test temperatures that is termed a failure envelope. This envelope can represent failure under more complex conditions, eg, creep and stress relaxation.

A Generalized Theory of Fracture Mechanics.   Fracture mechanics has provided a satisfactory starting point for discussing the failure both of glassy brittle polymers and elastomeric materials, but most polymers are not perfectly elastic. A more generalized theory of fracture has been proposed in which the requirement of perfectly elastic deformation is removed (166,167); accordingly, loading and unloading are represented by different functions. Suitable analysis yields the energy available for forming a surface crack after deduction of the energy losses throughout the bulk of the material $\mathcal{T}_0$, which can be compared with $\mathcal{T}$, the total energy expended by the system to produce unit area of crack growth. The comparison involves a loss function Φ that varies with external loading, temperature, and strain rate.

This theory has been tested by measuring the propagation of an edge crack in parallel-sided strips of four polymers. The values of $\mathcal{T}_0$ obtained approach the theoretical values calculated on the basis of the minimum energy of fracture being that which is required to break the interatomic bonds across the fracture plane. These preliminary results suggest that this theory may provide a starting point for a fundamental understanding of polymer fracture at the molecular and structural level.

Fatigue Failure.   Relatively low cyclic stresses can initiate microscopic cracks at centers of stress concentration, which on continued cycling can propagate and lead to eventual failure. Early studies with polymers followed the practice familiar with metals of cycling unnotched samples to produce plots of stress vs number of cycles to failure (N). Unless the frequency is kept low enough, adiabatic heating may be a problem, as it can lead to failure through melting. Tests with unnotched samples do not differentiate between crack initiation and crack propagation, and very sharp initial cracks must be introduced in order to examine crack propagation using the concepts of fracture mechanics.

The first quantitative studies on rubbers applied the tearing-energy concept of fracture to fatigue crack propagation (168); the crack-growth rate is expressed by the empirical relation

$$\frac{dc}{dN} = AT^n$$

where c is the crack length and T, the surface work parameter, which is analogous with the strain–energy release rate in linear fracture mechanics; A and n are constants that depend both on the material and the test conditions, with n normally between 1 and 6; T varies throughout the cycle and has a limiting value T_0 below which a fatigue crack does not propagate, corresponding with the minimum energy required per unit area to extend the rubber at the crack tip to its breaking point.

The generalized form of fracture mechanics previously discussed was applied to the analysis of the fatigue of low density polyethylene, a viscoelastic material (169). Fatigue characteristics were predicted from crack-growth data using the intrinsic flaw size c_0 as a fitting constant. It was suggested that the intrinsic

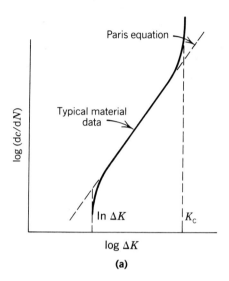

(a)

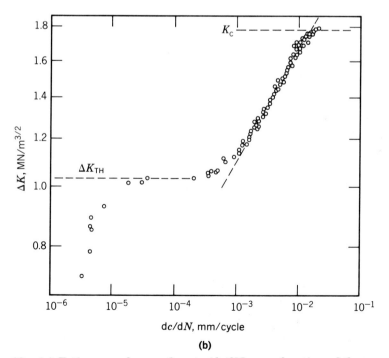

(b)

Fig. 62. (a) Fatigue crack growth rate (dc/dN) as a function of the range of the stress-intensity factor ΔK; (b) fatigue crack growth characteristics for a vinyl urethane polymer (171).

flaws were interspherulite boundary cracks and that c_0 corresponded to spherulite dimensions.

For glassy polymers an empirical relation

$$\frac{dc}{dN} = A'(\Delta K)^m$$

is usually adopted, where ΔK is the range of the stress-intensity factor, ie, $K_{max} - K_{min}$. This expression is also the most general form of a law for predicting fatigue crack-growth rates in metals (170). When $K_{min} = 0$ the form is identical with that of the relation used for rubbers: for plane strain the strain–energy release rate $G = 2T = K_{max}^2/2E$. Some typical results for polymers (Fig. 62) show divergence from the above law. There is a threshold value ΔK_{TH}, below which no crack growth is observed, and the crack accelerates as the critical stress-intensity factor K_C is approached. The empirical equation also ignores the important influence of the mean stress as distinct from the stress range. Further information is given in Ref. 172 in which physical variables such as crystallinity and molecular weight are discussed.

Recently several modifications of the above equation have been discussed, but in no case is it possible to assign a physical significance to the parameters of the equation. A fundamentally different approach is required before any physical insight can be gained (see FATIGUE in FRACTURE AND FATIGUE).

BIBLIOGRAPHY

"Mechanical Properties" in *EPST* 1st ed., Vol. 8, pp. 445–516, by V. A. Kargin and G. L. Slonimsky, Akademii Nauk SSSR.

1. R. S. Rivlin, *Philos. Trans. R. Soc. London Ser. A* **240**, 459, 491 (1948); **241**, 379 (1949).
2. M. Mooney, *J. Appl. Phys.* **11**, 582 (1940).
3. R. S. Rivlin and D. W. Saunders, *Philos. Trans. R. Soc. London Ser. A* **243**, 251 (1951).
4. R. S. Rivlin and D. W. Saunders, *Trans. Faraday Soc.* **48**, 200 (1952).
5. G. W. Becker, *J. Polym. Sci. Part C* **16**, 2893 (1967).
6. L. R. G. Treloar, *The Physics of Rubber Elasticity,* 3rd ed., Clarendon Press, Oxford, UK, 1975, Chapt. 11.
7. L. R. G. Treloar, *Proc. R. Soc. London Ser. A* **351**, 301 (1976).
8. K. C. Valanis and R. F. Landel, *J. Appl. Phys.* **38**, 2997 (1967).
9. R. W. Ogden, P. Chadwick, and E. W. Haddon, *Q. J. Mech. Appl. Math.* **26**, 23 (1973).
10. L. R. G. Treloar, *The Physics of Rubber Elasticity,* 2nd ed., Clarendon Press, Oxford, UK, 1958, p. 21.
11. T. Alfrey, *Mechanical Behavior of High Polymers,* Interscience Publishers, New York, 1948.
12. B. Gross, *Mathematical Structure of the Theories of Viscoelasticity,* VEB Hermann Haack, Paris, 1953.
13. G. W. Becker, *Mater. Plast. Elast.* **35**, 1287 (1969).
14. P. R. Pinnock and I. M. Ward, *Polymer* **7**, 255 (1966).
15. K. W. Hillier and H. Kolsky, *Proc. Phys. Soc. London Ser. B* **62**, 111 (1949).
16. H. Kolsky in N. Davide, ed., *International Symposium on Stress Wave Propagation in Materials,* Interscience Publishers, New York, 1960, p. 59.
17. H. Kolsky, *Structural Mechanics,* Pergamon Press, Oxford, UK, 1960, p. 233.
18. R. S. Marvin in *Proceedings of the 2nd International Congress on Rheology,* Butterworth & Co. (Publishers) Ltd., Kent, UK, 1954.
19. R. S. Marvin and H. Oser, *J. Res. Natl. Bur. Stand. Sect. B* **66**, 171 (1962).
20. A. B. Thompson and D. W. Woods, *Trans. Faraday Soc.* **52**, 1383 (1956).

21. S. Glasstone, K. J. Laidler, and H. Eyring, *The Theory of Rate Processes,* McGraw-Hill Inc., New York, 1941.
22. J. D. Hoffman, G. Williams, and E. Passaglia, *J. Polym. Sci. Part C* **14,** 173 (1966).
23. J. D. Ferry, *Viscoelastic Properties of Polymers,* 1st ed., John Wiley & Sons, Inc., New York, 1961, Chapt. 11.
24. M. L. Williams, R. F. Landel, and J. D. Ferry, *J. Am. Chem. Soc.* **77,** 3701 (1955).
25. F. Bueche, *J. Chem. Phys.* **21,** 1850 (1953).
26. G. Adam and J. H. Gibbs, *J. Chem. Phys.* **43,** 139 (1965).
27. G. Williams, *Trans. Faraday Soc.* **60,** 1556 (1964).
28. P. E. Rouse, *J. Chem. Phys.* **21,** 1272 (1953).
29. F. Bueche, *J. Chem. Phys.* **22,** 603 (1953).
30. B. H. Zimm, *J. Chem. Phys.* **24,** 269 (1956).
31. P. G. de Gennes, *J. Chem. Phys.* **55,** 572 (1971).
32. M. Doi and S. F. Edwards, *J. Chem. Soc. Faraday Trans.* **74**(2), 1789, 1802, 1818 (1978).
33. K. E. Evans and S. F. Edwards, *J. Chem. Soc. Faraday Trans.* **77**(2), 1891, 1913, 1929 (1981).
34. J. Heijboer in J. A. Prins, ed., *The Physics of Noncrystalline Solids,* North-Holland, Amsterdam, The Netherlands, 1965, p. 231.
35. M. F. Drumm, C. W. H. Dodge, and L. E. Nielsen, *Ind. Eng. Chem.* **48,** 76 (1956).
36. L. E. Nielsen, *Mechanical Properties of Polymers,* Van Nostrand Reinhold Co., Inc., New York, 1962.
37. L. Mandelkern, G. M. Martin, and F. A. Quinn, *J. Res. Natl. Bur. Stand.* **58,** 137 (1959).
38. K. Wolf, *Kunststoffe* **41,** 89 (1951).
39. N. G. McCrum, *J. Polym. Sci.* **34,** 355 (1959).
40. M. Takayanagi, *Mem. Fac. Eng. Kyushu Univ.* **23,** 1 (1963).
41. K. Schmieder and K. Wolf, *Kolloid Z.* **134,** 149 (1953).
42. M. Takayanagi in *Proceedings of the 4th International Congress on Rheology,* Wiley-Interscience, New York, 1965, Part 1, p. 161.
43. J. D. Hoffman, G. Williams, and E. A. Passaglia, *J. Polym. Sci. Part C* **14,** 173 (1966).
44. Z. H. Stachurski and I. M. Ward, *J. Polym. Sci. Part A-2* **6,** 1083 (1968).
45. *Ibid.,* p. 1817.
46. Z. H. Stachurski and I. M. Ward, *J. Macromol. Sci. Phys.* **3,** 445 (1969).
47. S. Turner, *Polym. Eng. Sci.* **6,** 306 (1966).
48. Y. H. Pao and J. Marin, *J. Appl. Mech.* **19,** 478 (1952).
49. Y. H. Pao and J. Marin, *J. Appl. Mech.* **20,** 245 (1953).
50. W. N. Findley and G. Khosla, *J. Appl. Phys.* **26,** 821 (1955).
51. D. J. Plazek, *J. Colloid Sci.* **15,** 50 (1960).
52. I. H. Hall, *J. Polym. Sci.* **54,** 505 (1961).
53. E. Guth, P. E. Wack, and R. L. Anthony, *J. Appl. Phys.* **17,** 347 (1946).
54. A. V. Tobolsky and R. D. Andrews, *J. Chem. Phys.* **13,** 3 (1945).
55. H. Leaderman, *Trans. Soc. Rheol.* **6,** 361 (1962).
56. H. Leaderman, *Elastic and Creep Properties of Filamentous Materials and Other High Polymers,* Textile Foundation, Washington, D.C., 1943.
57. B. Bernstein, E. A. Kearsley, and L. P. Zapas, *Trans. Soc. Rheol.* **7,** 391 (1963).
58. B. D. Coleman and W. Noll, *Rev. Mod. Phys.* **33,** 239 (1961).
59. P. H. de Hoff, G. Lianis, and W. Goldberg, *Trans. Soc. Rheol.* **10,** 385 (1966).
60. W. Goldberg and G. Lianis, *J. Appl. Mech.* **37,** 53 (1970).
61. K. C. Valanis and R. F. Landel, *Trans. Soc. Rheol.* **11,** 243 (1967).
62. R. A. Schapery, *Polym. Eng. Sci.* **9,** 295 (1969).
63. I. M. Ward and E. T. Onat, *J. Mech. Phys. Solids* **11,** 217 (1963).
64. A. E. Green and R. S. Rivlin, *Arch. Ration. Mech. Anal.* **1,** 1 (1957).
65. A. E. Green and R. S. Rivlin, *Arch. Ration. Mech. Anal.* **4,** 387 (1960).
66. A. E. Green, R. S. Rivlin, and A. J. M. Spencer, *Arch. Ration. Mech. Anal.* **3,** 82 (1959).
67. D. W. Hadley and I. M. Ward, *J. Mech. Phys. Solids* **13,** 397 (1965).
68. I. M. Ward and J. M. Wolfe, *J. Mech. Phys. Solids* **14,** 131 (1966).
69. F. J. Lockett, *Nonlinear Viscoelastic Solids,* Academic Press, Inc., London, 1972, Chapts. 5 and 6.
70. J. S. Y. Lai and W. N. Findley, *Trans. Soc. Rheol.* **12,** 243 (1968).

71. C. J. Morgan and I. M. Ward, *J. Mech. Phys. Solids* **19,** 165 (1971).
72. M. G. Brereton, S. G. Croll, R. A. Duckett, and I. M. Ward, *J. Mech. Phys. Solids* **22,** 97 (1974).
73. S. S. Sternstein and T. C. Ho, *J. Appl. Phys.* **43,** 4370 (1972).
74. C. P. Buckley and N. G. McCrum, *J. Mater. Sci.* **9,** 2064 (1974).
75. O. D. Sherby and J. E. Dorn, *J. Mech. Phys. Solids* **6,** 145 (1958).
76. M. J. Mindel and N. Brown, *J. Mater. Sci.* **8,** 863 (1973).
77. M. A. Wilding and I. M. Ward, *Polymer* **22,** 870 (1981).
78. M. A. Wilding and I. M. Ward, *Plast. Rubber Process. Appl.* **1,** 167 (1981).
79. E. L. V. Lewis, I. D. Richardson, and I. M. Ward, *J. Phys. E.* **12,** 189 (1979).
80. O. K. Chan, F. C. Chen, C. L. Choy, and I. M. Ward, *J. Phys. D.* **11,** 481 (1975).
81. L. Holliday in I. M. Ward, ed., *Structure and Properties of Oriented Polymers,* Applied Science Publishers, London, 1975, Chapt. 7.
82. I. M. Ward in I. M. Ward, ed., *Advances in Oriented Polymers,* Vol. 1, Applied Science Publishers, London, 1982, Chapt. 5.
83. A. Peterlin in A. Ciferri and I. M. Ward, eds., *Ultra-high Modulus Polymers,* Applied Science Publishers, London, 1979, Chapt. 10.
84. I. M. Ward, *Proc. Phys. Soc. London* **80,** 1176 (1962).
85. S. W. Allison and I. M. Ward, *Br. J. Appl. Phys.* **18,** 1151 (1967).
86. D. W. Hadley, P. R. Pinnock, and I. M. Ward, *J. Mater. Sci.* **4,** 152 (1969).
87. M. Takayanagi, K. Imada, and T. Kajiyama, *J. Polym. Sci. Part C* **15,** 263 (1966).
88. J. Hennig, *Kolloid Z.* **200,** 46 (1964).
89. I. M. Ward, *Mechanical Properties of Solid Polymers,* 2nd ed., John Wiley & Sons, Inc., Chichester, UK, New York, 1983, Chapt. 10.
90. R. E. Robertson and R. J. Buenker, *J. Polym. Sci. Part A-2* **2,** 4889 (1964).
91. H. Wright, C. S. N. Faraday, E. F. T. White, and L. R. G. Treloar, *J. Phys. D.* **4,** 2002 (1971).
92. G. Raumann and D. W. Saunders, *Proc. Phys. Soc. London* **77,** 1028 (1961).
93. G. Raumann, *Proc. Phys. Soc. London* **79,** 1221 (1962).
94. V. B. Gupta and I. M. Ward, *J. Macromol. Sci. Phys.* **1,** 373 (1967).
95. E. L. V. Lewis and I. M. Ward, *J. Mater. Sci.* **15,** 2354 (1980).
96. V. B. Gupta and I. M. Ward, *J. Macromol. Sci. Phys.* **2,** 89 (1968).
97. Z. H. Stachurski and I. M. Ward, *J. Macromol. Sci. Phys.* **3,** 427 (1969).
98. G. R. Davies, A. J. Owen, I. M. Ward, and V. B. Gupta, *J. Macromol. Sci. Phys.* **6,** 215 (1972).
99. G. R. Davies and I. M. Ward, *J. Polym. Sci. Part B* **7,** 353 (1969).
100. A. J. Owen and I. M. Ward, *J. Macromol. Sci. Phys.* **19,** 35 (1981).
101. M. Kapuscinski, I. M. Ward, and J. Scanlan, *J. Macromol. Sci. Phys.* **11,** 475 (1975).
102. E. L. V. Lewis and I. M. Ward, *J. Macromol. Sci. Phys.* **18,** 1 (1980).
103. E. L. V. Lewis and I. M. Ward, *J. Macromol. Sci. Phys.* **19,** 75 (1981).
104. R. G. C. Arridge, P. J. Barham, and A. Keller, *J. Polym. Sci. Polym. Phys. Ed.* **15,** 389 (1977).
105. P. J. Barham and R. G. C. Arridge, *J. Polym. Sci. Polym. Phys. Ed.* **15,** 1177 (1977).
106. A. G. Gibson, G. R. Davies, and I. M. Ward, *Polymer* **19,** 683 (1978).
107. W. Whitney and R. D. Andrews, *J. Polym. Sci. Part C* **16,** 2981 (1967).
108. R. I. Thorkildsen, *Engineering Design for Plastics,* Van Nostrand Reinhold Co., Inc., New York, 1964, p. 322.
109. P. B. Bowden and J. A. Jukes, *J. Mater. Sci.* **3,** 183 (1968).
110. S. Rabinowitz, I. M. Ward, and J. S. C. Parry, *J. Mater. Sci.* **5,** 29 (1970).
111. G. Biglione, E. Baer, and S. V. Radcliffe in P. L. Pratt, ed., *Fracture,* Chapman and Hall, London, 1969, p. 520.
112. A. W. Christiansen, E. Baer, and S. V. Radcliffe, *Philos. Mag.* **24,** 451 (1971).
113. K. Matsushigi, S. V. Radcliffe, and E. Baer, *J. Mater. Sci.* **10,** 833 (1975).
114. K. D. Pae, D. R. Mears, and J. A. Sauer, *J. Polym. Sci. Part B* **6,** 773 (1968).
115. D. R. Mears, K. D. Pae, and J. A. Sauer, *J. Appl. Phys.* **40,** 4229 (1969).
116. J. S. Harris, I. M. Ward, and J. S. C. Parry, *J. Mater. Sci.* **6,** 110 (1971).
117. I. M. Ward, *J. Mater. Sci.* **6,** 1397 (1971).
118. D. A. Zaukelies, *J. Appl. Phys.* **33,** 2797 (1961).
119. M. Kurakawa and T. Ban, *J. Appl. Polym. Sci.* **8,** 971 (1964).
120. N. Brown and I. M. Ward, *J. Polym. Sci. Part A-2* **6,** 607 (1968).
121. R. Hill, *The Mathematical Theory of Plasticity,* Clarendon Press, Oxford, UK, 1950, pp. 50, 318.

122. N. Brown, R. A. Duckett, and I. M. Ward, *Philos. Mag.* **18**, 483 (1968).
123. C. Bridle, A. Buckley, and J. Scanlan, *J. Mater. Sci.* **3**, 622 (1968).
124. C. Bauens-Crowet, J. A. Bauens, and G. Homès, *J. Polym. Sci. Part A-2* **7**, 735 (1969).
125. R. N. Haward and G. Thackray, *Proc. R. Soc. London Ser. A* **302**, 453 (1968).
126. R. E. Robertson, *J. Chem. Phys.* **44**, 3950 (1966).
127. R. A. Duckett, S. Rabinowitz, and I. M. Ward, *J. Mater. Sci.* **5**, 909 (1970).
128. A. S. Argon, *Philos. Mag.* **28**, 839 (1973).
129. J. C. M. Li and J. J. Gilman, *J. Appl. Phys.* **41**, 4248 (1970).
130. I. Marshall and A. B. Thompson, *Proc. R. Soc. London Ser. A* **221**, 541 (1954).
131. D. C. Hookway, *J. Text. Inst.* **49**, 292 (1958).
132. Y. S. Lazurkin, *J. Polym. Sci.* **30**, 595 (1958).
133. S. W. Allison and I. M. Ward, *Br. J. Appl. Phys.* **18**, 1151 (1967).
134. S. W. Allison, P. R. Pinnock, and I. M. Ward, *Polymer* **7**, 66 (1966).
135. G. Capaccio, T. A. Crompton, and I. M. Ward, *J. Polym. Sci. Polym. Phys. Ed.* **14**, 1641 (1976).
136. I. D. Richardson, R. A. Duckett, and I. M. Ward, *J. Phys. D* **3**, 649 (1970).
137. G. R. Irwin, *J. Appl. Mech.* **24**, 361 (1957).
138. J. J. Benbow and F. C. Roesler, *Proc. Phys. Soc. London Sect. B* **70**, 201 (1957).
139. J. P. Berry in B. L. Auerbach and co-workers, eds., *Fracture,* John Wiley & Sons, Inc., New York, 1959, p. 263.
140. R. P. Kambour, *Macromol. Rev.* **7**, 1 (1973).
141. N. Brown and I. M. Ward, *Polymer* **14**, 469 (1973).
142. G. L. Pitman and I. M. Ward, *Polymer* **20**, 895 (1979).
143. P. J. Hine, R. A. Duckett, and I. M. Ward, *Polymer* **22**, 1745 (1981).
144. J. P. Berry, *J. Polym. Sci. Part A* **2**, 4069 (1964).
145. S. S. Sternstein, L. Ongchin, and A. Silverman, *Appl. Polym. Symp.* **7**, 175 (1968).
146. A. S. Argon and J. G. Hannoosh, *Philos. Mag.* **36**, 1195 (1977).
147. A. S. Argon, J. G. Hannoosh, and M. M. Salama in D. M. R. Taplin, ed., *Fracture,* Vol. 1, Pergamon Press, Oxford, UK, 1977, p. 445.
148. E. H. Andrews and L. Bevan, *Polymer* **13**, 337 (1972).
149. N. Brown in R. A. Fava, ed., *Methods of Experimental Physics,* Vol. 16, Academic Press, Inc., New York, 1980, Part C, p. 233.
150. H. H. Kausch, *Polymer Fracture,* Springer-Verlag, Berlin, 1978.
151. S. N. Zhurkov and E. E. Tomashevsky in *Proceedings of the Conference on Physical Basis Yield and Fracture,* Oxford, UK, 1966, p. 200.
152. E. Orowan, *Rep. Prog. Phys.* **12**, 185 (1949).
153. K. E. Puttick, *J. Phys. D.* **11**, 595 (1978).
154. J. A. Kies and A. B. J. Clark in *Proceedings of the 2nd International Conference on Fracture,* Brighton, UK, 1969, paper 42.
155. H. R. Brown, *J. Mater. Sci.* **8**, 941 (1973).
156. R. A. W. Fraser and I. M. Ward, *J. Mater. Sci.* **9**, 1624 (1974).
157. P. I. Vincent, *Polymer* **15**, 111 (1974).
158. G. B. Bucknell, *Br. Plast.* **40**, 118 (1967).
159. I. Wolock, J. A. Kies, and E. B. Newman in Ref. 139, p. 250.
160. R. S. Rivlin and A. G. Thomas, *J. Polym. Sci.* **10**, 291 (1953).
161. A. M. Bueche and J. P. Berry in Ref. 139, p. 281.
162. F. Bueche, *Physical Properties of Polymers,* Wiley-Interscience, New York, 1962, p. 237.
163. T. L. Smith, *J. Polym. Sci.* **32**, 99 (1958).
164. T. L. Smith, *Soc. Plast. Eng. Tech. Pap.* **16**, 1211 (1960).
165. T. L. Smith and P. J. Stedry, *J Appl. Phys.* **31**, 1892 (1960).
166. E. H. Andrews, *J. Mater. Sci.* **9**, 887 (1974).
167. E. H. Andrews and E. W. Billington, *J. Mater. Sci.* **11**, 1354 (1976).
168. A. G. Thomas, *J. Polym. Sci.* **31**, 467 (1958).
169. E. H. Andrews and B. J. Walker, *Proc. R. Soc. London Ser. A* **325**, 57 (1971).
170. P. C. Paris in J. J. Burke, ed., *Fatigue, An Interdisciplinary Approach,* Syracuse University Press, Syracuse, N.Y., 1964, p. 107.
171. J. S. Harris and I. M. Ward, *J. Mater. Sci.* **8**, 1655 (1973).
172. J. A. Manson and R. W. Hertzberg, *Crit. Rev. Macromol. Sci.* **1**, 433 (1973).

General References

J. J. Aklonis, W. J. MacKnight, and M. Shen, *Introduction to Polymer Viscoelasticity,* John Wiley & Sons, Inc., New York, 1972.

E. H. Andrews, *Fracture in Polymers,* Oliver & Boyd, Edinburgh, UK, 1968.

A. Cifferi and I. M. Ward, eds., *Ultra-high Modulus Polymers,* Applied Science Publishers, London, 1979.

J. D. Ferry, *Viscoelastic Properties of Polymers,* 3rd ed., John Wiley & Sons, Inc., New York, 1980.

D. W. Hadley and I. M. Ward, *Rep. Prog. Phys.* **38,** 1143 (1975).

R. W. Hertzberg and J. A. Manson, *Fatigue of Engineering Plastics,* Academic Press, Inc., New York, 1980.

H. H. Kausch, *Polymer Fracture,* Springer-Verlag, Berlin, 1978.

H. H. Kausch, ed., *Crazing in Polymers, Advances in Polymer Science 52/53,* Springer-Verlag, Berlin, 1983.

A. J. Kinloch and R. J. Young, *Fracture Behavior of Polymers,* Elsevier Applied Science Publishers, Ltd., Barking, UK, 1983.

F. J. Lockett, *Nonlinear Viscoelastic Solids,* Academic Press, Inc., London, 1972.

N. G. McCrum, B. E. Read, and G. Williams, *Anelastic and Dielectric Effects in Polymeric Solids,* John Wiley & Sons, Inc., Chichester, UK, New York, 1967.

P. C. Powell, *Engineering with Polymers,* Chapman and Hall, London, 1983.

B. E. Read and G. D. Dean, *The Determination of the Dynamic Properties of Polymers and Composites,* Adam Hilger, Institute of Physics, Bristol, UK, 1978.

L. R. G. Treloar, *The Physics of Rubber Elasticity,* 3rd ed., Clarendon Press, Oxford, UK, 1975.

S. Turner, *Mechanical Testing of Plastics,* 2nd ed., Godwin, Harlow, and Longman, New York, 1983.

P. I. Vincent, *Impact Tests and Service Performance of Materials,* Plastics Institute, London, 1971.

I. M. Ward, *Mechanical Properties of Solid Polymers,* 2nd ed., John Wiley & Sons, Inc., Chichester, UK, New York, 1983.

I. M. Ward, ed., *Structure and Properties of Oriented Polymers,* Applied Science Publishers, London, 1975.

I. M. Ward, ed., *Developments in Oriented Polymers,* Vol. 1, Applied Science Publishers, London, 1982.

J. G. Williams, *Stress Analysis of Polymers,* 2nd ed., Ellis Horwood, Chichester, UK, 1983.

J. G. Williams, *Fracture Mechanics of Polymers,* Ellis Horwood, Chichester, UK, 1984.

A. E. Zachariades and R. S. Porter, eds., *Mechanics of High Modulus Fibers,* Marcel Dekker, Inc., New York, 1983.

D. W. HADLEY
University of Reading, UK

I. M. WARD
University of Leeds, UK

A

Abradability, 12
Abrasion and wear, **1**
ABS. See *Acrylonitrile–butadiene–styrene polymers.*
Acetal copolymer
 wear rate, 15
Acetal resins, 232
 chemical resistance, 55
 high modulus development, 461
 impact toughness, 500
Acrylic fibers, 346
Acrylonitrile–butadiene–styrene polymers (ABS)
 impact strength, 500
 wear rate, 15
Adhesion and wear, 8
Adhesive bonds
 heat-resistant polymers for, 449
Advanced composite materials, 328
Advanced composites, 62
Aerospace applications
 heat-resistant polymers in, 448
Aerospace industry
 composites in, 87
Aging, physical, **36**
Agricultural applications
 of fibers, 319
Air Force Ordered Polymer Program, 473
Alfrey approximation, 587
Alkyd resins
 chemical resistance, 53
Aluminum
 tensile modulus, 456
Amidel, 210
Amino resins
 chemical resistance, 53
Amorphous polymers, 601
 aging of, 36
Annealing, 36
Antiplasticizers, 171
Aramid
 hollow-fiber membranes, 322
Aramid fibers, 453
 from liquid crystalline solution, 549
 in composites, 63
Aramid pulp fibers, 346
Ardel, 210
Arenka, 453
Arnite, 224
Asbestos fibers, 346
ASTM standards
 polymer friction and wear, 11
Astrel, 210
Automobile bodies, 83

B

Ballistic applications, 349
Bauschinger effect, 636
Bayblend, 210
Bearings, 59
Beetle (PET), 224
Biomedical applications
 of fibers, 325
BKZ theory, 610
Blends
 of engineering resins, 237
Block copolymers
 high modulus, 465
BMC. See *Bulk molding compound.*
Boats, 91
Boltzmann superposition principle, 180, 581
Boron fiber
 in composites, 63
Breakage, 365
Brinell hardness, 424
Bulk molding compound (BMC), 127
Bulletproof glazing, 231
Bulletproof vests, 350
Butyl rubber
 chemical resistance, 54

C

Cables, 342
Cadon, 55
Capron, 210
Carbon
 lubricating fillers, 25
Carbon fibers, 346
 in biomedical applications, 327
 in composites, 63
 tensile modulus, 456
Carodel, 224
Celanex, 225
Celcon, 210
Cellulose
 liquid crystalline solutions, 530
Cellulosic fibers, 346
Charpy impact test, 395, 478, 646
Chemically resistant polymers, **52**
Chloroprene polymers
 chemical resistance, 54
Chlorotrifluoroethylene polymers
 chemical resistance, 55
Cleartuf, 224
Coatings, 60
 abrasion and wear testing, 11

T

U

V